AF761083

Information Systems for Managers in the Digital Age

WITH CASES

Edition 5.0

Gabriele Piccoli
Louisiana State University
and the University of Pavia

Federico Pigni
Grenoble École de Management

Prospect Press

Copyright © 2022 Prospect Press, Inc. All rights reserved.

No part of this publication may be reproduced, stored in a retrieval system or transmitted in any form or by any means, electronic, mechanical, photocopying, recording, scanning or otherwise, except as permitted under Sections 107 or 108 of the 1976 United States Copyright Act, without either the prior written permission of the Publisher, or authorization through payment of the appropriate per-copy fee to the Copyright Clearance Center, Inc. 222 Rosewood Drive, Danvers, MA 01923, website www.copyright .com. Requests to the Publisher for permission should be addressed to the Permissions Department, Prospect Press, 47 Prospect Parkway, Burlington, VT 05401 or e-mail to Beth.Golub@ProspectPressVT .com.

Founded in 2014, Prospect Press serves the academic discipline of Information Systems by publishing innovative textbooks across the curriculum including introductory, emerging, and upper-level courses. Prospect Press offers reasonable prices by selling directly to students. Prospect Press provides tight relationships between authors, publisher, and adopters that many larger publishers are unable to offer. Based in Burlington, Vermont, Prospect Press distributes titles worldwide. We welcome new authors to send proposals or inquiries to Beth.golub@prospectpressvt.com.

Editor: Beth Lang Golub
Production Management: Scribe Inc.
Cover Design: Annie Clark
Brick graphic on cover: Sunwand24/Shutterstock.com

References to internet websites (URLs) were accurate at the time of writing. Neither the author nor Prospect Press is responsible for URLs that may have expired or changed since the manuscript was prepared.

EDITION 5.0, With Cases:
eTextbook ISBN:
978-1-943153-83-1
Available from Redshelf.com and VitalSource.com

Printed Paperback ISBN:
978-1-943153-84-8
Available from Redshelf.com

For more information, visit
https://www.prospectpressvt.com/textbooks/piccoli-information-systems-for-managers-5-0

To Margaret, the finest teammate I could have asked for.
To Laura Jean and Erik, who always help put it all in perspective.
G. P.

To Paola, who teaches me how life is beautiful every day.
F. P.

Contents

Preface

Welcome to edition 5.0! The new subtitle, "In the Digital Age," reflects the new direction of this transformed edition. While staying true to our philosophy, in this new edition we focus on managers operating in a world where digital disruption, digital transformation, digital innovation, and digital entrepreneurship have become the norm. These trends permeate the new edition, and we have thoroughly updated coverage of cloud computing, systems development, value creation, and much more. Where appropriate we have removed discussion of obsolete technologies and concepts and condensed other material to maintain a streamlined, dynamic presentation for a one-term course (see below for more details). Of course, the book maintains its distinctive style, with rich descriptions, real-world examples, and quotes from executives. We hope you and your students will enjoy using it in your class as much as we do in ours.

The Philosophy of This Book

This book is inspired by the notion that today, half a century into the "information age" and in the midst of a "digital revolution," general and functional managers must be able to *actively* and *knowledgeably* participate in discussion and decisions about information systems and digital technology. We designed this book from the ground up, based on what has worked and proven useful to advanced undergraduate, master of management, and executive management students we have taught over the past 25 years in the United States and in Europe. Thus the topics this book covers, its writing style, and the examples, minicases, and full-length cases we use are all carefully chosen to be both *relevant* and *engaging* to an audience of students who may or may not plan to become information systems specialists.

In interactions with executives of large and small firms, we find that both general and functional managers need to be able to do two things when it comes to information systems decisions:

1. Evaluate the plethora of information technology and trends—from a strategic, not a technical, standpoint—in order to be able to identify and use the digital technology/technologies that will generate value for the organization.
2. Become effective partners of the information systems' function. To this end, they need to be familiar with those areas where they will have to come in contact with information systems professionals.

The main foci of this book are therefore the strategic role of information systems in the contemporary firm and the design and implementation of strategic initiatives that depend on digital technology to be successfully carried out. Over the years, we have come to believe that master's and executive MBA students, as well as undergraduates with an interest in information systems, don't need a "breadth book" that offers an overview knowledge of technology and technology issues. They are best served by an "analytics tools–focused" book that offers them frameworks and tangible guidance on how to ensure that their firms benefit from information systems and digital technology efforts. We designed this book, and we write every revision to provide such a toolkit for them.

Thus this is not a text *about IT* and *how IT works*. This is a book about the information system and digital technology resource and how it should be optimally deployed to achieve an organization's objectives. In other words, this book treats digital technology like any other organizational resource—an asset that general and functional managers alike need to understand enough to be able to plan for, select, deploy, and effectively manage with a keen eye to its strategic potential. The ultimate goal is to be able to create and appropriate value for their organizations.

How This Book Is Organized

The book follows a progression designed to engage skeptical students and "believers" alike while creating a sound long-term basis for decision making. We are particularly sensitive to the audience's needs with both the expository approach and content selection. The book uses a more colloquial and engaging writing style than traditional textbooks, with many examples and quotes centering on decision making by both general and functional managers. Our students, as well as many colleagues, told us that this book has the colloquial style of a consulting report rather than the dry writing style of a textbook. This is important feedback, as it confirms that once the "dry reading" obstacle is removed, students can see what the course is really all about: critical subject matter for managers, delivering a set of tangible and practical skills that will help them be assets in their respective organizations. One of our adopters once wrote to us, "The writing style is outstanding (lots of examples, which is critical). My nontechy MBA students (many of whom work full time as business managers) commented that they now see the relevance of taking my course!" This is what Gabe set out to do when originally designing the book, and it is the goal we both continue to target with each revision. The opening minicases, written from the perspective of functional or general managers, reinforce our pedagogical approach by clearly showing the relevance of the chapter's content while fostering discussion and generating momentum at the start of every class.

The book's content is organized into four sections, four stepping-stones that build upon one another in a sequential manner. Part I covers essential definitions and provides the conceptual foundation for the following chapters. Part II describes how digital technologies have changed, and continue to change, the competitive landscape and social environment, thus creating both opportunities and new threats for established firms and start-ups. Part III carries the defining content of this book, offering actionable frameworks to help managers envision how to develop value-adding strategic initiatives and gauge whether they can be protected so as to reap benefits in the long term. Part IV concludes the text by focusing on how to create and implement information systems at the core of any initiative. It also discusses common systems and major trends, as well as security, privacy, and ethical considerations.

Finally, the "With Cases" version of the book provides a number of full-length cases (included in Part V), written by us and colleagues, that were expressly designed to use the frameworks and debate the issues covered in the chapters. This tight parallel between content and cases has worked very well in our classes because students immediately see the applicability of the theories and concepts covered in the book.

New for Edition 5.0

The fifth edition was the most comprehensive revision since the first draft of this book. As the COVID-19 pandemic showed, the role of digital technology is only increasing. In the words of Marc Andreessen, "software programs the world." The world continues to change rapidly, and it has never been so compelling for students to understand the opportunities that digital technology provides for creating new strategic initiatives. While all chapters underwent a careful update, and several new and current examples are introduced to engage students in critically reviewing the discussed concepts, the fifth edition is a major rethinking and update of the content of the book. Staying true to its philosophy, very applied and focused on what general and functional managers need to know to extract value from digital technology investments, this edition fully captures the current technology and competitive landscape. The length of the text stayed the same, thanks to some condensing and streamlining, with the following major additions:

- Provided in-depth discussions of digital disruption, digital transformation, digital innovation, and digital entrepreneurship.
  - These concepts are introduced early (Chapter 1).
  - As with any concept in this book, we pay particular attention to definitional precision so as to ensure that our students know what they are talking about. Thus we ground these buzzwords in a clear explanation of the "digital technology" concept.
  - We also ensure students understand why it is warranted to use the term *digital* as separate from the more traditional *IT-enabled* (e.g., digital transformation vs. IT-enabled transformation).
  - Once these popular concepts are firmly established, we thread them throughout the book, thus ensuring the book is and feels up-to-date.
  - Reflecting the importance of these concepts, we have now titled Chapter 5 "Digital Transformation, Innovation, and Entrepreneurship." The chapter devotes a section to each. Keeping with the tradition of the book, we provide clear definitions and practical guidance and keep the writing flowing with examples and quotes from industry leaders.
- Provided in-depth discussion of digital platforms, marketplaces, and aggregators.
  - These concepts form a cornerstone of Chapter 4, where they are introduced, defined, and explained. They are related to foundational concepts we have already discussed, such as network and information economics, as well as the general thread of digital disruption informing the fifth edition.
  - From their introduction in Chapter 4, they are used where applicable in minicases and in the remaining chapters. They feature prominently in Chapter 8, with sections focusing on value creation as, or within, digital platforms and aggregators.
- Introduced important emerging concepts, such as the creator (or passion) economy, digital resources, and the algorithmic economy. These concepts are introduced, defined, and explained through examples. They are also discussed in relevant chapters—for example, in relation to value creation (Chapter 8) or system design and deployment (Chapter 11).
- Introduced the opportunities of emerging digital innovations, including 5G networks and quantum computing (Chapter 12).
- Expanded and updated the discussion of value creation and appropriation.
  - The section on value creation with data has been modernized to reflect recent attention in the area. Specifically, we now introduce, explain, and provide examples of various data monetization strategies.
  - We added a new section on value creation via digital platforms and aggregators for those organizations that are in a position to establish a dominant platform or become an aggregator.
  - We included a new section on value creation within digital platforms and aggregators for the many organizations that cannot establish a dominant platform or become an aggregator and must instead find ways to create value and thrive as part of the ecosystem of dominant platforms and aggregators.
- Further raised the profile of cloud computing throughout the book. This was already an important topic in earlier editions. In this revision, given the continued rise to prominence of the public cloud, we thread cloud concepts throughout. The following are some highlights:
  - When introducing digital resources, we show how they are by design servitized and thus could not be feasible without pervasive adoption of cloud computing.

  - In Chapter 10, we expand IS governance concerns to the cloud and introduce the issue associated with self-provisioning of computing resources by business units. We also discuss specialized governance approaches, like cloud centers of excellence.
  - We introduce and discuss the opportunity of serverless computing (Chapter 12).
- Significantly revised the discussion about systems design and deployment (Chapter 11) raising the profile of agile methodologies and providing an in-depth explanation of DevOps (with sidebar based on a real case).
- Updated key concepts in the book where research has furthered our understanding of the phenomena. The following are a couple of examples:
  - a broader focus on disruptive innovation rather than disruptive technology
  - a nuanced discussion of network effects, the different configurations they can assume, and the implications of such differences
- Revised all minicases to ensure they are timely and interesting. Where applicable, we created new minicases to help students engage with novel concepts—for example, Chapter 5 on the creator economy and digital platforms and Chapter 8 on competition within aggregators.
- Included concepts requested by adopters in the relevant chapters (e.g., business process automation [BPA] and business process improvement [BPI]).
- Made general updates throughout to statistics, trends, and quotes from industry leaders that keep the material feeling fresh, timely, and engaging for the students.

As usual, we maintain two editions of the book: "With Cases" and "Without Cases." The two versions are identical except for the full-length cases in Part V that the latter omits. The lower-priced "Without Cases" version gives you the opportunity to use your own selection of cases or to still support your pedagogy when you plan not to use the case method.

Instructor Resources

We have developed an extensive set of instructor resources, now all available on the publisher's website (https://www.prospectpressvt.com/textbooks/piccoli-information-systems-for-managers-5-0). You will find the standard set of materials, including the following:

- PowerPoint presentations for each chapter, updated for Edition 5.0 by Cheri Noteboom and Matthew Noteboom
- A test bank in Word and Respondus formats, updated and expanded by Cheri Noteboom and Matthew Noteboom
- an instructor's manual containing teaching notes for each of the opening minicases, along with tips and suggestions on how to use them to jump-start class with a high-energy discussion
- the full teaching notes of the cases, including the case analysis and the teaching plan

Information on pricing and ordering the different versions of the text can be found at the publisher's website: https://www.prospectpressvt.com/textbooks/piccoli-information-systems-for-managers-5-0.

Acknowledgments

While there are only two names on the cover of this book, and while we take responsibility for any errors or inaccuracies, a book like this one is really a joint effort. The ideas and concepts, teaching philosophy, pedagogical approaches, tools, and techniques covered came about through observation of and discussion with the many colleagues we have had the good fortune of interacting with over the years. Among the most influential, we want to acknowledge Roy Alvarez, Kathryn Brohman, Blake Ives, Dick Moore, Erica Wagner, and Rick Watson. Each of them has helped us in shaping the courses we have taught and ultimately the content of this book.

We would also like to acknowledge the contribution of the many students over the last 15 years who have brought a positive attitude and an open mind to the Information Systems Management course, those who have challenged our own understanding of the concepts in this book, and those who pushed us to find better ways to introduce and discuss them. We would also like to acknowledge the many companies and their management and employees who recognize that we can't have relevant education without their full engagement and support. These people have kindly volunteered their time to discuss many of the ideas explored in this book. Many have also agreed to allow their companies' examples and experiences to serve as case studies and welcomed my colleagues and us to share their knowledge.

We would also like to thank the colleagues who took time from their busy schedules to review the manuscript of the earlier editions and to offer valuable and constructive feedback. **From the first edition,**

Louis Beaubien, Providence College
Ed Christensen, Monmouth University
Blake Ives, University of Houston
John Kohlmeier, DePaul University
Paul Licker, Oakland University
Richard McCarthy, Quinnipiac University
Roberto Mejias, Indiana University
Paul Pavlou, Temple University
Graham Peace, West Virginia University
Cynthia Ruppel, Nova Southeastern University
John Scigliano, Nova Southeastern University
Michael Wade, IMD Lausanne
Erica Wagner, Portland State University
Ted Williams, University of Michigan, Flint

From the second edition,

Patrick Becka, Webster University, Louisville
Leida Chen, Creighton University
J. Michael Cummins, Georgia Tech
F. J. DeMicco, University of Delaware
David Dischiave, Syracuse University
Rassule Hadidi, University of Illinois, Springfield
Raymond Henry, Cleveland State University
Jeff Howells, University of Georgia
Shin-jeng Lin, Le Moyne College
Rodger Morrison, Troy University
Barbara Ozog, Benedictine University
Jeffrey Pullen, University of Maryland
Malu Roldan, San Jose State University

From the third edition,

Michael Bliemel, Dalhousie University
Sophie Buer-Chemin, Grenoble École de Management
Mark Cecchini, University of South Carolina
Mitchell Church, Coastal Carolina University
Michael Cummins, Georgia Tech
Pierre Dal Zotto, Grenoble École de Management
Greg Dawson, Arizona State University
Patrick Fan, Virginia Tech
Alberto Francesconi, University of Pavia
Robert Gordon, Molloy College
Iris Lui, Ming Chuan University
Brandi N. Guidry Hollier, University of Louisiana at Lafayette
Alex Heckman, Franklin University
Paul Licker, Oakland University
Efrem Mallach, University of Massachusetts, Dartmouth
John Muraski, University of Wisconsin, Oshkosh
Mark Rosso, North Carolina Central University
James Stewart, Syracuse University
Andrew Walker, Grenoble Ecole de Management

For Edition 4.0 we would like to thank the following:

Pierre Dal Zotto, Grenoble École de Management, for constant feedback and support in collecting the primary data for the opening minicase of Chapter 4.

Michael Rauchs, Cambridge Centre for Alternative Finance, for the input his work provided on the development of the blockchain section, and of course all of the book adopters who joined the IS4Managers.com community, asked questions, and provided us with feedback.

We also wish to thank the colleagues who provided input for the development of Edition 4.0. These include the following:

Robert Boncella, Washburn University
Jason Black, Northeastern University
Laura Brazie, Virginia Tech University
H. M. Chung, California State University, Long Beach
Geoff Dick, Northern Arizona University
John Drake, East Carolina University
Harold Fraser, California State University, Fullerton
Biswadip Ghosh, Metro State University of Denver
Robert Gordon, Molloy College
Shaoyi He, California State University, Long Beach
Derek Holbert, Franklin University
Carolyn Holton, Southeastern University
Esther Klein, Hofstra University
Gerald Lancia, Northeastern University
Diane Lending, James Madison University
Steve Nelson, Butler University
Sandra Newton, Sonoma State University
Alexander Pelaez, Hofstra University
John Pendergrass, Northern Illinois University
Hindupur Ramakrishna, University of Redlands
Mark Rosso, North Carolina Central University
Kevin Scheibe, Iowa State University
Yide Shen, Rowan University
William Spangler, Duquesne University
Nelson Velez, University of Massachusetts, Boston
Chris Zimmer, West Texas A&M University

Finally, we are grateful to the colleagues who provided extensive input for the development of the new Edition 5.0. Reviewers and User Survey Responders include the following:

Elizabeth White Baker, Virginia Commonwealth University
Brian Bender, Northern Illinois University
Jason Black, Northeastern University
Thomas Case, Georgia Southern University
Tung Cu, College of Business and Technology, Northeastern Illinois University
Pierre Dal Zotto, Grenoble Ecole de Management
Jerry DeHondt, Associate Professor, Ball State University
Geoffrey Dick, St John's University
Jeremy D. Ezell, James Madison University
Jason Ferguson, Bellevue University, College of Science and Technology
Jerry Fjermestad, New Jersey Institute of Technology
Peter Goldschmidt, The University of Western Australia Business School
Robert Gordon, Molloy College and Hofstra University
Gina Harden, Stephen F. Austin State University
Philip G. Lawrence, University of Redlands
Heejin Lee, Graduate School of International Studies, Yonsei University
One-Ki Daniel Lee, College of Management, University of Massachusetts Boston
Diane Lending, James Madison University
Ted Mikell, Baker College
Alanah Mitchell, Drake University
Mahesh Raisinghani, TWU College of Business, U.S.A.
Richard Sessions, School of Information Systems and Technology, San Jose State University
Marcos P. Sivitanides, McCoy College of Business, Texas State University
William E. Spangler PhD, Duquesne University
Manouchehr Tabatabaei, Georgia Southern University
Hari Stirbet, Wilfrid Laurier University
Shouhong Wang, University of Massachusetts Dartmouth
Erik Westgard, Metropolitan State University
Roy Wood, Northeastern State University, Tahlequah and Broken Arrow, Oklahoma

Last, but certainly not least, we would like to acknowledge Beth Lang Golub, who originally saw the opportunity for this book to be written and constantly supported its development. We also thank Andy Golub, Beth's husband, for his enthusiasm and indefatigable support of our work.

Foreword

An obvious takeaway from the COVID-19 pandemic crisis is the essential role played by digital technology and information systems. Online meetings, contactless purchases, and food delivery services—to name a few—proved to be absolutely essential components of economic survival. Moreover, the epidemic significantly accelerated the pace of digital transformation and disruption.

Helping students understand these technology-induced changes and develop the managerial skills to leverage digital resources in their organizations is the overarching objective of Gabe and Federico's book. While the text demonstrates the coauthors' deep understanding of both established and leading-edge technologies, it is their emphasis on applying and leveraging these digital technologies to drive business value that sets it apart. As such, in today's digitally transforming world, this book provides essential knowledge for all students of business.

While I have authored, edited, and reviewed textbooks written to teach students and executives how to use information technology to innovate and transform business and society and have also authored over 300 Harvard Business School case studies, Gabe and Federico did something that authors rarely succeed at: packing a lot of content in the book while ensuring the text is engaging and lively. The book is filled with rich descriptions and real-world examples, minicases, and quotes from executives. And as with previous editions, the option of a case-enhanced version is also available.

Undergraduate and MBA students will appreciate the concrete examples and cases and the very approachable writing style. Executive-level audiences will immediately see how to apply what they learn and the value that can be created by harnessing the power of digital innovation in their own organizations. This is anything but dry reading, and your students will thank you for using it!

Lynda M. Applegate
Baker Foundation Professor
Sarofim-Rock Professor of Business Administration, Emerita
Harvard Business School

There tend to be two major approaches to teaching information systems. The technology perspective instructs students about hardware and software, and the focus is very much on information technology. The transformative approach assists students in identifying how information systems can transform an enterprise's relationships with its key stakeholders, such as customers and suppliers. Under the transformative approach, the emphasis is on what the information systems as an enterprise can create and the value they can then unleash. Anyone can buy information technology, but it takes skill, vision, and persistence to create a transformative information system that can radically change an enterprise and even a society.

Gabe is a key member of the transformational school of information systems. His academic research and work with practitioners, the results of which are incorporated in this book, are directed at understanding and exploiting the transformative power of information systems. He has studied many examples of transformation, as you will find when you read some of the insightful cases in this book and the highly useful frameworks he presents. As you read this book, you will discover why information systems have been the major change engine for many enterprises and economies over the last five decades. Nothing else has had anywhere near the same influence on the way we live today.

As you develop your managerial skills, it is important that you realize that to become an organizational leader, you will need to demonstrate that you can design, lead, and execute transformational projects. Most of the morphing assignments you take on will require an intimate understanding of technology in organizations. Irrespective of which particular field is your area of concentration, you will find that at some point you will need to become deeply steeped in understanding how you can exploit one or more information technologies to create a new information system to generate a competitive advantage or superior customer experience. How well you integrate people, procedures, and information technology to create an information system will determine your success. This book is about developing your skills to successfully participate in, and possibly lead, an information systems–enabled project to create new opportunities for customer service and organizational growth. Read the book carefully, reflect on the frameworks and cases, incorporate the key insights into your model of the world, and above all, apply the knowledge gained to improve your organization's performance.

Richard T. Watson
Rex Fuqua Distinguished Chair for Internet Strategy
University of Georgia

As we all know, the required master's-level information systems course is a very difficult one to teach. I always admire the few faculty teaching this course who can transcend the reluctance, prejudice, and general disinterest too many of us confront from our students. Gabe is one of the talented few. He has enjoyed significant success teaching the required information systems course over the years, receiving teaching awards for his work with both master's and executive MBA students. In my foreword to the first edition, I said, "Hopefully, this book will arm the rest of us with some of Gabe's teaching magic." Now that we are in edition 5.0, I can report that I have talked to a number of colleagues who teach required MBA, executive MBA, and high-level undergraduate courses who have confirmed that this book has worked quite well for them.

I can now be confident in saying that this book will be a great addition to your arsenal, allowing you to leverage the enthusiasm of students already interested in the material and energize those who come in the door with a negative bias toward an "IT course." This book can make your course more compelling to your students thanks to Gabe's very approachable writing style, the wealth of examples he uses, the opening minicases that quickly create excitement and buzz, and the unique full-length cases in the full version of this textbook (several of which we wrote together). Most helpfully, Gabe has identified both the foundational and cutting-edge content that is most relevant to management students. With this book, you will find it much easier to demonstrate the relevance of information systems to your students and to create a positive learning environment in your classes.

Blake Ives, PhD
C. T. Bauer Chair in Business Leadership
C. T. Bauer College of Business
University of Houston

PART I

Foundations

In Part I, we lay the foundations for the study of information systems (IS). Although the press and commentators devote much attention to information technology (IT) and the (often substantial) IT investments that organizations make, modern managers don't have the time, or often the inclination, to become IT experts. After all, that's why organizations hire and pay IS and IT professionals.

Yet with IT becoming pervasive in both business and society at large, today's general and functional managers can no longer abdicate their obligation to make decisions about this crucial organizational resource. The good news is that you can be an effective manager without knowing a huge amount about IT and without knowing in detail how technology works. Such knowledge will surely help, and it will help you recognize opportunities associated with the barrage of new technologies that are constantly being commercialized. To be an effective general or functional manager, a proficient user of IT resources, and a productive partner of the firm's information systems and technology professionals, you need a strong grounding in the fundamentals of IS management and decision making.

As we describe in Chapter 2, information systems are sociotechnical organizational systems that encompass technology, the people who will be using such technology, and the business processes they execute to accomplish their daily tasks and carry out business activities. User-managers can rely on IT professionals when it comes to choosing among programming languages or the appropriate structure of a new database being implemented, but business managers must be able to design the appropriate information systems for their organization, plan and budget for the use of IT resources, and analyze whether a given information system creates a competitive advantage that can be protected.

This is not a book about IT and how it works. This is a book about information systems and the IS decisions that general and functional managers are routinely called on to make. In Part I, we lay the foundations upon which you will build your IS knowledge.

- *Chapter 1: Information Systems and the Role of General and Functional Managers.* The first chapter defines some basic terms and makes the case for why general and functional managers must be intimately involved in IS decision making.
- *Chapter 2: Information Systems Defined.* The second chapter defines what an information system is (the central concept in this book), places this definition in an organizational context, and draws the crucial distinctions between IT and IS.
- *Chapter 3: Organizational Information Systems and Their Impact.* The third chapter categorizes the different information systems found in modern organizations and provides the vocabulary you need in order to communicate with other managers and the IS professionals in your firm. This chapter also provides you with a foundation to consider the impact of various types of IT on the organization.

Chapter 1

Information Systems and the Role of General and Functional Managers

What You Will Learn in This Chapter

This chapter focuses on the role that general and functional **managers** play in the organizational use and management of information systems (IS). The chapter also describes the rise to prominence of information technology (IT) and the recent focus on **digital transformation, innovation,** and **entrepreneurship.**

MINICASE: Facing Termination?

The silence was beginning to become uncomfortable as you searched for words to answer the question from your chief executive officer (CEO). The boardroom had never looked so big, and it seemed as if her words were still echoing: "How could it get to this? You sat here telling us how this new application would dramatically improve our marketing efficiencies and customers' repurchase frequency. It has been over two months, and the bloody thing isn't even working!"

As you searched for the right way to respond, the events leading up to this moment flashed through your mind. It was more than two months ago when you sold the board on the benefits of a new sales force software as a service (SaaS) application. You had just been promoted to vice president of marketing, taking over from Tom Vecchio. Tom was an old-fashioned salesperson with a huge personality and an incredible memory. He was employee number four when the company launched back in 1982, and he had been instrumental in its early growth via personal networking—phone calls, rounds of golf, and birthday calls. He had surrounded himself with very similar people. You understood that culture, you had been one of the young guns a few years ago, and now you had replaced the master.

But things had changed in your industry, the competition was much tougher, and markets were now global. *How could a firm the size of this one run sales and marketing still just using a mess of "cutting-edge" Excel sheets?* you wondered once promoted. How ironic that you'd be the leader in your firm's "digital transformation." You had managed to never concern yourself with all that techie stuff. You were a pretty good user: email, web, Word, PowerPoint, Excel, some Python you learned at school but mostly forgot by now . . . the usual. But now your bonus depended on the performance of the whole function, not just the number of contracts you closed, and it seemed as if you had been getting all the heat about efficiencies that they could not put on Tom . . . they could scream all they wanted—he was untouchable. But you weren't!

It all seemed to have fallen into place when you went to the National Convention of the Sales Executives Association. At one of the booths, you had seen VelcroSoft and the sales force application VelcroSFA. There was a lot of buzz around the product both at the conference and in the press. The attendant at the booth told you about all the great features of VelcroSFA: cloud-native programming, mobile integration, automated recording of information at each stage in the sales process, automated escalation and approval, contact management, lead sharing for team selling, and in-depth reporting. It could even integrate with leading human resource (HR) systems for immediate computation of commissions, reduced data entry, and increased speed.

After you returned to the office, you read some more material about VelcroSFA, looked at the online demo, and called a couple of friends who used it in their organizations. It seemed to be the right application. You showed the online demo to some of the best-performing salespeople. They did not seem impressed, and they raised a bunch of issues. Joe, one of the old-timers, said, "The Excel sheet is heavy enough, and the Rolodex did wonders for me throughout my career; what do I need this 'cloud app' for?" Joe never liked you anyway since you had taken Tom's spot, you thought. Amanda, a younger associate, seemed more positive: "I'm willing to give it a shot, but it seems quite convoluted. I'm not sure I need all those functionalities." You recall thinking that they would change their minds once they saw their commissions go up because VelcroSFA would allow them to spend more time with customers. You liked the software more as you learned about it. They would too.

Jenny Cantera, the IT director, had pledged her help with the integration and, after looking at the brochure, had said, "Should take a weekend to have this application running and write the interface to the HR management system. I'm busy with the implementation of the new accounting system for the next three or four weeks, but I should be able to do this afterward." You had some doubts about Jenny. She was very smart and technically gifted, but she was a newcomer in your firm, and she had little experience in the position.

The board had been sold pretty easily on the purchase, a relatively low monthly fee of $175 per user and no "computer stuff" to buy. You had used the return-on-investment (ROI) calculations provided by VelcroSoft that included things like the initial investment for moving to the cloud, the cost of migration, and support. Granted, VelcroSoft personnel were very aggressive with assumptions underlying their calculations, but with a bit of effort on everyone's part, you truly believed you could achieve strong results. As soon as you got the go-ahead, you contacted the vendor and obtained the setup instructions and interface specifications to seamlessly integrate with the HR management system. Everything had gone perfectly up to that point, but your fortune seemed to turn right after.

First, you had the administrator package on your desk for more than a month. Jenny was running into unexpected trouble with the legacy accounting application. *This thing has been running since the early 1990s*, you chuckled to yourself. *When are these guys gonna follow me into the digital era?* Once she finally got around to integrating your product, she took one weekend to complete the migration and finally created the user accounts. The interface to the HR application was not operational yet—something about an "Apex JSON" malfunctioning . . . which you did not quite understand. However, you pressed on. Over the following week, you had encouraged your sales rep to "play around with the applications." You had even sent an email with the subject line "Up and running in the brave new world of digital salesforce!" But the response had been cool at best. Only a few accounts had been accessed, only one mobile app installed, and overall, the people you spoke to said they were too busy and would look at the software once the quarter closed.

Last weekend, when Jenny wrote the interface to the HR management system, all hell broke loose. On Monday (yesterday), the HR database was locked, and the HR management system was down. Apparently, the wrong sequence in the data migration procedures resulted in the reinitialization of the HR database . . . or so Jenny says. She was now scrambling to bring it back up and saying she might have to reinstall the legacy HR software application. She had already removed VelcroSFA, and at this point, it looked like the application would not be a priority for a while. You did not really mind; you had bigger fish to fry . . . you were concerned about getting fired.

Discussion Questions

1. Who do you think is to blame for the current state of affairs?
2. What do you think is the most critical mistake you made over the last two months? What were the principal mistakes made by others involved?
3. How could these mistakes have been avoided, if at all?
4. Should you take this opportunity to say good-bye to everyone and resign now? If not, what should you say in response to the CEO's question?

1.1 Introduction

It is hard these days to escape the hype and publicity surrounding IT and its business implications for transformation and innovation. IT has become more affordable and pervasive than ever before; just think about the wide array of technologies that you use for work and pleasure on any given day. You may own an iPad, a Microsoft Surface, or a Galaxy Tab; most likely you own a laptop and a smartphone; and perhaps you even have a game console, such as the Microsoft Xbox or Sony PlayStation.

At work, you are likely a heavy user of word processing and spreadsheet programs as well as some more specialized applications such as customer relationship management or business

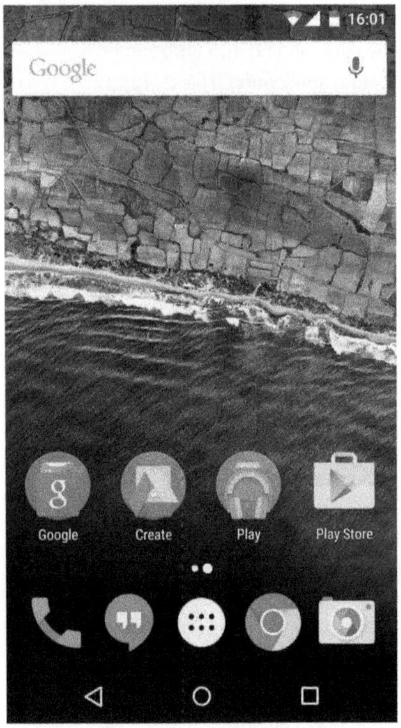

Figure 1.1. The Android operating system

Source: Copyright © 2015 by the Android Open Source Project, licensed under the Apache License, Version 2.0

intelligence software. You may even have been talented enough to create your own Android apps (Figure 1.1) or launch your own start-up.

More generally, for business, not-for-profit, and governmental organizations, IT has become a critical resource that draws significant investments. As Gartner reported in their Q4 2019 IT Spending Forecast, from a cross-industry perspective, IT spending is increasing, driven by software projects (e.g., cloud, security, blockchain, Internet of things [IoT], artificial intelligence) and a realization that every organization must fully embrace the "digital revolution" if it wants to thrive or even just survive.[1] The 2020 COVID-19 pandemic slashed budgets due to the global economic crisis it created.[2] However, it also impressed further acceleration to trends that were already underway and dispelled any doubt that digital transformation is imperative for all organizations. Microsoft CEO Satya Nadella captured this sentiment by saying, "We've seen two years' worth of digital transformation in two months."[3] Inspired Capital cofounder Alexa von Tobel had an even more memorable quote: "We started the year in 2020 and ended it in 2030."[4] In short, we live in the "digital age," and business solutions are increasingly digital first. If you want to start a bank today, your main imperative is not where to locate branches or automated teller machines (ATMs).[5] Rather, it is how to develop an omnichannel digital strategy (Chapter 5) that enables you to have a digital relationship with your customer.

It is now clear that the success of every enterprise, department, function, and employee depends increasingly on digital technology and IS. In 2011, Marc Andreessen, the cofounder of Netscape and the major venture capital firm Andreessen Horowitz, wrote a *Wall Street Journal* article in which he said, "My own theory is that we are in the middle of a dramatic and broad technological and economic shift in which software companies are poised to take over large swathes of the economy."[6] The article was prescient: for the first time in 2016, and even more as 2020 drew to a close, the largest U.S. companies by market capitalization were Apple, Alphabet (i.e., Google), Microsoft, Amazon, and Facebook (Figure 1.2). In fact, only three companies in the history of the U.S. markets passed the $1+ trillion (with at *T*) valuation: Apple, Amazon, and Microsoft. What do these companies have in common? They are the firms that underpin the digital revolution, and their value proposition is firmly rooted in software.

Not one to shy away from opinions, Andreessen later expanded his claim to "software programs the world." The key claim in his new thesis was that not only will technology companies continue to

1 Lovelock, J. 2019. "IT spending forecast, 4Q19 update: Emerging technologies in 2020." Gartner, retrieved from https://www.gartner.com/webinar/38061.

2 Lovelock, J. 2020. "IT spending forecast, 1Q20 update—view from the peak." Gartner, retrieved from https://www.gartner.com/webinar/3983792/.

3 Waters, R. 2020, April 29. "Microsoft growth accelerates as pandemic boosts cloud business." *Financial Times*, retrieved from https://www.ft.com/content/ac054397-eb9a-4198-a050-18961f39feb9.

4 Peers, M. 2020, October 1. "51% of subscribers disagree with Coinbase's Armstrong: The Information's tech briefing." The Information, retrieved from https://www.theinformation.com/articles/51-of-subscribers-disagree-with-coinbases-armstrong-the-informations-tech-briefing.

5 McCaffrey, O. 2020, June 7. "People aren't visiting branches. Banks are wondering how many they actually need." *Wall Street Journal*, retrieved from https://www.wsj.com/articles/people-arent-visiting-branches-banks-are-wondering-how-many-they-actually-need-11591531200.

6 Andreessen, M. 2011, August 20. "Why software is eating the world." *Wall Street Journal*, retrieved from https://www.wsj.com/articles/SB10001424053111903480904576512250915629460.

digitally disrupt traditional industries, but software is now the key ingredient of any value proposition—the key driver of digital transformation[1] and new value creation. The main reason for this shift has been the meteoric adoption of digital technology by consumers around the world, particularly after the advent of mobile platforms (Chapter 5). Customer relationships are now digital, and the main channel of interaction between consumers and the brands they love is a digital interface—often on their smartphone or tablet. This rise to prominence of digital relationships has forced all firms, not just software companies, to learn to build software and differentiate themselves from the competition by way of superior digital relationships with customers (but also suppliers and employees). Advances in cloud computing and SaaS, machine learning (ML), artificial intelligence (AI), IoT, and a host of other digital technology innovations have enabled forward-looking firms in any industry to digitally transform and use digital technology to innovate their value proposition. Thanks to their innovative mind-set, or under pressure from digital entrepreneurs seeking to disrupt their industry, incumbents have engaged in digital transformation to remain relevant. While perhaps not qualitatively different from other historical waves of digital disruption, the current era is certainly different in magnitude. It is safe to say that no industry is immune to digital disruption. These trends are of critical importance to you as a future business leader. If "software programs the world," it is important to be able to "code the programs." But it is even more important to know "what programs to write." So while computer scientists will continue to push the boundaries of the capabilities of digital technologies, understanding how these digital resources contribute to the creation of new economic value is the responsibility of business managers and IS professionals. There are no two ways about it: to be a successful manager in the digital age, you must understand information systems and the role they play in organizations. This book is your first step in that direction.

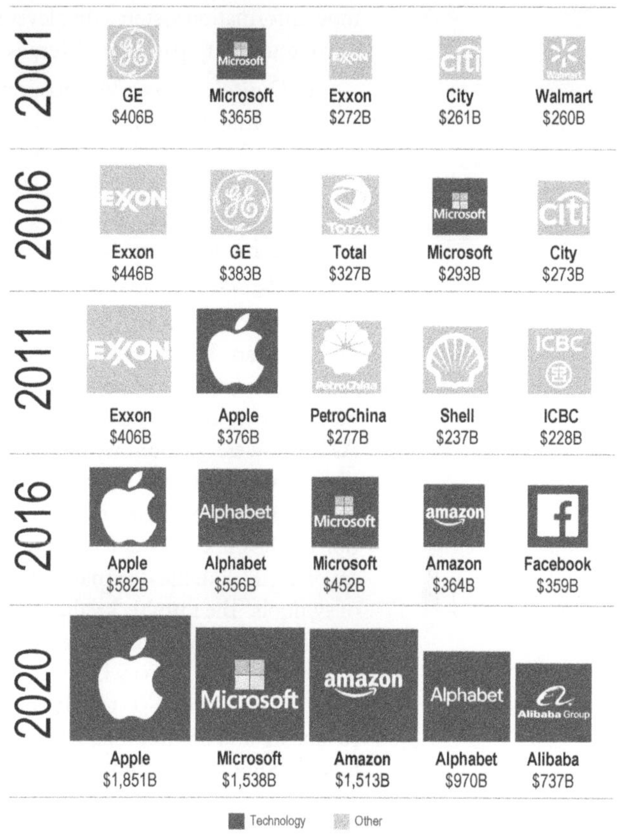

Figure 1.2. Largest companies by market cap and industry

Source: Based on Visual Capitalist, https://www.visualcapitalist.com/chart-largest-companies-market-cap-15-years/

1.2 Digital Technology

Two authors, Leavitt and Whisler, popularized the term *information technology* in the business literature in 1958.[2] They used it to refer to techniques and technologies designed to record and process data. In theory, IT does not need to be digital (e.g., a ledger and an abacus are both forms of IT). However, in this book, we adopt a stricter definition of IT, one that includes hardware, software, and telecommunication equipment.[3] We do so because our focus is on organizations and

1 Andreessen, M., et al. 2016, July 10. "Software programs the world." Andreessen Horowitz. https://a16z.com/2016/07/10/software-programs-the-world/.

2 Leavitt, H. J., and Whisler, T. L. 1958, November. "Management in the 1980's." *Harvard Business Review 36*(6): 41–48.

3 The exact definition of *information system* is a bit more complex, and we devote all of Chapter 2 to it.

those information systems that leverage the power of digital technology. The term *digital* is pervasive in contemporary parlance.[1] Here is a sample of definitions from the *Oxford English Dictionary*:

1. *Digital economy*: "An economy which functions primarily by means of digital technology, esp. electronic transactions made using the Internet."
2. *Digital money*: "Money represented and held in a digital form which can be transferred electronically from one party to another."
3. *Digital divide*: "The gulf between those who have ready access to current digital technology (esp. computers and the Internet) and those who do not."
4. *Digital footprint*: "A trace or the traces of a person's online activity which can be recovered by electronic means."

The Merriam-Webster dictionary provides a definition that captures the current use of the term *digital* as "relating to, or using calculation by numerical methods or by discrete units." The key term here is *discrete units*. Discrete means individually separate and distinct. It is the opposite of *continuous*. As if you were to count with your fingers, using a digital system to represent information implies the use of discrete values, as opposed to continuous units used in *analog* systems.[2]

In a digital computer, the specific digital system used to represent signals, data, and computations is the binary system. It is a positional numeral system that represents values using an ordered set of digits, like the decimal system you are most familiar with. Each digit can take a limited set of symbols. The number's value is determined by the combination of the symbol and its position. In the decimal system, we use 10 symbols [0, 1, 2, 3, 4, 5, 6, 7, 8, 9]. The binary system uses only 2 symbols [0, 1]. Thus while the decimal system is called "base 10," the binary system is called "base 2." The binary system is the abstraction at the heart of digital computers. Its application to storing and communicating data was first conceptualized by a 32-year-old researcher named Claude Shannon in 1948. In the paper describing his theory, Shannon proposed to use the term *bit*, short for *BInary digiT*, as the smallest unit for measuring information. As a testament to the importance of computing, the bit has joined the meter, the gram, the liter, and the minute as a determinate quantity—a fundamental unit of measure that enables our lives. Interestingly, the bit is atomic—you can't split it. Unlike time, which is continuous, digital data are, by definition, discrete. Thus you can't have a millibit or microbit.

Creating the binary system as the abstraction necessary to store data and perform computation was a breakthrough in computing. But it is only half the battle to create a working digital computer. In order for a machine to do the work for us, we need a physical device that can represent a binary numeric system. A transistor is the two-state device that enabled the implementation of the binary system inside digital electronic devices. Now you understand how simple, yet brilliant, the advent of digital technology was. Using electronic devices that could process electrical impulses (normally a positive or negative charge), the *conceptual* binary system could be *physically* implemented in machines that could perform computations and carry out instructions. Complex combinations of transistors wired together, called *integrated circuits*, enable computers to perform specialized basic operations. Such operations are combined to make complex software programs. You can think of these basic operations like notes in music. The notes are only seven, or 12 if you include all the tones. Yet in the hands of talented musicians (e.g., Chopin, Hendrix, Drake), those

1 While the general definition of *information technology* is broader than *digital technology*, our focus is on organizational information systems. In recognition of the fact that modern organizational information systems are enabled by digital technology, we will use both terms interchangeably throughout the book.

2 Indeed, the term *digital* derives from the Latin *digitus*, which means "finger." Since the 15th century, the term has also referred to "a whole number less than ten." The numerals we use for computations [0–9] are called *digits*—perhaps because our fingers were the first computation devices we ever had.

few notes become beautiful music (e.g., Piano Sonata no. 2, "Purple Haze," "Hotline Bling"). Integrated circuits typically combine billions of transistors storing and processing digital data to enable countless innovations, from music streaming to autonomous driving vehicles.

1.3 Digital Transformation, Innovation, and Entrepreneurship

The preceding discussion provides you with a solid conceptual foundation to understand the unique characteristics and potential of digital technology. Building on such a foundation, the following common terminology will be immediately clear to you:

- *Digital*: Digital refers to the discrete representation of signals and data in a binary format that is therefore readable and processable by a digital computer.
- **Digitization**: The process of digitizing, or converting signals from analog into the digital form (i.e., encoding into bits). More formally, digitization is any action that occurs when physical events, objects, content, and processes are encoded into digital form, therefore creating a digital representation of those events, objects, content, and processes.
- *Digital technology*: Electronic equipment and components that manipulate digital signals. A digital computer is the primary example of digital technology. While such computers take many physical forms (e.g., an iPhone, a Tesla car, an Amazon Echo smart speaker), they all have the same basic structure and components (Figure 1.3).

  It is the pervasiveness and unique nature of digital technology that lead to the powerful impacts on businesses and organizations described by Andreessen and others. As you enter the workforce, you will find that increasingly, for any organization, "digital business" is, in fact, business as usual. Business leaders now routinely speak about (and pour money into) digital transformation, digital innovation, and digital entrepreneurship efforts. We provide brief definitions here, and these concepts will be with us throughout the book. We will devote particular attention to them in Chapter 5 to ensure that you understand how digital transformation and innovation unfold in the firm.
- *Digital transformation*: The application and exploitation of digital technology to radically rethink organizational business and operational models. Digital transformation is underpinned by the redesign of existing organizational information systems. A much-publicized example of digital transformation, one that in hindsight shows how difficult it is to successfully implement such a large-scale effort, is provided by General Electric (GE). The venerable company, founded by Thomas Edison in 1892, makes large-scale industrial equipment like jet engines and train locomotives. In 2011, then CEO Jeffrey Immelt sought to radically transform the organization from a seller of hardware to a seller of solutions. The vision included charging customers for the service provided by GE equipment (e.g., a jet engine's flight hours) rather than the hardware itself. To pull off this transformation, GE would need to become a software company.[1] It did not really work, and GE changed course, along with its CEO, in 2017.

Figure 1.3. Digital technology everywhere
Source: CC BY 2.0, Evan-Amos / public domain

1 Immelt, J. R. 2017, September 1. "How I remade GE." *Harvard Business Review*, retrieved from https://hbr.org/2017/09/inside-ges-transformation.

- *Digital innovation*: The application of digital technologies by existing organizations to create new and improved products, processes, or services. A great example of digital innovation is the creation of a new service, iTunes (now Apple Music), underpinning a new process for purchasing and owning music. The iTunes model is the digital innovation credited with drastically reducing music piracy by offering customers a simple, affordable, legal process for owning high-quality digital music. As Steve Jobs liked to say, iTunes helped "keep honest people honest."[1] The system required music owners, the major labels, to agree to unbundle the CD and sell individual songs at a price perceived as affordable by most (i.e., 99 cents). It necessitates that music lovers have access to devices that could organize these individual songs into a music-owning experience that was superior to the existing one. Apple had already created the iPod for playing MP3 files. However, the digital innovation that revolutionized the music industry was the radical rethinking of the concept of music ownership and the music industry business model—a rethinking that took advantage of, rather than resisted, the unique characteristics of digital technology.[2]
- *Digital entrepreneurship*: The creation of new economic activities embodied in or enabled by digital technologies.[3] Many of the recent high-profile start-up ventures you are familiar with are apt examples. Uber and Lyft in transportation, Deliveroo and DoorDash in food delivery, Discord and Snapchat in messaging all have built new value propositions by leveraging the unique characteristics of digital technology as it became part of the fabric of society. A smartphone can, in real time, geolocate a person looking for a ride and a driver, message these individuals and process their inputs (e.g., a driver accepting a ride offer), track the device and show movement on a detailed map, and compute the appropriate charge for a ride and process payment. As the number of people carrying smartphones grew, the potential for the disruption of traditional taxi companies became real. Uber, Lyft, and a number of other similar ventures around the world were created to take advantage of this entrepreneurial opportunity.

The boundaries between the above three concepts are not clear-cut, as is the case often with organizational and business concepts. However, there is a clear difference in emphasis. Digital transformation is about rethinking existing processes and operating models of existing organizations. Digital innovation is about inventing and crafting new products, processes, or services, generally to create novel value propositions. Digital entrepreneurship is similar—however, while digital innovation typically resides with existing organizations (i.e., the industry incumbents), digital entrepreneurship revolves around creating new ventures focused on commercializing and exploiting the innovation.

1.4 General and Functional Managers

With the relentless pace and magnitude of business changes induced by digital technology, it is important that you reflect on the role of general and functional managers. A manager is a knowledge worker in a business or not-for-profit organization who is in charge of a team, a functional area (i.e., functional managers), a business unit, or an entire organization (i.e., a general manager). These

1 Tam, P.-W., Orwall, B., and Mathews, A. W. 2003, April 25. "With Apple stalling, Steve Jobs looks to digital entertainment." *Wall Street Journal*, retrieved from https://www.wsj.com/articles/SB10512259219370900.

2 Have a look at Steve Jobs's essay on the music industry and DRM: https://web.archive.org/web/20070207234839/http://www.apple.com/hotnews/thoughtsonmusic/.

3 von Briel, F., Recker, J. C., Selander, L., Hukal, P., Jarvenpaa, S., Yoo, Y., . . . and Fuerstenau, D. 2020. "Researching digital entrepreneurship: Current issues and suggestions for future directions." *Communications of the Association for Information Systems: CAIS* 48. https://doi.org/10.17705/1CAIS.04833.

individuals are typically trained in management schools and, particularly in larger organizations and entrepreneurial ventures, often hold a master's degree in management or business administration.

General and functional managers have, through schooling and on-the-job experience, developed strengths in business areas such as operations, marketing, finance, HR, accounting, and the like. Those of you who plan on a career in business consulting will develop a similar set of functional expertise. However, with the unrelenting pace of IT innovation and its widespread adoption in organizations, the personal success of managers and their areas of responsibility increasingly depend on these managers making optimal decisions about the use and management of information. Information is a critical organizational resource. As Thomas Watson Jr., the legendary chairman of IBM, recognized more than 50 years ago, "All the value of this company is in its people. If you burned down all our plants, and we just kept our people and our information files, we should soon be as strong as ever."[1]

To manage information effectively, the firm must adopt and use information systems and digital technology resources. Selecting, designing, and managing these resources are no longer exclusively the job of the technology professional but are now the shared responsibility of all general and functional managers. Bill Gates neatly captured this point when he stated, "Information technology and business are becoming inextricably interwoven. I don't think anybody can talk meaningfully about one without the talking about the other."[2] As a consequence, general and functional managers must work in partnership with information systems and technology professionals—those individuals with substantial technical training and education in management information systems, computer science, and engineering—when it comes to carrying out these decisions. *Partnership* is the key word here; it is no longer acceptable for general and functional managers to claim ignorance of IT and delegate all IS decisions to technologists. As Douglas Merrill put it when he was serving as the **chief information officer (CIO)** of Google, "The distinction between technology and business is antediluvian—it's gone."[3] The notion that the IT function needs to cater to its internal customer—"the business" (as if the IT function was not part of "the business"!)—no longer represents reality. So many managerial and organizational decisions depend heavily on information systems to be successfully implemented, and new information technology constantly opens up new possibilities for innovation in strategy, products, processes, and organizational design. The success of an organization in today's environment depends on the interplay of the information systems and digital technology with the other organizational resources (e.g., labor, brand, capital).

Let's look at examples of what happens when this partnership works well:

- Consider Zara, the world's largest apparel retailer. The root of its success can be traced back to the shared vision of Inditex founder Amancio Ortega Gaono and José Maria Castellano, a former IT manager. Both believed that computers were critical for building a responsive **supply chain**. The company profited from this unique blend of IT competencies and business visions for making "fast fashion" a reality—a sought-after chimera since the 1980s, consisting of the continuous design of affordable and fashion-inspired apparel. Zara gained the flexibility required to adapt its designs to the ever-changing preferences of customers by putting IT at the core of its store and supply chain coordination efforts. The 6,500 stores in 88 different countries

---

1 Quinn, S. B. 1992. *Intelligence enterprise: A knowledge and service based paradigm for industry.* New York, NY: Free Press: 244.

2 Gates, B., and Hemingway, C. 1999. *Business @ the speed of thought.* New York, NY: Warner Books.

3 Weiss, T. R. 2008, March 10. "Google CIO: Online success rests on talking with, not to, customers." *Computerworld*, retrieved from http://www.computerworld.com/article/2537599/internet/google-cio--online-success-rests-on-talking-with--not-to--customers.html.

receive new deliveries twice a week, and new products just take between two and three weeks to reach the store from the design table. For competitors, this generally takes months. The ability of the company to sense the market starts in stores, where sales data are generated and complemented with store managers' accounts of shoppers' preferences. Again, IT supports the supply chain coordination effort of the company to determine the best-performing products in each region, manufacture them, and deliver them as quickly as possible. Interestingly, Zara spends less than one-quarter of other brands in the industry.

- As it became clear that the Internet would be a viable channel for economic transactions, Michael Dell, founder and CEO of Dell, challenged his team to take advantage of the opportunity. The Internet offered Dell the chance to sell directly to the consumer (not just to business accounts) without having to compromise its direct model. In charge of the efforts was 30-year-old Scott Eckert, a recent master of business administration (MBA) graduate. Working closely with the information systems and technology group, Eckert developed Dell Online, the online store, and an application that allowed customers to configure their machines. As they say, the rest is history. Fueled by the staggering growth rates of the Internet itself, the online store generated one million dollars in revenue per day in six months, two million dollars per day in nine months, and three million dollars per day in a little more than a year since opening for business. Dell is now one of the largest computer makers in the world, and the Dell Online initiative was instrumental in helping the firm achieve this result.
- The Hotel Lugano Dante is a four-star hotel in Lugano, Switzerland, that is part of a small, two-property chain. While the Lugano Dante is a nice hotel, there is nothing intrinsically unique about it in terms of location, facilities, or any other structural characteristic. In the age of the Internet, where competition is only a click away, Carlo Fontana, the hotel's general manager, feared that his property could quickly become a commodity—an undifferentiated good that would be dragged into one price war after another. Believing in the power of differentiation through service, Fontana set out to enable his staff to create superior and highly personalized experiences for guests by providing attention to detail and catering to the unique needs of individual customers—far superior to the standard four-star service. Working closely with Davide Bernasconi, the director of IT, Fontana conceptualized and built a comprehensive system that provides accurate, real-time operational and guest data and disseminates it to each guest-facing employee. Staff members have personal iPads that allow them to produce and receive the necessary information to provide outstanding service. After a year of operation, the system has contributed to make the Lugano Dante the number-one hotel in its market based on customer reviews on Tripadvisor, with a 99% approval rate by customers that helps both differentiate the hotel and ensure above-average loyalty.

The preceding examples highlight a few critical lessons for prospective general and functional managers:

- The general and functional managers in the examples, while not IT experts, were making educated decisions about the deployment of information systems in their organizations.
- The objective of IT deployment was business driven. In other words, the new digital technology was brought in to serve the growing or changing needs of the organization. The people who perceived and understood the need for this new way of doing things were the general and functional managers.

- These general and functional managers were not **IT professionals** and had no specific training in technology, but they worked in close partnership with the IT professionals in their organizations to ensure the successful deployment of the needed information systems and technology functionalities.
- The general and functional managers did not use the new digital technologies firsthand once introduced. Rather, other employees, the **end users**, and the customers had direct contact with the hardware and software programs that had been introduced.

The last of the preceding points is important, as there is a clear difference between the roles of general and functional managers and end users.

1.5 General and Functional Managers versus End Users

End users are those individuals who have direct contact with software applications as they use them to carry out specific tasks. For example, I am an end user as I use word processing software (i.e., Microsoft Word) to write this book. You, as a student, are an end user as you engage in spreadsheet analyses using Microsoft Excel in your statistics, operations, or finance courses.

Most general and functional managers in organizations are also end users. They use software programs to improve their own personal productivity—from the very basic, such as email and calendars, to the most advanced, such as mobile management dashboards designed to keep a real-time pulse of the business's performance. Although being a sophisticated end user is an important asset for managers because effective use of software programs can lead to increased productivity, it is far more important for managers to have the skills and knowledge to make appropriate IS decisions at the organizational level.

It should now be clear why this book is not a training book on digital technologies and IT and why our focus is not on end-user skills. This book is expressly designed for current and future managers, for those individuals who have general management and/or functional management responsibility, and for those who serve, or will one day serve, on the board of directors of an organization. The premise of this book is that to be an effective manager, you need not know an inordinate amount of information about IT or how technology works, even though some of this IT-specific knowledge undoubtedly helps.

Rather, you really need to have and feel confident with a knowledge of information systems and the role that digital technology plays in them. (See Chapter 2 for definitions.) You need to know how to identify opportunities to use digital technology to your firm's advantage, how to plan for the use of information systems resources, and how to manage the design, development, selection, and implementation of information systems. These are the skills that separate effective managers (those who can be productive partners of the information systems function) from ineffective ones (those who delegate critical IT decision making and, more often than not, live to suffer the negative consequences of their decisions). In the words of Satya Nadella, Microsoft's CEO,

> Tech intensity comes down to three things: first, how you adopt the latest technology and integrate it into your own organization. Second, how you build your own unique digital capabilities. All organizations in all sectors of the economy will need a culture where all people are a part of the innovation process and are able to learn, test, and act on hypothesis. And then, finally, trust, both trust in technology, and trust in business model alignment.[1]

There should be no question in your mind that tech intensity requires savvy general and functional managers that understand the role of digital technology in their business.

1 Microsoft. 2019. "Satya Nadella at 2020 MS Ignite." https://videopress.com/embed/osQYkgA8.

1.6 Information Systems Professionals

One of the most enduring results of research in IS has been the degree of discomfort that executives exhibit when it comes to making decisions about information systems and digital technology. As a consequence, we often hear executives wonder why it isn't enough to hire "good" IT professionals and let them worry about all the IT stuff.

This stance immediately raises the question "How do you know if an IT professional is indeed good without knowing a minimum amount about what IT professionals do?" More important, "How can you establish a good partnership with your firm's IT group if you are not equipped to prove useful in the relationship?" As a simple example, imagine being the vice president of marketing for a retail chain (e.g., IKEA) that wants to be able to measure customer spending patterns precisely and rank shoppers based on the customer lifetime value they contribute to the firm. Business intelligence technology and techniques (Chapter 3) will enable this strategy. However, how can your IT group build the appropriate infrastructure, track all the relevant data, perform accurate analyses, and best segment the customer base unless you and your marketing team are intimately involved in the design and development of this technology and analytical infrastructure? The only recipe for success, in this case, is to have a productive partnership. You contribute the marketing-specific knowledge and make decisions about the critical requirements and capabilities of the initiative and the information systems. The information systems and technology professionals make decisions about platforms, interfaces, programming languages, and hardware performance. Here we detail some of the most common jobs you will find in any organizational IT function—the group of people who oversee the design, development, acquisition, implementation, and maintenance of the firm's IS resources (Table 1.1).

1.7 Fundamental IT Trends: The Staying Power of Moore's Law

Information systems—those organizational systems that enable the processing and management of an institution's information—are the backbone of organizational operations. These information systems are powered by more or less advanced digital technology at their core. Understanding the drivers and trends that shape the evolution of digital technology is important for you as a general or functional manager because new technologies constantly enable new strategies, new initiatives, and the effective management and use of greater and greater amounts of data and information.

The popularity and growing importance of IT-based information systems are self-evident in the 21st century. For those in need of further convincing, we only need to show statistics that capture the stunning rates at which IT has become, and continues to grow into, a critical tool for organizations, consumers, and individuals alike. For example, in 2019, the number of people with access to the Internet was estimated to have passed the four-billion mark—exceeding 53% of the world's actual population, many of them using smartphones to connect (Chapter 5).

Another source of evidence is the precision that has characterized Moore's law over the years since its announcement by Dr. Gordon Moore. In 1965, Dr. Moore, who three years later cofounded Intel Corporation, commented at a conference that if the current rate of improvement in the production process of transistor-based microprocessors continued, the number of transistors that could be etched on an integrated circuit (i.e., a microchip) would double every one to two years. What is remarkable about Moore's law is not so much how Dr. Moore arrived at it but rather how precisely it has held true over the last 50 years (Figure 1.4). While sustaining Moore's law is getting harder, researchers are looking into new ways for improving performance, including software and algorithm improvements, specialized chips and architectures, and whole new fields like quantum computing and biology-inspired neuromorphic computing. The bottom line: you can expect computational power to continue to increase.

Table 1.1. Information systems professionals

Position	Description
Executives	
Chief information officer (CIO)	The individual in charge of the information systems function. In organizations, the CIO reports directly to the CEO. They need to be well versed in business while not losing sight of the delivery of information services that enable the organization to operate effectively.
Chief digital officer (CDO)	The individual in charge of digital transformation and digital innovation in established organizations, he or she is the "transformer in chief."* Enlightened CIOs have historically covered this role. Firms that create the CDO role do so to free the CIO up to focus on technical and compliance issues (i.e., keeping the lights on) while the CDO can push digital innovation and change.
Chief data officer (CDO)	The individual who oversees all aspects of data use in an organization, from collection to compliance to value extraction. The emergence of CDOs is a testament to the strategic importance of data in organizations today as well as the increasing need to manage compliance with privacy and data management regulations.
Chief information security officer (CISO)	The individual in charge of digital risk management and cybersecurity. As the digital transformation of existing firms continues, cybersecurity threats and digital risk require executive-level oversight.
Technical Staff	
Architect	An architect is an individual in charge of developing a framework for the development of a system. An architect has a strong technical background, excellent technical skills, and significant experience with the design and development of IT systems.
Developer	A developer is an individual who builds high-quality, innovative, and performing software that complies with coding standards, technical designs, and the framework provided by architects.
Administrator	An administrator is an individual who is charged with the day-to-day maintenance of a system or collection of systems. *Database administrators* maintain data repositories; *network administrators* support data transmission systems; *system administrators* oversee software programs including operating systems and applications.
Analysts and Managerial Staff	
Analyst	*Analyst* is a general term referring to an individual who performs analysis in a specific field or topic area. ***Business analysts*** study the organization's business processes to identify opportunities for improvement. *System analysts* work directly with users to define the functionalities of new systems being designed. Their primary objective is to gather user requirements, validate them, and codify them in a precise enough manner to enable developers to write code to implement those functionalities.
Project manager	A project manager is the individual ultimately responsible for delivering an information systems project on time, within budget and scope. A good project manager has a blend of managerial skills (e.g., team leadership) and business skills (e.g., budgeting) and a good understanding of the technologies used in the project.
Information systems manager	An IS manager is an individual in charge of a team within the IT function of an organization. Teams can be in technical areas, like developers, or administrative areas, like the outsourcing team that manages the relationship with vendors or administrative teams planning and coordinating the activities of the IT function.
Data Science and Analytics	
Data scientists	A data scientist is an individual in charge of analytics efforts who has the full view of the analytics process end-to-end. Beyond technical skills, they have strong business acumen and communication abilities.
Data engineers	A data engineer is an individual who has distinctive skills in data access and data preparation
Subject matter experts	A subject matter expert is a general or functional manager with a deep understanding of the business and functional domain of analysis.
Source system expert	A source system expert is an individual with a deep understanding of the technology underpinning the business domain of analysis. They understand which business processes created the data and how the data are stored.
Other	
IT consultant	An IT consultant is an individual assuming roles like the ones described in this table. However, unlike the in-house IT function, consultants move from client to client as they take on different projects. Moreover, they expend substantial efforts in selling projects to potential clients.

* https://www.mckinsey.com/business-functions/organization/our-insights/transformer-in-chief-the-new-chief-digital-officer.

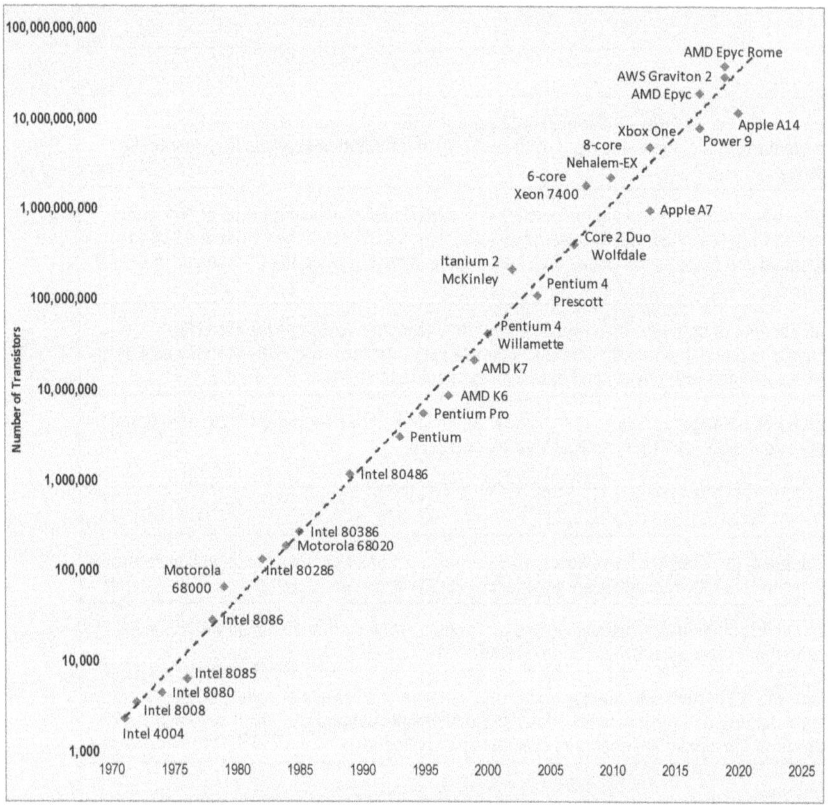

Figure 1.4. Number of transistors in different generations of commercial microprocessors
Source: Data from Wikipedia, https://en.wikipedia.org/wiki/Transistor_count

The unrelenting pace of performance improvement in microchip design and production processes has a number of important managerial implications, which we discuss in detail below. At a more general level, it is this unflagging progress that enables continued innovation and the reinvention of processes, strategies, products, and services. It is critical to recognize, however, that the responsibility for taking advantage of this continuous technological progress in organizations falls squarely on the shoulders of general and functional managers.

Processing Power and Memory Have Increased

As stated by Moore's law, the processing power of microprocessors has experienced exponential growth. Based on our early discussion of the digital representation of data and its processing, you understand the relationship between the number of transistors in a microchip and its power. Because a transistor is the basic unit of computational ability of a microprocessor, more transistors equate to greater computational power in any device that uses a microchip—your personal computer (PC), of course, but also your smartphone, your digital camera, your car, and even your microwave and your refrigerator. Not only has the programmable computational capacity available to us increased exponentially over the last two decades, but it also has proliferated to a multitude of devices, from gaming consoles to mobile devices.

The storage capacity of memory chips has also increased exponentially. With transistors also serving as the basic component of memory chips, a higher density of transistors makes for increased memory capacity and performance, thus providing the necessary complement to increasing the computational ability of microprocessors (i.e., solid-state device, random access memory). A similar pattern of evolution has occurred with respect to secondary storage capacity (i.e., hard disks, tapes), which has grown exponentially, shifting decisively to the digital format (Figure 1.6).

Costs of Computing Power Have Declined

The cost of computing power and storage has declined at astounding rates as Moore's law has shown its effects. By some estimates, all the power of a mainframe that cost more than $10 million in 1965 had been incorporated into a $7 chip by 2002.[1] Optimizations and specialization of

1 McFarlan, F. W. 2003. "The new strategic weapon: Information technology." Faculty Seminar Series, Boston, MA: Harvard Business School Press. CD-ROM.

processing systems have continuously sustained this trend. In other words, since the 1990s, the price-performance ratio has grown by an order of magnitude every four years. It is this amazing combination of power and affordability that has led to the spreading of "intelligence"—here defined as computational ability—in products ranging from the more advanced (e.g., smart cameras that can recognize human faces) to the more mundane (e.g., a hotel minibar that can recognize when items are consumed).

While the number of transistors on microchips has steadily increased, chip manufacturers have been able to contain energy requirements through innovative designs. Following Moore's law, the energy efficiency of computation improved more than 10 billion times from the mid-1940s to the year 2000. Today, processors' energy efficiency is still improving relentlessly, doubling approximately every three to four years (Figure 1.5).[1]

Figure 1.5. Performance, expressed in GFLOPS (gigaFLOPS, or billion floating-point operations per second), per watt

Source: Sun, Y., Agostini, N. B., Dong, S., and Kaeli, D., 2020, "Summarizing CPU and GPU design trends with product data," *ArXiv:1911.11313 [cs.DC]*, http://arxiv.org/abs/1911.11313

This trend, coupled with continued improvements in battery technology, has enabled the development of countless new portable devices based on digital computer architecture. Familiar examples are iPads and other tablets, smartphones, smart watches, Fitbits, and other wearable devices. In parallel, we have seen the computerization of products that exhibited little or no intelligence before, such as cars and appliances, and the transformation of plastic cards and labels into smart cards and radio-frequency identification (RFID) tags. This trend, sometimes called *pervasive computing* or the *Internet of things*, is poised to spur significant innovation in the years to come (Chapter 12).

Computers Have Become Easier to Use

One remarkable characteristic of the evolution of digital technology is that as computers become more powerful and more internally complex, they also become easier to use. This is an interesting side effect of Moore's law and one of the most intriguing characteristics of digital technology. Because software is extremely malleable, as computers become more powerful and are able to process more operations in a unit of time, they can take on more work on behalf of users. If you had the (sometimes painful) experience of using computers before the arrival of graphical user interfaces, then you know how difficult it can be to memorize a command (with perfectly correct syntax!) in order to have the computer perform a task. For example, in MS-DOS, in order to copy a file from the CD-ROM to the hard disk, you had to issue a command with the following structure:

```
COPY [/Y|-Y] [/A][/B] [d:][path]filename [/A][/B] [c:][path][filename] [/V]
```

Failing to remember the appropriate syntax would yield a cryptic error message and, typically, few hints as to how to go about fixing the error. Conversely, when using graphical user interfaces, copying a file from one location to another is as simple as identifying the right icon and dragging it to the target location. Voice-recognition interfaces like Siri or Cortana are also becoming more

1 Sun, Y., Agostini, N. B., Dong, S., and Kaeli, D. 2020. "Summarizing CPU and GPU design trends with product data." *ArXiv:1911.11313 [cs.DC]*, retrieved from http://arxiv.org/abs/1911.11313.

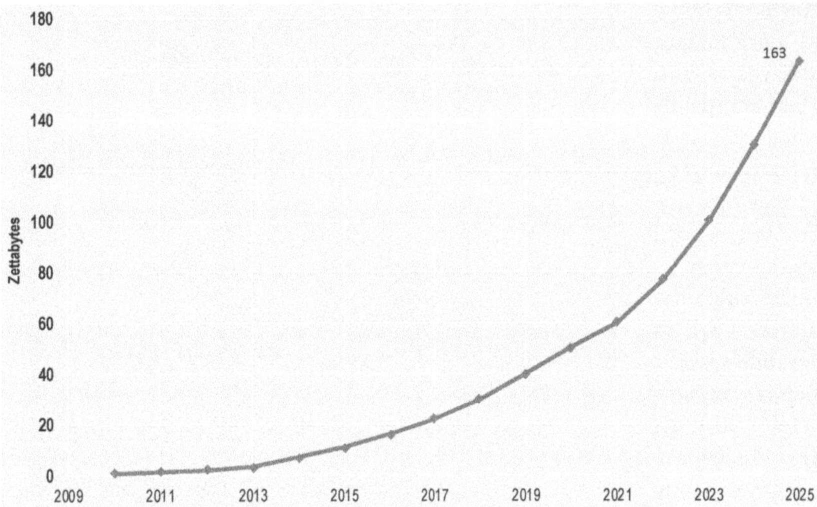

Figure 1.6. Estimated worldwide storage capacity per capita in zettabytes

Source: IDC, 2017, ***Data Age 2025***, retrieved from https://www.import.io/wp-content/uploads/2017/04/Seagate-WP-DataAge2025-March-2017.pdf

common, and we can increasingly issue commands to a computer by simply telling it what to do or, as with multitouch interfaces, by manipulating objects to convey our intentions to the machine.

The simplicity of user interfaces is just one example of how digital technology, as it becomes more sophisticated and complex, becomes easier to use. Another example is offered by the many software-enhanced objects we come into contact with regularly. Modern cars are quite difficult to service and quite powerful in terms of the number of software instructions that their controllers and control units can process. However, while all this software and hardware makes the car more internally complex, automobiles are easier and safer than ever to drive, with features like lane navigation, night-vision systems, rearview cameras, automatic brakes that activate if you get too close to the vehicle in front of you, and alert systems that can detect if you fall asleep at the wheel.

As more powerful computers can process more sophisticated software code, they are able to interact with humans in ways that are closer to our natural behavior—such as the interpretation of visual cues and the use of voice commands (e.g., Apple's Siri, Google Now, Amazon's Alexa, or Microsoft's Cortana)—and support us in unprecedented ways in an incredibly varied array of tasks. It is not a coincidence that senior citizens represent a sizable, and growing, portion of the online population. Note that these are individuals, age 65 and older, who often claim to be unable to program a cable box or TV recorder, yet they are connected to the Internet; emailing, chatting, and videoconferencing with their grandkids; sharing pictures; and organizing trips and cruises. Why? Because computing devices offer the things they want (e.g., keeping in touch with family), are affordable, and shelter them from much of the complexity of the machine's inner workings.

1.8 Other IT Trends of Managerial Interest

Beyond the lasting effects of Moore's law, there are other IT trends that are critical for general and functional managers to recognize.

Declining Storage Costs

The cost of computer storage has plummeted at ever-increasing speeds. It took 35 years from the shipping of the first hard disk by IBM to the introduction of the 1 gigabyte (GB) hard disk. It took 14 years from there to reach 500 GB and only two more (in 2007) for the 1 terabyte (TB) hard disk to be introduced by Hitachi. By some estimates, 1 GB of storage costs more than $600 in 1993, about $5 in 2003, $0.39 in 2007, and reached a cost of $0.02 in 2017 (Figure 1.7).

While the exact numbers are of secondary relevance, as a manager, you should realize that this trend has enabled the emergence of a whole host of strategic initiatives predicated on the collection and analysis of significant amounts of data. Consider the example of Kayak .com, the travel search engine that has the ability to make suggestions about *future* airline prices. While many sites can tell you what the lowest fare is today for a Paris to New York flight, Kayak.com can tell you whether you should purchase today or wait for fares to decrease. How do they do it? They use a proprietary forecasting model based on the analysis of years of multiple daily searches for airline tickets and hotel reservations to identify patterns of fare and hotel room rate changes. Cheap storage is a critical enabler for Kayak.com's service.

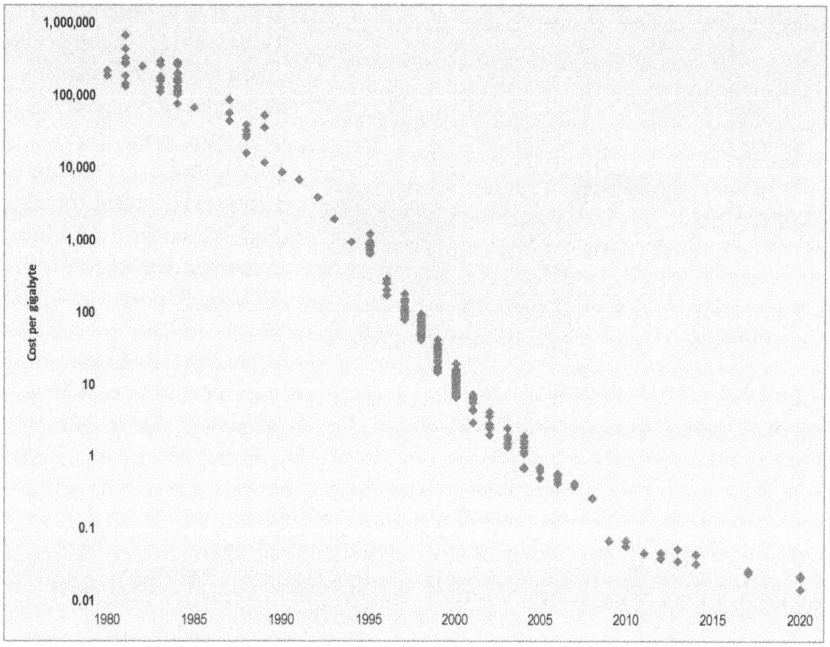

Figure 1.7. The decreasing cost per GB

Source: Adapted data (2017, December) from mkomo.com (2014, March 9), ***A history of storage cost: Update***, http://www.mkomo.com/cost-per-gigabyte-update

Ubiquitous Network Access

Ubiquitous networks are now a reality. In the early 1990s, the Internet, and its most visible services, such as the **World Wide Web** and electronic mail, took the world by storm. "The Internet changes everything" was the rallying cry that spurred the emergence of new business models and sent billions of dollars through the hands of venture capitalists and into those of high-tech entrepreneurs all over the world (Chapter 5).

The *dot-com era*, as it is sometimes called, has now come and gone. What remains is the Internet: a global network of networks relying on distributed ownership and openly available standards and communication protocols. Access to the Internet is increasingly pervasive, with wireless and cellular network access having eliminated the final obstacle to "always on" connections. We are approaching a world in which we won't wonder whether we'll have access to a high-speed Internet connection when we travel; we will assume so—just like we do today with access to electricity in many parts of the world.

Not only is network access becoming ubiquitous, but the costs of data transmission are becoming negligible. The convergence of these trends is spurring an amazing array of initiatives. Consider the recent trends of enabling cell phones to receive notifications when your bus is approaching your stop (no more waiting in the cold!) or enabling apps to pay bus fares and vending machines through NFC (no more being thirsty due to a lack of correct change!).[1] Some digital entrepreneurs, along with established players engaged in digital innovation, are positioning themselves to capitalize on ubiquitous network access.

Ubiquitous Computing and Digital Data Genesis

As IT costs and storage costs have plummeted, and as global data networks have become widely available, more devices are now intelligent (i.e., able to store and process data) and connected

1 Near-field communication (NFC) refers to a set of technologies enabling close-range data exchange between two devices. Contactless payment systems are available for Android and iOS smartphones.

Figure 1.8. Smart appliances

Source: Photo by LG Electronics / CC BY 2.0

(i.e., able to communicate with other devices), giving birth to IoT (Chapter 12). The smart home is now a reality, with smart appliances like a refrigerator that can tell you what's inside, alert you when food is going bad, suggest recipes that use ingredients you have on hand, and even order food (Figure 1.8); smart heating systems that you can control over the Internet from your office; and bathtubs that will allow you to start running a bath before you get home. For the aging population, the availability of cost-effective sensors enables people to remain independent in their homes longer than ever before.

Contemporary automobiles often pack as much computing power as your PC, and through satellite networks and telemetry technology, some can self-diagnose and email you alerts about preventive maintenance and potential trouble you may experience down the road. Moreover, the presence of intelligent devices enabling or monitoring events and human activity implies that data are generated by those activities in digital form. New devices like robots, drones, and autonomous vehicles increasingly generate new streams of data. This digital data genesis dramatically reduces the barriers to data collection and analysis. While there are serious implications to be considered—for example, privacy (Chapter 13)—digital data genesis enables further innovation (Chapter 12).

The trends discussed in this chapter, summarized in Figure 1.9, have had a few direct and indirect effects. First, computing devices are smaller and increasingly embedded in everyday products (Figure 1.10). As a consequence, digital content and services have increased dramatically and will continue to increase over time, along with the convergence of devices. Chances are you already carry a smartphone that doubles as a planner, a digital still camera, a digital video camera, an email client, a music player, a GPS, a compass, a personal assistant, and who knows what else.

Second, the rapid proliferation of easy-to-use computing equipment has spurred more digitization—the method by which content and processes become expressed and performed in digital form. Digital data genesis is a clear example of digitization, as is this book if you are reading it on an e-reader or listening to it in the form of a podcast. Finally, as computing devices become increasingly easier to use and more interconnected, they become pervasive. These trends have important implications for the manager of the future.

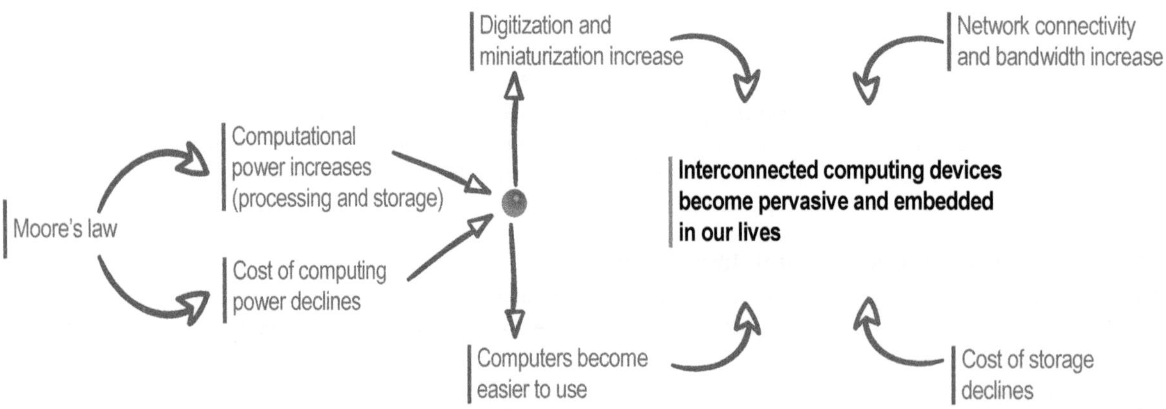

Figure 1.9. Interconnected trends

1.9 Managerial Takeaways

The discussion in the previous section depicted the world in which you, as a general and functional manager, will operate. You will increasingly be challenged to devise strategies and implement processes that enable your organization to take advantage of these trends and better serve your increasingly IT-savvy customers.

Figure 1.10. A stark example of miniaturization

But perhaps the most important implication of the pervasiveness of computing for today's management is to be found in the sheer size of organizational investments in digital technology, software, and services. The ever-increasing amount of money being spent on IT is largely due to one of the most interesting characteristics of software. Software is extremely malleable and can be molded into almost anything. Through the infinite combinations in which software code can be written by talented engineers and programmers, digital devices can morph into an infinite number of applications, many still left to invent! Vinton Cerf, credited as one of the "fathers of the Internet," said in a 2007 interview,

> Specifically for young people pursuing careers, software is an endless frontier. There isn't any limit to what you can do. I can confidently argue that maybe only 1% or 2% of the possible applications have already been implemented or thought of.[1]

And considering the proliferation of smart devices (Chapter 12), you may consider that number conservative. Cerf said, more recently,

> I think we are going to be surrounded by smart devices. There's something really magic, to be able to assume, that any device you have that has some programmability in it could be part of a communications network, and could communicate with any other random, programmable device. When you assume that's the norm, you have almost no limit to the kinds of ideas you can come up with.[2]

As a consequence, the number of opportunities for innovation and business success that depend on the appropriate use of digital technology is literally skyrocketing. If you think for a moment about the different types of software you encounter daily, you'll see that they support an almost endless variety of tasks.

For example, we use email, instant messaging, and Voice over IP (VoIP) applications such as WhatsApp to communicate with our friends, family, and colleagues. With these applications, digital technology acts as a communication device, replacing our traditional telephone or penned letters. We listen to streaming radio and watch videos on our PCs or mobile phones; we play video games and organize our pictures using Facebook or Google Photos. With these applications, our digital technology behaves like an entertainment device, replacing our radio, stereo, television, and scrapbooks. We use word processing, spreadsheets, and presentation software to carry out our work. With these applications, our digital technology takes on the role of a productivity tool for the office, replacing typewriters, calculators, and flip charts. Finally, and most important for you as a manager, digital technology can embed "intelligence" and decision-making capabilities previously required of employees or organizational functions. For example, check-in kiosks used by airlines and hotels replace the work of frontline employees, forecasting

1 King, J. 2007, July 30. "The grill: Google's Internet evangelist Vint Cerf on the hot seat." *Computerworld*, retrieved from http://www.computerworld.com/article/2552965/networking/the-grill--google-s-internet-evangelist-vint-cerf-on-the-hot-seat.html.
2 Park, B. 2014. "Wisdom from the father of the Internet—Vint Cerf." Startup Grind, retrieved from https://www.startupgrind.com/blog/wisdom-from-the-father-of-the-internet-vint-cerf/.

software and automatic inventory reorder software used in retail embed in algorithms some of the intelligence and decision-making abilities of skilled inventory managers, the electric grid is increasingly capable of automatically sensing and adjusting power consumption, and automated sprinkler systems in greenhouses monitor and analyze the state of the soil and decide when to turn on or off and even what and how much fertilizer to add to the water.

Beyond replacing earlier ways of doing business through digital transformation, a constant stream of new possibilities emerges through digital innovation and digital entrepreneurship. There is no better time in history to join the workforce as a digitally savvy manager.

Summary

This chapter laid the groundwork for this book by identifying the managerial and technology trends that make it imperative for managers to get involved in decision making pertaining to information systems and information technology. In this chapter, we learned the following:

- The development of digital technologies, their pervasiveness, and the characteristics of digital information—replicable at zero cost and with perfect precision, easily manipulated and compressed—are the enabling conditions of recent phenomena, including digital transformation, innovation, and entrepreneurship.
- General and functional managers, those individuals in organizations who have the responsibility to lead a functional area or a business, can no longer abdicate their right, and duty, to be involved in IS and IT decisions. They should act in partnership with the firm's information systems and technology professionals.
- The IS skill set required of managers pertains to decisions about identifying opportunities to use IT to the firm's advantage, planning for the use of information systems resources, and managing the design, development, selection, and implementation of information systems. While end-user skills (i.e., the ability to use computers proficiently) are an important asset for any knowledge worker, the critical skills for managers relate to the organizational, not personal, uses of information technology.
- CIOs, the leading figures in the information systems and technology function, are increasingly being selected from the functional and managerial ranks rather than from the technology ranks.
- Managers must increasingly partner with IS professionals to realize the full potential of the emerging IT-enabled value creation opportunities stemming from digital transformation.
- The enduring effects of Moore's law have led to increasingly powerful yet cheaper computing strength, declining costs of computer memory, and a dramatic improvement in the ease and breadth of the use of digital devices. Moreover, increasingly available network connectivity and storage capacity, improved battery life for portable devices, and the proliferation of intelligent devices have contributed to dramatically change the business and social landscape.

Study Questions

1. Explain the difference between the following terms: *digital transformation*, *digital technology*, *digital innovation*, and *digital entrepreneurship*.
2. Define the following terms: *general manager*, *functional manager*, *CIO*, and *end user*. Explain how these roles differ and the skill set each role requires.
3. Explain why it is critical to the success of firms that general and functional managers be directly involved in information systems and technology decision making.
4. Explain why firms are increasingly selecting CIOs from the managerial ranks rather than from the technology ranks.
5. Describe Moore's law and its direct and indirect effects on organizations.

Glossary

A glossary of key terms can be found at the book's website. Visit https://www.prospectpressvt.com/textbooks/piccoli-information-systems-for-managers-5-0, then click on "Student Resources" in the horizontal red menu bar.

Chapter 2

Information Systems Defined

What You Will Learn in This Chapter

This is one of the most important chapters of this book because it defines the foundational concepts that we will use throughout. We will define key terms, such as ***information system (IS)*** and ***information technology (IT)*** and will discuss *information system success* and *information system failure*.

MINICASE: Digital Innovation at McDonald's

As you get settled in your office for another day of work at WizConsult, the consulting firm you have recently joined, you sip your morning coffee and start your computer. Surprisingly for this early hour, your boss walks in and asks you to follow him immediately. You take one last sip of your coffee and go. Your boss is on the phone with a large McDonald's franchisee who operates 42 restaurants in the upstate New York area. They quickly bring you up to speed: McDonald's has been quietly evaluating whether to join the iPad menu bandwagon. The corporate office is investigating the option of rolling out iPad-based menus that would allow patrons to walk in and immediately sit at a table and then order using the device available at the table. They could place orders, customize them to their tastes, and then submit them. When their numbers are called, guests could then pay for and pick up their orders as they normally would.

Such iPad menus are a staple of popular restaurant chains: "iPads and their ilk are now a common sight on tabletops at numerous restaurants, from fast-casual burger chains to pillars of the casual-dining sector like Olive Garden and Chili's."[1] The approach that McDonald's is considering is very similar to the one introduced by Delta Air Lines in its terminals at JFK and Minneapolis airports. Coincidentally, you were at JFK last week (Figure 2.1)! At these restaurants, patrons sit down and use the devices to order directly from a graphically pleasing and interactive menu. As they wait for their food to be delivered, they can play games, check flight information, read the news, and surf the web. As Rick Blatstein, the chief executive officer (CEO) of OTG, the management company behind the concept, put it, "We are giving travelers a one-of-a-kind experience. . . . By combining cuisine with innovative seating and ordering technology, we are offering a truly unique airport travel experience."[2]

Your client explains that McDonald's doesn't expect to earn money initially from this service. He quotes a conversation he

Figure 2.1. iPads available for ordering and surfing in concourse G at the Minneapolis airport

Source: Photo by Andrea Pokrzywinski / CC BY 2.0

1 Whitney F., W. 2017. "Why tablets on restaurant tables are here to stay major chains can't get enough of the sales-boosting devices." Eater, retrieved from https://www.eater.com/2017/10/5/16428750/tablet-technology-restaurants-applebees-outback-steakhouse.

2 Robbins, K. 2010. "At JFK airport eateries, iPad will be your server." Delish, retrieved from http://www.delish.com/food/recalls-reviews/at-jfk-airport-eateries-ipad-will-be-your-server/.

had with Ron Jonson, president of McDonald's Northeast Division: "What we're banking on is that more customers will visit McDonald's. Moreover, your customers may stay longer as they can surf the Internet and read the news right on the iPad and therefore increase their consumption." The program seems to be gaining support at corporate, but your client has heard mixed feedback on the idea from his fellow franchisees attending a recent conference. He does not want to miss out on making extra revenue, as every bit helps. However, he does not want to waste money on the latest high-tech gizmo just for the sake of staying on trend. As a knowledgeable and enthusiastic early adopter of technology, you are now on the hot seat as your boss and the franchisee turn to you for a recommendation.

Discussion Questions

1. Drawing on your own experience at JFK airport last week, do you think that iPad menus will work well at McDonald's restaurants? Justify your answer.
2. Given your answer to Question 1, can you see exceptions? Or do you believe your answer applies to all McDonald's restaurants?
3. What do you suggest WizConsult's client should do tomorrow?

2.1 Introduction

Despite the well-documented challenges associated with achieving a satisfactory return on the IT investments, organizations around the world continue to spend significant amounts of money on digital technology, lured by its promise to yield efficiencies and improved competitive positioning. With technology spending on the rise, there is little doubt that being able to wring value from these investments is critical for organizations in the digital age (Figure 2.2). However, a narrow focus on IT investments alone is problematic. Instead, you should focus on IS and their design. To do so, we must first cover some important background information and introduce some key definitions.

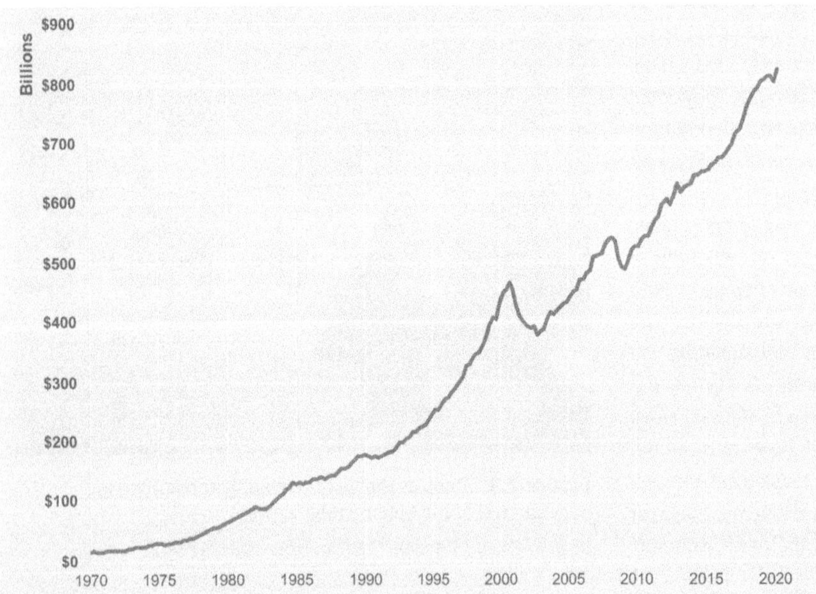

Figure 2.2. The rising investment in digital technology

Source: Data from Federal Reserve Economic Data, ***Private fixed investment in information processing equipment and software***, Q2 2020, A679RC1Q027SBEA, available at https://fred.stlouisfed.org/graph/?g=wpiT

2.2 Information Systems: Definition

In order to refocus our attention from a narrow IT investment perspective to a more comprehensive IS design perspective, we need to first define what an information system is and how it differs from IT.

IS, Not IT

Without a doubt, digital technology engenders a plentitude of confusing lingo, technical terms, and acronyms—a problem compounded by the many half-prepared, fast-talking individuals using terminology incorrectly.

Technical terms often turn into buzzwords once they are popularized. You have surely heard about Big Data, artificial intelligence, machine learning, cryptocurrencies, blockchain, augmented and mixed reality, quantum computing, and the Internet of things (IoT). Two quick examples will drive this point home:

- *Drones everywhere.* Strictly speaking, in computing and mechanical engineering, a *drone* is an unmanned aerial vehicle (UAV). Its defining characteristic is the use of machine learning to enable autopilot features. A drone without the ability to autopilot is not a drone (Figure 2.3); it is a radio-controlled aircraft. However, with the explosion of interest in quadcopters for recreational and commercial uses, the *drone* label is quickly losing its original meaning.
- *Holograms in the box.* Strictly speaking, a *hologram* is a three-dimensional rendering of an image that humans can perceive with the naked eye, without the use of special glasses or any other apparatus. With the increasing interest in augmented reality (AR), more vendors are referring to superimposed objects viewable by users through their headsets as holograms. Most recently, Microsoft even launched a product, HoloLens (Figure 2.4), that bakes this misguided use of terminology right into the product name.

Figure 2.3. A U.S. military drone

Source: Photo by Marion Doss / CC BY-SA 2.0

Figure 2.4. Microsoft HoloLens headset

Source: Photo by Microsoft Sweden / CC BY 2.0

Figure 2.5. The Ricasoli estate: The Castle of Brolio

Source: Photo by Darold Massaro

Of all the potentially confusing terms, none is more insidious than the term *information system*, usually abbreviated as IS. In this context, *information system* is often used as a rough synonym for *information technology*. But there is a critical difference between IT and IS!

Consider a simple example. The famous Ricasoli Winery is the oldest family-owned winery in Italy, producing wine in the heart of the Chianti region since obtaining ownership of the Brolio Castle (Figure 2.5) in the year 1141. It is to one of the members of the Ricasoli family, in fact—the baron Bettino Ricasoli—that we owe the formula of the "sublime wine," the Chianti. Bettino Ricasoli perfected the blend in 1872 after years of studies and experiments at a time when his estate was exporting wine all over Italy and beyond. Did the Ricasoli estate have an information system when the baron perfected the Chianti recipe?

The answer is yes, of course. The Ricasoli Winery's information system allowed the firm to take orders, track payments, organize activities around the farm, manage its inventory of aging wines, and help the baron collect information about the blends and treatments that enabled the wine to maintain its organoleptic characteristics when shipped far from the Brolio Castle. Using books and ledgers, the information technology of the time, the estate was able to keep track of all data and information necessary to its proper operations and longevity.

Yet the first known implementation of digital computing did not occur until World War II, and digital technology did not begin to enter business organizations until the postwar years. However, all kinds of organizations—from car manufacturers to laundry services, from banks to soft-drink makers—had been conducting business for decades (in some cases, as the example of the Ricasoli estate shows, even centuries!). Clearly, while IT is a fundamental component of any *modern* information system, we can see from these examples that IT (whether digital or not) and IS are separate concepts.

Information Systems as Sociotechnical Systems

Information systems are formal, sociotechnical, organizational systems designed to collect, process, store, and distribute information. Within this book, we primarily concern ourselves with formal organizational information systems—those that are sanctioned by a company, not-for-profit endeavor, or government entity. As students, you are likely very familiar with various *informal* information systems. For example, when you and your friends in this class use GroupMe, Snapchat, or WhatsApp to discuss topics or exams or to exchange course material, you are creating an informal information system (Figure 2.6). Such informal systems are beyond the scope of this book.

The key aspect of our definition is the notion of a sociotechnical system. Sociotechnical theory has a long tradition of research dating back to work done at the Tavistock Institute in London, England, in the early 1950s.[1] Sociotechnical theory questioned overly optimistic predictions about the potential benefits of new technology and suggested that the impact of new technologies on work systems was not a direct one but depended on the interplay of technology with other aspects, or components, of the work system.

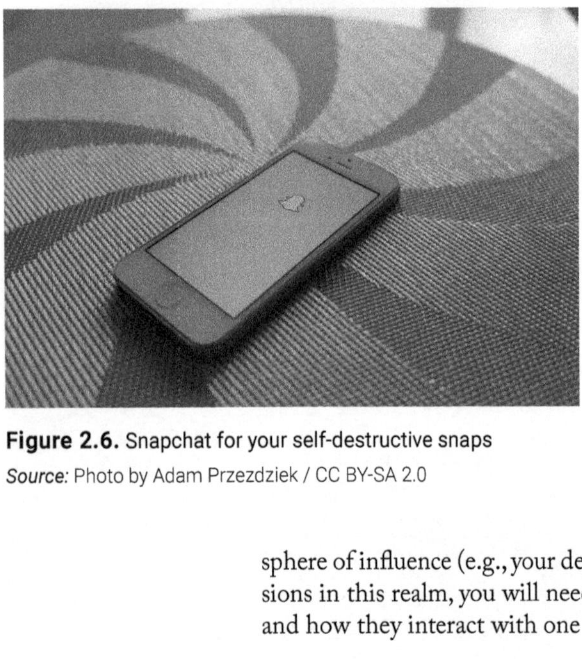

Figure 2.6. Snapchat for your self-destructive snaps
Source: Photo by Adam Przezdziek / CC BY-SA 2.0

The Four Components of an Information System

While sociotechnical theory is general in nature and applicable to any work system, it has been successfully applied to IT-based information systems. Specifically, we can represent any formal organizational IS as having four fundamental components that must work together to deliver the information processing functionalities that the firm requires to fulfill its information needs.

The four components of an IT-based information system are IT, people, processes, and structure (see Figure 2.7). They can be grouped into two subsystems: the technical subsystem and the social subsystem. The technical subsystem, comprising IT and processes, is that portion of the information system that does not include human elements. The social subsystem, comprising people and people in relation to one another (i.e., structure), represents the human element of the IS.

As a general and functional manager, you will be called on to make information systems decisions as they impact your sphere of influence (e.g., your department or functional area). In order to make appropriate decisions in this realm, you will need to have a solid understanding of each of the four components and how they interact with one another.

Information Technology IT is defined here as hardware, software, and telecommunication equipment. While we use a general label, *IT*, it is clear that today the technology element of information systems is almost exclusively digital technology. The IT component is a cornerstone

1 Trist, E. L., and Bamforth, K. W. 1951. "Some social and psychological consequences of the longwall method of coal-getting: An examination of the psychological situation and defences of a work group in relation to the social structure and technological content of the work system." *Human Relations* 4(1): 3–38. doi: 10.1177/001872675100400101.

of any modern IS, enabling and constraining action through rules of operation that stem from its design. For example, if you choose to collect and analyze data using spreadsheet software such as Microsoft Excel rather than using a relational database management system such as MySQL or Microsoft SQL Server (Figure 2.8), you are limited by the design of the spreadsheet software. Microsoft Excel cannot create meaningful associations between separate data elements. The result is substantial duplication of data, leading to redundancy, inconsistencies, and inefficient data management. This is because the design of Microsoft Excel is focused on computations and formulaic calculations, not on efficient data management.

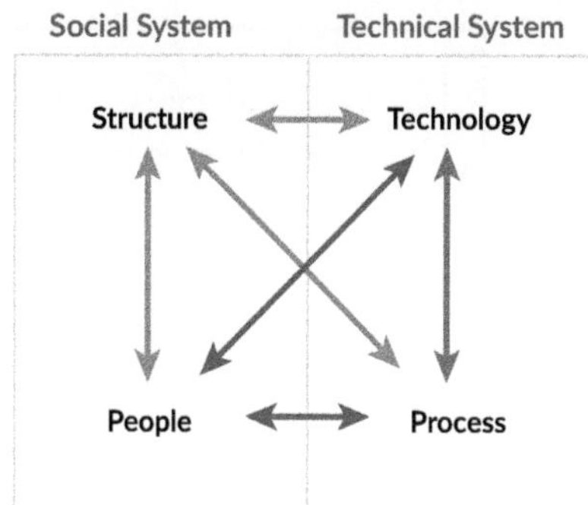

Figure 2.7. Information system components

In addition, it is important to remember that software design, still as much an art as a science, is driven in large part by the choices and opinions of the developers and programmers who create it. InfoWorld columnist Bob Lewis put it well when he said, "Every piece of software is an opinion." Software, particularly custom-developed applications, represents the developers' viewpoints on how the data should be represented, organized, and manipulated.

To relate to Lewis's quote, you need only think about the last time, on the phone with a customer service representative, you heard the phrase "The system won't allow that." In such a circumstance, the software design team, knowingly or unknowingly, restricted the functionality of the software to enforce a given behavior. Note, however, that this ability to enforce rules through software is a double-edged sword. For example, some car rental companies do not allow a rental contract to print unless the customer provides two distinct phone numbers. The design of the software seeks to ensure that the company can easily contact the renter and that valuable customer contact data can be tracked. However, those renters who do not have (or do not want to provide) more than one number and rushed employees trying to move quickly through the queue of waiting customers find it necessary to produce "phantom" phone numbers—thus defeating the very purpose of creating the restriction in the software.

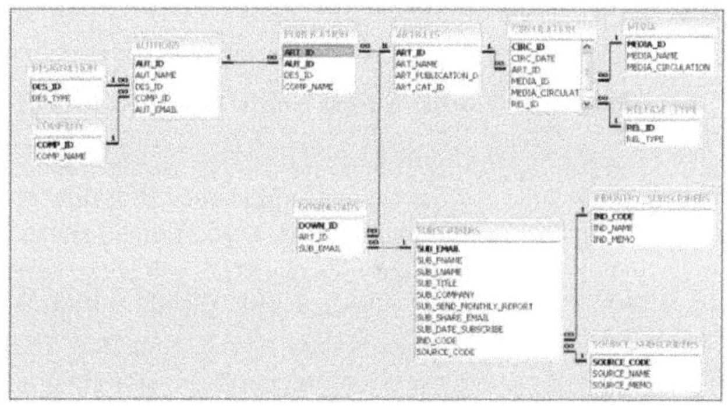

Figure 2.8. A relational database management system enables connections among data elements

Process The **process** component of an information system is defined here as the series of steps necessary to complete a business activity. Consider the job of a small, family-owned grocery store manager and the process he engages in when restocking inventory. The store manager must (1) check the inventory and identify the needed items; (2) call individual suppliers for quotes and delivery dates; (3) compare the various quotes; (4) select one or more suppliers for each of the needed items based on the terms of the agreement (e.g., availability, quality, delivery); (5) call these suppliers and place the orders; (6) receive the goods upon delivery, checking the accuracy and quality of the shipped items; and (7) pay the suppliers. Business processes can become very complex, spanning multiple individuals or organizational entities. For this reason, organizations resort to graphical aids such as business process maps when designing or evaluating processes (Figure 2.9).

Note that the same activity may be performed using a variety of different business processes. Note as well that gaps can exist between the official business process that forms the basis of training programs and customer service protocols and the informal ways in which these processes

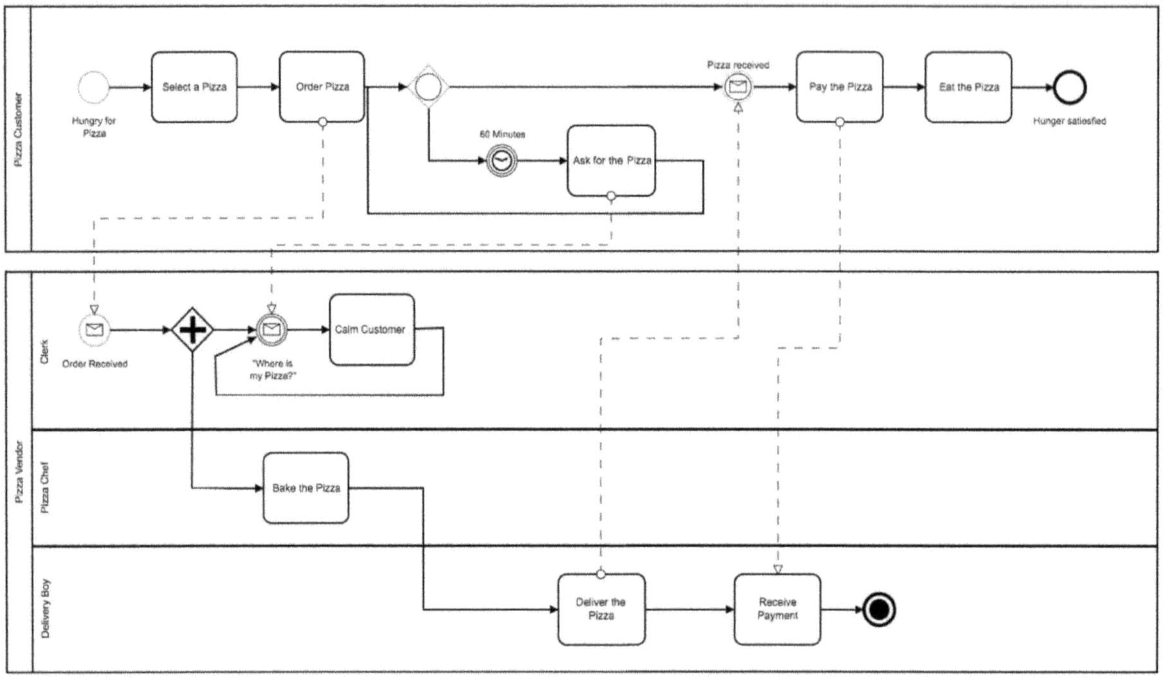

Figure 2.9. Business process map for the pizza-ordering process

are actually performed. This discrepancy is due to the fact that while many business processes are codified in procedure manuals and training materials, they are ultimately enacted by people. Consider again the case of the rental car company mentioned previously. While the stated business process calls for the collection of two separate customer phone numbers, you can imagine an employee—who sees a long line and is attending to a customer who is unable or unwilling to provide two numbers—typing in a fictitious number rather than wasting time arguing.

This potential discrepancy between the business processes as designed by the organization and the manner in which they are actually enacted is often the root cause of IS failure. When designing a new IS or when confronted with IS failure, it helps to think about what possible obstacles exist that may make it difficult for employees to accurately follow the business process.

People The people component refers to those individuals or groups directly involved in the information system. These individuals—whether they are end users, managers, or IT professionals—have their own set of skills, attitudes, preconceptions, and personal agendas that determine what they are able to do and what they will elect to do as part of the IS. A genuine understanding of the people involved, including their skills, interests, and motivations, is necessary when designing and implementing a new IS or when troubleshooting an existing IS that is not performing as expected.

Consider the example of a national government intent on rationalizing and improving communication between the local administrations (e.g., school districts) and the central institutions (e.g., ministry of education). As part of this initiative, email addresses are produced for the school district superintendents and each school's principal and assistant principal. You quickly realize that this simple initiative is more likely to be successful today than it was 10 or 15 years ago, when computer skills were scarce and computer anxiety was high even among highly educated individuals.

Structure The **organizational structure** component (or "structure," for short) refers to the organizational design (hierarchy, decentralized); reporting (functional, divisional, matrix); and relationships (communication and reward mechanisms) within the information system. Structure is difficult to identify sometimes, but you can think of it as the implicit or explicit rules that govern relationships between the people involved in the information system. Understanding the structure component is crucial because user resistance, incentive systems, and relationships are often silent enemies of IS success that go undetected before, and sometimes even after, IS failure becomes apparent.

Consider the famous case of a large consulting firm. The firm, with global operations, recognized the potential for knowledge sharing among its consultants. "If we introduce a knowledge management system," the thinking went, "we can create repositories of knowledge to which our consultants will contribute upon completing a project. This will enable us to surface and share best practices rather than having to reinvent the wheel with similar projects just because they are in different regions and involve a different team of consultants. Moreover, we will be able to identify subject matter experts to whom we will direct questions on specific topics. The outcome of this IS implementation will be increased turnaround time on projects, better quality results, and more satisfied clients."

A few months after the rollout of the knowledge management system, it became clear that usage was spotty at best. Upon careful analysis, the firm realized that there was little incentive for consultants to contribute to the knowledge base. In the fast-paced world of digital technology consulting, the road to success was the "billable hour," which was gained by spending productive time working on client projects. Moreover, the existing reward mechanisms were such that employees were incentivized to prioritize individual behavior and superior skills over teamwork and knowledge sharing. The inevitable conclusion was that in order for the knowledge management systems to reap the expected benefits, the tangible reward structure and the traditional mentality of the organization would need to change.

Systemic Effects

It should be clear from the preceding discussion that all four components are necessary to ensure that the information system is successful and delivers the functionality it was intended to provide. Imagine "breaking" any one of the four components in any of the preceding examples: the system as a whole would stop working. More subtly, the four components of an information system don't work in isolation but instead interact with one another—as noted by the arrows in Figure 2.7. This notion of the interdependence of the components goes by the name ***systemic effects***, indicating that changes in one component (e.g., the introduction of a new software application, a process redesign, a new organization chart, or turnover among employees) affect all other components of the system and, if not properly managed, its outputs. Mark Hedley, the former chief information officer (CIO) of global hotel chain Wyndham International, neatly captured this idea:

> Many companies rush out, buy software solutions, install them quickly, and then can't understand why the system failed. You have to look at what business issues exist, what people and processes pertain to that business issue, and what those people do. Technology won't solve [a problem] by itself—other components have to be part of the solution.[1]

Because of systemic effects, when called upon to design a new IS or to modify an existing one, you should focus not on optimizing the technology (i.e., adopting the most innovative and cutting-edge digital technology) or any other component individually; rather, you should

---

1 Piccoli, G. 2007, April. "Career isn't over: How CIOs are reaching new heights." *Cutter Benchmark Review*, retrieved from https://www.cutter.com/journal/career-isnt-over-how-cios-are-reaching-new-heights-486016.

optimize the IS as a whole (i.e., selecting components that create the best chance to deliver the needed information processing functionality when working simultaneously with the other components). This focus on information systems design rather than IT investment decisions also suggests that there are multiple ways to achieve the same information systems goal—as demonstrated by the many different ways in which similar organizations meet their information processing needs.

Understanding the importance of systemic effects is critical not only when designing a new system but also when troubleshooting an existing one that is underperforming in order to diagnose the root causes of the failure and devise the appropriate intervention. Consider a large telecom firm that introduces an IS to improve service quality at their call center. Instead of having call-center managers listen in on a small sample of calls to offer coaching to the associates in individual one-to-one meetings, the new system applies speech recognition and machine learning to automatically analyze every call and produce a standard report, with automated recommendations, that each associate is required to read. Call-center managers must review all reports but only meet face-to-face with severely underperforming associates. The system is met with discontent and rejection by the call-center managers, who complain about the fact that the new digital technology is not intuitive, is difficult to use, and limits their ability to make good decisions. How would you solve this problem if you were the head of the analytics project? There may be a number of options here:

- You could deem the new system a failure and cut your losses by reverting to the previous face-to-face approach.
- You could ascribe the failure to the quality of the user interface of the reporting application and then negotiate with the provider to improve the reports in order to make them more intuitive. This solution focuses on the IT component of the IS.
- You could ascribe the failure to the quality of the user interface of the reporting application but choose a different solution. You could work with both call-center managers and associates to convince them that the system is not as awkward as it seems and that a bit of training will solve all their issues. This solution focuses on the people component of the IS.
- You could ascribe the failure to the fact that your staff comprises mostly old-fashioned managers who rose through the ranks of the call center at a time when personal relationships, motivational cheers, and team spirit were the norm to improve service quality. They are not equipped to make the transition to data-driven decision making and evidence-based management. You could then turn over your staff, replacing all management with younger, tech-savvy, data-loving types. This solution also focuses on the people component.
- You could ascribe the failure to misplaced incentives. You could then call a meeting and inform both management and associates that the speech-recognition system is going to stay and that the next person who complains is going to be fired. You may, of course, offer motivation rather than threats if you are more of the positive-reinforcement management type. This solution focuses on the structure component of the IS.
- You could ascribe the failure to the fact that call-center managers lose the ability to engage directly with the associates, who instead spend time reading reports. You may then suggest that managers listen in on calls and review aggregate reports to offer coaching to all associates—not just the "severely underperforming" ones. This solution focuses on the process component of the IS.

2.3 Why Do Organizations Build Information Systems?

Now that we know what an information system actually is, we should step back and question why organizations build them in the first place. Simply put, a firm's objective when introducing information systems enabled by digital technology is to fulfill its information processing needs. Sometimes external requirements, such as financial reporting, safety, or tax regulations, mandate the introduction of a new IS. More typically, an organization introduces information systems in an effort to improve its **efficiency**[1] and **effectiveness.**[2]

In order to fulfill its information processing needs, an organization must capture relevant data that are then manipulated, or processed, to produce an output that will be useful to the appropriate users, either internal or external to the firm (e.g., customers). These data and information are typically accumulated, or stored, for future retrieval and use (see Figure 2.10). Thus while not a component of the information system per se, information plays a critical role in modern organizations.

Note that while the focus of this book is on information systems that leverage the power of contemporary digital technology, the processing of information does not necessarily require IT support in order to happen. Consider the last time you contributed to the in-class discussion of a case study. In order to produce valuable comments for the class, you gathered substantial information (i.e., you read the case, you read assigned articles and other class materials, you actively listened to the professor setting the context for the discussion, and you listened attentively to your classmates' comments). You then processed the information (i.e., during preparation, you thought about the case issues in the context of the information presented in the readings, and during class discussion, you thought about how your classmates' comments supported or contradicted your point of view). Finally, you produced some output—your insightful analysis for the whole class to hear. In other words, you engaged in significant information processing without ever using a computer.

Modern IT is not to be downplayed, however, as all but the most trivial of today's organizational information processing requirements cannot be fulfilled in a reliable and cost-effective manner without digital technology. But digital technology can only be successfully leveraged as an integral part of an information system, not in isolation. Moreover, while managers often focus on the financial results produced by the design and introduction of a new IS, there are nonfinancial outcomes, both planned and unexpected, that you need to be aware of.

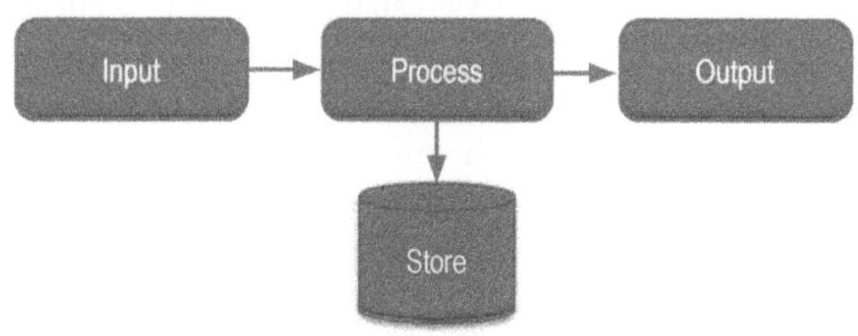

Figure 2.10. Information processing in an IS

Successful Information Systems

Any information system should be built according to an explicit goal (or a set of goals) designed to fulfill the specific information processing needs of the implementing organization.[3] Examples of typical IS goals include the following:

1 *Efficiency* is defined as the ability to limit waste and maximize the ratio of the output produced to the inputs consumed. In other words, a firm is more efficient when it produces more with the same amount of resources, produces the same with fewer resources, or produces more with fewer resources.

2 *Effectiveness* is defined as the ability to achieve stated goals or objectives. Typically, a more effective firm is one that makes better decisions and is able to carry them out successfully.

3 Note that while the IS goals should fit with the firm's strategic goals, the IS cannot ensure that the correct business objectives have been chosen. In other words, an IS is deemed successful when its information-processing goals are achieved—even in the face of business failure due to the pursuit of a flawed strategy.

- For a large retail store (e.g., Walmart, Carrefour)—To increase the efficiency and speed of customer checkout, perhaps using self-checkout stations.
- For a high-end car manufacturer (e.g., BMW)—To improve customer service by allowing individual customers to select the finish and accessories on their cars and quote real-time price changes and delivery date changes both at the dealership and online via the firm's website or mobile app.

Consider the choice faced by McDonald's franchisees who had to decide whether to offer Wi-Fi connectivity in their restaurants in 2003—taking a page out of the successful approach of Starbucks coffee shops. While a cursory examination may suggest that Wi-Fi simply does not fit with McDonald's strategic positioning, a more careful analysis would indicate that for some customers (e.g., busy parents, truck drivers) and in some locations (e.g., tourist areas), connectivity may be highly valued. The lack of a clear fit, however, led to McDonald's failing to develop a clear Wi-Fi strategy. After testing the program in 2003 with a $4.95 charge for two hours of use, the firm lowered the charge to $2.95 in 2004, allowing franchisees to decide whether they wanted to join the program. In 2009, Wi-Fi connectivity became a standard, available in every one of the more than 32,000 restaurants in the world, free of charge.

One of the difficulties that McDonald's executives faced was in clarifying for franchisees what the primary objectives of the program were. While at first management suggested that the program would increase consumption, the benefits weren't very clear. At the very least, if customers felt that McDonald's plastic seats were too uncomfortable to sit on for more than 15 minutes and that the food was too messy to fumble with their laptops during or after the meal, the systems would be largely unused, hence delivering little benefit to the franchisee who paid for its installation. When the program became a brand standard, it became clear that the value of Wi-Fi at each McDonald's was in attracting more customers to the stores with the added benefit that they could quickly check their email while on the road. Whichever the best objective is in this particular case, it is important to note that establishing a clear aim for the system is a critical prerequisite to a successful implementation. Table 2.1 highlights some famous high-profile information systems failures.

Information Systems Success: Definition

We can unequivocally deem an information system a failure if it is abandoned (i.e., the design and implementation phase is never concluded) or not used (i.e., the information system is completed only to be rejected by its intended users). In each case, the system is a failure because, due to nonuse, it will yield none of the promised benefits (e.g., efficiency improvements).

Beyond nonuse, discriminating successful information systems from failed ones requires an evaluation of whether, and to what extent, the system has delivered the expected results. This is why it is so important to articulate information system goals during the design and justification phase. Yet for as thorough a job you may do in defining the intended IS goals, there are many situations when unintended results, both positive and negative, will emerge. The celebrated Sabre reservation systems pioneered by American Airlines (AA), the first example of a successful strategic information system, was originally introduced simply to enable inventory control in response to a very tangible operational problem: AA found itself unable to manage and sell the increasing number of seats it supplied in response to mounting consumer demand for commercial air flights.

Information Systems Outcomes

Beyond efficiency and effectiveness improvements and the associated financial considerations, information systems have other direct and indirect effects on people within and outside of the firm (e.g., employees, customers, suppliers). These effects can be positive, including empowering

Table 2.1. Systems failures in business computing history

Year	Company	Outcome
2020	Heathrow Airport	A software failure, affecting T5's check-in services, resulted in the cancellation of more than 200 flights, delays, and thousands of passengers in "utter chaos."
2019	Boeing	An overaggressive autopilot and the lack of adequate pilots' training proved disastrous, resulting in the 737 Max fleet's total grounding. The resulting two deadly crashes cost the firm $19 billion and put the firm in serious trouble.
2018	Facebook and Cambridge Analytica	Eighty-seven million users' personal data were leaked without consent and then sold for political advertisement. The scandal cost Facebook allegedly five billion dollars.
2017	Amazon	An employee's mistyped command generated a cascade of failures that resulted in the outage of Amazon Web Services (AWS) cloud services in Northern Virginia, USA. The two-hour outage caused an estimated $150 million losses for S&P 500 companies, $160 million for U.S. financial services.
2016	Delta Air Lines	A computer outage in the airline main data center caused the cancellation of about 2,300 flights. The firm reported that the system failure resulted in $100 million less revenues in the month of August alone.
2015	Paderborn Baskets	A second-division German basketball team, the Paderborn Baskets, was relegated to a lower division for starting an official game 25 minutes late. The cause of the delay? An automatic Windows update that froze the digital scoreboard and prevented the start of the game.
2013	French Ministry of Defense	Expected to enter service in 2017, Louvois—Logiciel unique à vocation interarmées de la solde—was supposed to simplify and unify the payment system for the French army's 130,000 members. The system was scrapped in November 2013 after consuming €346 million.
2013	NHS Connecting for Health (UK)	Originally expected to cost £2.3 billion over three years, in June 2006, the total cost was estimated to be £12.4 billion over 10 years. The system was abandoned in September 2013 despite already costing taxpayers £10 billion.
2012	Royal Bank of Scotland Group (UK)	A failed upgrade to the payment processing system left customers without access to their accounts for two weeks and disrupted customers' payments either incoming (i.e., wages) or outgoing (i.e., bill payments). The company was later fined £56 million over the accident.
2011	Allied Irish Banks (AIB)	AIB sued Oracle Financial Services Software for €84 million, plus damages and lost profits, claiming it wasted the money on the failed implementation of a new retail banking system.
2011	U.S. Federal Bureau of Investigation (FBI)	In 2001, the FBI started work on a criminal case management system called Virtual Case File. The project was scrapped four years (and $170 million) later. The project was restarted under the name Sentinel and, as of May 2011, was unfinished after five years (and $400 million) in development.
2010 (2007)	AXA Rosenberg, a privately owned investment management company (U.S.)	A "coding error" led to underrepresented investing risk factors and resulted in investors losing a total of $217 million.
2010	Electronics retailer Dixons (UK)	Reiterated difficulties with its new eCommerce system were blamed for £15 million in lost revenue.
2009	Government of Victoria (Australia)	A smartcard ticketing system (called Myki), contracted in 2005 with AU$500 million budgeted, was rushed into operations (current cost: AU$1.3 billion) with limited functionality and still causes significant operational problems.
2009	Britain's National Health Service (NHS)	NHS scaled down its "Connecting for Health" program, designed to create the national electronics health record system, after investing an estimated £12 billion since the project began in 2002.
2008	Centrica, the largest utility supplier of gas to domestic UK customers	Centrica sued Accenture for £182 million in damages stemming from a collapse of customer service levels and loss of more than one million customers attributed to the failure of a "best-of-breed" customer billing system.

(continued)

Table 2.1. Systems failures in business computing history *(continued)*

Year	Company	Outcome
2008	J. Crew (U.S.)	Shares of the company fell more than 7% after the announcement of persistent website performance, order fulfillment, and call-center performance problems leading to slowing sales trends, lower gross margins, and three million dollars in unanticipated costs.
2007	LA Unified School District (LAUSD; U.S.)	LAUSD discovered that, due to ongoing payroll system problems, it overpaid 36,000 employees by a total of $53 million.
2007	Palm Beach County (U.S.)	Palm Beach County evaluated scrapping a $13.6 million upgrade to its computer systems that took three and a half years to develop (six months originally budgeted) due to an inability to operate with it.
2006	Child Support Agency (CSA; UK)	The CSA was shut down in part due to the problem-ridden deployment of a £456 million IT system built by Electronic Data Systems.
2005	Hudson Bay Co. (Canada)	Problems with their inventory system contributed to a $33.3 million loss.
2005	UK Inland Revenue	Software errors contributed to a $3.45 billion tax-credit overpayment.

Source: Adapted from ***IEEE***, ***BusinessWeek***, ***CEO Magazine***, ***Computerworld***, ***InfoWeek***, ***Fortune***, the ***New York Times***, the ***Guardian***, and the ***Wall Street Journal***

employees' and widening the scope of their responsibility, or negative, resulting in deskilling (i.e., reducing the scope of an individual's work to one, or a few, specialized tasks), loss of responsibility, and the creation of a monotonous working environment.

Another important outcome of information systems use pertains to their effect on future opportunities available to the firm. The introduction of a new information system may enable or constrain future information systems and strategic initiatives available to the organization. This is due to the fact that future systems typically rely on, or connect with, preexisting ones. Consider the example of social networking sites like Facebook or Instagram. As the user base grows larger and each individual member has more connections within the social network, Facebook and third-party developers have the opportunity to develop applications that leverage these data. Social games like *Candy Crush Saga* or *Farmville*, and applications such as *Where Have I Been*, could not have been easily created without the underlying Facebook social graph.

Beyond an appreciation of their goals and outcomes, successfully designing and implementing information systems requires recognition that any information system that you envision, or propose, exists in a unique and very specific organizational context.

2.4 Information Systems in the Organizational Context[1]

Consider the cultural differences between the retail giant Amazon and Whole Foods Market, the supermarket chain that focuses on natural and organic foods and sustainable agriculture. Amazon's control and measurement culture was the opposite of Whole Foods' individual high-empowering and decentralized one. While Amazon has been continuously seeking efficiency and productivity gains through tight control, performance management, and centralized systems, Whole Foods Market relied on local control throughout its history. Each shop manager was given wide latitude to adjust offerings based on local customers' preferences for fresh produce. Store managers also worked on building strong personal relationships with their customer base. Imagine both firms introducing an inventory management system enforcing a strict and centralized system. Will the same system yield comparable results in each of these firms? Where will

1 The following discussion is largely influenced by Silver, M. S., Markus, M. L., and Beath, C. M. 1995, September. "The information technology interaction model: A foundation for the MBA core course." *MIS Quarterly 19*(3): 361–90.

the "centralized inventory system" vision most likely come to fruition? Where is a culture clash most likely to occur when the new system becomes operational?

These questions are far from trivial. In this specific example, Amazon actually acquired Whole Foods Market for $13.7 billion in 2017. How would you handle such an acquisition? Would you look for synergies by integrating the successful inventory control systems used by Amazon into Whole Foods? Would you forego potential efficiencies and let Whole Foods run as a separate entity with its traditional organizational structure and business processes? Amazon ended up pushing its vision to introduce an order-to-shelf system optimizing deliveries, so inventory could directly move from truck to shelves, minimizing backroom overstock. The system set purchasing rules, displaying, and shelves replenishment. Scorecards were introduced to keep track of employees' performances. Whole Foods employees were frustrated and stressed by the new reward and demerit systems.[1] Interestingly, while Whole Foods dropped out of the *Fortune*'s 100 Best Companies to Work For list, the acquisition spurred an unprecedented digital transformation of the entire grocery industry.[2]

Every Organization Is Unique

The simple example just described should clarify that organizations are unique in many respects. This is true not only of companies in different industries but also of otherwise similar firms that compete head-to-head. Microsoft and Apple Computers were the two most recognizable names in the software industry throughout the 1990s, vying to establish their respective operating systems as the dominant platform. Yet the two firms had dramatically different images and cultures. Even starker is the difference between two of today's London-based airlines—British Airways and EasyJet. At the highest level of abstraction, a firm is characterized by its strategy, its culture, and its current infrastructure, stemming from the organization's history, size, product line, location, values, and so on.

Firm Strategy A firm's ***strategy*** represents the manner in which the organization intends to achieve its objectives. In other words, understanding a firm's strategy tells us what the firm is trying to do and what course of action it has charted to get there.

Consider two head-to-head competitors: Dell Computers and Hewlett-Packard (HP), who battled for supremacy of market share in the PC industry during the late 1990s. At the time, Dell and HP were the number-one and number-two makers of PCs, respectively. Dell focused on highly customizable, made-to-order devices that were assembled upon receipt of orders directly from consumers and business clients. HP focused on producing standardized devices to be sold through a wide distribution channel (e.g., Best Buy, Media World). Similar examples of direct competitors in the same industry with very different strategies abound.

Firm Culture A firm's ***culture*** is defined as the collection of beliefs, expectations, and values shared by the organization's members. The firm's culture, a broad representation of how the firm does business, is an important characteristic of the organization because it captures the often unspoken and informal way in which the organization operates. Practices that are deemed appropriate in one organization may not be in another one.

---

1 Blanding, M. 2018, May 14. "Amazon vs. Whole Foods: When cultures collide." HBS Working Knowledge, retrieved from http://hbswk.hbs.edu/item/amazon-vs-whole-foods-when-cultures-collide.

2 Begley, S., Marohn, E., Mikha, S., and Rettaliata, A. 2020, February 14. "Digital disruption at the grocery store." McKinsey & Company, retrieved from https://www.mckinsey.com/industries/retail/our-insights/digital-disruption-at-the-grocery-store.

Consider the 2004 merger between software titans Oracle and PeopleSoft. Announcing the merger, the *San Francisco Chronicle* posed the question, "What do you get when you combine a company run by an Armani-clad executive known for take-no-prisoners tactics with a firm led by a fatherly founder who hands out bagels and lets his workers wear flannel to work? . . . For the merger to succeed, Oracle faces the tough task of creating a cohesive company out of two firms with distinct, even contradictory, cultures." Quoting a Forrester Research analyst, the article concluded, "Pretty quickly, the PeopleSoft employees are going to divide themselves into those who say: 'I think I can work for Oracle' and 'I don't ever want to work at Oracle.'"[1]

Infrastructure When it comes to making information systems decisions, it is also important to consider the current digital technology infrastructure of the firm, generally called the **IT infrastructure.** The existing IT infrastructure, defined as the set of shared IT resources and services of the firm, constrains and enables opportunities for future information systems implementations.

The example of checkout scanners at grocery stores highlights this point. Once the infrastructure is in place, the grocery store can consider future initiatives that rely on it, such as automatic inventory reordering, but also checkout coupons (i.e., the ability to print coupons at checkout based on the items the customer bought), frequency shopper cards, and basket analysis (i.e., the ability to identify correlations among items purchased by the same customer).

External Environment Organizations themselves don't exist in a vacuum but instead are embedded in the **external environment** that encompasses regulation, the competitive landscape, and general business and social trends (e.g., outsourcing, customer self-service).

Consider three competitors, such as the Sony Corporation, Philips Electronics, and Samsung Group, headquartered in Tokyo, Amsterdam, and Seoul, respectively. While these three firms compete, at least in part, in the world market for consumer electronics, they have to contend with widely different local labor and taxation laws, governmental incentives, and so on. The external environment is fairly removed from day-to-day operations, yet these factors have an influence on the firm and, as a consequence, on the type of information systems the firm will need to introduce.

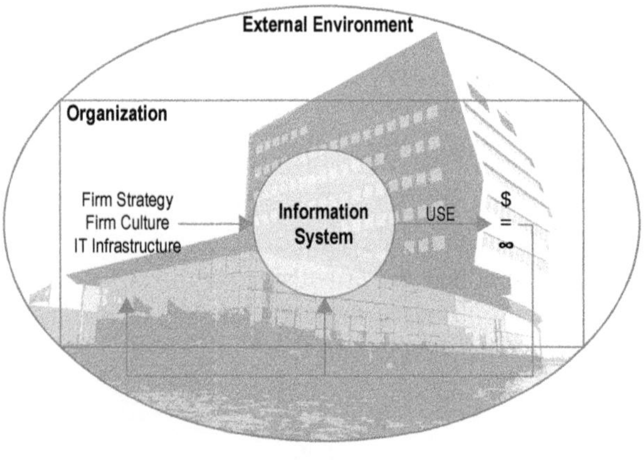

Figure 2.11. Information systems in an organizational context

Bringing It All Together

The previous discussion of the outcomes associated with information systems use and information systems in context is summarized in Figure 2.11.

The model in Figure 2.11 indicates that the immediate effect of information systems is whether they are used. If they are used, then intended and unintended outcomes ensue, including financial results, effects on people, and effects on the future opportunities and constraints available to the firm. The model also shows that information systems do not exist in a vacuum but are embedded in a specific organizational context, defined by the firm's strategy, culture, and IT infrastructure. Moreover, the organization itself does not exist in isolation but is embedded in the external environment, including social and competitive forces. The feedback loops represented by the solid, bold

1 Pimentel, B. 2004, December 15. "When firms merge, a clash of cultures: Oracle, PeopleSoft managing styles couldn't be more different." *San Francisco Chronicle.*

line remind us that the outcomes produced by the information system, whether positive or negative, will affect organizational characteristics and future information systems decision making.

This model is important for you as a manager because it draws to your attention all the external influences that will aid or undermine information system success.

2.5 Information Systems and Organizational Change

An understating of organizational change and the issues associated with it completes the discussion of information system design and successful implementation. As a manager, you must pay close attention to organizational change. With the widespread adoption of digital technology by today's organizations, increasingly this organizational change is brought on by the introduction of new IT. Defining information systems as sociotechnical systems can be instrumental in helping you better manage organizational change. Specifically, we can identify three levels of organizational change brought about by the introduction of new digital technology.

First-Order Change: Automate

The simplest order of change ensuing from the deployment of technology is automation. This level of change involves the IT and processes components of the information system but does not affect the sphere of the social subsystem (Figure 2.12). First-order change occurs when a digital technology innovation is introduced that modifies how an existing process is performed.

Consider those online banking tasks previously executed through a touch-tone phone interface—such as checking balances or transferring money between accounts. Many of us now perform those same tasks through a web interface or a smartphone. What has changed with the move to the web is the manner in which these processes are performed, not the tasks themselves or the individuals involved (e.g., customers). With the introduction of the web or the mobile phone as a customer interface, the process has become simpler and more intuitive.

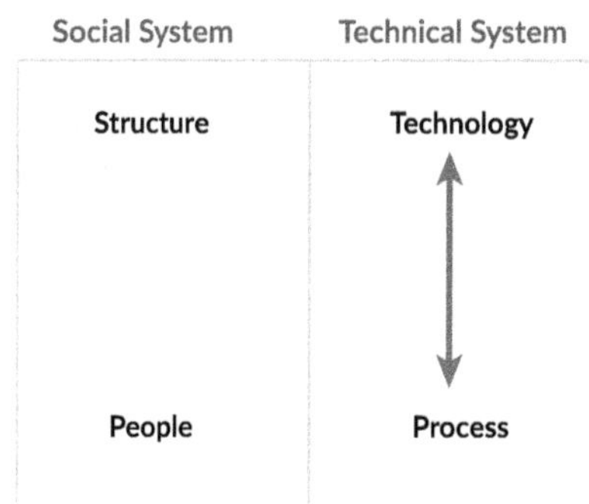

Figure 2.12. First-order change

Managing First-Order Change The limited scope of first-order change makes it relatively easy to envision, justify, and manage. General and functional managers understand how the new technology impacts the firm's operations, and the project is straightforward to justify because the financial benefits of the change, in terms of the return on investment, can be estimated with some precision. Thus first-order change requires little executive sponsorship and involvement.

Second-Order Change: Informate

Second-order change has major implications for the people component of the information systems as well as the IT and processes elements (Figure 2.13). With second-order change, not only will the manner in which the process is performed changes; those individuals who perform it will also differ—either their role changes or a new set of people is now involved. Moreover, the manner in which people interact with the technology also undergoes modification. This level of change typically occurs when the information intensity of the process being performed changes substantially due to the introduction of new digital technologies. For this reason, this level of change is called *informate.*

A good example of second-order change is customer self-service via digital technology. Consider airline check-in kiosks or apps. Traditionally, as an airline traveler, you'd have to go to the airport, line up, and interact with an

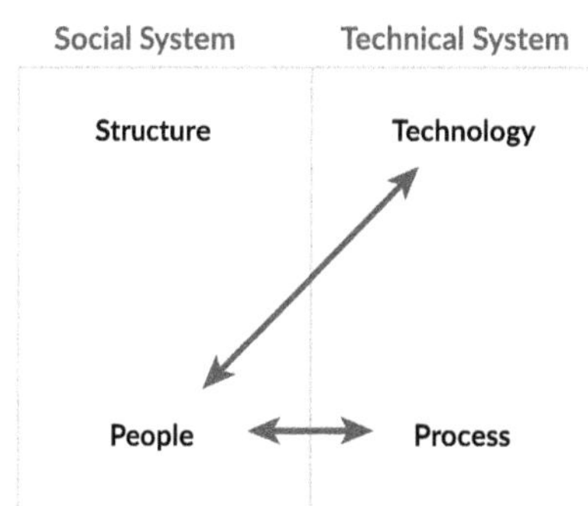

Figure 2.13. Second-order change

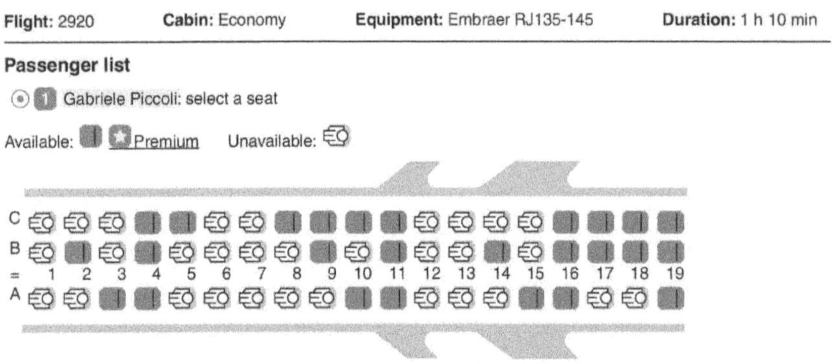

Figure 2.14. Choosing a seat during web check-in

agent, who would authenticate you by checking your ID card and then provide you with a seat and boarding passes. The advent of check-in kiosks, then web-based and finally mobile-enabled online check-in, has dramatically changed this process. First, it is now the machine that authenticates you—using your reservation number, a credit card, or a frequent flier card. Second, you can proceed to view the plane's seating chart, upgrade or change your assigned seat, select amenities like food or Wi-Fi, and retrieve the boarding passes, all independently of any airline agent (see Figure 2.14).

For those of us who are very tall and used to implore the agent for an exit-row seat, the kiosk has opened up a wealth of new possibilities. More important, the kiosks have had a dramatic impact on the agents themselves, who now serve in more of a training and troubleshooting role, helping travelers solve any problems they encounter with the system or working on special cases, rather than completing the routine check-in process for all travelers.

Managing Second-Order Change Since the primary impact of second-order change is on the people dimension of the sociotechnical system, second-order change provides much more of a challenge to managers than does first-order change. Those affected may include employees as well as customers. Thus providing appropriate training and overcoming the human tendency to resist change are key challenges.

The objectives of second-order change are typically far reaching as well. While first-order change is focused on automating existing tasks, informate-level change often seeks to take advantage of available market opportunities. Thus justifying this level of change is more difficult and requires a more speculative analysis.

Third-Order Change: Transform

Third-order change represents the most pervasive and radical level of change—as such, it is often referred to as *transform*. Third-order change subsumes first- and second-order change while also causing organizational structure disruptions (Figure 2.15). The interaction between structure and technology is substantiated by a change in the way the organization selects, uses, and manages technology. The interaction between the organizational structure and the people generally results in a change in the reporting and authority arrangement of the organization. A flatter or more permeable organizational structure usually emerges after the technology implementation. The interaction between the organizational structure and tasks manifests itself in a novel way of task accomplishment or a new set of tasks.

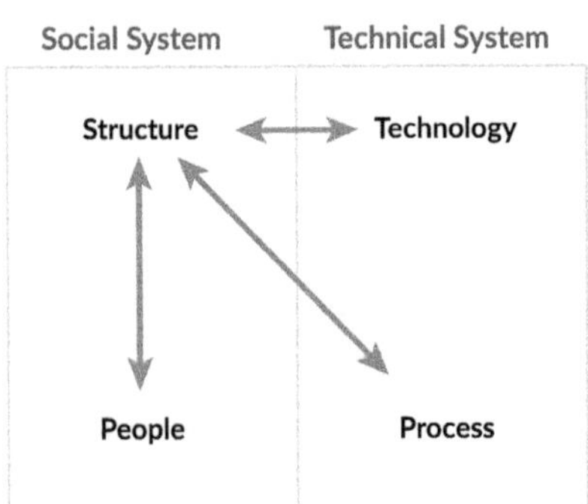

Figure 2.15. Third-order change

Consider a familiar example: Wikipedia. Wikipedia, the free, online, user-written encyclopedia, transformed the way we access and produce knowledge. Traditional encyclopedias have a strong top-down approach to content creation. Editors commission articles from individual experts who write an entry. Editors then review the entry and might share it with one or two other experts for peer review before finalizing and including it in the next release of the encyclopedia. We then buy the book and read the experts' entries. In contrast to this, thanks to the ability to easily connect any of the two billion people currently online across the globe and thanks to the capability of Wikis to allow easy, multiple editing of the same document, Wikipedia uses a very different knowledge codification and dissemination process. Anyone can contribute to an article, and the quality assurance is performed by the community of readers through a discussion page associated with each entry. As such, the people involved in content generation and the process by which knowledge is produced change dramatically. But the organizational structure of the online encyclopedia is also significantly different from its traditional printed counterparts. In fact, the very notion of an expert on a subject is lost and expertise is instead contributed by the population of readers at large.

Managing Third-Order Change Third-order change requires significant managerial and executive involvement. Championship by the top management team is necessary both for signaling purposes and to provide the political impetus to complete the transition. Changes in organizational structure are, in fact, likely to engender political battles and resistance by those whose authority and political influence is diminished. The history of the business process reengineering (BPR) movement (Chapter 3) provides countless examples of the perils associated with third-order change and the difficulties of managing dramatic change enabled by digital technology.

2.6 Managerial Takeaways

From the preceding definitions of information systems and IS success and the discussions of the roles of systemic effects and organizational context, a number of implications of interest to managers follow.

Don't Put the Cart before the Horse

We often receive inquiries from former students or other managers asking our opinion about various "digital technology solutions" (e.g., what is the best social media software for my firm to use for digital marketing?). However, you may now recognize that asking such questions is equivalent to letting technology drive decision making.

While it is common for strategy to be inspired by the functionalities of a powerful new software product, the selection of a specific digital technology should never be the point of departure of your information system design effort. When asked to express an opinion about a software program or other technology, you should always start by asking, "Why are you investigating this software program (i.e., what is the firm's strategy)? What are you attempting to do with the new digital technology (i.e., what is the IS goal)?"

In our opening minicase, the firm's intention was to improve its volumes (i.e., increase the number of customers and their level of consumption). This can be classified as the strategic goal. From this goal, the firm should derive a precise set of IS goals that detail what information-processing functionalities are needed to achieve the strategic objective (e.g., to create a comfortable space for restaurant patrons to more conveniently place orders at the table). Once the goals are set, the IS design team can identify, shape, and deploy the appropriate components of the system, of which IT is one.

Manage Systemic Effects

As in any other system, the components of an IS mutually influence one another (systemic effects). Similar to the ripples resulting from the act of throwing a rock into a pond, changes to one or more IS components impact, sooner or later, all other components. This influence may be significant or limited. However, when you are called upon to participate in IS design and implementation, you should try to contribute to your team's successful introduction of a new IS by anticipating these ripple effects and proactively managing them before they become a cause for concern.

Let's return to the example of airline check-in kiosks. When kiosks are installed, the business process of identifying travelers and assigning them seats must change for those customers who prefer this method. The role, and the skills required, of check-in agents is modified as a result of the change in the business process—at a minimum, they must be able to explain the kiosks' operation and provide helpful support as customers learn to use them. In this case, the new technology significantly impacts both process and people, while the effect on structure is limited. The more precise you can be a priori in estimating what the effects will be, the better able your firm will be to proactively manage the changes.

Information Systems Are in Flux

An information system is not designed "once and for all" as if it were a static artifact. Business strategy and the external environment evolve continuously. This evolution calls for a constant reevaluation of IS goals and needed information processing functionalities. In turn, this reevaluation will at times engender the need for changes to the design of an existing information system.

The design and use of an IS should be seen as an iterative process involving the cyclical evaluation of individual IS components and the assessment of how different organizational systems work together to support the business. The synergy among IS components, as well as among discrete organizational information systems, can be maintained over time only if there is a willingness to modify aspects of this IS configuration as needed. Any time a major change occurs, the current IS design must be reevaluated and the system must be optimized once again.

Optimize the Whole

As mentioned earlier, you should never lose focus of your objective: to optimize the information system rather than any of its constituent parts. Optimizing the system as a whole often requires that one or more components be de-optimized (i.e., they are not as powerful or cutting edge as they could be). This fundamental insight is one that is often forgotten when managers get caught up in the IT investment mentality—the most effective information system need not comprise all the best parts.

To the contrary, examples abound of firms that deployed cutting-edge digital technology where there was no need for it to deliver the desired information processing functionalities. In some cases, the adoption of digital technology in fact reduces the effectiveness of the IS as a whole, making the achievement of the needed information processing functionalities more difficult.

2.7 Conclusion

Like all frameworks, those presented in this chapter are valuable because of their ability to support a systematic and disciplined analysis of specific issues (e.g., software selection, IS design, system failure diagnosis). They don't offer cookie-cutter answers, but they provide the support necessary for a disciplined and thorough analysis. By using them to guide your thinking, you can be sure to complete a comprehensive analysis rather than being tempted to stop upon identifying the first or second most obvious explanations—perhaps those communicated by the most vocal members of your organization.

Furthermore, working with the models challenges conventional wisdom and management fads and fashions that may lead you to reach simplistic conclusions that, while perhaps true, often only reveal part of the story (e.g., "You must train your employees when new software is introduced"). The alternative—being driven by a best-practice mentality, benchmarking your organization against the performance of competitors—is a risky strategy (and you should always be wary of so many buzzwords in one sentence!). As a manager, you must always consider how new digital technology fits within the context of your own company. Does it suit the unique people, processes, and structures of your firm? If not, is it a wise strategy to make changes to these components in order to fit with the technology? Or do you need different digital technology? All too often, when selecting a system, considering change initiatives, or troubleshooting underperforming information systems, the focus is on digital technology and software functionality. The frameworks presented here challenge you to think in terms of overall IS design instead—perhaps a more difficult task, but certainly a more appropriate and productive one.

Summary

This is a critical chapter in the book because it provides fundamental definitions and sets the stage for the discussion in the next chapters. The chapter defined information systems as sociotechnical systems composed of four components: IT, people, processes, and structure. This definition and its implications provide the basis for this book.

Specifically, in this chapter, we learned the following:

- Information systems are designed and built with the objective of improving the firm's efficiency and effectiveness by fulfilling its information processing needs. Successful information systems are those that are used and that achieve their intended goals.
- Information systems exist in an organizational context, characterized by the firm's strategy, culture, and IT infrastructure. The organization itself is subject to the influences of its external environment, including regulatory requirements, social and business trends, and competitive pressures.
- Information systems are subject to *systemic effects* because the different components of a system are interdependent and changes in one component affect all other components of the system. Thus when designing a new information system or troubleshooting an underperforming one, you have a portfolio of options for achieving the system's goal.
- Increasingly, in modern firms, organizational change stems from the introduction of new information technology. Depending on the objectives and reach of the new system, we identify three levels of change—first-, second-, and third-order change—each requiring different levels of commitment and sponsorship to be successfully managed.

Study Questions

1. Describe the difference between information systems and information technology. Provide an example of each.
2. Provide an example of two organizations in which you think a similar information system would engender two very different outcomes. Explain why.
3. Provide two examples, from your personal experience, of information systems that generate positive and negative unintended results.
4. Define the concept of systemic effects. Explain why it is important for you as a general or functional manager to be aware of this concept.
5. Describe first-, second-, and third-order organizational change induced by the adoption of new IT. Provide an example, real or imagined, for each of these three levels of change.

Glossary

A glossary of key terms can be found at the book's website. Visit https://www.prospectpressvt.com/textbooks/piccoli-information-systems-for-managers-5-0, then click on "Student Resources" in the horizontal red menu bar.

Chapter 3

Organizational Information Systems and Their Impact

What You Will Learn in This Chapter

This chapter completes our introduction to the foundations of information systems (IS). In Chapter 1, we made the case for this book's value to current and future managers. In Chapter 2, we provided important definitions. In this chapter, we discuss the vocabulary and concepts that will enable us to categorize different types of information systems and to communicate with other managers and IS professionals. You may feel at times that this chapter has a lot of "historical stuff" in it. Yet one of the enduring characteristics of organizational information systems is that new systems seldom fully replace older generations. The result is that a large organization today will have an infrastructure that includes mainframes, large integrated **enterprise systems**, personal computers (PCs) running client/server applications alongside mobile phones and tablets running cloud-first applications, cutting edge **NoSQL databases** for **Big Data** analytics, and a real time, end-to-end product traceability solution based on the blockchain (Chapter 12). This chapter will help you organize and situate the wide array of organizational systems you will encounter when you join the workforce.

MINICASE: Integration at BigPharma Inc.

As you walk out of the boardroom, still shaking your head in disbelief, you mumble to yourself, "Boy, that was fun! They were really going at it today!" As you get to your desk, though, it hits you—you have to make sense of what just happened in there.

As the executive assistant to the chief executive officer (CEO) at BigPharma Inc., the second-largest pharmaceutical firm in the United States, you have had the luxury of attending all the executive team meetings and sometimes participating in decision making with your analyses without any of the responsibilities that come with making those decisions. However, far from what you had imagined the glamour of boardroom discussions to be, most meetings of the executive team were pretty boring. Not today!

Surprisingly, you had predicted this one to be a real snoozer. A pretty safe bet given the topic: the need to gain efficiencies by better integrating across functional areas. The meeting took a turn toward the exciting right out of the gate when Laura Jean Polly, your boss, announced that at PharmaMed (the premier industry trade event of the year), she had drinks with the senior VP of business development of BigCoSoft, the second-largest vendor of enterprise systems.

She said that BigCoSoft was interested in breaking into the pharmaceutical market and was seeking to sign up a high-profile customer. They looked at this contract as a mutually beneficial partnership that would lead to lots of press and advantages for both firms. The client would be a "showcase customer," featured on the website and in case studies. BigCoSoft was willing to waive licensing fees for the first three years. "But the biggest advantage," the senior VP had said, "is that with you on board, we will attract more customers. With critical mass, we can put huge development resources into this product."

At this point, you were thinking, "Yep, snoozer! I was right." Two seconds later, the first salvo was fired. Jane Pinket, the senior VP of finance, said, "Everyone knows that BigCoSoft's strength is manufacturing. Their financial package stinks. They will surely want to reuse that code, and I am going to have to take the hit. We can cut the same deal with LargeCoSoft. They already have an enterprise system for pharmaceutical firms, and their financial module is top notch."

"Another option could be to write a bolt-on," chimed in Erik Dino, the chief operations officer (COO). "That should take care of the missing finance functionalities." "But the human resource module of BigCoSoft also leaves much to be desired," interjected Joe Cole, the senior VP of human resources (HR). "Plus, we just spent $12 million on the overhaul of the benefits management system; am I going to get hit with more information systems service charges for an upgrade I don't need?"

This is about the time confusion set in, and the story became fuzzy as you got lost in the ping-pong volleys of comments, questions, and responses. With a heated topic on the agenda, it was painfully clear that the people in the room were more used to being listened to than to listening.

You were snapped back to attention when Ms. Polly closed the meeting by calling your name. She said, "Well, it looks like I underestimated how much my staff cared about systems! I will need a report with an investigation of the top three most viable options; the need to integrate is not going away, so we have to do something. I told BigCoSoft I would get back to them in three weeks."

Discussion Questions

1. Even as the lowly executive assistant to the CEO, it was apparent to you that there was some groupthink going on here. Was buying an enterprise system the only option?
2. You vaguely recall this idea of **system integration** from your information systems class two years ago. Could that approach work here?

3.1 Introduction

We formally define IS as those sociotechnical organizational systems designed to collect, process, store, and distribute information. We identified the four components of an IS and introduced the notion of systemic effects to represent the mutually interdependent relationships among the four components (Chapter 2). We also discussed how information systems fit within the organization in which they are embedded and within the larger external environment.

In this chapter, we rely on those definitions to explore the different information systems typically used in contemporary firms and the organizational impacts of those information systems. This knowledge is important for two reasons. First, information systems pervade the modern organization, so understanding how they are classified and organized is a prerequisite to being able to navigate the infrastructure of today's firm. Second, because you are ultimately responsible for the success of your company or the organizational function that you oversee (e.g., finance, marketing), it is critical that you manage organizational change when it occurs. Increasingly, the impetus behind organizational change comes from the introduction of IS enabled by digital technology, in the form of software programs and "intelligent" devices. It is therefore paramount that you have the appropriate vocabulary to join the conversation and that you have a solid understanding of what classes of software programs underpin organizational information systems today.

3.2 Computing Architectures and the Cloud

Understanding computing architectures will help you appreciate how digital services and software applications are delivered to enable your work and your decisions making. We cover the four architectures you may encounter in organizations today and devote particular attention to the dominant one: **cloud computing**.

To appreciate the distinct characteristics of different computing architectures, you need to know the logical structure of software applications. No matter how complex a software application is, you can think of it has having the following three components:

- User Interface Layer: It contains instructions specifying how the user interface should look and operate. For example, the instructions determining where to place menus on the screen or how to handle speech recognition belong in this layer.

- Application Logic Layer: It contains instructions specifying the functionalities of the software program. This layer uniquely defines the application's characteristics. For example, while Microsoft Word and PowerPoint share many of the elements of the user interface, their functionalities are significantly different.
- Data Management Layer: It contains instructions specifying where and how to store the application's data as well as how to search, retrieve, and secure them.

Mainframes and Terminals

Early computers did not fit in your pocket. In fact, they generally took up room and were housed in large cabinets referred to as *mainframes*. The term stuck, and to this day, a large digital computer supporting multiple users and multiple peripherals is called a mainframe. Mainframes are expensive; they support hundreds of simultaneous users and are purchased by organizations that require high performance computing, such as research centers, governments, large corporations, and universities. But how do all these users access the mainframe? They use terminals or, as they were called without much grace early on, "dumb terminals." In this context, a terminal is essentially an input/output device with no processing power used exclusively to access a mainframe (Figure 3.1). Today, if you are using your powerful PC to access a mainframe remotely, your computer is simply acting like a terminal to take advantage of centralized data and processing power. So where would the three layers of any software application you wanted to use have to reside? On the mainframe, of course. The terminals have no computational ability and cannot execute any instruction. One telltale sign that your computer is acting as a terminal is when you cannot use your mouse and have to rely only on keyboard commands. This is a centralized architecture.

Figure 3.1. The Digital Equipment Corporation VAX 11/780 mainframe with its terminals

Stand-Alone Personal Computing

What would you call the opposite of a centralized architecture? A distributed architecture, of course. But what would it look like? In a centralized architecture, you have one machine able to perform computations, the mainframe, accessed by many input/output devices, the terminals. In a distributed architecture, however, those terminals must be able to perform their own computations. PCs are full-fledged digital computers (Figure 3.2) and, as such, are able to execute instructions and run software applications independently.

As Moore's law continued to improve the computational power of microchips and the cost of computers plunged, Microsoft introduced the dominant software application of the modern era: Microsoft Office. Their vision of "a computer on every desk and in every home, running Microsoft software" is today a reality, and every knowledge worker (like you) uses a PC. The first PCs run their own software programs without connecting to a network, in a distributed computing architecture. A computer network, however, offers many advantages, even in a distributed computing environment. A network enables sharing—for example, the sharing of data and hardware resources (e.g., a printer, a shared disk).

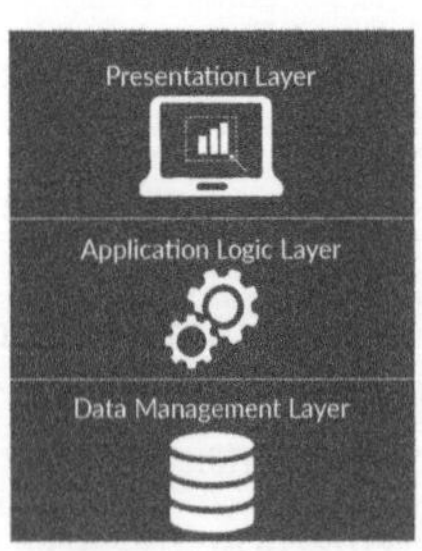

Figure 3.2. A stand-alone PC

The Client-Server Model

Shared processing, or the client-server model as it is generally called, is based on the idea of letting two or more machines share the load of executing the instructions in a software application. In other words, different computers execute the three elements of the same software application (Figure 3.3) instead of just one machine as in the previous two architectures (i.e., the mainframe or the PC). You intuitively understand that the client-server model enables the optimization of computations across the network as a whole rather than for each machine independently. The first implementation of such an approach entails separating the user interface layer from the application logic and data management layer—as it often happens in web applications. The local machine executes the instructions to render and manage the user interface layer. Yet by accessing a remote logic and data management layer over the network, the application can use centralized, easily sharable data.

The client-server model introduced two terms that are very familiar to you: *client* and *server*. In this context, a client is any software program that can make structured requests to a server in order to access resources that the server makes available. A server is a software program that makes resources available to clients. These definitions may surprise you, since you may imagine hardware, not software, when thinking of clients and servers. That's formally incorrect, and even though it is common colloquially to not make the distinction, you need to be clear on the difference.

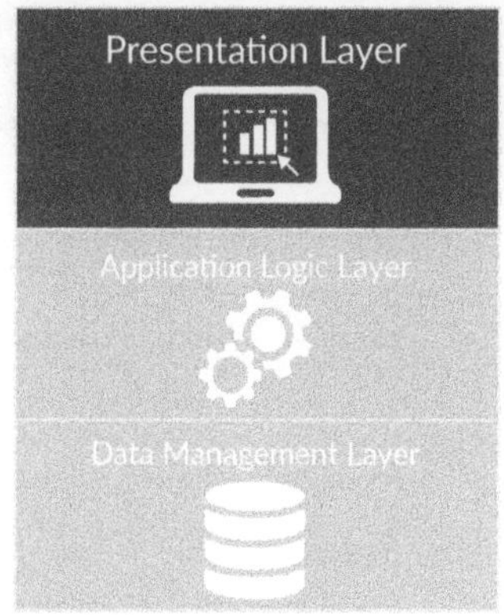

Figure 3.3. Elements of software applications

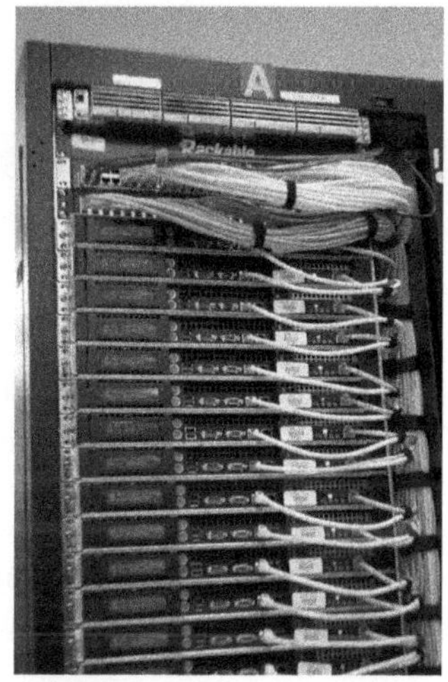

Figure 3.4. Rack of commodity servers

Source: Photo by D Coetzee / CC BY 1.0

Cloud Computing

Cloud computing uses the Internet to pool **digital resources**. Applications, computational and storage resources—the building blocks of digital technology solutions—"reside online, in the cloud" and are accessed by clients through the Internet infrastructure.[1] Consider Dropbox, a service you surely have used in the past. Dropbox provides users with the ability to back up, store, and share files. Until recently, Dropbox was running and offering its service as a cloud solution based on Amazon's infrastructure—yes, the same Amazon that sells books! Streaming services like Netflix or Spotify, social services like Reddit, and companies like Airbnb are all running on Amazon's servers.[2]

But what's all the hype? From a purely technical standpoint, the cloud is based on the client-server architecture with all forms of digital computers, such as laptops, tablets, or even cars, accessing remote servers (Figure 3.4). What is unique about the cloud-computing approach, however, is the notion that the utilization of, and payment for, the computing resources accessed in the cloud is dynamic and agile. An organization that sees a growing demand for its applications can rapidly scale the service. Moreover, cloud resources are flexible, allowing customers to acquire different services, like storage or processing capacity. This is significantly different from what happens in owned or outsourced data centers, where dedicated hardware is purchased for, or assigned to, the client and configured to run its applications. You should be familiar with cloud-computing services available at three levels of the technology stack: the application (SaaS), the platform (PaaS), and the infrastructure (IaaS; see Figure 3.5).

Software as a service (SaaS): when an application runs fully in the cloud—typically via a web browser. Salesforce, Gmail, and Dropbox are classic textbook examples of SaaS. Microsoft Office 365 is the SaaS version of the popular suite of personal productivity apps. The provider hosts the application in its data centers, and the customer accesses the needed functionalities over a computer network. Thus instead of licensing the application and requiring the customer to install it, maintain it, and generally support it, in the SaaS model, the software vendor performs these tasks.

1 The cloud is so named after the icon traditionally used to diagram the telecommunication network.

2 You can access some of Amazon's customers' stories at https://aws.amazon.com/solutions/case-studies/, and you can establish a free account yourself at https://aws.amazon.com.

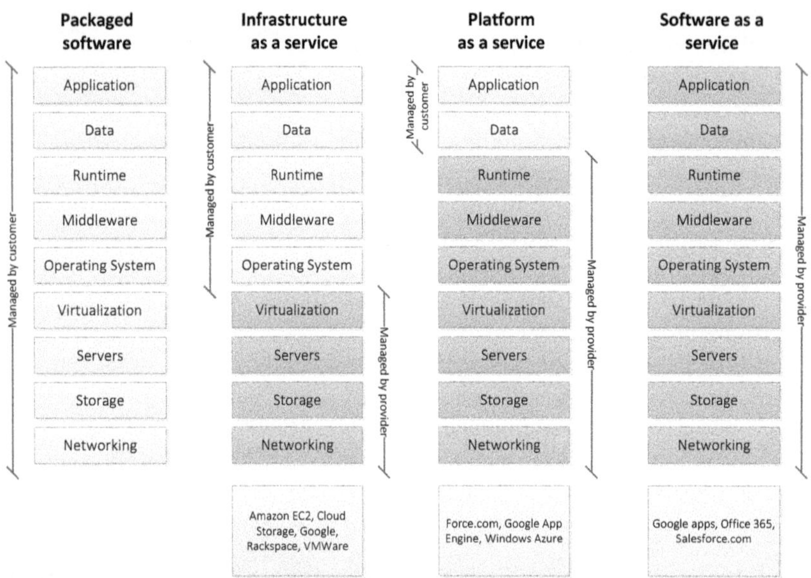

Figure 3.5. The elements of the cloud: IaaS, PaaS, and SaaS

Source: adapted from Microsoft, "Deploy Microsoft Dynamics AX 2012 R3 on Azure using Lifecycle Services [AX 2012]," retrieved from https://docs.microsoft.com/en-us/dynamics365/fin-ops-core/dev-itpro/lifecycle-services/ax-2012/deploy-2012-r3-azure-lcs

Platform as a service (PaaS): when the provider offers a platform on which the client builds its own applications, rather than offering a complete working application. In that case, you would be renting the use of hardware along with the functionalities of operating systems and utilities (i.e., storage, security, backup and recovery, application serving, and the like). This model is generally used for development, testing, and deployment of applications, as components are already in place. For example, companies can use the platform components from Force.com—the developers' side of Salesforce.com—to create and deploy their own cloud applications.

Infrastructure as a service (IaaS): when the client purchases the use of hardware functionality—in essence, computational power, storage capacity, and network connectivity—without acquiring the hardware. In the IaaS model, you rent just the virtualized hardware. All software—including the operating system, backup and recovery, and the like—are your responsibility. The IaaS provider takes care of running and maintaining the infrastructure for a fee. Like all other cloud models, IaaS platforms offer scalable resources that can be adjusted on demand. The Amazon service that powers Netflix and Airbnb, called Amazon Web Services (AWS), is another example.

The cloud market has greatly changed in the last few years. While the cloud-computing concept has been around for some time, a decade ago, firms were just starting to experiment heavily with it and develop growing confidence in the stability of the services and the providers. The focus was first on private cloud implementations—cloud-computing architectures executed within organizational boundaries. Today, the situation has radically changed. Analysts agree that cloud services will sustain a two-digit year-over-year growth in the next few years (Figure 3.6). Cloud services are now mainstream components of the digital infrastructure of large organizations, and the major software companies dominate the scene (Figure 3.7).

This growth is built on the cloud solutions' key advantages compared to on-premises implementations:[1]

- *Lower entry barriers.* Firms with limited investment capacity have access to dynamically priced enterprise-class digital resources, shifting digital technology costs from capital expenditures to operational expenditures. Start-ups can realize their digital innovations with initial investments an order of magnitude lower than in the past. Just two decades ago, a tech start-up would have to do a series A round of funding just to buy the servers and hire the staff to launch its product.

1 Marston, S., Li, Z., Bandyopadhyay, S., Zhang, J., and Ghalsasi, A. 2011. "Cloud computing—the business perspective." *Decision Support Systems 51*(1): 176–89. doi: 10.1016/j.dss.2010.12.006.

- *Faster innovation.* The immediate access to digital resources reduces time to market. Without an up-front investment, firms can deploy solutions faster, thereby facilitating innovation.
- *Higher scalability.* Solutions can easily grow proportionally with the firm's needs, and new digital resources can be allocated or reduced depending on the actual client requirements.

Today, organizations rely on mixed environments where functionalities are delivered through a combination of on-premises and cloud solutions.

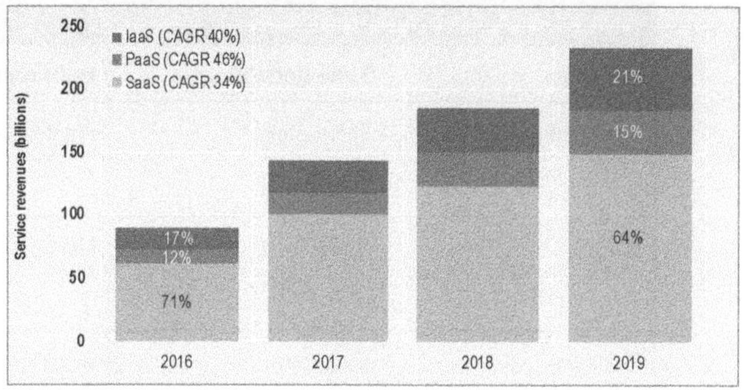

Figure 3.6. The evolution of cloud platform delivery models (compound annual growth rate [CAGR])

Source: IDC, 2020, ***Worldwide Semiannual Public Cloud Services Tracker*** (2017, 2018, 2019), retrieved from https://www.idc.com/tracker/showproductinfo.jsp?containerId=IDC_P29737

3.3 Categorizing Organizational Information Systems

Since Leavitt and Whisler popularized the term *information technology (IT)* in the business literature in 1958, researchers have advanced a number of approaches to classify and describe the role that digital technology plays in organizations.[1] These efforts focus on categorizing the software applications as the point of departure for understanding the function and functionality of the information systems built around them.

Classification models are useful for two reasons. First, they provide you with a vocabulary to interact with your colleagues and with IS professionals—a vocabulary that today in any contemporary organization you are assumed to know. Second, the models described below provide the basis for you to develop your own thinking about the role that technology plays in your organization and on how to best manage its impacts.

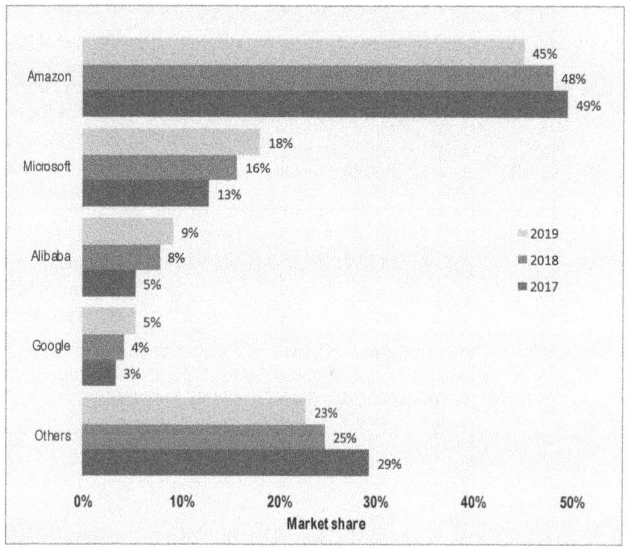

Figure 3.7. Major cloud infrastructure providers by market shares (IaaS, PaaS)

Source: Gartner, September 2017, Data from market share analysis: Public cloud services, Worldwide, 2016

Hierarchical Perspective

The hierarchical perspective recognizes that decision making and activities in organizations occur at different levels. At each level of the hierarchy, the individuals involved have different responsibilities, make different types of decisions, and carry out different kinds of activities (see Table 3.1).

The type of information systems introduced to support each level must take the differences into account (Table 3.2).

Evolution of the Hierarchical Perspective The hierarchical perspective has proven very useful over the years in enabling managers and IS professionals to easily identify the main characteristics and purpose of information systems and the digital technology products designed to support them. However, this perspective is becoming increasingly less representative due to

1 Leavitt, H. J., and Whisler, T. L. 1958, November. "Management in the 1980s." *Harvard Business Review*: 41–48.

Table 3.1. Activities by hierarchical level

Activity	Time horizon	Hierarchical level	Characteristics
Operational	Short-term	Frontline employees	Low discretion Highly structured Transaction focused
Tactical	Mid-term	Middle management	Repeatable Semistructured Recurrent
Strategic	Long-term	General management Functional management	Externally focused Ad hoc Highly unstructured

Table 3.2. Systems by hierarchical level

Level	System	Description	Example
Operational	Transaction processing systems (TPS)	Automate recurring activities and structure day-to-day operations, ensuring that they are performed with speed and accuracy	Scanner checkout system in a grocery store
Tactical	Decision Support Systems (DSSs)*	Provide the information needed by functional managers to engage in tactical decision making. DSSs focus on internal operations and rely on TPSs as data sources	Inventory monitoring and planning system for a grocery store
Strategic	Executive Information Systems (EISs) and Executive Dashboards	Enable rapid evaluation of highly aggregated organizational and trend data while still providing drill-down features that enable executives to view detailed information.	Monitoring system enabling comparison across stores and planning for new locations

* There is some confusion about terminology at this level, with some sources referring to information systems in support of middle management as management information systems (MISs). The term ***MIS*** can be misleading, as it is often used to refer to the collection of all the information systems used by the firm. We therefore use the term *DSS*.

the recent organizational trend toward the adoption of flatter hierarchies with fewer layers between frontline operations and strategic decision making. Moreover, we witnessed a trend toward the empowerment of operational personnel who increasingly enjoy decision-making discretion. You are probably already familiar with applications such as Slack that enable real-time communication and document sharing in organizational and cross-organizational teams. Slack's main competitor, Microsoft Teams, has it right in the name. These applications recognize that the traditional organizational hierarchy, while still important, is increasingly coupled with flatter and more agile organizational structures—like cross-functional teams.

Functional Perspective

The functional perspective categorizes information systems on the basis of the organization function they support (see Figure 3.8). **Functional systems** rely on the principle of local optimization, which suggests that information processing needs are unique and homogeneous within a functional area. Thus systems are tailored to those highly specific needs and use a language that is familiar to the professionals in the area. As a result, today there is a vast software industry catering to the information processing needs of every functional area in almost any industry sector.

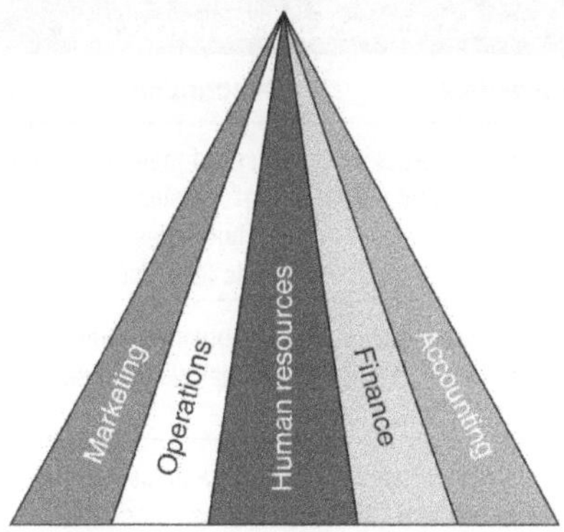

Figure 3.8. Functional perspective

Process Perspective

The primary limitation of information systems designed according to the functional or hierarchical perspectives is their lack of **integration** across organizational functions and levels resulting in redundancy and "siloed" data. Like silos used in farms to store and keep different grains separate (Figure 3.9), applications that only serve a functional area or hierarchical level segregate data and limit decision-making quality and process integration.

Figure 3.9. Silos on a farm

Photo by Arjan / CC BY 2.0

Business Process Reengineering **Business process reengineering (BPR)** emerged in the early 1990s as a way to break down organizational silos in recognition of the fact that **business processes** are inherently cross-functional. BPR is a managerial approach that employs a process view of organizational activities seeking internal **business integration** and the rationalization of activities across separate functions and units. As Michael Hammer put it in the *Harvard Business Review* article that first popularized the term *BPR*, "We should 'reengineer' our business: use the power of modern information technology to radically redesign our business processes in order to achieve dramatic improvements in their performance."[1] Such redesign has a process focus—a way of organizing work that centers on the steps necessary to create value for customers (Chapter 2)—without regard for what functional areas would traditionally be responsible for the process steps. Only with this approach can the firm "stop paving the cow paths" and achieve drastic performance improvements.[2] Today few people explicitly refer to BPR, generally *older folks* like us, but the process perspective and the idea of using digital technology to completely redesign business processes are now standard managerial practice. Digital transformation projects, for example, imply a substantial amount of BPR (Chapter 5). There are also a number of contemporary managerial practices that leverage the core BPR ideas (Table 3.3).

The Dark Side of Reengineering As with any major managerial intervention, BPR has drawbacks. First, the radical third-order change (Chapter 2) required by BPR efforts engenders significant resistance by those involved. People tend to be very comfortable with the way they operate, and changes in their job role, job scope, or responsibility require the development of new sets of skills and training that is often met with resistance. Second, BPR initiatives are very expensive because they require the firm to retire its legacy systems and develop a costly integrated technology infrastructure. Applications that had been developed to enable a functional perspective rarely can be adapted to support a process perspective. Finally, the BPR methodology developed a bad reputation after its initial popularity because of its complexity and the fact that for many organizations, BPR led to significant downsizing and layoffs.

---

1 Hammer, M. 1990, July–August. "Reengineering work: Don't automate, obliterate." *Harvard Business Review*: 104–12.

2 Ibid.

Table 3.3. Business process perspectives

Practice	Description	Example
Business process management (BPM)	A set of methods and techniques for improving the firm's business processes and performance using digital technologies both as an element of change and as enabling tools (e.g., for modeling and analyzing business processes).	The methodology followed by a U.S. bank to continuously improve and automate its business processes.
Business process automation (BPA)	The business practice focused on using digital technologies to improve business processes and workflow efficiency.	The automation of the customers' application process within the bank's loan process, so that 80% of its 1.75 million applications did not require human intervention anymore.*
Business process improvement (BPI)	The business practice focused on continuously improving business processes.	The automation of the loan process of the bank and its iterative improvement.
Robotic process automation (RPA)†	An enabler of BPA and BPI focused on using software agents to perform repetitive and tedious knowledge work tasks.	The use of an RPA robot to perform data entry from customer loan applications for documentation and analysis.

* Forrester. 2019, March. "The Total Economic Impact™ of the IBM automation platform for digital business." Retrieved from https://www.ibm.com/account/reg/us-en/signup?formid=urx-38452.

† Lacity, M., and Willcocks, L. P. 2016. "Robotic process automation at Telefónica O2." MIS Quarterly Executive 15(1): 21–35.

Table 3.4. The locus of integration

	Description
Internal	The locus of integration is internal to one organization. The firm is seeking to unify and coordinate owned assets that reside within the boundaries of the firm.
External	The locus of integration is external to one organization. Two or more organizations engage in interorganizational efforts to integrate business processes, data, or applications.
Business integration	The object of integration is one or more business activities carried out by individuals, groups, and departments within one firm or across multiple organizations. The result is streamlined business processes that encompass previously separate activities.
Systems integration	The object of integration is digital technology, resulting in unified information systems and **databases**. The outcome of system integration is a collection of compatible systems that regularly exchange information (i.e., data integration) or the development of integrated applications that replace the former discrete ones (i.e., application integration).

3.4 Enterprise Systems

The focus on breaking down barriers between siloed applications, started with the process perspective and continuing to this day, introduced a class of software solutions focused on integration: *enterprise systems*. The goal of integration is to organize, streamline, and simplify a process or an application. We categorize integration efforts on two dimensions: their locus and object. Object refers to what is being integrated, whether business activities or technology. Locus refers to the position or location of the objects of integration (Table 3.4).

Over the last three decades, the software industry has introduced a number of enterprise systems. We discuss the most important, Enterprise Resource Planning (ERP), and briefly introduce the others in the remainder of this section.

Enterprise Resource Planning

ERP systems find their roots in the material requirement planning (MRP) approach used by manufacturing firms to write software designed to automatically translate master production schedules into requirements for subassemblies, components, and raw materials. German-based

SAP extended this integrated approach to manufacturing processes that spanned other functional areas of the enterprise, such as HR, engineering, and project management. Logically, they called the resulting system ERP, a confusing label that only makes sense when you know its genesis.

While much early development in this area was focused on functionalities (i.e., what the software applications could do), the parallel development of the BPR methodology led to increasing attention to changing the way activities in the enterprise were performed. Today ERP vendors incorporate "best practices" into their applications with the objective of offering a ready-made set of menus of business processes native to the application (i.e., "best-practice software"). It is clear from the genesis of enterprise systems that their defining features are native integration and a goal to support all components of the firm's IT infrastructure. Thus ERPs are modular, integrated software applications that span (all) organizational functions and rely on one database at the core (Figure 3.10). Their principal characteristics are modularity, application and data integration, and configurability.

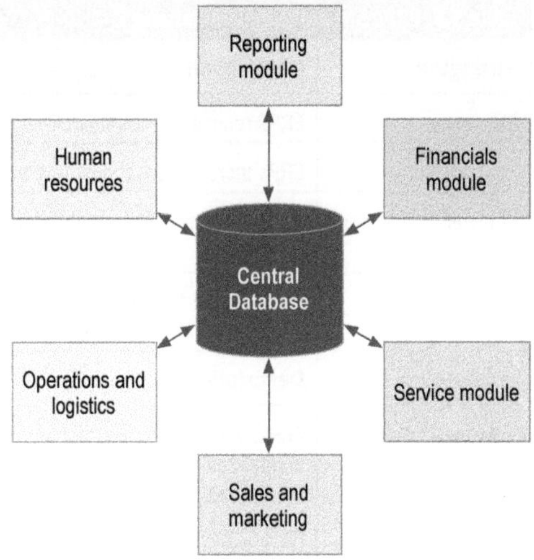

Figure 3.10. Enterprise system modules

- Modularity: ERPs are organized in separate elements, typically separated by functional area (e.g., financials, HR, logistics). Such modularity enables customers to decide which modules they need.
- Application and data integration: With application integration, an event that occurs in one module (e.g., a sale) automatically triggers events in one or more of the other separate modules (e.g., manufacturing scheduling, commission credit to the sales rep). ERPs also implement data integration by relying on one logical database at the core. Thus while there may be multiple physical data stores and locations where information resides, they will be treated as one.
- Configurability: ERPs come with configuration tables that allow the adopting firm to choose among a predefined set of options during the implementation of the application. Such options are configured by selecting parameters available in the application (e.g., the last-in, first-out [LIFO] or the first-in, first-out [FIFO] method for inventory accounting).

ERPs also allow the firm to extend the capabilities of the standard application by creating "bolt-on modules." Bolt-on modules, typically written using a programming language that is native to the ERP (e.g., ABAP in SAP), are used to further tailor the software to the specific needs of the organization. ERPs are mature solutions, and there is considerable knowledge about their advantages and disadvantages (Table 3.5).

Other Enterprise Systems

While the ERP represents the prototypical example of enterprise systems, there are a number of more specialized solutions in use in the large and midsized corporations you may soon join (Table 3.6).

3.5 Analytics and Big Data

One of the trends characterizing the last decade has been the emergence of business analytics and Big Data (Chapter 1). The *Oxford English Dictionary* defines the verb *analyze* as to "examine methodically and in detail the constitution or structure of (something, especially information), typically for purposes of explanation and interpretation" and to "discover or reveal (something)

Table 3.5. ERPs strengths and limitations	
Strengths	**Description**
Efficiency	ERPs reduce direct and indirect costs by streamlining business processes and operations.
Responsiveness	ERPs increase the ability of the firm to address customer needs thanks to application and data integration.
Knowledge infusion	ERPs, like any off-the-shelf application, enable the infusion of knowledge into the adopting firm by embedding industry best practices.
Adaptability	ERPs enable customization during installation thanks to their configurability and ability to write bolt-on modules (even though such adaptability is not comparable to that of custom-made applications).
Limitations	**Description**
Standardization	ERPs, despite their adaptability, benefit from standardization. Limiting customization ensures that the organization capitalizes on the development economies of scale of the vendor and that it will find it easiest to transition when upgrading to new versions over time. For this reason, ERPs are jokingly called "software concrete"—adaptable and moldable while being poured yet very inflexible and difficult to modify after it has solidified.
Strategic clash	ERPs force the firm to choose among the set of business processes supported by the software (i.e., the best practices). When a firm has developed unique practices that give it an advantage, it will lose them or have to engage in significant customization if it adopts the ERP.
High costs and risks	ERPs are large software implementations that entail costly projects carrying a considerable amount of risk.

Table 3.6. Other enterprise systems	
System	**Description**
Supply chain management (SCM)	SCM systems integrate the set of logistical and financial processes associated with the planning, executing, and monitoring of supply chain operations.
Customer relationship management (CRM)	CRM systems enable a strategic orientation that calls for iterative processes designed to turn customer data into customer relationships through active use of, and learning from, the information collected. They help the organizations use customer data to make inferences about customer behaviors, needs, and value to the firm so as to increase its profitability.
Knowledge management (KM)	KM systems integrate processes and data the firm uses to create, codify, gather, and disseminate knowledge within the organization.

through detailed examination." Thus generally speaking, analysis is concerned with extracting relationships, and therefore insight, from data. From a purely definitional standpoint, business analytics is the examination of business data in an effort to reveal useful insights that enable superior decision making. The continuing effects of the digital revolution and the changing nature of business data have created an unprecedented opportunity to (creatively) extract value from data (Chapter 8). Depending on the type of insight being generated, we categorize analytics efforts as:

- **Descriptive Analytics** The focus of descriptive analytics is to give an account of *what has occurred*. Descriptive analytics is substantiated in reports and visualization of data through executive dashboards, **online analytical processing (OLAP)** tools, or scorecards (see below).
- **Predictive Analytics** The focus of predictive analytics is to estimate *what will occur*. Predictive analytics is substantiated in statistical modeling and machine learning algorithms that extrapolate current trends. Political polling is an example of predictive analytics.

- **Prescriptive Analytics** The focus of prescriptive analytics is to state *what should occur*. Prescriptive analytics is substantiated in optimization algorithms that seek to identify targets to guide decisions or monitor current operations. Revenue management is an example of prescriptive analytics.

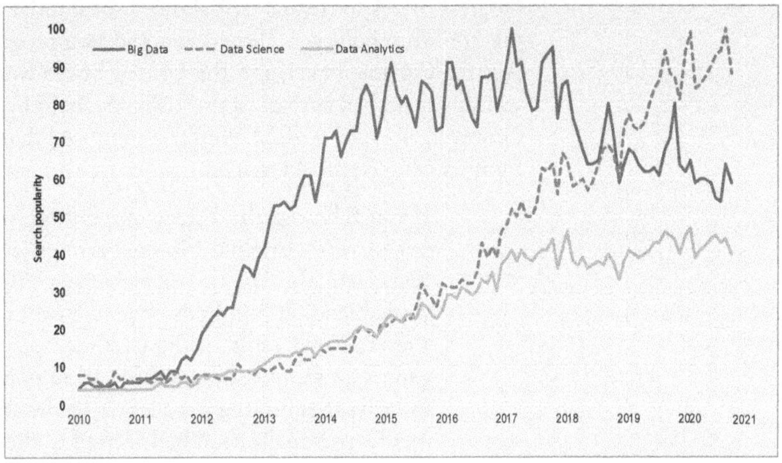

Figure 3.11. Google Trends results for search terms "Big Data," "Data Science," and "Data Analytics"

While the buzz about Big Data, analytics, and data science is inescapable, there are well-documented challenges associated with achieving a satisfactory return on investments when it comes to these projects. Recent data from the consulting firm IDC indicates that global spending on Big Data initiatives exceeded $160 billion in 2018.[1] With interest in analytics projects mounting (Figure 3.11), the survey found that "45% of organizations are still at a low level of maturity for data excellence—either Level 1 or Level 2. Only 19% of organizations reach the highest Level 5." Those of us who have been around the business and organizational applications of information for a while are used to seeing similar "disappointing" results. In fact, commenting on the survey results, Capgemini's vice president of North American business information management remarked, "If we look at it in the analogy of other technologies that have come along the way—a website, then digital presence, digital ecommerce, digital store, payments, and so forth—we saw the same type of errors in the beginning of those technology trends."[2]

Advanced analytics projects, while idiosyncratic in many respects, are a subset of a more general class of projects: business IT projects. After all, just like any other major organizational initiative dependent on digital technology, such as ERP, **CRM**, or **KM**, analytics and Big Data projects

- Require the implementation of new digital technology at their core,
- Require varying degree of process change,
- Focus (or should focus) on fulfilling organizational or business objectives, and
- Must be successfully deployed and used in organizations in order to produce returns.

Consider an early illustrative example: Netflix and its million-dollar analytics competition in 2006. The early Netflix value proposition was delivering DVDs via the postal service. It was movie rental without the stores. The Netflix competition promised the million-dollar prize to the first developer of an algorithm that, using all data collected by Netflix across its full customer base, could surpass by at least 10% the performance of its Cinematch algorithm. Cinematch, a predictive analytics algorithm, computed the expected rating each renter would assign to a movie, and the result was used for suggesting future rentals. The competition also provided a series of

1 Vesset, D., Maddox, M., Hamel, J., Olofson, C. W., Schneider, L., Kusachi, S., Bond, S., Ward-Dutton, N., Gopal, C., Fleming, M., and Marshall, C. L. 2019. *IDC FutureScape: Worldwide data, integration, and analytics 2020 predictions* (#US44802519; p. 19). IDC, retrieved from https://www.idc.com/getdoc.jsp?containerId=US45599219.

2 Bertolucci, J. 2015. "Big data success remains elusive: study." InformationWeek, retrieved from https://www.informationweek.com/big-data/big-data-analytics/big-data-success-remains-elusive-study/d/d-id/1318891.

progress prizes for teams that showed substantial predictive improvement but could not reach the 10% threshold yet. Three years and two progress prizes later, a team with members from Austria, Canada, Israel, and the United States won the big prize on September 21, 2009. The winning team had surpassed the 10% threshold by combining outputs of about 800 algorithms. However, on April 6, 2012, Netflix announced that it would not be implementing the "winning formula." The Netflix blog read,

> If you followed the Prize competition, you might be wondering what happened with the final Grand Prize ensemble that won the $1 million two years later. This is a truly impressive compilation and culmination of years of work, blending hundreds of predictive models to finally cross the finish line. We evaluated some of the new methods offline but the additional accuracy gains that we measured did not seem to justify the engineering effort needed to bring them into a production environment. Also, our focus on improving Netflix personalization had shifted to the next level by then.[1]

This example delivers an important lesson: analytics projects are, first and foremost, IS projects (Chapter 11). As such, their success depends on more than just superior algorithms, efficient software programs, and reliable hardware infrastructure. That is, they depend on more than just technology (Chapter 2). Remember this lesson when becoming involved in a business analytics project in your career.

There is little debate about the fact that extracting insight from observations, what today we call analytics, has been a fundamental business activity since humans started organized economic endeavors.[2] So why is this concept "trending" now? What is all the publicity and hype surrounding data science and Big Data all about? While it may appear that the analytics trend exploded on the scene within the last five years, its growth has paralleled that of the digital technology industry. The industry has always sought to leverage the digital computer to aid human decision making, from military applications enabling the precise computation of ballistic trajectories, to the **decision support systems** and expert systems of the 1970s and 1980s, through the **business intelligence** revolution of the 1990s. The real catalyst for the current attention to analytics is the proliferation of data generated by sensors (e.g., humidity, light), machines (e.g., GPS in cars), and increasingly, humans (e.g., Tweets). The trend is so apparent that a popular business publication even referred to humans as "walking data generators."[3] While boiling down humanity to data generation may make you cringe, the idea is that through the devices we carry (e.g., smartphones), wear (e.g., fitness trackers), and use (e.g., televisions, computers), we are leaving an unprecedented amount of "digital traces." Life is increasingly computer mediated, and computers track data comprehensively and precisely.

In the remainder of this section, we discuss the evolution of business and organizational analytics over time. We do so to provide you with needed background and to frame the development of the technologies that today comprise the analytics professional tool kit. The section is organized around four main eras of data processing in business (see Figure 3.12). While the boundaries between them are not clear-cut, each era is defined by the predominant functionalities of the

---

1 Netflix Tech Blog. 2012, April 6. "Netflix recommendations: Beyond the 5 stars (Part 1)." Medium, retrieved from https://medium.com/netflix-techblog/netflix-recommendations-beyond-the-5-stars-part-1-55838468f429.

2 Wixom, B. H., and Piccoli, G. 2020, October 5. "Actioned analytics pave the way to new customer value." MIT Sloan Management Review, retrieved from https://sloanreview.mit.edu/article/actioned-analytics-pave-the-way-to-new-customer-value/.

3 McAfee, A., and Brynjolfsson, E. 2012, October. "Big Data: The management revolution." *Harvard Business Review* *90*(10): 60–68.

digital technology that emerged at the time. However, as we mentioned earlier, you will see technologies from each era still operating once you join the workforce.

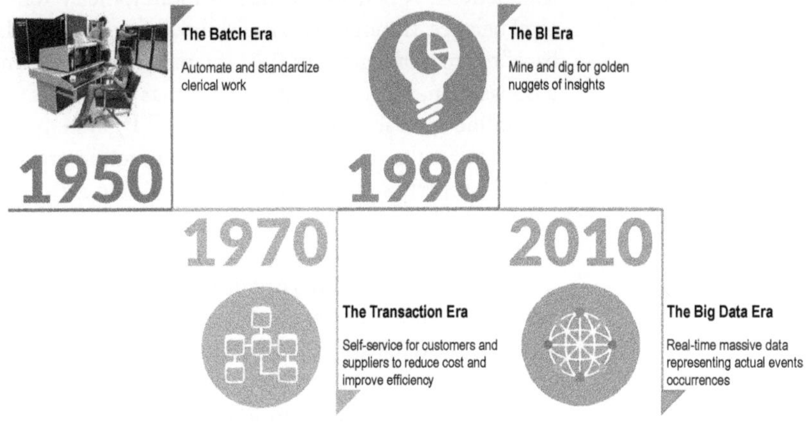

Figure 3.12. Digital data processing eras

Source: Pigni, F., Piccoli, G., and Watson, R., 2016, "Digital data streams: Creating value from the real-time flow of Big Data," ***California Management Review 58***(3): 5–25

The Batch Era

At the dawn of digital business data processing in the 1950s, transactions were batched and processed by a mainframe at periodic intervals—daily, weekly, or even monthly—one at a time. Computer "cycles" were scarce and extremely expensive. Computer "time" was shared among projects and users. Each job was queued, waiting for a system's availability. To get an idea of how this operates, consider that both the program and the data had to be written on punch cards (Figure 3.13) before being dropped in a job.[1]

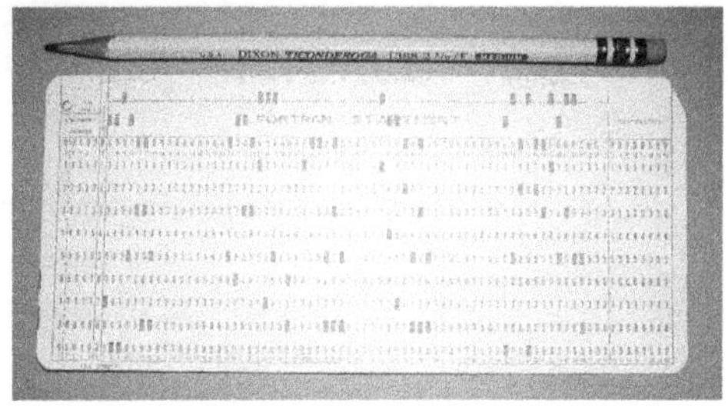

Figure 3.13. A punch card from the 1970s

Source: Photo by Wesley Fryer / CC BY-SA

Leveraging the technology of the day, the main opportunity for value creation was in the automation of clerical work and the standardization of routine decisions. Batch processing is still commonly used because it is an efficient way to deal with business process workloads, automated tasks, or processing-intensive tasks. For example, Netflix, a modern technology enterprise, uses a cloud-based batch architecture to transcode[2] the original digital content it receives from the studios into multiple, and often device-specific, streamable formats.[3] Jobs may be limited to a single movie or episode or may concern entire libraries of movies (170 terabytes [TB] of data!). Consider that when the iPad was launched back in April 2010, Netflix had to transcode its entire catalog to a new format and have it ready for "day one," even though Apple approached Netflix a few months before launch.

The Transaction Era

The introduction of computer terminals, PCs, and networks enabled the processing of online transactions. In computing terminology, *online* means in real time, and it is the opposite of *batch*. When referring to IT-enabled data management (e.g., in a database structure), the term *transaction* identifies a single logical operation on the data. A transaction is allowed only to complete successfully or fail entirely. By definition, a transaction cannot be partially completed. Withdrawing cash from an automated teller machine (ATM) or booking a flight are examples of

1 You are familiar with hard disks and solid-state drives, the hardware used for storing data today. Punch cards were the first storage media. They consisted of paper cards or paper tapes on which holes were punched to represent data in a digital format. A hole is a 1 and no hole is a 0 in binary form . . . pretty cumbersome!

2 Transcoding is the process of converting from a digital format to a different one.

3 Pearce, R. 2012, December 4. "How Netflix has the cloud do the heavy lifting for video transcoding." *Techworld Australia*, retrieved from https://www2.cio.com.au/article/443660/how_netflix_has_cloud_do_heavy_lifting_video_transcoding/?fp=4&fpid=40000.

transactions. When you withdraw money, the system has to debit your account and reduce the total balance. If money is not disbursed from the machine, your account balance must be "rolled back" to credit the amount (i.e., the money you never received). A withdrawal must be treated as a transaction, and you easily can see that customers would not be happy otherwise!

Online transaction processing (OLTP) systems are designed to provide real-time or near real-time results. To ensure reliable transaction processing by software, OLTP must respect the ACID properties of every transaction (Table 3.7).

Database Management Systems In technical parlance, a database is a self-describing collection of related records. Today's organizations manage their databases using a **database management system (DBMS)**—the software program (or collection of programs) that enables and controls access to the database. A DBMS equips a database administrator with the tools to maintain and administer the data (e.g., protect it through authentication, schedule backups) and enables application/data independence. When using a DBMS in a client/server architecture, applications need not store the data themselves but rather issue requests to the DBMS. The database can therefore be shared among multiple applications, and upgrades to one of the applications or the database itself can be made independently.

You are perhaps most familiar with personal DBMSs, such as Microsoft Access, that allow individuals or small groups to create and manage very small databases. Such a system can be confusing because it typically embeds both the DBMS and the database applications (Figure 3.14). Now consider Sabre Holdings Corporation, the parent company of Sabre Travel Network and Sabre Airline Solutions. Sabre is anchored by a database that today supports tens of thousands of users, ranging from travel agents and individuals seeking to make airline and hotel reservations, to airline check-in agents issuing boarding passes and seat assignments, to airline employees routing planes and managing their maintenance schedules. The Sabre database was once estimated to perform 85,000 transactions *per second* concerning 70 airlines and more than 100,000 hotels. For such large operations, and for much smaller ones like grocery stores and eCommerce websites, you will need a multiuser DBMS (e.g., MySQL, Oracle Database 19c, Microsoft SQL Server). Making requests to the DBMS is a set of separate database applications (Figure 3.15) typically configured in a three-tier client-server architecture.

Table 3.7. ACID principles for reliable transaction processing

Atomicity	Transactions are atomic. If one part of the transaction fails, the system must cancel the transaction. For example, if while withdrawing money at the ATM, the cash dispenser jams, your balance should not be debited.
Consistency	Transactions are consistent. In other words, only valid data are committed to long-term memory and stored in the system. For example, if the airline seat assignment system requires only letters in the first-name field, no transaction with numbers in that field is accepted.
Isolation	Transactions are nonconcurrent. If the system has yet to store the results of a transaction while writing the results of a second transaction, its database may end up holding invalid data. For example, if you are withdrawing money from an ATM while your sister is at home moving money electronically from the same account, the resulting balance may be invalid unless the system maintains isolation of the two transactions.
Durability	Transactions are durable when they can be recovered in the face of system failure. In other words, once the system has successfully processed the transaction, it will no longer lose it. For example, once the agent has changed your seat, the change is recorded in a transaction log. If anything were to go wrong with the database, the current state could be recreated by reprocessing the transactions from the log.

The Relational Model This is the dominant approach to organizing data in DBMSs today. When first introduced, it was a distinct departure from existing flat file systems that required programmers to worry about data management directly. With the flat file system, a programmer has to write instructions to handle all three layers of software applications (Figure 3.2). This no longer happens thanks to the relational model and DBMS.

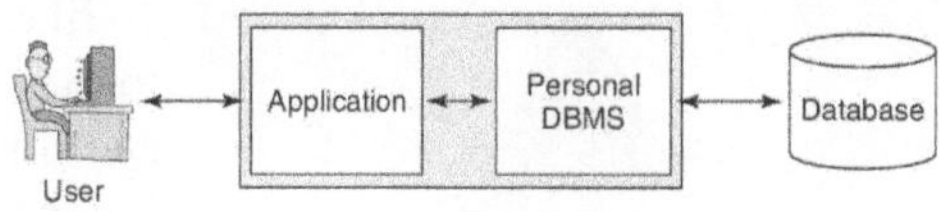

Figure 3.14. Individual application and DBMS

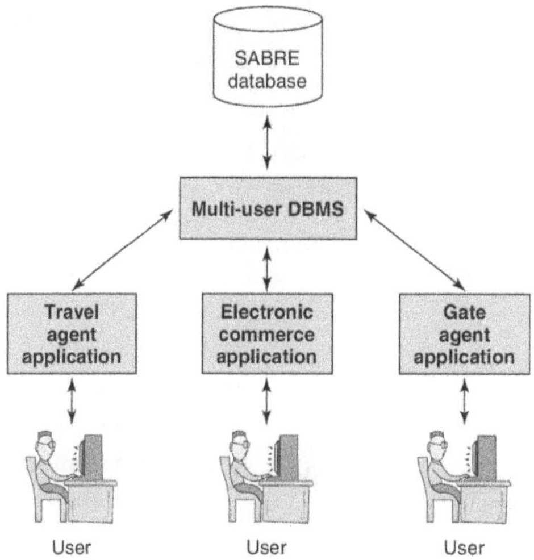

Figure 3.15. Multiuser database structure

The basic idea of the relational model is that data and the relations among them can all be organized in tables, providing a logical view of the data abstracted from the hardware. As such, relational databases provide application/data independence, tasking the DBMS with the data management layer and leaving the applications programmer to worry about only the logic and user interface layers. In a relational DBMS, the structure of the data, the metadata, is called the *schema*. Duplicated records are not allowed, so each row needs one or more attributes (columns) that uniquely identify it. These fields are the *primary key* of the table. The reason is simple: duplicate elements would generate ambiguous results. Consider the funny example of a hacker, code-named Droogie, who registered a personalized license plate in California with the text *NULL*.[1] NULL in a relational database identifies a missing or undetermined value. As Droogie reasoned, if his license plate is *NULL*, he will not receive any parking ticket because the database will treat his car as nonexistent. In a funny twist of fate, however, he started receiving the tickets of all vehicles for which a record was treated as NULL (e.g., the license was missing or not recorded correctly). By choosing the *NULL* primary key, Droogie ended up tricking the system into resolving the ambiguous results . . . making him the target of all previously ambiguous tickets!

Records in various tables are related, and relational DBMSs use a standard language, Structured Query Language (SQL), to create meaningful queries for reporting purposes. The database and application logic can be separated and differently optimized. While database programmers focus on implementing and optimizing *how* the database works, application programmers concentrate on *what* to do with the data stored in the database. This intuition was remarkable, and relational DBMSs are arguably the most successful class of enterprise software. They "run operations" in all but the smallest firms around the world. Some estimates put the industry over the $50 billion mark in 2019.[2]

Business Intelligence Era

In organizations, data are the typical by-product of daily transactions that the firm completes as it handles business in the present. Examples include websites and system activity logs, goods moving from warehouses to stores, shoppers' purchasing through apps, and fans walking through the turnstile at a stadium. **Transaction processing systems (TPSs)** are concerned with automating these day-to-day activities, and their implementation is typically justified by efficiency improvements and business process support. The business intelligence (BI) era began when firms focused their attention on the use of transaction data for decision support.

1 Barrett, B. 2019, August 13. "How a 'NULL' license plate landed one hacker in ticket hell." *Wired*, retrieved from https://www.wired.com/story/null-license-plate-landed-one-hacker-ticket-hell/.

2 Palanca, T., and Menon, S. 2020. "Market share analysis: Database management systems, worldwide" (no. G00717751). Gartner, retrieved from https://www.gartner.com/document/3988592.

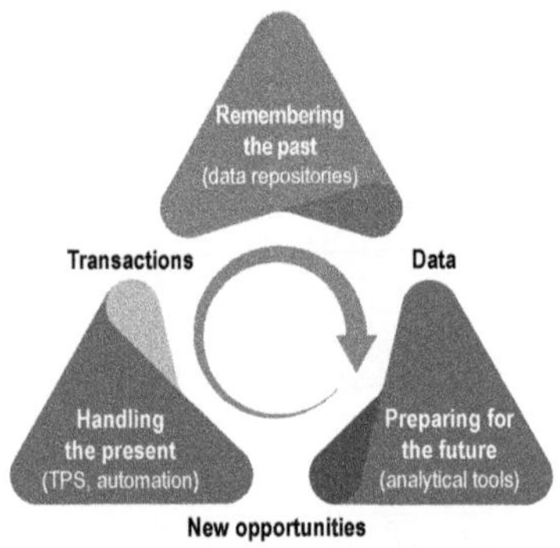

Figure 3.16. The information systems cycle

BI encompasses the set of techniques, processes, and digital technologies designed to gather and interpret data about the business in order to improve decision making and advance the organization's interests. The **information system cycle** helps better frame the defining characteristics of BI and shows how BI differs from transaction processing systems (Figure 3.16). Business data progress from their inception in transaction processing systems, to their storage in data repositories and increasingly in **data warehouses**, and finally to their use for insight extraction through analysis.

This model illustrates how transactional data, when not disposed of, create records of the past (i.e., the data are used to *remember the past*), thus becoming the raw material for valuable business intelligence and knowledge about business operations. Consider the example of Anheuser-Busch, the parent company of beer brands such as Budweiser and Michelob. Anheuser-Busch's distributors carry handheld devices rather than the traditional clipboard when they visit the stores they supply. They use the device to take orders and also to gather data about competitors' products and strategy (e.g., pricing, placement, promotions). The data are immediately uploaded to Anheuser-Busch's data warehouse, where they are joined with demographic, marketing, and other external data to become available for analyses using an application called BudNet. Mapped to the IS cycle, this application shows how data progresses through the BudNet application, from collection as orders are taken (handling the present), to long-term storage in the data warehouse (remembering the past), to its employment for analytical purposes (preparing for the future; Figure 3.17).

As the example above shows, business intelligence encompasses transaction processing, since these transaction processing systems generate the data. Traditional BI focus is on structured, mostly internal, data, and it is performed with "enterprise IT tools" in integrated suites (e.g., Oracle Business Intelligence, SAP Business Objects, and IBM Cognos, to cite some of the major industry players). However, the defining characteristic of BI is a conscious focus on the analysis and communication of the data generated.

Components of the **Business Intelligence Infrastructure** From our definition of business intelligence, it is clear that BI is not a single digital technology or set of technologies. In fact, as was

Figure 3.17. Business intelligence at Anheuser-Busch

the case in the Anheuser-Busch example, in order to engage in business intelligence, the firm must develop an information system paying particular attention to each of the four components (Chapter 2). Yet given the sheer volume of data and information that a firm needs to manage as part of its BI initiatives, digital technology is a critical component. As a consequence, business intelligence applications now represent a thriving segment of the software industry.

We use the term *business intelligence infrastructure* to refer to the set of digital technologies designed to create, manage, and analyze large repositories of data in an effort to extract value from them. Beyond the transaction processing systems that generate the needed data, the main components of a BI infrastructure are data warehouses, **data marts**, query and reporting tools, OLAP, and **data mining**.

Data Warehouse A data warehouse, or, more precisely, an enterprise data warehouse, is a data repository that collects and consolidates data from multiple source systems, both internal to the organization and external, with the purpose of enabling analysis. A data warehouse typically has the following characteristics:

- *Large in size.* Data warehouses easily span into the scale of terabytes—the rough equivalent of all the content of an academic library.
- *Large in scope.* Data warehouses draw information from a wide variety of source systems.
- *Enabling data integration.* Data warehouses compile and collect data from multiple source systems, ensuring that data are accurate and current.
- *Designed for analytics.* The defining characteristic of a data warehouse, what makes it different from a large transactional database, is its focus on analysis and analytics.

A data warehouse is typically the cornerstone of a BI infrastructure, but the repository is only valuable insomuch as the data it contains are accurate—a condition that IT professionals like to call GIGO: garbage in, garbage out. Thus the primary driver of the complexity and cost of building a data warehouse is the need to gather and clean the data to be compiled—the extracting, transforming, loading (ETL) process, as it is known. What may appear to be a trivial process at first glance is in fact a lengthy and complex undertaking designed to ensure that redundancy, data integrity violations, and inconsistencies are kept to a minimum.

As mentioned above, a data warehouse is optimized for analysis. While traditional transactional databases enable analysis and reporting, their structure is optimized for fast data retrieval at the atomic level. For example, if you asked a gate agent to change your seat assignment as you prepared to board an American Airlines flight, she would have to retrieve your individual record, make the change, and store it before issuing the new boarding pass. Such a transaction is focused on the present, accesses one record, and addresses a specific item in the record (Table 3.8).

Now consider our Anheuser-Busch example. The firm is not interested in any one of its target customers individually (i.e., how much beer you purchase). Rather, it focuses on large groups of individuals who share some characteristics in an effort to identify patterns and draw conclusions about their collective behavior (i.e., how much, when, and how people in your neighborhood purchase a given brand of beer). Such transactions seek to access multiple (i.e., thousands of) historical records and aggregate them on several dimensions of interest (Table 3.8).

While aggregation is possible with transactional databases—and indeed it is performed every day in organizations to extract reports—as the size of the database grows, aggregating data in transactional databases degrades performance. Thus by using specialized techniques such as

Table 3.8. Transactional versus analytical databases

Transactional database	Analytical database
Atomic level	Aggregate level
Current data	Historical data
Individual record access	Multiple record access

multidimensional representations and preaggregation, an analytical database is optimized for enabling complex querying and the analysis of large amounts of data with an acceptable response time.

Data Mart A data mart is a scaled-down version of a data warehouse that focuses on the needs of a specific audience. Like a data warehouse, a data mart is a repository built explicitly to enable analysis. Unlike a data warehouse, though, a data mart is designed for the specific needs of a narrowly defined community of knowledge workers (e.g., marketing group, accountants). The advantages of a data mart over a data warehouse are that the data mart is smaller in scope, and thus easier to build, and uses audience-specific data classifications and language. In very large organizations that have already created an enterprise data warehouse, data marts may be introduced to simplify and focus analysis by a department or function.

In many cases, a firm will develop data marts in order to take an incremental approach to its BI strategy. In this case, the firm will introduce one or more data marts before creating a data warehouse. This incremental strategy can help generate buy-in from senior executives and momentum behind business intelligence initiatives. The drawback of this approach is that a proliferation of data marts creates the potential for replicating the problem that centralized data storage was designed to eliminate—data redundancy and a lack of data consolidation.

Online Analytical Processing The term *online analytical processing* (OLAP) refers to a class of software programs that enables a knowledge worker to easily and selectively extract and view data from analytical databases. The defining characteristic of OLAP tools is that they are user driven. In other words, an analyst must issue a query that specifies what data items the user is interested in. Note that OLAP users need not be IT specialists; in fact, if you elect to become an analyst for a financial institution or a marketing organization, you stand a very good chance of using OLAP tools yourself.

Data Mining Data mining is the process of automatically discovering nonobvious relationships in large databases. The recent popularity of data mining is due to the availability of powerful computer systems that can quickly search through large volumes of data contained in data warehouses. Using years of compiled data, for example, Walmart analysts sought to identify what the best-selling items were in areas under threat of an approaching hurricane. Much to everyone's surprise, the most important item needed to prepare for a hurricane was not water, wood, or nails. Walmart found that strawberry Pop-Tarts were the top sellers! While it is relatively easy to make sense of this finding once we are told about it (i.e., Pop-Tarts have a long shelf life, they need not be cooked, kids like them), it is a nonobvious and largely unexpected finding beforehand. Walmart capitalized on similar analysis and leveraged the data collected during past disasters to forecast and stock useful products during an emergency.[1]

Like OLAP tools, data mining is used to analyze historical information. Unlike OLAP, though, data mining is more akin to a brute force approach, enabling the software to identify significant patterns by analyzing all possible combinations rather than relying on an analyst to structure a specific query. Of course, analysts are still heavily involved in data mining, as they must interpret the results. Yet as in the Walmart example, data mining is used to seek unexpected (i.e., nonobvious) relationships among data items. The following is a list of possible patterns a data mining application may identify:

---

1 See, for example, Walmart. 2018. "2018 global responsibility report summary" (p. 232). Retrieved from https://corporate.walmart.com/media-library/document/2018-grr-summary/_proxyDocument?id=00000162-e4a5-db25-a97f-f7fd785a0001: 42–45.

- *Associations.* Associations occur when one event can be correlated to another event (e.g., beer purchases are highly associated with chips purchases in the week leading up to the World Cup finals).
- *Sequences.* Sequences occur when one event leads to another subsequent event (e.g., a rug purchase is followed by a purchase of curtains).
- *Anomalies.* Anomalies occur when data deviate from an expected pattern (e.g., when a credit card is used in a physical store in Grenoble, France, and later that day in another store in Baton Rouge, Louisiana).
- *Classification.* Classification occurs when categories are generated from the data (e.g., customer profiles based on historical spending).
- *Forecasting.* Forecasting occurs when patterns in the data can be extrapolated to predict future events (e.g., sales forecasts based on weather patterns).

Big Data Era

As software continues relentlessly to "eat the world," the pervasiveness of digital technologies and the ubiquity of the Internet allow organizations to collect and store increasing amounts of structured and unstructured data. This phenomenon is generally called *Big Data.* While the exact genesis of the term is unclear, by some accounts, it was first used in the IT industry in the mid-1990s during a lunch conversation at Silicon Graphics (SGI).[1] At the time, the now defunct SGI was the leading authority in computer graphics and high-performance computing and a leading supplier for Hollywood studios and NASA. The term was used to convey the idea that a new digital infrastructure was emerging to cope with the growth of both data storage and processing needs. As often happens in the digital technology industry, the idea slowly grew in labs and IT departments, then caught the attention of early adopters pushing the limits of their current infrastructure, and at one point burst onto the mainstream business scene.

The current definition of *Big Data* is credited to Doug Laney, a consultant with the META Group, who wrote a two-page 2001 report outlining emerging trends for the next five years.[2] The report was organized along three lines of data evolution: volume, velocity, and variety (the three *V*s):

- *Volume* is the amount of digital data that organizations have to store and manage. This dimension is best thought of as a trend rather than a threshold. It is not clear what would constitute "Big Data volume"—is it terabytes, petabytes, exabytes? What is clear, however, is that organizations have to manage increasing amounts of data (Chapter 1) and that the new data storage architectures are required to do so. As an example, YouTube's own statistics showed that in 2020, its two billion users uploaded a total of 500 hours of video footage . . . per minute![3]
- *Velocity* is the speed of creation and use of new digital data. This dimension relates to the need to gather, process, and communicate information in real time. New data sources like sensors, radio-frequency identification (RFID) tag readers, video streams, clickstreams, and tweets—to cite just some of them—require modern IT architectures capable of handling streaming data. By contrast, traditional data warehousing environments were designed for batch processing and the support of

---

1 Lohr, S. 2013, February 2. "The origins of 'Big Data': An etymological detective story." *New York Times,* retrieved from http://bits.blogs.nytimes.com/2013/02/01/the-origins-of-big-data-an-etymological-detective-story.

2 Laney, D. 2001, February 6. "3D data management: Controlling data volume, velocity and variety." *Application delivery strategies: META Group.*

3 https://www.omnicoreagency.com/youtube-statistics/.

online analytical processing within the predetermined confines of internal organizational data. Data velocity enables firms to sense and respond to relevant events (e.g., a market change, a competitor's move) and fuel the trend of accelerated business operations that proprietary networks created in the 1980s and the Internet consolidated in the late 1990s and early 2000s.

- *Variety* is the kind of digital data that organizations have to store and manage. You can roughly think about variety in terms of data types (e.g., text, image, video, sensor readings). More important, you should think about variety in terms of the structure of the data. If BI was the realm of structured data (i.e., data that are configured within a predefined organizing schema), a Big Data infrastructure must handle unexpected formats and "schema-less" data. A defining characteristic of the Big Data paradigm is that both the problem definition and the data model are (often) not completely known in advance.

Big Data: Technology We are in the midst of the Big Data era: both technology and management practices are evolving rapidly. However, there are some consolidated technologies and practices that form the cornerstone of an organization's Big Data infrastructure. Relational DBMSs are not going away any time soon and still power the great majority of organizational systems and business applications. The market is maturing with the presence of notable commercial (e.g., Oracle, Microsoft SQL, IBM DB2) and open source solutions (e.g., PostgreSQL, MySQL). Data warehouses are also going to remain very relevant, with a number of active cloud solution providers. One of the leaders in this space, Snowflake Inc. (Figure 3.18), recently completed the most successful initial public offering (IPO) of 2020.

The volume, velocity, and variety of data many organizations are attempting to collect as part of their "Big Data strategies" has forced them to seek new data storage solutions, and vendors are of course happy to fill the need. Most prominent are data lakes, a new class of storage solutions expressly designed to serve as the centralized repository of all structured and unstructured data collected by the organization, with the ability to grow at any scale. Unlike a data warehouse, typically housing only structured data and requiring the schema design up front, the data lake prioritizes data collection. Thus data are stored "as-is," without having to first structure them. As you can guess, if you have been paying attention in this chapter, data lakes replace neither relational databases nor data warehouses—rather, they complement these earlier technologies having different characteristics and uses (Table 3.9). On the analytics side, large-scale data-driven applications like Google's search engine or Instagram's social graph ran up against some of the intrinsic limits of traditional solutions. NoSQL databases represent the best attempt at overcoming the limits of analytical database, providing

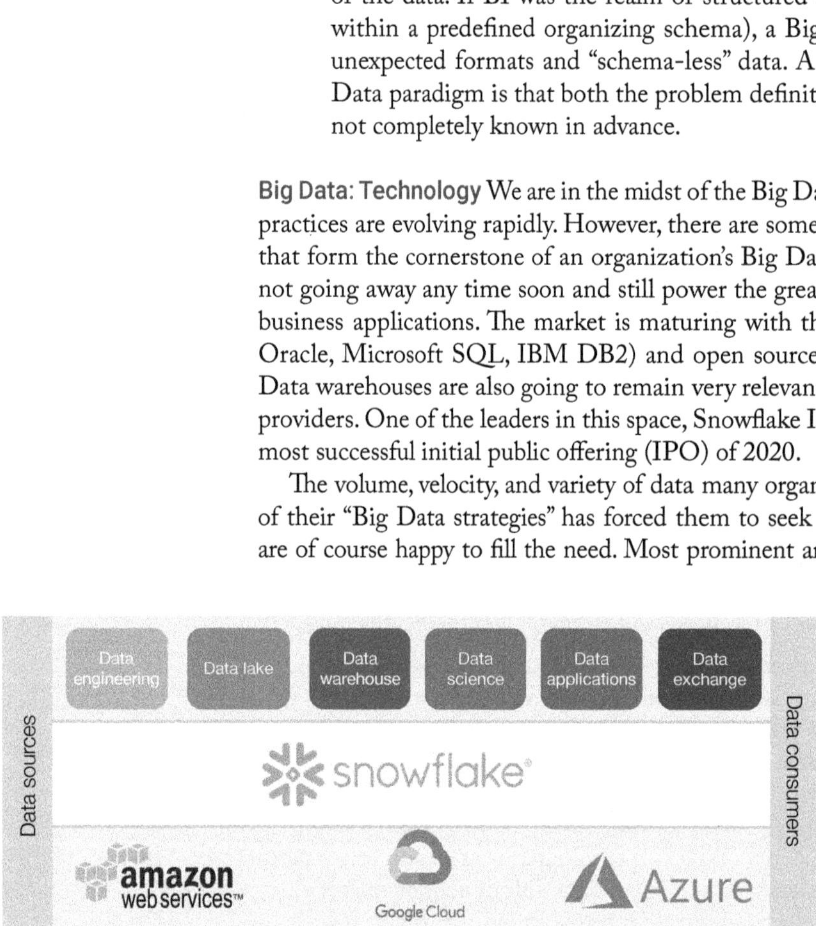

Figure 3.18. Snowflakes' data cloud architecture
Source: Author's redrawing

- a simple and flexible schema (or schema-less) structure to accommodate heterogeneous data,

Table 3.9. Activities by hierarchical level

Characteristics	Data warehouse	Data lake
Data	Long-term	General management Functional management
Schema	Relational from transactional systems, operational databases, and a line of business applications	Nonrelational and relational from Internet-of-things devices, websites, mobile apps, social media, and corporate applications
Price/performance	Designed prior to the data warehouse implementation (schema-on-write)	Written at the time of analysis (schema-on-read)
Data quality	Highly curated data that serves as the central version of the truth	Any data that may not be curated (i.e., raw data)
Users	Business analysts	Data scientists, data developers, and business analysts (using curated data)
Analytics	Batch reporting, business intelligence, and visualizations	Machine learning, predictive analytics, data discovery, and profiling

Source: What is a data lake?, Amazon Web Services, retrieved from https://aws.amazon.com/big-data/datalakes-and-analytics/what-is-a-data-lake/

- horizontal scalability adapted to distributed environments composed of commoditized servers, and
- high availability.

As the volume, velocity, and variety of data continue to increase, NoSQL database systems complement earlier solutions. To understand the reasons NoSQL databases have emerged as a solution to the scalability problem of relational DBMSs, we have to understand how distributed data management systems, like Hadoop or MongoDB, operate. Systems can scale vertically (scale in) or horizontally (scale out), depending on how they grow to accommodate more computational capacity. The term *horizontal scalability* refers to the capacity of a distributed system to improve its performance by adding new nodes (e.g., more computers are added to the network). In contrast, vertical scalability means that a single node is made more capable (e.g., using a faster processor or increasing the system memory). One of the main advantages of horizontal scalability is that it distributes the workload on clusters of commodity servers and hardware. Hardware can be added incrementally as needed. The typical architecture before the emergence of Big Data favored vertical scalability, as individual servers provided enough resources to manage the workload. While horizontal scalability is the optimal approach to handle massive amounts of data, it comes at the cost of higher software complexity, latency, and network coordination due to the fact that the application now needs to manage this distributed structure. Moreover, not all tasks can be parallelized, and proper scaling in distributed environments is still challenging.[1]

To intuitively understand the fundamental difference between SQL and NoSQL databases, consider a simple case where data are distributed on two different servers. Imagine that the database system locks both copies of a table on the two servers when handling a transaction. The database will always remain consistent, but it will not be available for other transactions. Conversely, if one or both tables accept write requests at all times, the database will be always

1 Amdahl's law, established in 1967 by Gene Amdahl, an IBM scientist at the time, prescribes that for fixed-size problems, the performance improvement of parallelized computations (the scaling) is limited by the quantity of operations that in any case need to be serialized, the problem size, and the type of processing nodes. However, as computing power increases, programs tend to increase in size, too, making the case for better scaling.

Table 3.10. Relational versus NoSQL databases

	Relational databases	NoSQL databases
Optimal workload	Relational databases are designed for transactional and strongly consistent OLTP applications and are good for OLAP.	NoSQL databases are designed for a number of data access patterns that include low-latency applications. NoSQL search databases are designed for analytics over semistructured data.
Data model	The relational model normalizes data into tables that are composed of rows and columns. A schema strictly defines the tables, rows, columns, indexes, relationships between tables, and other database elements. The database enforces the referential integrity in relationships between tables.	NoSQL databases provide a variety of data models, such as key-value, document, and graph, which are optimized for performance and scale.
Performance	Performance is generally dependent on the disk subsystem. The optimization of queries, indexes, and table structure is often required to achieve peak performance.	Performance is generally a function of the underlying hardware cluster size, network latency, and the calling application.
Scale	Relational databases typically scale up by increasing the compute capabilities of the hardware or scale out by adding replicas for read-only workloads.	NoSQL databases typically are partitionable because access patterns are able to scale out by using distributed architecture to increase throughput that provides consistent performance at a near boundless scale.
Application program interfaces (APIs)	Requests to store and retrieve data are communicated using queries that conform to an SQL. These queries are parsed and executed by the relational database.	Object-based APIs allow app developers to easily store and retrieve data structures. Partition keys let apps look up key-value pairs, column sets, or semistructured documents that contain serialized app objects and attributes.

Source: Adapted from ***What is NoSQL?***, Amazon Web Services, retrieved from https://aws.amazon.com/nosql/

available for writing, but it will likely fall into an inconsistent state at times, and the two tables may have different data.

To overcome relational DBMS limits in a distributed environment, NoSQL databases had to accept consistency limitations. But how can one trade something as fundamental as data consistency? The solution is to accept inconsistencies but only for a limited period. Consider Amazon's DynamoDB, a NoSQL database supporting the firm's shopping cart operations. A stringent requirement for the shopping cart is its availability. The cart has to always allow customers to put products in, even when a network node or a server fails. The shopping cart is a critical part of Amazon's customer service experience, and any downtime has a direct financial impact. In this scenario, customers can always order, but Amazon accepts the risk of not being able to ship all items purchased, as availabilities and orders may be inconsistent for short periods of time. In other words, Amazon does not lock a product availability record every time a customer adds a product to her shopping cart despite the fact that, in theory, if the sale goes through, that item may not be available to other customers. If they did, the probability of missing a sale would increase (consider that Amazon's shopping cart service handles tens of millions of requests and millions of checkouts per day).[1] As service downtime is clearly not an option, Amazon trades consistency for availability.

It is not critical for you to understand all trade-offs associated with different database structures (that's a course in and of itself). What you should grasp, however, is that different database systems are architected differently and are designed to optimize performance on different requirements (Table 3.10).

1 DeCandia, G., et al. 2007. "Dynamo: Amazon's highly available key-value store." *ACM SIGOPS Operating Systems Review 41*, ACM, retrieved from http://dl.acm.org/citation.cfm?id=1294281: 205–20.

NoSQL databases are designed to take advantage of the scale-out architecture of the cloud environment, maintaining performances when transactions grow. Compared to SQL databases, they can better handle large volumes of data and semi- and unstructured data and are more flexible in the definition of their structure. Additionally, NoSQL databases store data in forms that are closer to the way applications handle them, allowing developers to store data based on the characteristics of the application. However, no one approach can be superior on all dimensions. The relational DBMS has been, and continues to be, the workhorse for operational systems and transactional processes. But the Big Data era calls for new approaches as well. Being aware of the options and having a sense of the inherent trade-offs they entail will make it easier for you to partner with database experts in your advanced analytics projects.

3.6 Managerial Takeaways

The material in this chapter tends to polarize our readers. Some love to learn how the infrastructure of organizations evolve over time, to understand the genesis of systems they have encountered at work, and to understand how they work. Others don't enjoy this chapter and wonder why a manager should need to hear all this jargon and this "historical stuff." In our view, there are two managerial takeaways in this chapter. First, we tried to clarify the landscape of digital technologies and applications you will encounter when you go to work. Having understood their genesis and their respective strengths and limitations, you will be better able to understand the value each system in your organization delivers. You will also be better able to identify mismatches and underoptimized solutions. Our favorite example, albeit on a small scale, is the many organizations that wring Microsoft Excel to do the work of a database. Just because you may have a hammer, not everything you encounter is a nail. This chapter should also lay to rest the myth that organizations can easily "update their systems" when "new solutions" arise in order to "stay ahead of the curve."[1] You now know that this is a naïve view because new technologies often complement, rather than supersede, earlier systems. Moreover, some older solutions stand the test of time because they are optimal for specific tasks that have not gone away (e.g., batch processing). Second, while we recognize that there is a lot of jargon in this chapter, this is now the language of business. As the digital revolution continues to transform legacy organizations, you are expected to be conversant in it. Organizing the information around general frameworks (e.g., computing architectures) and long-term trends (e.g., business data processing eras), will help you organize the vocabulary and, more importantly, gain credibility with your technology counterparts at work.

Summary

This chapter completed our introduction to the foundations IS by explaining the vocabulary and concepts you need in order to categorize different types of information systems and communicate with other managers and IS professionals. Specifically, in this chapter, we learned the following:

- Organizational applications are delivered to users through different computing architectures and structures, either centralized or distributed.
- Cloud computing has been a revolution in the software market, lowering the barriers to entry and giving access to the benefits of enterprise-class IT services to even start-ups and small organizations.
- SaaS, PaaS, and IaaS are cloud delivery approaches in which a provider manages the application, the platform, or only the infrastructure through its data centers and the customer accesses the needed functionalities over the network.
- Different organizational information systems can be characterized through a hierarchical, functional, or

1 We hope you noticed this sentence is packed with buzzwords to make a point!

process perspective. The functional perspective underpins many of the most recent managerial trends, including BPR and business systems integration efforts.

- BPR, defined as a managerial approach calling for a process view of organizational activities, was one of the principal management trends of the mid-1990s. While its popularity has faded somewhat, you should not forget the key lessons of BPR: firms evolve over time, as do technologies; old assumptions may no longer be valid today; and you have an opportunity to use IT to dramatically improve the efficiency and effectiveness of the firm's business processes.
- ERPs are modular, integrated software applications that span (all) organizational functions and rely on one database at the core. The defining characteristics of enterprise systems are their large scope (seeking to support all aspects of an organization's IT infrastructure in an integrated fashion), their modularity (enabling adopting firms to select which components they need), and their configurability (allowing adopting organizations to choose among a predefined set of options during the implementation of the application).
- While enterprise systems offer much promise in terms of efficiency improvements, increased responsiveness, knowledge infusion, and adaptability, they have some significant limitations as well, including the trade-off between standardization and flexibility, the limitations of best-practice software, the potential for strategic clash, and the high costs and risks of the implementation process.
- Seeking to understand the structure of information and extracting insight from observations are decidedly human activities, not a new trend. Analytics has been a key concern of IS professionals throughout the four main eras of data processing in business since the early applications of IT to business in the 1950s. Most recently, analytics was embedded in BI and Big Data efforts. BI encompasses the set of techniques, processes, and technologies designed to gather and interpret data about the business in order to improve decision making and advance the organization's interests.
- BI has been one of the dominant trends in organizational computing over the last decade. It encompasses the set of techniques, processes, and technologies designed to enable managers to gain superior insight into and understanding of their business and thus make better decisions. A firm that intends to engage in business intelligence needs to create a business intelligence infrastructure that typically is centered on a data warehouse. Internal transaction processing systems and external sources feed the data warehouse. Once the data have been structured for analysis in the data warehouse or a data mart, they can be examined using analytical tools such as OLAP and data mining.
- *Big Data* is the umbrella term under which the major recent trends converge. The impacts on decision making and knowledge extraction are a paradigm shift in the way management and organizations have traditionally made decisions. NoSQL databases represent a new breed of technologies developed to overcome the limits of current data management approaches when dealing with Big Data.

Study Questions

1. Describe the principal differences between transaction processing systems, management information systems, and **executive information systems.**
2. Provide an example of functional systems. What is the defining characteristic of these systems?
3. Provide two examples of transaction processing systems.
4. What is batch processing? Provide two current examples of batch processing.
5. Define the concept of BPR. Can you provide an example, real or imaginary, of a company operating under old assumptions that are no longer valid? How would you propose to redesign the firm's business processes? What information technology would you expect the firm to adopt to enable the redesigned process?
6. What are the key limitations and drawbacks of BPR?

7. How have we defined the concepts of business and systems integration? What is the relationship between the two? Can you provide examples of each?
8. What is an enterprise system? Can you describe its genesis? Identify the main advantages and risks that a firm should consider when evaluating the opportunity for the installation of an enterprise system.
9. What is an ERP system? After presenting its principal characteristics, discuss its main advantages and limitations.
10. Define the following terms: *business intelligence*, *business intelligence infrastructure*, *data warehouse*, *data mart*, *OLAP*, and *data mining*. What is the relationship among the various elements of BI infrastructure?
11. Define the term *Big Data*. What are the main implications of the three *V*s—volume, velocity, and variety—on current managerial practices?
12. What is the relational model? What are its main limitations in the current context?
13. What is the difference between data mining and Big Data?
14. Describe the client-server model. What are the main differences compared to a centralized model?
15. Present the main differences and discuss the main advantages and disadvantages of peer-to-peer and three-tiered architectures.
16. What are the main advantages and disadvantages of a cloud solution compared with an on-premises solution?

Glossary

A glossary of key terms can be found at the book's website. Visit https://www.prospectpressvt.com/textbooks/piccoli-information-systems-for-managers-5-0, then click on "Student Resources" in the horizontal red menu bar.

PART II

Competing in the Digital Age

The rallying cry of enthusiastic engineers, entrepreneurs, venture capitalists, investors, and just about everyone else during the late 1990s (or the dot-com era, as it became known) was "The Internet changes everything." In such a statement, there was certainly quite a bit of FOMO (fear of missing out), and while most people were unsure as to how exactly the Internet was going to change everything, they did not want to be left behind . . . in case it did. The frenzy took the NASDAQ—the tech-focused electronic equity security market—past 5,000 points in March 2000 before seeing it tumble down to a 1,400-point low. The market crash notwithstanding, the dot-com era ushered in what some have named the "network economy." It is evident now, two decades after the crash, that the Internet, and the many digital technologies and digital innovations that are built on the Internet infrastructure, dramatically changed the competitive landscape for almost every company. Today we can see a resurgence of the positive mentality that drove the growth of the Internet. The driving force of this renewed optimism has been cloud computing, the mobile platform, digital transformation, and entrepreneurship.

The most apparent changes brought about by the network economy are the ones at the interface between companies and their clients, in what is called the *business-to-consumer* space. You could not imagine running an airline or a hotel today without a digital relationship with your customers, enabling them to learn about your offer, book reservations, and even check in. The same goes for banks, retailers in the widest variety of sectors, publishers, movie studios, and even celebrities! However, while the front end grabbed all the headlines, the bulk of the "digital revolution" took place behind the scenes, within company walls and in their relationships with other businesses: the *business-to-business* space. This trend is captured by a quote attributed to Satya Nadella, the current chief executive officer (CEO) of Microsoft. Nadella, referring to the potential for efficiencies ushered in by digital technologies, said, "Now information technology is at the core of how you do your business and how your business model itself evolves."[1]

Whether the Internet and modern digital technologies do indeed "change everything" is really not the issue.[2] There is no doubt that a global, affordable, digital network infrastructure for communication where all objects can potentially interconnect is a critical business enabler. There is also no doubt that the pervasive adoption of digital technology by consumers and companies alike has created unprecedented opportunities to rethink business operations and business models. You, as a general or functional manager, must be able to harness digital technology to benefit your organization. In order to do so, you must be able to answer two broad questions:

1. What impacts do modern digital technologies have on the competitive landscape? How do they change the environment your firm is, and will be, competing in?
2. How have modern digital technologies been used by organizations to transform their business? How can they be used by your firm to take advantage of the changing competitive landscape and innovate to stay relevant?

---

1 Rosoff, M. 2016. "Satya Nadella on why you'll love Cortana, how cars are like data centers, and what's spurring all these global startups." *Business Insider*, April 4, 2016. https://www.businessinsider.com/satya-nadella-microsoft-interview-with-business-insider-2016-4.

2 Fingar, P., and Aronica, R. C. 2001. *The death of "e" and the birth of the real new economy: Business models, technologies and strategies for the 21st century*. Tampa, FL: Meghan-Kiffer Press.

Part II of this book is devoted to answering these two questions.

- *Chapter 4: Digital Disruption and the Competitive Environment.* This chapter focuses on the first question and discusses how digital networks differ as economic entities from traditional and physical goods. With this backdrop, the chapter discusses how Internet, mobile, and related digital technologies have changed the modern competitive landscape and the implications this has for strategy in the modern firm.
- *Chapter 5: Digital Transformation, Innovation, and Entrepreneurship.* This chapter tackles the second question. It provides a background on the technologies shaping the current digital transformation and innovation trends. It then introduces a vocabulary with which to understand digital business trends past, present, and future.

Chapter 4

Digital Disruption and the Competitive Environment

What You Will Learn in This Chapter

This chapter focuses on the revolutionary changes that have occurred in the global economy since the advent of the commercial **Internet** in the mid-1990s and the successive waves of digital innovation that are continuing to this day. The digitally networked world is widely different from the prenetworked one because networks have peculiar economic characteristics. Moreover, pervasive networks are the basis for many recent trends, such as social media, the gig economy, and the unbundling of work from unemployment. The successful firm, and the effective manager, must be able to design and implement strategies to take advantage of, rather than suffer from, these changes. The concepts and examples discussed in this chapter will help you do so.

MINICASE: Airbnb and the Ghost of Start-Ups Past

"You always had an entrepreneurial streak," you tell yourself with a chuckle as you reflect over a hot cappuccino at your favorite coffee shop. With a technical undergraduate degree and a soon-to-be-granted master of management degree, a start-up sounds like a perfect way to jump back into the real world. Despite the COVID-19 pandemic crisis, venture funding was plentiful, and you were convinced that the right idea would get funding. Cameo (https://www.cameo.com/) raised $6.5 million from heavy hitters like Kleiner Perkins to enable people to book personalized celebrity videos delivered directly to their smartphone . . . for a fee. Clubhouse (https://www.joinclubhouse.com/), which billed itself as "a new type of social network based on voice—where people around the world come together to talk, listen and learn from each other in real-time"[1] received funding by elite venture capital firm a16z at a valuation rumored to be around $1 billion. Airbnb went public in late 2020, quickly surpassing the $100 billion valuation mark. Airbnb was one of those rare companies whose brand had become a verb. While you may "Google" **information**, people were increasingly "Airbnbing" for vacation and work.

As a testament to its real or perceived digital disruption of the lodging industry, Airbnb was worth more than Marriott International, founded in 1927 ($48 billion), and Hilton, founded in 1919 ($33 billion), combined. As you ponder the issue, you reflect on lessons from tech ventures of the past. One that you studied deeply was eBay. You recalled an interesting article from back in 2004 drawing a parallel between eBay and Amazon. At the time, the two firms were, respectively, 60th and 66th in *BusinessWeek*'s Top 100 Brands.[2] While *digital disruption* had yet to become a popular term, eBay and Amazon had clearly begun to disrupt retail:

> eBay and Amazon.com, the Internet's top two eCommerce sites, are taking opposite approaches to growth. eBay raised its prices this month for the fourth year in a row, while Amazon renewed its pledge to keep cutting prices even if it means lower profits.[3]

You recall Meg Whitman, at the time eBay's chief executive officer (CEO), saying, "The eBay **marketplace** is a powerhouse. . . . We continue to enjoy ever-bigger, ever-faster cycles of success,

1 Monroy, L. 2021, April 9. "Clubhouse's allure and pitfalls as the new social media sensation." *San Jose Inside*. https://www.sanjoseinside.com/news/clubhouses-allure-and-pitfalls-as-the-new-social-media-sensation/.

2 "The global brand scorecard." 2004, August. *BusinessWeek* 72: 2.

3 Walker, L. 2004, January 29. "A study in e-commerce opposites." *Washington Post*.

fueled by the unlimited opportunity of our huge addressable market."[1] At the time, eBay was reaching the peak of its financial achievement and growth. You recall the same article quoting Amazon's founder and CEO, Jeff Bezos: "We will, for years and years and years, consistently give back the gains we get in lower operating costs to our customers in the form of lower prices."

Airbnb seemed to be on a similar trajectory, destined to become synonymous with digital disruption. As you ponder your next move, you cannot help but think that replicating Airbnb's success is predicated on understanding its dynamics and those of similar successful disruptors.

Discussion Questions

1. Reflect on what you have read and your knowledge of the impact of new technology on the competitive landscape and ask yourself, Why would Amazon and eBay act so differently?
2. Apply that thinking to the lodging industry. Is Airbnb in the same position as eBay? What about traditional incumbents, such as Marriott and Hilton?
3. What general rules can you extrapolate from your understanding of these cases? Rules that could help you position your own start-up?

1 Ibid.

4.1 Introduction

It is now clear that the Internet and the mobile **platform** served as the basis for an unprecedented number and variety of digital transformations and innovations. We discuss the Internet itself as well as the many innovations it has enabled in the next chapter, while devoting the present one to the foundational concepts that underpin those innovations.

The commercialization of the Internet and the digital technology services built upon it in the early 1990s ushered in an era of "digital disruption" driven by a powerful economic shift: the elimination of distribution costs for information goods (see Section 4.4) and the drastic reduction of transaction costs for most economic activities. The recent history of the newspaper industry provides a stark example of this disruption. Prior to widespread adoption of the Internet, newspapers were a primary means of delivering information to individuals in separate geographic locations. But in order to do so, a newspaper publisher needed to own, or at least control, the means of distribution of such information. That is, it was impossible to deliver the news without presses that would imprint them on paper and trucks that would move those papers to newsstands within reach of readers. By controlling these physical assets—the presses and the trucks—publishers could bundle the political commentary, sports results, business news, and editorial opinions that readers wanted with classified advertisements. The former was what people looked for, and the latter was what paid the bills and brought in the profits. This bundle of information controlled by those with access to the prerequisite physical assets that provided access to customers was the norm in many industries: music labels controlled the printing and distribution of CDs, television and cable channels controlled access to programming in the homes and to viewers. Each medium had its dedicated distribution channel and specialized complementary assets.

A similar dynamic occurred with physical goods and services. Supermarkets controlled access to groceries and packaged goods by consumers, thanks to their considerable investment in stores. Shopping malls offered one-stop-shopping convenience by bringing a wide variety of brands to one physical location. Such concentration of supply reduced transaction costs—the sum of all costs a buyer (or seller) incurs when making any economic trade. One-stop-shopping convenience is a perfect example of lower transaction costs, since buyers need to invest time to drive to only one location in a physical space to fulfill all their needs, rather than physically driving to multiple stores in different parts of town. But what if each of the stores in the mall could be reached with one click and the goods could be delivered to your door? Transaction costs are slashed, and the value of controlling mall real estate drops dramatically.

Let's return to the recent history of newspapers. As more newspaper readers connected to the Internet and could access content directly without waiting for the newspaper to be delivered to their home the next day, the value of controlling printing presses and trucks decreased. More importantly, there was no longer any economic reason for the main money-making elements of newspapers—classified advertisements—to be bundled with the news and opinion sections of a traditional newspaper. Sites like Craigslist and eBay in the United States and similar local players around the world (e.g., Subito.it in Italy, Leboncoin.fr in France) emerged to allow people to list their goods and reach buyers directly. Compared to classified ads in the local paper, these websites offered superior selection to buyers and a greater market reach to sellers. Everyone was better off . . . except the newspaper publishers, of course (Figure 4.1).

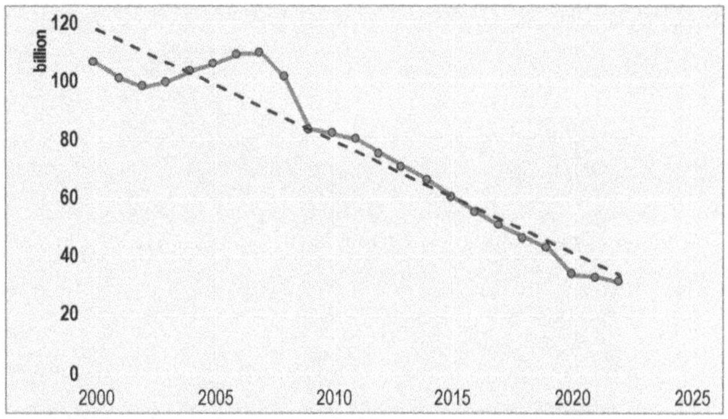

Figure 4.1. Global newspapers' advertising spending, 2000–2022 (billion USD)

Source: Zenith, 2020, "Advertising expenditure forecasts June 2020," https://www.zenithmedia.com/product/advertising-expenditure-forecasts-december-2020/

As content began to proliferate on the Internet, a new value proposition came to the fore—organization and curation. As Nobel Prize winner Herb Simon once put it, "A wealth of information creates a poverty of attention."[1] If everyone can publish what they like thanks to the Internet, how can readers find what they are looking for? When content is plentiful, what is the valuable service that will attract people? It is the ability to organize the content and provide easy access to the most useful content (i.e., curation). In this world, the search engine becomes the aggregator of attention, with Google being the dominant player thanks to the technical superiority of its PageRank algorithm.[2] As more consumers gravitated to Google to start their online journey, content providers increasingly tried to make their material easily discoverable by Google's algorithms. They organized content catered to the specifications of the Google crawler and invested considerable time and money optimizing for the PageRank algorithm. New professions were born (e.g., search engine optimization, search engine marketing). With Google increasingly aggregating consumer attention and gaining visibility into user interest simply by collecting their queries, it was natural for advertisers to focus their attention, and budgets, away from newspapers. Shifting advertisement dollars to Google had the added advantage that the marketing message could be tailored to the specific intent of the consumer, as expressed explicitly in the search queries (Figure 4.2). Later, with the emergence of the mobile platform as the dominant digital channel for reaching consumers (Chapter 5), Facebook figured out how to catalyze and aggregate attention. The endless algorithmic feed of news and updates from friends and acquaintances became the place where consumers would spend moments of idle time—on the

1 Simon, H. A. 1971. "Designing organizations for an information-rich world." In *Computers, communication, and the public interest*, edited by Martin Greenberger, 40–41. Baltimore, MD: Johns Hopkins Press. https://digitalcollections.library.cmu.edu/awweb/awarchive?type=file&item=33748.

2 The PageRank algorithm is the core of the Google search service—it is the program that determines how to rank web pages related to a given search query. Stibel, J. 2008, June 23. "The brain behind Google." *Harvard Business Review*, retrieved from https://hbr.org/2008/06/the-brain-behind-google.

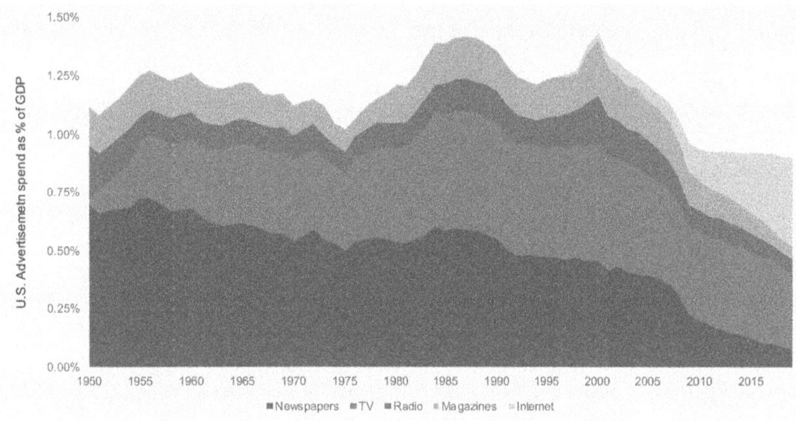

Figure 4.2. Seventy years of U.S. advertising

Source: Adapted from Evans, B., 2020, June 15, "News by the ton: 75 years of US advertising," retrieved from https://www.ben-evans.com/benedictevans/2020/6/14/75-years-of-us-advertising

bus, waiting at the doctor's office, even in the bathroom. As user attention migrated to Google and Facebook, so did advertisers' money.

Conceptualizing and implementing digital initiatives now that the Internet is a cornerstone of business infrastructure requires a basic appreciation of what the Internet is and how it works (Chapter 5). More importantly for general and functional managers, however, the ability to leverage the Internet, mobile platforms, and related digital technologies requires an understanding of the economic characteristics of networks and information. In the remainder of this chapter, we cover those topics.

4.2 Platforms, Marketplaces, and Aggregators

Digital Platforms

Strictly speaking, the dictionary defines a platform as "a raised level surface on which people or things can stand." In computing terms, a platform is a software foundation on which other software is built. The operating system is a platform for software applications. A software program written to run on the Windows platform will likely work on various versions of Windows—Windows 7, Windows Vista, or Windows 10. However, it will not run on a Macintosh or a Linux machine. The operating system exposes so-called application program interfaces (APIs) to the applications built on top of it. Each API offers a set of functions. There are APIs for sending documents to a printer, APIs for reading and writing from external memory, and so on. This approach enables application developers to leverage the APIs to build their applications efficiently. As application development moved to the Internet, and more recently to the mobile platform (Chapter 5), the dominant platforms running on smartphones and tablets are iOS and Android. Messaging systems, such as WeChat in Asia, have themselves become platforms. WeChat users, for many of whom the smartphone is the only computer they own, use the messaging app to do all kinds of things—from transacting business with the many thousands of stores that have a WeChat presence, to exchanging money or sending gifts to each other, to playing games.

With the increasing strategic prominence assumed by digital technologies in recent years, the term *platform* has broadened from its original technical definition. A digital platform is an evolving sociotechnical system (Chapter 2) that exposes resources that enable digital innovation and competitive actions by other firms. Digital platforms facilitate the design and implementation of competitive actions by exposing resources and a design architecture that suppliers can leverage to envision and carry out their competitive initiatives. In this context, a *resource* is any asset or capability that the platform makes available via a software interface. The prototypical example of a digital platform is the iPhone operating system, iOS. iOS exposes, via its APIs, thousands of assets and capabilities, from access to hardware components (e.g., GPS) to advanced digital capabilities such as image recognition. Developers who want to create applications for the iPhone or iPad use these APIs to access the digital resources exposed by iOS and combine them with their custom software to deliver the apps' functionalities. In other words, Apple designed iOS to help developers employ a set of standardized methods to access the different capabilities of the Apple iPhone (or iPad). Uber is a great example. It geolocates you and nearby drivers in real time, by making an API call to iOS to retrieve the current location from the GPS in the smartphones. By doing so, it allows you to call a ride with the tap of a button while knowing the charge in advance and being able to track a driver coming to pick you up. Through such digital innovation enabled by the iOS platform, Uber was able to offer a far more convenient service than that offered by traditional taxi companies. Another example is Shopify. Created in 2006, Shopify emerged to help inexperienced merchants gain an online presence. The idea behind Shopify matured two years earlier, when founder Tobias Lütke started building the online shop for Snowdevil, his snowboarding business. He quickly realized that marketed solutions were inadequate and the skills required for opening even a small eCommerce presence were beyond the reach of most small business operators. Shopify over time became an all-in-one eCommerce platform supporting all facets of the online transaction, including building the shop and managing payments, shipping, and customer relationships. Instead of just selling the eCommerce service, Shopify decided to expose the platform's resources via API. Thus software developers for larger and more sophisticated operators could directly access the underlying services and customize them to their needs. Exposing Shopify's resources through API had the additional advantage of allowing independent developers to contribute software add-ons to the Shopify app store. By 2020, Shopify had become the leading eCommerce platform, counting more than 1,000,000 stores, and the pandemic rocketed the firm's market capitalization beyond $120 billion.[1]

Note that digital platforms are sociotechnical systems, not just technology. Along with technical resources, they provide a complete set of design rules, a common architecture that includes interfaces, and a set of standards and governance rules. Consider the iOS platform again (Figure 4.3). Apple, the owner of the platform, doesn't just enable access to digital resources like GPS on the iPhone. It also produces guidelines and rules for inclusion in the App Store.[2] Disputes over one of these rules, the requirement for in-app purchases being subject to a 30% fee paid to Apple, was at the basis of the recent complaints by major apps like Fortnite, Basecamp, and Spotify.[3]

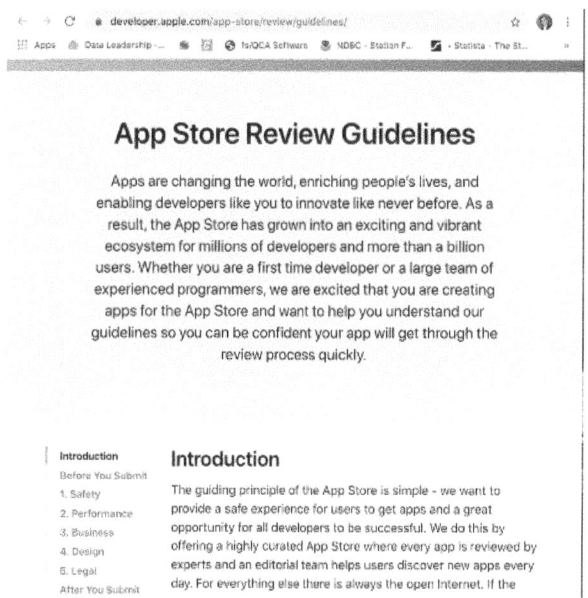

Figure 4.3. Apple's guidelines and rules for inclusion in the App Store

Digital Marketplaces

Digital marketplaces are communities of buyers and sellers who transact via digital technologies. The prototypical examples of marketplaces are Craigslist, eBay, or SeatGeek, the self-proclaimed leading mobile-focused ticket platform that enables fans to buy and sell tickets for sport, concert, and theater events. But there are many other examples of

1 Bradshaw, T. 2020, May 6. "Shopify surges as retailers rush online." *Financial Times*, retrieved from https://www.ft.com/content/c6298fb5-48f2-4bf6-996b-b64bb26525a1.

2 Apple. 2021. "App Store Review Guidelines." Retrieved from https://developer.apple.com/app-store/review/guidelines/.

3 Carman, A. 2020, June 17. "Epic Games and Match Group join Spotify in protesting Apple's App Store fees." The Verge, retrieved from https://www.theverge.com/2020/6/17/21294230/epic-games-match-group-spotify-eu-apple-app-store.

marketplaces where supply and demand meet. Match.com, the dating site, is one example, as is Cargo.one.[1] Marketplaces have three main objectives:

- They exercise control over the products and services that can be listed by sellers. The marketplace owner thus determines the acceptable product categories that can be sold and sets quality expectations for the listings.
- They function as enablers of trust, as they can guarantee customer protections (e.g., refunds), reliable payment transfers (e.g., escrow services), or provide mechanisms to decrease information asymmetry (e.g., ratings and reviews).
- They facilitate the discovery of products and services by providing resources for sellers to improve the visibility of their products and by taking direct action to curate and organize product listings.

Aggregators

Aggregators are organizations that manage an integrated digital platform and a marketplace. In other words, these organizations aggregate, on their marketplace, customer demand for the innovations that are built upon their digital platform. A prototypical example of an aggregator is Amazon. Starting as an online retailer, Amazon was able to attract consumer demand for a wide range of products, from books to lawnmowers. But in 2010, Amazon introduced the Amazon Marketplace (originally launched as zShops in September 1999), to enable other retailers to reach the consumer demand it had successfully aggregated. In order for suppliers to create and manage their Amazon store, they rely on the resources that Amazon makes available through its Amazon Marketplace Web Service (MWS), including product presentation, inventory management, payment, advertisement, and the like. Thus as an aggregator, Amazon owns and controls the marketplace in which customers can access the online store of suppliers. Doug Herrington, senior vice president for Amazon's North American Consumer Business, aptly summarized Amazon's aggregator strategy: "We had the insight early on that we had all these customers and all this traffic and why not think of a way to allow other people who have things to sell to come on and take advantage of this" . . . for a fee, of course!

Another example is the dominant food delivery player in Europe, Deliveroo. Deliveroo exposes a set of digital resources via its digital platform, known as the Restaurant Hub. Restaurants on Deliveroo leverage those resources to set up their menus and processes (e.g., what kind of food customizations are available) on the platform. Deliveroo, like DoorDash or Uber Eats in North America, is an aggregator because it also controls the digital marketplace where patron demand meets restaurants' offerings. Aggregators are unlike marketplaces without a platform, such as Cameo, Craigslist, or Instacart, or platforms without a marketplace, like Windows, the Microsoft operating system, and Stuart, the Paris-based provider of geolocalized independent couriers who deliver goods for retailer and restaurants.

Some of today's most powerful organizations are aggregators—such as Apple, which integrates the iOS platform with the App Store. An aggregator is in a unique position of power. By way of its control of the marketplace, it aggregates customer demand (e.g., restaurant's patrons), thus attracting to its platform those suppliers who want to capture such demand (e.g., restaurants). Thanks to the zero-distribution cost and negligible transaction costs enabled by the Internet, aggregators attract both demand and supply into their marketplace at almost zero cost. The result is their ability to gain an unprecedented global scale. Note that a firm becomes an aggregator when it integrates a digital platform with a marketplace. As such, the product or service being

---

1 Cargo.one is a European website enabling freight-forwarding companies with 24–7 online booking for the cargo capacity of multiple major airlines such as Lufthansa and Ethiad.

commercialized by suppliers is a combination of their original products or services and resources made available by the aggregator. Consider Deliveroo again. Those restaurants that join Deliveroo do so in order to be able to deliver food. They do so by integrating their original products (e.g., the food, the customization options, the service) with resources exposed by Deliveroo's digital platform (e.g., the drivers, the ordering experience). The same argument can be made for Amazon retailers and iOS developers. Contrast this with a traditional online marketplace like Craigslist or Expedia. If you want to sell your old bicycle on Craigslist, you simply list it the way we used to do in the classified ads of newspapers in the pre-Internet days. A hotel that lists its rooms on Expedia simply commercializes its product, room nights, on an alternative distribution channel.

The integration of a digital platform and a marketplace enables aggregators to influence the suppliers' strategic choices regarding product design and product commercialization. By controlling the platform, suppliers' product design choices are dependent on the architecture of those resources exposed by the aggregator, their technical requirements, and their associated governance rules. For example, iOS applications requiring access to users' geographical locations are constrained by the availability of iOS location APIs (i.e., the Core Location framework), their technical specifications (e.g., the framework's programming language such as Swift and Objective-C), and their usage rules (e.g., the requirement to prompt the user to grant or deny the location request). By integrating a marketplace, the commercialization of suppliers' products is influenced by the marketplace's rules and limitations. For example, when commercializing their applications on the App Store, developers cannot charge customers for software updates and other transactions, limiting their business models to one-time payments or subscriptions. As aggregators become more prevalent, you, as modern managers, will need to be able to understand their dynamics and navigate competition within them (Chapter 8).

4.3 Network Economics

Consider this question: How can a smartphone app of fewer than 10 megabytes (MB) of code, as useful and innovative as it may be, enable the firm that owns it to fetch one billion dollars in less than two years since it was released? This is what happened to Instagram, a firm that sold its business—the Instagram app—for one billion dollars to Facebook when it had only 13 employees.[1] Valuations are a monetary indication of the potential of the firm to generate profits. How could Instagram achieve such a large valuation so quickly? Much of the answer is to be found in the economics of networks.

As basic economics teaches us, water is cheap, and diamonds are expensive. Yet water keeps us alive, and diamonds have few practical uses. The reason is that typical resources are valued as a function of their scarcity. Skilled labor and managerial talent, the resources you sell to your employer, are an example. While we can debate the morality of sky-high executive compensations, it is clear that their proponents justify them with the argument that few people in the world have the talent and experience to run large, complex business operations. The same argument is used for professional athletes in popular sports—there is only one LeBron James who can fill seats in an arena, make people tune into the games he plays, and have kids clamoring for his shoes and jerseys. The likes of Cristiano Ronaldo, Lionel Messi, and Neymar Jr. fill soccer stadiums and help earn lucrative television contracts for their teams. Like scarce diamonds, these people command a premium price for their services. Value is typically found in scarcity, and the heart of business strategy is about being unique in a positive way (Chapter 7).

---

1 One billion is an eye-popping amount for a 13-employee company. Yet if you think about how much value the Instagram product created for Facebook shareholders, it's easy to argue they got a bargain when purchasing Instagram!

Figure 4.4. Crazy fans make for great entertainment value

Figure 4.5. Networks find value in plentitude

Networks Are Different: Value in Plentitude

We suspect that you get significant value from your Instagram use. It is not monetary value, unless you are an influencer, but it is in the form of social relationships, news, and probably often entertainment (Figure 4.4). How much would you have been willing to pay for the very first copy of the Instagram app in October 2010? If value is found in scarcity, would you have paid hundreds of thousands of dollars for that uniquely scarce application? Probably not!

In fact, you probably would take a pass and pay nothing for it.[1] The very first copy of Instagram is worthless. The same goes for the very first fax machine, telephone, or copy of Zoom, TikTok, WeChat, or WhatsApp. On the other hand, if a business associate of yours had already downloaded Zoom, you might consider purchasing it as well if the price were low enough given the amount of real-time videoconferencing you needed to engage in with this associate. As more of your friends download the app, you begin to see significant value in it, and you would consider paying more for it.

So where do Instagram, fax machines, and Zoom draw their value from? Not scarcity, but the opposite, plentitude (Figure 4.5). In fact, the value in a network is proportional to the number of connected nodes.[2] Similar arguments can be made for network technologies, like the telephone, instant messaging, video sharing, railroads, and the telegraph. While traditional goods obey the law of diminishing marginal utility—stating that the more of a good someone has available to consume, the less value one sees in each incremental unit of that good[3]—for networked products, the relationship is reversed. The insight underlying these examples is that networks differ dramatically from most other goods, as their value is tied to how many other nodes are in the network (plentitude) rather than how few (scarcity).

Physical and Virtual Networks

In the previous section, we talked about computer networks and the Internet. There are, of course, other types of networks, like the telephone network or the railroad network. These are ***physical networks***, where the nodes are connected by physical links (i.e., railroad tracks, telephone wires). However, a critical insight for you as a manager is that network economics also applies to **"virtual" networks**.[4] The defining characteristic of a virtual network is that connections between **network nodes** are not physical but intangible and invisible. The nodes of a virtual network can be either people or devices.

Whether tangible or intangible, network connections enable network nodes to share and communicate. In a virtual network, the people in the

1 We are aware of the fact that Instagram is a free app for users in the sense that you don't pay any money to download and use it. But don't fool yourself. You are indeed paying with your time and attention, but this discussion is beside the point at this stage.

2 In the language of networks, a *node* is any connected entity. Nodes can be people, organizations, computers, telephones, or fax machines.

3 Recall the last time you were very thirsty. You would have gone out of your way for a glass of water. When you found a fountain, you started drinking, and with each sip of water, the value of the water decreased until you wanted no more.

4 We adopt the term *virtual networks* following the definition used in Shapiro, C., and Varian, H. R. 1999. *Information rules*. Boston, MA: Harvard Business School Press.

network can share information (i.e., share files of the same format) with other members of the same user network (e.g., BitTorrent file-sharing users), expertise (e.g., information on how to use a given software program within a certain community of practice), or just images of their friends dressed up for the game (Figure 4.4). Note that a virtual network is generally sponsored by an organization or technology that enables it, controls access to it, and manages its evolution. Apple, for instance, sponsors the App Store network, while ByteDance controls the TikTok network of short video–sharing users and Facebook runs the Instagram network of image-sharing friends. Maintaining control of the network puts the sponsor in a position of advantage even in terms of its capacity to monetize it.

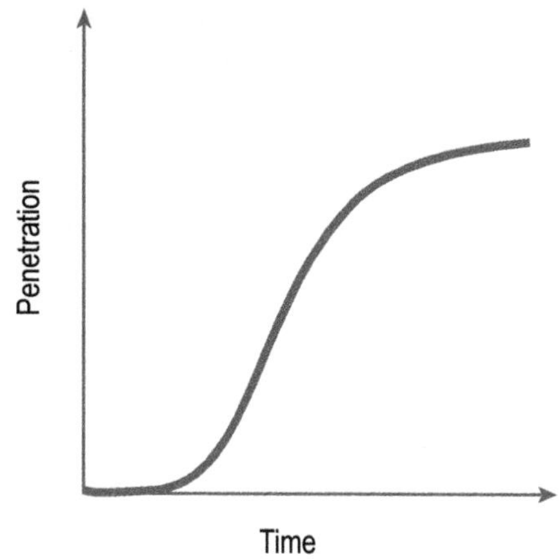

Figure 4.6. Classic technology adoption curve

Key Concepts and Vocabulary

In order to understand how networks operate and to explore their potential for firm strategy beyond an intuitive level, we need to introduce some vocabulary and some fundamental concepts.

Positive Feedback Adoption of a new technology product or service typically follows the pattern represented by the S curve (see Figure 4.6). **Positive feedback** is simply defined as that self-reinforcing mechanism by which the strong gets stronger and the weak gets weaker. It is very similar to the process by which return in a microphone quickly becomes louder and louder until it reaches a deafening volume (i.e., a phenomenon known as the Larsen effect).

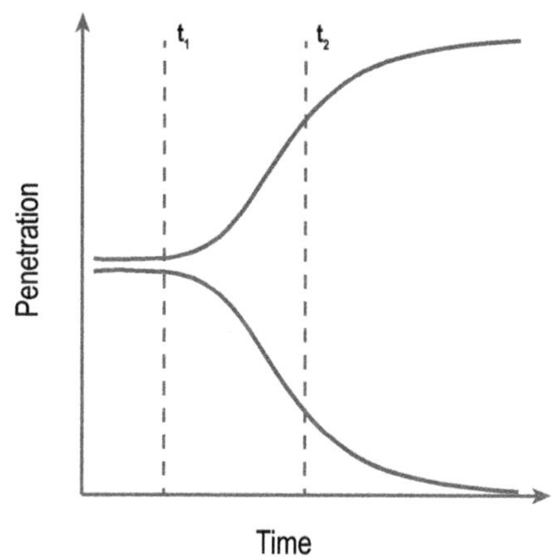

Figure 4.7. The dynamics of positive feedback

Positive feedback is a well-known economic phenomenon at the heart, for example, of economies of scale. In industries with strong economies of scale—say, automobile manufacturing or cloud-computing infrastructure—there is a significant advantage stemming from size. Imagine two car makers, one larger than the other. The larger manufacturer can spread its fixed cost across a larger volume of cars, thus being able to reduce its prices—assuming everything else is the same, including expected profit margins. With lower prices, the larger manufacturer will sell more cars, thus being able to further spread its fixed costs. With even lower prices, the cycle begins again, allowing the dominant manufacturer to further reduce unit cost and increase volumes (see Figure 4.7). Note that the smaller manufacturer, losing market share to the larger one, will see its unit cost increase as it loses market share, thus having to raise prices and seeing even lower sales. While the losing firm may still have a chance at time t_1, things look compromised at time t_2.

Positive feedback sets in motion a virtuous cycle, benefiting the larger firm, and a vicious cycle, penalizing the smaller one. Thus the stronger gets stronger and continues to grow while the weaker gets increasingly weaker. Unless the smaller firm is able to identify a profitable niche or somehow differentiate its product, it will likely fade into oblivion, unable to sustain itself as a viable business.

Negative Feedback The above discussion should clarify that there is nothing inherently positive (i.e., good) about positive feedback—particularly for the firms on the losing side of it! In other words, positive feedback simply means that the process is self-reinforcing, not that it is beneficial.

Negative feedback is the term used to refer to the opposite dynamic. If negative feedback is at play, the stronger gets weaker and the weaker gets stronger. Negative feedback typically characterizes economies of scale and takes effect when the dominant firm has reached a significant size. After a certain size, economies of scale no longer reduce unit cost and, due to coordination costs and increasing overhead, further growth is hampered. In other words, past a certain size, the dominant firm encounters difficulties that limit further growth.

Network Effects Positive and negative feedback play a crucial role in physical and virtual networks because the value of a network to its members is a function of the number of nodes in the same network. Positive feedback dynamics that occur in networks go under the name of ***network effects***, *network externalities*, or *demand-side economies of scale*.

Network effects occur when a new node (e.g., a new WhatsApp or Zoom user), while pursuing his or her own economic motives, creates value for all the other members of the network by making the network larger and thus more valuable. Network effects have the characteristic of economic externalities—hence the name *network externalities*.[1] That is, they create spillover effects that have an impact on other individuals: positive for those members of the growing network and negative for the members of the other ones.

Consider once again the example of Instagram, and ask yourself how the firm reached 300 million monthly active users in just four years from its inception and today tops 500 million daily active users.[2] Instagram enables those who download it to share images and important moments of their lives. If you are an Instagram user, a friend or colleague who had already downloaded it probably alerted you to its existence. Once you downloaded the application and started using it, you probably began to recruit your own friends and colleagues. The reason is that any one of them who downloads Instagram makes your use of the app more valuable. In other words, since you can now interact with more people, Instagram is more useful to you. The term *evangelist effect* describes this dynamic and the incentive that current members of the network have to "spread the word" and convince others to join it. While the evangelist or *viral effect* generally co-occurs with the network effect, it serves to speed up users' adoption. A similar dynamic has fueled the growth of many other applications you may use today: Snap, Waze, TikTok, Clubhouse, and the like.

Positive feedback associated with traditional economies of scale typically exhausts itself well before one firm can achieve market dominance, but this is not the case for network effects. Positive feedback associated with network effects can play out, without limit, until one firm dominates the network and all others disappear—a situation typically called "winner-take-all" (see Figure 4.7). Moreover, the network effect is a powerful barrier to entry, protecting the winner from competitors (Chapter 9).

A firm that finds itself on the losing side of network effects can survive under two conditions:

1. It can become compatible with the dominant player, thus being able to connect to the dominant network and tap into its value. When Apple found itself on the losing side of the battle for dominance of the personal computer (PC) platform, it was forced to seek compatibility with Windows products—a sponsor of the dominant

---

1 As you may recall from your introductory economics courses, an externality occurs when an economic actor, pursuing his or her own economic motives, affects other actors' economic positions. Consider the example of a large resort able to charge $500 a night. If a pig farm is opened upwind of the resort, with no intention to hurt the resort but simply to pursue a business opportunity, the resort will no longer be able to charge premium prices . . . or perhaps any price at all! This spillover effect is called an *externality*—a negative externality, in this case.

2 Murphy, H. 2019, January 31. "Facebook seeks to knit Instagram and WhatsApp with core app." *Financial Times*, retrieved from https://www.ft.com/content/96c131ba-2524-11e9-8ce6-5db4543da632.

network. However, the sponsors of the dominant network will often resist this move toward compatibility.
2. It can find a niche that is different enough from the broader market and big enough to sustain the firm. Before Apple became compatible with the dominant Wintel platform, it was able to survive by offering a far superior product for designers and publishers, who found the Macintosh computer and software much better than Windows machines for their needs (Figure 4.8).

Figure 4.8. The original Apple Macintosh

Source: Photo by Luc Legay / CC BY-SA 2.0

The dominant network sponsor may react by trying to either block or limit compatibility. It may also try to take over the available market niches. Its ability to do so will depend on the characteristics of the market it competes in and the demand for product variety that characterizes it.

Tipping Point and Tippy Markets A ***tippy market*** is one that is subject to strong positive feedback, such that the market will "tip" in favor of the firm that is able to reach critical mass and dominate it. A tippy market is therefore a market with "winner-take-all" tendencies. A ***tipping point*** is the watershed of dominance, that moment in the evolution of a market where one organization or technology reaches critical mass and goes on to dominate it. You can think of it as the point of no return where winners and losers are defined. In Figure 4.7, the tipping point occurred sometime between times t_1 and t_2.

The lower the cost of production and distribution of a product, and the stronger the network effect, the quicker the onset of the tipping point (Figure 4.9). Apps such as WhatsApp represent a good example. The daily active users have now passed the two billion mark (Figure 4.10)—not bad for a company in business for only twelve years.

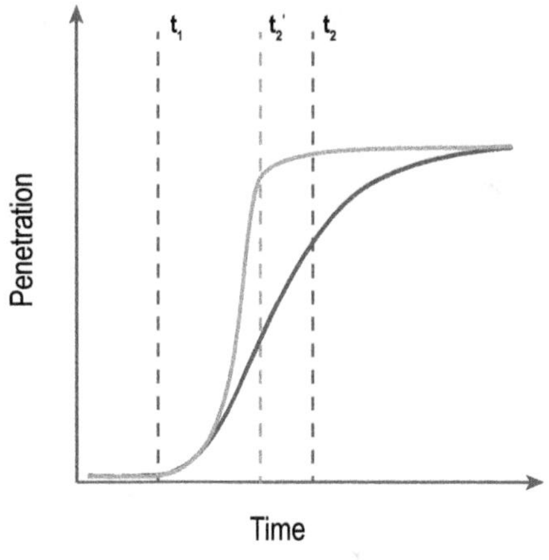

Figure 4.9. Adoption curve in the presence of a network effect

Not All Markets Tip "Not all markets tip" and winner-take-all dynamics are more the exception than the rule. Consider Garmin, the maker of navigation and communication equipment (Figure 4.11). Garmin is a vertically integrated company that directly controls the design, manufacture, marketing, and sales of its products. Its automotive GPS navigators help customers plot their course, identify points of interest along the road, and receive real-time driving directions. Is Garmin in a tippy market?

No, the automotive GPS navigation market is not a tippy market. If you own a Garmin device to help you find your way (i.e., you are in the Garmin network) and I purchase my own

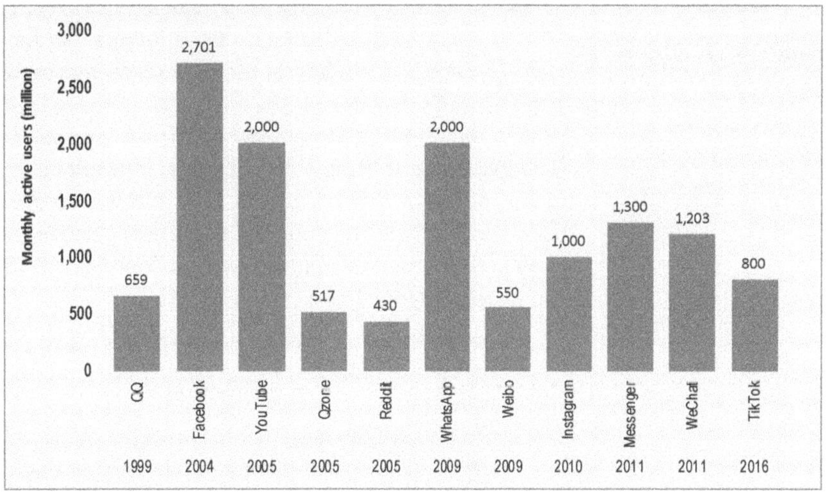

Figure 4.10. Top 10 social networks by number of active users and year of creation

Source: Wikipedia, 2020, "List of social platforms with at least 100 million active users," ***Wikipedia***, retrieved from https://en.wikipedia.org/w/index.php?title=List_of_social_platforms_with_at_least_100_million_active_users&oldid=985032735

Figure 4.11. The Garmin Fenix, maps at your wrist

Source: Photo by Jan Capek / https://i.redd.it/parfebjb52z31.jpg

Figure 4.12. RCI's website

GPS navigation device, you don't stand to benefit in any way from my purchase. We both use the receiver independently to find our own way. Now imagine if the navigation device automatically uploaded the location and speed of each user to a database that computed the average speed of cars on different routes. When I seek to find the quickest way home during rush hour, the more people that are in the Garmin network uploading their speed and position, the more accurate directions I would get—thus creating strong network effects and a tippy market.

Contrast the above example with that of Resort Condominiums International (RCI). RCI is a well-established timeshare company founded in 1974 to enable "exchange vacations." In the timeshare model, customers buy the right to one or more vacation weeks in a condominium unit in a specific property and a specific location (e.g., unit 235 at the Outrigger Palms at Wailea, on the isle of Maui, Hawaii). Those who own a timeshare can enjoy it during their allotted time. Alternatively, when seeking variation in their vacation, timeshare owners can exchange the right to use their unit with others. RCI and other companies that support exchange vacations create the market and facilitate the process of finding suitable trading partners, managing credits (e.g., a week in a high-end Maui resort may be worth two weeks in a midscale Florida property), and providing the many other services necessary to enable the exchange.

RCI has a membership base of 3.7 million timeshare owners, with more than 4,200 affiliated resorts in over 100 countries. RCI now has a website that supports many of its interactions with its members and prospective timeshare owners (Figure 4.12), but it is certainly not a "network business." Is RCI's industry characterized by strong network effects? The answer here is yes. Every new member who joins the RCI network increases the network's value for the entire current (and prospective) membership because the new member's unit increases the pool of available options. Moreover, it increases the potential pool of people interested in current members' own units—thus making current RCI members' units more likely to be requested for exchange. It follows that for prospective timeshare buyers, joining the largest timeshare exchange network offers the highest value.

How to Recognize a Tippy Market The two examples above show that "being on the Internet" is no guarantee of being able to harness network effects and that even non-Internet businesses can benefit from network effects. How can we recognize a tippy market a priori? How can you tell if your firm has the potential to harness positive feedback?

Whether a market will tip toward a dominant technology or a dominant player depends on two factors (Figure 4.13):

- *The presence and strength of economies of scale.* Strong economies of scale, whether traditional economies of scale or network effects, provide an advantage to larger firms.
- *The variety of customer needs.* Customer demand for variety creates the potential for the development of distinct market niches that the dominant player may be unable to fulfill.

When economies of scale are significant and customer needs are fairly standard, then the conditions for market tippiness are strongest. Consider one of our first examples, the fax machine. There are strong network effects in the faxing industry—and by now, you should have no doubt about how and why they occur. Moreover, the need for fax machines is a very standardized one—customers don't need much variety in the service. It follows, then, that one dominant fax network would emerge. While more than one fax standard may have been vying for dominance in the early days, the market eventually tipped toward one. The others disappeared. A similar dynamic has played out for modems, videocassette recorders (VCRs), and, more recently, high-definition DVDs.

When economies of scale are limited and the market has a wide range of different needs, the potential for market tippiness is the weakest. This is because not only is there a slight advantage associated with increasing size, but there are also a number of smaller niches that can support multiple smaller, focused players. Consider sports cars. Typically, the top manufacturers, such as Ferrari or Lamborghini (Figure 4.14), make a limited number of cars. Thus economies of scale are small. Moreover, those who seek to purchase high-end sports cars do so partly to differentiate themselves from the crowd. Those who purchase exclusive goods seek variety—at times, even uniqueness. This is therefore a market that is likely to sustain a large number of relatively small players.

When economies of scale are significant and demand for variety is high, the potential for market tippiness depends on the number and size of the available market niches. The ability to tap into a sizeable market niche seeking a product with unique specifications (e.g., graphic designers) is what allowed Apple to survive prior to ensuring compatibility with Microsoft-dominant products. When economies of scale are limited, even if the demand for variety is low, the potential to create positive feedback is small and the market is unlikely to tip.

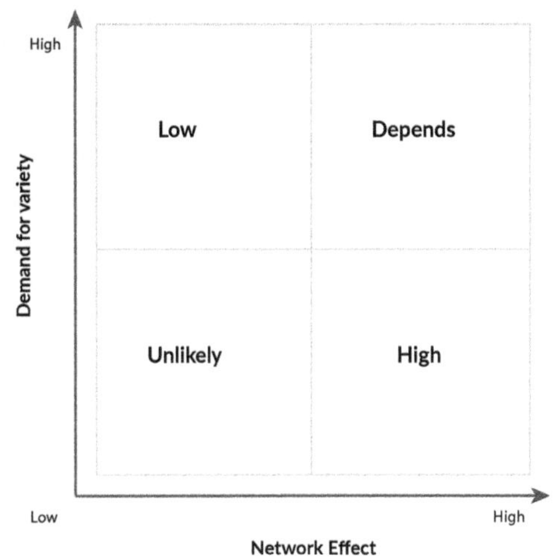

Figure 4.13. Likelihood of market tippiness

Figure 4.14. The Lamborghini Aventador

Source: Photo by Norbert Aepli / CC BY 3.0

Two-Sided Networks and Digital Marketplaces

Now that we have discussed the dynamics of networks in their purest sense, we can complicate the picture a bit. When network effects are present, the addition of one node to the network directly creates value for all the existing members. However, positive feedback can also occur in what we term ***two-sided networks***—that is, networks that have two types of members, each creating value for the other.

Digital marketplaces such as Craigslist or eBay, where sellers of goods encounter buyers online, are perfect examples of two-sided networks. But the nature of the marketplace exchange is not limited to buy/sell transactions. Consider an early example: Adobe. In 1990, the firm invented the now ubiquitous Portable Document Format (PDF) standard. Documents that are converted to PDF are guaranteed to look exactly the same on any platform. Chances are that you are a heavy user of PDF documents, reading them with Adobe's Acrobat Reader, for which you paid exactly nothing. In a similar fashion, Microsoft gives away its Windows Media Player, as do all other streaming audio/video makers. Why is this so? Does it make sense for a company to give away its product?

The strategy described above is sensible, particularly as the firm tries to establish its product as the standard. As you realize when you move from wanting to read PDF files to wanting to create them, Adobe Acrobat is not free. In other words, Adobe created a market for software programs that would ensure documents' cross-platform accuracy of display. Then it established its technology as the standard by creating a market of users, all those who downloaded and used the free reader. As the number of users grew, fueled by some early adopters of the authoring software (i.e., PDF document makers), organizations that produce and publish documents (e.g., companies publishing manuals for their products online) decided to adopt Adobe Acrobat. In turn, the increasing number of PDF documents available for download created even more demand for the reader software, aided by the "Free: Download Acrobat Reader" link that would be placed next to it.

You immediately recognize this process as an example of positive feedback. If you tried to break into the cross-platform document maker market as of today, you would be taking on a next-to-impossible feat. Adobe today provides a whole family of content creation tools dubbed the Adobe Creative Cloud, for which yearly licenses run in the hundreds of dollars. More generally, in a two-sided network, the value of the network to one type of member depends on the number of members from the other side who take part in the network. Consider again digital marketplaces such as Alibaba.com or Match.com. In these exchanges, whether catering to consumers or businesses, buyers are interested in the number of suppliers they will be able to reach (i.e., selection), while sellers are interested in the number of buyers they will be able to reach (i.e., potential sales volume). In this case as well, the firm that enables the marketplace (i.e., the sponsor of the network) and is first to reach critical mass (i.e., the first to pass the tipping point) will dominate the industry, leaving little room for any competitor (i.e., any competing marketplace).

Network effects are of central importance to digital platforms as well, whether they control a marketplace or not. Consider Windows, one of the first and most dominant digital platforms in history. Those who purchase a Windows machine are not so much interested in the operating system itself (i.e., Windows) but on the applications they can install on their computer to do their work (e.g., Excel). Computer buyers choose Windows when the applications they need are available for the platform. But application developers have finite resources, and they will write software for the platform that ensures the greatest potential for sales—in other words, the one with the greatest number of users.

This is the so-called chicken-or-the-egg problem that digital platforms, digital marketplaces, and aggregators must solve. You may use Expedia to look for airfare or hotel rooms, but you would not use the site if you could not find a variety of providers. The providers, hotels and airlines, would not list on Expedia if there was not a critical mass of travelers like you using it. In order to solve the chicken-or-the-egg problem, platforms and marketplace owners typically subsidize one side of the network in order to build network membership that they hope will attract members on the other side. We saw an example of this approach earlier when we discussed Adobe's strategy of giving away the PDF reader for free. A more recent example is the iPhone. When Apple first launched the iPhone in 2007, it had a number of Apple-built apps that made the device useful to consumers. When Apple opened up the App Store a year later, developers had an incentive to write software for the iOS platform because there was an installed base of users. As more apps became available, the iPhone became an increasingly compelling value proposition for consumers, who purchased more iPhones. The App Store started with 800 apps in 2008, crossed the one million mark in 2013, and by 2020, it counted almost two million. While the iPhone is a beautifully engineered piece of hardware, it is the variety of things that users can do with it that has been driving its adoption. As you recognize, this is a clear example of positive feedback based on network effects in a two-sided market. The video gaming market provides an even starker example, with console makers like Sony, Microsoft, and Nintendo often selling the devices at a loss. By some accounts, Sony was losing $60 for each PlayStation 4 it sold, which was a substantial improvement from the launch of the PlayStation 3, when the firm was losing more than $250 per unit.

Multisided Markets

Things can get even more complex in so-called multisided markets. These are networks that have more than two types of participants involved in transactions. An example is Instacart, the dominant grocery delivery player in the United States. Instacart orchestrates a complex market involving traditional supermarkets, independent shoppers, and grocery buyers and promises to deliver your order within two hours (Figure 4.15).

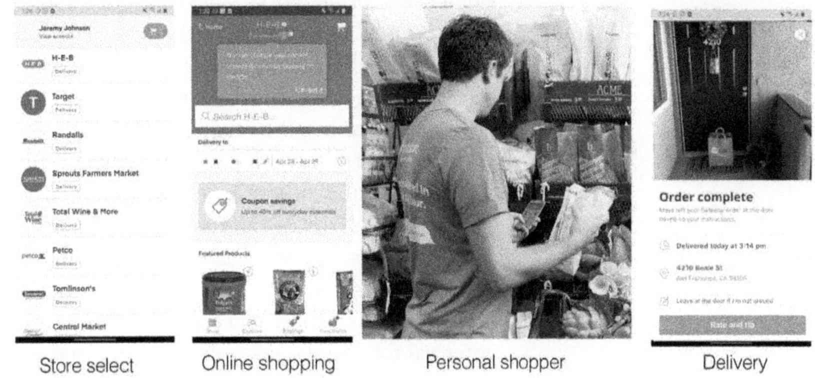

Figure 4.15. Instacart shopping process
Photo composition from the Instacart website

Supermarkets partner with Instacart to provide faster checkout and preferential lanes to its shoppers. The shoppers work as independent contractors, like Uber drivers, to fulfill and deliver orders that customers place via the Instacart app. As you quickly recognize, each of the three sides benefits from a critical mass in the other two sides. If there are too few supermarkets to choose from, buyers will not join the Instacart network, and shoppers will not receive enough work. If there is no critical mass of shoppers, the service will be slow, and customers will not receive a level of service that entices them to return. Finally, unless there is a critical mass of customers, supermarkets cannot justify the investment of time and money required by the Instacart partnership, and shoppers cannot rely on Instacart for a stable flow of work. While there can be even more complex arrangements, the dynamics in multisided networks, and the strategic trade-offs, are similar to those in two-sided markets. The challenge for the platform owner or the marketplace sponsor is to ignite the positive feedback inherent in the network effects that characterize their markets.

Network Properties

Following the discussion to this point should convince you that network effects are a dominant force for business strategy in the digital age. As Scott Cook, founder of Intuit, once put it, "When network effects are possible, it is the most important thing in the world to follow, to understand, and to make happen."[1] But it is important for you to recognize that not all networks are structurally the same, and their characteristics have deep implications for value creation and value capture.[2] The following are four characteristics of networks that you should evaluate and monitor when confronted with network effects:

1. **Network Effect Strength:** The strength of positive feedback may vary by network and may vary over time for the same network. When the sheer number of users has a direct relationship to the perceived value of joining the network, as in the case of TikTok, where a growing number of members makes for more varied and entertaining content, then network effects are very strong. As you may notice, the strength of the positive feedback varies in time. While the early users are essential for reaching a critical mass and allowing enough content to be generated, past a certain size, every new user addition has less impact. This is even more evident

1 Blitz, Amy. 2001, March. "Interview with Scott D. Cook." HBS Baker Library Special Collections. https://www.library.hbs.edu/Entrepreneurs/resources/persons/scott-d.-cook.

2 Feng, Z., and Iansiti, M. 2019. "Why some platforms thrive . . . and others don't." *Harvard Business Review* (January/February): 118–25; Coolican, D., and Jin, L. 2018, December 14. "The dynamics of network effects." Andreessen Horowitz, retrieved from https://a16z.com/2018/12/13/network-effects-dynamics-in-practice/.

when users (or network nodes) are highly similar in terms of their contribution to the network. For example, in the case of Waze, the first or the most active users in a specific location make the largest contribution to the network value. However, past a specific network size, any additional "wazer" adds increasingly less valuable signals for "sensing the road." In other situations, the quality rather than the number of nodes determines value in the network. For example, in gaming, it is not only the sheer number of games available that attracts gamers to a specific platform but also the availability of the few games they really care about. For this reason, gaming console makers invest considerable amounts of money to sign exclusives with highly anticipated games.

2. **Network Clustering:** The topology of a network (i.e., its "shape") determines its potential for winner-take-all dynamics to emerge (Figure 4.16). Among the first organizations to experience this was OpenTable, the online restaurant reservation provider. The firm famously noticed that it did not matter so much to restaurant patrons how many total restaurants joined the OpenTable network. The key was how many restaurants were in the person's own city or even neighborhood in metropoles with bad traffic like New York or Los Angeles. The firm figured they needed about 10% of the restaurants in a neighborhood for the service to become valuable to diners in the area.[1]

   Food delivery platforms like Deliveroo and Uber Eats have similar dynamics. In fact, the different platforms dominate in different markets (Figure 4.17). In contrast, a messaging service like WhatsApp will benefit from user growth, no matter where such growth is located.
3. **Disintermediation Risk:** Network effects are weaker when participants can easily transact outside of the network (i.e., disintermediate the network owner). Consider a marketplace that connects dog walkers to dog owners, such as Rover. If you find a dog walker your dog loves, you will want to hang on to his or her services. It is very simple for you to do so by exchanging phone numbers with the dog walker and connecting directly. But in doing so, you have disintermediated Rover, who no longer has a role in the transaction. If you were in charge of the marketplace, like Rover, you'd want to recognize this structural characteristic of your service and find ways to counter this trend and continue to capture transactions—if possible. The challenge here is to do so without degrading the service or making it cumbersome for network participants to join and stay in your network (e.g., onerous legal terms may dissuade participants). Other networks do not suffer from this risk. For example, you have to be on TikTok to discover the latest TikTok dance craze. TikTok cannot be disintermediated.
4. **Multihoming Potential:** This characteristic stems from the nature of positive feedback that we described earlier. *Multihoming* is the technical term used to describe the simultaneous participation of network nodes in multiple competing networks. You are multihoming when you download the Uber and the Lyft apps and call on the service of the two competitors at different times. When the costs of multihoming are low, then the strength of the network effects are diminished for each of the competing network owners. A classic example of high multihoming costs is provided by network effects that are rooted in hardware—such as those

1 D'Arcy, C. 2019, July 29. "Hidden networks: Network effects that don't look like network effects." Andreessen Horowitz, retrieved from https://a16z.com/2019/07/29/hidden-networks-effects/.

for console-based games—or those rooted in the need to migrate significant **data**—such as a music library.[1] Sometimes the cost of multihoming is artificially created, relying on exclusives. Netflix, Amazon's Prime Video, Hulu, and Disney+ try to attract users following a similar playbook as gaming consoles. They produce and distribute exclusive content in order to attract customers based on both the quality and quantity of their catalogs. While Netflix is the leading streaming platform service, 67% of Netflix subscribers also use Amazon Prime Video, and 50% subscribe to Hulu.

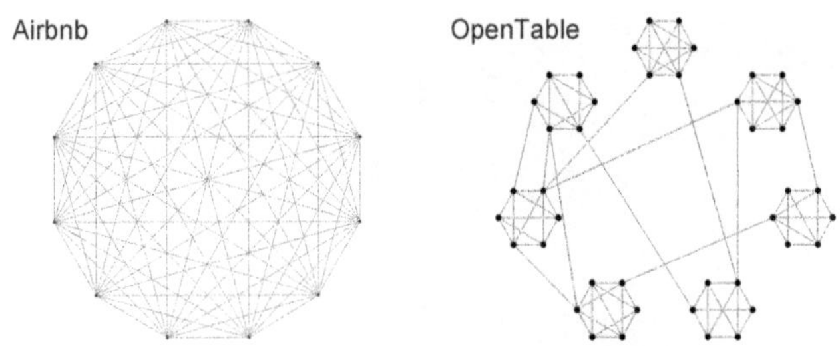

Figure 4.16. Global and locally clustered network topologies

Source: Adapted from Feng, Z., and Iansiti, M., 2019, "Why some platforms thrive . . . and others don't," ***Harvard Business Review***

By now, you have developed a good understanding of network effects and their prevalence in the digital age. The examples of successful organizations are a good source of inspiration, as is the post hoc analysis of the managerial decisions taken by firms that successfully leverage network effects. Understanding how Sony won the high-definition storage format war with Blu-ray by losing $250 per PlayStation 3 they sold (subsidy strategy) is a good exercise.[2] Now that you are familiar with the theory, we challenge you to pay attention to the strategies of established and emerging digital innovators and study how they leverage network effects. Ultimately, as you will notice, both theory and creativity guide their decisions.

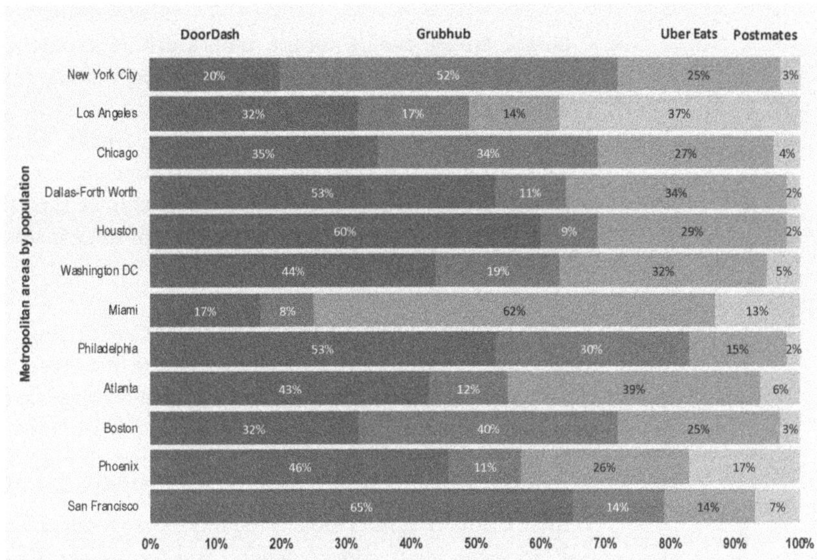

Figure 4.17. Food delivery platforms dominate different markets

Source: Yeo, L., 2020, October 19, "Which company is winning the restaurant food delivery war?" Second Measure, retrieved from https://secondmeasure.com/datapoints/food-delivery-services-grubhub-uber-eats-doordash-postmates/

Managerial Takeaways

Network economics have substantial implications of managerial interest.

Network Effects, Not Just Networks As we have seen in the many examples provided above, network effects and, more generally, positive feedback create the precondition for winner-takes-all dynamics. Network effects occur in the presence of technology standards, like the fax machine or a computer platform, but are not restricted to the technology arena. They also occur in the presence of virtual networks and communities of interest.

1 Tucker, C. 2018, June 22. "Why network effects matter less than they used to." *Harvard Business Review*, retrieved from https://hbr.org/2018/06/why-network-effects-matter-less-than-they-used-to.

2 Michiyo Nakamoto. 2006, January 5. "Sony and Toshiba set for DVD format war." FT.com, retrieved from https://www.ft.com/content/ddc91238-7dd2-11da-8ef9-0000779e2340.

Figure 4.18. The Ashley Madison website

Consider, for example, dating communities, such as Match.com, or employment search sites, such as Monster .com. While these communities do not use or require their members to purchase any specific technology or buy into any technical standard, their value is directly proportional to the number of users they can attract, and as a consequence, they enjoy strong network effects. This dynamic became apparent when a famous hacker attack exposed the list of members of the Ashley Madison website. Men and women join the community, allegedly, to have an affair (Figure 4.18). When the list was exposed, hackers demonstrated how the site handled its two-sided network effects strategy. There were only 4.5 million accounts held by women—about 13% of the total.[1] But this is somewhat predictable, and in fact, the firm itself claimed to have 5.5 million female members. What is more interesting is that the true number of accounts held by "live" women was much smaller—with a great proportion of profiles being fake accounts operated by bots.[2] This hack makes for a great example of a two-sided network subsidizing strategy. When one side of your network is limited in numbers and prevents your market from starting network effects dynamics (i.e., the chicken-or-the-egg problem), find a way to develop the other side. If you are not too concerned with ethical behavior, you can do like Ashley Madison and "manufacture" the entities you need.[3]

The Threshold of Significance We defined the tipping point as the watershed of dominance—that moment in the evolution of a market where one organization or technology reaches critical mass and goes on to dominate it. Traditionally, the onset of the tipping point would take some time, as the dominant technology slowly gained acceptance and became adopted by more users. For example, in the classic battle for dominance of the VCR market, Sony and JVC struggled to establish their standards—Betamax and VHS, respectively (Figure 4.19).[4] The battle between these competing standards lasted more than a decade and, as we know today, was won by JVC's VHS technology. Today's general and functional managers do not have a decade to monitor competitors before the onset of the tipping point. As a case in point, the more recent battle in the DVD market between two technologies vying for dominance—Sony's Blu-ray and Toshiba's high-definition DVD (HD DVD)—played out much more quickly . . . and Sony came out on top this time. The market can tip even quicker for digital products delivered over the Internet. In some cases, by the time a competitor realizes that a technology is emerging (i.e., the new technology reaches the threshold of significance), it is too late to react. This is arguably what happened with TikTok, which went from practically unknown to dominant in a matter of months. In these markets, being the innovator and the first mover is critical.

---

1 BinaryEdge. 2015, August 31. "Ashley Madison: A final analysis of the data." *BinaryEdge* (blog), retrieved from http://blog.binaryedge.io/2015/08/31/ashley-madison-a-conclusive-analysis.

2 A bot is a computer program designed to respond and interact with users in a humanlike manner. More details on this alleged fraud can be obtained here: Newitz, Annalee. 2015, September 8. "How Ashley Madison hid its fembot con from users and investigators." Gizmodo. http://gizmodo.com/how-ashley-madison-hid-its-fembot-con-from-users-and-in-1728410265.

3 Disclaimer: we do not advocate for fraud, but this is too good an example not to mention it. Newitz, A. 2015, August 31. "Ashley Madison code shows more women, and more bots." Gizmodo, retrieved from http://gizmodo.com/ashley-madison-code-shows-more-women-and-more-bots-1727613924.

4 You are surely familiar with the VHS standard for videocassettes. A competing standard was developed by Sony, called Betamax. You don't know it because it lost the standard wars to VHS and faded into oblivion.

Users Select a Network One of the most important implications of the above discussion is that customers will pick a network, not a product or a service provider. Let's return to your decision to buy a PC. If all your friends are in the Microsoft Windows network, you are most likely to join them in that network. Yet while many of your friends may be using Dell laptops, you could choose a Lenovo machine or even a Mac—which now enables you to be a member of the dominant network. In other words, you will not care about what hardware or software you use, as long as it enables you to tap into the resources of your network of interest.

Figure 4.19. Betamax and VHS tapes

Controlling the Network Provides Competitive Advantage Firms are willing to engage in standards battles and invest significant resources to achieve critical mass because sponsoring a dominant network provides the firm with a position of competitive advantage. Consider the court battle involving Facebook and Power Ventures, owner of the domain Power.com, which offered a single sign-on interface to access one's social networks, including LinkedIn, Twitter, MySpace, and AOL or Yahoo! Messenger. When Power.com, in an attempt to make good on its motto "All your friends in just one place," cached Facebook content to add the most popular social network to its list of sites, Facebook's lawyers sprang into action and sued.[1] They won the 2009 copyright infringement proceedings, and Power.com quickly faded away. Why was Facebook the firm that sued Power.com, while other social networks did not? Being the dominant network, Facebook had the most to lose from participating in an arrangement that enabled compatibility across networks. In the case of Apple and Google control of iOS and Android, their dominance of mobile platforms has enabled them to gross $120 billion in 2019.[2]

The Importance of Mutual Exclusivity Up to this point, we have treated competing networks as mutually exclusive. This is true for many of them. Consider the high-definition standards battle once again. Very few individuals were willing to purchase two DVD players, one using the Blu-ray standard and the other using the Toshiba HD DVD standard. This is because they not only needed to purchase two devices that cost money and take up space but would also have to maintain separate movie collections with some titles in one format and some in the other.

Simply put, there are costs associated with multihoming, trying to be a member of both networks. The steeper these costs, the more valuable it is to be able to control and retain ownership of the network. Consider Zoom and the videoconferencing market. Zoom is a small app that can be installed quickly and requires little configuration. If a competitor with very similar characteristics were to emerge, it would not cost users much to run both applications simultaneously. In fact, Zoom was not a first mover, and many of us also have Skype, Microsoft Teams, and Google Meet running on our computers. It is a bit of a nuisance to have to install and run multiple videoconferencing applications, but they are not mutually exclusive, and the power associated with controlling the dominant network (i.e., Zoom) is therefore limited. Similarly, when joining the network depends on the purchase of a physical device, the lower its cost and the more multihoming is likely to occur. Despite all the efforts of console makers, more than one-third of

---

1 Wikipedia. n.d. "*Facebook, Inc. v. Power Ventures, Inc.*" *Wikipedia*, last updated March 28, 2018, retrieved from https://en.wikipedia.org/wiki/Facebook,_Inc._v._Power_Ventures,_Inc.

2 "The state of mobile in 2020: The key stats you need to know." 2020. App Annie, retrieved from https://www.appannie.com/en/insights/market-data/state-of-mobile-2020-infographic/: 49.

PlayStation owners have an Xbox, and almost half of Xbox owners enjoy their PlayStation.[1] Why should gamers own two systems, each one requiring a specific investment in software? Because the "entry" ticket is sufficiently low and spread over a console's life span (typically seven years). In fact, one of the authors of this book is convinced that exclusive titles are worth this additional cost! Value is in the eye of the customer (Chapter 7).

4.4 The Economics of Information

In the introduction to this chapter, we hinted at the unique characteristics of information goods. As networks of interoperable digital devices have continued to expand, one of the most important results of managerial interest has been the unprecedented amounts of data and information that are being captured, stored, processed, and distributed by modern organizations. In order to be able to wring value from data and information, you must understand their economic characteristics. Information, like networks, has some interesting traits that differentiate it from physical goods. These unique characteristics of data and information have significant implications for firm strategy and competition.

Data and Information

Information systems (IS) researchers typically draw a distinction between the terms *data* and *information*. *Data* are codified raw facts—things that have happened (e.g., a customer has lodged a complaint) are coded as letters and numbers and stored, increasingly, by way of digital devices (Figure 4.20). *Information* is data in context (Figure 4.21). In other words, data become information when they have been given meaning and can therefore be interpreted by individual users or machines. It follows, then, that information is audience dependent; one person's data are another person's information.

Classic Information Goods and Digital Goods

The unique characteristics of information are best understood by first looking at products where information is at the heart of the value proposition. **Classic information goods** are those products that a customer purchases for the sole purpose of gaining access to the information they contain. Examples include software, books, music, TED Talks, and the news.[2]

For example, this book (and any other book) is a classic information good. The only reason you purchased it was to be able to gain access to its content.[3] The same goes for movies. The only reason you go to the theater or pay to stream a movie is to gain access to the experience that a film provides. The news represents yet another example. Whether you read the newspaper, watch television, or visit a news agency website, the only reason you do so is to acquire the information provided. Other classic information goods are stock quotes, class lectures, and the like (Figure 4.22).

A simple test for recognizing information goods is to verify whether the product can be digitized (i.e., can be encoded into bits and stored in a digital format). If so, the product is an information good, and once digitized, it is a digital good. You may be holding

7975593065 HAKE4H Th, Dec 30, 2010 6:45 Th, Dec 30, 2010 18:38 DL1072 ATL MSY DR. GABRIELE PICCOLI

Figure 4.20. A string of data

Ticket number	Record Locator	Departure	Arrival	Flight #	Departure	Arrival	Passenger
7975593065	HAKE4H	Th, Dec 30,2010 6:45	Th, Dec 30,2010 18:38	DL1072	ATL	MSY	DR. GABRIELE PICCOLI

Figure 4.21. Contextualized data become information

1 Harding-Rolls, P. 2020. "Next-generation games consoles." Ampere Analysis, retrieved from https://www.ampereanalysis.com/report/next-generation-games-consoles.

2 See the TED Talks website: https://www.ted.com/talks.

3 While it can convincingly be argued that we can use a book in many other ways (e.g., as a status symbol), we are focusing here on the use of books and other information goods in their primary function—that of conveying the author's message.

a paper copy of this book, but the same content is available in e-book format, and it could be rendered as sound if we decided to publish an audiobook version. I have a version of this same book on my computer in Microsoft Word format, while the publishers have a version on their computers in Adobe InDesign.

Now consider the chair on which you are sitting as you read this book. Could that be digitized? The answer is no, of course. The plans and drawings for making the chair could be digitized (another example of an information good), but the chair itself is a physical product that you purchased not for its information content but for its ability to support you in a comfortable reading position. You may send schematics for a chair anywhere in the world and have a 3D printer produce the chair, but again, the schematics are an information good, whereas the chair is not.

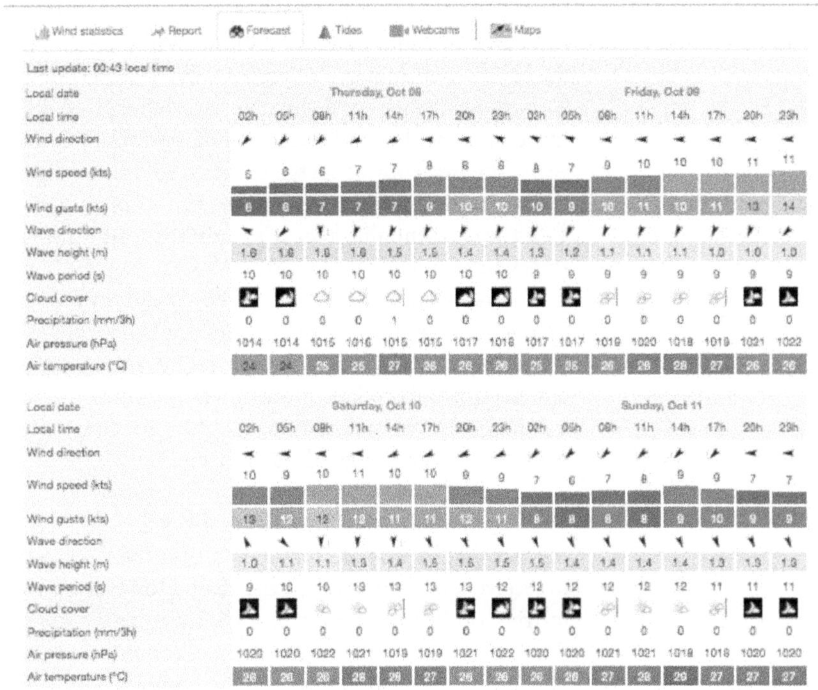

Figure 4.22. Wind and waves forecast: A classic information good

Source: Windfinder.com

The Economic Characteristics of Information

The fairly specific definition of classic information goods provided above is important because information—and as a consequence, digital goods—has some unique and interesting characteristics.

Information Has High Production Costs The first copy of an information good is very expensive to create in terms of time and money. Consider this book once again. Writing this text required a substantial amount of time in front of a computer typing a first draft, editing it multiple times, selecting appropriate examples, responding to reviewers' comments, and performing further editing as necessary. More subtly, before the writing process even began, substantial time was invested in studying these ideas, learning frameworks proposed by other authors, developing the unique frameworks and analytical models that are original to this text, doing the interviews and writing up the case studies, teaching the material, and taking notes on what seemed to work well in the classroom and what did not. After the draft of the book was completed, editors from the publishing house revised it and typeset it, assistants fact-checked the information and obtained the required copyright permissions for images and quotations, and so on. While this book was a big undertaking, you can envision projects that are even more costly and time consuming. Think about the latest big-budget film you have seen, a project that easily cost hundreds of millions of dollars and involved a large number of people who spent years developing their craft. All that work went into creating a less than two-hour-long entertainment experience for you to enjoy. In summary, information goods are very costly to produce. This is particularly true in relative terms, when the cost of producing the first copy is compared to the cost of producing the second one.

Information Has Negligible Replication Costs This is where information goods begin to differ drastically from physical goods. For as long as it took to create the first copy of this book or the blockbuster movie you last saw, the second copy could be produced at a fraction of the cost. Consider software—say, the copy of Microsoft Word we are using to write this book. Word is made up of millions of lines of code written by hundreds of Microsoft programmers over the years. The first copy of such a complex software program takes a significant amount of time and money to produce, but what about the second copy? Producing the second copy of Microsoft Word was essentially free. It simply took up a few gigabytes of storage space on a hard disk somewhere at Microsoft Corp.

For many information goods, the second copy, and all subsequent ones, has such a low cost of production that it is essentially free. This is not true of physical goods, such as the chair discussed above, or a car, or a meal. In a restaurant, for example, food is the second largest component of cost, second only to labor. Thus no matter how many steaks au poivre the restaurant cooks one evening, each one will require roughly the same amount of ingredients. The second copy of a steak (i.e., a physical good) is not free.

The Information Is Not the Carrier Consider the way you access your favorite magazines. Despite the fact that you can go online and read the *Economist* or *TIME* magazine, you might still like to receive a physical copy in the mail. However, you should realize that the paper is simply the "carrier" of the information, not the information itself. This becomes obvious once the information has been digitized and you access content through a browser on a PC or with a smartphone. The authors are old enough to remember a time when pictures were physical and had to be developed by specialists, music was carried by vinyl records and cassette tapes (we are now dating ourselves), and software came on CDs and DVDs (and even before, floppy disks). Today, in the post-PC era (as Steve Jobs used to call it), apps are downloaded onto your smartphone through wireless channels, and the DVD is quickly going the way of the floppy disk (Figure 4.23) and the tape cassette.[1]

A stark example of this inexorable transition is the end of Blockbuster, the place that we as kids visited to get movies—first on VHS tapes and later on DVDs. At its peak in 2004, Blockbuster operated more than 9,000 stores worldwide, whereas the last store recently ceased operations and became available for rent on Airbnb.[2] Today, if you are into movies, you get them streamed through your cable provider or you have an account on Netflix and/or the other streaming services available on the Internet.

Figure 4.23. 8", 5.25", and 3.5" floppy disks

Information Has Negligible Distribution Costs As with replication costs, the distribution costs associated with information goods are very low. Distribution costs are defined here as the expenditures associated with delivering the information good to the customers so that they can access its content.[3]

Consider once again the example of a big-budget movie. How does the movie get to the theater for your enjoyment? Traditionally, the studios

1 Ulanoff, L. 2010, June 2. "Steve Jobs ushers in post-PC era." *PC Magazine*, retrieved from http://www.pcmag.com/article2/0,2817,2364545,00.asp.

2 Garrand, D. 2020, August 12. "The last Blockbuster in the world will soon be available to rent on Airbnb—for just $4 a night." CBS News, retrieved from https://www.cbsnews.com/news/last-blockbuster-airbnb-4-dollars-night/.

3 A point of confusion here is that the term *distribution cost*, for information goods like movies and music, often refers to marketing expenditures (which can be significant for information goods). In this book, the term *distribution cost* is used in its strictest sense and does not include marketing expenditures.

copied films onto reels—a typical movie fits on five or six reels—and shipped them in film cans to the theaters. However, strictly speaking, the cost of distributing the cans is the cost of distributing the carrier of the information, not the information itself. In fact, modern delivery systems for in-room entertainment in hotels, for example, rely on digitized movies that are downloaded onto servers that in turn stream them to the TV sets in the rooms. High-definition movie theaters also have done away with the reels.

Where the infrastructure for digital distribution has been created (e.g., the App Store for software applications in the iOS ecosystem, Netflix for movies and TV series, Spotify for music), the distribution cost of the information goods is indeed negligible—free in practice. Information goods are therefore characterized by high fixed costs and very low marginal costs. The cost of producing the first copy is steep, whereas the cost of making and delivering incremental copies is near zero.

Costs Are Sunk Unrecoverable costs—those expenses that the firm has incurred to create its product or service but cannot be recuperated—are termed *sunk costs*. For example, if you are remodeling your kitchen and purchase some new flooring only to find out that your significant other hates it and vetoes your installing it, you can return the material and recover the expense, but you can't recuperate the costs (in terms of time, effort, and gasoline, in this case) of selecting and transporting the flooring. Your time and the gas you wasted are sunk costs. Information goods are unforgiving. If nobody is interested in reading our book after it becomes available for sale, we will be unable to recover all the expenses associated with writing and publishing it. If you have dreams of making it big in the music industry, all the time and money you invested in making your first tracks are lost if nobody cares for your form of artistic expression.

What's unique about information goods is that the bulk of their production costs are sunk costs. It follows, therefore, that there is significant risk involved in producing information goods, and consequently, a good deal of attention and research needs to be devoted to gauging and creating demand for them.

Information Has No Natural Capacity Limits While the creation of new information goods entails the significant risk that the investment will not be recovered, the upside is also significant. Information goods face almost no constraints to reproduction. Let's return to the example above and imagine that your songs struck a chord with the executives at Shady Records, who see in you the next big hip-hop star. When your songs make it into the iTunes music store, there is no limit to how many times they can be downloaded or streamed (i.e., how many digital copies can be generated for next to zero cost).

Information Is Not Consumed by Use Perhaps the most intriguing characteristic of information is that it can be reused multiple times. Physical goods, like an apple or one night in a hotel room, are destroyed through their use. That is, if you eat the last apple in the kitchen, there is nothing left for anyone else. If you occupy room 235 at the Ritz in Paris on March 19, 2021, that room, that night, will not be available for others to reserve and use. Conversely, information goods are not consumed by use. When you are watching the FIFA world cup, there are millions of other fans around the world sharing the same experience All the people in the theater with you can enjoy the movie you are "consuming." The fact that you read the news this morning does not preclude me from learning the same facts from the nightly news . . . or even from the very same newspaper you left on the subway on your way to work (assuming newspapers were still a thing).

Figure 4.24. Acqua Di Giò by Armani

Source: Photo by Net Sama / public domain

Information Goods Are Experience Goods Experience goods are those products or services that need to be tried (i.e., experienced) before their quality can be assessed. All new products and services are experience goods; in fact, perfume trials and samples of shampoo have been used for decades to entice people to buy. However, information goods are experience goods every time. If you are happy with the scent of the Acqua Di Giò by Giorgio Armani perfume (Figure 4.24) you sampled, you can make it your perfume of choice and purchase it over and over with confidence (or at least until the manufacturer changes the formula).

But how do you know if next week's copy of *TIME* magazine will be as good as today's? Can you be sure that the next book by Dan Brown is worth reading (assuming the first one was) or that the next release of *Grand Theft Auto* is worth playing?

Managerial Takeaways

The unique economic characteristics of information and classic information goods described above have some important implications for you as a general or functional manager:

- *Information is customizable.* Information goods can often be modified with relative ease. For example, movies are typically edited and remastered for different showings or different audiences, or for leveraging new audio and video features (e.g., 4K or 8K video). Bonus cuts and extra material are often included in DVD releases or collector's editions as a somewhat desperate attempt to entice movie lovers to own their own copies. Physical goods are typically much more difficult to customize before or after they are produced. Imagine realizing that your kitchen is too small after purchasing a new house.
- *Information is reusable.* Because information is not consumed by use, it is reusable multiple times and, because it is customizable, in multiple forms.
- *Information is often time valued.* The value of information is tied to the user's ability to employ it. Often timely use of the information is necessary to reap the potential value. Stock quotes represent a perfect example of this. Stock quotes on a 15-minute delay are useless information to a stock trader. Another example is represented by book publishers, who often release hardcover versions of popular novels and business books before releasing paperbacks that sell for much less. The cost of production of hardcovers is not the reason for the price difference. The same applies to movies and video games. Their price decreases after the release. Publishers are simply "versioning" their product to capitalize on the fact that some customers are willing to pay a premium to read the book, watch the movie, or play the game as soon as it is released.
- *Information goods can achieve significant gross profit margins.* Because of their economic characteristics—high production costs and low replication and distribution—firms that produce successful information goods can enjoy vast profit margins. Microsoft Corporation has enjoyed legendary profits over the years thanks in large part to its two cash cows: Microsoft Windows, the dominant operating system software for PCs, and Microsoft Office, the dominant suite of productivity tools. It is only with the recent rise to dominance of the mobile computing platform (Chapter 5) that Microsoft has de-emphasized Windows and the PC to reinvent itself by focusing on becoming a cloud service provider.

Information-Intensive Goods

As you read the above discussion about classic information goods, you may have wondered what applicability it has to industries that don't deal directly with these goods. That is, how useful is the above discussion to executives in industries such as restaurant franchising, car manufacturing, cruise ship operations, or health care? In each of these industries, you may reason, the value proposition customers seek is a tangible product or service, not information. While this is true, a quick look "under the hood" will reveal that information plays a critical role in these businesses as well.

Authors Evans and Wurster, from their vantage point at the media and convergence practice of the Boston Consulting Group, claimed way back in 1997 that "every business is an information business." As an example, they cited health care, an industry that offers a very "physical" service, but where one-third of its cost is "the cost of capturing, storing and processing such information as patient's records, physicians' notes, test results, and insurance claims."[1] Most industries, while not dealing directly with information goods, rely on information to create and bring to market their product or service; from research and development, to logistics, to distribution, to sales and marketing, information is the "glue" that holds together business operations. This means that most products and services are **information-intensive goods.** For information-intensive goods, while information is not exclusively what the customer seeks when purchasing the products or service, information is either one of their critical components or a necessary resource during their production process.

The role that information plays could be at the periphery of the product or service (e.g., informational material about the features of a product, the brand) or could be embedded in the product itself as knowledge (e.g., R&D and product development research). Consider, for instance, McDonald's Corporation, the franchiser of the popular fast-food restaurants. Is McDonald's in the "restaurant business" or is it in the "information business"? While you may opt for the first answer, a careful analysis reveals that the second is a more accurate label. What McDonald's Corporation sells to its franchisees around the world is sales volume through customer traffic due to its strong brand (i.e., information in the consumer's mind), management know-how (i.e., information about optimal practices, ranging from pricing, to purchasing, to human resource management), and various other support services (e.g., training, bulk purchasing contracts). Thus much of what the franchisor offers is in the form of information and knowledge.

Information is also embedded in the production and organizational processes that enable the transformation of inputs into products and services to be sold. Moreover, products and services in today's economy are increasingly "augmented" by information services. As computers have become increasingly embedded in products and services and customer relationships are more computer mediated, the value proposition customers receive from these products and services is increasingly dependent on their information content. Information is also embedded in the production and organizational processes that enable the transformation of inputs into products and services to be sold.

Consider a modern example: self-driving cars (Figure 4.25). The value proposition of a self-driving car is to get you to your destination safely without you having to devote attention to the road. In short, then, the value proposition of a self-driving car is to create extra time for the user. How does the car "create time," then? Is a self-driving car an information good? Probably not, but it is certainly an information-intensive good, as its value proposition depends on the car's ability to obtain and analyze massive amounts of sensor data in order to make appropriate decisions on the road. Its "analytics engine" is as

Figure 4.25. Waymo's self-driving Chrysler Pacifica
Source: Photo by Daniel Lawrence Lu / CC BY-SA 4.0

1 Evans, P. B., and Wurster, T. S. 1997, September–October. "Strategy and the new economics of information." *Harvard Business Review*: 70–82.

important as its internal combustion or electric engine. The latter uses gasoline or electricity as its fuel, while the former uses digital data streams.

Because of its pervasiveness, information has become a clear source of competitive advantage. Many of the most admired modern organizations draw their advantage from a superior ability to capture, manage, and distribute (or use) information (Chapter 8).

4.5 Information in Networks

As we have seen, information has unique economic characteristics. However, it has traditionally been constrained to rely on physical carriers in order to be delivered. Film reels carry movies to the theater, books carry text and images to readers, and professors carry lecture content to a class of students. The fact that information has had to rely on a physical carrier has acted as a constraint, limiting its ability to behave according to its inherent characteristics and benefiting those who controlled the complementary physical resources as we showed in our opening example on the newspaper industry.

It should be clear then how revolutionary the advent of the Internet was. It was the global network that allowed information, once digitized, to flow unrestricted by legacy carriers. Consider the process of organizing your honeymoon prior to 1993, the date of the commercialization of the Internet. Back then, you would likely visit a travel agency with your spouse and, after waiting in line for your turn to speak with an agent (the carrier of the information you were seeking), you would tell the agent your likes and dislikes and receive some suggestions. Based on these suggestions and your reaction to them, a skilled travel planner would narrow down his or her suggestions, asking increasingly specific questions and offering advice based on his or her superior knowledge of destinations and even individual resorts. This example suggests that when information is constrained by a carrier, such as the travel agent, it is not allowed to behave like information. That is, while information is not consumed by use and is cheap to reproduce and distribute, since it has to be delivered by a person, it has to follow the economics of the carrier—only one person can speak with the agent at a time. The travel agency could hire and train new agents to reduce the lines, but this would be a costly proposition.

This example is representative of a phenomenon known as the trade-off between **richness** and **reach**. *Richness* represents the amount of information that can be transmitted, the degree to which the information can be tailored to individual needs, and the level of interactivity of the message. *Reach* represents the number of possible recipients of the message. Traditionally, as information has been constrained by its physical carrier, a firm would have to make a trade-off decision between the number of people it wanted to communicate a message to (i.e., reach) and the depth of the message (i.e., richness). With the advent and widespread adoption of a cheaply and easily accessible information infrastructure, from the Internet to the mobile platform, this trade-off is disappearing. This is why "influencers" and digital entrepreneurs in the "creator economy" (Chapter 5) can reach a global audience through Instagram or TikTok, with zero distribution cost. Note that the trade-off between reach and richness has not been eliminated. There are still compromises to be made between reaching a large audience and offering a very rich exchange. However, with each new wave of digital technology, the trade-off is less pronounced. Innovations like virtual reality (Chapter 12) and machine-learning powered chatbots promise continued digital disruption of traditional industries and business models.

Process Virtualization

Our discussion of information-intensive goods demonstrates the growing importance of information in products and services. The trade-off between richness and reach described earlier is, in fact, a special case of a larger trend: the increasing number and variety of activities performed in the digital space. For example, travelers check in on flights using a smartphone app, suppliers automatically replenish warehouses by monitoring pallets of goods at their clients' site with radio-frequency identification (RFID) transponders, movies are delivered to your television through the Internet—this list is endless and encompasses a larger set of processes than just those within organizational boundaries (i.e., business processes). Process virtualization theory provides a general framework for understanding this trend.[1]

For example, let's consider dating as a process. At the core, dating is the series of steps individuals perform in order to test their compatibility (i.e., looking for similar interests, hobbies, or beliefs and values) with the ultimate goal of becoming a couple. Until recently, dating demanded physical proximity and face-to-face interaction between the interested parties, limiting the options to individuals in the local community. As you probably know, online dating has dramatically increased in popularity and social acceptability since Match.com launched in 1995 (Figure 4.26). Today, online dating is an industry worth more than two billion dollars in the United States alone, and by some accounts, the majority of U.S. heterosexual couples now meet online (Figure 4.27).

Despite its success as a maturing industry with increasing segmentations (e.g., farmersonly.com, tallfriends.com, and eurodate.com), online dating clearly has some shortcomings compared to its traditional counterpart.[2] For example, a number of surveys suggest that the majority of profile pictures on dating sites are at least three years old. That said, online dating makes for a great case to explain process virtualization theory. The adjective *virtual* is used in computing to connote something that does not physically exist, but it is made to appear to exist through software. Virtual reality (Chapter 12) is the perfect example. When appropriately executed, virtual reality is immersive, and while the users know that what they are sensing (seeing, hearing, and feeling) does not physically exist, they perceive it as real. In computing, the term *virtual* also refers to activities that are carried out, accessed, or performed by means of a digital computer. With this precise understanding of the term *virtual*, we define *process virtualization* as the performance of

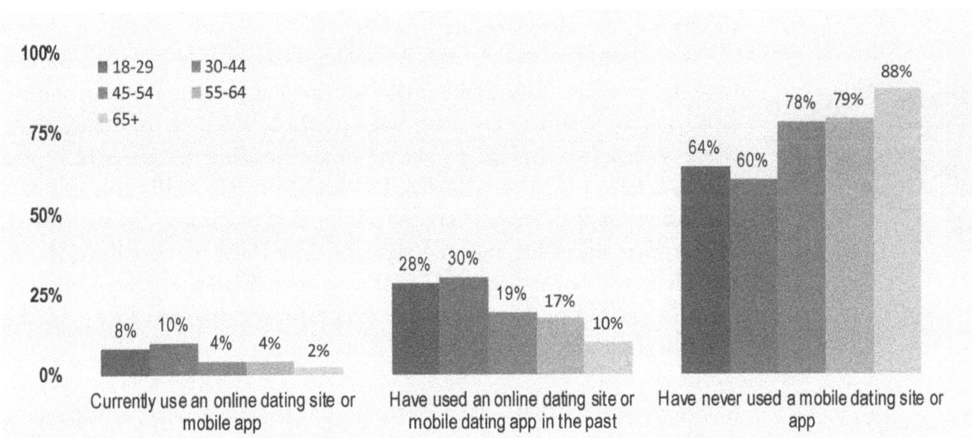

Figure 4.26. Online dating site and mobile dating app use among U.S. Internet users as of April 2020

Source: Morning Consult, 2020, table CMS15, p. 244, National Tracking Poll #200473, retrieved from https://morningconsult.com/wp-content/uploads/2020/04/200473_crosstabs_CONTENT_CORONAVIRUS_Adults_v2_JB.pdf

1 This section draws its conceptual foundations from Overby, E. 2008. "Process virtualization theory and the impact of information technology." *Organization Science 19*(2): 277–91. doi: 10.1287/orsc.1070.0316.

2 We made up only one of these three domain names, and you'd be surprised to see the degree of segmentation of dating sites.

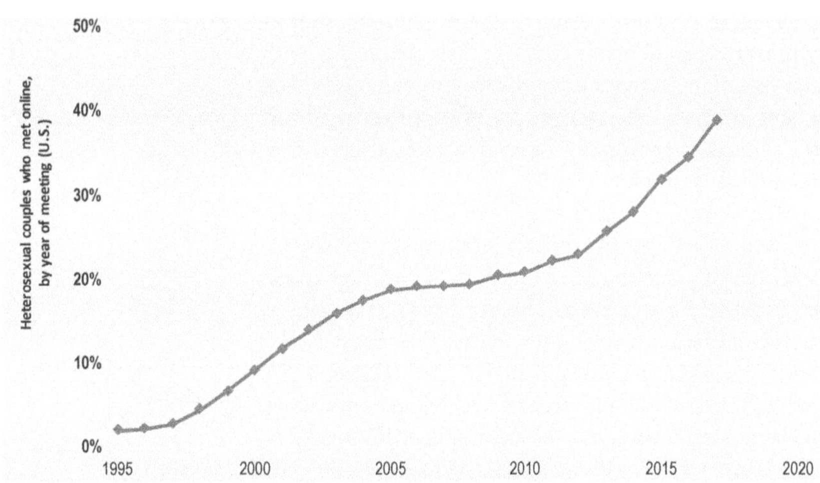

Figure 4.27. Heterosexual couples who met online in 2017

Source: Data from Rosenfeld, Michael J., Thomas, Reuben J., and Hausen, Sonia, 2020, October 3, "How couples meet and stay together 2017 fresh sample," in ***The end of the American Internet***, edited by Benedict Evans, Stanford, CA: Stanford University Libraries, retrieved from https://www.ben-evans.com/benedictevans/2020/10/3/the-end-of-the-american-internet

all or some of the steps in a process by software instructions executed by a digital computer.[1] Thus ***process virtualizability*** is the degree to which a goal-oriented series of activities (i.e., a process) can be performed without the physical interaction among participants or among participants and objects involved in the process. The virtualization of a process can be complete or partial depending on whether all or only some of the steps in the process are virtualized. Examples abound. Shopping is the process of browsing through and evaluating products in a physical space such as a store or open-air market. Online shopping is the act of browsing or evaluating digital representations of products via a software application such as a browser or mobile app (Figure 4.28). In the case of dating, the activity consists of a physical, colocated meeting of two individuals for the purpose of getting to know each other better. When virtualized, dating becomes online dating. The interaction takes place at a distance and becomes mediated by communication software applications running on a digital computer (e.g., PC, smartphone, or virtual reality headset).

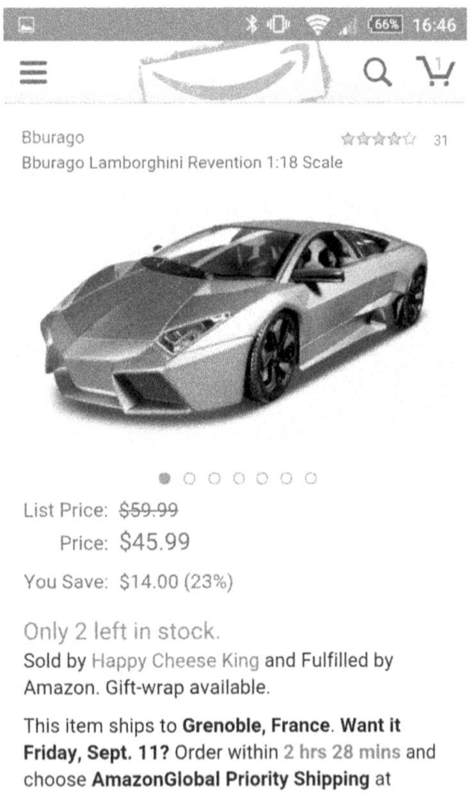

Figure 4.28. Shopping online with the Amazon Underground app

As you intuitively understand, not all processes are equal with respect to their virtualizability. Some processes, such as looking for information or buying an airline ticket, have been performed online for a long time, while others, such as getting a haircut or appreciating the aroma of a good espresso, have yet to be virtualized. Process virtualizability theory helps determine which processes have the potential to be successfully virtualized. The framework posits that the virtualizability of a process depends on four elements, or requirements, of the process: the sensory requirements, the relationship requirements, the synchronism requirements, and the identification and control requirements.

Sensory Requirements These represent the need for process participants to be able to experience a range of sensory stimuli in order to engage in the process, including tasting, seeing, hearing, smelling, and touching the other process participants and/or objects. If you have ever been to a wine tasting, you

1 Strictly speaking, process virtualization does not need to occur through digital computers and has been happening for centuries without IT. For example, distance learning effectively removed the physical interaction between teacher and students. Mail ordering removed the need for interaction between customers and clerks through catalogs. Pragmatically, in this book, we focus only on process virtualization that occurs through IT.

know that there is more than just . . . tasting (Figure 4.29). With taste, you appreciate the flavor, the structure, or the profile of the wine, but sight (i.e., color and "wine legs") and smell (first, secondary, and tertiary aromas) are critical senses engaged by the experience. For online wine tasting to work, you would need to virtualize not only each of these sensations—and that's beyond current technological state of the art—but the combination of all of them.

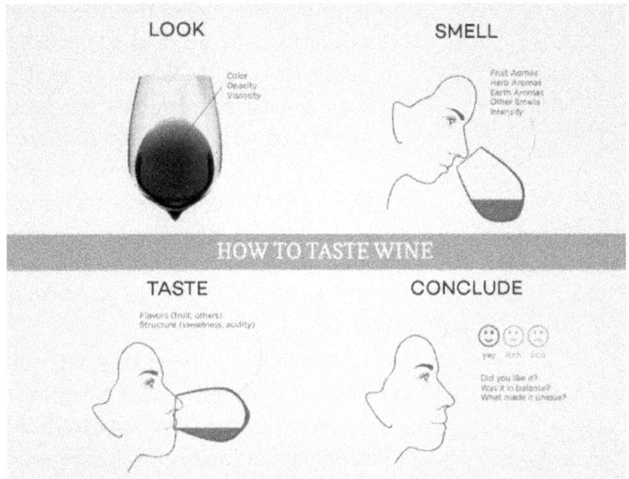

Figure 4.29. The wine-tasting process

Source: WineFolly, "How to taste wine," ***WineFolly*** (blog), retrieved from https://media.winefolly.com/how-to-taste-wine-method-tasting-winefolly.jpg

Relationship Requirements These encompass the need for process participants to interact in a social or professional context so as to acquire knowledge or develop trust and friendship. The idea is that physical, face-to-face interaction transmits a broader range of nonverbal cues such as gestures, posture, tone of voice, and facial expressions that are capable of communicating the warmth and attentiveness necessary in relationship development. Those who cyclically predict the death of the business travel industry when new technology emerges, such as videoconferencing in the 1980s, Internet desktop video conferencing in the 1990s, and telepresence most recently, seem to severely underestimate the importance of those cues. Another example comes from mentorship. During their doctoral studies, students engage in apprenticeship over many years of close contact, both professional and personal, with mentors and fellow doctoral students. Replicating that experience, even if possible, requires specifically addressing the need to build the relational bonds that characterize the mentorship process.

Synchronism Requirements These represent the degree to which the activities that make up a process need to occur in real time or with minimal delay. Physical processes are indeed naturally synchronous, as they happen through the interaction of colocated participants. In a virtualized process, this natural condition is not met. The telephone first, and Internet videoconferencing more recently, well illustrates how synchronism plays a role in communication processes. For example, Skype asks users about the delays disrupting the call to assess the quality of experience of its services (Figure 4.30).

In a different context, the grocery shopping industry, experience showed that fruit and vegetable sales are harder to virtualize than the sale of standardized packaged goods. This is not only because of the need for greater sensory requirements but also because of the delays introduced by online sales between the ordering and delivery of perishable goods.

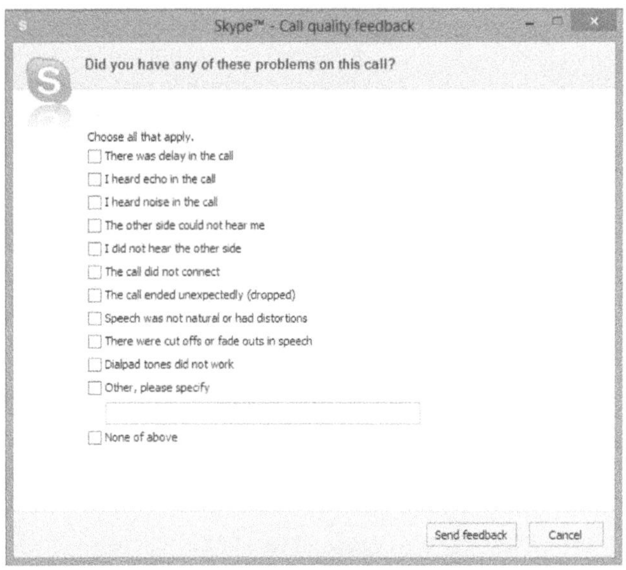

Figure 4.30. Skype's call quality feedback form

Identification and Control Requirements These concern the degree to which the process requires the unique identification of all participants and the ability to influence or exert control over their behavior. Virtual processes are intrinsically mediated by technological means and are then subjected to identify spoof, as participants cannot physically ascertain another's identity. Not surprisingly, fake profiles and old or "photoshopped" images are common in online dating. Similarly, the need to ascertain authorship of students' work and to control the access to tests is still an open issue in online education. Coursera, the leading massive open online courses (MOOCs) platform, which is pushing the boundaries of what's possible in digital education, complemented webcam verifications with Signature Track, a typing style analysis system, to reduce cheating. Online transactions face a similar challenge. The pioneers of the electronic commerce revolution, Amazon and eBay, had to design innovative reputation systems based on user feedback to increase participants' trust in their digital marketplaces. Online reviews are now ubiquitous.

While, strictly speaking, virtualization is possible without computers (e.g., mail-order shopping), it is the relentless innovation of networked digital technology that enables all but the most trivial of modern process virtualization initiatives. But what specifically about digital technology and its evolution fosters greater process virtualization? Three capabilities of digital technology aid in virtualizing more physical processes: representation, reach, and monitoring.

Representation This is the capability of digital technology to effectively simulate actors, objects, their properties and characteristics, and the manner in which users interact with them. More specifically, representation provides a sensory and relational representation of the process elements. The senses of sight and sound are more easily represented by modern digital technology, whereas smell and touch are not yet mainstream despite the advancement in olfactory and haptic interfaces. Online music stores provide a simple example. Because sound has been digitized with lossless compression (Chapter 1), shoppers can perfectly sample music before purchasing it. Since the sensory experience of music listening is fully captured by digital sound files, music retail selling was one of the first successful cases of virtualized retail.

Reach This is the capability of digital technology to overcome both time and space constraints. In essence, it allows the flexible participation of users in processes. Videoconferencing, for example, enables people to meet and discuss synchronously at a distance. In gaming, *Fortnite* and *League of Legends* are examples of massively multiplayer online games (MMOGs) that allow the concurrent participation and interaction of millions of players worldwide. *Fortnite* hit 10.8 million concurrent players during the in-game live concert of Marshmello.[1] In education, Coursera has experienced similar success. Partnering with 227 institutions, the platform offers more than 6,000 courses and reaches 43 million students worldwide.

Monitoring and Identification These are the capability of digital technology to authenticate process participants and objects and track their activity. The advancements in authentication through innovative systems (e.g., Coursera's Signature Track) or biometrics like the iPhone's TouchID and FaceID are examples of the relationship between digital technology evolution and process virtualizability as they apply to identification and control requirements. Another example of increased monitoring capability is provided by the Global Entry border protection program in the United States, where selected and approved individuals are allowed to enter the country by simply scanning their fingerprints at a computer kiosk.

---

1 Phillips, T. 2019, February 4. "Ten million players attended *Fortnite*'s Marshmello concert." *Eurogamer*, retrieved from https://www.eurogamer.net/articles/2019-02-04-10m-players-attended-fortnites-marshmello-concert.

It is critical for you to realize that process virtualizability depends on both the characteristics of the process expressed by its requirements and the potential of new technology to affect sensory, relationship, synchronism, or identification and control requirements. As a consequence, process virtualizability is a moving target. As digital technology continues to evolve, you can expect more processes to be virtualized, leading to continued waves of digital transformation, innovation, and entrepreneurship.

Managerial Takeaways

As new technology eases the trade-off between richness and reach and the progress of digital technology leads to the virtualization of more processes, a number of implications are critical for you as a manager to consider.

Continued Questioning of Traditional Business Models

A number of "traditional" business models were predicated on the fact that information was constrained by its carrier. The newspaper industry, with which we opened this chapter, is the prototypical example. Such business models have been facing, and will continue to face, challenges from focused organizations that exploit digital technology. Consider the traditional travel agent business model one more time. Many travel agents do an outstanding job of providing valuable consultations and advice to prospective travelers. In the past, others based their value proposition on the fact that most travelers didn't have the ability to book their own travel. As airlines, hotels, and car rentals created their own bookable websites and travel intermediaries such as Kayak.com and Skyscanner made it easy to quickly query multiple providers, traditional travel agents faced significant pressure, and many closed up shop (Figure 4.31). With technology advancing relentlessly, even those travel agents who offer valuable information and expertise will come under increasing pressure.

Figure 4.31. A travel agency gone out of business

The Importance of Customer Interface If information is increasingly allowed to travel independently of its carriers, it becomes feasible to unbundle traditional products and services and bundle products that could never be brought together before. Consider the retail banking industry. Traditionally, a customer would purchase a bundle of services from her retail bank—say, a checking account, a savings account, certificates of deposit, mutual funds, a car loan, and a mortgage. An important value driver of this bundle would be the convenience of one-stop shopping (i.e., being able to visit one branch to address a number of financial needs).

As we discussed earlier in this chapter, aggregators excel at owning the customer interface and aggregating demand. Today, for example, the same customer who had all of her financial needs fulfilled by the same bank may use Quicken or Mint to manage her finances (Figure 4.32). She might have checking and savings accounts with the local bank, a car loan with another local provider, a mortgage through LendingTree, mutual funds

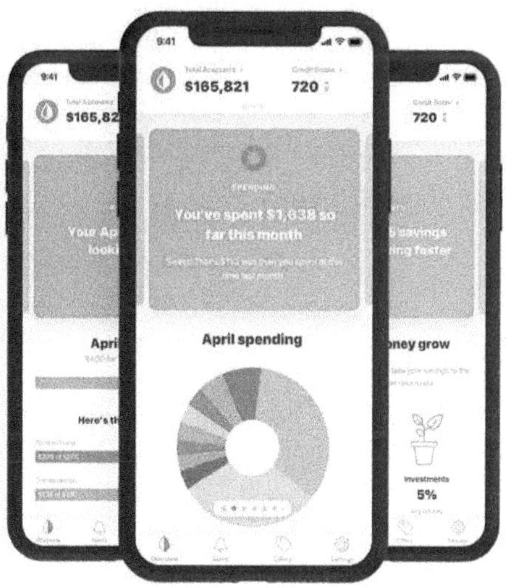

Figure 4.32. Mint on the iPhone

and retirement planning with Schwab, and some stocks holdings with TD Ameritrade. Being able to easily download monthly statements to Mint, she can keep the pulse of her financial standings and easily switch providers, as long as the new one also enables downloads to Mint. The old adage about the importance of "location, location, location" is in many industries based on information friction and is being challenged by digital technologies that eliminate transaction costs. Having a direct relationship with the customer, or owning the customer interface, is increasingly critical. In the banking case, it is Mint, a software company, which is in the strongest position.

The Decreasing Value of Asymmetric Information Perhaps the most evident implication of the emergence of technologies that ease the trade-off between reach and richness is the amount of information modern customers have available to them. This has put significant pressure on organizations that benefit from asymmetry of information.[1] If an organization bases its value proposition on the inability of individuals to obtain and use information at low costs, that position is increasingly untenable. Today, for example, within a few minutes, you can shop for a car online, find out the factory price from Edmunds.com, research various dealer packages, find out the value of your used car trade-in, and walk into a dealership ready to negotiate.

Obstacles

While there are many examples of industries where the effects of the easing trade-off between reach and richness are being felt, there are a number of obstacles that have been slowing and will continue to slow down this process.

New Technology Must Replace All Characteristics of the Old One Consider again the example of newspapers. Newspapers do not offer the best platform for consuming the news; they are not as timely as television or the Internet, they support neither video nor high-quality images, they offer limited space, and they have many other drawbacks. Why do we still buy them and read them (albeit in fewer and fewer numbers)? Why don't we all read the news online? Most newspapers have websites, anyway.

The answer is simple: with all the drawbacks newspapers have due to the constraints of printed paper, the broadsheet is still the most portable, most convenient, and easiest-to-read device for accessing the news. While e-books and tablets are increasingly challenging the dominance of paper, the newspaper will not disappear until new technology is able to supplant its advantages. The lesson is clear: old technology goes away only when the new one has replaced all of its relevant characteristics. Until then, the new and old technologies tend to coexist—as do newspapers and online news today.

Retaliation from Incumbents As we attempt to envision how new technology changes society and the competitive environment, seeking ways to create value in the new environment, it is easy to commit a critical fallacy: ignoring incumbents' retaliation. This was one of the main mistakes many observers made during the early days of the Internet. Retaliation can come in a number of forms:

- *Legal means*, such as those used by the music industry in reaction to the advent of digital music and the MP3 compression standard
- *Legislative means*, such as the lobbying efforts of taxi drivers to stave off Uber operations

---

1 Asymmetry of information is the condition where one party to a transaction has more information than the other party. Purchasing a used car is a classic example, where the seller knows the history of the car, while the buyer has to rely on the seller or use other imprecise means to gauge the quality of the car (e.g., take it to a mechanic).

- *Hybrid offers*, such as those provided by retailers with physical stores and online operations so as to leverage their existing infrastructure
- *Heightened competition*, such as that started by hotels companies in reaction to the offering of Airbnb accommodations

While it is enticing to think about the promise of the new digital technology and the opportunities it offers, you always need to remember that the road from today's status quo to tomorrow's reality is paved with competitive battles.

Human Resistance to Change Perhaps the most powerful bottleneck to some of the changes discussed above is human inertia. New technologies and new ways of doing business, while offering advantages, entail costs in the form of learning to use the new technology or simply stopping the old routine. While easily dismissed, these considerations can spell the difference between success and failure in a fledgling business. The history of digital technology is full of great ideas that fell prey to the "if we build it, they will come" mentality and failed to address and manage human resistance to change. Sometimes a crisis or discontinuity will impress an acceleration to slow-moving trends. In a few months, the COVID-19 pandemic forced a lot of people to overcome their resistance to change and accept remote work, online learning, food delivery, and a host of other virtualized processes.[1] "There are decades where nothing happens; and there are weeks where decades happen"—this observation, paraphrasing Vladimir Lenin,[2] seem to describe the effects of the pandemic on the acceleration of digitization trends fairly accurately.

Attention Challenges A by-product of the unprecedented availability of information is the increasing difficulty people encounter in keeping up with it. Customers' attention is not only required for your product or service to be considered; it is also required to educate customers about its advantages and how they can best use it. People's time and attention is perhaps the scarcest resource an organization has to deal with. The scarcity of attention leads to slow adoption rates for all but the most revolutionary of innovations.

Consider, for instance, online grocery shopping, an industry until recently only remembered for having produced the largest failure of the dot-com era. Webvan, the poster child for online grocery shopping, burned through $1.2 billion in funding before it closed up shop, much to the dismay of its few but enthusiastic customers (Figure 4.33). Yet online grocery shopping remained alive and well in many locales,[3] and the COVID-19 disease helped Instacart, the leader in the space,

Figure 4.33. Webvan stock certificate

Source: Scripophily, 2018, "Webvan—Bankrupt dot com brought home the bacon," retrieved from http://scripophily.net/webvancom.html

1 Watson, R., Ives, B., and Piccoli, G. 2020. "Guest editorial: Practice-oriented research contributions in the Covid-19 forged new normal." *MIS Quarterly Executive 19*(2). AIS eLibrary, retrieved from https://aisel.aisnet.org/misqe/vol19/iss2/2.

2 O'Toole, G. 2020, July 13. "Days into which 20 years are compressed—quote investigator." *Quote Investigator* (blog). https://quoteinvestigator.com/2020/07/13/decades-weeks/.

3 Peters, K. 2003, July 2. "Online grocery showing fortified strength, study says." *InternetRetailer*, retrieved from http://www.internetretailer.com/internet/marketing-conference/27475-online-grocery-showing-fortified-strength-study-says.html; McKinsey Insights. 2014, December. "How to win in online grocery: Advice from a pioneer." McKinsey & Company, retrieved from https://www.mckinsey.com/industries/retail/our-insights/how-to-win-in-online-grocery-advice-from-a-pioneer.

reach a valuation of over $17 billion in late 2020.[1] The lesson of the Webvan story is not so much that online grocery shopping was an ill-conceived idea but rather that the consumer adoption rate of this radically new way of performing a task that is thousands of years old was much slower than the adoption rates Webvan needed to survive as an online-only grocery operation.

4.6 A Note about Disruptive Innovation

In the introduction to this chapter, we used the term *digital disruption* to refer to the dramatic changes that digital technology and the Internet spurred in countless industries, from media to retail. Beyond the role of the Internet in changing the competitive landscape, as a general or functional manager, it is important that you are aware of the theory of **disruptive innovations**. The theory explains how some types of innovations can blindside managers, with devastating effects on their organizations. It is best understood by first drawing a distinction between sustaining and **disruptive technologies**.[2]

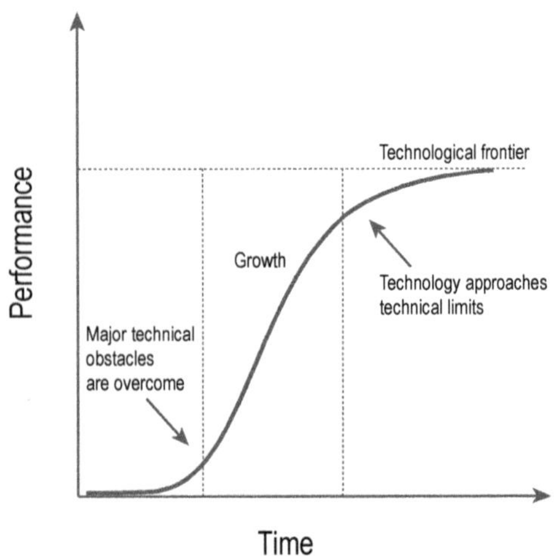

Figure 4.34. Product performance improvements over time

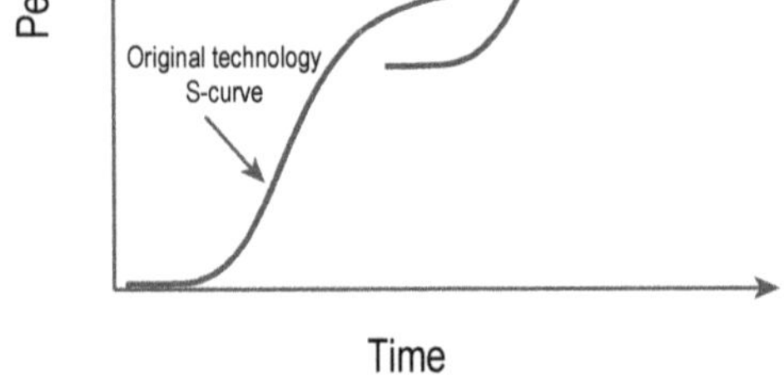

Figure 4.35. Performance improvements over the established performance trajectory

Sustaining Technology

The defining characteristic of **sustaining technologies** is that they maintain or rejuvenate the current rate of performance improvement of the products and services that use them. The performance trajectory of a technology (e.g., capacity of electric car batteries) is typically captured visually by the use of the S curve (Figure 4.34). The S curve suggests that as the product is first introduced, its performance is limited. With design refinements comes a growth period where substantial improvements are achieved until the technology underpinning product performance plateaus and further performance improvements become marginal.

Sustaining technologies are those new technologies that enable a product's performance to continue to grow—in other words, sustaining technologies extend the useful life of the product in its existing market with its mainstream customers who demand further and further improvements along the traditional performance dimensions (Figure 4.35). A sustaining technology will therefore be a good candidate to replace a previous generation because it offers the same set of attributes, but it yields superior performance. Thus firms that are using the existing technology in their product will find it appealing to switch to the sustaining technology as they seek to improve their products along the established performance trajectory.

Consider, for instance, the mechanical excavation industry. At the turn of the century, the tool of choice in this industry was the steam shovel. Steam shovels used a steam engine to generate the power required to pull the cables that would lift buckets full of dirt to be moved. In the early 1920s, the steam shovel began to be replaced by gasoline-powered shovels that offered superior performance on the critical performance dimension: the ability to move dirt in a fast, reliable, and cost-effective

1 Wilhelm, A., and Dickey, M. R. 2020, October 8. "Instacart raises $200M more at a $17.7B valuation." TechCrunch, retrieved from https://techcrunch.com/2020/10/08/instacart-raises-200m-more-at-17-7b-valuation/.

2 Christiansen, C. M. 1997. *The innovator's dilemma: When new technologies cause great firms to fail.* Boston, MA: Harvard Business School Press.

manner. This is an example of sustaining technology, since the new technology (i.e., gasoline engines) enabled manufacturers of dirt-moving equipment to improve the performance of their product on critical performance dimensions.

Disruptive Technology

Unlike sustaining technologies, disruptive technologies cannot be used in products aimed at the mainstream market because they are inferior with respect to the critical performance dimensions valued by mainstream customers. However, disruptive technologies have two characteristics that make them deceptively dangerous:

- The technology offers a different set of attributes than the technology used in products currently aimed at the mainstream market.
- The performance improvement rate of the technology on critical performance dimensions is higher than the rate of performance improvement demanded by the market (Figure 4.36).

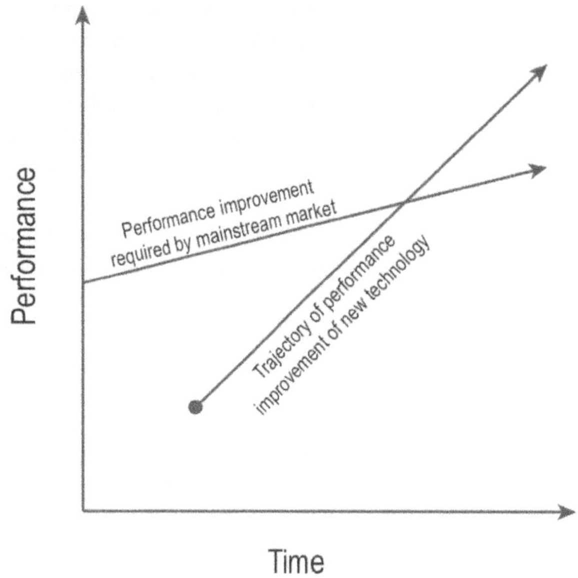

Figure 4.36. Market expectations for performance and new technology performance improvement over time

Thus while a disruptive technology delivers inferior performance when compared to the current technology used in products aimed at mainstream customers, it enables products that offer a different bundle of characteristics. Consider the prototypical example: the hard disk industry. The critical performance dimension for customers of hard disk drives (i.e., computer manufacturers) is the storage capacity of the disk. With remarkable precision, hard disk drive manufacturers have been blindsided by the emergence of an architecture design that enabled smaller and smaller drives to be produced—from the original 14-inch drives to the 8-, 5.25-, 3.5-, 2.5-, and 1.8-inch architectures. This is because these changes were disruptive in nature.

Each generation of smaller disk drives did not at first offer the same storage capacity as the established one. Yet each generation offered a performance rate of improvement on this dimension (the critical performance dimension) far superior to the speed with which computer manufacturers required greater storage capacity. As a consequence, when the storage capacity of the smaller drives reached market needs, the incumbent's mainstream customers switched to the new entrants. No longer having storage capacity concerns, they now valued the other characteristics offered by the new technology (i.e., smaller size, reduced power consumption).

Disruptive Innovation

After the original formulation of the theory, which centered on the difference between sustaining and disruptive technology, its author broadened it.[1] The shift enabled a wider applicability of the insight and clarifies some confusion about what constitutes a technology. Disruptive innovations are those novel value propositions (i.e., new products, services, solutions) that start as inferior to the dominant solution on the primary performance dimension but offer a different bundle of characteristics and a rate of improvement on key performance metrics that outstrip the rate of improvement expected by the mainstream market. Whether they are anchored in a clearly identifiable new technology or whether they stem from innovative rearchitecting of existing components or even a

1 Christensen, C. M., Raynor, M. E., and McDonald, R. 2015, December 1. "What is disruptive innovation?" *Harvard Business Review*, retrieved from https://hbr.org/2015/12/what-is-disruptive-innovation.

new business model is not the discriminant. This broader formulation of the theory points to two possible situations where emerging disruptive innovations will survive early in their life: the low end and a new market. The reason is that a disruptive innovation, which by definition is launched by a small new entrant in the market, needs time and revenue to grow to challenge the incumbent. As a consequence, it is only possible for it to exist in a void in the market that the incumbent has decided not to capture, typically because it does not offer attractive profits.

Low-End Market Start A disruptive innovation can have enough space to grow when it starts at the low end of the mainstream market, where large incumbents are unwilling to compete. In this market segment, there are usually customers who are unable to pay premium prices and are therefore willing to accept "good enough" products that are cheaper. Rather than service these low-end customers by making low-margin, stripped-down versions of its solution, the incumbent will typically move upmarket, where customers are willing to pay a premium for products at the high end of the performance spectrum. To do so, the incumbents will readily invest in sustaining technologies that enable them to keep pushing their product's performance. For example, while interchangeable lens cameras are still preferred by professionals, as they offer more control and superior picture quality, smartphone cameras have improved dramatically with each new generation. The old joke "the best camera is the one you have with you" is a testament to the power of "good enough" technology. Even in the early days of smartphones, for most people, the camera was good enough. Apple used to boast at every keynote that the iPhone was the most popular camera in the world. As quality improved, however, smartphones started to win photography awards, increasingly taking market shares from dedicated photo camera producers (e.g., Canon, Sony, or Nikon), who had no choice but to continue to move upmarket.[1] As the theory of disruptive innovation would predict, sales of dedicated cameras plummeted (Figure 4.37). The remaining customer base supporting the incumbents is now only hard-core photographers and professionals . . . but for how long?

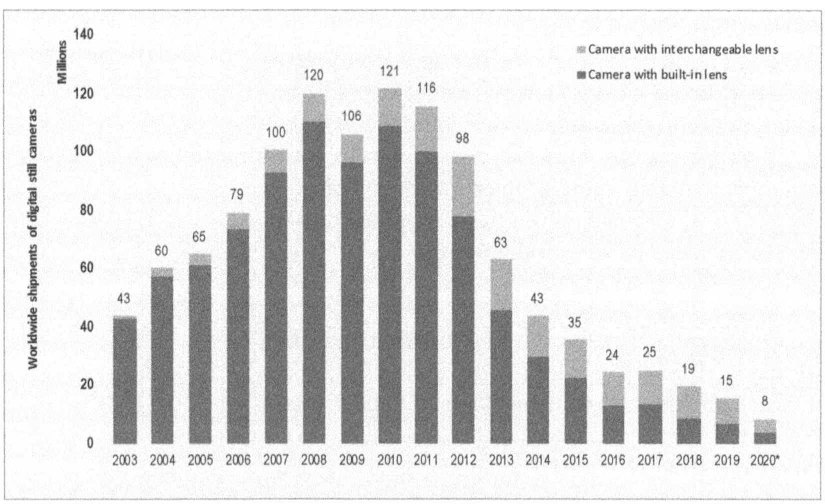

Figure 4.37. Worldwide digital still camera shipments

Source: CIPA, "Statistical data," retrieved from http://www.cipa.jp/stats/dc_e.html

New Market Start A disruptive innovation can also have enough space to grow when it pioneers a completely new market. This is perhaps the most common occurrence, where the innovator identifies a new unserved need and targets it. This is, by definition, an unproven market with uncertain potential and no guarantee of success. Moreover, these markets typically value the different bundle of characteristics offered by the disruptive innovation. Thus the incumbent finds it unattractive on financial grounds and more difficult to target because it would require the development of a new solution. The recent launch of the Apple M1 chip powering the Mac lineup of PCs is an example.

1 You can check see for yourself by checking the latest iPhone Photography Awards™ (IPPAWARDS) at https://www.ippawards.com/gallery/.

As with the disruptive technologies that originally inspired the theory, the superior rate of improvement of disruptive innovations on the critical performance dimension leads to a "cross-over event" when the disruptor reaches the performance level expected by the mainstream market (Figure 4.38). While this moment typically occurs years after the time when the disruptor first entered the market, it is at this time that the incumbent is in trouble. At this point, the different bundle of characteristics offered by the disruptor becomes a positive differentiator for the mainstream customers of the incumbent, who is typically unable to match the disruptor's solution and loses its dominant position. A very recent example is offered by the CPU business, historically dominated by Intel. As the story goes, Apple, an Intel customer for its Mac family of PCs, approached Intel in 2005, asking the firm to design and produce the CPUs for its future lineup of smartphones. As Paul Otellini, Intel's CEO at the time, tells the story,

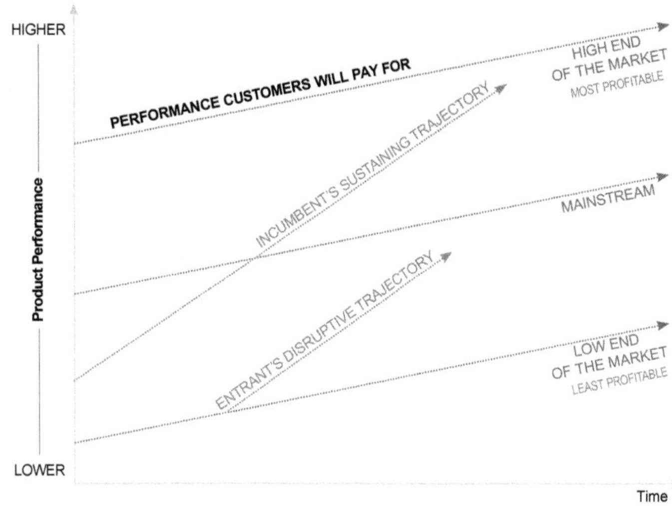

Figure 4.38. The disruptive innovation model

Source: Adapted from Christensen, C. M., Raynor, M. E., and McDonald, R., 2015, December 1, "What is disruptive innovation?" ***Harvard Business Review***, retrieved from https://hbr.org/2015/12/what-is-disruptive-innovation

> We ended up not winning it or passing on it, depending on how you want to view it. And the world would have been a lot different if we'd done it. . . . The thing you have to remember is that this was before the iPhone was introduced and no one knew what the iPhone would do . . . At the end of the day, there was a chip that they were interested in that they wanted to pay a certain price for and not a nickel more and that price was below our forecasted cost. I couldn't see it. It wasn't one of these things you can make up on volume. And in hindsight, the forecasted cost was wrong and the volume was 100× what anyone thought.[1]

Thanks to "fabless chip design" enabled by manufacturers like Samsung and TMSC, Apple decided to design its own chips for the iPhone.[2] The design imperative for these chips was power consumption, since they had to work in a small portable device where battery life was a critical performance measure. For this reason, they could not compete with Intel's designs on computational power—the key performance dimension for use in PCs.

However, as is clear now, the A series of Apple chips was a disruptive innovation. It was characterized by a different bundle of characteristics, chief among them its superior energy efficiency and a rate of improvement on the dimension of computational power that exceeded not only that of the PC market but even that of the incumbent, Intel (Figure 4.39).

The disruption was complete in 2020, when Apple stopped using Intel CPUs in its PCs and introduced a variation of the A chip for the Mac line, called the M1 (Figure 4.40).

---

1 Madrigal, A. C. 2013, May 16. "Paul Otellini's intel: Can the company that built the future survive it?" *Atlantic*, retrieved from https://www.theatlantic.com/technology/archive/2013/05/paul-otellinis-intel-can-the-company-that-built-the-future-survive-it/275825/.

2 In the semiconductor industry, a fabless company is a firm that just designs the chip and then partners with a specialized manufacturer to produce it. Chip manufacturing is an extremely capital-intensive business and a concentrated market.

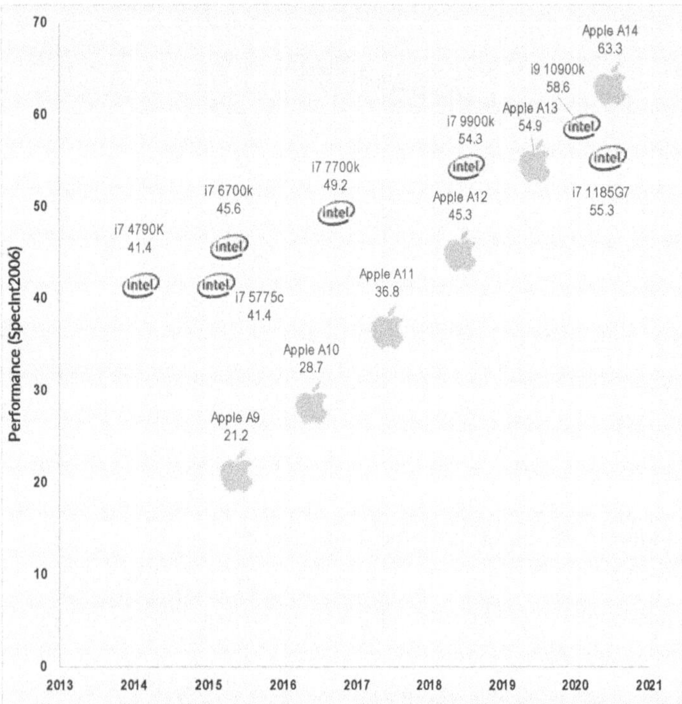

Figure 4.39. Intel and Apple chip performance compared

Source: Adapted from Frumusanu, A., 2020, November 10, "Apple announces the Apple Silicon M1: Ditching x86—What to expect, based on A14," Anandtech, retrieved from https://www.anandtech.com/show/16226/apple-silicon-m1-a14-deep-dive

Recognizing a disruptive innovation is a lot easier in hindsight, and we should not think we are smarter than Intel's executives. But there are many examples of potential disruptive innovations enabled by modern digital technology, and as a manager, you must be very aware of them. Many suggest that 3D printing and digital manufacturing (Chapter 12) are disruptive innovations (Figure 4.41). The industry recently experienced two-digit growth reaching four billion dollars in 2020, and it is expected to grow at a compound rate of 18% until 2025.[1] By bridging the physical and the digital worlds, 3D printing is making a new business ecosystem emerge in which digitally designed objects are shared online and locally manufactured. The technology has been around since the 1980s, but only recently has the massive drop in the cost of 3D printers made the technology affordable and personal. The story may repeat itself, and personal 3D printing may disrupt the manufacturing industry just as home and personal computers changed the information technology (IT) industry forever.

Managerial Takeaways

Familiarity with the dynamics of disruptive innovations is important for today's general and functional managers because disruptive innovations typically blindside leading firms (i.e., the incumbents in the market), who see their position of dominance lost to those upstarts that were able to ride the disruptive technology wave.

Differential Rates of Improvements The deadliest characteristic of disruptive innovations is their rate of evolution on the currently established performance metrics. As shown in Figure 4.36, a disruptive innovation begins with performance that is well below the needs of the firm's mainstream customers. Moreover, the disruptive innovation will likely not improve at a rate sufficient to overcome the existing or sustaining innovation available. However, this is misleading information for a manager! In truth, it is irrelevant whether the disruptive innovation will ever outstrip the current one on key performance metrics. Rather, you should estimate whether, in the foreseeable future, the disruptive innovation will catch up to market needs on the critical performance dimensions (i.e., become good enough for mainstream customers).

It is often the case that a firm, focused on its most demanding and most advanced customers, will push the performance of its products using new generations of sustaining innovations. However, such a relentless focus on the most demanding customers may end up pushing the firm to increasingly overshoot the needs of its mainstream customers. When this is the case, the incumbent firm becomes particularly vulnerable to disruptive innovation.

1 ReportLinker. 2020. "COVID-19 impact on 3D printing market by offering—global forecast to 2025." Retrieved from https://www.reportlinker.com/p05949870/COVID-19-Impact-on-3D-Printing-Market-by-Offering-Global-Forecast-to.html.

Different Sets of Attributes Become Relevant As disruptive innovations close the gap between the performance level they offer and mainstream customer needs, the novel set of attributes they offer may become increasingly attractive to potential customers. In other words, as the disruptive innovation closes the gap on the primary performance metrics, the technology's other characteristics may become a source of positive differentiation. At this point, customers defect from established suppliers offering the standard products and begin to adopt the new innovation—but it is typically too late for established players to make the switch and regain their dominant position.

Figure 4.40. Apple M1 key features and specifications

Source: Apple keynote event, 2020, October 13, retrieved from https://www.apple.com/apple-events/october-2020/

Listening Closely to Customers Might Spell Trouble Conventional business wisdom suggests that a firm is well served by listening closely to its customers in an effort to develop the products and services that best serve their needs. While you should not ignore this suggestion, we add a word of caution: listening attentively to your most aggressive customers will create a bias toward prompt adoption of sustaining innovation and a reluctance to buy into disruptive innovation.

Your best customers are constantly pushing the envelope of your product performance. Consider again the classic case of the hard-drive industry. High-end computer manufacturers, seeking to outdo each other, will seek larger and faster hard disks to be included in their machines. As a consequence, they will create an incentive for you to adopt those innovations that offer improved performance on the accepted set of performance metrics. A technology that enables the development of smaller hard disks is not valued, even though these smaller devices have other interesting characteristics, such as compact size, lower energy consumption, and less need for heat dispersion. They simply are not good enough on the "important" dimensions that your best customers are clamoring for. However, if the new, smaller hard disks are a disruptive innovation, their performance will soon meet the needs of your mainstream customers. When this happens, all those other characteristics may become valuable, and you'll be left with a rapidly shrinking market.

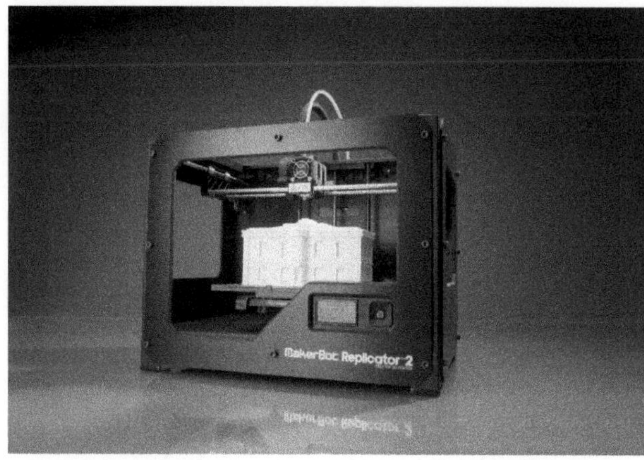

Figure 4.41. 3D printer

Source: Photo by Creative Tools / CC BY 2.0

What to Do?

Those studying disruptive innovation change suggest the following approach to managing organizations that face the emergence and development of disruptive innovations:

- Monitor market developments for the emergence of new technologies or innovations and determine whether they are of the sustaining or the disruptive kind.
- When disruptive innovations emerge, envision the new market they would likely be best suited for. One of the greatest challenges faced by the incumbent firm is to identify which customers will likely appreciate the new blend of features and

functionalities that the disruptive innovation supports. While a producer of large mainframes does not care much about the power requirements and physical size of hard disks, these are critical characteristics for laptop manufacturers.

- Spin off a new division that focuses exclusively on the commercialization of products based on the disruptive innovation. Separating the group that is blazing the trail for the new innovation may be necessary to create the appropriate financial incentives. Disruptive innovation starts off serving the needs of a small, niche market. As such, it is difficult for large companies to get excited about—and, more important, divert resources to—such small markets. A separate entity, focusing on that business and competing with the other small firms in the new market, would have no difficulty creating the appropriate incentives.

Summary

This chapter provides you with a framework to understand how new technologies are shaping the competitive landscape you will encounter as you enter the job market. Specifically, in this chapter, we discussed three broad topics: network economics, information economics, and disruptive technologies.

- Value in networks—physical ones such as the telephone network and virtual ones such as eBay's online community of buyers and sellers—is created by plentitude. This value driver is the opposite of the principal value driver of most other goods and services: scarcity.
- Because the most valuable networks are the largest ones, an individual's act of joining a network creates value for the other members of the network—a phenomenon termed *network effects*. In industries subject to strong network effects, particularly when the demand for variety is low and networks are mutually exclusive, winner-take-all dynamics ensue, and the market is dominated by one organization.
- When confronted with network effects, four characteristics of networks should be evaluated and monitored:
  1. *Network Effect Strength*: The strength of positive feedback may vary by network and may vary over time for the same network.
  2. *Network Clustering*: The topology (the "shape" of a network) determines its potential for winner-take-all dynamics to emerge.
  3. *Disintermediation Risk*: Network effects are weaker when participants can easily transact outside of the network.
  4. *Multihoming Potential*: Network nodes (e.g., users, devices) simultaneously participate in multiple competing networks.
- Digital platforms, markets, and aggregators represent the integration of a digital platform, and a marketplace enables aggregators to influence the suppliers' strategic choices regarding product design and product commercialization.
- Information, a prevalent resource in the modern competitive landscape, has unique economic characteristics. In its purest form, information has high production costs, which are sunk, and negligible replication and distribution costs. The production of information faces no natural capacity limits, and information is not consumed by use. As a consequence, information is infinitely reusable, highly customizable, and often time valued.
- When discussing information as an organizational resource, it is important to distinguish the information itself from the carrier of the information. Historically, information as a resource or product had been constrained by the economics of the carrier. The advent of the Internet, a global infrastructure for information exchange, has in many cases separated the two. New technology continues to push the frontier of the richness/reach trade-off and, in the process, threatens established business models in information industries and beyond.
- The process virtualization theory provides an overall framework to analyze the likelihood for an activity to be performed in the digital world. The framework identifies four main characteristics of the process, called requirements, that determine its "virtualizability." As technology improves, processes that today are impossible to virtualize may find their digital transposition. Indeed, process virtualizability depends on both the characteristics of the process and the potential of new

technologies to affect process requirements. Monitoring process virtualizability opens up the opportunity to anticipate new forms of interaction leveraging new uses of existing technologies.

- New technologies can be characterized as sustaining or disruptive. Sustaining technologies are those that maintain or rejuvenate the current rate of performance improvement of the products and services that use them. Conversely, disruptive technologies are those that offer a different set of attributes than the technology the firm currently uses in its products, and their performance improvement rate is higher than the rate of improvement of market needs. Disruptive technologies are particularly dangerous for established firms, which typically tend to underestimate their potential impact on the firm's current business. Proper monitoring and management of disruptive technologies by the incumbent are necessary because, due to the rate of performance improvement and the different set of features they offer, once a disruptive technology has achieved acceptable performance improvements on the traditional dimensions of performance, customers quickly defect to products that use it.

Study Questions

1. Define the term *Internet* and offer examples of its principal services. What is the difference between the Internet and the World Wide Web?
2. What do you see as the likely evolution of the Internet in the near future?
3. Explain each of the following concepts: positive feedback, network effects, and tippy markets. Explain how the three concepts relate to one another.
4. Offer an example of a tippy market and an example of a market that does not tip.
5. Differentiate physical and virtual networks and provide examples of each.
6. Explain the defining characteristics of a two-sided network and provide an example.
7. Define *networks*, *marketplaces*, *aggregators*, and *platforms* and discuss their differences.
8. What is the defining characteristic of classic information goods? How do they differ from information-intensive goods? Provide examples of each.
9. Information, as an economic entity, behaves quite differently than traditional goods. Identify the principal economic characteristics of information and describe the primary implications for strategy.
10. Explain what we mean by the richness/reach trade-off. Why is this concept important today for general and functional managers? Provide examples of recent technologies that have pushed the richness/reach trade-off frontier farther out. What industries or organizations are under pressure as a consequence of this development?
11. Do you believe that "the Internet changes everything," or is it "just another technology"? Be sure to defend your position.
12. What is process virtualization theory? Discuss the case for the virtualization of the "ordering food" process.
13. Use the process virtualization theory to discuss the likelihood for two processes of your choice to be performed digitally.
14. What is the difference between sustaining and disruptive technologies? Offer one example of each.
15. What would you advise an incumbent firm to do in the face of the emergence of new technology? For example, if you were an executive for American Airlines, what would you do about the recent introduction of very light jets (VLJ)—such as the 2,000 preordered Eclipse 500 VLJs from Eclipse Aviation, which are able to reach 300 miles per hour and have a range of 1,125 nautical miles?[1]

Glossary

A glossary of key terms can be found at the book's website. Visit https://www.prospectpressvt.com/textbooks/piccoli-information-systems-for-managers-5-0, then click on "Student Resources" in the horizontal red menu bar.

1 CNN. 2006, March 31. "Light jets promise air revolution." Retrieved from http://edition.cnn.com/2006/TRAVEL/03/31/private.jet/.

Chapter 5

Digital Transformation, Innovation, and Entrepreneurship

What You Will Learn in This Chapter

In the previous chapters, we have introduced the economic forces underlying the waves of digital disruption that have taken the business world by storm since the commercialization of the Internet in the 1990s. The chapter provides you with the background and history of two decades of digital disruption, and it will help you understand key ideas such as digital transformation, digital innovation, and digital entrepreneurship. The objective is to support your development of a solid grounding in the concepts and vocabulary of **digital business**—a vocabulary that is no longer the exclusive province of Silicon Valley insiders but an integral part of the language of all modern organizations.

MINICASE: Chasing Waves and Changing Lives

Your Instagram notification fires off. There is a new video from your childhood friend Federico. You two grew up together and enjoyed sports. One summer, Federico had a new contraption he wanted you to try—a windsurfer. It was cool to try to master the thing (while mostly falling off). You liked the challenge, but it did not change your life. Federico, on the other hand, was hooked, and he started traveling to different windsurfing destinations with his father. Upon returning, he would show you pictures and talk about new moves he was trying. As you were starting college, Federico told you he secured a sponsorship and was going to try to make it on the pro tour. As he chased his dream, you finished college and got a job with McKinsey's digital strategy practice.

As you were watching the new video, Federico DMs you: "What's up, old pal! How is the high-pressure world of consulting going? Do you have a few minutes to help your old beach bum friend? :-)" You immediately replied that you were free, and two minutes later, the phone rang. Federico was in Sardinia, training for the upcoming season, even though the COVID-19 pandemic crisis led to most events being canceled. With budgets being squeezed, many of the pro windsurfers on tour were losing sponsors or seeing their already limited contracts being slashed. Windsurfing was not a major sport. For all but the very top sailors, a sponsorship meant technical equipment support but no money. Yet windsurfing was not a cheap sport. It required athletes to visit windy destinations to train and compete, bring bulky equipment (and pay airline excess baggage fees), and invest in repairs and new material when the inevitable wipeout damaged it.

Federico told you times were tough, and it was increasingly difficult to self-sustain while trying to be a professional athlete in such a demanding sport as windsurfing. "Moreover" he confessed, "I also want to have a broader impact, motivating people to live their best life and to live sustainably. You should see how humans are damaging the environment. In the most beautiful beaches around the world, you will see microplastics. And I am sure you know about the Great Pacific Garbage Patch." Federico's current "revenue model" was based on advertising and branding. He had technical sponsors that supplied him the basics in terms of equipment. Beyond the sport, he was a brand ambassador and influencer for Christopher Ward watches, Seek Discomfort clothing, and Milo, a communication device specially geared for action sports. Federico had over 25,000 followers on Instagram, where he posted every day, and over 7,000 on Facebook. He also had a YouTube channel hosting his vlogs and action videos totaling tens of thousands of views. His followers and viewers came from all corners of the earth, from the United States, to Italy, to Japan. Yet with rough estimates of payout for YouTube videos at $3 to $5 per 1,000 views, it was difficult to sustain a pro athlete career on a pure ad strategy.[1] A typical influence marketing campaign could net a few hundred dollars per post.[2]

Federico concluded, "Ultimately I want to be one of the top ten windsurfers on tour and inspire people as I do it. Since I get equipment from my technical sponsors, I estimate I need about $50,000 a year to have the resources to focus on training and really give it a shot. Do you have any brilliant ideas? Digital tech moves so quickly, maybe there are opportunities I could pursue that are better than what I've been doing so far." Seeing your old buddy inspiring people from the top of the podium would be really cool. Perhaps the recent explosion of digital platforms and resources for "creators" could be the solution.

Discussion Questions

1. What do you think of Federico's current strategy? What do you see as the pros and cons of his current "influencer" strategy?
2. Do you think Federico could develop a direct relationship with an audience? Is the windsurfing community big enough to help him reach his $50,000 a year target?
3. What specific suggestions do you have for Federico as he tries to generate enough revenue to continue chasing his dream?

1 https://influencermarketinghub.com/how-much-do-youtubers-make/.
2 https://influence.co/go/rates/instagram_followers/25k-50k.

5.1 Introduction

In the previous chapter, we introduced the concepts and techniques you need to understand the successive waves of digital disruption that have taken the business world by storm since the commercialization of the Internet in the 1990s. We introduced the notions of network economics and information economics, and we discussed process virtualization theory and the role of disruptive technologies in shaping the competitive landscape. These concepts are critical for general and functional managers, who increasingly find themselves managing in the digital economy. With this theoretical background as the basis, we now define and provide examples of digital transformation, digital innovation, and digital entrepreneurship. We then formally discuss the digital technologies that have spurred repeated waves of digital disruption, starting with the Internet through the mobile platform to the most recent: digital resources. Our objective here is to help you become familiar with the history, language, and current manifestations of the digital business landscape in order for you to be a more effective manager in the digital age.

5.2 Digital Disruption and Its Manifestations

Elon Musk, the celebrity chief executive officer (CEO) of Tesla and SpaceX recounts a story from his early days in Silicon Valley. A 24-year old Musk, with younger brother Kimbal and friend Greg Kouri, founded Zip2. After some iterations, Zip2 became the software backbone powering newspapers' efforts to create online business directories. As Musk tells the story, he and his brother were pitching their services to publishers of the Canadian Yellow Pages. Dismissing their ideas, one of the executives tossed one of the big yellow books at them and said, "You ever think you're going to replace *this*?"[3] Well, we know how this story ends—that's exactly what happened. Internet services made the Yellow Pages obsolete (as we write this, we wonder if you even know what the Yellow Pages are . . . if not, ask your parents!).

3 Locke, T. 2020, April 4. "Elon Musk tried to pitch the head of the Yellow Pages before the Internet boom: 'He threw the book at me.'" CNBC, retrieved from https://www.cnbc.com/2020/04/03/elon-musk-pitched-the-head-of-yellow-pages-before-the-internet-boom.html.

The above example is a perfect illustration of digital disruption, a disruption so significant that the industry that was affected no longer exists in its pre-Internet form. In today's parlance, we would say that through their digital entrepreneurship efforts, the Musk brothers and many other Internet enthusiasts disrupted the paper directory listings industry. As we have been saying throughout this book, such digital disruption has already changed many industries, from publishing, to music, to entertainment, and is in the process of changing all remaining industries. No organization is immune from the transformative power of digital technology.

There are three manifestations of digital disruption that you should be aware of: digital transformation, digital innovation, and digital entrepreneurship. The three are distinct but related, with the commonality that digital technology plays a central role in each of them. However, you'd be surprised at how difficult it is for managers, consultants, and even academics to define such seemingly common and intuitive terms. In this section, we explain how digital phenomena are different in magnitude and, sometimes, in quality from more traditional information technology (IT)-enabled phenomena.

Digital Relationships and Digital Business Modeling

Before discussing each of the three manifestations of digital disruption, consider this question: When is the last time you physically visited a bank, or a travel agency, or even a bookstore? You still have financial, travel, and reading needs, but increasingly, the default option we resort to is fulfilling those needs through a digital technology interface—a web browser, a smartphone, even a smart speaker. If you do visit a physical store, you are increasingly using technology as part of that transaction. Many grocery stores allow you to scan items in your basket while you go through the shop using a dedicated barcode reader or your personal smartphone. Amazon Go stores don't even have check-out lanes—you grab the items and . . . go.[1] As a manager, now consider the other side of that question. How should a bank, a travel agency, or a bookstore maximize its ability to profit from customers shifting to digital relationships? They need to change their business processes and often, more dramatically, their **business model** to take advantage of the idiosyncratic characteristics and potential for value creation enabled by digital technology. Note that this is a constant state of affairs, not a one-and-done occurrence. As new digital technologies lift previous constraints and enable new opportunities, the potential for business process and business model change is constantly renewed.

Business Model A business model is an abstraction that captures the firm's concept and value proposition while also conveying what market opportunity the company is pursuing, what product or service it offers, and what strategy it will follow to seek a dominant position. The business model may also identify what organizational capabilities the firm plans to leverage to turn the concept into reality. In short, the business model tells us who the firm's customer is, what the firm does for its customers, how it does it, and how it is going to be compensated for what it does.

The term *business model* acquired prominence with the advent of the Internet and the emergence of **electronic commerce.** Up until the commercialization of the Internet, with few exceptions, it was clear what a firm did and what its value proposition was (i.e., its business model) once we knew its industry. The statement "I produce beer," or "I sell groceries," or "I am a real estate agent" clearly conveyed what the firm did, what its cost structure was likely to be, and how the firm would create and sustain a revenue stream. In other words, the underlying business model was implicitly but clearly defined.

1 "Amazon Go." n.d. Amazon. https://www.amazon.com/b?ie=UTF8&node=16008589011.

Table 5.1. The business model canvas nine building blocks

Question	Building block	Description
1. Who is the business designed to serve?	Customer segments	The groups of people (i.e., consumers) or organizations for whom your firm is creating value
2. What will the firm do for those customers?	Value proposition	The specific set, or bundle, of products and services that create value for customers
	Channels	The physical or digital touch points the firm utilizes to deliver value to its customers
	Customer relationships	The tangible and emotional connections the firm establishes with the customer
3. How will the business create value?	Key resources	The tangible and intangible assets within the control of the firm, without which the value proposition could not exist in the current (or planned) form
	Key activities	The actions that the firm must be able to perform well to offer its value proposition
	Key partnerships	The upstream relationships the firm must be able to leverage to deliver its value proposition
4. How will cash flow in and out of the business?	Revenue streams	The "monetization" of the firm's value proposition through a specific pricing mechanism
	Cost structure	The outflow of cash to upstream partners and resource providers the firm sustains while operating the business model

However, the ability to use malleable digital technology enabled the creation of a seemingly unending stream of new "digital solutions" to customer problems. Complementarily, the emergence of the network economy provided the vehicle for scaling these solutions on a global basis. Consider the business model of aggregators (Chapter 4), like Deliveroo in the food delivery business. Deliveroo aggregates demand for food delivery by reaching, theoretically, all consumers in a specific area (e.g., London). Anyone who has a smartphone can download the Deliveroo app and start ordering. The more customers are aggregated, the more restaurants in the area have an incentive to join. Restaurants can participate by simply configuring their digital menus through the Deliveroo Restaurant Hub, an online set of tools for restaurant partners. Riders can also join "digitally" by downloading the Deliveroo rider app, set up an account, and start taking orders. A business model predicated on intermediating a large-scale, three-sided marketplace (Chapter 4) of self-provisioning customers, restaurants, and riders was simply not feasible prior to the pervasive adoption of interconnected digital devices (Chapter 1), such as smartphones, that encompass multiple sensors (e.g., GPS) and are always connected to the Internet. There are a number of tools that can help entrepreneurs and managers alike think through their business model in a disciplined fashion. The **business model canvas** is one such tool that has received considerable attention since its publication in 2010 (Table 5.1).

The Elements of the Business Model Canvas The business model canvas is a handy "thinking tool" that can help managers conceptualize and crystallize the design of a new business model. It also provides a language for describing and communicating the model. While a thorough discussion of the business model canvas is beyond the scope of this book, we describe here the rationale for the model, and we provide a brief definition of its nine building blocks.[1] At the highest level of abstraction, it answers four fundamental questions (Table 5.1).

---

1 The business model canvas was conceptualized by an IS PhD student, Alex Osterwalder. Thus it should not surprise you to find out that rather than simply writing a book, Alex and his team leveraged digital technology and the Internet to produce a number of digital tools and content. We encourage you to visit their repository of material at https://strategyzer.com/.

In the late 1990s and early 2000s, as the Internet became a stable platform for commerce, entrepreneurs and organizations seeking to profit in the "network economy" designed and introduced a number of novel business models. Ever since the emergence of the Internet and the mobile platform as a reliable infrastructure for commerce, business model evolution spurred by digital technology has continued to accelerate. Consequently, as you enter the workforce, you may join one of the many established corporations engaged in digital transformation. You may find employment in a consulting firm helping its clients with their digital innovation efforts, or you may engage in digital entrepreneurship by launching your own start-up seeking to digitally disrupt an entire industry.

Digital Transformation

We defined digital transformation as the application and exploitation of digital technology to radically rethink organizational business and operational models (Chapter 1). We added that digital transformation is underpinned by the redesign of existing organizational information systems. But how is digital transformation different from business transformation in general? If it is the use of digital technology that differentiates the two, what's new here? How is modern digital transformation unlike previous trends, like business process reengineering (BPR) and business process management (Chapter 3)? While we are the first to admit that the "digital technology hype machine" is always thirsty for new and exciting buzzwords, we would argue that the term *digital transformation* is necessary to emphasize that business transformation today is radically impacted by the pervasive nature of modern digital technology (Chapter 1). Experts point out that "a digital transformation involves rethinking the company's value proposition, not just its operations. A digital company innovates to deliver enhanced products, services, and customer engagement. Digital is exciting, thrilling—and a bit unnerving!"[1] Of course, any rethinking of the company's value proposition will require business process change (Chapter 2). However, unlike IT-enabled business transformations of the past, such as BPR (Chapter 3), the emphasis of digital transformation initiative is on embracing digital relationships with customers through digitally enhanced products or services. Forward-looking chief information officers (CIOs) who understand the transformative power of digital technology have always sought to create (Chapter 8) and appropriate value by leveraging digital technology (Chapter 9). For them, digital transformation is nothing new. But the sheer magnitude and pervasiveness of the changes required now that digital technology is the primary infrastructure for business has catalyzed general and functional managers' attention in every industry. Everyone is doing digital transformation, not just the most forward-looking CIOs anymore.

When business executives today realize the need to transform their business, both the impetus for change and the solutions available depend heavily on digital technology. This is sometimes a radical change that calls into question the value of existing resources, as in the case of newspapers and their printing presses (Chapter 4). Such seismic change forces a fundamental rethinking of the firm's business model. Jeff Lawson, CEO of Twilio, recounts a personal incident about famed retailer Nordstrom (Figure 5.1). You probably are not old enough to remember, but Nordstrom was the gold standard in customer relationships for decades.[2] Their sales agents were renowned for getting to know customers personally, recognizing them when they walked in the door, and striking up conversations about past purchases (i.e., "Hi Jill, how was your daughter's baby shower? Did she like that cute baby dress you gifted her?").[3]

---

1 Ross, J. 2017, September 29. "Don't confuse digital with digitization." MIT Sloan Management Review, retrieved from https://sloanreview.mit.edu/article/dont-confuse-digital-with-digitization/.

2 NRF. 2019, March 4. "Nordstrom sets the standard for customer experience. Again and again." National Retail Federation, retrieved from https://nrf.com/blog/nordstrom-sets-standard-customer-experience-again-and-again.

3 Spector, R., and McCarthy, P. D. 2005. *The Nordstrom way to customer service excellence: A handbook for implementing great service in your organization.* Hoboken, NJ: John Wiley & Sons.

Lawson tells the story about a wedding he had to go to a few years ago and, as busy tech executives do, he had forgotten and was looking for a suit at the last minute. He had no time to go shopping for suits and, having lost a lot of weight recently, he didn't have a suit he could reuse. What saved him was a Nordstrom exchange he had six months prior. At the time, after a purchase, he was prompted to receive a text receipt rather than the typical paper printout. Reluctant to provide his number, he declined. The sales agent checking him out told him he could use the number he received the text from to text back directly to him if ever needed. So Lawson agreed to receive the receipt via text. Six months later, in a time crunch, he remembered the exchange, dug out the text message, and asked for help. The sales rep responded with pictures of three suites in his size range, along with his own suggestions. The rep also said that if Lawson picked one of the options, he could swing by at 6 p.m. to do a fitting in five minutes—in and out. The suit would be tailored and ready for pickup the next day.[1]

Figure 5.1. W. Nordstrom, cofounder of the Nordstrom department store chain, in front of his shoe store in Seattle, Washington, opened in 1901

Reflect on what happened here. A retailer that had historically made personal relationships between its outstanding sales representatives and customers a cornerstone of its success had to confront a new world of busy consumers who are increasingly used to shunning **brick-and-mortar** retailers for online transactions. How could Nordstrom transform its business and the resources that defined its competitive advantage before the digital age to remain relevant to its digital consumers? How could the firm reinvent personal relationships for the digital age? While the helpful suggestions of a highly skilled sales representative are still valuable to customers, how could the firm provide them if customers no longer visited the store as often? We don't know if these efforts will be enough to save retailers such as Nordstrom from the digital disruption tsunami that digital technology—and the recent COVID-19 pandemic—has unleashed on the physical retail industry. Clearly, however, those who will survive are the ones that will be able to navigate the required digital transformation and change their operations and business models to suit the digital age:

> To become digital, leaders must articulate a visionary digital value proposition. This value proposition must reassess how digital technologies and information can enhance an organization's existing assets and capabilities to create new customer value. Being digital is *not* just introducing mobile apps for customers. It is taking advantage of the opportunity to redefine a business—and possibly even an industry.[2]

In the Nordstrom case described above, this meant redefining the traditional personal relationship between sales associates and customers and extending it to changing customer needs by taking advantage of the characteristics of available digital technology (e.g., multimedia text messaging).

1 Thompson, B. 2020, October 15. "Twilio acquires Segment, what is Segment, an interview with Twilio CEO Jeff Lawson." Stratechery, retrieved from https://stratechery.com/2020/twilio-acquires-segment-what-is-segment-an-interview-with-twilio-ceo-jeff-lawson/.

2 Ross, J. 2017, September 29. "Don't confuse digital with digitization." MIT Sloan Management Review, retrieved from https://sloanreview.mit.edu/article/dont-confuse-digital-with-digitization/.

Digital Innovation

The above discussion of digital transformation shows you how the concepts are interrelated. We defined digital innovation as applying digital technologies by existing organizations to create new and improved products, processes, or services (Chapter 1). There is certainly quite a bit of digital innovation in a digital transformation, but the scope is different. With digital innovation, the change engages an existing organization impacting products, processes, or services, generally to extend or reinvent the firm's existing value propositions. In an increasingly digital economy, digital innovation becomes an essential process for incumbents to remain relevant. The effects of digital innovation may be subtle and sometimes even difficult to forecast. Consider Starbucks, the global chain of coffee shops. At the forefront of digital innovation, the company began testing a mobile app in 2009, launching a wallet-like mobile payment program. Customers could buy rechargeable cards in stores, purchase virtual cards online, or configure the app to automatically recharge when the balance was low. The app was a success from the start. As Adam Brotman, senior vice president and general manager of Starbucks Digital Ventures, put it,

> The customer response to our mobile apps has been phenomenal. Not only are they using their phone as a wallet, but as a connection point to Starbucks as they manage their Starbucks Card accounts, send eGifts and get into the holiday spirit with the Starbucks Cup Magic app.[1]

However, the breakthrough digital innovation was mobile order-ahead from the app, introduced in 2014. Instead of just simplifying payment, the process innovation allowed line-heating coffee addicts to preorder their drink, pay for it in the app, and collect it as soon as it was ready at the shop—no delays, no lines. In a funny twist of fate, the innovation was so successful that a line of mobile app users was now forming at the stores. Further innovation was needed, in an example of the digital world shaping physical spaces: "We're going to redesign new stores and existing remodels to reflect the fact that Mobile Order & Pay, although it's in its nascent stage, is obviously going to be a significant part of the morning business."[2] In 2019, Starbucks opened an experimental pickup-only store to serve digital orders and provide relief to traditional shops. When the COVID-19 pandemic struck, Starbucks leveraged these new store formats to survive, closing around 400 cafés and building 300 new pickup-only stores.[3]

The Starbucks story well illustrates the main characteristics of digital innovation. Its focus is on a product, service, or process, the scope is limited, and the objective is clear. In this case, their objective is creating value for customers by streamlining the ordering process using digital technology, thereby increasing convenience and eliminating a process pain point (i.e., waiting in line to order and then again to pick up the coffee). Like all successful improvements, digital innovations are far from a one-time investment. They are characterized by recurrent changes to enable organizations to anticipate and react to market dynamics.

---

1 Starbucks. 2011, December 5. "Starbucks mobile transactions exceed 26 million within first year." Starbucks Stories & News, retrieved from https://stories.starbucks.com/stories/2011/starbucks-mobile-transactions-exceed-26-million-within-first-year/.

2 "Starbucks Corp. (SBUX): Q1 2017 Earnings Call." 2017, January 26. Starbucks Investor Relations, retrieved from https://investor.starbucks.com/files/doc_financials/quarterly/2017/transcripts/SBUX_Q1-2017-Earnings-Call-Transcript.pdf.

3 Jiang, I. 2020, June 20. "Starbucks is shutting hundreds of cafes and opening futuristic pickup-only stores in their place. Here's what it's like to buy coffee from a pickup-only Starbucks store." *Business Insider*, retrieved from https://www.businessinsider.com/starbucks-pickup-only-store-review-2019-11.

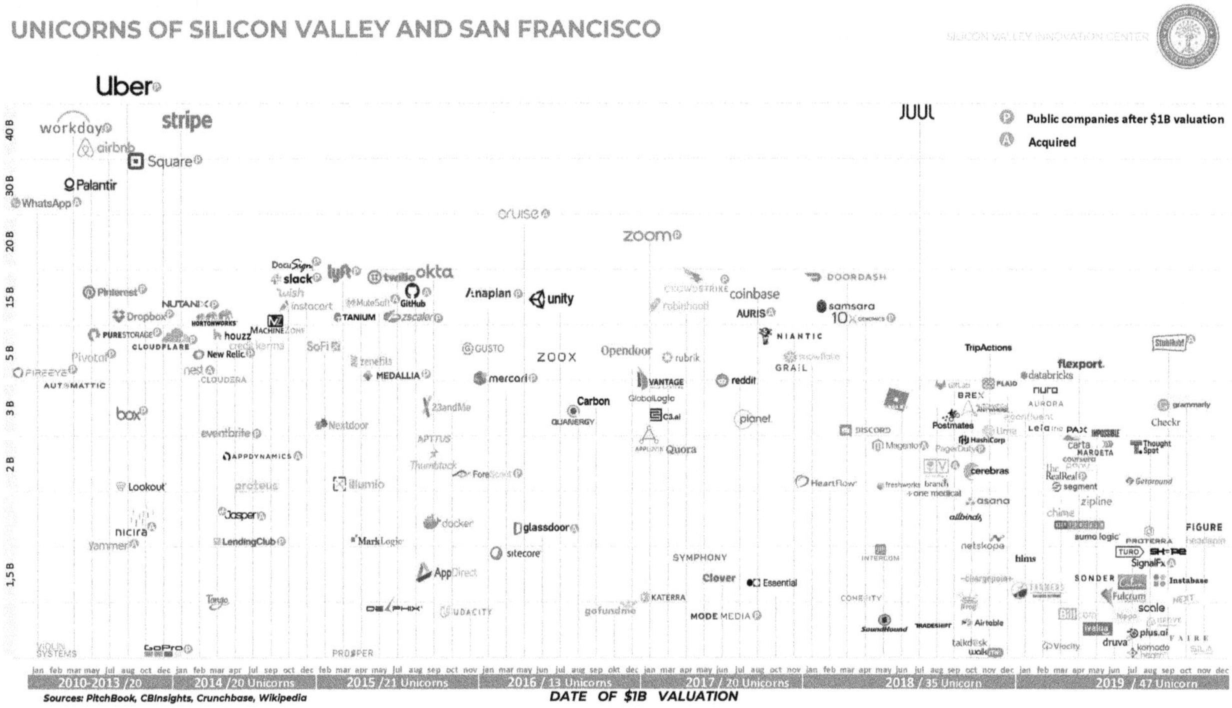

Figure 5.2. The unicorns of the Silicon Valley, firms valued over one billion dollars

Source: Silicon Valley Innovation Center, 2020, February 18, ***Infographic: Who are the unicorns of Silicon Valley?***, retrieved from https://siliconvalley.center/blog/infographic-unicorns-of-silicon-valley

While digital innovation may not look new to the most tech-savvy among you, you should consider that in the early 2000s, its strategic value was still questioned.[1] So what has changed? With digital technologies becoming so pervasive (Chapter 1), the scope of opportunities for digital innovation has exploded. There's no other way around it. When every business is a digital business, digital innovation is the "engine" that keeps a firm competitive.

Digital Entrepreneurship

We defined digital entrepreneurship as the creation of new economic activities embodied in or enabled by digital technology (Chapter 1).[2] Thus digital entrepreneurship is a form of digital innovation that is generally enacted by a new venture expressly created to pursue it—as with Silicon Valley start-ups (Figure 5.2).

As with the opening story about Elon Musk's Zip2, digital entrepreneurs are typically those responsible for disrupting entire industries. Companies like Revolut and Monzo, so-called digital banks, are set on disrupting the banking industry. Their business model hinges on consumers' embracing digital relationships for something as sensitive as financial transactions. Owning no physical branches and no brand, they compete by making digital onboarding seamless and keeping

1 Carr, N. 2003. "IT doesn't matter." *Harvard Business Review 81*(5): 41–50.

2 von Briel, F., Recker, J. C., Selander, L., Hukal, P., Jarvenpaa, S., Yoo, Y., . . . and Fuerstenau, D. 2020. "Researching digital entrepreneurship: Current issues and suggestions for future directions." *Communications of the Association for Information Systems: CAIS.*

costs low while also providing superior customer experience via their mobile applications. A person in good standing can open a Revolut account in two minutes from the comfort of their home or on the train on their way to work. They can then immediately start using the digital debit card associated with the account and receive and send transfers to other bank accounts. Through the app, customer can access numerous financial services, such as commission-free stock trading or travel insurance policies. The application also automatically analyzes spending and enforces budget goals aimed at promoting a healthy and responsible management of personal finances. The complete Revolut value proposition is digital, as is the firm's relationship with the customer.

The advent of digital banks is forcing legacy providers to acknowledge the threat and respond to it through digital transformation efforts. Some large legacy banks, such as BBVA in Spain, have fully embraced digital relationships and provide sophisticated app functionalities alongside their branches and traditional services.[1] As in any other industry under the threat of digital disruption spearheaded by digital entrepreneurs, the question is whether digital banks can gain the legitimacy and trust that legacy banks have in the eyes of mainstream consumers faster than the incumbents can complete their digital transformation and establish solid digital relationships with their customers. As with digital transformation, digital entrepreneurship is nothing new for tech-savvy entrepreneurs. What is different, however, is the magnitude and reach of modern opportunities for digital entrepreneurs. Digital technology is so pervasive and imbued in our lives that it is now possible for talented individuals with minimal IT skills to launch their own independent ventures and become digital entrepreneurs in the creator economy.

The Creator Economy The creator economy (sometimes called the "passion economy" or the "unbundling of work from employment") offers perhaps the clearest example of how the Internet, the mobile platform, and modern digital technologies have democratized digital entrepreneurship. In the 1980s and 1990s, a substantial investment in money and time was needed to launch a digital technology start-up that had any chance of success. Expensive specialized hardware had to be purchased, secured, and maintained and software had to be written by highly skilled software engineers. The rise to prominence of commodity servers in the 2000s and even more so the recent mainstreaming of cloud computing reduced the barriers in terms of monetary investment and time to market, shifting CAPEX to OPEX (Chapter 3).[2] However, software innovation remained the province of highly skilled specialists. Moreover, once a digital innovation had been created by digital entrepreneurs, the start-up needed to reach and educate customers. Often the only way to do so was to create a sales organization staffed with highly skilled and expensive technical sales executives.

What is drastically different in the creator economy is that, armed with a direct digital relationship to a global market of potential customers and a set of easy-to-use and continually improving digital tools to create content, manage the business, and nurture digital relationships with fans, talented solopreneurs around the world are becoming digital entrepreneurs without any deep technology skills.[3] The economics of it are surprisingly good for creators of interesting things.[4] As venture capitalist Li Jin put it,

> The top-earning writer on the paid newsletter platform Substack earns more than $500,000 a year from reader subscriptions. The top content creator on Podia, a platform for video courses and digital memberships, makes more than

1 BBVA. 2017, July 13. *NEWS BBVA*, retrieved from https://www.bbva.com/en/bbva-best-mobile-banking-world/.
2 *CAPEX* stands for capital expenditure, like the cost of a data center with racks of servers, and *OPEX* stands for operating expenditure, like the cost of ongoing services (e.g., computing resources from an IaaS provider).
3 The *Oxford English Dictionary* defines *solopreneur* as "a person who sets up and runs a business on their own."
4 Kelly, K. 2008, March 8. "1,000 true fans." *Technium*, retrieved from https://kk.org/thetechnium/1000-true-fans/.

$100,000 a month. And teachers across the US are bringing in thousands of dollars a month teaching live, virtual classes on Outschool and Juni Learning.[1]

Figure 5.3. A Postmates robot delivery
Source: Postmates, http://www.postmates.com

Creators, as these digital entrepreneurs are typically called, are very different from workers in the so-called gig economy. In the gig economy, a solopreneur contributes undifferentiated inputs, like a ride on Uber or a grocery store run on Instacart. For this unit of "independent" work, the individual is compensated by the aggregator (Chapter 4) that is managing the marketplace in which those services are required (e.g., Uber, Instacart). However, and crucially, there is no direct relationship or loyalty between the customer ordering the food and the Deliveroo's rider who is delivering it. Even if you really like the rider who just delivered your burger, you could not request his services in the future. The aggregator controls the customer relationship and the flow of work, and gig workers are interchangeable. As a consequence, gig economy workers command very little pricing power and are often just a stand-in until full automation can be achieved. The main cost of an Uber or Lyft ride is the compensation of the driver. Imagine the unit economics of these ride-hailing platforms when self-driving cars become the norm. When that happens, the incentive for Uber to stop using gig workers will be impossible to resist. As an early example of things to come, in Los Angeles, Postmates is experimenting with local deliveries using a robot (Figure 5.3).

Compare the above scenario for gig workers with that of solopreneur and "creator" Ben Thompson. He is the founder and sole content producer for the Stratechery.com website, to which we subscribe.[2] Ben, a business school graduate, had a passion for strategy and technology. Fusing his passions into Stratechery, he launched a blog and a newsletter conveying his analysis of current technology strategy news. One of his weekly posts is free on the website, while for $120 a year, you get two more weekly analyses in your email in-box. Ben is clearly very good, and we are happy customers (by some accounts, he has a lot of happy customers).[3] We don't always agree with his conclusions, but his articles always make us reflect. Ben is not substitutable. What we are paying for by subscribing to his service is the expectation of quality content and useful analysis of the kind he provides, not a commodity service that many can deliver. Therefore, Ben has a direct digital relationship with us, his customers, rather than being a supplier to an aggregator that retains control of the customer relationship. In fact, Ben works hard to foster such a direct relationship, acknowledging feedback from readers and engaging directly with those who reach out via email or on Twitter. Historically, the major limitations of this model were the partial reach of the message without a sales force and the fact that running a company takes time. In the case of Stratechery, solid technology strategy analysis of the kind that Ben does is a full-time job. But there are a number of activities that are necessary when running a company, even a small one like Stratechery: marketing, member management, payments, refunds, content creation, publishing, newsletter mailing, online forum management, payment of suppliers, taxes. For these

---

1 Jin, L. 2019, October 8. "The passion economy and the future of work." Andreessen Horowitz, retrieved from https://a16z.com/2019/10/08/passion-economy/.

2 https://stratechery.com/.

3 Stegmann, A. 2020, August 28. "Ben Thompson's Stratechery should be crossing 3 million in profits this year." Medium, retrieved from https://medium.com/hyperlinked/ben-thompsons-stratechery-should-be-crossing-3-million-in-profits-this-year-c146433f0458.

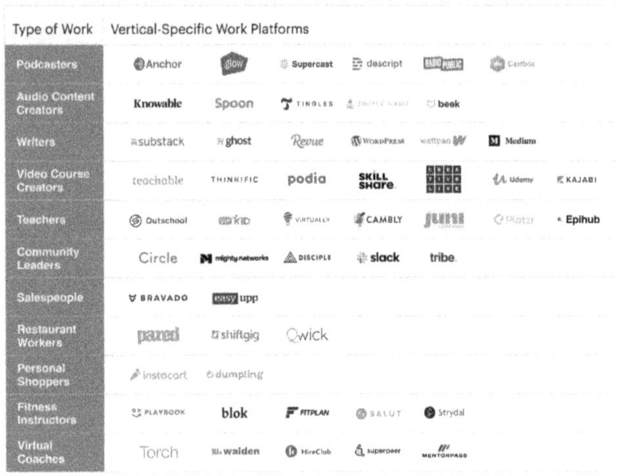

Figure 5.4. Platforms enabling creators in specialized verticals

Source: Jin, L., 2020, July 29, ***Unbundling work from employment***, Li's Newsletter, retrieved from https://li.substack.com/p/unbundling-work-from-employment

reasons, until recently, a writer would prefer to work for a magazine, an artist would prefer to work for a studio, and a professor would have to work for a university. These institutions provided the reach and the management, allowing the creators to focus all their energy on creating content. The Internet, with its global reach at zero marginal cost (Chapter 4), and social media solved the first problem. The latter is being solved by a growing host of general and specialized digital tools designed to enable creators. For example, Stratechery uses the following for content creation, mailing, and business management: Memberful (membership management), WP Engine (to host WordPress), ChartMogul (subscription management), Discourse (discussion management platform), Mailchimp (mailing), Stripe (payments), Gandi.net (hosting), Algolia (search functionalities), and Jetpack (WordPress optimization and security). Given the potential size of the creator economy, there are also digital platforms emerging with the purpose of enabling creators to be digital entrepreneurs. These platforms offer integrated solutions ranging from content creation to running the business (Figure 5.4).

5.3 Technologies for Digital Disruption

Now that you understand digital disruption and its three main manifestations (digital transformation, digital innovation, and digital entrepreneurship), it is important to be familiar with the waves of digital technologies that enable such digital disruption. We will cover them in chronological order and focus on ensuring that you can differentiate among them and be conversant in their unique characteristics.

The Internet

Simply put, the Internet is "a network of networks." In other words, the Internet is broadly defined as a collection of networked computers that can "talk to one another." This simple definition points to a fundamental issue: the Internet is an infrastructure upon which *services*—such as email, the web, instant messaging (IM), and many others—are delivered. Wikipedia, the free, web-based encyclopedia, provides a more complete definition:

> The Internet is the global system of interconnected computer networks that uses the Internet **protocol** suite (TCP/IP) to communicate between networks and devices. It is a network of networks that consists of private, public, academic, business, and government networks of local to global scope, linked by a broad array of electronic, wireless, and optical networking technologies. The Internet carries a vast range of information resources and services, such as the inter-linked hypertext documents and applications of the World Wide Web (WWW), electronic mail, telephony, and file sharing.[1]

From this definition follow a number of observations, but before discussing them, you should carefully note that the definition of the Internet is very general and "device agnostic." In other

1 Wikipedia. 2020. "Internet." *Wikipedia*, last updated October 31, 2020, retrieved from http://en.wikipedia.org/wiki/Internet.

SIDEBAR 5.1. The TCP/IP Protocol Suite

The TCP/IP protocol suite governs data transmission on the Internet. Early in the development of TCP/IP, the designers introduced the principles of encapsulation and layering to simplify complex end-to-end transmission. Instead of trying to design a single monolithic protocol, they opted for using multiple layers—or *stack*, as it is called today—of independent protocols. With encapsulation, each upper layer in a transmission could access only what was needed from the lower layer. The most important layers of the TCP/IP protocol stack are described below:

- Application layer: It is the uppermost layer, the one used by the applications that end users interact with. This is the domain of the higher-level protocols, such as HTTP, FTP, SMTP, SSH, that determine the application-to-application rules of communication.
- Transport layer: The Transmission Control Protocol (TCP) is the layer responsible for ensuring that a communication transaction occurs between the correct sender and the correct receiver. It is the protocol that specifies the node-to-node rules of communication. For example, it requests the retransmission of packets that may have been lost, rearranges packets that arrive out of sequence, and makes sure that the complete message as intended by the sender is passed up the stack.
- Internet layer: The Internet Protocol (IP) is the layer that addresses the data packets and ensures that they reach their destination. While logically, TCP manages the message, IP manages the single packets. This is the layer that enabled the internetworking of different networks.

Another one of the original design principles of the Internet was the end-to-end principle. According to this principle, "intelligence" should be located at the edges of the network with communication equipment between nodes only responsible for routing packets. One of the reasons behind this principle was the need for network resilience (such that if failure occurred anywhere on the network, communication could automatically route around the problem) and scalability (such that any network or node could be easily added to the Internet as long as it subscribed to the TCP/IP protocol). In short, the TCP/IP protocol family "abstracts" all the details of the underlying physical networks by defining the address format, the packet format, and the transfer and handling mechanisms regardless of the physical medium (cable, fiber optics, wireless) or type of network being used.

words, as a collection of computer networks, the Internet can connect any device based on the digital computer architecture—such as a laptop, a smartphone, a face-recognition digital camera, and so on. Today, there are more devices connected to the Internet than there are people with access, and analysts predict that by 2025, Internet-connected devices will outnumber humans five to one.[1] Samsung's president and CEO of their consumer electronic division H. S. Kim echoed this trend in his 2020 speech at the International Consumer Electronics Show in Las Vegas: "Today, the majority of our devices are IoT-enabled, and we'll continue to make them as connected and intelligent as possible in a way that makes sense for our consumers."[2] The extreme flexibility of the Internet protocol (see Sidebar 5.1) is what allows such scalability and a variety of devices to be connected to the Internet.

Internet Services

As the Wikipedia definition above suggests, the nodes of the Internet "carry an extensive range of information resources and services." A common misconception is that the terms *Internet* and *World Wide Web* (or *web*) are synonymous. This is incorrect, and it is important to differentiate the two. The Internet is the infrastructure upon which many services are made available. Typically, you will connect to the Internet, the infrastructure, to access the services you want to use (e.g., IM).

1 IDC. 2019, June 18. "The growth in connected IoT devices is expected to generate 79.4ZB of data in 2025, according to a new IDC forecast." Retrieved from https://www.idc.com/getdoc.jsp?containerId=prAP46737220.

2 Kim, H. S. 2020, January 2. "Samsung in the 'Age of Experience.'" Samsung Newsroom U.S., retrieved from https://news.samsung.com/us/samsung-age-of-experience-hs-kim-ces2020/.

Figure 5.5. A Dropbox transfer

The web is a service available on the Internet and, alongside electronic mail and messaging, it is the most popular. However, there are many other services that we use daily—for example, Voice over IP (VoIP) with WhatsApp, streaming music with Spotify, synchronous electronic discussion with Skype, and even file synching and transfers with Dropbox (Figure 5.5) or the old trusty File Transfer Protocol (FTP; Figure 5.6).

Distributed Ownership The Internet is "publicly accessible," meaning that no single entity owns it, regulates its use, or otherwise controls it. In fact, the Internet has many owners but no one who centrally controls it. In other words, different portions of the Internet (i.e., different networks connected to other networks) are owned by different entities—literally millions of them. For example, your university network, while connected to the public Internet, is privately owned by your university. Your university manages and pays for it. Distributed ownership has been perhaps the main strength of the Internet, limiting regulation, fostering experimentation, and ensuring widespread access leading to significant growth.

Multiplicity of Devices The Internet is a digital network consisting of millions of smaller digital networks. Each of these smaller digital networks encompasses a collection of digital devices, called nodes. The simplest digital network to visualize is perhaps a home network (Figure 5.7). Your home network may be composed of a couple of personal computers and a smartphone, a smart TV, a console, and a printer to which both computers and smartphones can send documents. Using a home router, wired or wireless, and a broadband modem (e.g., 4/5G, cable, fiber, or DSL), you connect to the Internet. Each of these digital devices—the two computers, the printer, and the router—are nodes on your home network.

Your home network is a tiny contributor to the larger Internet. The fancier ones among us may have more cutting-edge devices, such as a connected thermostat, a wireless media center to stream music and videos from a computer through the living room stereo or to listen to Internet radio stations, a wireless webcam to monitor the front door, smart speakers to set timers while cooking, a video game console, and even cooler stuff! A connected car—equipped with a GPS device, the OnStar system, or a permanent Internet connection (Figure 5.8)—is another example of a networked node, as is a modern smartphone. As the price of microchips and bandwidth keeps dropping (Chapter 1), the number and types of devices that become nodes of a network will continue to increase. In other words, the Internet is in continuous expansion.

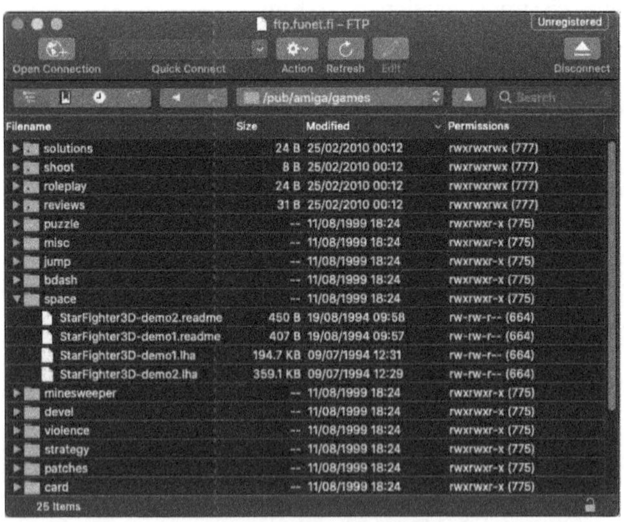

Figure 5.6. FTP client accessing a public server

Open Standards The Internet relies on open technology standards and protocols. A protocol is an agreed-upon set of rules or conventions governing communication among the elements of a network (i.e., network nodes). For example, case study discussion in your class follows a (stated or implicit) protocol. Typically, the professor sets the stage for the case discussion. You listen, and when the floor is open, you might raise your hand. When you are called on, you contribute your perspective. You may disagree with classmates, but you do so in a polite manner, addressing ideas rather than individuals. Respect of the protocol by all is necessary to enable constructive communication rather than unproductive classroom chaos.

Similarly, in order to communicate, network nodes need to follow an agreed-upon set of rules. On the Internet, such a set of rules is the TCP/IP protocol mentioned in the Wikipedia definition. Nobody owns the TCP/IP protocol; as such, it is an open (i.e., freely available) standard, as opposed to a proprietary one. The same holds true for the other technologies that enable the Internet and its services, such as HTML (the language used to write web pages). While there are standard-setting bodies that decide how each of these technologies should evolve (in the case of HTML, it is the World Wide Web Consortium, or W3C), no entity can charge for their use. Thus anyone who wants to embed support for these standards in his or her applications can do so . . . and innovation continues to thrive on the Internet!

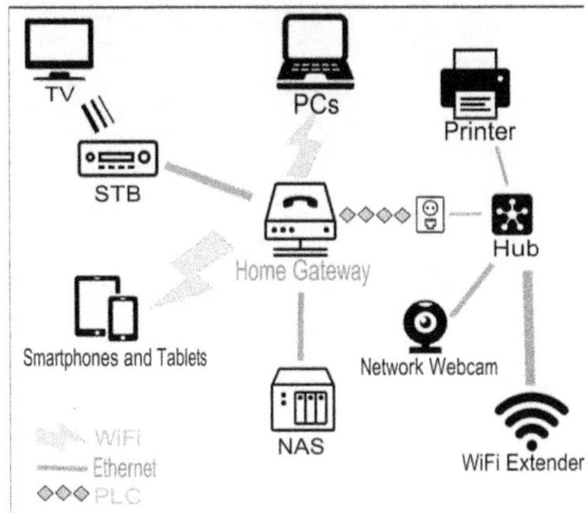

Figure 5.7. A simple home network

Source: Figure by Norrellstrange / CC BY-SA 4.0

As a final note, we offer a word of caution. The definition of the Internet as a network of networks can be very misleading. It typically conjures up a vision of computers of various shapes and sizes hooked together by a maze of cables of different shapes and colors—the Internet as a bunch of interconnected computers. While this image more or less correctly captures what the Internet looked like when it first became mainstream, it is more confusing than helpful today. For example, while it may not look like it at first glance, modern smartphones are full-fledged digital computers, and the cell phone network, while not a cable, is a data transmission channel. In fact, modern smartphones are way more powerful than the computers of just a few years ago (Figure 5.9).

Figure 5.8. The Tesla Model S connected dashboard

Source: Photo by Steve Jurvetson / CC BY 2.0

Web 2.0

With the success and widespread adoption of the Internet and the technologies associated with it came the need to categorize and conceptualize its evolution. The underlying infrastructure and its defining characteristics (Chapter 4) have not changed dramatically since its inception. Moreover, many of the technologies at its core (e.g., the TCP/IP protocol, HTML, JavaScript) are still the bedrock of Internet operations and the pillars upon which innovation happens. However, over the last two decades, we have witnessed a significant evolution of the front end and the way in which the Internet is utilized by both organizations and individuals. The term ***Web 2.0***, popularized in 2004, labels the second wave of innovation and evolution occurring on the

Figure 5.9. New and old portable devices versus old desktop computers

Figure 5.10. Yahoo.com website (1999)

Gmail

Figure 5.11. Gmail logo

Internet after the shakeout following the original thrust of mainstream Internet innovation (1993–2001). We can roughly date the rise to prominence of Web 2.0 between 2001 and the dawn of the mobile platform era spurred by the launch of the iPhone in 2007.

While skeptics considered *Web 2.0* a hollow marketing term, proponents of the label drew a distinction between the first incarnation of the World Wide Web (Web 1.0) as made of web pages populated with text and static images (Figure 5.10) and the dynamic nature of Web 2.0 sites (e.g., YouTube) (Figure 5.11). They considered *Web 2.0* a useful umbrella term for categorizing both emerging technologies and business innovations that represent a significant departure from the paradigm of Web 1.0. More specifically, the proponents of Web 2.0 pointed to a number of defining features that differentiate it from its predecessor.

Two-Way Conversations The metaphor for the web, when it first became widely utilized in the early 1990s, was the printed publication. The web was made of pages (not a coincidence in terminology) filled with text and, later, images. One key difference between traditional printed publications and web publishing, however, was presented by the information navigation possibilities offered by links in hypertext documents. Despite its navigational interactivity, however, in the early days, the web was a one-way broadcasting medium. Organizations and individuals would *create* web pages, and an audience of visitors would *visit* those pages. Today, the metaphor for individual and organizational web presence is a two-way conversation. Blogs, a prototypical example of Web 2.0 technology, are a vehicle for individuals to communicate with the (potentially) huge audience of web surfers. However, a staple of blogs is the possibility for that "audience" to comment and respond to the original post. Those responses are themselves public, searchable, and linkable, making them an intrinsic and valuable component of the blog itself. The bloggers, in turn, can comment on the comments of the "audience," thus turning the blog into a full-fledged asynchronous conversation.

Interactive User Experience If the metaphor for the early web was the printed page, relatively static and unchanging over time, Web 2.0 sites were designed to be more akin to desktop applications than documents. Using a set of programming technologies centered on the Ajax framework, Web 2.0 sites dynamically respond to user behavior and to other events.[1] Such applications are generally referred to as Rich Internet Applications (RIA). Consider, for example, the popular Google email client, Gmail (Figure 5.11). Unlike previous web-based email clients, where the in-box was a static page of text and images, the Gmail in-box behaves like a local email client (i.e., a software program that resides on the user's own computer). If a new email arrives, the page is automatically updated with the new information without any user intervention (i.e., you don't need to refresh the page).

User-Generated Content If the "surfers" of the early web were consumers of content, modern web surfers could be better labeled "prosumers."[2] This evolution is in large part due to Web 2.0

1 Ajax (Asynchronous JavaScript and XML) represents a group of web programming technologies designed to enable the development of interactive client-side applications running in a web browser.

2 *Prosumer* is a term that represents the combination of *producer* and *consumer*—thus indicating that individuals both create and utilize content and resources available on the web.

technologies that dramatically lowered the barriers to the production of content by the general population of web users. The most enduring example of user-generated content is online reviews and engagement tools like likes, upvotes, and retweets. As a testament to the power of user-generated content and the crowds of users contributing to it, as the Web 2.0 phenomenon was taking hold, *TIME* magazine recognized the generic "you" (i.e., the mass of individuals cooperating and communicating through the web) as the "person of the year." In the explanation, *TIME*'s editors wrote,

> Who are these people? Seriously, who actually sits down after a long day at work and says, I'm not going to watch *Lost* tonight. I'm going to turn on my computer and make a movie starring my pet iguana? I'm going to mash up 50 Cent's vocals with Queen's instrumentals? I'm going to blog about my state of mind or the state of the nation or the steak-frites at the new bistro down the street? Who has that time and that energy and that passion? The answer is, you do. And for seizing the reins of the global media, for founding and framing the new digital democracy, for working for nothing and beating the pros at their own game, *TIME*'s Person of the Year for 2006 is you.[1]

The above quote reads a bit naïve and quaint in 2021, after a year dominated by "fake news" and troll farms. However, we are unrelenting optimists, and we believe that humanity will eventually find a positive balance in its use of these powerful communication technologies.

Emergent Structure The publishing metaphor of the early web, along with the technical restrictions imposed by the technology of the time, made it so that content on the web had to be structured by a designer and users would accept it and work within its limits. However, as technology progressed, making it easier for users to interact with and customize online applications to suit their needs, it became possible to allow structure to take form dynamically.

There is no clear date marking the transition from Web 1.0 to Web 2.0. However, the value of discussing Web 2.0 for today's managers lies in developing an appreciation for how Internet technology has evolved and for the implications that this evolution has had, and continues to have, for organizations and businesses.

Crowdsourcing

The connections enabled by the Internet and the interactivity introduced by Web 2.0. made it possible to "monetize the Internet of people" through **crowdsourcing.** Crowdsourcing is an online activity in which an individual or organization proposes the voluntary undertaking of a task to a heterogeneous "crowd" of individuals via an open call. Crowdsourcing approaches can be broadly categorized on the basis of the class of problems they address:

- *Knowledge discovery and management.* This is used to solve information gathering, organization, and reporting problems, such as the creation of collective resources. The basic idea is that the necessary knowledge is already available and that the crowd can be leveraged to discover and organize it. The typical example is Wikipedia, where articles are the result of the collaborative writing and research efforts of multiple contributors.
- *Distributed human intelligence tasking.* This approach is used for tasks where human intelligence is required to process data or information instead of computers. Generally consisting of simple repetitive tasks, it is ideal for processing large batches of data that only humans can handle (e.g., tag images, filter image content, verify

1 Grossman, L. 2006, December 13. "Time's person of the year: You." *TIME*.

catalogs, fill surveys, edit transcripts, write reviews). At Amazon's Mechanical Turk, requesting firms submit a project (called HITs—human intelligence tasks) that workers can choose from and be compensated for upon completion.
- *Broadcast search.* This focuses on finding a single specialist, generally outside of the field of expertise of the organization. The broadcast-search approach is appropriate for problems where the "right" answer exists but is not yet known by the organization. Kaggle and Innocentive.com are examples of this category. Kaggle runs data science competitions commissioned by paying organizations, and Innocentive.com seeks innovation solutions from talented individuals who compete to provide ideas and solutions to the submitted business, social, policy, scientific, and technical challenges.
- *Peer-vetted creative production.* This is adapted for ideation problems where an organization resorts to the crowd for supporting the creative phase and receives multiple design proposals. This approach is ideal for design, aesthetics, or policy problems where contributors have direct interests or are potential customers. Threadless.com, for example, acts as a platform where users can both propose and vote on T-shirt designs that are then made available for purchase on the website.[1]

The Mobile Platform

While the Internet provides the technical infrastructure for communication, increasingly, the services available on the Internet are not accessed through a desktop computer but through a mobile device—a smartphone, tablet, or **Internet of things (IoT)** device (Chapter 12). In 2010, *Wired* published an article with an intriguing title: "The Web Is Dead, Long Live the Internet." The main point of the article was that "the World Wide Web is in decline, as simpler, sleeker services—think apps—are less about the searching and more about the getting . . . As much as we love the open, unfettered Web, we're abandoning it for simpler, sleeker services that just work."[2] Much of this transition from the web to apps was driven by the growth of the mobile platform and the two competing ecosystems: Apple iOS and Google Android.

It is hard to appreciate just how quickly the mobile platform has risen to prominence, becoming the personal computing infrastructure of the world. The chart below shows how quickly Microsoft's dominant control of the operating systems market faded. In the span of 10 years, the giant from Redmond went from controlling more than 92% of global machines running its operating system (Windows) to less than 40% (Figure 5.12).

For many people, particularly in the developing world, a smartphone represents their first (and only) digital computer, and unlike a desktop or laptop computer, almost every adult on the planet owns one (Figure 5.13).

Smartphones and tablets have the ability to run an operating system and therefore software applications—just like any other digital computer. But smartphones trace their roots to the efforts started in the early 1990s by the likes of IBM and Nokia to pack functionality into the portable phone, thus rendering it a multipurpose device rather than just a telephone. The history of the smartphone intersects with the rise, during the same years, of palmtop devices. These were handheld computers known as personal digital assistants (PDAs), like those produced by Palm Inc. (Figure 5.14) and later Handspring (Figure 5.15). Those devices were conceived as computers with an operating system, and as a consequence, software developers recognized opportunities to build applications for the mobile platform (Chapter 4).

1 Brabham, D. C. 2013. *Crowdsourcing*. Boston, MA: MIT Press, 44–50.

2 Anderson, C., and Wolff, M. 2010, August 17. "The web is dead: Long live the Internet." *Wired*, retrieved from https://www.wired.com/2010/08/ff_webrip/.

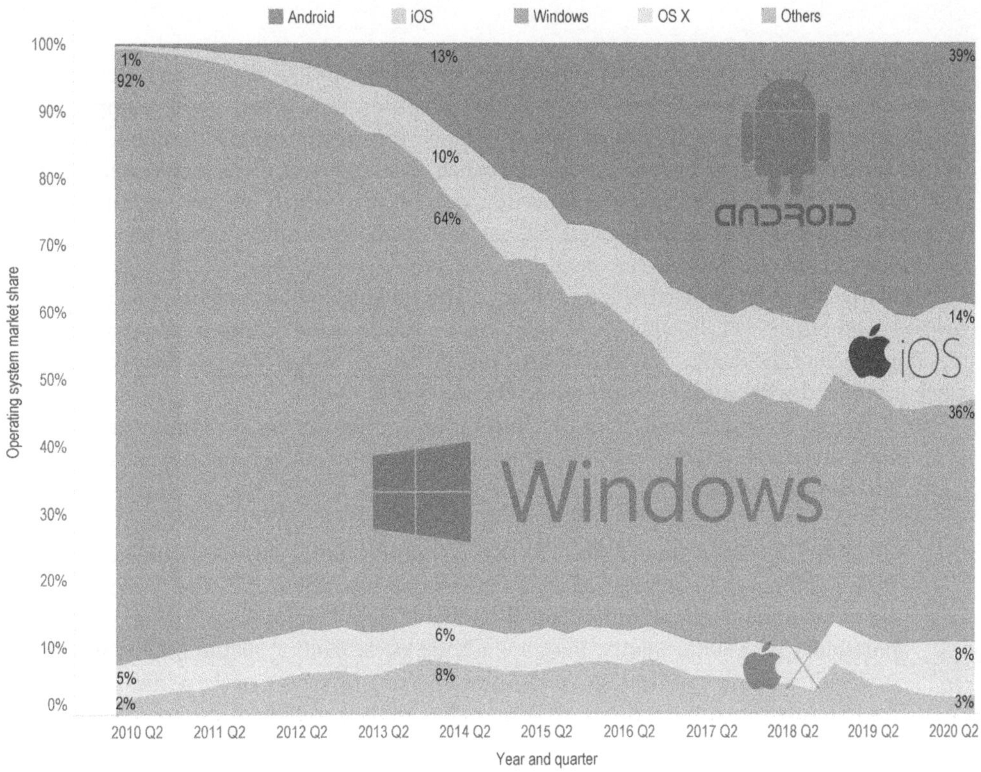

Figure 5.12. Operating systems market share

Source: StatCounter.com, retrieved from http://gs.statcounter.com/os-market-share

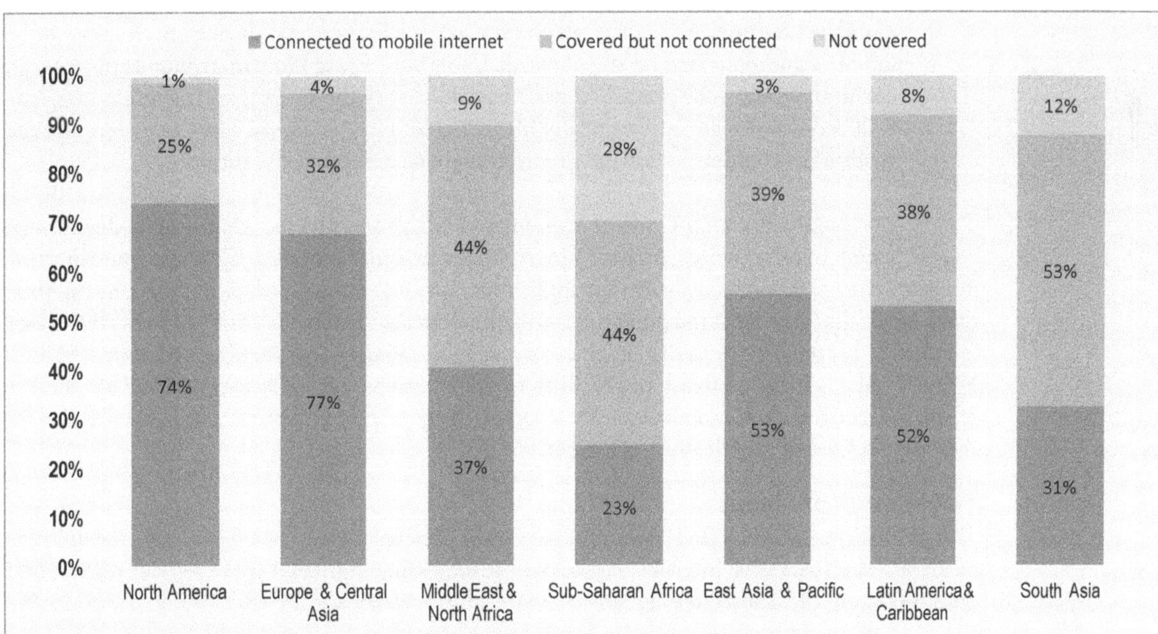

Figure 5.13. Global population and mobile penetration

Source: Bahia, K., and Suardi, S., 2019, ***The state of mobile Internet connectivity 2019***, Connected Society, p. 60, retrieved from https://www.gsma.com/mobilefordevelopment/wp-content/uploads/2019/07/GSMA-State-of-Mobile-Internet-Connectivity-Report-2019.pdf

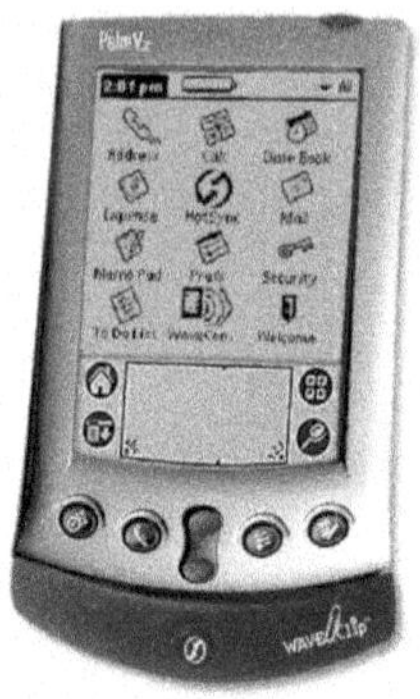

Figure 5.14. The Palm Vx

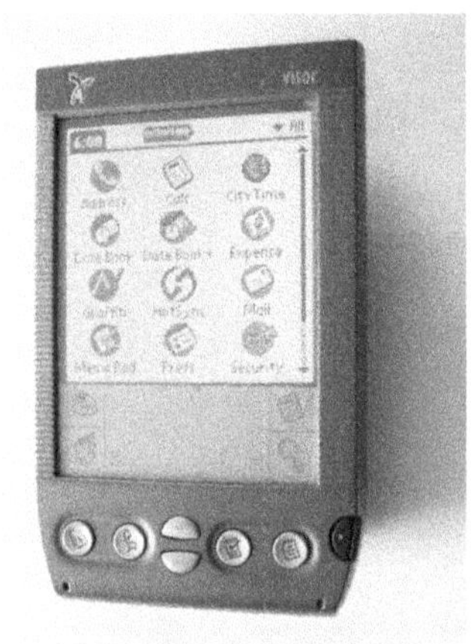

Figure 5.15. The Handspring Visor

Source: Photo by Waldohreule / CC BY-SA 3.0

As people were carrying around cell phones and PDAs and hardware kept increasing in power, combining the two devices was a natural extension (and a very good example of the polymediation, the convergence of multiple digital devices into one introduced in Chapter 1). While it is not our intent here to produce an exhaustive history of the smartphone, this brief discussion of its genesis helps you appreciate the context in which we have seen the rise of portable computers that look like cell phones and application stores where you can easily download software for your mobile device.

Similarly, the success of tablets has affected the existing market of both notebooks and portable computers. While tablets also trace their roots to earlier computing innovations, it was the launch of the Apple iPad in 2010 that established the category as it is defined today—a category created with unprecedented speed (Figure 5.16).

Smartphones are digital computers. Architecturally, mobile devices are like any other digital computer; however, they do have the peculiar characteristics of ubiquity, identifiability, and context awareness.

Ubiquity Ubiquity represents the idea that users of the device can access needed resources from (in theory) anywhere.[1] Mobile devices and wearables offer the highest level of potential ubiquity among commercially available digital technology because they marry portability with connectivity. Imagine a hotel guest landing at the airport without a reservation simply geolocating her business meeting by typing in the address, searching for hotels within a five-minute walking radius from that address, evaluating rates, and booking a room.

Identifiability Identifiability represents the idea that mobile devices uniquely identify their user. In order to access the data grid, both smartphones and tablets utilize the cellular network and use a subscriber identification module (SIM) card. Each SIM card has a unique identifier, and since telephones are usually personal devices, each user can be uniquely authenticated on the network. Using your Pizza Hut smartphone app, you can create and order your pizza. You don't have to specify the address and payment information because you are automatically identified. In fact, you could even save your favorite custom pizza orders and simply reorder them as desired in the future.

Context Awareness Context awareness is enabled by the fact that mobile devices can be geolocated. In other words, modern smartphones that incorporate a GPS receiver can communicate their position to any software application running on them. Such applications can then make use of the location of the person carrying the device and infer the context in which the user is embedded at the time or in the vicinity to other geolocated entities (e.g., a restaurant, a friend). You can quickly imagine many possibilities to craft business initiatives around the availability of this information. You can also quickly imagine many privacy violations and other abuses that these capabilities of mobile devices engender!

The Internet of Things

The label *Internet of things* (IoT) attempts to capture the convergence of a few trends: the ubiquity of the Internet, pervasive Internet access, and the increasing digitization of everyday devices and objects (Chapter 1). These "intelligent" or "smart" objects utilize the TCP/IP protocol and are able

1 We specify "in theory" because, despite technical feasibility, many users are constrained in their access to resources by their data plan.

to communicate using the existing Internet infrastructure.[1] Having been trained as academics, we are not very fond of buzzwords, but we must recognize that IoT is a reality. There are many more devices connected to the Internet infrastructure than there are people. The Internet itself had to accommodate a larger addressable space for opening to the projected 40 billion "things" connected by 2025. The move from IPv4 to IPv6 was a necessary precondition to accommodate all the new connected devices.[2]

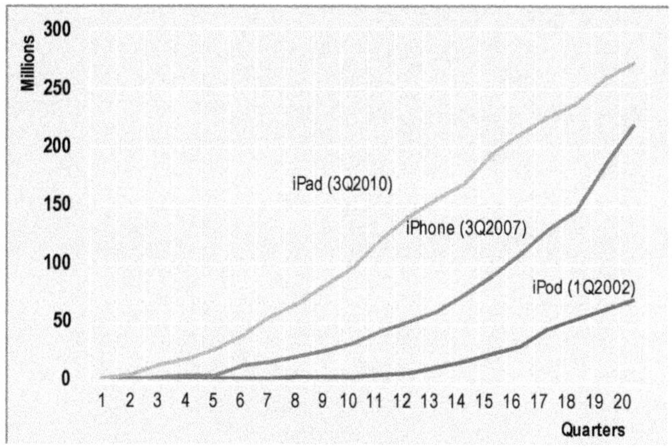

Figure 5.16. Cumulative units shipped by quarters since launch (with launch quarter appearing in parentheses)

Source: Data from AAPLinvestors.net, 2018, "iPad versus iPhone versus iPod," retrieved from http://aaplinvestors.net/stats/iphonevsipod/

Underpinning the rapid development in IoT are two trends. First is the commoditization, and consequential drastic price decline, of sensors brought about by the smartphone (Chapter 3). You may not often think about it, but besides being a full-fledged computer, the smartphone you have in your pocket is packed with miniaturized and sophisticated sensors. Your phone has one or more of the following: a digital camera, microphone, GPS receiver, accelerometer, gyroscope, ambient light and proximity sensor, compass, and barometer, along with speakers, a touch-sensitive screen, and all kinds of wireless communication devices. As global sales of smartphones took off in the last decade, the marginal cost of production of all these sensors drastically declined (see Figure 5.17). As they became increasingly cheaper, it was cost effective to embed these highly sophisticated sensors into more everyday objects.

Second is the emergence and consolidation of the cloud-computing architecture (Chapter 3) based on reliable Internet networking. As you recall, the cloud enables local devices to tap into powerful remote computing services. Therefore, simple and cheap IoT devices appear very powerful by offloading much of the computation to the cloud. Smart IoT objects bridge the physical world with cloud-computing services, offering new product functionalities and determining new opportunities for value creation. It is their numbers and the need to make sense of the data they produce that make the IoT relevant to organizations and managers. Consider the example of Google's Nest thermostat. When introduced, it was a pioneering instrument that offered Internet connectivity for remote heating/cooling systems management and programming. A major feature of the product was its ability to learn automatically from user behaviors, user interaction, and motion sensors. If you ever had to deal with a traditional programmable thermostat, you know the frustration users endure to set a program they are happy with. This same thought spurred Tony Fadell, founder of Nest, to find a solution in 2009.[3] Formerly at the head of Apple's iPod and iPhone divisions, Fadell was concerned by the high levels of carbon dioxide that were pumped into the atmosphere because of the lack of an easy and automatic way to control heating and cooling systems in homes and offices. The Nest thermostat had to have an intuitive interface, a color display, and a set of convenient features, setting the product apart from its legacy counterpart. But above all, the thermostat had to automatically learn from users' preferences and habits to efficiently manage the temperature, also

1 Most discussions of the IoT use the term *intelligent* or *smart* very loosely to refer to any device that is able to perform computations. IoT devices incorporate a digital computer and are therefore able to process and execute instructions coded into software.

2 McMilian, R. 2015, May 13. "Coming this summer: U.S. will run out of Internet addresses." *Wall Street Journal*, retrieved from http://www.wsj.com/articles/coming-this-summer-u-s-will-run-out-of-internet-addresses-1431479401.

3 Levy, S. 2011, October 25. "Brave new thermostat: How the iPod's creator is making home heating sexy." *Wired*, retrieved from https://www.wired.com/2011/10/nest-thermostat/.

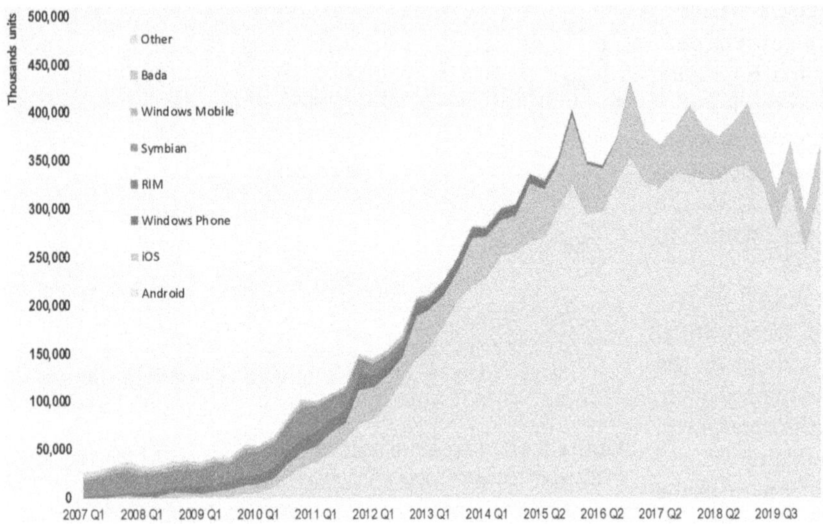

Figure 5.17. Global smartphone sales to end users broken down by operating system from 2009 to 2020

Source: Wikipedia article on Gartner data and analysis from authors, https://en.wikipedia.org/wiki/List_of_best-selling_mobile_phones

allowing remote user control through the Internet.

Consider now the entire series of Google Nest home products. When multiple Nest products such as the smoke detector, smart speaker, or the surveillance camera are in range, the devices can exchange data through a local wireless network and coordinate their operations. For example, the camera can start recording when smoke is detected, or, in a less dramatic example, you can just ask the Google Assistant to lower the temperature.

This case well exemplifies the general architecture behind IoT. Smart devices are digital computers in the shape of everyday objects that are capable of interconnecting to exchange data. They can process instructions, but they often have sophisticated software that enables them to perform substantial analysis and modify their operation over time. Furthermore, interconnectivity allows the devices to coordinate to create novel applications that each of them individually could not perform—for example, increase security and safety by providing a video feed during a fire to potentially improve the effectiveness of rescue operations. Generally, IoT devices pack limited computational power and storage capacity but rely on cloud service's almost unlimited digital resources. Consider the example of the Amazon Echo smart speaker (Figure 5.18). The device has a microphone array (several microphones working in parallel) that is constantly listening to conversations. When a user utters the appropriate keyword ("Amazon" or "Alexa"), it awakens and begins to interact with the person, using natural speech. The speech processing does not happen on-site—the sound is transferred to a cloud service managed by Amazon, it is processed (i.e., understood), and a voice response is generated. For this reason, when Internet connectivity is down, voice recognition won't work. Interestingly, Amazon provides an application program interface (API) to this cloud service—called the Alexa Skills Kit (ASK)—that enables independent developers to build applications leveraging speech processing as a digital resource.

Figure 5.18. Amazon voice-activated assistant—the Echo Dot

Source: Amazon.com

It should be clear now that the complexity behind IoT is centralized in cloud-based powerful systems that providers make accessible to their devices (Figure 5.19). Consider the example of Google Nest devices again. Camera features, like facial detection or advanced motion analysis, are too computationally taxing to be embedded in devices and are therefore cloud based—at least for the moment. Recent developments, both in hardware and software, are taking machine learning and artificial intelligence (AI) digital capabilities (Chapter 12) closer to the user—at the "edge" of the cloud—enabling IoT devices to limit the need for constant connectivity to provide their services. IoT digital capabilities, as any cloud service, have the added advantage of being highly scalable, always up to date, and accessible through a multiplicity of devices (e.g., smartphones with the appropriate apps).

Smart-device capabilities can be grouped into four main areas:[1]

- *Monitoring* is the ability of smart objects to detect and sense the physical world. They can measure data about themselves (e.g., their state of operation) and the surrounding environment. Our running example, the Google Nest family of products, demonstrates this concept. The thermostat, camera, speakers, and smoke detector all capture environmental information. The thermostat, for example, can monitor temperature, humidity, near and far field activity with a 150-degree-of-view angle, and ambient light.
- *Control* refers to the possibility of setting product functions and personalizing user experiences either directly from the device interface or remotely through the Internet. The Nest thermostat, for example, provides a convenient interface for setting user preferences (i.e., the temperature) and an extended set of features accessible via a web browser or through the mobile app.
- *Optimization* consists of a product or service performance enhancement enabled by monitor and control capabilities such as predictive diagnostics, service, and repairs. In this sense, Nest claimed that the use of their learning thermostat saved U.S. customers about 10% to 12% on their heating bills and about 15% on their cooling bills.[2]
- *Autonomy* is the combination of previous capabilities that allows for autonomous product operations, enhancement and personalization, self-coordination of operation with other devices, and self-diagnostics. The case where the camera turns on when smoke is detected is an example of the planned coordination of objects.

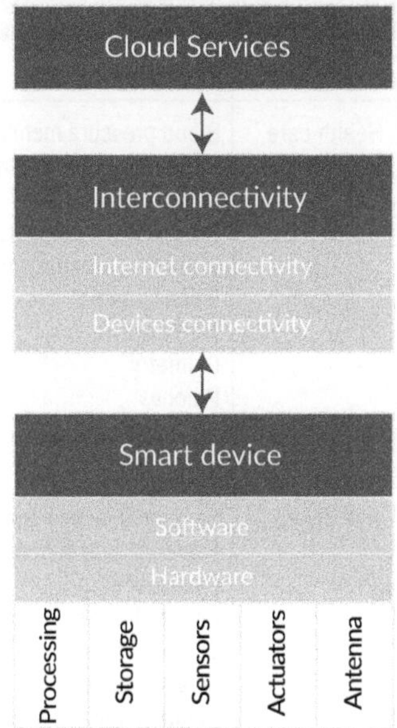

Figure 5.19. IoT layered architecture

Source: Adapted from Porter M. E., and Heppelmann, J. E., 2014, November, "How smart, connected products are transforming competition," ***Harvard Business Review 92***

IoT implications for business are disruptive, changing how value is created for customers and the competitive perimeter. New business ecosystems will arise, compete, and eventually coexist. Having been through the eCommerce and eBusiness revolutions, we see many parallels. IoT is a natural extension of the Internet, linking together the physical and digital worlds.

Wearable Devices

Wearable devices, or wearables, are smart clothing items or accessories worn by individuals. The concept has been around since the early days of the Internet, but the recent innovations in digital and sensor technologies are the enablers that made them feasible as mainstream products.

Just like IoT objects, wearables look and behave just like their "dumb" counterparts. Thanks to embedded IT and connectivity, they provide functionality designed to facilitate or support their owners' activities. Wearables are a diversified category of digital devices ranging from bulky head-mounted displays or backpack systems to watches and fitness trackers or futuristic subcutaneous devices (see Table 5.2).

Wearables represent an important subset of IoT, but their peculiarity is that they are physically worn by people. As such, they provide a more intimate interaction with their users. If you own a smart watch (Figure 5.20), a fitness band (Figure 5.21), or smart earbuds, you are already experiencing examples of the new user interaction and engagement opportunities offered by wearables. You are easily aware of the calories you burn per day, the steps you take, your heartbeat, or the hours you slept last night. The digital capabilities of wearable devices are making it trivial to collect

1 For a more in-depth analysis of them, see Porter, M. E., and Heppelmann, J. E. 2014, November. "How smart, connected products are transforming competition." *Harvard Business Review 92*: 11–64.

2 Nest Whitepaper. 2015. "Energy savings from the Nest Learning Thermostat: Energy bill analysis results." Nest.com, retrieved from https://nest.com/downloads/press/documents/energy-savings-white-paper.pdf.

Table 5.2. Wearables applications and product categories

Application	Product category
Health care and medical	Blood pressure monitors Continuous glucose monitoring Defibrillators Drug delivery products ECGs that monitor pulse Insulin pumps Medical alert Oximetry Patches Smart glasses
Fitness and well-being	Activity monitors Emotional-measurement fitness and heart-rate monitors Footpads and pedometers Sleep sensors Smart glasses Smart clothing Smart watches Audio earbuds Heads-up displays
Entertainment	Bluetooth headsets Heads-up displays Imaging products Smart glasses Smart watches
Industrial	Hand-worn terminals Heads-up displays Smart clothing Smart glasses
Military	Hand-worn terminals Heads-up displays Smart clothing

Source: Walker, S., 2013, "Wearable technology—Market assessment," IHS Electronics & Media, retrieved from http://cdn2.hubspot.net/hub/396065/file-2568104498-pdf/Blog_Resources/IHS-Wearable-Technology.pdf?t=1427903372862

and leverage people analytics at scale. Fueled by consumers interested in health monitoring, the wearables market has rapidly developed, and it is set to maintain double-digit growth rates through 2024.[1]

In the corporate space, wearables are used for hands-free operations, maintenance, tracking productivity, accessing customer data in real time, receiving alerts, health and safety monitoring, and training.[2] Google Glass, for example, long considered a failure in the consumer space, has been reimagined in the Google Glass Enterprise Edition project. The device, mounted on safety work glasses (Figure 5.22), enables workers to access training videos, images annotated with instructions, or quality assurance without having to stop their activity to walk over to a computer or grab a tablet—a capability that enabled several firms to maintain operations and guarantee remote assistance during the pandemic.[3] As Brian Krzanich, CEO of Intel, put it during the 2015 Consumer Electronic Show's keynote speech, "The rise of new personal computing experiences, intelligent and connected devices, and the wearable revolution are redefining the relationship between consumers and technology."[4] Seeing kids conversing with Alexa and receiving the heartbeat report from our smart watch at the end of a run suggest this prediction is becoming reality.

Digital Resources and the Algorithmic Economy

As smartphones became the device of choice for digital consumers, digital mobile platforms took on the role of essential infrastructure for commerce. The owners of the two dominant mobile platforms, Apple with iOS and Alphabet with Android, have profited handsomely as reflected in their stock prices (Figure 5.23). But a rising tide lifts all boats, and the mobile platform engendered a huge economy in which developers could profit from making apps for those platforms (Figure 5.24).

As we discussed earlier in the chapter, consumers expect to be able to have a digital relationship with organizations of all sorts—for-profit companies, nongovernmental organizations, and public institutions. The interface for such digital

Figure 5.20. An example of a smart watch

Source: Photo by LG Electronics / CC BY 3.0

1 IDC. 2020, September 25. "Worldwide wearables market forecast to maintain double-digit growth in 2020 and through 2024, according to IDC." Retrieved from https://www.idc.com/getdoc.jsp?containerId=prUS46885820.
2 Wallin, L.-O., and Jones, N. 2019. "Use wearable technology to enhance frontline worker productivity: Field service management" (no. G00434217; p. 8). Gartner.
3 Castellanos, S. 2020, May 29. "Coronavirus pandemic brings new use cases for augmented reality." *Wall Street Journal*, retrieved from https://www.wsj.com/articles/coronavirus-pandemic-brings-new-use-cases-for-augmented-reality-11590777284.
4 Business Wire. 2015. "Intel CEO outlines the future of computing." *Business Wire*, retrieved from http://www.businesswire.com/news/home/20150106006877/en/Intel-CEO-Outlines-Future-Computing.

relationships is the mobile app; no wonder that there are about two million apps currently available on the Apple App Store!

The success of mobile platforms and the imperative for all kinds of organizations to build digital interfaces for interacting with their customers is planting the seed for the new evolution of digital technologies. This evolution requires digital transformation and creates fertile ground for digital entrepreneurship. Twilio is a great example of this emerging trend, along with other digital entrepreneurial ventures such as Stripe or PubNub (see Sidebar 5.2). These are firms that do not sell apps. Rather, they provide critical building blocks for those who have to build apps—whether start-ups like Instacart (Figure 5.25) or century-old banks like BBVA.

Consider Instacart, the popular grocery shopping service. As we described earlier, Instacart manages a multisided network that enables you to order groceries from your favorite supermarket and have an independent shopper pick up and deliver them to your house within a two-hour window. Key elements of the Instacart value proposition are its ability to process complex payments involving the supermarket, the shopper, the customer, and, of course, a fee for Instacart. A fundamental feature of Instacart is real-time communication. When placing an order, customers are advised to be ready to respond to shoppers' messages: "While your shopper is preparing for your order, be ready to respond to questions she or he may have about replacement options by using our in-app chat functionality."[1] You intuitively realize how the ubiquity of smartphones and connectivity are critical enablers for Instacart's success. The lack of such enablers was at the heart of the failure of Webvan and the idea of large-scale home grocery delivery in the late 1990s (Chapter 4). What you may not know, however, is that Instacart engineers did not build any payment or communication functionality. Rather, Instacart orchestrates a number of digital resources made available by specialists like Stripe or Twilio. Ravi Gupta, the former COO and CFO of Instacart, explained,

> We tried to build the things that we are uniquely suited to build . . . we are uniquely suited to build a great front-end ecommerce experience, and a great enterprise experience for grocers, we are uniquely able to fulfill, we are

Figure 5.21. An example of a fitness band

Source: Photo by LG Electronics / CC BY 3.0

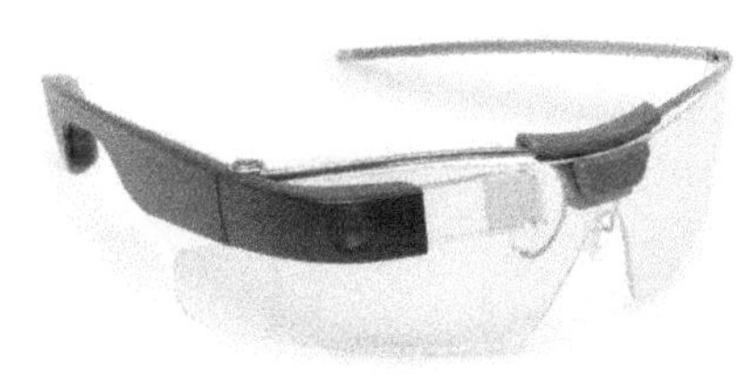

Figure 5.22. Google Glass Enterprise Edition

Source: https://www.x.company/glass/

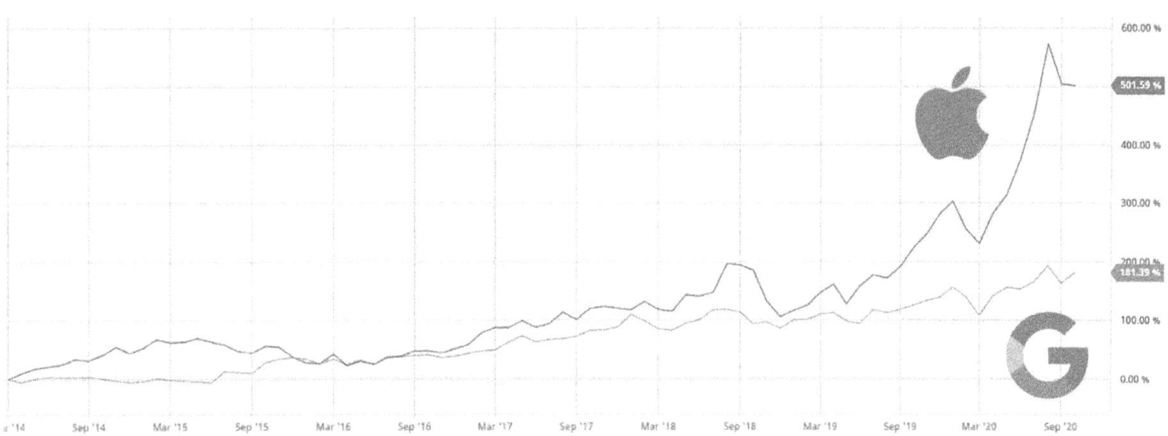

Figure 5.23. Six years of Apple and Google stock prices

Source: Data from barchart.com

1 Instacart. 2020, March. "Tips for using Instacart during this busy and uncertain time." Instacart, retrieved July 26, 2020, from https://news.instacart.com/tips-for-using-instacart-during-this-busy-and-uncertain-time-914d15625684.

> uniquely able to onboard shoppers and give them a great experience. We are not as well suited as other people [like Stripe] to have the payment infrastructure for the entire thing.[1]

Unlike Webvan, who had to build, secure, and maintain its own IT resources (e.g., data centers), Instacart relies on digital resources such as Amazon Web Services (AWS), RDS, and EC2, for relational database storage and virtualized computing resources, respectively. On the payment side, Instacart uses Stripe, and on the messaging side, Instacart uses Twilio and never considered insourcing a messaging infrastructure or messaging capabilities, as Max Mullen, one of the cofounders of Instacart, admitted: "I don't think we could have started Instacart without Twilio. When we started out, dispatching orders to shoppers was a key problem, and Twilio was an enabling technology for us."[2]

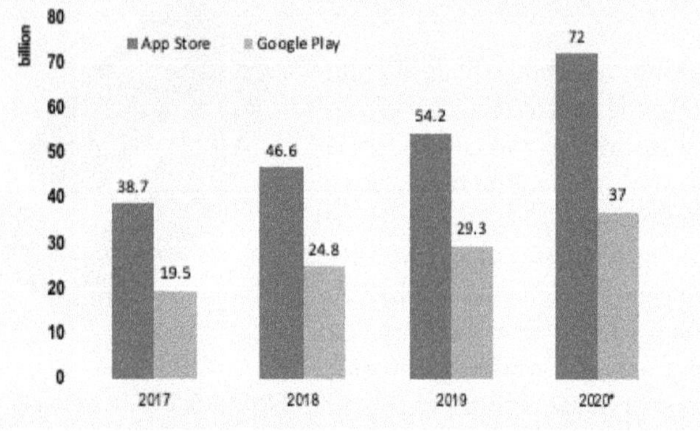

Figure 5.24. Global consumer spending in mobile apps and games

Source: SensorTower, 2020, "total 2020 estimates are authors' forecasts based on 1H 2020 and Q3 2020 values," retrieved from https://sensortower.com/reports

Instacart is a great example of the increasing prominence of digital resources. It exemplifies how digital innovation and digital entrepreneurship increasingly revolve around the ability to solve customer problems via software applications. More importantly, the Instacart example shows how this shift toward digital relationships has created an ample opportunity for firms that are able to provide the building blocks that organizations need to create their digital relationships with customers. We call these building blocks digital resources.[3] They are a specific class of digital objects that (a) are *modular*, (b) *encapsulate objects of value*, assets and/or capabilities; and (c) are accessible by way of a *programmatic bitstring interface*. Without getting too technical, this means that digital resources offer a useful capability (e.g., payment processing)—they can be used as modules or components in a new application (e.g., Instacart using Stripe payments) and are plugged into the new application via a software interface, typically an API.[4] Think of this software as Lego bricks that you can use in your new app without having to reinvent a whole set of functionalities. Using such resources via APIs, a company like Instacart can create an app without having to develop its communication elements, payment processing functionalities, or the code to manage notifications. It can instead rely on digital resources exposed by Twilio, Stripe, and PubNub, respectively (Figure 5.25). A new "industry" of

Figure 5.25. Instacart and enabling digital resources

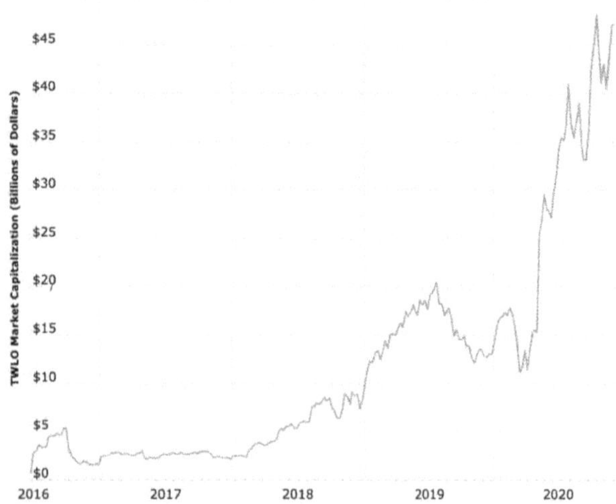

Figure 5.26. Twilio's market capitalization

Source: Macrotrends.net, available at https://www.macrotrends.net/stocks/charts/TWLO/twilio/market-cap

1 Gupta, R. 2019. "Delivering scale: Lessons from Instacart's hypergrowth." Stripe 2019 Annual Conference, retrieved from https://www.youtube.com/watch?v=tNwkOBW5Lfw.

2 Mullen, M. 2020. "Instacart's on-demand grocery service brings home the bacon for thousands of happy customers." Twilio, retrieved from https://customers.twilio.com/1049/instacart/.

3 Piccoli, G., and Rodriguez, J. 2020. "Digital strategic initiatives: Conceptualization and illustration." *BLED 2020 Proceedings, 39.* AIS eLibrary, retrieved from https://aisel.aisnet.org/bled2020/39.

4 See Chapter 4.2: "Platforms, Marketplaces and Aggregators."

SIDEBAR 5.2. Twilio and Its Digital Resources

Digital platforms expose resources to enable developers to contribute to the overall application ecosystem (Chapter 4). Digital resources are made available by way of digital interfaces, typically an API. Before the emergence of the **algorithmic economy**, firms used to acquire complete software solutions from vendors to capitalize on the economies of scale associated with off-the-shelf software applications (Chapter 3) and applied customization only when necessary. Organizations like Twilio don't provide complete applications. Rather, they offer "application components," or digital resources, as we call them.[1] If this difference seems trivial to you, consider that Twilio is a company worth around $40 billion (Figure 5.26).

Twilio "encapsulates" into digital resources the communication functionalities that are standard in modern apps. For example, sending short message service (SMS) message requires developers to orchestrate several software pieces and maintain or connect to an SMS gateway (and pay for it) to route messages on the phone network.[2] Therefore, just to enable SMS functionalities in its app, a firm like Uber would have to incur development, maintenance, and service costs.

An alternative solution to sending an SMS message is to rely on Twilio and accomplish the objective with a few lines of code (Figure 5.27). The text could be a promotional code, a confirmation that your package is on its way, or a notification to a driver that a pickup has been scheduled. You immediately recognize that the firm capable of exposing a digital resource to cleanly and efficiently enable text messaging has a huge addressable market for its services—and the more apps will be developed, the larger the customer base will be.

Note that in the text example, a developer incorporating text messaging using Twilio's digital resources need to only have a rudimentary understanding of messaging. As long as she knows how to make requests via the API, Twilio takes care of processing those requests. Twilio exposes the messaging digital resource, along with a plethora of other communication capabilities, including voice synthesis and recognition for call-center automation, chatbots, video, and email (Figure 5.28).

```
import os
from twilio.rest import Client

account_sid = os.environ['TWILIO_ACCOUNT_IDENTIFIED']
auth_token = os.environ['TWILIO_AUTHENTICATION_TOKEN']
client = Client(account_sid, auth_token)

message = client.messages.create(
body='Hello Gabe! Here's the Twilio example! Fede',
from_='+653525150',
to='+15558675310'
)
print(message.sid)
```

Figure 5.27. An example of Python code used to send an SMS message leveraging Twilio—no more 400-page manual to study!

1 Piccoli, G., Rodriguez, J., and Grover, V. 2020. "Digital strategic initiatives and digital resources: Construct definition and future research directions." In *Proceeding of Forty-First International Conference on Information Systems*, Hyderabad, India; Piccoli, G., and Rodriguez, J. 2020. "Digital strategic initiatives: Conceptualization and illustration." In *Proceedings of the 33rd Bled eConference, enabling technology for a sustainable society*, Bled, Slovenia.

2 One of the authors still remembers with trepidation when, in 2002, his start-up was asked to integrate SMS capabilities for a client. He was given the 400-page manual from the telecommunication company explaining the specification of the SMS protocol and gateway instructions the telecom company used to process text messages.

digital resource creators is emerging, and IT advisory firms such as Gartner have taken notice. They label digital resources as "packaged business capabilities" (PBCs) for the "composable" enterprise:

> PBCs are delivered in product style by SaaS vendors, central IT and other application 'factories'... They are then assembled in the organization, by central or business unit IT teams, to deliver uncluttered fit-for-purpose support for specific job roles' or individuals' practices.[1]

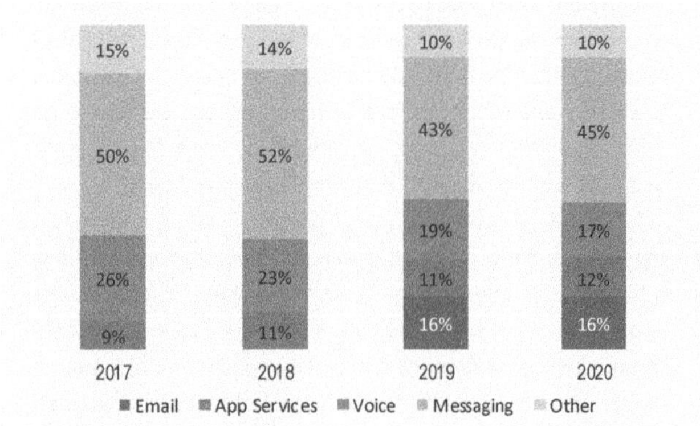

Figure 5.28. Twilio revenue contribution by digital resource

Source: Twilio Investor Day 2020 presentation, https://investors.twilio.com/events-and-presentations/default.aspx

The Instacart example is not unique. By some accounts, the average app you use on your smartphone or tablet leverages 17 different APIs to access digital resources that help them deliver their value proposition.[2] This trend is spawning an algorithmic economy—some call it the API economy, where digital entrepreneurs are identifying those components that are general and widely used across applications, like payment.[3] They then focus on building world-class digital resources and making them available, for a fee, to firms who use them in the innovative digital products and services they create. Stripe, the leading provider of payment digital resources, captures this vision neatly on its website: "We believe that payments is a problem rooted in code, not finance. We obsessively seek out elegant, composable abstractions that enable robust, scalable, flexible integrations."

At the forefront of this trend are digital entrepreneurs and start-ups like Stipe or Twilio. However, in an algorithmic economy, any organization that has valuable resources that could be exposed through digital interfaces can commercialize them. For example, large established corporations are increasingly seeking ways to monetize their digital assets. Imagine a large credit card company that could expose, opportunely anonymized and aggregated, the spending data of their customer base. You could imagine local tourism guides or newspapers around the world using the data to create innovative services like a list of restaurants, bars, or other venues that are popular with the local crowd.

If you are tempted to believe that the above is too technical because you are a manager and not an "IT guy," consider this: Digital innovation is increasingly enabled by the creation and orchestration of digital resources. More generally, the subsequent waves of digital disruption that began with the commercialization of the Internet created a specific vocabulary that is now taken for granted in business. As a manager in the digital age, you must be conversant with the language if you plan to contribute to your firm's digital innovation efforts. There is no escaping the reality that in the world you will be managing, innovation is digital innovation.

5.4 Categorizing Digital Business Initiatives

A number of definitions of the terms *electronic commerce* (eCommerce) and ***electronic business*** (eBusiness) have been proposed over the years. Perhaps the simplest definition of the term *electronic*

---

1 Natis, Y., Gaughan, D., and Alvarez, G. 2021. "Strategic roadmap for the composable future of applications." ID: G00433984.

2 Casado, M. 2018, March 10. "The world through an API." Andreessen Horowitz, retrieved from https://a16z.com/2018/03/09/api-world-summit/.

3 van der Meulen, R. 2016, January 20. "The algorithm economy will start a huge wave of innovation." Gartner, retrieved from http://www.gartner.com/smarterwithgartner/the-algorithm-economy-will-start-a-huge-wave-of-innovation/.

commerce is the broadest one: an online exchange of value. A more specific one, adopted in this book, is as follows: the process of distributing, buying, selling, marketing, and servicing products and services over computer networks such as the Internet.

The term *electronic business* originally referred to the digital enablement of internal organizational business processes, such as logistics. However, recognizing the increasingly interconnected nature of business operations upstream and downstream in the value chain, the term rapidly evolved to encompass interorganizational processes spanning such areas as electronic purchasing and supply chain management. Thus we broadly define the term *electronic business* as the use of digital technologies to enable business processes and operations. While the terms *eCommerce* and *eBusiness* are still in use, they are subsumed under a more general label: *digital business*.

The most immediate way to classify different types of digital business ventures and innovations is to identify the parties involved in the transaction (Table 5.3).

Another valuable categorization approach is by company structure (Table 5.4). The last two decades of experimentation led to the consolidation of some established business models for digital business (Table 5.5). While small organizations focus on one business model, a large organization might have portfolios of business models. Consider Amazon, for example. They are a retailer when they sell and deliver books

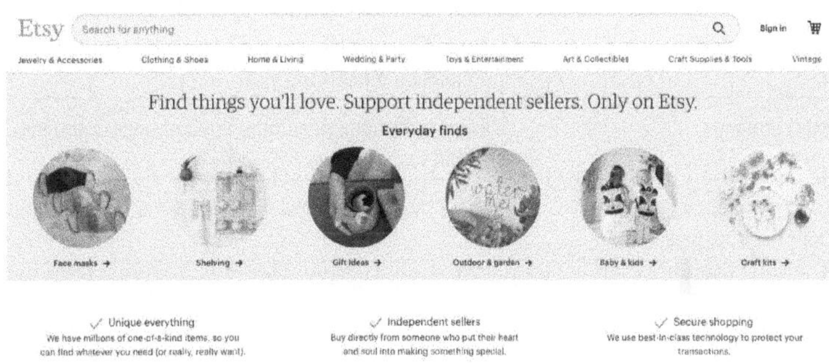

Figure 5.29. Etsy eCommerce website

Table 5.3. Digital business ventures categories by transaction type

Type	Description	Example
Business-to-Consumer (B2C)	Enable transactions involving a for-profit organization on one side and an end consumer on the other	Online retailers (e.g., Amazon)
Business-to-Business (B2B)	Enable transactions involving two or more business entities	Wholesalers (e.g., Alibaba)
Consumer-to-Consumer (C2C)*	Enable transactions involving consumers transacting online with each other	Marketplaces (e.g., Craigslist or Etsy; Figure 5.29)
eGovernment	Enable transactions involving legislative and administrative institutions	Electronic tax filing

* You may be tempted to lump our previous discussion about the "creator economy" in the C2C category. However, this would be a mistake. If, thanks to digital technology and the direct digital relationships it enables you to have with your followers, your passion becomes your main activity, you are now a business. For the purpose of this classification, then, you'd be engaging in a B2C transaction—even if you are a "one man" organization.

Table 5.4. Digital business venture categories by company structure

Structure	Description	Example
Brick and mortar	Firms that have physical operations and locations (e.g., stores) and don't provide their services exclusively through the Internet.	Oink & Moo BBQ, the most popular food truck in America in 2017*
Bricks and clicks	Firms that have hybrid operations. Typically, brick-and-mortar operations that embraced digital transformation.	Walmart, Tesco, or any major retailer with physical and digital operations
Pure play	Firms "born online." Organizations that have no stores and provide their services entirely through a digital interface with the customer.	Google, TikTok, Stripe, or any other online-based organization

* https://www.cnbc.com/2017/04/07/10-most-popular-food-trucks-in-america-.html.

Table 5.5. Main business models for digital businesses

Type	Description	Example
Online retailing	Firms that take control of inventory to resell it at a profit	BestBuy.com, Fnac.com, Mediaworld.it
Information intermediaries or infomediaries	Firms that use digital technologies to provide specialized information services	Google search, Skyscanner.com, Trovaprezzi.it
Content providers	Firms that develop and publish new information or new materials	Reuters.com, Netflix, creators in the passion economy
Social networking and online communities	Firms that intermediate the online congregation of individuals, social groups, or communities of interest	Instagram, LinkedIn, TikTok, Stack Overflow (Figure 5.30)
Digital platforms	Firms that expose resources to enable digital innovation and competitive actions by other firms	Android, Shopify, WeChat
Marketplaces	Firms that manage communities of buyers and sellers who transact via digital technologies	Expedia.com, eBay, Etsy
Digital resources and cloud service providers	Firms that offer digital technology resources or services via the Internet	Stripe, Dropbox, Salesforce

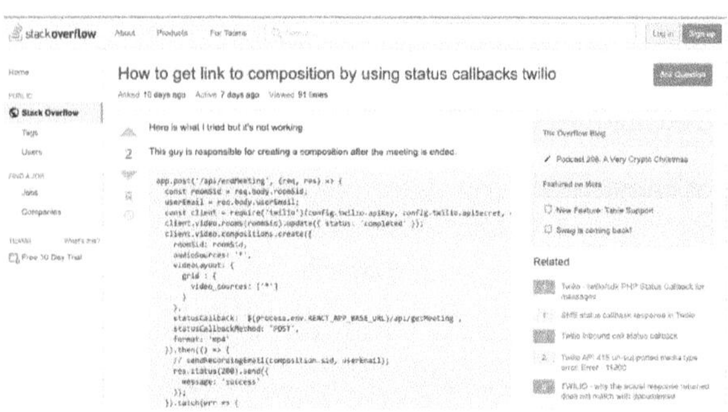

Figure 5.30. A request for help addressed to the Stack Overflow community

that they have purchased and stored in their warehouse; they are a retail service provider when they expose services such as listing or fulfillment by Amazon through their digital platform to other retailers; they are a computing service provider when they expose digital resources via the AWS platform.

A key element in a business model is the specification of how the firm plans to make money—that is, the firm's **revenue model**. With many new business models adopted by digital businesses leveraging traditional revenue models (e.g., **pay for service**), business model innovation has spurred some pioneering ways to produce revenue (Table 5.6).

5.5 Managerial Takeaways

As we discussed in the introductory chapters, the emergence of new technologies often has dramatic impacts on organizations. The rapid adoption of the Internet, the mobile platform, and the relentless pace of digital technology innovation is no exception. As a manager, you will be called upon to manage how your firm can benefit from digital innovation while resisting digital disruption of its traditional business model. In order to do so, you should be familiar with some of the main issues and opportunities brought about by the digitization of business.

Disintermediation

The hallmark of the Internet is connectivity. As such, its emergence and widespread adoption by consumers enabled any organization that so chooses to establish (at least technically) a direct relationship with its customers. All of a sudden, it was practical for companies as diverse as hotels and computer manufacturers, automakers, and insurance companies to reach customers directly rather than through one or more middlemen.

The term ***disintermediation*** refers to the process by which a firm's distribution chain is shortened through the elimination of one or more intermediaries. Disintermediation has a direct impact

Table 5.6. Main revenue models for digital businesses

Type	Description	Example
Pay for service and pay per use	The firm offers a product (e.g., books) or a service (e.g., insurance) for sale, and it is compensated much like a traditional store or service provider	Dropbox, LegalShield.com, Twilio,* Airbnb†
Subscription	The firm offers its product for a time-based fee, and customers pay for the right to access the content	*New York Times*, Netflix
Advertisement support	The firm sells "access to its audience" to interested advertisers	Facebook, Instagram, *Huffington Post*
Affiliate	The firm generates revenue from a third party by referring customer traffic to it	Digital businesses or influencers linking to products on Amazon
Freemium (contraction of free-premium)	The firm gives away its product or service for free and sells premium versions	Dropbox, Mailchimp
In-app purchasing	The firm provides free apps or games that have premium content or features for sale	*Fortnite*

* Increasingly, the computing resources offered by cloud providers are billed as a function of usage. This pay-per-use revenue model is the standard for the new breed of companies whose business model consists of providing digital resources. This model is quite interesting because it ties directly the success of the client—WhatsApp, in this case—and the revenue of the digital resource provider (i.e., Twilio).

† When adopted by marketplaces such as Airbnb or eBay, this revenue model entails charging for the service of enabling transactions.

on those organizations that find themselves being . . . well, disintermediated—such as travel agents and car dealers. Those organizations that were caught in the middle had to recast their value proposition in order to survive. For example, many travel agents have found it difficult to stay in business after airlines and, increasingly, hotels have eliminated commissions. Many others, though, have been able to leverage their superior knowledge about travel products and rules and are now prospering using a consulting rather than commission model (i.e., they receive fees from travelers who value their service and knowledge). In the worst-case scenario, a firm facing disintermediation may be forced to harvest and close the business.

Reintermediation

As managers and observers tried to make sense of the Internet as a business opportunity, many thought that disintermediation would lead to the demise of distribution channels in most industries. While disintermediation failed to eliminate traditional intermediaries, the Internet created opportunities for new intermediaries to exist alongside their brick-and-mortar counterparts—a process known as ***reintermediation***. Consider the insurance industry, for instance. Today, insurance companies reach consumers directly (e.g., Progressive.com), through traditional insurance brokers and through independent online insurance brokers (e.g., Insure.com).

Another example is offered by many of the infomediaries discussed above. While it is true that traditional travel agents have been forced to reinvent their value proposition, a number of Internet travel agents (e.g., Booking.com, Expedia, Opodo) have emerged and are thriving due to their ability to help travelers gather information and uncover low prices. Similar dynamics have occurred in traditional retail and many other industries.

Market Efficiency

Since its advent, IT has contributed to reduced search and transaction costs, thus improving the efficiency of markets. The Internet and digital technologies that leverage it continued and perhaps accelerated this process, empowering customers with the instruments and technologies they need to sift through large amounts of product and service data.

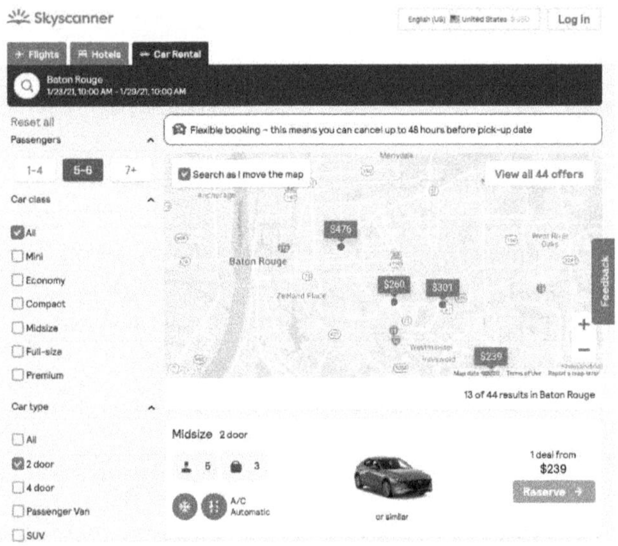

Figure 5.31. Subset of results from metasearch infomediary Skyscanner

Prior to the arrival of the Internet, customers faced significant costs when searching for products and services. They would have to either visit physical stores or call multiple outlets to describe what they were looking for and inquire about availability and price. In either case, the process would be fairly time consuming and therefore costly. The outcome of this process has been heightened competition and an increasing difficulty in profiting from strategies rooted in asymmetry of information or high search costs. Perhaps the best illustration of market efficiency is provided by travel metasearch infomediaries such as Skyscanner (Figure 5.31).

Channel Conflict

The emergence of the online channel created a conundrum for many organizations that had an established distribution chain. Should they disintermediate, following the promise of reduced distribution costs and a direct relationship with customers, or should they work with the channel in an effort to identify mutually beneficial Internet-enabled initiatives? At the heart of this dilemma is the difficulty of moving distribution from the traditional channel (the one currently producing the revenue stream the organization needs to survive) to the online direct channel (the one that promises the highest profitability in the long run). The term ***channel conflict*** captures this dilemma.

Two examples highlight the difficulty faced by organizations confronting channel conflict. When Dell began selling computers through its website, it faced no objections from distributors because it had no distributors. When Compaq, Dell's principal competitor at the time, sought to respond by creating its own direct-sale website, it faced significant resistance from electronics store chains carrying its devices (e.g., Circuit City).

Renaissance Cruises had an even more traumatic encounter with the channel conflict dilemma. Taking a page out of the airline and lodging industries, the company decided to embrace the Internet channel in the late 1990s and drastically reduce travel agent commissions. In retaliation, the travel agent community boycotted the Renaissance product, and the firm quickly encountered financial difficulties. A public apology in the pages of the *Wall Street Journal* notwithstanding, these difficulties culminated in bankruptcy once the events of September 11, 2001, severely hampered the travel industry.

Figure 5.32. Self-service kiosks
Source: Photo by Gallowolf / CC BY-SA 3.0

Customer and Employee Self-Service

Another important implication of the widespread adoption of electronic commerce and electronic business has been the emergence of customer and employee self-service. Aided by easy-to-use websites and the increasing degree of comfort that the general public has developed with information technology of all kinds, digital self-service is a growing trend requiring managerial attention (Figure 5.32).

Examples of this trend abound, from kiosks at airline counters, at post offices, in hotel lobbies, and in fast-food restaurants to self-checkout counters at grocery stores, to web-based software that allows you to compute fairly complex tax returns without ever speaking to a professional. With the advent of the mobile platform and the proliferation of apps that let digitally savvy individuals transact directly with the organizations of their choice, digital self-service is increasingly becoming the norm in customer relationships.

Long-Tail Strategies

In statistics, the notion of a **long tail** represents a power law relationship between two quantities. A power law between two quantities occurs when the frequency of an object occurs as a power function of certain characteristics of that object (Figure 5.33). The best-known example of this phenomenon in economics is the so-called Pareto principle, or the 80–20 rule (e.g., 80% of your sales will come from 20% of your product catalog). More generally, the long tail represents a phenomenon by which the frequency of an event is related to some characteristic of that same event. For example, Chris Anderson in his 2004 *Wired* magazine article on the subject noted that, according to the Recording Industry Association of America, only the top 10 major label music CDs were actually profitable.[1]

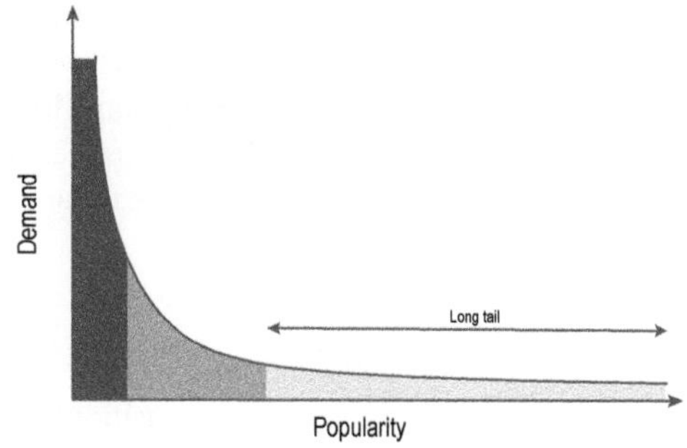

Figure 5.33. The long-tail distribution

However, Anderson observed that online stores could capture sales from "not-so-popular" items as well. More specifically, he documented that the online retailer Amazon carried, at the time, 2.3 million books and could carry inventory that dwarfed those of a typical physical bookstore (about 130,000 books). Importantly, he also documented that the total sales of books at Amazon from the 130,000 books one would also find in traditional stores (i.e., the high-volume "hits") accounted for fewer than 50% of total sales. In other words, the long tail of "not-so-popular" products delivered a greater share of sales than the *fat head* of popular items. While one could argue that this is partly due to the fact that Amazon faced significantly higher competition for the popular items, the findings suggest that customer interests are highly varied and that there is some demand for niche products. When this is the case, a firm has the opportunity to craft a strategy that leverages the dis-homogeneity in customers' demands. Such a strategy should focus on minimizing inventory holding and distribution costs—easiest to do with information goods such as digital songs or e-books (Chapter 4)—as well as enabling customers to search for the more obscure items and even be alerted about previously unknown items that would be of interest to them based on previous purchases through collaborative filtering or similar technologies.[2]

Online-to-Offline

Online-to-offline (O2O) refers to the use of digital technology (i.e., online) to spur transactions in physical stores (i.e., offline). There are a number of O2O techniques aimed at engaging online customers in the physical retail space, and new approaches emerge as new information technologies are introduced and adopted by customers. Typical examples of O2O are services such as product pickups, free shipping, returns, and payments. As explained by Joe Tsai, Alibaba's executive vice chairman, "You can imagine a customer going into a store using an Alibaba app, experiencing the product firsthand, but operating the whole purchase online so he doesn't have to wait at the checkout counter."[3] In this scenario, having found a product online, the customer goes to the store to physically evaluate it but then completes the transaction online. Another example is Amazon, a

1 Anderson, C. 2004. "The long tail." *Wired 12*(10), retrieved from http://www.wired.com/2004/10/tail/.

2 Collaborative filtering is the technique of alerting customers about products of potential interest to them based on an analysis of their pattern of behavior relative to people who exhibit similar tastes. A famous example of the use of collaborative filtering is Amazon.com's purchase recommendations.

3 Reporters, F. T. 2015, August 11. "Alibaba places big bet on bricks and mortar." *Financial Times*, retrieved from http://www.ft.com/intl/cms/s/0/c305c71a-4010-11e5-9abe-5b335da3a90e.html.

digital business trailblazer since its founding in 1994, which has recently begun opening physical bookstores. As we write this book, Amazon has 13 total locations in various U.S. cities, and it is pioneering some intriguing O2O techniques, including prominently displaying books that have received more than 10,000 online reviews on Amazon.com or books devoured in three days by customers reading the e-book version on the Amazon Kindle (Figure 5.34).

Summary

The years between 1993 and 2001 saw the dawn of the network economy. In the 2001–7 time span, innovation revolved around Web 2.0 and focused on transitioning from static web pages to dynamic web applications. With the launch of the iPhone in 2007, we entered the current period of digital transformation, dominated by the Internet and mobile platforms and the apps that leverage them. In this chapter, we sought to provide you with the tools to make sense of past developments and understand future trends.

- The Internet, traditionally thought of as a network of computer networks, is evolving into a global information grid enabling ever-changing devices, and the people who use them, to easily connect and disconnect from it. The rapid pace of evolution and innovation on the Internet is enabled by its characteristics: distributed governance and the reliance on publicly available open standards supporting a multiplicity of compatible devices and offering a number of services.
- Digital business finds its roots in the development of information technology and networking over the last 40 years. But the recent acceleration of innovation in this area has been enabled by affordable computing equipment, widespread access to the Internet, the increasing ease of use of information technology, and the availability of open standards.
- As smartphones became the device of choice for digital consumers, digital mobile platforms took on the role of essential infrastructure for commerce. Their success and the imperative for all kinds of organizations to build digital interfaces for interacting with their customers seeded the evolution of new digital technologies that require digital transformation and create fertile ground for digital entrepreneurship. New firms like Twilio, Stripe, or PubNub do not sell apps. Rather, they provide digital resources, critical building blocks for those who have to build applications.
- We have categorized digital business initiatives on two dimensions. By looking at the type of transaction taking place, we classified electronic commerce as **business-to-consumer (B2C)**, **business-to-business (B2B)**, **consumer-to-consumer (C2C)**, **consumer-to-business (C2B)**, and **eGovernment.** Focusing on the company structure of the organizations involved, we classified concerns involved in electronic commerce such as brick and mortar, **bricks and clicks**, and **pure play**.
- We defined a business model as the document that captures the firm's concept and value proposition while also conveying what market opportunity the company is pursuing, what product or service it offers, and what

Figure 5.34. Amazon O2O e-book strategy

strategy the firm will follow to capture a dominant position. The dominant business models that have emerged in the network economy are online retailing, infomediaries, content providers, social networking, crowdsourcing, marketplaces (or exchanges), and cloud computing.

- A key feature of a business model is the revenue model—the firm's plan for building a revenue stream. The dominant revenue models that have emerged are pay for service, **subscription**, **advertisement support**, **affiliate**, and **freemium**.
- The rapid adoption of the Internet and the emergence of the network economy have had some significant implications for both established organizations and upstarts. Disintermediation (the process by which a firm's distribution chain is shortened through the elimination of one or more intermediaries), reintermediation (the process by which new online intermediaries carve a niche for themselves alongside their brick-and-mortar counterparts), market efficiency (the main outcome of the lower search costs enabled by Internet technologies), channel conflict (the dilemma faced by organizations deciding whether to disintermediate their legacy distribution channels), the emergence of widespread IT-enabled self-service, long-tail strategies (the opportunity to craft business strategies leveraging the dis-homogeneity in customers' demands), and O2O techniques (the engagement of online customers in the physical retail space) are the most relevant.

Study Questions

1. What is the Internet? Are the Internet and the web the same thing? How do they differ?
2. What is the difference between the terms *electronic commerce* and *electronic business*? Why has the distinction largely faded in recent years?
3. Discuss the main features of Web 2.0 that differentiate it from its predecessor.
4. Define the term *mobile platform* and discuss its impact on today's business environment.
5. What are the idiosyncratic main capabilities of *smart devices*? How may they challenge current business models?
6. Explain the difference between *smart devices* and *wearables*, giving specific examples.
7. Define the term *API* and discuss the business model of those firms leveraging them as their core value proposition.
8. Define the business model canvas and provide a brief description of its main building block. Which critical business questions does it aim to answer?
9. Define each of the following terms and provide examples: *business-to-consumer (B2C)*, *business-to-business (B2B)*, *consumer-to-consumer (C2C)*, *consumer-to-business (C2B)*, and *eGovernment*.
10. Define each of the following terms and provide examples: *brick and mortar*, *bricks and clicks*, and *pure play*.
11. Explain what we mean by the terms *business model* and *revenue model*. How do the two differ? What are the principal business models and revenue models adopted by modern organizations?
12. What problems can be effectively addressed through crowdsourcing? Define the terms and discuss, with proper examples, how crowdsourcing may effectively address each class of problems.
13. Define the concept of "long tail." Which strategies does it enable?
14. Provide an example and critically assess the impacts on the customer experience of an O2O approach you recently experienced.

Glossary

A glossary of key terms can be found at the book's website. Visit https://www.prospectpressvt.com/textbooks/piccoli-information-systems-for-managers-5-0, then click on "Student Resources" in the horizontal red menu bar.

PART III

The Strategic Use of Information Systems

The potential for the strategic use of digital technology and information systems (IS) has been a source of debate since it became clear that information technology (IT) had important business applications. Given the continuous evolution of digital technology (Chapter 1) and its effects on the competitive environment (Chapter 4), there is no simple recipe for extracting value from digital technology investments. There are stark examples of companies that have successfully harnessed the potential of ever-more-powerful digital technology, grabbing headlines, market share, and profits and others who have famously squandered large sums of money with little visible benefit.

The ability of megastore operator Walmart to manage information for competitive advantage is the stuff of legends. The firm built its own satellite-based telecommunication network in the 1970s to support real-time communication with its stores located in rural areas of the United States. As a testament to his faith in the potential of information systems pervading the company, founder Sam Walton declared in 1992, "We've spent almost $700 million building up the current computer and satellite systems we have. . . . What I like about it is the kind of information you can pull out of it on moment's notice—all those numbers."[1]

So why are some firms able to exploit information systems for sustained competitive advantage while others cannot? Perhaps the most enduring research result that can help in answering this question is offered by executives' surveys, which show remarkable consistency with the finding that the average senior manager feels that digital technology and IS decisions are well outside of their comfort zone. To this day, many business managers and chief executive officers (CEOs) are still uncomfortable with planning for the use and management of digital innovations.[2]

Compounding the above problem is the fact that the IS function has traditionally been led by technologists. Because of the vastly different background and knowledge base of business executives and technology executives, the result has often been failed communication and a delegation of "all IT issues" to technologists. More recently, however, we have witnessed a trend reversal, with the IS function being led by many "new school" chief information officers (CIOs), who are well versed in the inner workings of the business and may act as the driving force for their firms' digital transformation. A number of organizations also introduced chief digital officer (CDO) positions. While this is a step in the right direction, it is hardly enough because, as talented as today's CIOs are, they are not spending their time addressing operations problems the way chief operating officers (COOs) do, marketing problems the way chief marketing officers (CMOs) do, or financial problems the way chief financial officers (CFOs) do.

The above call to action is particularly important when it comes to using information and digital technology to underpin value-adding strategic initiatives. You, as future general and functional managers, must feel comfortable with planning and setting a direction for the use of IS resources, with identifying opportunities to use technology to create and appropriate economic value, and with deciding under what circumstances these initiatives can be protected against competitive retaliation.

1 Walton, S., with Huey, J. 1992. *Sam Walton: Made in America*. New York, NY: Bantam: 271.

2 Kark, K., Briggs, B., Phillips, A. N., Lillie, M., Tweardy, J., and Buchholz, S. 2020. 2020 global technology leadership study. Deloitte Insights, retrieved from https://www2.deloitte.com/us/en/insights/topics/leadership/global-technology-leadership-study.html.

Part III speaks to general and functional managers and covers the key IS decisions that all modern managers must be comfortable making.

- *Chapter 6: Strategic Information Systems Planning.* This chapter provides an overview of the strategic IS planning process, from the definition of an overall information vision to the identification of strategic initiatives.
- *Chapter 7: Value Creation and Strategic Information Systems.* This chapter sets the background for analyzing the use of information systems and technology to create and appropriate value. We define key terms and explain the framework used to analyze value creation and appropriation potential of specific strategic initiatives.
- *Chapter 8: Digital Value Creation.* This chapter discusses a number of frameworks and analytical models that have been advanced over the years to help managers envision how to use information systems and technology to create and appropriate economic value.
- *Chapter 9: Digital Value Capture.* This chapter completes the puzzle by focusing on sustainability. Once a firm has successfully created value with information systems and technology, it must be able to defend its position of advantage and appropriate the value created over time.

Chapter 6

Strategic Information Systems Planning

What You Will Learn in This Chapter

This chapter focuses on the strategic information systems (IS) planning process and the role that general and functional managers need to play in it. Strategic IS planning is a fundamental aspect of IS management because it ensures that information systems and digital technology decisions are not made in a haphazard fashion. Rather, they are made with a clear understanding of business strategy and an overall sense of direction with respect to what the firm is trying to achieve with its use of IS resources.

MINICASE: Strategic Information Systems Planning at CFCU

As you return to your hotel room in Athens, Georgia, tired from your first day at the new client site, you start thinking about the challenge ahead. Your consulting firm was called in by a regional bank—Campus Federal Credit Union (CFCU). The small regional bank has long tradition in the community, having started in 1967 to serve the retail banking needs of local residents through 21 branches, a capillary network of automated teller machines (ATMs), online banking facilities since 2001, and more recently, a mobile baking app.

CFCU's operation centered on personal banking with a wealth of products, ranging from standard checking accounts, to certificates of deposit, to mortgages. While CFCU was a conservative player, offering the pledge of being a safe place for its customers' money, digital transformation was sweeping the industry, creating opportunities and pressure to innovate and reach new customers. In early client calls, CFCU leadership told you they were looking for something "disruptive" to demonstrate their ability to succeed at digital transformation. As you suspected and your meeting today confirmed, however, the bank's main market was local residents with stable financial needs who would benefit from new services—such as straightforward online banking tools or flexible personal loans.

Your consulting firm had been engaged by the client to "develop a strategic information systems plan." Yet the series of meetings you had today left you with the distinct impression that the client had already formulated some ideas. In a meeting with you, the senior vice president of marketing said, "We're missing the boat. Large national banks like HSBC target our customers with precise marketing offers. We must invest in Big Data analytics and online-to-offline (O2O) customer experience management (CEM) tools including AI-driven superior customer service. We need to introduce digital innovations pioneered by FinTechs. For example, the blockchain ledger can secure our transactions and reassure our customers frightened by data breaches. The return on investment (ROI) for us is that our bank will still exist in five years!" From the director of technology services, you heard mostly frustration: "They want me to deliver a world-class infrastructure on a shoestring. The chief financial officer (CFO), to whom I report, makes me justify every digital technology investment. ROI and net present value calculations are all he cares about! Yet two-thirds of my budget goes to keeping the lights on, and the rest goes to various pet projects for one or the other of the executives. How am I going to modernize our archaic infrastructure under these conditions?"

The chief executive officer (CEO) seemed to recognize some of these issues and thought your help would be instrumental in changing things for the better. In the opening meeting, he said,

"Your firm is the best at this, and you know our industry well. We need you to assess our current operations and draft a strategic plan; we will follow it at once!" As you organize your notes, a nagging feeling overcomes you—the firm does not seem to have great unity of purpose at the moment, so how does it expect you to be able get all members of the firm on the same page? "Well," you tell yourself, "one step at a time." You have certainly been in more difficult situations.

Discussion Questions

1. What do you see as the major pitfalls of the current manner in which the information systems budgeting and prioritization process is run?
2. What do you believe are going to be the major challenges you will encounter on this assignment?
3. What should be your next step as you get to work tomorrow morning?

6.1 Introduction

With digital technology increasingly embedded in all aspects of business operations, the most successful organizations are those that are able to establish a productive partnership between information technology (IT) executives and their functional counterparts. Results of the 2020 Deloitte and *Wall Street Journal* "Intelligence Global Survey" showed that CEOs see the chief information officers (CIOs) and technology leaders "as their primary strategy partner, more than all other C-suite roles combined" (Figure 6.1).[1]

Being able to establish a productive partnership between your organization's technology professionals and the executive team is predicated on recruiting the right people and devoting significant attention to the development of the relationship. As Samir Dalya, Motorola Solutions CIO, put it, "Tech leaders need an informed opinion, and that comes from really understanding the business. Having a dialogue with business leaders is table stakes."[2] However, whether or not you are able (and lucky enough) to establish a productive partnership with the IS professionals in your organization, you must be involved in the strategic planning and management of information systems in your firm. Failing to do so, and leaving all IS decisions to IT professionals, will simply result in your joining the ranks of unsatisfied and disappointed general and functional managers scattered throughout the business world. Moreover, as cloud computing (Chapter 3) becomes the standard approach for running the digital technology infrastructure of modern organizations, general and functional managers are increasingly able to provision digital technology independently of the IT function. When provisioning a new sales force automation application required the IT function to purchase and configure servers in the company's data center, install them, and manage the software application and the underlying databases, general and functional managers could not act independently of IT professionals. The

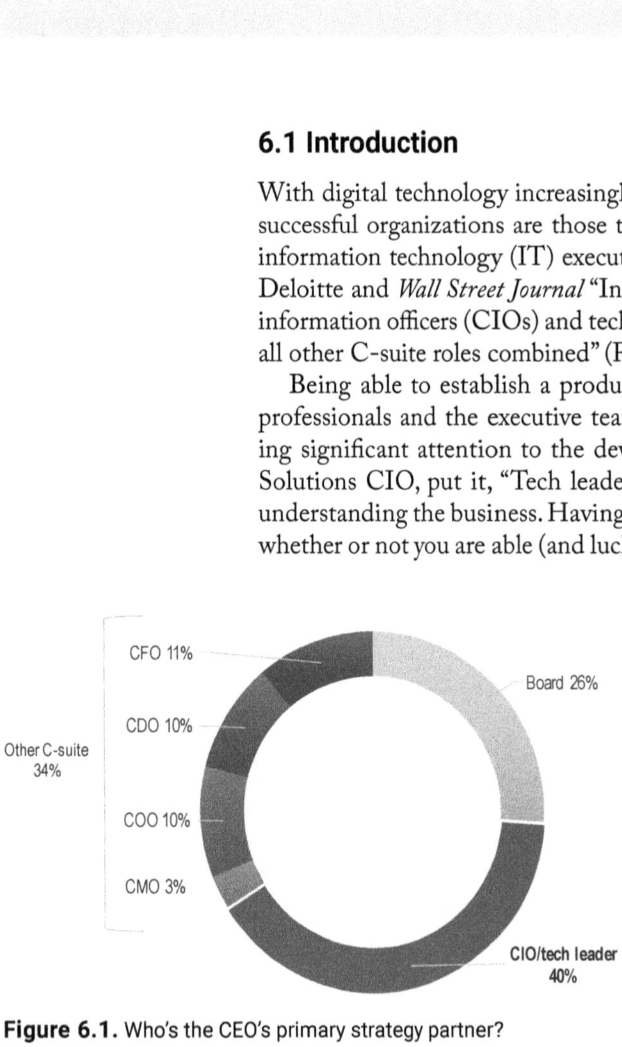

Figure 6.1. Who's the CEO's primary strategy partner?

Source: Adapted from Deloitte and WSJ Intelligence, 2020, ***The new CIO: Business-savvy technologist***

1 Deloitte and WSJ Intelligence. 2020. "The new CIO: Business-savvy technologist" (p. 16). https://www2.deloitte.com/content/dam/insights/us/articles/6550_wsj-new-cio/DI_WSJ-New-CIO.pdf.

2 Ibid.

software as a service (SaaS) delivery mode (Chapter 3) changed all that. Today, marketing and sales executives can simply contract with Salesforce and provision the most popular sales force automation suite independently. Human resource (HR) managers can contract directly with Workday and have the leading HR management suite of tools running without ever needing a server. While this chapter clarifies why we would not advocate it, such a haphazard approach to application provisioning is possible and does happen in modern organizations. More importantly, you realize that this level of (at least theoretical) independence of general and functional managers gives them more power in IS decisions. But with power comes responsibility. Thus the imperative for current and future general and functional managers to be deeply involved in the IS planning process has only grown stronger.

A Word about Strategic and Operational Planning

The realm of strategy pertains to the decisions that an organization makes with respect to how it will develop and deploy its resources over time in an effort to achieve its long-range objectives. Consider for a moment the Roman proconsul Julius Caesar. When he set forth to conquer Gaul (Figure 6.2), he had to decide how to best create and deploy Roman military resources in order to succeed in the conquest—in essence, he had to devise a strategy for the conquest of Gaul. Caesar and his generals had to decide how many legions to raise, how to equip them, how to train them, where to locate camps, and how to organize the logistics of supplies and provisions to sustain the campaign.

As we see here, the realm of strategic thinking and, as a consequence, the realm of strategic planning are concerned with long-range decisions about resource development and deployment. Conversely, operational decision making is concerned with local decisions in the present. For example, as the famed tenth legion took camp at Bribacte (near modern-day Autun in Bourgogne), the camp had to operate with day-to-day routines for mounting guard, for cleaning, for provisioning, and so on. All of these were operational, not strategic, decisions.

As this simple example from a historic military campaign clarifies, there is a clear difference between strategic and operational decision making. While both strategic planning and operational effectiveness are critical to the success of a venture, the remainder of the chapter focuses exclusively on strategic IS planning—that is, how the firm intends to build and deploy its IS resources in pursuit of its long-range business objectives.

Strategic Alignment

A firm that has been able to achieve a high degree of fit and consonance between the priorities and activities of the IS function and the strategic direction of the firm has achieved **strategic alignment**. Research in this area has consistently shown that alignment has a direct impact on firm performance, and alignment is perennially on the top 10 list of CIO priorities. Yet strategic alignment is very difficult to achieve and maintain, particularly in those highly competitive environments where opportunities arise and fade quickly and strategic priorities change constantly. Thus

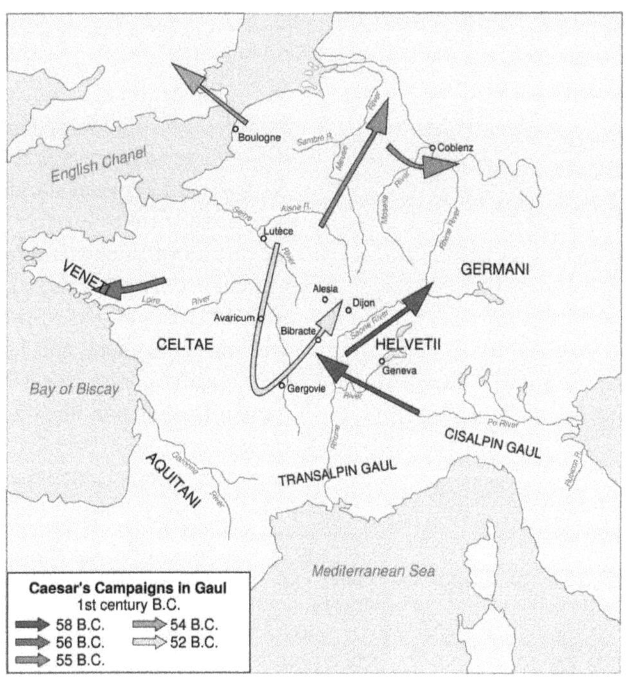

Figure 6.2. The Roman Gallic campaign
Source: Photo by Sémur / CC BY-SA 2.5

ensuring a high degree of strategic IS alignment requires as much improvisation as careful planning. Data from McKinsey provides some comfort, suggesting that IT executives have become adept at aligning IT strategy with the needs of the organization[1] and that high-performing IT organizations present an active CIO involvement in the business.[2] However, the IS function still struggles to systematically propose new ways of creating value using technology resources (Figure 6.3), and executives' current perceptions of IT performance are still negative.[3]

6.2 Six Decisions Requiring Managerial Involvement

It is clear to you by now that senior management must be involved in IS management. One way for them to do so is by addressing six key IS decisions:[4]

1. *How much should we spend on IT?* This is perhaps the most critical question that you, as a general manager or as a member of the executive committee, will be called on to ask because the answer informs all subsequent IT-related decision making. As we saw in Chapter 2, organizations are widely different with respect to their strategy and objectives, their culture, their history, their current infrastructure, and so on. It follows that even two head-to-head competitors may need very different investments in IT—different in terms of both quantity (i.e., how much to spend) and quality (i.e., what initiatives should be the recipient of the IT budget). This question is designed to force senior executives to discuss and decide on the role of information systems and technology in the organization. Failing to ask and answer this question puts the firm's IT department in a reactive mode, having to decide on individual projects with little guidance as to the overall picture.
2. *Which business processes should receive the IT dollars?* This question requires executives to decide which business processes are most important to the firm at a given point—clearly a business decision—and, as a consequence, should attract IT funding. In other words, it is within the purview of business managers to decide on the allocation of resources among the many possible projects the firm can pursue—information systems and technology projects included!
3. *Which IT capabilities need to be company-wide?* Large organizations constantly battle the trade-off between standardization and flexibility. Increasing standardization enables the firm to manage operations efficiently, while flexibility enables the firm to react more quickly and more effectively to local needs. Because information systems often enable the standardization or flexibility of operations, IS professionals are often left with the responsibility of making this decision. Yet this is a business decision that general and functional managers should not hand over. After all, the executive committee is in the best position to weigh the costs/benefits of standardization and flexibility.
4. *How good do our IT services really need to be?* The quality of service provided by an organization's IS department is measured by the reliability and uptime of the

---

1 Craig, D., Kanakamedala, K., and Tinaikar, R. 2007. "The next frontier in IT strategy: A McKinsey survey." McKinsey on IT, retrieved from https://www.mckinsey.com/business-functions/mckinsey-digital/our-insights/a-rising-role-for-it-mckinsey-global-survey-results.

2 Arandjelovic, P., Libby, B., and Naufal, K. 2015, February. "Why CIOs should be business-strategy partners." McKinsey & Company, retrieved from https://www.mckinsey.com/business-functions/mckinsey-digital/our-insights/why-cios-should-be-business-strategy-partners.

3 Ibid.

4 Ross, J. W., and Weill, P. 2002, November. "Six IT decisions your IT people shouldn't make." *Harvard Business Review*: 84–92.

IT infrastructure, data accessibility and flexibility, responsiveness to user needs, and the like. This question forces executives to make conscious decisions about the degree of service the firm needs and what they are willing to pay for. With the rise to prominence of cloud services, every organization should evaluate how its core services are provisioned, asking questions like, What resources and services can migrate to the cloud? How should our internal IT function change? How should the IT skills of internal staff change as more services are provisioned on the public cloud?

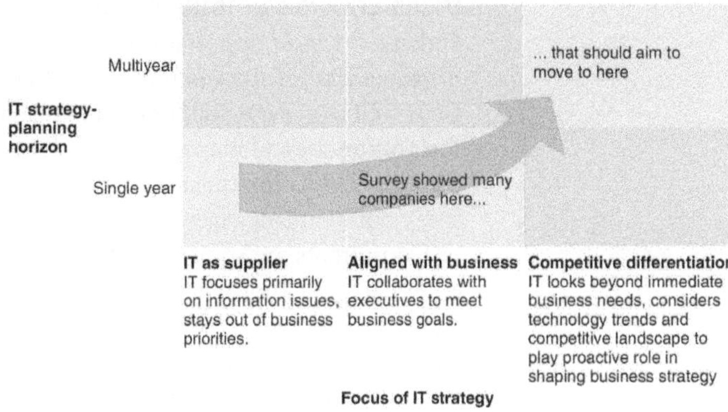

Figure 6.3. McKinsey IT strategy maturity matrix
Source: Adapted from ***McKinsey Quarterly***, 2007

5. *What security and privacy risks will we accept?* Remember the old adage "You get what you pay for"? When it comes to computer security and privacy risks, this proverb is more accurate than ever. Security and privacy decisions are about managing, not eliminating, risk (Chapter 13). Different organizations face different threats and have different degrees of risk aversion. For example, while a hospital cannot afford to skimp on redundancy and uptime of life-support systems, a small independent restaurant may not need to worry about (and pay for) more than basic security and backup procedures. The critical point is that security and risk management decisions are not IS decisions. Rather, they are business decisions that can have a dramatic impact on the future viability of the business and, as such, need the full attention of the executive team.
6. *Who do we blame if an IT initiative fails?* This question draws attention to the need to identify business sponsors for most IS projects. Any project that involves users and organizational departments beyond the IS group should have a clearly identified business sponsor who is responsible for its successful implementation. It is the job of senior executives to allocate resources (e.g., create time for business managers to be involved with relevant IS projects) and assign responsibility.

Note that these are wide-ranging business decisions, necessitating senior executives' input, but they require a technical understanding of the various alternatives, costs, and implications. Thus they need to be made in partnership with the IS professionals and the IS function. The planning process helps structure this partnership.

6.3 The Purpose of Strategic Information Systems Planning

The planning process has a number of objectives, chief among them that of clarifying how the firm intends to use and manage IS resources to fulfill its strategic objectives. Note that we used the term *planning process*, not *planning document*. As most executives will attest, the *process of planning* is as important as, if not more important than, the final documents that represent its output. The planning process requires discussion, clarification, negotiation, and the achievement of a mutual understanding. While these documents can serve to bring new hires, consultants, and vendors up to speed on what the company is looking to achieve with its adoption and use of IS resources, the time spent discussing and writing the documents is what cements mutual understanding for the individuals involved. The planning process offers a number of advantages, discussed next.

Plans Enable Communication

Perhaps the most important outcome of the IS planning process is that it enables and supports intraorganizational communication. As the **planning team** members, composed of IS professionals as well as general and functional managers, assess current IS resources and set guidelines for their future use and management, a shared mental image of the role of each member of the team emerges. This communication is critical, since individuals typically have different expectations, speak different languages, and often have different objectives and priorities.

Consultants are sometimes brought into the planning team because they bring significant experience and knowledge of the planning process and they can serve as catalysts for discussion and facilitators of the communication process. Yet their firm-specific knowledge is limited, and it will be easy for the organization to dismiss a plan that is formulated by a consulting firm. Thus it is critical that consultants serve as members of the planning team rather than as delegates of the firm.

Plans Enable Unity of Purpose

Organizations achieve their best results when a clear strategy and clear goals have been identified and lead to concerted efforts from the organizational units and employees. The IS plan serves as a contract of sort, wherein the objectives of IS deployment are specified and clear responsibilities are agreed upon. When this happens, coordinating action and achieving a unity of purpose become simpler.

Plans Simplify Decision Making over Time

When a firm has not developed an IS plan, it has failed to create a context for decision making. Under these circumstances (all too common, unfortunately), the firm will find itself selecting projects to fund as part of the yearly budgeting process with little sense of overall direction and purpose. As a consequence, projects will be funded year to year in a haphazard fashion, resulting in an uncoordinated infrastructure and redundancy of systems and efforts and leading to a heightened risk of missing opportunities and wasting resources.

6.4 The Strategic Information Systems Planning Process

While **strategic IS planning** can be a lengthy and complex process, particularly for large organizations, its basic structure is fairly straightforward. It consists of gathering information about the current availability and performance of IS resources. It also involves a series of decisions, increasingly specific, designed to provide a road map for decision making about information systems. The strategic IS planning process typically evolves in five phases:

1. *Strategic business planning*. A strategic business plan consists of an organization's mission and future direction, performance targets, and strategy. **Strategic plans** are a prerequisite to IS planning. Note, however, that strategic planning is itself informed by available IT and current IS trends. As discussed in Chapter 2, digital technology is a critical enabler for modern firms, often determining the strategic opportunities available to them.
2. ***Information systems assessment***. An IS assessment consists of taking stock of the firm's current IS resources and evaluating how well they are fulfilling the needs of the organization.
3. ***Information systems vision***. An IS vision consists of a concise statement that captures what the planning team believes should be the role of IS resources in the firm. It provides an articulation of the ideal state the firm should strive for in its use and management of resources.

4. ***Information systems guidelines***. IS guidelines represent a set of statements, or maxims, specifying how the firm should use its technical and organizational IS resources.
5. ***Strategic initiatives***. Strategic initiatives are long-term (three- to five-year) proposals that identify new systems and new projects or new directions for the IS organization.

Note that while we are presenting the strategic planning process in a sequential manner, this is a simplification of reality, and the process is really an iterative one (Figure 6.4). In other words, downstream analyses may lead to a reevaluation and change in decisions made earlier. We now will discuss each step in greater detail.

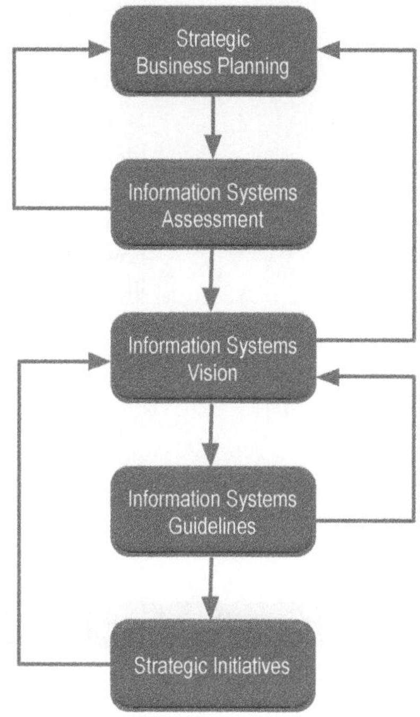

Figure 6.4. The iterative strategic IS planning process

Know Who You Are: Strategic Business Planning

Information systems are enablers of business strategy and operations. They allow a firm to achieve its stated goals while also creating opportunities for new strategic directions and initiatives that new technologies make possible. Thus effective IS planning can only occur in concert with business planning. In other words, unless the planning team has developed a clear understanding of the firm and what makes it successful as well as a deep understanding of the business strategy and its future goals and objectives, planning for the use and management of IS resources is an exercise in futility. IS planning without business planning is equivalent to trying to decide what car you should rent for your upcoming vacation before you have decided where you will go, with how many friends, and at what time of the year.

Know Where You Start: Information Systems Assessment

Once the planning team has a clear grasp on the strategic direction the firm intends to pursue, more research is needed. The team needs to perform an information systems resource assessment that includes taking an inventory of the IS resources the firm is currently using and critically evaluating them in terms of how well they are meeting the business needs of the organization. The planning team should assess the firm's current use of, and satisfaction with, these resources. The objective is to understand what resources are available and whether they are currently satisfying organizational objectives. Note that we refer here to an information systems assessment, not an information technology assessment, as including technical resources, data and information resources, and human resources:

- *Technical resources* are composed of hardware, software, and networking components that make up the firm's digital technology infrastructure. Inventorying of these resources can be done by examining documents such as IT schematics and speaking with selected IS professionals.
- *Data and information resources* are composed of databases and other information repositories. An assessment of these resources can be done by examining documents, such as database structure and data schemas, and by speaking with informants, including technical personnel and the customers of the data resource.
- *Human resources* are composed of IS professionals (those individuals who are responsible for creating and managing the IT resources) and the user community (including general and functional managers as well as end users). An assessment of these resources requires an examination of individuals and their skills, attitudes, and preconceptions as well as an examination of reporting structures and incentive systems. This can be done by examining documents, such as the firm's organization

chart, and speaking with informants from the various hierarchical levels of the IS function and the business. (See Table 6.1 for some suggested questions to ask at these three stages.)

During the assessment stage, the planning team reviews company documents and public literature and interviews key informants. The documents analyzed and individuals interviewed depend on the size and structure of the organization. Note that obtaining the needed information requires skilled questioning. Often it is not enough to ask informants a direct question (e.g., how is the IS function performing?); they may not be willing to share the information, or more likely, they may be unable to answer a question posed in this way. But skilled questioning and probing from a number of different angles usually will uncover the needed information (e.g., what IS services do you need that are lacking today?).

The output of the assessment stage should be a snapshot, using both text and graphics, of the current "state of IS resources" in the organization. A well-developed assessment document should clearly convey to the reader what IS resources are currently available and how well they are serving the needs of the organization. It should also inherently suggest potential areas of concern.

Table 6.1. Suggested questions for . . .

Technical resources	Human resources	Data and information resources
What hardware comprises the organization's IT infrastructure?	How many full-time IS professionals are currently employed by the organization?	What data are currently collected?
What platforms are currently in use?	How is the IS function organized?	Where and how are the data collected?
What is the current application portfolio?	Who does the head of the IS organization report to?	Where are the data stored? In what format?
Are there any redundant systems?	What role does the IS function serve?	Are data shared across applications? How?
What networking infrastructure is currently in place?	What is the current skill set of the in-house IS professionals?	What applications access the data?
Does the IS organization provide any shared services to the business?	What is the IS sophistication level of the end users and user-managers?	Who owns the data (e.g., the IS organizations, local departments)?
Is the IT infrastructure centralized or decentralized?	What are the industry performance benchmarks?	Who is in charge of maintaining the accuracy of the data?
What systems are on-site, and what systems are off-site?	How does the organization compare against industry benchmarks?	Who is in charge of ensuring the security and backup of the data?
Are any components of the IT infrastructure outsourced?	Who are the leading competitors?	What rules are followed to determine data ownership and responsibilities?
What is the age of the current application portfolio?	What performance levels have leading competitors attained?	
How do the existing applications relate to one another as a system?	What are the user-managers' opinions of the current IT infrastructure and application portfolio?	
How are applications normally obtained (in-house development, acquisition)?	What are the users' perceptions of the current IT infrastructure and applications?	
Who owns the IT infrastructure?		

Know Where You Want to Go: Information Systems Vision

With a clear understanding of the organization's business strategy, an inventory of the current resources available, and a solid assessment of their present performance, the planning team begins its real work—looking forward. The first step at this point is to spell out the role that information systems should play in the organization. In some organizations, IS operations and technology resources are critical to the firm's survival and success. For other firms, IS operations are not so critical to their survival and continued success.

Consider the case of the 2017 Amazon Web Services (AWS) S3 outage, one of the most expensive digital service outages of all times. In February, a mistyped command from an Amazon employee working on the cloud-storage billing system generated a cascading failure of S3 services. The resulting four-hour outage of the popular cloud platform cost $150 million to S&P 500 companies and $160 million to U.S. financial service firms using the infrastructure.[1] Obviously, flawless IT operations at Amazon AWS are a must, as it powers the technology infrastructure of large swaths of the modern digital economy!

Contrast the above case with that of Morton's Restaurant Group Inc., the world's largest owner and operator of company-owned upscale steakhouse restaurants, with upscale steakhouses in the North American, Singapore, and Hong Kong markets. The company has several applications, ranging from unit-level point-of-sale to corporate procurement and financial systems. Disruptions to the performance of these applications, even if protracted, do not endanger the viability of the organization. Imagine, for example, a four-hour outage of the network or of the point-of-sale systems. Morton units can revert to a manual system and continue with operations. While this is certainly not a scenario that Morton's executives would like to experience, the impacts are much less severe than those that Amazon AWS and its clients experienced. To get a rough estimate of IT downtime costs, market research firm Gartner reported that, on average, a network outage costs $5,600 per minute, or roughly $300,000 per hour.[2] And this cost will continue to rise in the foreseeable future as organizations increasingly invest in digital transformation.[3]

With more organizations relying on information systems enabled by digital technology and cloud computing for running their operations, protracted disruptions to the firm's IT infrastructure are certainly going to create problems. However, the impact of these disruptions can vary dramatically from organization to organization.

Aside from the impact on day-to-day operations, information systems play a more strategic role in some firms and a more tactical one in others. Some organizations' success in the marketplace depends heavily on their ability to introduce digital innovations and manage information systems strategically. Google and Apple come easily to mind, as do UPS and FedEx.

For these firms, information systems play a strategic role, and the organization must constantly look for applications of IS that enable it to be more competitive. For others, typically those organizations in more mature and less digital industries, being cutting edge is less important. In fact, in some firms, information systems are nothing more than a "necessary evil"—a resource the company needs to have and use but not one that will provide a leg up on the competition. As software continues to "eat the world" (Chapter 1), the number of firms in this last category is shrinking.

---

1 Stevens, L. 2017, March 2. "Amazon finds the cause of its AWS outage: A typo." *Wall Street Journal*, retrieved from https://www.wsj.com/articles/amazon-finds-the-cause-of-its-aws-outage-a-typo-1488490506.

2 Lerner, A. 2014, July 16. "The cost of network downtime." Gartner, retrieved from http://blogs.gartner.com/andrew-lerner/2014/07/16/the-cost-of-downtime/.

3 ITIC. 2020. "ITIC 2020 global server reliability report." Information Technology Intelligence Consulting, retrieved from https://www.lenovo.com/us/en/resources/data-center-solutions/analyst-reports/itic-2020-global-server-reliability-report/.

SIDEBAR 6.1. Information Systems Vision of a Cruise Line

Sidebars 6.1 to 6.4 refer to the same company, a large cruise line operator with multinational operations. The examples are adapted from the actual firm's information systems planning document from 2000. We refer to this company as the "Large Cruise Line."

The IS function will assume more of a leadership role within the corporation. While the IS function will continue to service the organization by providing a solid IT infrastructure and supporting and maintaining existing systems, the IS function will no longer be considered exclusively a support arm to the rest of the organization.

In order to maintain our leadership position, we must use information to do the following:

- Set the customer service standard in the industry for consumers and business partners by using customized and personalized information, beginning with the first contact and continuing throughout the relationship.
- Enable the company to be the employer of choice in our industry by empowering a workforce with accurate, timely information and thus accelerating change and innovative decision making.
- Assume a leadership role as innovators in the use of the Internet as an enabling technology that drives business growth, internal and external communications, operating efficiencies, and new sources of revenue.

Information Systems Vision Whether information systems are crucial to the firm's success or merely useful, whether they are strategic or a necessary evil, it is important that the planning team is able to clearly articulate what the role of IS *should be*. We refer to this statement as the information systems vision (see Sidebar 6.1).

The IS vision is a concise statement that captures what the planning team believes should be the role of IS resources in the organization. It provides an articulation of the ideal state the firm should strive for in its use and management of IS resources.

The IS vision must be aligned with and reflect the firm's business strategy and, as a consequence, will be unique and highly specific to your company. While the industry your firm competes in, the product or service it offers, and the competitive environment will have some influence on the role that information systems should play in your organization, the position and role of the IS function should ultimately depend on a conscious decision by senior management and the planning team. Thus companies that compete in the same industry, even head-to-head competitors, will have different IS visions that reflect their strategic posture.

For example, while the Ritz-Carlton and W Hotels compete in the luxury segment of the lodging industry, the former positions itself to offer traditional luxury, while the latter has a much more edgy image, catering to a younger and more tech-savvy customer base. We can therefore expect the two hotel chains to have developed very different IS visions. You can also imagine how the IS vision of a digital resource creator (Chapter 5) like Stripe reads. Cofounder and president John Collison captured the strategic role of digital technology in a 2017 interview: "I don't see any other payment companies thinking about payments as an application program interface (API) problem, as a software problem, or an infrastructure problem. We look much more like Amazon Web Services than any traditional payment company."[1]

---

1 We introduce a variety of examples throughout the chapter to demonstrate the practical applicability of the concepts we advance. For clarity, we also provide a running illustration focused on Stripe. Stripe is a San Francisco–based company incorporated in 2009 and funded by major venture capital firms (e.g., Sequoia, a16z). Stripe Payments, the firm's first and most successful product, is an online payment processor that affords the capability of making and receiving online payments to any organization seeking to transact online.

Deciding what the role of information systems in your organization should be, and developing a concise IS vision that encapsulates it, poses a difficult task. Two analytical tools that have been developed to help managers involved in this process are the **critical success factors** (CSFs) methodology[1] and the **strategic impact grid.**[2]

Critical Success Factors A technique that has been used over the years to help focus managers' attention to the firm's information needs is the critical success factors (CSFs) methodology. Critical success factors are defined as the limited number of areas, typically three to six, that executives must effectively manage to ensure that the firm will survive and thrive. CSFs represent those fundamental things that "must go right" for the business to flourish. At the risk of oversimplifying things, the CSFs methodology has the merit of focusing attention on fundamental issues and of helping ensure that the planning team is able to prioritize. Note that the CSFs methodology asks that managers focus not on information systems but on business objectives—that is, the CSFs methodology asks that you identify what the firm must do right (not what the IS department must do right) to ensure the ongoing success of the organization. With the CSFs identified, it becomes easier to think about the role of IS in achieving them.

Let's return to the AWS example. Given that AWS's revenue stream and reputation are fully dependent on their cloud services and digital resources being operational, and given the significant disruptions that follow protracted downtime, one of AWS's CSFs is likely to ensure the optimal performance (i.e., reliability and speed) of online store operations. Other CSFs we could imagine for AWS are as follows:

- Continue to grow the breadth of digital resources offered on the platform.
- Increase the customer base through aggressive pricing that enables the firm to maintain and grow its economies of scale.

The Strategic Impact Grid Another analytical framework that helps define the role of information systems in a specific company is the strategic impact grid. The main advantage offered by the strategic impact grid is its ability to enable simultaneous evaluation of the firm's current and future IS needs. This is achieved by plotting the firm on the following two dimensions: the current need for reliable information systems and the future need for new information system functionalities.

Current Need for Reliable Information Systems This dimension focuses on current day-to-day operations and the functionalities of the existing systems. Not all organizations, even fierce head-to-head competitors, need the same information systems and the same degree of reliability of their technology infrastructure. The planning team should achieve some consensus about where the firm falls on this dimension by determining the following:

- Is there a risk of a tangible loss of business if one or more systems fail for a minute or more?
- Are there serious negative consequences associated with even a small degrading response time of one or more systems?
- Are most core business activities online, and do they require real-time or near-real-time information processing?

1 Rockart, J. 1979, March–April. "Chief executives define their own data needs." *Harvard Business Review 57*: 81–93.

2 Nolan, R., and McFarlan, F. W. 2005, October. "Information technology and the board of directors." *Harvard Business Review 83*(10): 96–106.

- Will even repeated service interruptions of up to 12 hours, while troublesome, cause no serious consequences for the viability of the business?
- Can the company quickly revert to manual operations for the majority of transaction types when systems failure occurs? While unwelcome, do such disruptions not endanger the business as a viable concern?

Future Needs for New Information System Functionalities This dimension is forward looking and is concerned with the strategic role that new digital technology capabilities play for the organization. While the industry the firm competes in has some bearing on this question, different organizations pursuing different strategies will fall on different locations of the spectrum. The planning team should achieve some consensus about where the firm falls on this dimension by determining the following:

- Do new systems and new functionalities of existing systems promise major process and service improvements?
- Do new systems or new functionalities of existing systems promise major cost reductions and efficiency improvements?
- Do new systems or new functionalities of existing systems promise to close (or create!) major gaps in service, cost, or process performance with competitors?
- Is IS work mostly maintenance of current state-of-the-art functionalities? Does the firm foresee no major new systems that are crucial to business success within the current planning horizon?
- Do new systems promise little strategic differentiation, and do customers not expect any major new functionalities or services?

At the intersection of these two dimensions, we find four possible roles that information systems can play in the organization (Figure 6.5). A common error by managers is to use the strategic impact grid to separately map the firm's current and future position (i.e., where the firm is today and where it should be). The strategic impact grid simultaneously captures current operations and future impact. Therefore, if used correctly, it will show where the planning team thinks the firm falls and, as a consequence, what the use of IS resources should be going forward.

Support Quadrant The organization falls in the support quadrant when information systems are not mission critical for current operations and this state of affairs is not likely to change in the foreseeable future. Many building contractors fall into this quadrant, as does the Morton's Restaurant Group Inc., discussed above.

When a firm finds itself in the support quadrant, it should view information systems as a tool to support and enable operations, but one that offers little potential to significantly harm or benefit the organization. As a consequence, firms in this quadrant are typically cost conscious and conservative in their IS investments decision making, with the head of the IS function typically reporting to the CFO.

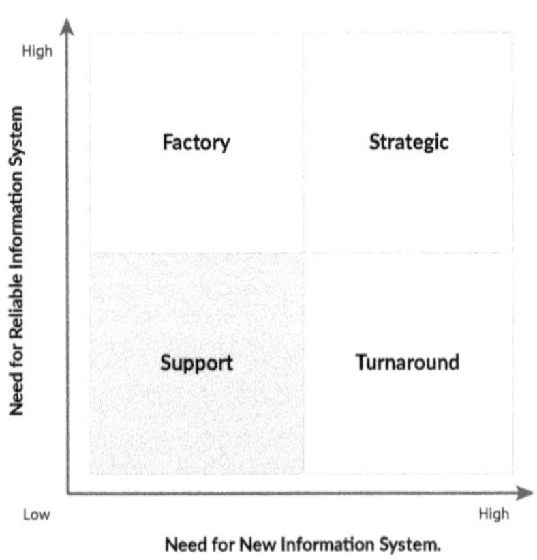

Figure 6.5. The strategic impact grid

Factory Quadrant The organization falls into the factory quadrant when it currently has a technology infrastructure that enables the business to operate with the needed degree of efficiency and effectiveness. Disruptions to this infrastructure, even small ones, can endanger the firm's well-being and future viability. Yet within the planning horizon under consideration, the firm appears to be in a stable state and the planning team foresees a limited potential for new systems and functionalities to make a substantial contribution. Like a factory working steadily, if the current state of IS affairs is maintained, the firm will be in good shape. NASDAQ, the company that runs the largest U.S. electronic stock market, is an example of a firm that must ensure flawless operation of its current systems. Airlines and large chemical plant operators represent other examples.

When a firm finds itself in the factory quadrant, it must closely monitor its current systems and must be willing to fund their maintenance and upgrade. Yet because of the minor future potential for the impact of new systems, the organization may take a conservative stance toward future investments.

Turnaround Quadrant The organization falls into the turnaround quadrant when information systems are not considered mission critical for current operations. Yet unlike firms in the support quadrant, the planning team believes that the current state of affairs is due for a change in the near future, and new information systems or new functionalities of existing systems will be critical for the business's future viability and continued (or expected) success. As the term *turnaround* suggests, the firm is (or should be) readying to change its IS posture. The recent emergence of food delivery companies, such as Uber Eats, Deliveroo, and DoorDash, have forced restaurants to react. Many chains are in the midst of significant digital transformation efforts. They are infusing digital technology into their operations, evaluating how to work effectively with aggregators (e.g., Deliveroo) and/or white-label delivery providers (e.g., as Stuart), and introducing their own proprietary smartphone apps. These firms are clearly in the turnaround quadrant, moving from a historical position where digital technology was not critical to their success to a world where their relationship with customers is increasingly digital. When a firm finds itself in the turnaround quadrant, it typically needs to engage in some reorganization (e.g., by reevaluating its organizational structure) and engage in third-order change (Chapter 2). When in the turnaround quadrant, the firm will also need to take an aggressive stance with respect to digital technology investments and the acquisition of necessary skills.

Strategic Quadrant The organization falls into the strategic quadrant when information systems are critical to the firm's current operations and the planning team foresees new information systems or new functionalities of existing systems to be critical for the future viability and prosperity of the business. In other words, outstanding IS operations and a relentless attention to IS innovation are a must for companies in this quadrant. For pure play organizations (Chapter 5) as well as the most recognized names in eCommerce, such as Amazon.com and eBay, survival depends on flawless IS operations. These organizations are constantly on the lookout for new digital technology and new information systems that can provide them with a competitive edge. Large banks find themselves perennially on this list as well, under attack from fintech start-ups and with blockchain (Chapter 12) as the latest wave of disruption keeping them on their toes.

When a firm is in the strategic quadrant, it must be very proactive with respect to information systems and digital technology investments. This is typically done by having a CIO with a strong voice on the executive team. For these organizations, information systems are part of the firm's DNA. For example, Amazon defines itself as a technology company that happens to be in the retail business rather than a retail company that uses technology.

Know How You Are Going to Get There: Information Systems Guidelines

While the IS vision articulates the destination—the ideal state the firm should strive for when it comes to using and managing IS resources—it provides little guidance as to how the firm should deploy its resources to achieve this goal. Thus the next stage in the IS planning process consists of identifying a parsimonious set of guidelines that, if followed, will enable the firm to achieve its information vision. This set of guidelines, sometimes referred to as the information systems architecture, is prescriptive in nature—identifying the guiding principles that the firm will follow when using and managing IS resources.

Why Develop Information Systems Guidelines? The building of a custom home offers a good metaphor for understanding the role of IS guidelines. While you have a vision for what you'd like your dream home to be like—including location, style, size, number, and types of rooms—you probably know very little about construction codes, permits, materials, and the like. Thus you engage an architect to whom you describe your vision and rely on her to formalize it in the form of blueprints and schematics (Figure 6.6).

When the architect delivers the drawings and floor plans, you will be able to see if she captured your vision and suggest any changes. A good architect will also make suggestions about what changes will make the house better, reduce cost, speed up construction, and so on. Once you sign off on the blueprints, they become binding—a sort of contract. Armed with them, you or your architect on your behalf will engage a builder who will actually create the house.

The blueprints enable the contractor to develop the house according to your vision, without having to interact much with you—or even knowing what your overall vision was. Contractors will know where they should locate the rooms, their sizes, and where to install plumbing, windows, doorframes, and electrical outlets. They will also know what materials they should use. If you change your mind during construction and request changes, you will be responsible for the cost. On the other hand, if the builder makes a mistake and forgets to create a planned window opening in the master bedroom, he or she will be responsible for the cost of the modifications. The blueprint serves as the guideline for what should happen and for dispute resolution.

The process and purpose of creating IS guidelines are very similar to those described above in the context of custom home building. The firm executives have a vision for what they want the business to achieve and how they want it to operate. Working in partnership with IS professionals, as part of the planning team, they will refine their vision and establish a blueprint that will enable communication, establish responsibility, and guide future decision making.

Communication The primary objective of the information systems guidelines is to simplify tactical and operational decision making and to ensure that future decisions are aligned with the IS vision. By establishing a coherent set of rules stemming from the IS vision, these guidelines ensure that future information systems and technology decisions are made in accordance with the overall objectives of information systems use and management in the firm rather than haphazardly and in an uncoordinated manner.

Imagine, for example, an organization that opts for a strictly supporting role of information systems. Such a company will likely strive to buy low-cost IT products (e.g., refurbished or older hardware) and will be conservative when it comes to digital innovation (e.g., waiting to buy applications until they have become a competitive necessity). Conversely, an organization in the strategic quadrant of the strategic impact grid that has identified flawless personalization of the customer experience as a critical success factor may decide to move to a centralized architecture with centralized servers that gather all customer data in one location for easy retrieval and analysis.

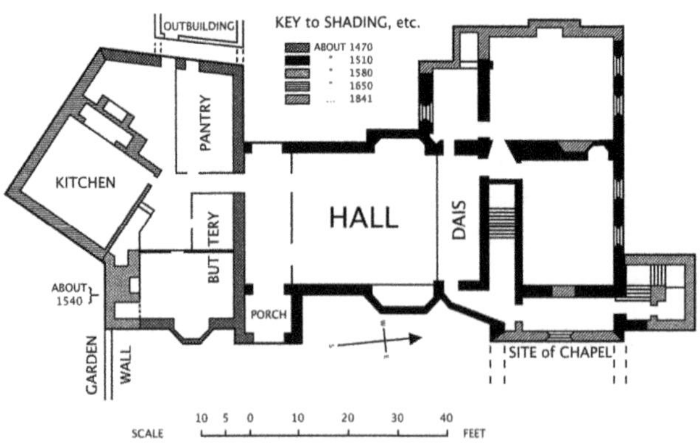

Figure 6.6. The blueprint of the floor plan of a custom house
Source: Photo by The Man in Question / CC BY-SA 3.0

Identify Responsibilities The IS guidelines also set expectations for behavior, serving a similar binding purpose as policy rules or the custom blueprints and schematics described above. Decisions that are made in accordance with the IS guidelines are in line with expectations and will typically be deemed appropriate. Decisions that are made outside of the guidelines are treated as exceptions and will typically need substantial justification. If the firm finds itself regularly making exceptions, it will need to reevaluate the quality and accuracy of the IS guidelines.

Long-Range Decision Support Because the firm will not engage in the IS planning process every year, the IS guidelines must be general enough to provide direction over a number of years. Yet it is crucial that they be actionable. Thus they need to be specific enough to clearly spell out what the firm should do and, as a consequence, what it should not do when it comes to the deployment of IS resources.

Imagine that you just got an internship with the Large Cruise Line (discussed previously). You are eager to make a contribution, and after reading the company's IS vision (see Sidebar 6.1), you remember your long-lost cousin Vinnie, who just launched a start-up. The brochure of his flagship product reads, "Our personalization digital solution enables your company to track customer preferences and offer outstanding service." A lightbulb comes on, and you run to your CIO and suggest, "We should buy and implement this digital innovation; it is perfectly aligned with our vision!" She ponders your suggestion for all of five seconds and then denies your request, adding, "I'm glad you read the planning document, but you seem to have stopped reading too early. The second technical guideline [see Sidebar 6.2] rules out relationships with new and not established vendors for core systems like those housing our customer data."

Technical and Organizational Guidelines IS guidelines address every aspect of IS decision making, both technical and organizational. While technical guidelines and organizational guidelines are deeply intertwined, it helps to separate them out during the planning process and in the planning documents.

Technical Information Systems Guidelines The IS guidelines that focus on the technical components of the firm's information systems must address future decisions pertaining to the hardware and software infrastructure, networking services, and storage and protection of organizational data and information. (See Sidebar 6.2 for an example.)

Technical guidelines will not typically specify the vendors to be used or particular platforms or applications. Instead, they are broad enunciations of direction for the technical components

SIDEBAR 6.2. Technical Information Systems Guidelines of the Large Cruise Line

1. The movement toward standardization will evolve over time, and we will remain flexible in our approach. However, our major objective is to achieve centralized and standardized products that are more easily managed and controlled, especially across multiple continents with limited staff to maintain them.
2. We will follow the trends of dominant vendors and be guided by these leaders rather than the niche market players for both hardware and software core systems.
3. We will buy software packages (rather than develop custom code) that provide generic solutions for business functions that are not part of our core competency and not part of what constitutes our competitive advantage.
4. We will not obtain monolithic packages that drive significant duplicate code and data.
5. We will store data centrally for all mission-critical applications.
6. Mission-critical systems will have complete fallback solutions to redundant systems to minimize the impact of any disaster.

of the infrastructure. As the example in Sidebar 6.2 shows, the statements produced by the planning team of the cruise line are aligned with the firm's IS vision (see Sidebar 6.1) and are both general, making them relevant and useful for years to come, and precise, thus specifying what decisions are legitimate and what decisions should not be made.

Organizational Information Systems Guidelines The IS guidelines that focus on the organizational components of the firm's information systems must address those decisions that pertain to human resources, the organization of the IS function, reporting and hierarchical structures, and the like. (See Sidebar 6.3 for an example.)

These statements focus on IT governance issues. IT governance pertains to the relationship between the IS function and the other departments in the organization. It specifies who is responsible for proposing and sponsoring application development and how maintenance and new purchases should be evaluated (Chapter 10). The statements also pertain to outsourcing and vendor relationships, human resources decisions (e.g., what type of individuals the IS function is looking to hire, the type of IS skills that the firm deems necessary), and the like.

Know How Well Equipped You Are to Get There: Information Systems SWOT

Having defined the IS vision and the broad maxims to achieve it—the IS guidelines—the planning team must now review how well equipped the firm is to achieve their vision in accordance with the stated guidelines. This step is the last piece of analysis before the team develops an action plan and proposes tangible initiatives. It consists of a strengths, weaknesses, opportunities, and threats (SWOT) analysis focused on the firm's current IS resources and capabilities.

The iterative nature of the planning process becomes clear at this stage. As the planning team members evaluate how well positioned the firm is to attain the IS vision, they may realize that they are attempting to do too much and that the vision, as stated, is not achievable given the current set of strengths and weaknesses and the landscape of opportunities and threats. In this case, the IS vision should be revised to be more realistic. Failing to do so will create an unattainable, not actionable, vision that will make people cynical and defeat the purpose of the planning process.

As shown in the example (see Sidebar 6.4), this stage of the analysis is designed to reveal the internal factors that can be exploited to achieve the vision as well as highlight the internal weaknesses that must be carefully managed. It also enables an externally focused analysis that

SIDEBAR 6.3. Organizational Information Systems Guidelines of the Large Cruise Line

- We will focus our expenditures on projects of strategic value and long-term importance over short-term fixes that deviate from our overall strategy.
- Outsourcing will be considered for IS operations and legacy applications where possible and feasible.
- Business-supported projects will be governed by the business case and will be evaluated by the full project costs and values in terms of people, process, and technology.
- Business-supported projects will require the participation of the business throughout the engagement.
- While the IS function will be developing systems, at all times, our mind-set will be that of a business professional first and will always consider the business opportunity and/or impact of systems that we develop or purchase.
- The IS function will create a mixed environment of both seasoned professionals and new, eager, recent graduates. The persona of our IS function will be that of a level-headed, technologically excited individual.
- We will strive to avoid silos of data and silos of skill sets within our company and thus enable our staff to grow and to minimize disruption when specialized staff are moved to other assignments and/or leave the company.

seeks to uncover new technologies and trends that generate opportunities for the firm as well as threats that may undermine the ability of the firm to achieve its IS vision.

In a well-developed plan, this section of the analysis is consistent with the previous ones and forms the basis for the next section. In other words, having read the SWOT analysis and given the proposed vision and the guidelines, it should become clear what the firm needs to do during the current planning cycle.

From Planning to Action: Proposed Strategic Initiatives

After so much discussion and analysis, it is time to move to action. The last component of the strategic IS plan is the identification of strategic initiatives. Strategic initiatives are long-term (three- to five-year) proposals that identify new systems and new projects (e.g., supply chain management) or new directions for the IS organization (e.g., create a CIO position reporting directly to the CEO). These initiatives need not be precisely articulated, but they do need to identify a set of future avenues for exploitation of the IS resources. They also must be tightly aligned with the IS vision and the proposed role of IS in the organization. For example, it would not be appropriate for the planning team to propose a change to the organizational structure, seeking to establish a new CIO position who reports to the CEO, after having decided that information systems play a support role in the organization and having crafted a defensive IS vision.

Managerial Takeaways

There is one clear message that emerges from this chapter. If you are a manager joining the workforce today, as digital technology is revolutionizing every industry, you must roll up your sleeves and join the table where decisions about digital technology in your firm are being made. The six decisions we discussed at the beginning of the chapter should convince you that digital technology decisions are business decisions. The six questions are based on the premise that general and functional managers need to be involved in the decision making that affects their organization's investment in, and use of, information systems and digital technology resources.

The strategic IS planning process is the table that you need to join. But in order to be an asset on the planning team, you must understand the planning process, its purpose, and the type of decisions to be made as it unfolds. Information systems are complex organizational systems

SIDEBAR 6.4. Information Systems SWOT at the Large Cruise Line

Strengths:

- The IS staff is competent in the implementation and maintenance of new technology.
- User-managers, on average, understand IS concepts and have a good relationship with the IS function.

Weaknesses:

- There are currently four nonintegrated systems housing customer data.
- The current IT infrastructure supports a silo, function-centric approach and does not support flexible, timely response to customer needs.

Opportunities:

- New technology, such as the XML standard and data warehousing applications, is now robust enough to warrant migration to integrated data repositories.
- No competitor is currently offering IS-enabled integrated customer solutions.

Threats:

- Our competitor is moving swiftly to establish itself as the customer service leader in our industry through the deployment of integrated, channel-independent customer service systems.
- Our preferred suppliers have developed the capability for electronic data communication, but we are currently unable to connect to their systems. This inability to communicate hampers our efficiency and may drive suppliers to the competition.

that exist at the intersection of business and technology. Setting a direction for their use and management requires a blend of skills, technical and organizational, that are rarely housed in one organizational function or department. The planning process must occur as a partnership among those with technical skills, the IS group, and general and functional managers. This chapter was devoted to helping you become a productive partner to your more technical colleagues. It is within the guidelines of the IS plan that the firm devises future strategic initiatives. A number of frameworks and techniques have been developed to support their identification and analysis. We will discuss them in the remaining chapters of Part III.

Summary

This chapter provides the basis for the ensuing chapters and describes the strategic IS planning process. Specifically, in this chapter, we discussed the goals of the strategic information system planning process and its components, with a focus on the role played by general and functional managers.

- Strategic IS planning is the process by which the firm, by way of the planning team, develops a shared understanding of the role of IS resources use in the organization.
- General and functional managers play a crucial role on the planning team, despite the fact that they typically lack technical knowledge. Their role is to help identify the firm's strategy and, in light of that business strategy, to help decide how IS resources should be used to achieve it.
- General and functional managers should also take the lead in answering questions, such as how much money should be spent on IT, to what business processes these funds should be directed, what IT capabilities should pervade the organization, what levels of IT service should be achieved, what degree of IT risk the firm will accept, and who is responsible for IT initiatives.
- As critical members of the planning team, general and functional managers will help craft the firm's IS vision and guidelines. The IS vision provides an articulation of the ideal state of IS resource use, while the guidelines offer a context for decision making.
- With the basic planning mechanisms in place, the firm moves to action and identifies strategic initiatives to be implemented in order to achieve the stated IS vision. These strategic initiatives often stem from what the organization believes are available opportunities as well as weaknesses that must be managed.

Study Questions

1. Why should general and functional managers be involved in information systems planning decisions despite their lack of technical expertise?
2. Jeanne Ross and Peter Weill of MIT's Center for Information Systems Research (CISR) suggest that senior managers be involved in six IS management decisions. What are these decisions? What is the guiding principle behind this need for senior executives' involvement?
3. What is the purpose of strategic IS planning? Who needs to be involved in this process? Why?
4. What are the key components of the strategic IS planning process? Can you define and describe each one?
5. What purpose do the CSFs methodology and the strategic impact grid play in the planning process? Can you provide examples of firms in each of the four quadrants of the strategic impact grid?
6. What is the purpose of the IS vision? Can you provide an example?
7. What is the purpose of IS guidelines? Given the IS vision you have proposed in response to Question 6, can you provide an example of guidelines that are aligned with it?

Glossary

A glossary of key terms can be found at the book's website. Visit https://www.prospectpressvt.com/textbooks/piccoli-information-systems-for-managers-5-0, then click on "Student Resources" in the horizontal red menu bar.

Chapter 7

Value Creation and Strategic Information Systems

What You Will Learn in This Chapter

This chapter focuses on the strategic role of information systems (IS) and the digital technology that enables them. The definitions, analytical frameworks, examples, and exercises in this chapter will help you develop a knowledge base that allows you to confidently identify and evaluate the added **value creation** potential of information technology (**IT**)**-dependent strategic initiatives.** In this chapter, you will become well versed in the language of **added value** analysis and **strategic information systems.**

MINICASE: Consulting for the Al Dente Restaurant Chain

Al Dente is a London-based small chain of four restaurants focused on fresh pasta dishes. The concept is based on quality of food and the mystique of its fresh pasta lab and kitchen. While housed in the basement of each restaurant, the pasta lab and kitchen are open to customers who want a tour to understand what makes fresh pasta so good. The concept has been very successful since its inception, seemingly hitting the target with a sophisticated clientele willing to pay a premium for authentic, fresh Italian pasta. Located in densely populated, affluent, residential London neighborhoods, each Al Dente restaurant offers dine-in options, with a very popular wine bar concept, but does a lot of takeout business. While it has yet to experiment with delivery, worried about the quality of the dishes when they get to destination, Al Dente's head chef was nonetheless experimenting with packaging to find the right solution, ensuring the preservation of Al Dente's food quality upon delivery.

The restaurant chain is in need of upgrading its Kitchen Display System (KDS) from Oracle, which is at the end of its useful life. The KDS is the unsung hero of back-of-the-house restaurant operations. The system delivers critical functionalities, such as meal coursing, delayed routing, and cook times, which ensure accurate order management. As orders are received from the point of sale (POS), the KDS routes them and displays them to the kitchen staff so as to ensure that all items are prepped and cooked for simultaneous delivery and optimal quality. The Al Dente KDS is at the end of its five-year useful life. After five years, Oracle stops guaranteeing support for legacy applications and requires customers to upgrade hardware and software to the current version. While some restaurants take the risk of running obsolete, unsupported software to save money, Al Dente is a dynamic, growing chain focused on quality food and service and has already decided to upgrade to the Oracle MICROS Express Station 400 (Figure 7.1) and sign a new five-year agreement. The Al Dente setup for each restaurant is a dedicated Windows machine based on commodity hardware, with two overhead 24-inch antiglare displays with an integrated soundbar, sealed against humidity, grease, and other airborne contaminants. The hardware/software bundle costs Al Dente $1,000 per restaurant and requires signing the Enterprise License, totaling $200 a year per restaurant.

WizTech, a California-based high-tech firm specializing in cloud solutions, has recently contacted Al Dente's general manager. This contact is very timely, given the imminent upgrade to Oracle MICROS Express Station 400. WizTech is beginning to commercialize a cloud-native KDS solution that delivers the exact same functionalities as the leading KDS. It also uses the same displays. However, being cloud native, it does not require any other hardware, including the dedicated Windows machine. Moreover, to ensure a degree of business continuity, WizTech offers, free of charge, a "work offline service" that enables temporary operations in case of unavailable Internet access; this service is secure and managed seamlessly by WizTech until the connection is back in operation. Thus the WizTech solution is a *perfect substitute* for the Oracle KDS. WizTech's premium solution has a subscription price of $500 per year for four restaurants. The hardware, setup, and installation for all four Al Dente restaurants costs WizTech $1,500 to produce and install.

Discussion Questions

1. What should Al Dente's general manager do?
2. Does WizTech enjoy a competitive advantage (or disadvantage) in this market?
3. Can you quantify such an advantage (or disadvantage)?

7.1 Introduction

Arguably a manager's greatest contribution to their organization is to help the firm attain and maintain a competitive advantage. Intuitively, a firm has a **competitive advantage** when it is able to outperform competitors thanks to some unique characteristic of its operations or its offerings.[1] More precisely, the primary role of managers in business organizations is to contribute to the *creation* and *appropriation* of economic value by their firm. But what does it mean to create value? Why would an organization want to engage in value creation? And perhaps most importantly, how can you ensure that your organization benefits from its value creation strategies and initiatives?

7.2 The Analysis of Added Value

Added value is one of those terms that we all too often hear being tossed around as a slogan in presentations and press releases to convey the vague idea that a firm is doing something worthy of attention. Consider the following:

- From the website of a state information technology agency: "We will work hard to make sure that everything we do is 'value added' for our customers and achieves [the agency's] vision."
- From the title of an article published by the *Washington Times*: "Wi-Fi going from an added value to an expected amenity."
- From Nike's vice president (VP) of digital innovation: "We have access to content that cannot be found elsewhere, and we bring this unique added value to our customers with this jersey."

Do you know what it means to "create value" and to "have added value," beyond buzzwords and slogans? Can you carry out a disciplined analysis or compute a number for the value added by an initiative you have envisioned? What would this number mean? What decisions could you make based on this analysis?

The analysis of added value is a formal mechanism that managers and analysts use to answer these questions and evaluate how much of the value created by the firm can be appropriated in the form of profits.[2] While the analysis of added value can be applied to any firm's initiative, we will constrain our focus to those projects that leverage digital technology at their core. This analysis is an essential step in the decision of whether you should go ahead with the initiative. It stands to reason that if the proposed initiative creates no tangible value, you should shelve it. More insidiously, if the

1 Later in this chapter, we will revisit this definition using the more precise concepts introduced by the analysis of added value.

2 The term *value appropriation*, sometimes expressed as *value capture*, denotes the ability of an organization or an individual (e.g., customer, employee) to capture some of the value created. For a firm, such appropriation of value ultimately results in profits.

initiative does contribute to the creation of value but your firm will be unable to appropriate such value created, then you should also not go on with it, in most cases.

This type of analysis is useful not only when you are innovating—in other words, when you are endeavoring to create value in novel ways and offering things that no competitor is currently offering—but also when you are evaluating how to respond to a competitor who took the leadership position. The analysis of added value can help you measure how much benefit your competitor is drawing from the innovation and what benefits are likely to accrue for you if you choose to replicate the initiative. While in many cases, you will only have limited information and you will not be able to create precise estimations for value created and added value, a disciplined analysis will nonetheless help you carefully analyze the potential of the initiative.

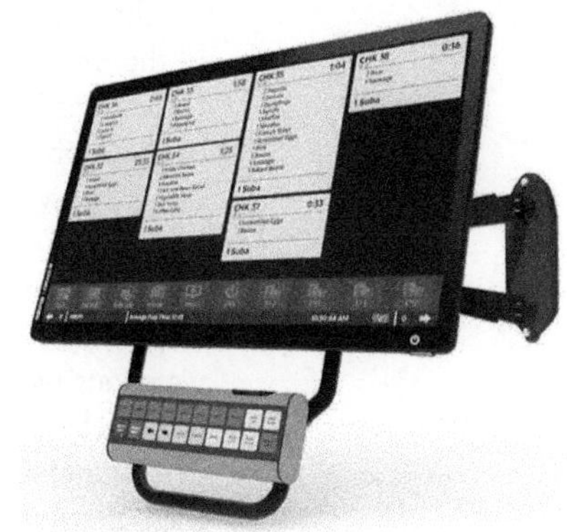

Figure 7.1. The Oracle MICROS Express Station 400

Source: Oracle, 2020, "Kitchen Display Systems," retrieved from https://www.oracle.com/dk/industries/food-beverage/restaurant-pos-systems/kitchen-display-systems/

The Definition of Value

Economic value is generated when *worthwhile things* that did not exist before are created. Thus value is generated not when something novel is done but only when this "something novel" is deemed worthwhile by someone else. As attentive observers have long recognized, this is the crucial difference between inventors and entrepreneurs. Inventors are those individuals who create new products and new technologies—in short, new things. These new technologies or products can be amazing, technically flawless, and beautifully engineered, but they will not create value until they fit an opportunity in some market. Entrepreneurs know this full well and focus on *market opportunities* and the development of *solutions* to customer problems or desires—that is, entrepreneurs look for new ways to create value rather than new technologies or new products. Often the novel solution being marketed relies on a new technology, but it is the invention that serves the market opportunity, not the other way around. As Amazon chief executive officer (CEO) Jeff Bezos put it, "Our job is to invent new options that nobody's ever thought of before and see if customers like them."[1] While the solution offered may not be technically superior or beautifully engineered, it will be commercially appealing precisely because it does contribute to the creation of economic value. Identifying these opportunities is certainly not simple, and tools like the business model canvas (Chapter 5) are designed to help perform a disciplined analysis of value creation opportunities.

The Firm's Transformation Process Economic value is created through a **transformation process** when some input resources that have a value of $\$x$ in their next best utilization are transformed into outputs for which customers are willing to pay $\$x + \$v$. When such a transformation process takes place, we conclude that value, in the amount of $\$v$, has been created. In other words, this new value was not there before and would not come to be unless the transformation process did occur, since the input resources would remain untransformed and simply maintain their original worth of $\$x$. Typically, a firm that enacts the transformation process seeks to monetize (at least) some of this value created in the form of profits.

Input resources are traditionally represented by any factor of production, such as raw materials, labor, equity and debt capital, managerial talent, support services like transportation and storage, and so on. In other words, anything that is used to generate the product or service and then market it, sell it, and support it is to be considered an input resource. For modern digital startups, the input resources are then specialized knowledge and skills in domains including software development, system design, digital relationships development, and the like.

---

1 Robischon, N. 2017, February 13. "Why Amazon is the world's most innovative company of 2017." Fast Company, retrieved from https://www.fastcompany.com/3067455/why-amazon-is-the-worlds-most-innovative-company-of-2017.

The output of the transformation process is the product and/or service that the firm engaging in the transformation process is seeking to sell and that a customer is interested in acquiring. In the case of Facebook, for example, the digital innovation that catapulted the company to the top of the mobile advertiser list was the infinite scrollable feed of news and friends' posts. The feed design, now ubiquitous in mobile app design, enabled Facebook to fill customers' idle minutes of boredom. Customers used it, thereby demonstrating that value was created, and advertisers paid to reach those users. A more traditional example is the physical retail industry. As a large retailer, Tesco (Figure 7.2) uses input resources such as labor (over 400,000 employees), physical stores (over 7,000), warehousing facilities, trucks and transportation equipment, energy sources (e.g., diesel fuel for the trucks, electricity for the stores), equity and debt capital, and so on. By employing these resources in its transformation process, which consists of acquiring products in bulk, warehousing them, and then distributing them to conveniently located stores, the firm is able to offer something that its customers are willing to pay for: convenient access to a large selection of mainstream products.

Defining the Components of Value Created

A formal analysis of added value requires some key definitions:

- ***Supplier opportunity cost*** *(SOC)*. SOC is the minimum amount of money suppliers are willing to accept to provide the firm with the needed resources.
- ***Firm cost*** *(FC)*. FC is the actual amount of money the firm disbursed to acquire the resources needed to create its product or service.
- ***Customer willingness to pay*** *(CWP)*. CWP is the maximum amount of money the firm's customers are ready to spend in order to obtain the firm's product.
- ***Total value created*** *(TVC)*. The TVC in the transaction is computed as the difference between CWP and SOC (TVC = CWP - SOC).

While FC is an accounting amount, resulting from the total of all costs incurred to create its offering, SOC and CWP determine TVC. We discuss them further below.

Figure 7.2. Entrance of a Tesco store in 1949

Source: Photo by Peter McGowan / CC BY 2.0

Supplier Opportunity Cost SOC, the lower bound of value creation, is an important figure, as it represents the value that the needed resources would have in their next best use. For this reason, it is defined as an opportunity cost. A rational supplier will only provide the firm with its services (e.g., labor, managerial talent, raw materials) if it receives at least the same sum of money it would have received from any other buyer (i.e., another firm seeking to use the resource the supplier offers).

Note, however, that suppliers will typically not be paid an amount equal to their SOC. This is because the firm acquiring the resources will not be able to precisely estimate this number—in fact, in most cases, the suppliers themselves may not have a precise estimate available—and the suppliers will happily accept any offer exceeding their opportunity cost.

For an example, think back to the latest job offer that you received and accepted. During the interview, you, the supplier of labor to the hiring firm, formulated some idea regarding your willingness to work for the company and how much you'd want to get paid. When the offer came through, it most likely exceeded this minimum requirement you had established, and you took the job . . . without returning the "excess salary" you received.

The simple example above addresses a very important issue. SOC is a theoretical minimum; the actual amount of money the firm will disburse to acquire the needed resources (i.e., the FC) is the outcome of a negotiation effort in the presence of an asymmetry of information between the negotiating parties. We will see that this important point resurfaces later when we discuss price considerations.

Customer Willingness to Pay The other end on the value continuum is represented by the CWP. As we noted above, an inventor is someone who generates a new idea and creates a new product or technology. An entrepreneur is someone who matches a novel product or service to a market opportunity. This difference should be clear now with the terminology of value creation. Unless some customer is willing to part ways with his money in order to acquire what the inventor has created and this amount is larger than the SOC, no value has been generated and no economically viable venture can develop. In other words, value is in the eyes of the customer. The most elegantly engineered and technically beautiful product is valueless unless a customer is willing to pay for it.

The history of IT products and services is littered with examples of innovations that, while perhaps were technically amazing, were met with cool customer response and dwindled into market oblivion as a consequence. Do you remember the Picturephone (Figure 7.3), the Apple Newton, Audrey (3Com's Internet appliance; Figure 7.4), the Philips CD-I, the Sony Glasstron (head-mounted displays for virtual reality), WebTV, the Amazon Phone (Figure 7.5), Webvan's online grocery service, and Snap Spectacles (Figure 7.6)? Unless you are a historian of sorts, the answer is most likely no.

Figure 7.3. The Picturephone: An innovation introduced 35 years before its time

Source: Photo by LabguysWorld.com / CC BY 3.0

Figure 7.4. 3Com's Audrey, the Internet appliance

Source: Wikimedia

Figure 7.5. The Amazon Fire Phone, unboxed

Source: Copyright © Romazur / Wikimedia Commons / CC BY-SA 4.0

Computing the Total Value Created

Simply defined, value is the difference between CWP and SOC—that is, value is created when resources that in their next best use would be worth a given amount are transformed into something that a customer is willing to pay more for.

Consider a simple fictitious example.[1] Your grandmother was famous for baking a great-tasting cake. In her will, she entrusts the secret recipe to you, so you decide to become an entrepreneur and start baking the specialty cake. You can bake a single cake using amounts of eggs, flour, sugar, and the secret grandma ingredients worth about $4.00 together. An hour of your time invested in making the cake is valued at $9.25, under the assumption that you only qualify for minimum-wage jobs (i.e., if you used the same amount of time to work for the best other job you qualify for, you would earn $9.25). Finally, you use electricity to bake the cake, and some gas and wear and tear on your car to deliver it, in the amount of $3.75. The local gourmet coffee shop, whose owner knew your grandma personally and who had tasted the cake, is willing to pay you as much as $25.00 for each one of your homemade cakes. This is because she thinks that she can get 12 slices out of every cake and sell each one for $3.00, thereby making a nice $11.00 profit.

This information allows us to precisely compute the TVC in this *cake-making and -selling* transaction. TVC = CWP - SOC (see Figure 7.7)—that is, TVC = $25.00 - ($4.00 + $9.25 + $3.75) = $8.00. Or, more formally, taking

1 While this example could be considered trivial, we encourage you to pay close attention, since it elucidates every aspect of the analysis of added value. The example is clearly fictitious in that it assumes perfect information by all parties involved.

resources valued at $17.00 in their next best use and producing a good valued at $25.00 by a customer, you have contributed to creating a value of $8.00 through a "cake-making transformation process."

Figure 7.6. The original Snap Spectacles

Source: Photo by Danny Robinson / CC BY 2.0, edited by the authors

Appropriating the Value Created

Up to this point, we have discussed the process of value creation—a process in which suppliers, the focal firm (i.e., you), and the customer partake. However, TVC only tells us if there is *an opportunity* to make a profit. That is, if the TVC is greater than zero, someone will benefit, but we still don't know who is going to capture the value created.

Value appropriation is the process by which the TVC in the transaction is split among all the entities who contributed to creating it (i.e., suppliers, the firm, and the customer). It is typically the outcome of a negotiation process between the firm and the suppliers to determine the FC and between the firm and the customer to establish product prices (see Figure 7.8). When a firm appropriates value, it does so in the form of higher profits. When customers appropriate value, they do so in the form of savings (i.e., paying less than what they would have been willing to pay).[1]

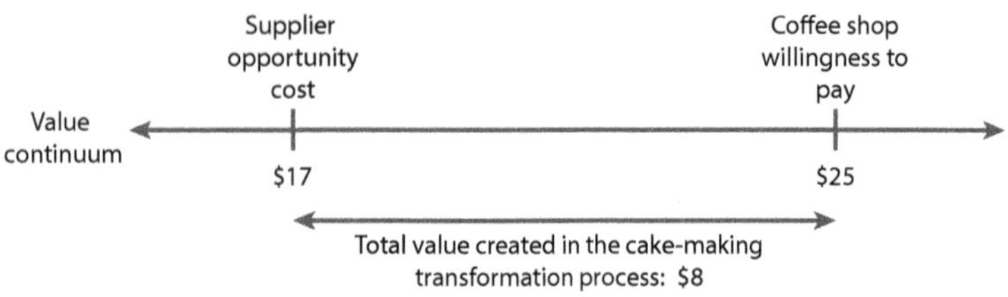

Figure 7.7. Total value created

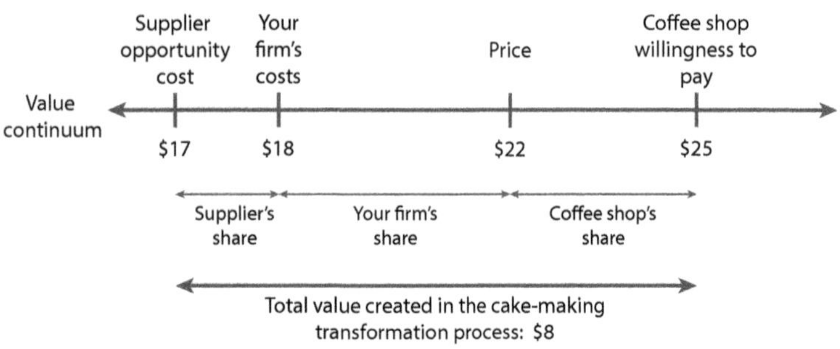

Figure 7.8. Value appropriation

Let's return to the example of your grandma-recipe cake-making venture. Note that at this point, we have said nothing about either your actual cost of making the cake (i.e., the FC) or the price you and the coffee shop have agreed upon for each cake you will deliver. In other words, at this point, we know how much total value has been created (a theoretical amount), but we know nothing about how this value is going to be realized and who is going to appropriate it.

Let's assume that while the store that provides you with the ingredients for your cake may be willing to drop the price to about $4.00 per cake because of your bulk buying, you are unaware of the store's SOC and do not negotiate hard. You simply pay the published price of $5.00 per cake in ingredients. Thus your FC equals ($5.00 + $9.25 + $3.75) = $18.00.[2]

Now since you do not know that the coffeehouse's actual CWP is $25.00 (they certainly are not going to tell you!), you accept the coffeehouse's first offer and do not negotiate hard . . . should you consider a negotiation class at this point? The price you agree on is $22.00 (Figure 7.8).

We now have all the information we need to compute value appropriation and determine how much of the TVC each of the entities involved will retain. Here is how the TVC, $8.00, would break down in this case:

---

1 As customers, we have all come across a "good deal" a few times in our lives. A good deal is nothing but a situation where the price we are asked to pay is lower than our CWP.

2 Note that in this simplistic example, you are playing two roles. On the one hand, you are a supplier of the labor resource. On the other hand, you are the entrepreneur—the firm—who will retain any profits the business is able to generate.

- The suppliers appropriate $1.00 in excess profits—that is, one dollar more than the minimum amount of money they would have been willing to accept to provide you with the needed resources (i.e., SOC).
- You appropriate $4.00 in excess profits—that is, four dollars more than the minimum amount of money for which you would have been willing to stay in this venture (i.e., your FC).
- The customer, the gourmet coffee shop, appropriates $3.00 in savings—that is, three dollars less than the maximum amount of money it would have been willing to disburse to acquire your specialty cake (i.e., their CWP).

The Definition of *Added Value*

With the terminology of value creation and value appropriation now in place, we are ready to discuss added value. A firm's added value is defined as *that portion of the TVC that would be lost if the firm did not take part in the exchange.* That is, the firm's added value is measured as that portion of the value created in the transaction involving the firm minus the total value that could be created if the firm did not exist. A firm's added value is therefore the unique portion of the TVC that is contributed by the firm itself. Added value depends on the effects of existing competition.

Recall the example of your cake-making venture. In that example, your firm was the only one in the market that could take the ingredients that were valued at $17.00 and, using grandma's secret recipe, transform them into a final product—this one-of-a-kind cake that customers would be willing to pay a maximum of $25.00 for. We established that the TVC in the transaction was $8.00.

What was your firm's added value then? It was the full $8.00. If you decide not to bake the cake, as the only person who has knowledge of the secret recipe (i.e., the unique transformation process), none of the potential value would be realized. That is, all $8.00 of the new value would be lost, as we would be left with the raw resources worth the original $17.00.

Added Value in a Competitive Market

Let's now assume that you were not the only one who had been given the secret recipe in your grandma's will. Much to your surprise, your estranged cousin Bettie also received the secret recipe, and you just found out that she is entering the cake-baking business in your area! In fact, she just contacted the gourmet coffee shop you have been in talks with and is attempting to undercut you. She has the exact same cost structure as you do and produces a cake that is no different in any respect than yours. In other words, you and Bettie produce two products that are perfect substitutes for one another.

Mapping this scenario to the added value model demonstrates that the TVC has not changed (see Figure 7.9), since both you and Bettie encounter the same SOC and produce a cake that the coffee shop owner would pay the same amount of money for.[1]

What has changed in this scenario is your firm's added value. In the first scenario, you were the only one who could create the cake. Now if your firm were to leave the exchange, Bettie's

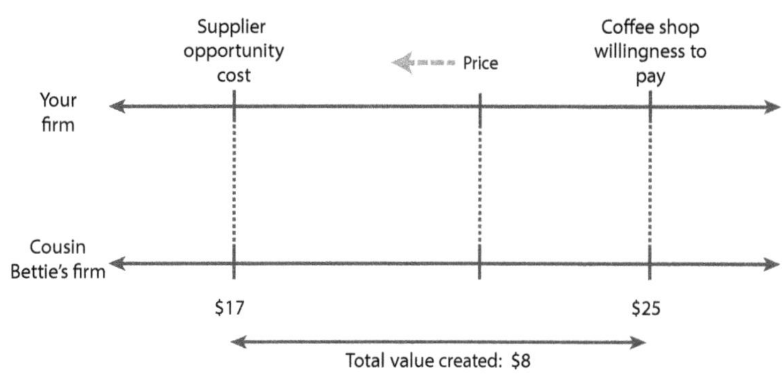

Figure 7.9. Total value created

1 This assumes that Bettie also has a minimum-wage job as the next best use of her time.

firm could step right in and, using the exact same resource, produce the exact same cake. You do *nothing unique* and, as a consequence, your added value is now $0.00.

Pricing Considerations

Again, while we don't need to know price to compute added value (a theoretical value), price becomes important to gauge what portion of the value created each entity partaking in the transaction can appropriate. In our first scenario, when you were the only person who knew the secret recipe for the cake, we assumed you'd agree to the first price the coffee shop owner proposed (Figure 7.8)—$22.00. It should be clear, though, that you were in a position of bargaining power and could have pushed the envelope. In fact, since we know the coffee shop owner's CWP, we can assert that you could have charged as much as $25.00.

Once your cousin Bettie enters the market and begins to offer her cake, a perfect substitute of yours, you are no longer in a position of bargaining power. We can therefore expect a price war to cause prices to drop as low as your FC since neither of you offers anything unique (your added value is zero). Under this circumstance, any price you quote that exceeds your FC would provide an incentive to the coffee shop and cousin Bettie's firm to strike a deal to cut you out. Cousin Bettie would quote a price lower than yours to get the business, and a lower price would be a better deal for the coffee shop. This would force you to underbid Bettie, and the process would only end when one of you is willing to quote a price equal to the FC. In this case, neither competitor will make any extra profit, while the customer, the coffee shop, will reap big savings.

The lesson is clear: no matter how much value your firm contributes to creating, unless you can be (at least in part) unique in your value creation, you will quickly compete this value away to customers. This is, of course, the essence of competition and a force that in your capacity as a manager you need to learn to manage.

The Relationship between Added Value and Competitive Advantage

The insight about pricing discussed above is critical, and it should clarify that a firm should focus on creating value, through innovation and the use of IT, not just for the sake of doing so but in order to be able to appropriate at least a portion of the TVC. In theory, the maximum amount of value that a firm can appropriate equals its added value. Thus the imperative when engaging in new strategic initiatives is for a firm to create *added value*. The firm's added value is also a measure of its competitive advantage because it measures the extent to which the firm is able to do something unique and valuable.

How Is Added Value Created?

Now imagine one last twist in the story of your cake-making venture. Imagine that, since you have some artistic talents, you are able to garnish the cake with some icing that makes it unique. The coffee shop owner thinks that she can sell each slice of your "personalized" cake at a premium, say $4.00, claiming that the cake is a "coffee shop exclusive." Thus she is willing to pay you $29.00 per cake.

Personalizing the cake takes you 12 minutes of extra labor ($1.85), and the value of the extra ingredients needed is 15 cents. In this case, your SOC has risen by $2.00. But the extra investment of resources leads to an increase in CWP to $4.00. Cousin Bettie has no artistic talents and is unable to personalize her cakes. Your firm's added value is now positive (Figure 7.10).

Perhaps a simpler way of analyzing this is to focus on the value of the different characteristics of your product. That is, with an incremental investment of $2.00, you increase the coffee shop's willingness to pay by $4.00 over what the competition can do, thus generating added value in

the amount of $2.00. Note that price will be once again determined as a consequence of a negotiation process, but you certainly now have a competitive advantage over Cousin Bettie's firm, and you can expect to appropriate as much as two dollars (e.g., charging the coffee shop no more than $20.00).

Two Ways to Create New Value

More important than the calculations produced above is the meaning of this analysis and the considerations that follow from it.

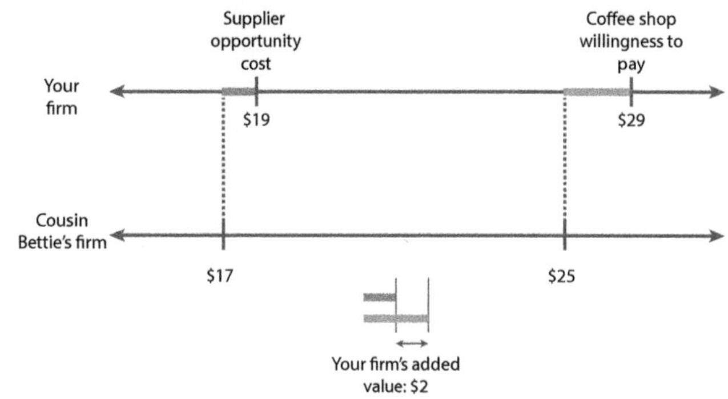

Figure 7.10. Added value under competition

Increasing Customer Willingness to Pay Added value is created by doing something of value for customers, thereby increasing CWP through an investment of resources that does not exceed CWP—otherwise we would be simply destroying value and wasting resources. Apple is a great example of value creation through increased willingness to pay. Steve Jobs, speaking at D in 2010, articulated his firm's success as focusing on making great products, pricing them aggressively, and going for volume.[1] Translated into the terminology of this chapter, this approach reads, "Our strategy is to drive CWP as high as we possibly can (i.e., great products) and then try to price low enough so that people perceive that they are appropriating significant value when they buy our products and therefore buy them in large volumes."

The most visible competitive battles take place on this end of the value continuum. The firms that have innovated use marketing to educate customers about their new products and services and drive up CWP (Figure 7.11). However, this focus on customer-facing initiatives can be misleading.

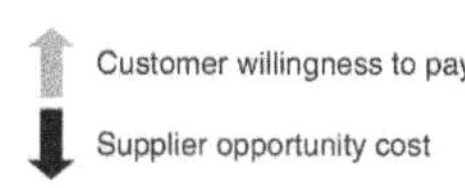

Figure 7.11. How value is created

Decreasing Supplier Opportunity Cost By definition, value can be created not only by increasing CWP over SOP but also by reducing SOC without a comparable reduction in CWP. In other words, there is great potential to create value by working with the firm's suppliers and creating incentives for them to supply you with needed resources for less money (Figure 7.11).

The prototypical example of this strategy is offered by Walmart, a firm that has made "everyday low prices" its slogan and the key driver behind most of its decisions. In the late 1980s, Walmart worked with one of its largest suppliers, Procter & Gamble (P&G), to find creative ways to lower its cost structure. Walmart pioneered a technique now called "continuous replenishment."

Continuous replenishment is based on the idea of "pulled logistics," whereby a supplier (e.g., P&G), not a customer (e.g., Walmart), is in charge of managing the customer's inventory. In the specific case of the Walmart/P&G partnership, leveraging its data network, Walmart provided P&G with real-time purchase data from each of its stores. That is, any time a box of Pampers or a carton of Tide detergent is sold at any Walmart store, P&G receives an electronic notification. This near-real-time knowledge of actual consumer demand enabled P&G to optimize its production schedules and its supply chain operations. As a consequence, P&G's own costs decreased, and P&G became willing to supply Walmart for less money than it charged other customers who could not engage in continuous replenishment. In other words, P&G's SOC for dealing with Walmart, and only Walmart, decreased thanks to this data-sharing initiative. On the other end of the value continuum, where Walmart serves the end consumer, no change occurred,

1 The All Things Digital Conference—or D for short—is an annual event for executives of the computer and software industries hosted by the *Wall Street Journal*.

as customers didn't need to know about continuous replenishment or see any changes in what Walmart offered, and Walmart still commanded the same CWP for its service.

Similarly, CareFusion, a global medical technology firm now part of Becton Dickinson, lowered supply chain partner costs by deploying a software as a service (SaaS) platform to foster collaboration. Thus they were able to gain greater supply chain efficiency with limited partners' specific investment. Lower supply chain costs emerged from better supply chain visibility and operations. For example, thanks to this platform, CareFusion partners could print product labels directly from the system. As all firms in the transaction use the same data and the same codes, CareFusion can process inbound deliveries of materials more efficiently. This approach brought process time down to one hour a day from the six it usually required. As CareFusion's vice president of operations commented,

> What suppliers get—and what we get—is complete transparency. Suppliers have the same information and data about our relationship as I have. They can see my inventory levels. They can see if I'm overstocked or about to run out. They can see their performance level against the delivery targets we set for them.[1]

Managerial Takeaways

The above examples and discussion of the analysis of added value suggest a few important concluding remarks of particular value to managers.

Value Is in the Eye of the Customer We define CWP as the maximum amount of money that a customer is willing to give the firm to acquire its products or services. Unless there is a market for the innovation—that is, customers who appreciate it—no value has been generated.

Customer Willingness to Pay Is Not the Same as Price Price is a consequence of a negotiation between the firm and the customer, depending on the available information and the degree of competition in the market. Note, however, that a firm that is able to create higher CWP does not need to charge higher prices to benefit. Imagine two firms vying for customer business and charging the same price. If one is able to command a slightly higher CWP—for example, by having a stronger brand—it will get the customer's business.[2]

Value Can Be Tangible or Intangible Our cake-making example suggests that the coffee shop is willing to buy your cake because they can resell it and turn a profit. We buy clothes because of the tangible outcomes they provide (e.g., warmth). However, we also buy them for looks, style, fashion, for the way they make us feel, to fit within a certain group, to support a particular manufacturer, and so on. All these "intangible" drivers of value are as important as the tangible ones. As much as possible, goodwill, brand effects, loyalty, and all other intangible drivers of value should be estimated and measured. Techniques such as focus groups and market research are typically used to do this.

---

1 Twentyman, J. 2015. "Cloud is new platform for supply chain partners to get together." *Financial Times*, retrieved from http://www.ft.com/intl/cms/s/0/c4c07ef8-f316-11e3-a3f8-00144feabdc0.html?siteedition=uk.

2 Note that in some industries where the cost structure involves significant fixed costs and small variable costs (e.g., the lodging or airline industries), attracting more customers, even if at the same price as the competition, can result in significant profitability gains. This is because any revenue from incremental business will mostly flow through to the bottom line.

Creation of Value Is Not Appropriation Beyond increasing CWP, there is much opportunity to create added value by focusing on suppliers' opportunity cost and by providing advantages to suppliers to work with the firm. Yet as a firm, it is crucial only to create value that you can appropriate (i.e., added value). Creating value can be done even in circumstances when this value can be appropriated by others (often customers). You must have added value (which is unique to your firm) to be sure to appropriate it.

Competitive Advantage and Added Value Are Closely Related While the framework for the analysis of added value is a simplification of reality, as it assumes perfect information of all entities involved and the absence of switching costs, it highlights what a firm must do to gain a competitive advantage. True competitive advantage is a function of added value (see Sidebar 7.1). Note, however, that the analysis of added value is focused on the short term and tells us nothing about the long-term resilience (e.g., resistance to erosion) of the firm's competitive advantage. An analysis of sustainability, the focus of Chapter 9, is the analytical tool needed to evaluate if any added value we create can be appropriated over time. As we learned in this chapter, it is not enough to be able to create value; we must be able to appropriate it. To do so, we must protect any added value we create from erosion by competitors.

7.3 Strategic Information Systems

Without a doubt, information systems and technology engender a plentitude of confusing lingo, technical terms, and acronyms—a problem compounded by the crowds of half-prepared, fast-talking individuals using terminology incorrectly. In this book, we were very careful in defining

SIDEBAR 7.1. How to Perform Added Value Analysis

Depending on the specific characteristics of the initiative under investigation, the analysis of added value can be more or less straightforward. You should try the following:

Clearly Define the Initiative and Understand What It Entails

The first step in the analysis requires that we are very clear about what the firm will do for customers or suppliers and what resources are necessary to create the product or perform the service being sold by identifying the intended value proposition.

Identify the Comparison

Because added value is defined in comparative terms, it is critical to identify a baseline comparison. This baseline can be the competitor's initiative or the firm's own offers—when the firm is innovating with products or services that improve on the current state of the art in the industry.

Estimate Customer Willingness to Pay

Estimating CWP can be a very complex process requiring substantial approximation and research. In order to simplify and focus this process, it helps to start by listing all the positive CWP drivers—defined as what the firm does as part of its offer to increase CWP. Note that any initiative has both positive and negative effects. That is, any initiative entails trade-offs; as the firm does some things of value for its customers, it also forgoes doing other things. It is therefore critical to also surface these negative CWP drivers and discount their effects.

Estimate Supplier Opportunity Cost

This analysis is similar to the one above and includes both positive and negative change. When the initiative's main contribution to value creation is on the SOC side, SOC must be used. When the main effect of the initiative is on CWP, then a simplifying assumption using FC as a proxy for SOC is acceptable.

Estimate Added Value

With the above information in hand, you can measure added value and begin to draw value appropriation considerations.

what IS is and in differentiating it from digital technology (Chapter 2). We defined an IS as a sociotechnical system that includes digital technology, processes, people, and organizational structure.

The distinction between digital technology and IS is a critical prerequisite to understanding the strategic potential of information systems and the role that information technology plays in the creation and appropriation of added value. This distinction also shows why the firm that focuses solely on digital technology investments to become competitive (i.e., blindly purchasing computer systems) is wasting its money. Digital technology investments are only appropriate within a larger IS design and only as components of information systems.

Definition: Strategic Information Systems

As you may intuitively expect, not all information systems that an organization seeks to design, develop, and use may be strategic. The foremost objective of strategy in for-profit business ventures is to gain a competitive advantage and sustain superior financial performance. To do so, the firm uses its resources to create value by either reducing SOC or increasing the customers' willingness to pay for its product and services, or both. A firm achieves a competitive advantage when it is able to generate added value by creating a unique and positive difference between CWP and SOC. At that point, the firm is in a position to appropriate, in the form of profits, the added value it has created. In short, competitive strategy can be defined as the art and science of *being unique*.

We define *strategic information systems* as those information systems used to support or shape the competitive strategy of an organization. More succinctly, with the terminology discussed in this chapter, we can define *strategic information systems* as those that enable the creation and appropriation of value.

Strategic or Not? Depends on the Purpose Strategic information systems are not defined by their functionality or the organizational function they support (Chapter 3) but are instead defined in terms of their objectives and the purpose they serve (e.g., improving the firm's competitive standing). Consider two well-established examples from American Airlines (AA):

1. *The Sabre reservation system*, typically considered to be the foremost example of a strategic information system, was originally created as an airline seat inventory system (i.e., a transaction processing system).
2. *The SMARTS system*, also considered a tool that enabled AA to gain a competitive advantage, was expressly designed to enable AA regional sales representatives to craft highly tailored incentive schemes for travel agents. SMARTS was an analytical tool (i.e., a decision support system).

No Need for Proprietary Digital Technology Contrary to conventional wisdom, strategic information systems do not have to rely on proprietary technology. eBay has provided the starkest example of this rule (Figure 7.12). eBay has dominated the U.S. online auction market since its inception using commonly available technology—namely, Internet technologies and the web.

A simple look at a competitor's auction site (Figure 7.13) shows remarkable similarities and comparable functionalities. eBay deploys little in the way of proprietary digital technology, and its technology has been duplicated by competitors. However, such replication of the technology is not enough, as it is the whole of eBay's initiative, enabling it to leverage network effects (Chapter 4), that underpins eBay's value-creating strategy and determines its added value. As we discuss more fully in Chapter 9, replicating the digital technology at the core of defendable strategic information systems is often a useless move.

Strategic versus Tactical Information Systems The definition of strategic information systems that we use is helpful in discriminating and identifying the many systems that are not strategic. These are systems that do not position the firm to create added value, even though they are important (often crucial) for the business's operations. We refer to these as ***tactical* information systems**. Consider the following examples:

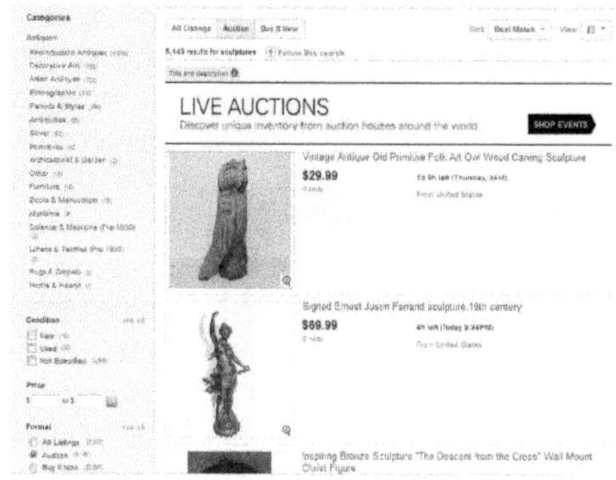

Figure 7.12. eBay's auction site

- All organizations that have salaried or hourly employees must pay them and maintain a complex set of records to compute tax withdrawals, accrued vacation time, sick leave, and so on. These software applications, typically called human resource management systems, are critical for the smooth operation of organizations. Can you imagine working for a company that consistently sent you a paycheck with errors? You'd be looking for another job fast! Human resource management systems are critical. Yet they are not strategic. They typically do not enable the creation of added value—rare is the firm that offers a unique value proposition based on its ability to correctly cut paychecks.
- A restaurant's primary information system, anchored by its POS software, is used to manage reservations, seating, order taking and delivery, and billing. Clearly, most modern restaurants could not operate as effectively without such a system. Yet POSs are generally tactical in nature, not strategic, because they rarely allow the restaurant to create unique value.
- Similarly, no matter how well it is run, an email system is unlikely to be the foundation of a strategic information system, and the same argument can be made for productivity software such as Microsoft Word and Excel, no matter how advanced their features may be.

Figure 7.13. eBid.net auction site

No modern organization could run without email and productivity software such as Microsoft Excel. Yet it is important to recognize that, as vital as they are, these systems are not strategic, and implementing or upgrading tactical systems will not create a competitive advantage.

IT-Dependent Strategic Initiatives

As future managers, you may often propose new initiatives that need information systems and digital technology to be enacted. You will also be called upon to help in the analysis, design, and development of strategic information systems early in your career. As a graduate of a management program, you will be paid for your analytical and decision-making abilities. This will likely include being involved with strategic information systems decisions.

In this capacity, you will be focusing on specific projects and initiatives. We use the notion of *IT-dependent strategic initiatives* in this book to refer to identifiable competitive moves and projects that enable the creation of added value and that rely heavily on the use of IT to be successfully implemented. IT-dependent strategic initiatives have three defining characteristics.

Initiative IT-dependent strategic initiatives consist of specific projects with clear boundaries that define what the initiative is designed to achieve as well as what it is and is not designed to do. For example, a freight shipper's package-tracking initiative has very clear boundaries. It

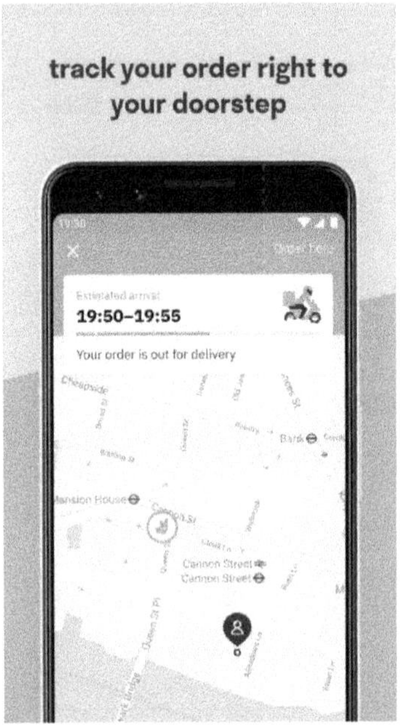

Figure 7.14. Deliveroo's app tracking an order in real time

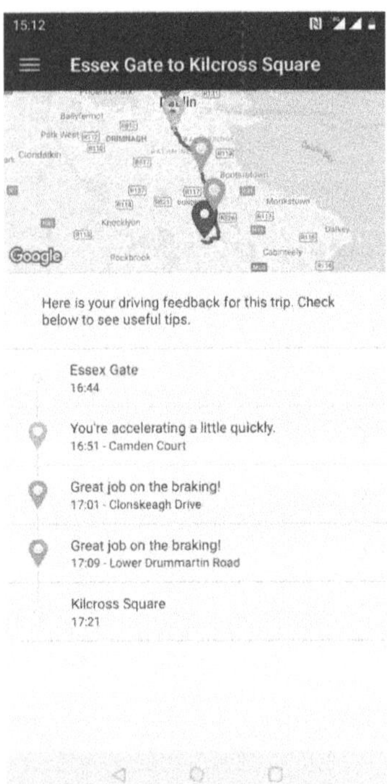

Figure 7.15. AXA's DriveSave Android application

is designed to allow customers to gain visibility with respect to the current location of their parcels by logging onto a website. Tracking tools for online transactions are now available in lots of industries—you can even track in real time your dinner as it is being prepared and delivered (Figure 7.14)!

Strategy The firm introduces IT-dependent strategic initiatives with the definite objective of producing a new value that the IT can appropriate. In other words, the firm seeks to create competitive advantage through the initiative. The freight shipper's package-tracking initiative, originally pioneered by FedEx in the mid-1990s, was primarily designed to improve customer service (and therefore CWP). Note, however, that as it happened, this initiative also had the potential to shift much of the current volume of tracking inquiries away from call-center operators and onto the web, thereby reducing FCs as well.

IT Dependency IT-dependent strategic initiatives cannot be feasibly created and executed without the use of digital technology at their core. Note that this IT core need not use cutting-edge or new breakthrough digital technologies. While the Internet and the web were relatively new to the shipping public in the mid-1990s, the technology at their core had been around for more than 20 years. FedEx's package-tracking initiative is clearly IT dependent, as it would not be feasible for the freight shipper to heavily promote a call-center-based package-tracking system due to the high costs of call-center operations versus online automated tracking systems.

Examples of IT-dependent strategic initiatives abound in modern businesses and include some of the most important and recent trends: cloud integration, supply chain management, Big Data and analytics, 3D printing, and Internet of things (IoT) initiatives. However, as we see in the FedEx example above, IT-dependent strategic initiatives can also have a much more limited scope, including projects such as introducing Internet cafés on cruise ships, insurance companies monitoring drivers' behaviors (e.g., AXA or Progressive; see Sidebar 7.2), or start-ups leveraging smartphones to promote a platform for health management (e.g., Propeller Health; see Sidebar 7.3).

IT-Dependent Strategic Initiatives, Not IT Investments Throughout this book, we have encouraged you to focus on information systems as a whole, not on digital technology investments in isolation. The notion of the IT-dependent strategic initiative keeps with this focus, and it is important to plan for the use and management of strategic information systems. Using this definition helps us shift attention away from investments in technology and recognize that digital technology investments can only pay off if they are part of a larger and cohesive information system design. It is the ability of the IT-dependent strategic initiative to create added value that we must focus on, not the uniqueness or innovativeness of the technology at its core. Thus IT-dependent strategic initiatives do not simply consist of building an app, a digital device, or a software program that, allegedly, generates competitive advantage until it is successfully replicated; rather, they consist of configuring an activity system, dependent on IT at its core, that fosters the creation and appropriation of economic value.

SIDEBAR 7.2. AXA's Insurance DriveSave

Initiative

DriveSave allows AXA car insurance customers to receive discounts on their car insurance premiums. The application (Figure 7.15) monitors and informs users on driving characteristics, including acceleration, speed, distance, turning, and stopping force. An overall driving behavior score ranging between 1 and 100 is then calculated. Scores greater than 70 allow customers to benefit from additional discounts. The initiative targets price-savvy, technologically prone, safe young drivers.

Strategy

The initiative has the objective to increase AXA's ability to correctly assess customers' profiles and associated risks. By increasing the visibility of customers' driving behavior, insurers can reduce their costs and transform these savings into competitive prices.

IT Dependency

On the customer side, the core technological components are embedded in smartphones. The initiative relies on AXA's IT capabilities for the development, management, maintenance, and integration of the new system.

SIDEBAR 7.3. The Propeller

Initiative

The Propeller (https://www.propellerhealth.com/) is a sensor for inhalers capable of recording use time, frequency, and location (Figure 7.16). The collected data can be used to better understand the severity and triggers of asthma and chronic obstructive pulmonary disease (COPD). The target customers are people affected by these pathologies desiring to keep a complete record of their respiratory health for improving their well-being.

Strategy

For patients, value is created in the form of a convenient, unobtrusive, precise, and complete record of Propeller uses. Additional value is created through the reduced hassle of managing asthma, personalized feedback, information, and medication reminders.

IT Dependency

The initiative is based on Bluetooth communication between the inhaler sensor and the patient's smartphone. The development of the application requires particular attention to interface design and data visualization to foster patients' access and understanding of information.

Summary

This chapter provides the background for understanding strategic information systems decisions by discussing fundamental concepts and analytical frameworks. Specifically, in this chapter, we introduced the notion of value creation and appropriation.

- Economic value is created when some input resources that have a value of $\$x$ in their next best utilization are transformed into outputs for which customers are willing to pay $\$x + \$v$.
- The value thus created is partitioned among those entities involved in its creation—a firm, its suppliers, and its customers—a process known as *value appropriation*.
- A firm is able to appropriate that portion of the TVC that would be lost if the firm did not partake in the exchange—a figure we termed *added value*.
- Strategic information systems are those that are designed and developed to create and appropriate value. They differ from tactical information systems, which are those systems that, while are often critical for the firm's operations, do not enable the creation of distinctive value.
- IT-dependent strategic initiatives consist of identifiable competitive moves and projects that enable the creation

Figure 7.16. Propeller Health iPhone application

of added value and rely heavily on the use of IT to be successfully implemented. They should be the focus of the search for new value creation.

- Managers, because of their understanding of the firm's processes and customer needs, take center stage in the identification and analysis of opportunities to create value with information systems. When doing so, they should focus on IT-dependent strategic initiatives rather than take a narrow focus on IT investments.

This chapter laid the foundation for the analysis of value creation and appropriation of IT-dependent strategic initiatives. In the next chapter, we examine frameworks and analytical tools designed to help you envision and take advantage of opportunities to deploy IT-dependent strategic initiatives.

Study Questions

1. Explain the value creation process. How does a firm contribute to the creation of economic value?
2. Provide two examples of firms that you think have been able to create value using information systems. The first one should be a firm that has done so mainly by focusing on CWP. The second one should be a firm that has done so mainly by focusing on SOC.
3. Think about the last time you bought something that you felt was "a great deal." Why did you think the product or service was such a great deal? Do you believe that the transaction was considered "great" by the firm from which you acquired the product or service? Why or why not? Explain using the framework of value creation and appropriation.
4. What is the difference between value creation and value appropriation? Why is this difference important?
5. Provide an example of a well-known firm that you think currently has added value. Explain your example using added value analysis.
6. Define the concept of strategic information systems and provide an example.
7. Define the concept of tactical information systems and provide an example.
8. Define the concept of IT-dependent strategic initiatives and provide an example.

Glossary

A glossary of key terms can be found at the book's website. Visit https://www.prospectpressvt.com/textbooks/piccoli-information-systems-for-managers-5-0, then click on "Student Resources" in the horizontal red menu bar.

Chapter 8

Digital Value Creation

What You Will Learn in This Chapter

In the previous chapter, we laid the foundations for our discussion of strategic information systems (IS) and information technology (IT)-dependent strategic initiatives by ensuring that you have clarity on key vocabulary (e.g., added value) and analyses (e.g., how value is created). In this chapter, we introduce analytical models that managers use to identify opportunities to create value with digital technology and to design and develop value-adding IT-dependent strategic initiatives.

MINICASE: Digital innovation at Domino's

On Thursday, October 8, 2020, Stu Levy, chief financial officer (CFO) and executive vice president of Domino's, the international pizza-delivery chain, was announcing their 38th consecutive positive quarter.[1] The tone was very optimistic. Domino's was outperforming competitors and growing its market share. There were many reasons for this success, but Rich Allison, chief executive officer (CEO), president, and director of Domino's Pizza Inc., was clear. The "catalysts" of their continued positive performance were their business strength and brand equity. He firmly believed that Domino's digital innovation was key to evolving the brand through a revolutionary customer experience. The results seemed to prove him right: 75% of domestic U.S. sales were generated via digital channels, and more than two-thirds of those were through mobile devices.[2]

The role of online ordering and innovation had a long tradition at Domino's, beginning with the web-based Pizza Tracker in 2008.[3] The company introduced its iPhone app in June 2011. The app automatically detected the Domino's location closest to the client (Figure 8.2) and, once logged in, consumers could select from either the standard menu or an available list of specials from the chosen location. They could also request a custom order by selecting pie size, crust type, toppings (Figure 8.3), and any other possible personalization (e.g., extra sauce). Frequent customers could store previous orders and customized pies in the app, making reordering quick and effortless. One of the advantages of the iPhone ordering application was the ability to avoid calling the restaurant. The app enabled customers to place orders, even during store closing times, and schedule delivery or pickup for a later convenient time.

Upon completing their order, customers would pay and then could track the progress of their pizza up to the delivery stage, through the preparation, baking, and quality assurance steps (Figure 8.4). The app was continuously improved and was also made available on Android. In October 2014, Domino's included an additional voice ordering feature. Similar to Apple's Siri, Microsoft's Cortana, or Google Now, "Dom" took orders, suggested additions, and guided customers through the purchasing process (Figure 8.5). As Dennis Maloney, head of multimedia

1 Seeking Alpha. 2020, October 8. "Domino's Pizza, Inc. (DPZ) CEO Ritch Allison on Q3 2020 results—earnings call transcript." Seeking Alpha, retrieved from https://seekingalpha.com/article/4378244-dominos-pizza-inc-dpz-ceo-ritch-allison-on-q3-2020-results-earnings-call-transcript?part=single.
2 Motley Fool. 2020, June 14. "Domino's Pizza Inc (DPZ) Q2 2020 earnings call transcript." Motley Fool, retrieved from https://www.fool.com/earnings/call-transcripts/2020/07/16/dominos-pizza-inc-dpz-q2-2020-earnings-call-transc.aspx.
3 https://hospitalitytech.com/pizza-tracker-tech-allows-dominos-customers-track-orders-store-door.

Figure 8.1. Domino's AnyWare digital channels

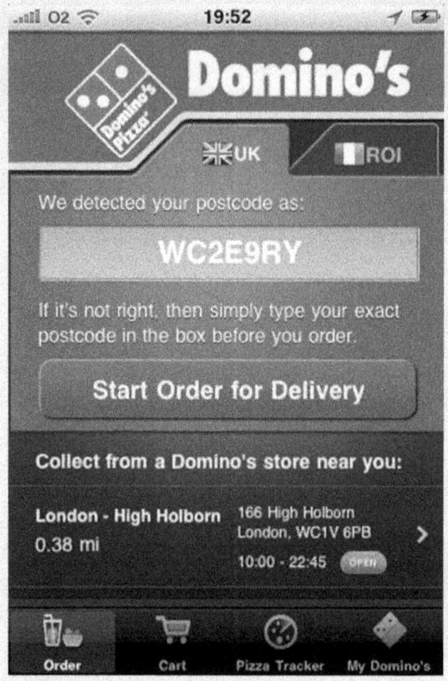

Figure 8.2. Domino's iPhone app welcome screen

Figure 8.3. Customers have playful ways to choose their toppings

Figure 8.4. Domino's order tracking feature

Figure 8.5. Domino's new interface and "Dom" integration

Source: Image courtesy of Domino's

marketing, said, "He's fun, but very focused on the pizza ordering experience."[1]

In 2015, the firm introduced ordering via Twitter, where customers could do so via text by sending a pizza slice emoji, and in February 2016, Domino's extended ordering and tracking capabilities to the Apple Watch.[2] Their strategy was clear: deepen the digital relationship with pizza lovers and continuously improve every aspect of the customer experience through relentless digital innovation. Or as executives put it, "We will continue to invest to grow our digital lead. We are just as committed to keeping our lead as we were to achieve it."[3] As of 2020, Domino's pizza lovers could place their order through any major digital channel, from WhatsApp to Google Home (Figure 8.1).

One area where the pizza delivery giant had taken a conservative approach was food delivery aggregators. In fact, the firm had shunned all the major aggregators of demand, from Deliveroo in Europe to DoorDash and Uber Eats in the United States. Allison explained, "When a customer orders from Domino's, I want a Domino's uniformed driver to show up and deliver that pizza . . . It helps us to control customer experience and also the quality of our product."[4] However, the competitive pressure from delivery aggregators forced Domino's to lower its long-term sales growth forecast.[5] Yet as of 2020, Domino's continued its commitment to independent delivery.

Discussion Questions

1. Do you believe that Domino's multichannel digital ordering is an example of an IT-dependent strategic initiative? Explain.
2. Do you believe that the multichannel digital ordering initiative has the potential to create added value? Substantiate your answer.
3. What would you do next if you were put in charge of the initiative?
4. How do you judge Domino's decision not to join food delivery aggregators?

1 Waters, R. 2015. "Artificial intelligence: A virtual assistant for life." *Financial Times*, retrieved from http://www.ft.com/intl/cms/s/0/4f2f97ea-b8ec-11e4-b8e6-00144feab7de.html.

2 https://www.eater.com/2016/2/17/11039872/dominos-apple-watch-app.

3 Seeking Alpha. 2020, October 8. "Domino's Pizza, Inc. (DPZ) CEO Ritch Allison on Q3 2020 results—earnings call transcript." Seeking Alpha, retrieved from https://seekingalpha.com/article/4378244-dominos-pizza-inc-dpz-ceo-ritch-allison-on-q3-2020-results-earnings-call-transcript?part=single.

4 Lucas, A. 2019. "Domino's won't be switching to UberEats anytime soon." CNBC, retrieved from https://www.cnbc.com/2019/01/15/dominos-wont-be-switching-to-ubereats-anytime-soon.html.

5 https://www.cnbc.com/2019/10/08/dominos-pizza-ceo-says-uncertainty-around-the-fate-of-third-party-delivery-led-to-slashed-outlook.html.

8.1 Introduction

As you know, the strategic information systems planning process is designed to create an overall context for IS decision making (Chapter 6). The planning documents typically end with the identification of strategic initiatives the firm should focus on. As the Internet, followed by smartphones and the digital technologies that leverage the pervasive information infrastructure, continues to revolutionize the competitive environment (Chapter 4), organizations must engage in digital transformation and innovation (Chapter 5). In this chapter, we introduce the frameworks and analytical models that IS professionals, as well as general and functional managers, use to contribute to value creation efforts in their firms (Chapter 7). Taking advantage of these frameworks requires an analytical mind-set, lots of discipline, and a good dose of creativity.

Digital technology has been a "competitive weapon" for organizations since the 1980s.[6] Early efforts modeled firms as **value chains** that transform inputs into outputs through a sequential

6 Ives, B., and Learmonth, G. P. 1984. "The information system as a competitive weapon." *Communications of the ACM* *27*(12): 1193–201.

set of value-adding activities (Chapter 7). Those organizations are embedded in a value network with other firms, and a unidirectional flow of activities typically occurs: suppliers selling to a firm, which in turn sells downstream to a distribution channel or a consumer. This is still the structure of many large organizations that provide substantial value to their customers, from Pizza Hut, to Delta Air Lines, to ExxonMobil. The first part of this chapter focuses on established frameworks to help you create value in these types of firms.

Leveraging the relentless digitization of activities and the rise to prominence of the Internet aggregators, digital platforms and digital resource creators have emerged (Chapter 4). They operate differently—they orchestrate the activity of suppliers and customers and don't focus on creating finished products. Rather, they enable the suppliers who leverage their resources to do so. Aggregators and platforms are increasingly dominating the competitive environment. When a firm has the chance to own a digital platform, it is imperative for it to seize the opportunity. Some of the largest firms in the world are aggregators and digital platform owners. Many fast-rising digital entrepreneurial ventures, from Shopify to Snowflake, embrace platform strategies. Moreover, as aggregators and digital platforms concentrate economic power, organizations find themselves forced to compete within their ecosystem. They simply cannot afford to not join. An apt example is provided by the restaurant industry. The 2020 COVID-19 pandemic dramatically accelerated a trend already underway, that of food delivery and ghost kitchens.[1] Restaurants have had to figure out how to compete for visibility and orders inside aggregators such as Deliveroo and DoorDash. The most innovative among them have begun experimenting with "virtual restaurants." Ordermark, for example, leveraged its online ordering platform capabilities to monitor popular items on the major U.S. food delivery platforms and created virtual restaurants selling those items. For food preparation, it relies on a network of prequalified restaurants that have excess capacity in their kitchen.[2] Virtual Dining Concepts (VDC), another example of clever digital entrepreneurship at work, uses the same strategy.[3]

The world in which you will manage increasingly includes both kinds of organizations, and sometimes the same firm configures both as a value chain and as a platform. Consider the company with the greatest capitalization in the world: Apple. Apple manufactures hardware, such as iPhones, iPads, and the less successful HomePod. That business is very much structured like a traditional value chain. While Foxconn physically manufactures the bulk of iPhones, it is Apple that controls the design, the inputs, and the transformation process. Once manufactured, the iPhone gets to you either directly from Apple or through a traditional distribution chain of online and brick-and-mortar retailers. However, Apple is also a dominant aggregator, integrating a platform (the iOS operating system) with a marketplace (the App Store). It is this combination that enables it to foster the ecosystem of app developers who continue to find ways to make your iPhone more valuable by developing new apps.

As the competitive landscape continues to evolve in response to waves of digital disruption, you will need frameworks to analyze the potential for your firm to develop and own its digital platform. You will also think about how to best compete if your organization has no opportunity to own a platform. How can you thrive as a supplier inside a powerful aggregator? The latter part of this chapter focuses on emerging frameworks to help you.

---

1 A ghost kitchen is a delivery-only food preparation facility—in other words, a restaurant stripped of the dining area used for welcoming customers.

2 Shieber, J. 2020, October 27. "Spying a pivot to ghost kitchens, SoftBank's second Vision Fund pours $120 million into Ordermark." TechCrunch, retrieved from https://techcrunch.com/2020/10/27/spying-a-pivot-to-ghost-kitchens-softbanks-second-vision-fund-pours-120-million-into-ordermark/.

3 https://www.virtualdiningconcepts.com/.

8.2 Traditional Models of Digital Value Creation

In this section, we explore the influential strategic models that have been and continue to be used to analyze the value creation potential of digital technology for those organizations embedded in traditional value networks. Those strategic models are as follows:

- **Industry analysis**
- Value chain analysis
- **Customer service life cycle (CSLC)** analysis

We present these models below, and using current examples, we show how they can still be fruitfully applied by managers seeking to create IT-dependent strategic initiatives (Chapter 7).

Industry Analysis

The industry analysis framework is grounded in the basic notion that industries offer different potential for profitability.[1] A simple analysis lends support to this assumption (see Figure 8.6). The industry analysis framework suggests that industry differences can be analyzed a priori by managers using an analytical model known as the five forces framework (see Figure 8.7). Armed with the results of this analysis, executives can decide whether to enter an industry or forgo investment. More importantly, for an organization that is already a player in a given industry, such analysis can offer guidance as to what to do to increase the appeal (i.e., average profit potential) of the industry. Thus from a simply analytical framework, the industry analysis model becomes a prescriptive one. It is in this capacity that it can be used to surface opportunities to introduce IT-dependent strategic initiatives.

Five Competitive Forces The industry analysis framework identifies five structural determinants of the potential for profitability of the average firm in a given industry. Each determinant is discussed next.

The Threat of New Entrants This force represents the extent to which the industry is open to entry by new

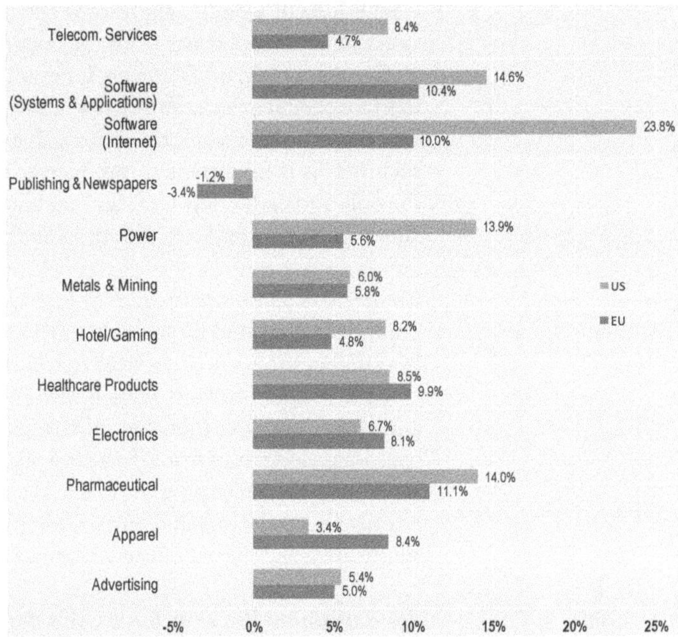

Figure 8.6. Industry differences in average profitability (cumulated values of net income on cumulated sales for the sector)

Source: Data from Damodaran, A., 2018, January, retrieved from http://www.damodaran.com

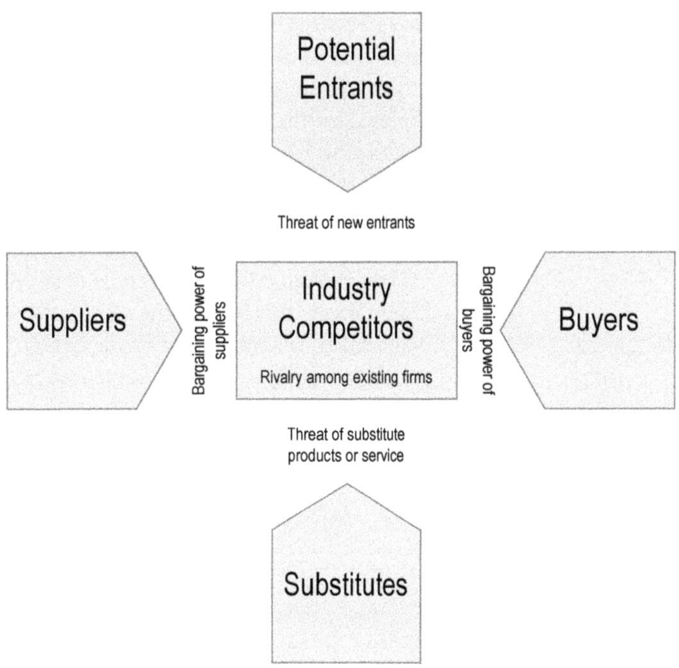

Figure 8.7. Industry analysis

1 Porter, M. E. 1980. *Competitive strategy: Techniques for analyzing industries and competitors.* New York, NY: Free Press.

competitors or whether significant barriers to entry make it so that the existing firms need not worry about competition from the outside.

Consider car manufacturing, for example. Car manufacturing is characterized by a substantial need for capital investments in research and development as well as a need for significant production capacity. Moreover, the automotive industry is characterized by strong economies of scale, such that it is important to produce a large number of vehicles to stay competitive—as Tesla, the only successful new U.S. car manufacturer in 70 years, has found out. For these reasons, the auto industry is characterized by a low threat of new competitors due to strong barriers to entry.

The Threat of Substitute Products or Services This force represents the extent to which the products or services marketed by the firm in the industry of interest are subject to potential substitution by different products or services that fulfill the same customer needs.

For example, products such as the iPod and digital music files, introduced by firms traditionally in the computer industry, substituted CD players and other devices traditionally offered by consumer electronics firms. In a clear sign of the times, Sony announced in October 2010 that it would stop manufacturing the iconic Walkman cassette player. While nostalgia brought back vinyl records for a collectors' and purist audiophile audience, music cassette players have now been fully substituted, and digital music streaming services have "eaten" 85% of the entire music industry![1]

The Bargaining Power of Buyers This force represents the extent to which customers of those organizations in the industry have the ability to put downward pressure on prices. Highly concentrated buyers and low switching costs (Chapter 9) typically conspire to increase the bargaining power of buyers.

Consider, for example, a company like Walmart, which, because of huge sales in its more than 11,500 stores worldwide, can purchase the bulk of a manufacturer's production capacity. With its size and focus on low prices, Walmart is famous for influencing—some would even say *dictating*!—prices set by its suppliers. During the COVID-19 pandemic, Amazon policed merchants accused of charging high prices for high-demand and low-supply products (e.g., sanitizers). It did so on behalf of customers in its marketplace (i.e., the buyers) and enforced pricing rules algorithmically. Interestingly, as with any proprietary algorithm owned by aggregators and platforms, it was impossible for suppliers to ascertain exactly what triggered the ban, and some merchants took no chances and simply pulled the monitored products to avoid risking a ban.[2]

The Bargaining Power of Suppliers This force represents the extent to which those individuals and firms who sell production inputs to the organizations in the industry have the ability to maintain high prices. This force is the same as the previous one, where the firms in the industry of interest have taken the role of the buyer rather than the seller.

As a future supplier of labor resources, you should pay significant attention to this force. If you can put yourself in a position of bargaining power toward the industry of your interest (e.g., by choosing to concentrate in a field of study that is highly sought after but in short supply), you stand to reap significant benefits in terms of salary.

---

1 Friedlander, J. P. 2020. "2020 mid-year music industry revenue report" (p. 3). RIAA, retrieved from https://www.riaa.com/wp-content/uploads/2020/09/Mid-Year-2020-RIAA-Revenue-Statistics.pdf.

2 Soper, S. 2020, June 11. "Amazon's crackdown on price-gouging of disinfectant products made shortages worse." *Fortune*, retrieved from https://fortune.com/2020/06/11/amazon-price-gouging-cleaning-supplies-disinfectant-sanitizer-wipes-shortage/.

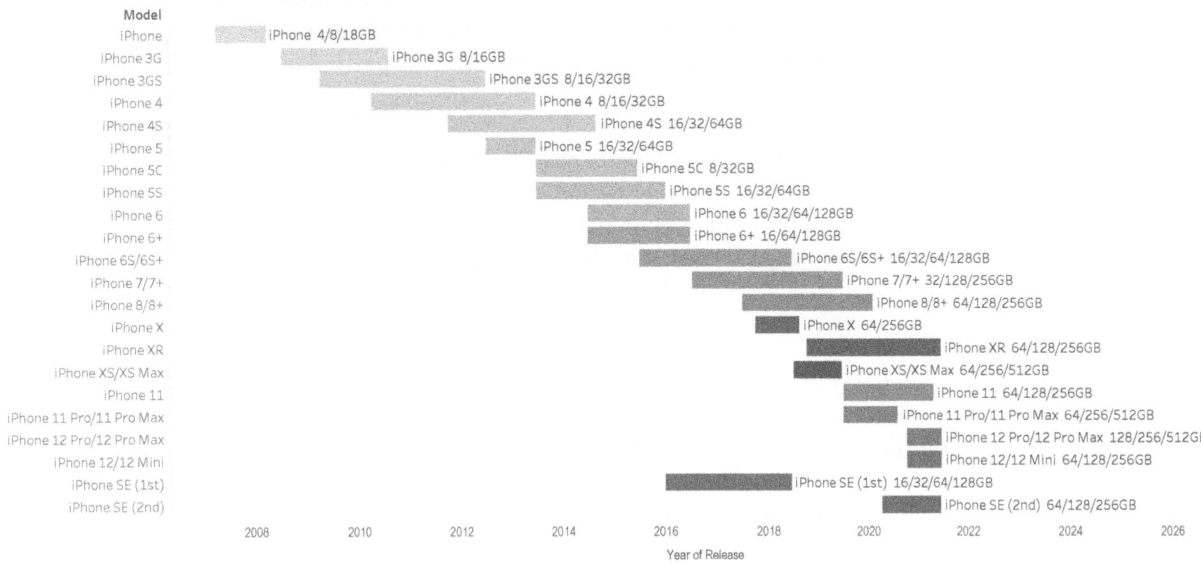

Figure 8.8. Consumer electronics: A hypercompetitive industry

The Rivalry among Existing Competitors This force represents the extent to which fierce battling for position and aggressive competition occur in the industry. The degree of competition in an industry can vary dramatically. The term *hypercompetition* refers to industries characterized by fierce rivalry among existing firms and a very rapid rate of innovation leading to fast obsolescence of any competitive advantage and a consequent need for a fast cycle of innovation. The search engine industry may be an example, as is the consumer electronics industry, where new products are constantly introduced (Figure 8.8).

Industry Analysis and the Role of Information Systems Researchers and consultants who have adapted industry analysis to the search for opportunities to introduce IT-dependent strategic initiatives suggest looking for ways to use information systems to affect one or more of the industry forces, thereby using IT resources to tip the balance of power to the firm's advantage. While the five forces framework is a *time-honored* tradition, it is still relevant, and its proponent, Professor Michael Porter, has justified its value in response to major technology evolutions like the Internet[1] and, more recently, the Internet of things (IoT).[2] The following are some of the questions that are typically asked during the analysis:

Can the Use of IT Create or Increase Barriers to Entry in the Industry? Investments in information systems may be such that they reduce the threat of new entrants. Consider, for example, the need for an automated teller machine (ATM) network or an online banking app in the banking industry. Entry into the banking industry nowadays requires access to a network of ATMs and online banking facilities. Likewise, access to the lodging industry requires access to a computerized central reservation system (CRS) and a substantial number of interfaces

1 Porter, M. E. 2001, March. "Strategy and the Internet." *Harvard Business Review 79*: 62–78.

2 Porter, M. E., and Heppelmann, J. E. 2014, November. "How smart, connected products are transforming competition." *Harvard Business Review 92*: 11–64.

to the plethora of traditional and emerging distribution channels. Entry into the gaming industry depends on the possibility of establishing a platform popular enough (e.g., PlayStation, Xbox, Switch) to attract developers producing games and ultimately users. All of these are barriers to entry.

Can the Use of IT Decrease Suppliers' Bargaining Power? With the emergence of the Internet and its related technologies as viable business tools, examples abound of digital innovations that have contributed to shift power away from suppliers and toward buyers. Consider, for example, Alibaba, the sourcing, procurement, and expertise provider that enables companies to easily access global suppliers. Alibaba enables organizations to dramatically increase their bargaining power with suppliers by reducing the search costs for identifying qualified suppliers.

Can the Use of IT Decrease Buyers' Bargaining Power? As much as the Internet has helped firms strengthen their bargaining position toward suppliers, it has also reduced their bargaining power toward customers. Just as companies can shop for alternatives when looking for production inputs, so can their customers. Yet some opportunities to strengthen relationships with customers, thus reducing their incentive to shop around, still exist.

Consider travel intermediaries like Orbitz.com. While competitors are literally one click away, by storing personal preferences (e.g., preferred airlines), personal data (e.g., frequent flier miles), and billing information, travel intermediaries can levy switching costs (Chapter 9) and reduce their customers' bargaining power. This approach was made popular by Amazon's one-click buying, an innovation for which the firm even received a process patent in 1999 and the "1-Click" trademark. The greater convenience of the shopping experience netted the firm billions and, more importantly, reinforced Amazon's position as a customer service leader.

Can the Use of IT Change the Basis of Industry Competition? The introduction of a new information system or a new digital innovation by a firm, whether an incumbent or a new entrant, sometimes spurs a revolution that forces competitors to take notice and react. A stark example of this dynamic was presented by the advent of online retailing in the mid-1990s.

When Amazon.com burst onto the scene, with its ability to offer huge selection and high levels of customer service without a single store, shock waves reverberated in the retail sector. Firms such as Amazon, which seized the opportunity presented by the Internet to sell directly to consumers, trained consumers to self-serve in ways unheard of before and changed forever the notion of what it means to be a retailer. Today, it would be a grave mistake for any large retailer to neglect the online channel of distribution. Examples abounds including Uber in transportation, Airbnb in lodging, Netflix in entertainment, Stripe in finance, and Coursera in education. Today, we can find similar "digital disruptions" in almost any sector.

Value Chain Analysis

While useful in identifying potential opportunities to improve the profitability of the industry and suggesting ways in which managers can deploy information systems and digital innovations to neutralize or minimize the unattractive features of an industry, much of the potential for the employment of strategic information systems concerns intraindustry competition. In other words, much of the time that you, as managers, will be spending analyzing opportunities to deploy strategic information systems will be with respect to the ability to create added value. Thus you will not worry so much about average industry performance; rather, given the industry that you are competing in, you will seek to outperform your competitors by using information systems to create added value and competitive advantages.

Value is created when a firm employs its transformation process to use resources that have a value of $\$x$ in their next best use and to generate a product or service that customers are willing to pay $\$x + \$v$ for (Chapter 7). If what the firm does is unique, such that no competitor is able to offer comparable value, the firm creates added value and has achieved a position of competitive advantage. But what does the firm's "transformation process" look like? How do these input resources become the final product or service that customers pay for?

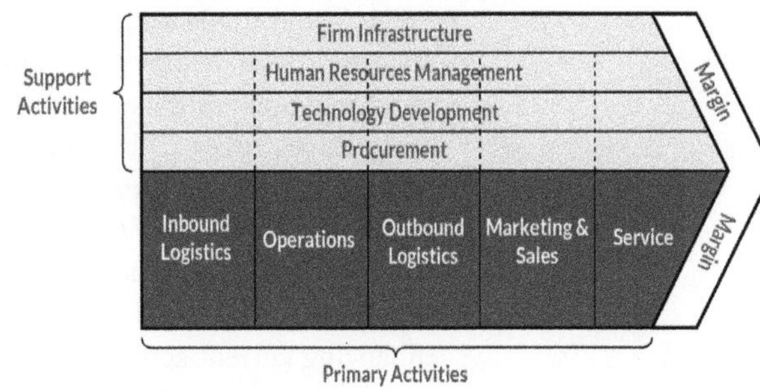

Figure 8.9. Value chain framework

Primary and Support Activities The classic framework to logically represent a firm's transformation process is the value chain (Figure 8.9).[1] The value chain model maps the set of economic activities that a firm engages in and groups them into two sets: primary activities and support activities.

Primary Activities Primary activities represent the firm's actions that are directly related to value creation. These are the activities that define the firm's unique transformation process such that they are typically performed by all firms engaged in the same or a similar line of business. The classic value chain identifies five primary activities: inbound logistics, operations, outbound logistics, marketing and sales, and service. It is the execution of these activities that enables the transformation of the input resources in final products or services—and, as a consequence, the creation of value.

Let's return to the automotive industry. We can think of the car manufacturing transformation process in terms of these five primary activities. A car manufacturer needs to procure and receive component parts (i.e., inbound logistics), have assembly lines that take these components and put them together in the various car models (i.e., operations), deliver the vehicles to the distribution channel and its dealership network (i.e., outbound logistics), create a demand for its make and models (i.e., marketing and sales), and ensure that any problems with its products can be addressed by a network of repair shops (i.e., service).

Support Activities Support activities represent the firm's actions that, while not directly related to the transformation process, are nevertheless necessary to enable it. These activities do not define the organization's unique transformation process, which is to say that they are typically performed by a wide range of firms offering diverse products and services. The classic value chain identifies four support activities: firm infrastructure, human resource (HR) management, technology development, and procurement.

Take HR management, for example. Any firm that relies on labor resources, from car manufacturers to hospitals, must be able to recruit, train, evaluate, pay, promote, fire, and generally manage the labor force. Yes, these activities do not directly impact the product (e.g., the car traveling down the assembly line). However, no firm without an HR function would be in business for long!

The competitive advantage of value chain organizations traditionally stems from the ability of the firm to transform input resources into a product or service that is both valuable and unique—thereby having created added value (Chapter 7). Thus in order to have a competitive advantage, the firm must perform a different set of activities than competitors, or it must perform the same set of activities but in a different (and value-adding) manner. Information systems and IT have a long tradition of enabling such unique transformation processes.

1 Porter, M. E. 1985. *Competitive advantage: Creating and sustaining superior performance.* New York, NY: Free Press.

Value Chain Analysis and the Role of Information Systems Using the value chain to identify opportunities to deploy IT-dependent strategic initiatives requires managers to recognize, understand, and analyze the activities the firm performs so that they can be enhanced or transformed using IS resources. This approach is grounded in the realization that a firm's value chain has both physical and information processing components and that information is a critical enabler of the firm's activities. As software continues to "eat the world" (Chapter 1), the information component of value chains is steadily increasing in importance. Thus the search for the strategic deployment of IT should focus on the role that information technology can play in evolving and enhancing current activities.

An example is offered by a classic IBM commercial. A malicious-looking character is seen walking the aisles of a supermarket stuffing various goods into the pockets of his trench coat under the suspicious eye of a security guard. As he walks out of the store, apparently without paying, he is called to by the guard, who, after some suspense-inducing delay, says, "Sir! You forgot your receipt."

This is an example of how IBM envisioned the digital innovation of the grocery store checkout process in the future using radio-frequency identification (RFID) chips embedded in everyday goods. Such a widespread adoption of RFID would have enabled grocery stores to no longer have checkout lines. In this example, the checkout process is radically transformed, leading to substantial efficiency improvements and creating the potential for new initiatives (e.g., tracking household purchases over time). The firm that is closest to realizing this vision is in fact Amazon with its Amazon Go stores. However, the technology Amazon is banking on for this go around in cashier-less stores is video and machine learning rather than RFID. Stay tuned to find out if they can scale this concept or whether we will still be queueing up at grocery stores in the future.

The Value Network Another insight that emerged from the introduction of value chain analysis is that a firm has relationships both upstream and downstream. In other words, the firm's own value chain exists in a larger value network that comprises the firm's suppliers upstream and the firm's customers downstream (Figure 8.10). The points of contact between these separate value chains are called **linkages**, and they offer significant opportunities for the deployment of IT-dependent strategic initiatives.

For example, the inbound logistics activities of your firm can be thought of as an extension of the outbound logistics activities of your suppliers. Recognizing these linkages enables firms to create partnerships and develop interorganizational systems that may benefit all parties involved. Continuous replenishments, the initiative pioneered by Walmart and Procter & Gamble (Chapter 7), is a classic example.

A Word of Caution Managerial frameworks are designed to map and simplify the complexity of real organizations in order to enable a disciplined analysis of complicated phenomena. In the case of the value chain model, the objective is to help you as a manager identify opportunities to change the transformation process your organization engages in and thus uncover ways to create new value. It is therefore critical to recognize that a general framework, like the value chain, often needs to be adapted to the specific realities of your firm.

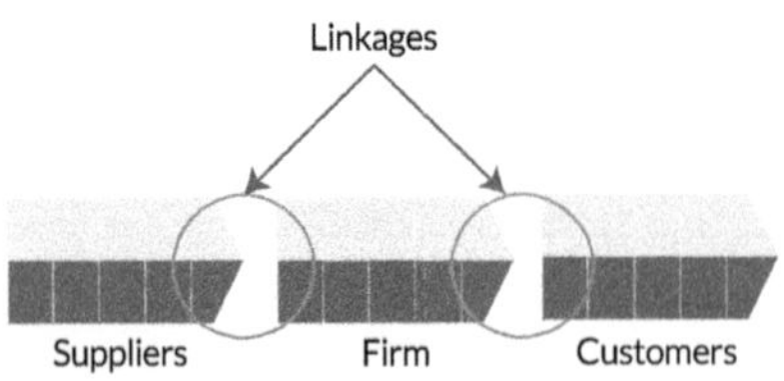

Figure 8.10. Linkages in the value network

The original model portrayed in Figure 8.9 is clearly best suited to represent the realities of manufacturing companies where raw materials and component parts are transformed into final products that need to be marketed, sold, and later serviced. Service businesses (e.g., consulting and law firms, banks, entertainment venues, laundry services, restaurants) work very differently, and while they also have to complete primary and secondary activities, the activities they perform and the sequence of events can be very different. Figure 8.11, for instance, portrays the value chain model mapping the sequence of primary activities as they occur in a hotel or resort.

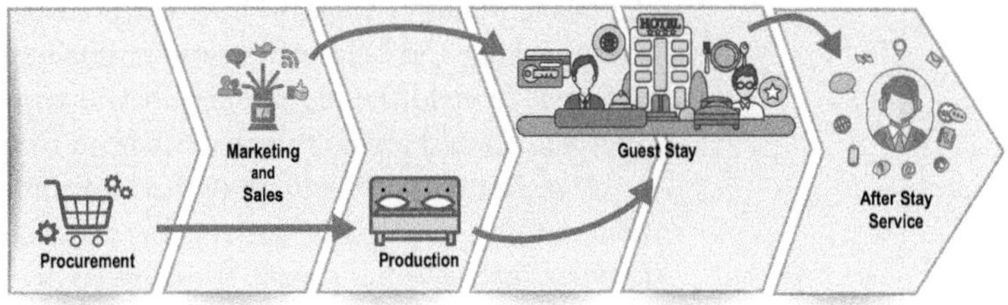

Figure 8.11. Sample value chain of a lodging outfit

New digital business models may even challenge your understanding of the value chain. Consider in this sense a platform like Twilio, a YouTuber like PewDiePie, or an aggregator like Deliveroo. It is imperative that when using the value chain, or any other managerial framework, you do not simply apply it "as is" but, using your in-depth knowledge of the specific firm and industry you are analyzing, adapt the model to your needs. After all, as a supplier of managerial talent, this is what you are paid to do!

Customer Service Life Cycle Analysis[1]

Where the value chain mostly focuses internally, the CSLC draws attention to the relationship between the firm and its customers. Its premise is that a firm can use information systems to create value through digital relationships with customers. The CSLC breaks down the firm-customer relationship into 13 stages, grouped into four primary phases; for each one, it shows how to address customer needs, enhancing the service offered and creating economic value. The objective is to envision how information systems can be deployed to improve digital relationships and customer experiences no matter when and how they interact with the firm.

Four Phases The CSLC framework maps the four major phases of customer interaction with the firm: requirements, acquisition, ownership, and retirement (Figure 8.12). During the requirements phase, the customer realizes the need for a specific product or service and begins to focus on its attributes. During the acquisition phase, the customer orders, pays for, and takes possession of the product or service. The next major phase is ownership; here the customer has the product or is receiving the service and must deal with issues regarding its efficient and effective use. The final phase is retirement, in which the customer may begin to think about repurchasing, trading in, or dismissing old products.

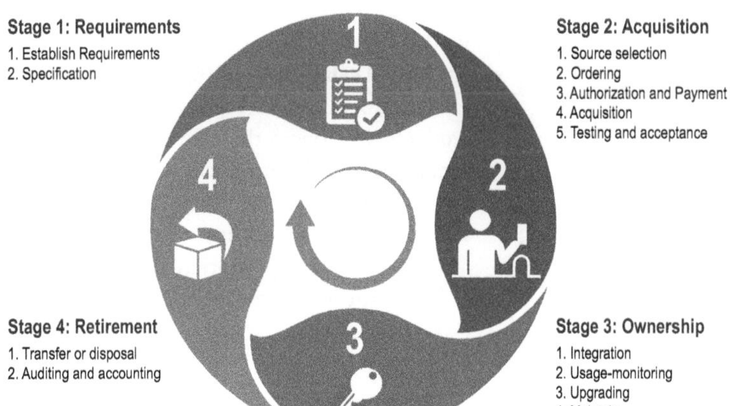

Figure 8.12. The customer service life cycle

Thirteen Stages Each of the four main phases is then further broken out into subphases, or stages helping you identify where your customers may be frustrated or underserved and where interaction can be improved with the help of

1 Portions of this section are adapted from Piccoli, G., Spalding, B. R., and Ives, B. 2001. "The customer service life cycle: A framework for Internet use in support of customer service." *Cornell Hotel and Restaurant Administration Quarterly 42*(3): 38–45; and Ives, B., Palese, B., and Rodriguez, J. A. 2016. "Enhancing customer service through the Internet of things and digital data streams." *MIS Quarterly Executive 15*(4).

digital technology (Table 8.1). As a creative planning framework, the CSLC is designed to stretch your thinking and help you view your business with a fresh perspective. The stages that have the potential to yield the highest payoff will vary by customer segment, by product, and over time. Thus you should repeat the analysis from time to time.

Organizations achieve competitive advantage through their ability to envision and implement value-adding strategic initiatives. The most innovative ideas are often not the most expensive or resource intensive but simply those based on the best understanding of how customer needs can effectively be satisfied. The CSLC provides you with a basis for evaluating a firm's relationship with customers, benchmarking against competitors, and uncovering opportunities to use digital technology to improve customer willingness to pay (CWP) through outstanding customer service.

8.3 Value Creation with Data and Information

The strategic role that information systems and digital technology can play in the modern organization was acknowledged more than 30 years ago. However, recent trends have provided a new impetus for the search for IT-dependent strategic initiatives. The catalysts were the commercialization of the Internet in 1993 and the growth of the mobile platform since the introduction of the iPhone in 2007 (Chapter 5). The Internet and the mobile platform ushered in an era of digital disruption, and today's consumers and businesses could not imagine working (and living?) without them. Parallel to this upsurge of global networking was the continuance of long-term trends, such as the declining cost/performance ratio of computing equipment and the declining cost of storage (Chapter 1). Affordable, powerful, interconnected computers and cheap storage are the engines behind another enduring trend, that of value creation with data and, more recently, Big Data (Chapter 3). The traditional strategic models described in the previous section were characterized by a view of information as a support resource to the production and transformation processes that the firm engages in. The two frameworks we cover in this section take as their starting point the realization that the data that modern organizations generate through their day-to-day operations have significant strategic value in their own right.[1]

Virtual Value Chain

The **virtual value chain (VVC)** model builds on the generally understood value chain model.[2] In a value chain, through a series of logically sequential activities, raw materials are transformed into products or services that the firm then distributes to customers. The process is a value-adding one, such that the products or services being distributed have a higher value (and command higher CWP) than the inbound materials and services. The VVC maps out the set of sequential activities that enable a firm to transform data in input into some output information that, once distributed to the appropriate user, has higher value than the original data (Figure 8.13).

The critical insight underpinning the VVC is that information is no longer just a support resource for physical activities, those described by the physical value chain model, but can itself be treated as an input of a productive

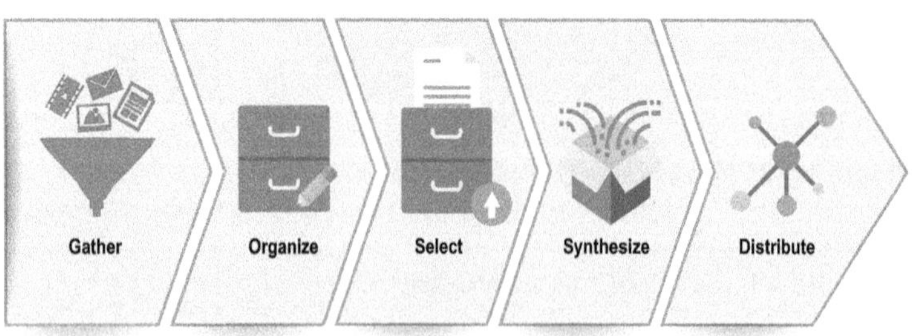

Figure 8.13. Virtual value chain framework

1 Wixom, B., and Beath, C. 2015. "Let's start treating data as a strategic asset!" (research briefing). MIT Center for Information Systems Research, retrieved from https://cisr.mit.edu/publication/2015_0901_DataStrategicAsset_WixomBeath.

2 Rayport, J., and Sviokla, J. 1995, November–December. "Exploiting the virtual value chain." *Harvard Business Review 73*: 75–85.

Table 8.1. Customer service life cycle stages

Stage	Description	Example
Requirements		
Establishing requirements	Establish a need for the product or service	• Netflix opening page with personalized recommendations for what you should watch • Carrefour's in-store beacons pushing messages suggesting products you may be interested in purchasing
Specification	Determine the product or service attributes	• Nike By You sneakers customization initiative • Hilton HHonors app room to select your room like a seat in an airline
Acquisition		
Source selection	Determine where to obtain the product or service	• ShopSavvy barcode scanner and comparison app to check for the best price around • Skyscanner to compare airfares and select the best combination for your travel
Ordering	Order the product or service from a supplier	• Starbucks app for ordering your breakfast or lunch on the go • Amazon's Alexa to make an order with your voice
Authorization and payment	Transfer funds or extend credit	• PayPal online payment solution for an added guarantee in online transactions • Apple Pay and Alipay contactless payment systems to forget about your Visa
Acquisition	Take possession of the product or receive the service	• Lime app to localize a scooter and start riding around the city • Streaming music and video services to immediately access all the content you may want
Testing and acceptance	Ensure that the product or service meets specifications	• Salesforce's free 30-day trial to test the CRM before the actual purchase • FedEx's SenseAware multisensory device to monitor a shipment's conditions during delivery
Ownership		
Integration	Add to an existing inventory or integrate with existing internal business processes	• My Disney Experience app to plan a visit to the amusement park, make dining reservations, secure access to attractions and shows, and set up event-related reminders • FedEx's Ship Manager to automate the shipping process from printing labels to package tracking
Usage monitoring	Control access and use of the product or service	• OTIS's remote elevator monitoring to constantly track their performance and immediately intervene in case of an accident or malfunctioning • Google's Nest cam to receive alerts and video clips when specific events are detected, like an unexpected movement or the presence of a family member
Upgrading	Upgrade the product or service if conditions change	• Tesla's over-the-air updates to add new features and increase car performance • Lufthansa app to exchange your miles for an upgrade or purchase a better seat
Maintain	Analyze, diagnose, and repair the product or service as necessary	• An operating system's automatic updates to download security fixes to counter the latest exploits discovered by hackers • iPhone's hardware diagnostic app to check your phone health
Retirement		
Transfer or disposal	Transfer, resell, return, or dispose of the product or service	• Dell asset recovery service to properly dispose of end-of-life machines, format their hard disks, and eliminate all the customers' data • The service offered by manufacturers of networked printers and copiers to automatically reorder toner and ink cartridges when they are running out
Auditing and accounting	Monitor evaluations and expenses related to the product or service	• American Express's solution to monitor travel expenses for their large customers • AXA's DriveSave app to log in real-time mileage, speed, and cornering to give customers a cashback refund on the insurance cost to reward their safe driving performances

transformation process.[1] The VVC uses the same logic of the physical value chain but recognizes information as the "raw material" being transformed through the chain of activities.

Five Activities The VVC identifies five sequential activities that must be completed in order to harness its power:

1. *Gather.* The firm collects information from transaction processing systems and any other sources—both internal to the organization (e.g., orders received) and external (e.g., census data).
2. *Organize.* The firm stores the gathered data in a way that makes later retrieval and analysis simple and effective.
3. *Select.* Users identify and extract the needed data from the data repository created in the previous step.
4. *Synthesize.* The firm packages the selected information so that it can be readily used by the intended consumer for the specific purpose to which it is directed (i.e., decision making, sales).
5. *Distribute.* The firm transfers the packaged information to its intended user or customer.

Three Classes of Strategic Initiatives The VVC model identifies three classes of strategic initiatives that firms can implement upon adopting VVC thinking: **visibility, mirroring capabilities**, and new customer relationships. These three classes of activities have an increasing order of complexity and uncertainty of results, with visibility being the most intuitively appealing and easy to justify financially and new customer relationships being the toughest to sell to senior management. Note that the five value-adding activities of the VVC are performed with each initiative. What changes is the level of complexity and departure from the traditional way of doing business for the organization.

Visibility The first application of the VVC is termed *visibility*. In this case, the firm uses the sequential activities in the VVC to "see through" organizational processes that it was previously treating as a black box. An enduring example of visibility is offered by online retailers. Because their customers' behavior is computer mediated as consumers shop and purchase online, online retailers collect significant amounts of individual and aggregate data. These data include what web pages a customer views, which web pages seem to go unused, what path through the site customers are taking as they shop, whether and where they abandon the transaction, how customers react to advertising and banners, and so on. In other words, customers' shopping and purchasing processes are significantly visible to online retailers, a degree of accuracy and detail that was unprecedented in brick-and-mortar stores. But digital transformation has enabled traditional retailers to leverage visibility as well. Zara, the world's largest fashion retailer, is one of the most successful users of digital technology for strategic advantage. The firm outpaced competitors by quickly adapting to changing demand conditions. The firm designed a flexible production and logistic system heavily reliant on IT, and it is capable of adapting their collections in real time based on information about customers' tastes gathered from their 2,259 stores located in 96 countries. Zara has introduced RFID technology to streamline logistic and point-of-sale operations, directly tackling stock visibility and replenishment. As Pablo Isla, chairman and CEO of Zara owner Inditex SA, said, "It

1 In this section, to avoid confusion, we use the term *physical value chain* to refer to the traditional value chain model.

gives us great visibility, knowing exactly where each garment is located. . . . It really changes how we operate our stores."[1] These early digital innovations created a degree of inventory accuracy and visibility that enabled further innovation (Chapter 2). Zara's customers can now check the available assortment in their preferred physical stores, place orders, and pick up the items within half an hour all via the firm's mobile app. Isla recently explained, "[Zara's] integrated store and online platform strategy . . . is a cornerstone of [Zara's] unique business model with three key pillars: flexibility, digital integration and sustainability."[2]

Inventory of parts in a manufacturing company is typically held to reduce uncertainty of demand and inbound logistics. Clearly this insurance against uncertainty comes at a cost—the cost of capital tied up as well as the cost of write-offs for obsolete inventory. A firm that is able to gather and use superior information, and thereby reduce uncertainty, can limit its inventory cost and run a leaner operation. Retailers are now going even further with digital innovation. Through microcameras embedded in store mannequins and shop windows, the video feed of surveillance cameras, the Wi-Fi signal strength, or Bluetooth and beacon technologies, digital innovators attempt to replicate the level of visibility available to online retailers.[3] These technologies help measure customer proximity to specific shelves, how long they linger by a product, and their path through the store.[4] Offline-to-online (O2O) techniques (Chapter 5) represent a good example of the impact that digitization and process virtualization (Chapter 4) are having on business firms in all sectors.

As devices become "intelligent" via embedded microchips and software, the firms that use them have more opportunities to harness the power of visibility. Imagine a firm that manages vending machines, which are increasingly computerized and network connected. What decisions could you make if the vending machines were able to maintain real-time communication with a central server about sales and inventory level? How much more efficient could your firm be with this information? What new strategic initiatives would you implement with this infrastructure in place?

Mirroring Capabilities A further application of the VVC, termed *mirroring capabilities*, consists of shifting some of the economic activities previously completed in the physical value chain to the information-defined world of the VVC. That is, some of the activities that were previously physical in nature (i.e., completed by employees) become completely information based. In one example of innovative thinking, Coca-Cola rolled out the Freestyle drink dispenser, an RFID-enabled vending machine that draws ingredients from 30 flavor cartridges and is able to mix over 160 different drinks (Figure 8.14). While we may debate whether any of these drinks are actually worth drinking, here lies Coca-Cola's strategy. The Freestyle machine is connected to central servers through the cellular phone network. By letting people mix and match their drinks and keeping track of which drinks are being consumed and when, the firm gains visibility in the soda-drinking process. More importantly, the firm is betting on being able to perform extremely

---

1 Bjork, C. 2014, September 16. "Zara builds its business around RFID." *Wall Street Journal*, retrieved from https://online.wsj.com/articles/at-zara-fast-fashion-meets-smarter-inventory-1410884519.

2 Inditex. 2020, September 16. "Inditex returns to profitability and generates €734 million net cash in the second quarter." Inditex, retrieved from https://www.inditex.com/en/article?articleId=653527&title=Inditex+returns+to+profitability+and+generates+%E2%82%AC734+million+net+cash+in+the+second+quarter.

3 Correa, A., Barcelo, M., Morell, A., and Vicario, J. L. 2017. "A review of pedestrian indoor positioning systems for mass market applications." *Sensors 17*(8). doi: 10.3390/s17081927.

4 Kirkpatrick, K. 2020. "Tracking shoppers." *Communications of the ACM 63*(2): 19–21. Retrieved from https://cacm.acm.org/magazines/2020/2/242338-tracking-shoppers/fulltext.

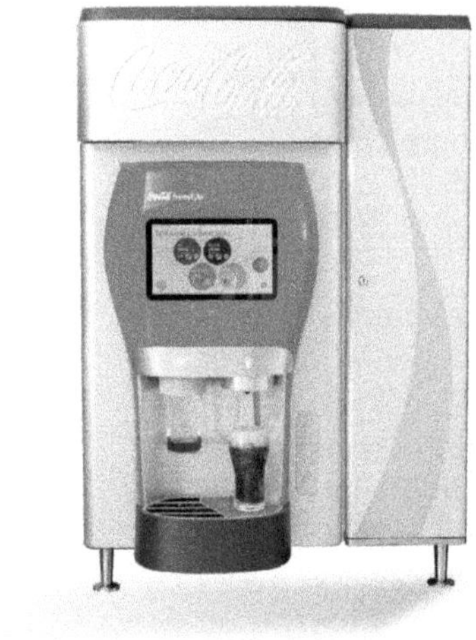

Figure 8.14. Coca-Cola's Freestyle drink dispenser

Source: Coca-Cola press kit

precise taste tests that will aid new product development and real-time business analytics.[1] In other words, Coca-Cola is taking the act of physical taste-testing in the lab and, at least in part, mirroring it into the digital world. Moreover, Coca-Cola leases the machines to fast-food outlets, enabling them to access real-time drink consumption data, all the while providing greater variety to those soft-drink-crazed individuals who need to have 100 different flavors!

Examples of mirroring capability are pervasive in firms that need to perform much testing and simulation. For example, a recent trend among drug manufacturers is to test the effect of new drugs using computer models rather than real patients, thereby drastically speeding up trials and reducing the cost of new drug development. Another great example is training in virtual environments. *Second Life*, a virtual reality massively multiplayer computer game, popularized this trend before the large availability of current virtual reality sets. Launched in 2003, after almost two decades, the platform still enjoys an average of 900,000 active users per month. Over the years, many organizations, from hospitals, to police departments, to the military, created virtual learning environments, but some pioneering simulations leveraged the immersive reality characteristics of *Second Life* and created situations that would be extremely expensive and time consuming to design in real life (Figure 8.15). Imagine staging a terrorist attack or a blackout in the emergency room to train people in crisis response skills! In fact, the traditional approach to this type of training is classroom cases or video scenarios—techniques that pale in comparison to the virtual world in terms of realism and skill acquisition potential.

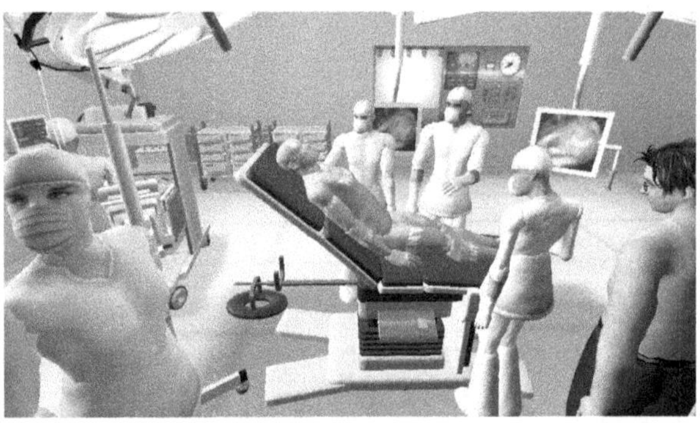

Figure 8.15. A simulated emergency room in *Second Life*

Note that mirroring capabilities are beneficial when transferring the activity to an information-based platform if they enable the firm to perform the activity more efficiently and effectively. They are also beneficial when the activity can be transformed with significant performance improvements. For this reason, a mirroring capability approach is different from a mere automation of the existing activity using computers. To fully grasp this concept, think about the Ergo Bike Premium 8i from Germany's Daum Electronic. The Premium 8i looks like a regular high-tech stationary bicycle. However, it allows you to race other bicycle enthusiasts across the world in some of the greatest bicycle race segments on the globe (e.g., Hawaii's Ironman Triathlon). The bike simulates the terrain, giving you the sensory feeling of being there. It allows you to monitor your heart rate and other statistics as well while talking to your racing partners over headsets. Imagine climbing the Alpe d'Huez along with your friends, following in the footsteps (or bike tracks) of the great racers in the history of the Tour de France—Gino Bartali, Fausto Coppi, Eddy Merckx, Greg LeMond, Miguel Induráin. Perhaps, then, a better term for this strategy would be *digitizing capabilities* rather than *mirroring capabilities*.[2]

---

1 The mirroring capability strategy is a precursor to the introduction of digital twins (Chapter 12). As real-time streams of digital data generated by devices like the Coca-Cola Freestyle drink dispenser become more common, they enable the creation and calibration of highly detailed representations of entities in the physical world. These representations are called digital twins.

2 We do maintain the original terminology in the framework to be consistent with the original authors.

New Digital Value The above two types of strategies, visibility and mirroring capabilities, are mainly concerned with internal operations and the creation of value within the confines of the organization. The third stage, **new digital value**, is instead concerned with the organization's relationship with the customer and the firm's ability to increase CWP (Chapter 7) using the information generated through the VVC to create new value in the form of new information-enabled products or services.

Consider a classic example of this strategy: personalization through suggestive selling initiatives. Using your individual purchase history as well as a technique known as *collaborative filtering* that compares your purchases to those of others with similar interests, online retailers are able to propose items that may be of interest to you in an effort to increase their sales. This approach, pioneered in retail, is now common in many industries. Current examples come from the music service Spotify and the video streaming service Netflix, which developed a proprietary recommendation engine based on customers' visual preferences further refined through consumption behavior.

Value Matrix When introduced to the VVC and the potential for value creation through the use of organizational data, it is natural to gravitate to this way of thinking. However, it is important to remember that most organizations today need to pay significant attention to the traditional environment and their established operations. For as relevant as the potential to create value through information is, much of the opportunity for value creation remains in the firm's physical transformation processes. Thus combining the traditional (physical) value chain and the VVC offers a cohesive framework, termed the ***value matrix***, that general and functional managers can use to seek and exploit opportunities for the deployment of IT-dependent strategic initiatives in their organizations (Figure 8.16). This is what happened with Domino's. In an effort to increase product quality, the firm invested in systems to track the pizza-making process in all its stages. This infrastructural investment in support of the main process in its value chain became the basis for gathering the data needed to engage in VVC activities. First, Domino's achieved visibility internally and, in 2008, exposed part of this information to hungry customers via the Pizza Tracker. As the firm continued to digitally transform its physical operations over the years, new opportunities for customer-facing digital innovation became available, and today, the firm is the recognized digital innovation leader in its industry with 75% of sales via digital channels in the United States.

By being mindful of the five steps of the VVC as they apply to data generated throughout the physical value chain, managers can uncover opportunities for new value creation and appropriation. The caveat expressed before about analytical frameworks holds here as well, and you need to map the value matrix that most accurately represents your firm's individual context (see, for example, Figure 8.17).

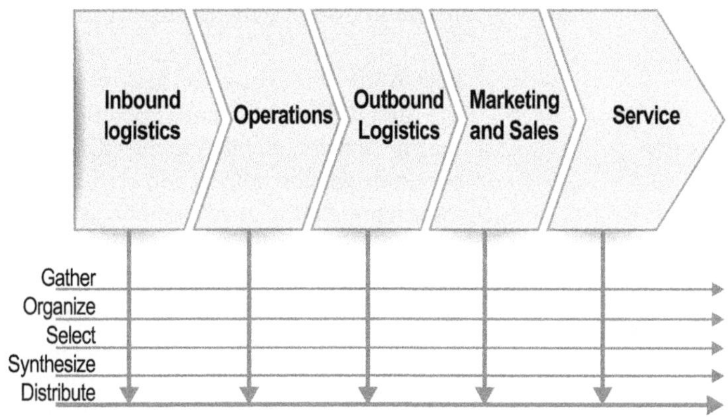

Figure 8.16. Value matrix

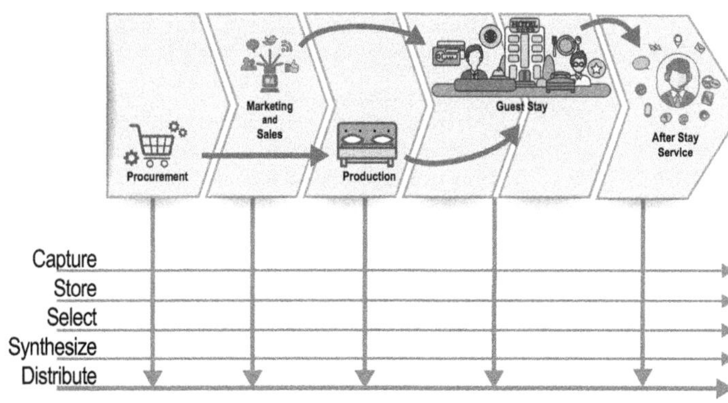

Figure 8.17. Sample value matrix for lodging

Data Monetization[1]

A more recent analytical tool for thinking about extracting value from data is the **data monetization** framework.[2] In this context, *data monetization* refers to the notion of leveraging digital data and analytics as the means of digital value creation. The model identifies three general classes of data monetization strategic initiative: **improving**, **wrapping**, and **selling**.

Before discussing each one, it is critical for you to reflect on what it means, in practice, to extract value from data. Producing "valuable" data is only the first step in the journey toward value creation. Unless the data and the resulting analytics are acted upon, no value is realized—as in the case of the Netflix 2006 competition (Chapter 3), where the winning ensemble algorithm was not deemed suitable for implementation in production.[3] The rule is simple: no action, no value creation. Moreover, in many cases, the value created with data decays over time with the delay, or latency, associated with extracting needed data from an event of interest (capture latency), transforming data into usable information (analysis latency), and taking action based on this new information (decision latency).

A useful way to think about the gap between data and action is the information offerings consumption path first suggested by SAS, one of the leading developers of analytics software (Figure 8.18).[4] The model shows how, in order for data to provide value, it must be generated (raw data) and organized for use or analysis (prepared data). Insight is extracted from the data by way of reporting, standardized recurrent aggregations and manipulations, or more in-depth and ad-hoc analytics. Finally, for the tangible economic value to be realized, the firm must design and execute specific processes. While perhaps intuitive, the model shows how data generation is only the first step in a chain of activities that will lead to some action. Again, no action, no value creation.

Improving This class of initiatives is inward looking and refers to projects that seek to leverage digital data to advance operational processes and to refine decision-making processes in the organization. While you may think that, a few decades into the digital age, there are no more opportunities for process and decision-making improvement with digital technology, in fact, these represent the most common avenue for value creation. For one, initiatives in the *improving* category are completely within the control of the organization, and they generally use internal data. Thus barriers to their implementation are minimal. More importantly, you should by now recognize that new opportunities emerge with each wave of new digital technologies that create novel digital data streams and more powerful data analyses. Consider the recent Big Data trend (Chapter 3) and the rapid emergence of machine-learning algorithms (Chapter 12). Their combination supports unprecedented improvements in operational processes. Alphabet, for example, runs some of the largest data centers in the world. Cooling a data center is an expensive and polluting process. Using machine learning, Alphabet has cut 40% of energy used for cooling and 15% of overall energy overhead.[5] The same combination of Big Data and machine learning fosters better

---

1 In literature, the term *data monetization* indicates "the act of exchanging information-based offerings for legal tender or something of perceived equivalent value." While the term may be confused with *revenue model* or *value appropriation*, we stick to its original meaning.

2 Wixom, B. H., and Ross, J. W. 2017, January 9. "How to monetize your data." MIT Sloan Management Review, retrieved from https://sloanreview.mit.edu/article/how-to-monetize-your-data/.

3 Wixom, B. H., and Piccoli, G. 2020, October 5. "Actioned analytics pave the way to new customer value." MIT Sloan Management Review, retrieved from https://sloanreview.mit.edu/article/actioned-analytics-pave-the-way-to-new-customer-value/.

4 Buff, A., Wixom, B. H., and Tallon, P. P. 2016, July. "Foundations for data monetization." Center for IS Research Working Paper #402, Sloan School of Management, Massachusetts Institute of Technology.

5 Google. 2019, September. "Data Centers." Google Sustainability, retrieved from https://sustainability.google/progress/projects/machine-learning/.

decision making in agriculture. Around the world, a number of "precision agriculture" start-ups support farmers in improving their yields sustainably. Bilberry, for example, leverages camera video feeds on sprayers to recognize and map weed species within crops. By specifically targeting those pests hindering crop growth, the French company claims an ability to reduce herbicide use by more than 80%.[1]

Note that organizations today are awash in internal data (Figure 8.19). Data are generated by sources such as raw material and inventory moving through the production process, customer inquiries, complaints and reviews, or app and website clicks, to name a few. Data are plentiful. But for improving strategies to bear fruit, firms need a change of operational process or a new approach to decision making (no action, no value, remember?). Organizations often struggle with the ability to recruit talent to transform data into action and with decisions about the responsibility (and costs) for the analytic, process design, and implementation activities that must follow data generation.

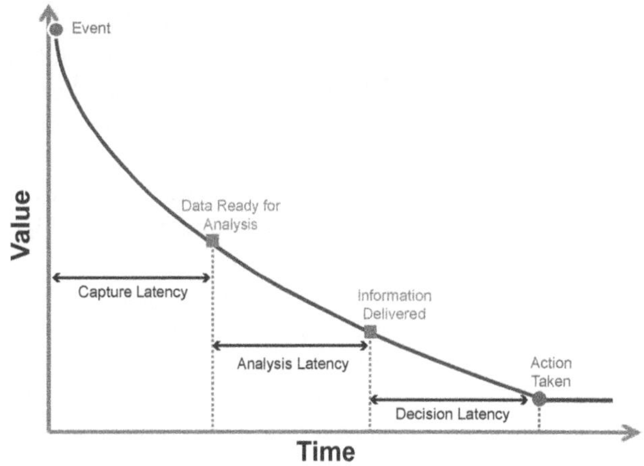

Figure 8.18. Drivers of latency for action based on data and analytics

Source: Pigni, F., Piccoli, G., and Watson, R. T., 2016, spring, "Digital data streams: Creating value from the real-time flow of Big Data," ***California Management Review 58***(3): 5–25, doi:10.1525/cmr.2016.58.3.5

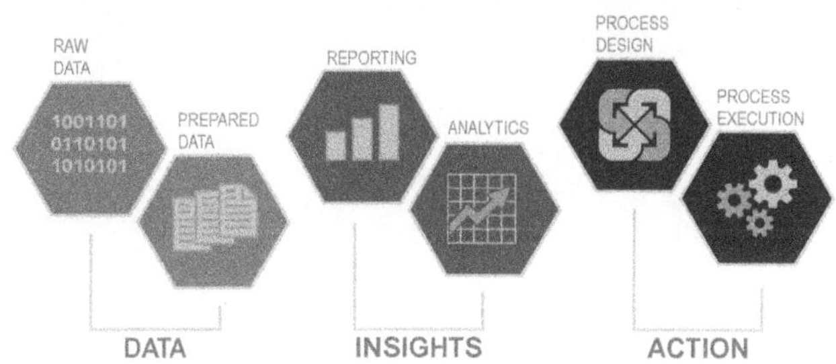

Figure 8.19. The information offerings consumption path

Source: Adapted from Buff, A., Wixom, B. H., and Tallon, P. P., 2015, July, "Foundations for data monetization," Center for IS Research working paper #402, Sloan School of Management, Massachusetts Institute of Technology

Wrapping These initiatives are customer facing, not internal, and consist of using data and analytics to "augment" the value proposition of a company's main products or services. With wrapping efforts, the firm increases the customer willingness to pay for its offerings, an increase "monetized" by charging a premium for the wrapping service, by increasing sales, or by deepening customer loyalty. Early examples of wrapping followed the advent of eCommerce, when delivery service companies exposed the data of their internal logistics system to introduce web-based shipment tracking.[2] More recently, the sensor readings collected through smartphones and wearable devices put firms like Apple or Fitbit in stiff competition to offer customers detailed and precise health and well-being analytics. The Fitbit Premium subscription, for example, leverages these data for providing more personalized fitness programs up to data-driven one-to-one coaching services.[3]

When engaging in wrapping, the sky is the limit, and a focused framework like the CSLC can be quite helpful in stimulating your creativity. However, having read the book up to this point, you realize that for data wrapping to create value, customer action is needed (again: no action, no value).

---

1 Pucci, J. 2019, July 2. "Twentysomethings of Bilberry lead 'green-on-green' race." PrecisionAg, retrieved from https://www.precisionag.com/in-field-technologies/imagery/twentysomethings-of-bilberry-lead-green-on-green-race/.

2 Lappin, T. 1996, December 1. "The airline of the Internet." *Wired*, retrieved from https://www.wired.com/1996/12/ffedex/.

3 Stoll, J. D. 2020, October 24. "Inside Fitbit's quest to be your health monitor." *Wall Street Journal*, retrieved from https://www.wsj.com/articles/inside-fitbits-quest-to-be-your-health-monitor-11603540800.

Selling Selling initiatives are also customer facing. Unlike wrapping efforts, however, they generate revenue as a stand-alone offering, not as a part of some other product. The selling strategy is the core business model of information businesses and infomediaries (Chapter 5). Consider STR Global as an example. The firm, incorporated in 1985, is the leading provider of hospitality industry market data, offering competitive benchmarks, industry analysis, trends, and profitability analyses. As more organizations undergo digital transformation, it is increasingly common to see firms that are not expressly in the information business attempting to directly exploit the data they generate through selling (Figure 8.20).

Consider the Finnish pharmaceutical distributor Tamro. The firm started to sell data about drug sales to pharmacies that could then benchmark their performance with that of competitors. The same data are then provided to drug manufacturers, who benefit from the granularity of sales information. Cleverly, the firm acts as both a physical product distributor and an information provider, up- and downstream in its value network.[1]

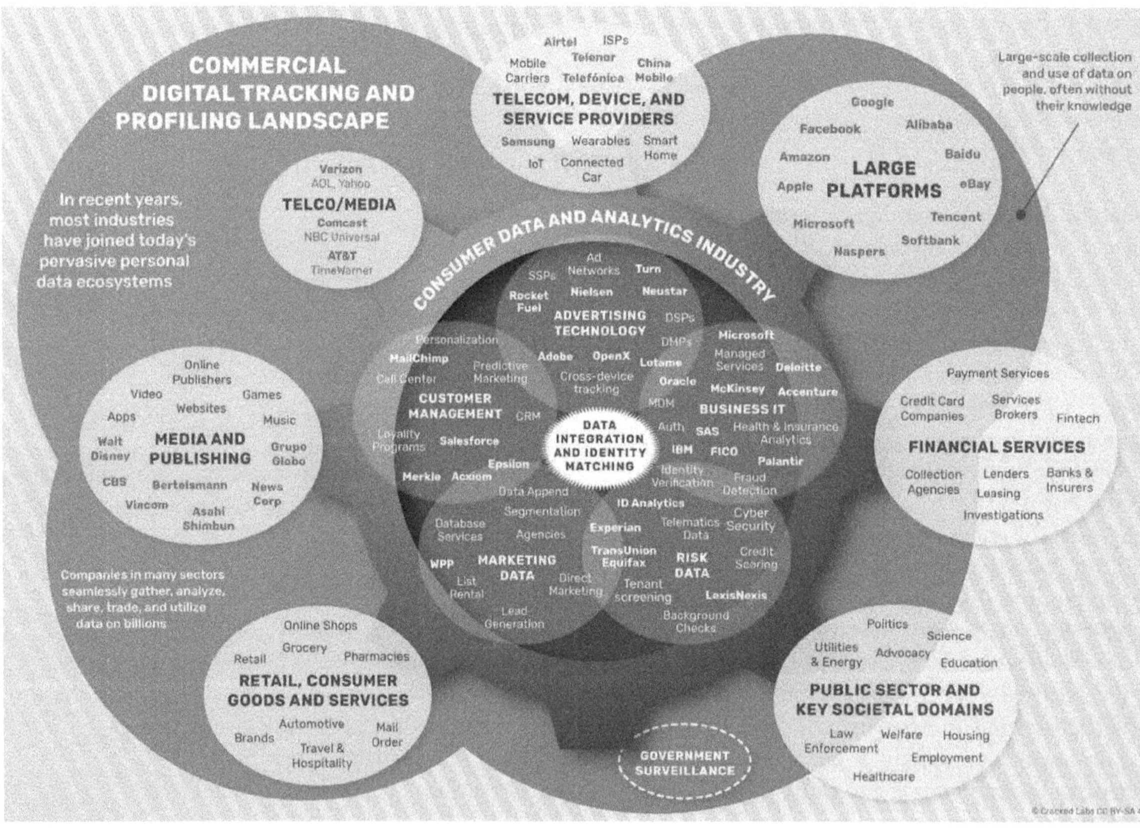

Figure 8.20. Commercial digital tracking and profiling landscape

Source: Christl, W., 2017, June 8, "Corporate surveillance in everyday life," Cracked Labs, retrieved from http://crackedlabs.org/en/corporate-surveillance

---

1 Parvinen, P., Pöyry, E., Gustafsson, R., Laitila, M., and Rossi, M. 2020. "Advancing data monetization and the creation of data-based business models." *Communications of the Association for Information Systems 47*. doi: 10.17705/1CAIS.04702.

When it comes to selling initiatives, remembering the "no action, no value" rule is particularly important. Selling initiatives can occur at each stage of the information offerings consumption path, from the outright trade of raw data to the development of digital resources (Chapter 5) that fully automate some action on behalf of the customer. On one end of the spectrum, consider credit card companies that sell their raw transaction data to hedge funds and advertising networks[1] that analyze them to inform their campaign and investment strategies. On the other, consider Stripe and its fraud detection functionality. Any developer that uses Stripe's digital capability to process payments can set a threshold for refusing transactions that are suspected of being fraudulent. All the analytics to make the determination, and even the processing or halting actions for each payment, are performed by Stripe. What is clear in both of these examples is that the data are not enough to create value—there must be a process in which an action is produced. No action, no value.

Regardless of what form selling takes, this data monetization approach requires a distinct business model.[2] For this reason, while intuitively appealing to any organization that generates data as a by-product of its core business, the selling strategy is not simple to execute. There are few examples of successful data monetization selling initiatives by firms that don't make information their core business (i.e., information businesses).

While related, the three classes of data monetization strategies differ on a number of dimensions, as summarized in Table 8.2.

The discussion up to this point has largely assumed that the data were collected and are available, at least in their raw form. However, you realize that the opportunity to collect data from operations is not uniformly distributed across companies. It depends on the industry that they are in and the degree to which they have digitized their operations. In other words, some firms are highly constrained when it comes to gathering customer data and may have no better way to obtain it than to pay a representative sample of customers to take the time to respond to surveys. Others have more data than they can ever hope to meaningfully put to use. We refer to this as the **degree of unobtrusive data capture**. Despite being a mouthful, it is a largely intuitive concept that indicates the extent to which, in the normal course of business, customer data are collected and stored in a readily usable format. As an example, consider the data sources that modern firms use to map their customers' journey (Figure 8.21). As you can see, aside from the use of customer surveys, all the techniques are unobtrusive.

An early analysis of practices in your industry can be illuminating in this regard. Imagine, for example, a fine dining restaurant. Fine dining is an industry where the personalization of the service is highly suitable, yet much of the data needed to carry it out are generated in fleeting customer-server

Table 8.2. Key differences between the three data monetization strategies

	Improve	Wrap	Sell
Value creation driver	Internal process improvement	Product/service differentiation	Data monetization
Primary metrics	Process/decision improvement	Existing product/service revenue	New revenue stream
Primary customer	Internal process owner	Existing customer	New customer / data marketplace

1 Helm, B. 2020, May 12. "Credit card companies are tracking shoppers like never before: Inside the next phase of surveillance capitalism." Fast Company, retrieved from https://www.fastcompany.com/90490923/credit-card-companies-are-tracking-shoppers-like-never-before-inside-the-next-phase-of-surveillance-capitalism.

2 Buff, A., Wixom, B. H., and Tallon, P. P. 2015, July. "Foundations for data monetization." Center for IS Research Working Paper #402, Sloan School of Management, Massachusetts Institute of Technology.

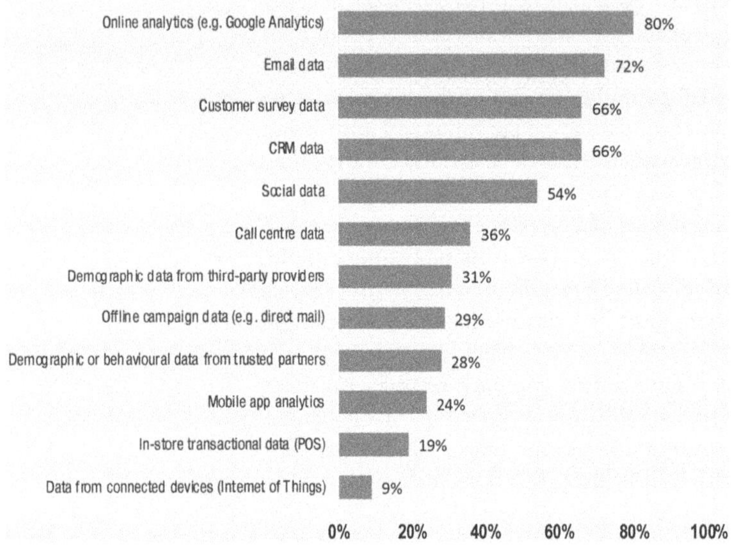

Figure 8.21. Data sources used to understand the customer journey
Source: Econsultancy, 2017, ***The new marketing reality***, p. 15, IBM, respondents: 806

exchanges that are difficult to capture and codify for easy storage and retrieval. Add to this mix the high employee turnover typical of the food service industry and it becomes clear why, for as much patronage as we give to our favorite restaurants, we generally don't receive a commensurate degree of personal service.

Compare the difficulty a restaurant has with collecting and storing its customer data in a readily usable format to the relative simplicity of the same task for a ghost kitchen. Granted, the potential depth of the relationship is lower, but the ease with which online-delivery-only restaurants can collect, store, and process the data you provide is much greater.

The degree of unobtrusive data capture for a firm is generally fixed at any point in time. However, the relentless pace of innovation in digital technology—for example, in sensors (Chapter 1)—constantly enables new digital data streams (Chapter 12) and new human habits. For example, while much of the information about customers' gambling behavior in casinos was traditionally left to busy and fallible casino hosts and pit bosses, the advent of electronic slot machines ushered in a new era. Casino executives realized that a modern slot machine is, in essence, a digital computer and that a computer records all the transactions it performs with great speed and accuracy. Tying these transactions to individual customers once they were convinced to use magnetic strip cards was a relatively small step. Today, the natural course of business in the casino industry is such that a company can have an accurate, real-time picture of each of its customers' slot-playing behavior. With the declining price point of RFID tags, this same level of precision is increasingly integrated to table playing as well, not just slots.

8.4 Value Creation via Digital Platforms and Aggregators

Up to this point, we have focused on traditional ways to create value from digital technology and digital data. However, as you know, aggregators, digital platforms, and digital resource creators operate differently. When a firm has the chance to own a digital platform, it is imperative for it to seize the opportunity. Those who are unable to do so must often compete within a dominant aggregator or digital platform—and must do so at their best. In either case, you as a manager should understand how to create value in both circumstances. In this section, we focus on digital value creation within and across digital platforms and aggregators ecosystems.

Value Creation as a Digital Platform

Increasingly, organizations are seeking opportunities to establish a digital platform. The "**platform play**" consists in developing resources that can be exposed to other organizations who will leverage them into their own digital innovations. In other words, the platform is a provider and orchestrator of resources, rather than a provider of finished products or services in a linear value chain. Resources can be assets (i.e., tangible or intangible things the firm controls) or capabilities (i.e., competencies the firm developed; see Sidebar 8.1).

SIDEBAR 8.1. DIGITAL RESOURCES

Resources are any tangible or intangible means of creating the firm's products, services, processes, or activities. They are formally defined as "assets and capabilities that are available and useful in detecting and responding to market opportunities or threats."[1] Assets are any tangible or intangible things the firm owns or controls. Capabilities are the repeatable patterns of actions a firm is able to execute to deploy its assets in creating and offering its products or services to the market. Less formally, assets are things the firm controls and capabilities are things the firm can do.

Traditional IT assets are associated with hardware and software (e.g., IT infrastructure, information repositories), whereas IT capabilities stem from human and organizational competencies (e.g., IS-business partnerships, software development skills). When digital technology is integrated with organizational resources, we have IT-enabled resources. These resources have properties that emerge from the combination of the technology and the organizational resource, properties that neither the digital technology nor the organizational resource alone would have. Consider the example of telemedicine. Remote surgery is the notion that a doctor can operate on a patient at a distance with the aid of local staff and robotic equipment that she controls remotely via a computer (Figure 8.22).

More common today, a remote surgery system is a sociotechnical system (Chapter 2) that displays emergent properties. For it to work, you need doctors with the competencies to perform the surgery and skills in operating the telemedicine equipment. These individuals are geographically separated from the patients. Therefore, you also need local medical staff to prepare both the patient and the telemedicine robotic equipment. The emergent property is the ability to perform remote heart or cancer surgery, a resource that neither the remote doctors nor the local equipment and local doctors would have taken independently. Only when the local technology and staff work in combination with the remote heart surgeons can the medical procedure successfully happen.

Finally, digital resources (Chapter 5) are modular assets or capabilities that can be incorporated into digital innovations via software interfaces. These assets and capabilities are embedded in software, such as Stripe or Twilio communication capabilities.

1 Wade, M., and Hulland, J. 2004. "Review: The resource-based view and information systems research: Review, extension, and suggestions for future research." *MIS Quarterly 28*(1): 109. doi: 10.2307/25148626.

Consider Shopify. The firm believes that "anyone, anywhere, can start a business," and it exhorts would-be digital entrepreneurs to "bring your business online. Create an ecommerce website backed by powerful tools that help you find customers, drive sales, and manage your day-to-day."[1] Shopify is a digital platform. It enables digital innovation in the form of online retailing for traditional stores and individuals. Unlike Amazon, the dominant aggregator in the retail space, Shopify does not manage a marketplace, and it does not maintain direct relationships with consumers. In fact, you may never have heard of the company even if you have used its services through the digital storefront of one of its more than one million customers in 175-plus countries. The "powerful tools" that Shopify makes available to digital entrepreneurs are the online global and multilanguage store, Shopify Shipping, Shopify Payments, and Shopify POS.[2] These resources enable online retailers to create, manage, and maintain digital relationships with their customers. Shopify is hailed by some as the "anti-Amazon"—the firm that helps independent retailers escape the control of the Amazon marketplace.

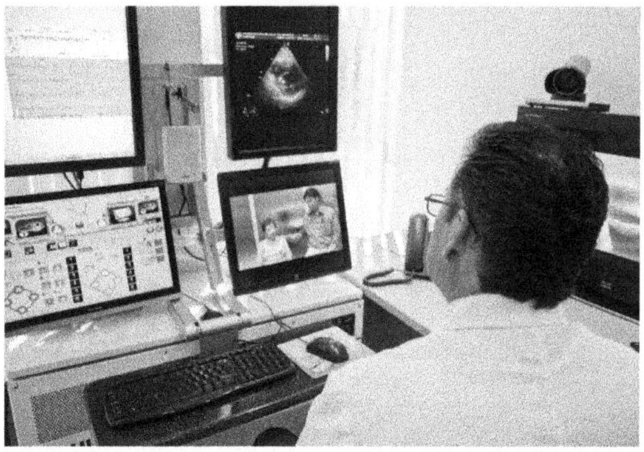

Figure 8.22. Telemedicine equipment
Source: Photo by IntelFreePress / CC BY-SA 2.0

1 https://www.shopify.com/.
2 https://www.shopify.com/pricing.

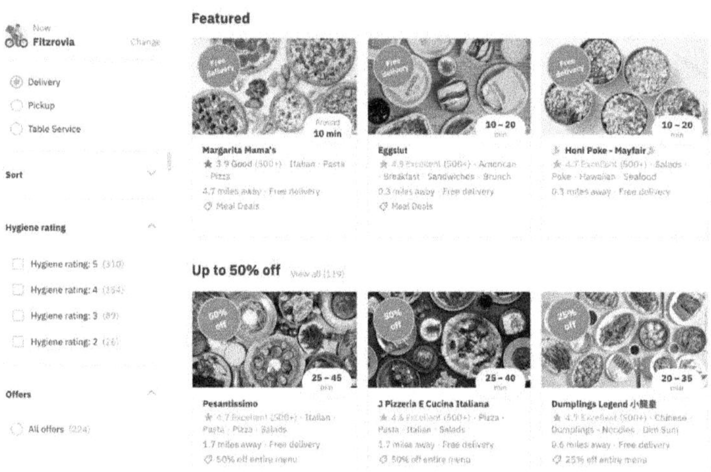

Figure 8.23. Deliveroo's local delicatessen promotions

The resources exposed by a platform may be digital or not. If the firm is a digital resource provider in the algorithmic economy (Chapter 5), like Stripe or Twilio, the resources it exposes are integrated by the digital innovator into its products (e.g., the Instacart app) via a programmatic interface (i.e., an application program interface [API]). However, a platform could expose more traditional resources that are not accessed programmatically. For example, Deliveroo enables restaurants to run promotions, allowing some of their menu items to appear in the top carousel (Figure 8.23). A restaurant must manually enable the promotions via a dashboard, as there is no way to access the promotion capability programmatically through APIs.

Creating value as a digital platform hinges on the firm's ability to create and market resources that other firms are willing to pay for. A platform can be a generalist or a specialist. A generalist develops resources that have a wide range of applications across many diverse industries. Examples include Stripe in payment and Twilio in digital communication. Service providers as different as Uber and OpenTable use Stripe to process payments, as do niche retailers like the Taittinger Champagne house and creators like Ben Thompson of Stratechery.com. When competing as a generalist, the platform must identify a vexing problem and provide effective solutions that are easily integrated into the user's own digital innovations. A platform may also position itself as a specialist in a particular vertical market. Teachable is one such example focused on educators interested in creating and marketing their own online courses. Teachable competes with other providers, such as Udemy or Kajabi, on its ability to create a set of specialized resources for designing, creating, and administering online teaching materials.

As you know already, products and services do not have intrinsic value. Value is created only when some user is willing to pay for what has been created (Chapter 7). Resources exposed by platforms are no different in that it is not enough to create the resource—the platform must ensure that a sufficient number of digital innovators are willing to pay for its use. A pivotal role is played by trust. Trust in vendors and in their ability to deliver quality products at the agreed-upon conditions has always been necessary in business relationships. But for digital innovators, choosing a platform to work with is even more important because the integration of digital resources into a digital innovation creates a permanent dependency on the resource provider. Instacart, for example, relies on Twilio's communication resources that are a prerequisite for the application to work. That is, any downtime or obstacle to the ability of Twilio to operate brings the Instacart service to a halt. This scenario is quite different from the traditional approach to software services, where the user licenses and installs a software application in its data center. In this scenario, even if the vendor ceases operations, the application does not stop working. Digital resources are "live" and within the control of the platform for as long as the digital innovation relies on them. The platform could change pricing terms; it could modify or even discontinue the resource that is in fact a critical input into the services of the innovator. Being able to convince innovators to tie their future to the platform requires a compelling value proposition and significant trust.

Digital platforms are enabled by the cloud (Chapter 3) and typically embrace "servitization" by offering software as a service (SaaS). This approach is similar to the servitization of music by streaming services. When you enroll in Apple Music or Spotify, you don't own the music, as you did when you purchased CDs or MP3s. Rather, you have a right to listen to a library of songs (i.e., use the service) for as long as you pay the monthly subscription. Servitized digital resources exposed by digital platforms are monetized through a subscription or a pay-per-use revenue model (Chapter 5). The digital platform business model is very different from the one of firms in a traditional value network, where suppliers would sell their product or service to a downstream customer (e.g., Intel selling microprocessors to Dell at a margin). In servitized businesses, revenue is received over an extended period of time—namely, the customer lifetime—but the main effort the firm makes in acquiring those customers is front-loaded. The ability to retain existing customers is also critical, with customer's satisfaction and continued delivery of customer value being paramount. When dissatisfied customers leave the firm, or *churn*, the organization foregoes the future revenue stream it worked so hard to initiate.

As you see from the discussion above, because of its novelty and uniqueness, the servitized model requires a number of specialized metrics to track and measure value creation. You must be familiar with the most important ones if you are to fruitfully join the conversation (Sidebar 8.2). The peculiarities of servitization are reflected in cash flow statements and projections. While revenue is received over time, the customer acquisition cost (CAC) is accounted for immediately.[1] This creates a cash flow problem that is typical for digital platforms and must be proactively and attentively managed by balancing CAC with ***expected lifetime value***. Interestingly, the more aggressive a firm is in terms of growth targets, the more cash it needs to support the cost of acquisition of new customers. This is the reason, for many digital entrepreneurial ventures, we witness very high valuations while the firm is losing a lot of money. The investors are simply recognizing that the significant up-front investment will be paid off over the lifetime of the digital relationship the firm is establishing with its customers.

Some platforms have the ability to leverage network effects (Chapter 4). This is the case when customers must "join the platform" by consciously adopting the platform's own product. The classic example is operating systems (e.g., Windows) and gaming consoles (e.g., Xbox). These products are platforms that expose resources to digital innovators (e.g., software application and video game developers). But unlike platforms like Shopify or Stripe, they require customers who want to access the digital innovations to own their product. You have to purchase Windows to use Adobe Reader on your PC, and you have to purchase a Nintendo Switch to play *Kirby Star Allies*. Conversely, you can purchase ketchup from the Heinz online store without being aware that the retailer is powered by Shopify or book a tour of the Taittinger Champagne house in Reims, France, without knowing that the payment is processed by Stripe.

Platforms that serve as the nexus of the relationship between digital innovators (e.g., game developers) and their customers (e.g., gamers) can leverage two-sided network effects (Chapter 4). For example, consumers evaluating games will have to decide which console to buy (i.e., which platform to join). As more consumers join a specific platform—say, the Nintendo Switch—the incentive for game developers to write games for that audience grows and, through positive feedback, the stronger gets stronger and the weaker gets weaker. Value creation through platforms requires the firm to create and market resources that digital innovators find useful in creating their product or service. For those platforms that also require customer adoption, it is even more critical to ignite network effects and take advantage of positive feedback.

---

1 The CAC includes all direct and indirect expenses accounted in order to acquire new customers. Among other items, we can find the following: advertisement, salespeople base and commissions, exhibitions and trade shows, travel expenses, demos, and free trials.

SIDEBAR 8.2. The Metrics of a Servitized Business

- **Monthly recurring revenue (MRR):** MRR measures the (normalized) monthly subscription revenue, and it represents the foreseeable revenue that a firm receives monthly. It is computed by summing all the monthly subscriptions or fees, eventually dividing multimonth payments by the number of months. MRR growth is a clear indication that the business is growing and serves as a primary metric of interest to investors.
- **Annualized run rate (ARR):** ARR represents the annualized revenues extrapolated from the data of a shorter period (typically a month). It is computed by identifying periods representative of the overall business and uses them as a baseline for the annual estimation. The choice of the representative months is critical. For example, taking November and December in retail as a base for your analysis would result in an overestimation, as those months are typically the most profitable.
- **Annual recurring revenue (ARR):** This consists of a "cloud" declination of the annualized run rate. The two are often confused, as both are referred to as *ARR*. The underlying logic is, in any case, the same. The only difference is that annual recurring revenue is specific for those business models based on subscriptions or recurring fees. It is used to project on a yearly base the current revenues, assuming the customer base remains the same. It is computed as annualized MRR, thus multiplying MRR by 12. The choice of ARR or MRR for the analysis depends on the type of subscription or contracts. Businesses relying on monthly payments may better use MRR, while those based on annual subscriptions should use the ARR.
- **Churn rate:** This is the rate at which your customers cancel their service subscriptions. It is computed by dividing the number of canceled subscriptions or customers in a period (e.g., monthly, quarterly, yearly) by the total number of subscriptions or customers at the beginning of the period. Thus (customers at the beginning of the month – customers at the end of the month) / customers at the beginning of the month. While the **churn rate** represents a key metric in platform businesses, the fundamental business question to answer is Why do customers churn?
- **Revenue churn rate:** Also called the *MRR churn rate*, this measures the rate at which revenue is lost because of customers canceling their subscriptions. It represents the monetary counterpart of the churn rate. It is particularly significant when the offer is articulated on different plans. It is computed as a percentage of the ratio between the revenue churned in a period and the revenue at the beginning. Thus (revenue lost in a month / revenue at the beginning of the month) * 100.
- **MRR retention rate:** This measure represents an estimate of the effect on the monthly revenues of the customers that churn. It is computed by dividing the MRR of the renewed subscriptions by the MRR supposedly to be renewed.
- **Customer retention rate (CRR):** CRR represents the proportion of customers who renew their subscription. In the case of a monthly base, it is computed, as a percentage, as (customers at the end of the month – new customers) / customers at the beginning of the month.
- **Lifetime value of a customer (LTV):** Also referred as customer lifetime value (CLV), this measure represents the cumulative revenue that the average customer will produce for the firm for the duration of the relationship. It is computed as average monthly revenue per customer * lifetime months. The time interval can vary (monthly, quarterly, yearly) depending on the objective of the analysis. It is customary to limit lifetime months to 36 for computation purposes.
- **Customer acquisition cost (CAC):** CAC indicates the average up-front cost the company incurred in order to acquire each new customer during a given year. It is computed by dividing the total annual sales costs by the number of annual new customers.
- **LTV CAC ratio:** This ratio indicates how much revenue the company will receive over the lifetime of a customer given the up-front cost of acquisition. It is an important measure of profitability, and it is desirable that the ratio exceed three.
- **Months to recover CAC (CAC / Monthly Revenue):** This metric measures the estimated time to profitability—meaning, how long before the firm recovers the CAC and starts to have a cumulative positive cash flow. A desirable time to recover is less than 12 months.

Igniting network effects is both art and science. While digital entrepreneurs and innovators are constantly devising new "growth hacks" for leveraging network effects, there are a number of approaches that have proven useful in the past:

1. **Create one side of the network.** Apple launched the iPhone in 2007. At the time, there was no App Store, since Steve Jobs could not be convinced about turning the iPhone from a device into a platform. Despite its closed nature, the iPhone sold one million units in just 74 days from its release and sold over six million units before being discontinued. In 2008, Apple executives managed to convince Jobs that exposing resources of the iPhone to external developers (i.e., creating a platform), and allowing them to contribute apps, would dramatically improve the value of the iPhone to consumers and, through network effects, lead to exponential sales growth (Figure 8.24).

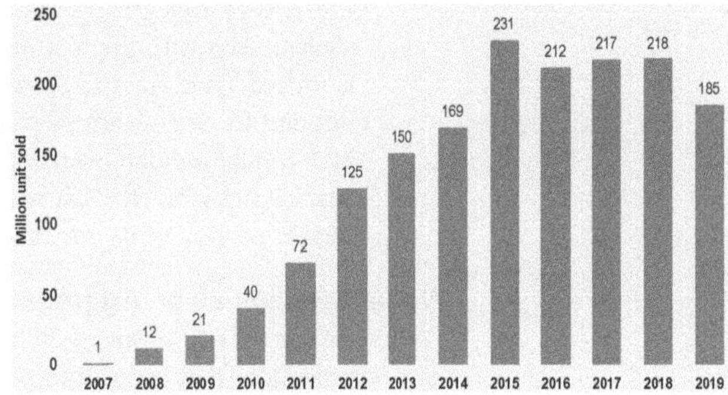

Figure 8.24. Apple iPhone total sales worldwide

Source: Apple's 10-K annual filings, retrieved from https://investor.apple.com/sec-filings/default.aspx

   The App Store was launched in July 2008, and the rest is history. Note, however, that Apple's decision to create the first apps internally was instrumental in igniting the network effect and attracting independent developers because it proved customer demand for the phone.

2. **Subsidize one side of the network.** A fairly standard practice for igniting network effects and solving the chicken-and-egg problem (Chapter 4) is to actively subsidize one side of a two- or multisided market. This is the strategy pioneered by firms like Adobe when pushing free downloads of Adobe Reader and currently pursued by food delivery aggregators (e.g., Deliveroo, Uber Eats) when running promotions such as free delivery to new customers.
3. **Come for the tools, stay for the network.** This strategy is available to organizations that provide tangible, direct benefits to adopters even when they use the service individually. Pinterest, the image-sharing and social media service, was able to use its "online scrapbook" function to attract customers. The firm made it easy for individuals to collect and organize images found on the web. As the community of "image collectors" grew, the value of sharing and communicating increased, and the network effect was ignited. An example in the business-to-business (B2B) platform context is offered by Snowflake. The recently public company is primarily a cloud-based provider of data-warehousing solutions (Chapter 3). As such, it attracts clients based on the strength and performance of its "data warehouse-as-a-service" software. However, one of the most innovative aspects of Snowflake's platform is its Snowflake Data Marketplace, which, as the firm states, "gives data scientists, business intelligence and analytics professionals, and everyone who desires data-driven decision-making access to live and ready-to-query data from your ecosystem of business partners and customers, and from potentially thousands of data providers and data service providers."[1] Given your understanding of network effects (Chapter 4), you quickly recognize what is going on here. As organizations join the Snowflake ecosystem to use its data warehousing products internally (i.e., the

1 Snowflake Data Marketplace. 2020. Snowflake Data Marketplace, retrieved from https://www.snowflake.com/data-marketplace/.

tools), they are creating value for all other Snowflake users who may be interested in sharing data with the firm (i.e., the network). To the extent that such data sharing is indeed value creating, you can see how the positive feedback loop would make it difficult for organizations to leave the ecosystem, even if they were attracted away by a competing data warehousing provider. If Snowflake is able to ignite the data-sharing network effect, it will require collective switching efforts (Chapter 9) by a sizable portion of its customer base to imperil Snowflakes' leadership position.

Value Creation as an Aggregator

When a digital platform successfully integrates a marketplace, it configures as an aggregator. The few organizations that can pull this off achieve a position of considerable power because they collect pay-per-use revenue from digital innovators using the resources they expose, leverage network effects to stave off competition, and retain control over the digital relationship with the end customer. Becoming a successful aggregator implies the ability to grow a platform by creating and marketing resources and, if available, to leverage network effects. It also requires the ability to create and manage a thriving two-sided or multisided marketplace. Once the marketplace is established, the aggregator's greatest asset is arguably the community of ecosystem participants, suppliers, and customers who choose to do business on the platform and transact in the aggregator's marketplace. Integrating a marketplace with a platform provides four advantages and "control levers" to the aggregator:

1. **Control of digital relationships with customers.** Through its owned and managed marketplace, the aggregator has direct control of the relationship with customers in its ecosystem. Such a direct relationship endows the aggregator with a position of power (Chapter 4) that tends to commoditize suppliers. Consider the Amazon Marketplace. If you are an Amazon Prime customer, you have implicitly signaled that you value your relationship with Amazon. If a supplier ceases to sell its products on the marketplace, the most likely outcome is that you will purchase a similar product from a competitor. For example, while notorious brands including Nike and Ikea stopped selling on Amazon, third parties filled this void on the marketplace, and most customers did not defect from Amazon.[1]
2. **Governance and curation of digital innovations.** The integration of a marketplace and platform that characterizes aggregators affords them substantial influence over the suppliers' digital innovations. Consider Apple and its control over the App Store. The terms of service for using the iOS platform specify, among many other rules, that pornography will not be allowed. Developers must submit their apps for approval before they become available to consumers for download in the App Store. Ownership of the marketplace gives Apple complete gatekeeping power. While any digital platform could craft such service terms (e.g., Windows), their enforcement is more difficult without control of the marketplace.
3. **Data visibility.** Control of the marketplace endows the aggregator with full visibility of customer preference and transaction data. Deliveroo, for example, can monitor customer preferences by tracking their browsing behavior. Doing so enables the aggregator to identify those restaurants that have the highest customer engagement and those that, while initially attracting attention, are unable to translate engagement into orders. Deliveroo knows the menu items that attract attention and those that are ignored. This level of visibility enables novel strategies, such as adjusting the ranking

1 Palmer, A. 2020, January 10. "More brands are leaving Amazon, but the strategy could backfire." CNBC, retrieved from https://www.cnbc.com/2020/01/10/more-brands-are-leaving-amazon-but-the-strategy-could-backfire.html.

of restaurants in the app depending on the time of the day, the day of the week, or customer past order history with the goal of maximizing customer conversions. Full visibility also enables the aggregator to estimate CWP through experimentation. For example, by dynamically and programmatically changing prices, Amazon estimates the demand curve for each of its products and sets prices that maximize revenue.

4. **Algorithmic control**. Aggregators have full control of the ranking algorithms that present choices to customers in the marketplace. Such algorithms are opaque to everyone else since they are an important and closely guarded intellectual property asset. But control of the algorithms enables the aggregator to shape competition inside the marketplace to its advantage. Google has notoriously leveraged its position as a search engine to increase the visibility of its other products, such as Google Reviews, Google Shopping, and Google Travel. There is ample speculation that Amazon favors its own products, Amazon Basics, in the product categories where it actively competes.[1]

Having described the strategies for value creation as a platform or an aggregator, it should be clear that achieving such a position is no small feat. In fact, the set of organizations that has the business model and capabilities to succeed in value creation as a platform or aggregator is a small one. But the continued proliferation of platforms and aggregators makes it all but certain that the firm you are working for finds itself having to leverage platform resources or finds itself competing within an aggregator. This is the case for online retailers selling their products on Amazon Marketplace and/or leveraging Shopify's resources to reach customers directly. It is the case for restaurants competing for visibility and orders in Deliveroo or DoorDash and game developers creating titles for the Xbox, PlayStation, or Nintendo Switch.

Value Creation within Digital Platforms

Creating value as a digital innovator or digital entrepreneur in a digital platform requires the ability to successfully leverage the platform's exposed assets and capabilities. While competitors have access to the same resources, it is the ability of the digital innovator to create differentiated products and/or services that ensure superiority. For example, while all PlayStation games leverage the same platform resources, games can vary widely in terms of how well they do so. Consequently, developers who invest in fine-tuning their games for a specific platform can provide superior gameplay. Similarly, online retailers that build deep integrations with Shopify's resources, such as order fulfillment, can create superior shopping experiences.

Value Creation within Aggregators

If value creation in a digital platform unfolds like the traditional value creation process (Chapter 7), aggregators add a layer of complexity. Aggregators are very dangerous because they control the customer interface. As we showed when we introduced the creator economy, aggregators commoditize their suppliers (e.g., drivers in Uber). The analysis of added value (Chapter 7) clarifies how being the provider of an undifferentiated commodity leaves the supplier with no added value and no pricing power. The most powerful aggregators tend to create this dynamic. So the best strategy to compete in an aggregator is . . . don't.

Return to the example of Shopify. The firm enables anyone to establish an online retail presence and manage a direct relationship with its customers. A similar value proposition is offered by

---

1 Edelman, G. 2019, November 24. "Amazon doesn't favor its own brands—except when it does." *Wired*, retrieved from https://www.wired.com/story/amazon-gating-private-labels-antitrust/; Mattioli, D. 2020, April 24. "Amazon scooped up data from its own sellers to launch competing products." *Wall Street Journal*, retrieved from https://www.wsj.com/articles/amazon-scooped-up-data-from-its-own-sellers-to-launch-competing-products-11587650015.

Stuart, the Paris-based provider of white label food delivery drivers. For those restaurants willing to forego the exposure offered by aggregators such as Deliveroo and able to take online orders independently, Stuart offers the resources for completing the delivery of the ordered items. In other words, with the exception of platforms that require reaching a customer via a physical device (e.g., iPhone, Xbox), in most industries, there are digital platforms and resource providers that enable digital innovators and entrepreneurs to transact directly with customers and forgo having to join an aggregator. Those organizations that are able to do so typically have a strong brand (e.g., the *New York Times*) or a highly differentiated product that they can leverage to persuade customers to engage directly with the firm (e.g., Nike).[1] In other situations, carefully crafting incentives for direct relationships, such as loyalty programs, is a viable solution. Large hotel operators, such as Marriott or Hilton, do so by strategically leveraging their frequent traveler programs.

But escaping the aggregator is not easy. There is a reason the aggregator is successful. In the case of Amazon, for example, while transaction costs on the Internet are low and "the competition is just one click away," the fact of the matter is that there is value in the convenience of shopping from a single reliable (and trusted!) retailer. Amazon has been laser focused on making sure that they are such a reliable source since introducing one of their first digital innovations in 1999: one-click shopping. Amazon relentlessly developed its superior customer service strategy with initiatives like a no-questions-asked return policy, Prime membership, quick refunds, competitive prices, and speedy services. The result is that Amazon today is a perennial member of the top-three list of most trusted and valuable brands[2]—a remarkable feat given the scrutiny the firm is under for its labor practices[3] and its ability to effectively police its marketplace.[4] If escaping the gravitational pull of the aggregator is impossible for your firm, it must devise a strategy to successfully create value *within* the aggregator's ecosystem. The following are critical elements of a successful value creation strategy:

1. **Visibility.** The aggregators' marketplace serves an important role as a discovery mechanism for customers. On Amazon.com, 78% of searches are for generic goods (e.g., "women's running shoes") rather than for specific brands or suppliers. Therefore, being visible in the marketplace becomes critical for attacking customers. Visibility is driven by the aggregator's ranking, discovery, and search algorithms. While no aggregator exposes such algorithms to scrutiny, the firms provide indications of the variables that are taken into account. For example, Amazon proclaims to optimize its shopping and discovery algorithms to feature the products customers will most likely want based on their search, another effort at increasing convenience.[5] General rules typically apply to any marketplace, such as having better product ratings and compelling product images and descriptions. Other variables

1 Novy-Williams, E., and Soper, S. 2019, November 13. "Nike pulling its products from Amazon in e-Commerce pivot." Bloomberg, retrieved from https://www.bloomberg.com/news/articles/2019-11-13/nike-will-end-its-pilot-project-selling-products-on-amazon-site.

2 Morning Consult. 2020. "Most trusted brands 2020." Morning Consult, retrieved from https://morningconsult.com/most-trusted-brands/; Winter, D. 2019, June 11. "Top 100 global brands 2019: The full ranking." *Financial Times*, retrieved from https://www.ft.com/content/3a3419f4-78b1-11e9-be7d-6d846537acab.

3 Greene, J. 2020, April 15. "Amazon fires two tech workers who criticized the company's warehouse workplace conditions." *Washington Post*, retrieved from https://www.washingtonpost.com/technology/2020/04/13/amazon-workers-fired/.

4 Moorhead, P. 2020, July 6. "Amazon establishes counterfeit crime unit to combat fake products." *Forbes*, retrieved from https://www.forbes.com/sites/moorinsights/2020/07/06/amazon-establishes-counterfeit-crime-unit-to-combat-counterfeiters-products/.

5 Mattioli, D. 2019, September 16. "Amazon changed search algorithm in ways that boost its own products." *Wall Street Journal*, retrieved from https://www.wsj.com/articles/amazon-changed-search-algorithm-in-ways-that-boost-its-own-products-11568645345.

are aggregator specific. Moreover, since algorithms are proprietary and constantly changing, supplier's experimentation is critical to improve visibility.

2. **Compliance.** Aggregators own the relationship with the customer, and they typically position themselves as the advocates of the customer. Their curation and control occur via rules that govern suppliers' participation in the marketplace. Suppliers that excel in complying with the aggregator's rules and guidelines stand to benefit. For example, Apple traditionally gives a design award and writes articles in the App Store about those apps that provide services aligned with the firm's stated priorities (e.g., health, privacy). Deliveroo and Uber Eats promote restaurants that excel in their ability to consistently fulfill customer orders rapidly and accurately.
3. **Conversion.** Aggregators receive revenue when customers transact on the marketplace, and suppliers aim to optimize their offerings to gain visibility and maximize conversion and thus sales. For a supplier, gaining visibility is critical to increase conversions, and aggregators' discovery algorithms typically reward high conversion rates. To jumpstart this virtuous cycle, a supplier must learn to leverage the full range of resources exposed by the aggregator. For example, creating appealing videos and descriptions and responding to the Q&A section of the listing are activities known to increase conversion in the Amazon Marketplace.
4. **Outside moves.** Thriving within an aggregator does not only depend on actions that leverage the aggregator's resources. "Outside moves" are strategic actions a supplier carries out without leveraging the aggregator's resources, outside of the aggregator's platform. For example, by directing customer traffic to the aggregator, a seller may improve its visibility. Restaurants often run marketing campaigns on social media to promote their listings on Deliveroo and other aggregators. Outside moves drive qualified traffic to the aggregators, increasing sellers' conversion and visibility. Perhaps the most compelling outside moves are those that improve sellers' products or services with the discovery engine's ranking algorithms. For example, restaurants competing in Deliveroo can benefit greatly by improving kitchen processes to reduce food preparation times—a key variable of the Deliveroo ranking algorithm. Some of the best Amazon sellers continuously scan the highest ranked products in each category to complete their listings.

8.5 Managerial Takeaway

As you know by now, the competitive landscape has changed dramatically. As a future general or functional manager, the context in which you will be called upon to make strategic decisions is as complex as it has ever been. And such complexity is likely to continue to increase as digital technology takes center stage and competition is increasingly global. Your role in the organization is to contribute to the creation and appropriation of value and to deploy digital technology to enable digital value creation and capture. To do so, you have to make sure you first understand what configuration your organization takes. Is it a value chain, like McDonald's, Hilton Hotels, Ferrari, or Epic Games, the maker of *Fortnite*? In this case, much of the early thinking about digital value creation can be applied. Traditional frameworks like the VVC or the CSLC can be very useful to you. Moreover, you should make sure that your firm can engage in data monetization through improving and wrapping initiatives. On rare occasions, these types of companies can also monetize their valuable data directly by selling it, even though it is typically hard for firms that are not information businesses to sell data effectively. Finally, you will need to be very attentive about how your organization competes within aggregators or digital platforms. This is dangerous ground where digital value creation is severely constrained. If, instead, your organization is an aggregator, a digital platform, or a digital resource creator, it will operate differently. Rather than resembling a value chain, these organizations orchestrate the activity of suppliers and customers and don't focus on creating finished products. Your role in this type of company is to ensure that the firm gives itself the best chance to seize the opportunity. As the competitive landscape continues to evolve in response to waves of digital disruption, the frameworks in this chapter should help you think about how to best compete.

Summary

In this chapter, we focused on IT-dependent strategic initiatives, discussing the frameworks and analytical models that have been proposed over the years to help you identify opportunities to create value with IT and to design and develop value-adding, IT-dependent strategic initiatives.

Specifically, we introduced the following:

- Industry analysis, focusing on the characteristics of the industry your firm competes in, seeks to help you identify opportunities to deploy information systems to improve the profitability of the industry.
- Value chain analysis focuses on the firm's own unique transformation process. It seeks to spur your thinking about how information systems and technology can be used to introduce new activities and/or change the way the firm's activities are currently performed.
- The CSLC suggests that there is ample opportunity to create value by using information systems and technology to enhance the relationship with customers and enable superior customer service. The CSLC identifies four major phases and 13 stages in which the relationship between the firm and its customer can be mapped. Each one offers opportunities for value creation.
- The VVC recognizes the importance of the wealth of information available to today's organizations in the search for value creation. It identifies five sequential activities that a firm can use to transform raw data input into information outputs that have more value than the inputs. Using this approach, a firm can develop one of three classes of strategic initiatives: visibility, mirroring capability, and new digital value.
- Data can also offer the potential to create value with different strategies. Data monetization consists of devising strategies for profiting from the use of digital data as the core asset and analytics as the means of value extraction. Three general classes of IT-dependent strategic initiatives can be used to monetize data: improving, wrapping, and selling. The viability of the chosen strategy depends also on the degree of difficulty the firm encounters in collecting and using the needed data and the value of data in time.
- With digital platforms increasingly dominating the strategic discourse, traditional models of value creation may fall short. Platforms are providers and orchestrators of resources that can be assets (i.e., tangible or intangible things the firm controls) or capabilities (i.e., competencies the firm had developed).
- Creating value as a digital platform hinges on the firm's ability to create and market resources that other firms are willing to pay for. Enabled by the cloud, digital platform revenues consist of customers' fees and subscriptions. While received over the customer lifetime, the CACs are front-loaded, making the ability of a platform to retain existing customers fundamental. Specific metrics are used to assess a platform's performance.
- In digital platforms, effectively managing the network effect requires firms to devise appropriate strategies. The examples of the past showed the importance of creating one side of the network, subsidizing it, or attracting members with a core feature and then leveraging the ecosystem value to lock them in.
- The objectives and strategies vary if you are creating the platform or if you are developing an aggregator. As an aggregator, the control exerted ranges from the relationship with customers to the "gatekeeping" of the digital resources and marketplace. The visibility gained on the transaction constitutes a unique opportunity for value creation. The additional possibility of controlling discovery engines and algorithms on the platform allows for the devising of appropriate strategies to maximize conversions and overall transaction revenues.
- Participating in a platform or within an aggregator requires compliance with the terms of the service and adapting to its processes. As aggregators and platforms live on customer transaction fees or subscriptions, sellers

will need to specifically aim at being easily identifiable, often devising search engine optimization (SEO)-like strategies to maximize customers' conversion. Visibility within an aggregator is critical for the seller—they should develop a deep understanding of the algorithms that may maximize their exposures and eventually leverage resources beyond those available on the platform.

- Once the firm identifies a potentially value-adding strategy, it must ensure that it can appropriate the value created over time. In other words, the firm that has created competitive advantage by way of an IT-dependent strategic initiative must ensure that the advantage is sustainable; this is the topic of the next chapter.

Study Questions

1. Describe the focus and principal objectives of industry analysis applied to information systems. Select one of the five competitive forces and offer an example of a firm that you believe has been able to influence it by way of an IT-dependent strategic initiative.
2. Describe the focus and principal objectives of value chain analysis applied to information systems. Why is it important to contextualize the value chain? Provide an example of a firm that you think has been able to create competitive advantage using information systems. Identify the primary activities most impacted by information systems in this firm.
3. Describe the CSLC and its primary objectives. Provide an example of a firm that, in your opinion, has created competitive advantage using information systems to enable superior customer service. What stages of the CSLC are mostly impacted by the firm's IT-dependent strategic initiative?
4. Describe the basic tenets of the VVC. How does it differ from the physical value chain? Can you identify an example for each of the three applications of the VVC?
5. Choose a console of your preference and discuss the strategies Sony, Microsoft, or Nintendo devised to support the success of their respective platforms. Are they acting as aggregators? Please specify your answer.
6. Think about your last job or the job you'd like to have once you graduate. Where would you place this firm's product or service on the dimensions of theoretical repurchase frequency and **degree of customizability**? How difficult is it for the firm to collect and use customer data? Is the firm engaging in a customer data strategy? If not, is it missing the boat?
7. Consider the following quarterly results, then calculate the missing cloud metrics with the help of a spreadsheet, and comment on them. Would you consider the situation promising?

		Q1			Q2			Q3		
		Jan.	*Feb.*	*Mar.*	*Apr.*	*May*	*June*	*July*	*Aug.*	*Sept.*
Number of customers at the end of the month	**Base ($50/month)**	500	535	540	560	612	803	497	457	428
	Premium ($299/month)	10	15	17	18	22	27	30	36	49
Metrics	**Churn rate**									
	MRR									
	Revenue churn rate									

Glossary

A glossary of key terms can be found at the book's website. Visit https://www.prospectpressvt.com/textbooks/piccoli-information-systems-for-managers-5-0, then click on "Student Resources" in the horizontal red menu bar.

Chapter 9

Digital Value Capture

What You Will Learn in This Chapter

Business managers historically had a love-hate relationship with digital technology. They recognize its potential to help the firm compete, but they often lack the analytical tools and experience to make sound decisions about its use. The plethora of pundits who comment on the strategic potential of information systems and digital technology only add to the confusion. As you recall from the analysis of added value (Chapter 7), value creation is only useful insomuch as the digital innovator can appropriate, or capture, some of it. Imitation by competitors is the primary challenge to value appropriation. In this chapter, we provide a set of concepts and an analytical framework that will help you establish whether specific information technology (IT)-dependent strategic initiatives can be defended against competitors' retaliation. In other words, we will explore under what circumstances a competitive advantage rooted in the strategic deployment of digital technology is sustainable. Given the nature of sustained advantage, many of the cases and examples we analyze will be historical. Don't make the mistake of thinking that these are "old" examples. Using historical cases ensures that the story has fully played out and that the insights we can draw from them are reliable.

MINICASE: Digital Innovation at Domino's—Should You Follow Suit?

Five years after graduating, your information systems (IS) consulting business is thriving. At 6:45 a.m., when you power up your computer to check email and start the day, you see a late-night message from Gregg Yves, the vice president of marketing and customer service initiatives at Regional Chain of Pizza Shops (RCPS), headquartered in Bristol, England. He is an old friend whom you met through your university alumni network when you were in school and with whom you have kept in touch. Why would he be emailing so late? As you read his message, it becomes clear:

> Do you remember when we discussed the Domino's smart-watch app, they use to enable customer ordering?[1] At the time, we were focused on rolling out our online web ordering solution, and we did not pay much attention. We later considered following suit, but we felt that the effort was not justified by the potential benefits. Now they pushed it further again. They introduced a new Alexa skill. These devices sell like hot cakes, and I just read that over one-third of the population has access to one.[2] And their Domino's AnyWare initiative? You can order from a bunch of digital channels—I mean, I can order from my TV!
>
> I've been thinking about this all night. Can we match them? Should we follow suit as well? Do we need to move rapidly? I'd rather take a wait-and-see approach here, as I don't know if it makes sense for my customers, but I'm afraid that if we miss the boat on this initiative, the repercussions could be significant. You're an expert on this strategic information systems stuff; can you give me your insight and direct my thinking a bit?

1 See the minicase in Chapter 8 for a description of the Domino's smart-watch ordering initiative.

2 Lee, P., and Calugar-Pop, C. 2020. "The connected home: Just getting started." Deloitte, retrieved from https://www2.deloitte.com/uk/en/pages/technology-media-and-telecommunications/articles/the-connected-home-just-getting-started.html.

Discussion Questions

1. Do you believe that the Domino's voice-ordering initiative is sustainable?
2. What are the pros and cons of the wait-and-see approach that Gregg prefers?
3. What is your recommendation? What should RCPS do next?
4. What else could Gregg do if he decides not to engage in digital innovation?

9.1 Introduction

When it comes to using IS and digital technology, the primary objective of managers is to create added value (Chapter 7). However, creating added value is just one facet of the job; the firm must be able to capture the value created over time to truly benefit. In other words, any *competitive advantage* the firm generates with the implementation of its IT-dependent strategic initiative must be defended over time to ensure that the firm will be able to reap the benefits of digital innovation. Failing to do so will quickly lead to a situation where competitors match the leader and customers rather than the innovator end up appropriating the value created (Chapter 7). Digital entrepreneurs often refer to this effort as "creating a moat"—like the moat of a castle (Figure 9.1)—to ensure the ability of the new venture to appropriate the digital value it created.

Figure 9.1. The Château de Chenonceau on the River Cher

Source: Public domain

Before analyzing how digital value is appropriated, however, it is essential to establish that it can indeed be captured. That is, Can a firm protect an advantage based on the innovative use of information systems and digital technology? Can IT-dependent strategic initiatives deliver ***sustained* competitive advantage**? You may be tempted to respond affirmatively. But as digital technology becomes increasingly pervasive and commoditized, some argue, it does not afford any potential for sustained advantage. However, not all digital technology is created equal or behaves the same—particularly when used strategically.

We describe two very different historical examples showing how two different IT-dependent strategic initiatives, based on two different digital technologies at their core, can produce opposite results when it comes to appropriating economic value.

High-Speed Internet Access in Hotel Rooms

During the dot-com days of the late 1990s, as the number of Internet users was increasing at a staggering rate, a host of organizations—ranging from airport operators, to coffee shops, to malls—began offering high-speed Internet access (HSIA) and wireless connectivity to their customers. Lodging operators offered in-room HSIA capabilities in their hotels. Soon, HSIA became one of the hottest technologies to come to the lodging industry in a while. Companies offered HSIA as a paid amenity, with $9.95 for unlimited daily use being the most popular pricing option. While take rates (i.e., guests' actual usage) were much lower than expected, HSIA quickly became a "must-offer" amenity, based on the assumption that business travelers, who were used to high-speed connections at home and in the office, would snub hotels that could not have them surfing in the fast lane.

The inevitable result was an increasing number of properties that offered HSIA, and quickly, the amenity became free of charge—as the analysis of added value would predict (Chapter 7). For example, in February 2001, the Sheraton Vancouver Wall Centre announced that it was offering HSIA free of charge to all guests. In the press release introducing the initiative, the

Figure 9.2. Free Wi-Fi is a standard in three-star hotels

Source: Photo by Wesley Fryer / CC BY 2.0

HSIA vendor declared, "Offering this service as an amenity with no charge to the guest will certainly differentiate the Sheraton Vancouver Wall Centre from its competitors."[1] Any hoped-for differentiation did not last long, however, as more operators had to join the "HSIA as a free amenity" trend. Soon, free HSIA moved from a property-level amenity to a brand-level, free amenity. Omni Hotels began offering HSIA system-wide to all guests starting in February 2003, and Best Western and Holiday Inn quickly followed suit, as did many other major chains (Figure 9.2). Following the trend of countless amenities before it, HSIA was rapidly becoming just another cost of doing business in the lodging industry and a competitive necessity. HSIA is valuable indeed, and more bandwidth is always better. However, because it could not be protected from rapid imitation, all the value it created flowed to hotel guests rather than the hotel companies that introduced it. History is replete with examples of technology innovations whose value was "competed away" to customers. But this is not always true, as the next section demonstrates.

Business Intelligence at Caesars Entertainment

Caesars Entertainment (Figure 9.3) has been widely celebrated for its innovative use of business intelligence (Chapter 3) in support of its efforts to better understand its customers. Caesars (which at the time was known as Harrah's Entertainment) invested heavily in digital technology.[2] The advent of electronic slot machines ushered the firm—and later, the entire industry—into a new era. But technology is only the beginning of this story. Harrah's did not simply buy a bunch of computer systems, flip on the switch, and watch the dollars roll in. Instead, the firm embarked on a large-scale reorganization, centralizing and focusing operations on the brand and away from individual property interests.

Consider how Harrah's use of data differentiated them from the industry average. They started collecting customers' data for every transaction made at their properties through the usage of loyalty cards. For example, customers were incentivized to swipe the loyalty card to earn points before playing a slot machine. The same mechanism was used for all customer transactions, including dining, entertainment, and shopping at any affiliated property. Harrah's leveraged digital technology to collect customers' transactional data and segment them based on their spending behaviors. They then customized the services, pricing, and promotions accordingly. By discovering the drivers of profitability, the firm was able to differentiate and personalize the customer experience in a way that was revolutionary to the gaming industry, and it did so across all its properties.

Figure 9.3. Harrah's Casino in Atlantic City

The important point in this discussion, however, is how the early advantage could be sustained. As part of the reorganization, Harrah's hired a new breed of analysts, known as decision scientists (Chapter 1). The 200-member analytics team had the mind-set and skills to gather and analyze data

1 Hospitality Net. 2001. "RoomLinX Inc., the Sheraton Vancouver Wall Centre and One Wall, takes lead to offer high speed Internet as an amenity at no charge to guests in every guestroom." Hospitality Net, retrieved from https://www.hospitalitynet.org/news/4007069.html.

2 In this chapter, we refer to Caesars Entertainment as Harrah's in reference to the historical name the firm had at the time. It is interesting to note that Harrah's success enabled it to acquire Caesars Entertainment, but the firm decided to retain the latter's name after the merger.

about gamblers' characteristics and activities. By carrying out scientific experiments, Harrah's was able to become both more efficient (i.e., spend less) and more effective (i.e., spend better) in its use of funds to attract and retain gamblers while also increasing customer satisfaction and share-of-wallet (i.e., the percentage of the gambling budget a player would spend with Harrah's rather than its competitors).

The returns on Harrah's use of technology were considerable. At the same time, the centralized IT infrastructure and the processes it developed enabled Harrah's to expand its distribution with relative ease and control. The expansion strategy that followed produced impressive results, making Harrah's the world's largest provider of casino entertainment following the acquisition of Caesars Entertainment in 2004 for $5.2 billion. Clearly, Harrah's was able to appropriate quite a bit of the value it created with its IT-dependent strategic initiative. Revenue and operating income grew to $10.8 billion and $1.7 billion, respectively, in 2007, growing more than three times from the $3.4 billion and $283 million of year 2000.

However, a series of creative financial maneuvers and the 2008 financial crisis brought Harrah's to its knees. Even though their loyalty program was still producing a higher fair share of customers and spending, their superiority in analytics was not enough to save the firm.[1] After 15 years of sustained competitive advantage, the firm's attempt at financial engineering sent Caesars toward bankruptcy. On January 15, 2015, the firm's largest business unit filed for Chapter 11, sinking under the weight of its buyout debt. Interestingly, the most valuable and sought-after individual asset of the company remained the 300-gigabyte (GB) database housing loyalty program data for about 45 million customers, with an estimated value of one billion dollars.[2] As a final testament to the centrality of the business intelligence initiative to Harrah's and Caesars' success, all the properties sold in the bankruptcy experienced decreasing revenues—having lost access to the analytics and reward program once the new owners took them over.[3] Interestingly, both Caesars Entertainment and the Total Rewards program survived to 2015's debacle. In 2019, Total Rewards was renamed "Caesars Rewards" months before the merger with Eldorado Resort, a merger that grew the loyalty program to over 55 million members.[4] Caesars Rewards was still the leading gaming loyalty program in the world, and executives hoped it would be a crucial asset to help them navigate out of a new crisis: the 2020 COVID-19 pandemic.

Tesla and the Strategic Value of Autonomous Driving

There is another example where technology may lead to a sustained advantage. We don't have the benefit of history in this analysis, as this example is yet to fully play out. We are talking about Tesla's Autopilot program. Tesla took the industry by surprise when, in October 2015, it released a software update that enabled its cars to use eight cameras, front-facing radar, and 12 ultrasonic sensors mounted on the car to drive in traffic without human intervention. Their release made all other solutions on the market look obsolete. A plethora of online tests and reviews hyped the performance of Tesla's Autopilot, welcoming it as the first demonstration of the semiautonomous driving systems' capabilities. While competitors including Waymo (part of the Google universe) and Cruise (General Motors) apparently intended to release self-driving solutions only when

---

1 Caesars Entertainment. 2014, April 10. "SEC filing form 8-K." Caesars.com, retrieved from https://investor.caesars.com/static-files/af580d2e-9e8d-468b-aa99-f3c866bebef2.

2 O'Keeffe, K. 2015, March 19. "Real prize in Caesars fight: Data on players; customer loyalty program is valued at $1 billion by creditors." *Wall Street Journal.*

3 Short, J. E., and Todd, S. 2017, March 3. "What's your data worth?" *MIT Sloan*, retrieved from https://sloanreview.mit.edu/article/whats-your-data-worth/.

4 Chin, K. 2019, June 24. "Eldorado resorts to buy Caesars in $8.58 billion casino deal." *Wall Street Journal*, retrieved from https://www.wsj.com/articles/eldorado-resorts-to-buy-caesars-for-8-58-billion-11561372166.

fully mature, Tesla opted for a more incremental approach. Through software updates, they promised to make existing cars progressively more autonomous. The strategic choice was attributed to Autopilot reliance on "fleet learning." At the car level, Autopilot complements information in the environment captured from the cameras with radar and sensor readings. These inputs are then processed by dedicated software that recognizes the surroundings, routes the vehicle, and then activates the appropriate car system (e.g., steering, brakes, shifts, accelerator, suspensions, windows). Tesla Autopilot leverages deep learning neural networks (Chapter 12) trained on the data generated by all cars equipped with Autopilot-ready hardware. In other words, the key element of the strategy is shared "learning," achieved by aggregating data from all the Tesla cars on the road. This sharing should enable even a new car to recognize its surroundings and self-drive on roads it has never driven before. Then through over-the-air software updates, smarter versions of Autopilot are sent to the fleet to improve the cars' self-driving capabilities. Elon Musk is prone to overpromising, stating during a press call in late 2016,

> I feel pretty good about this goal [by the end of 2017]. We'll be able to do a demonstration guide of full autonomy all the way from LA to New York. So basically, from home in LA to Times Square in New York. And then have the car go and park itself by the end of next year.[1]

New Year's Day 2018 has come and gone, but the Tesla did not make it to Times Square for the celebration. The vows were renewed every year, and in 2020, Musk remained "confident that we will have the basic functionality for level 5 autonomy complete this year. . . . I drive the bleeding edge alpha build in my car personally. Almost at zero interventions between home and work."[2]

While a healthy dose of skepticism is good anytime we talk about digital innovation (and Elon Musk predictions), the Tesla example does illustrate what could happen if the firm is able to keep up its innovation trajectory given the strategic choices it has made with Autopilot. Autopilot is a strategic component of Tesla's value proposition, and it may indeed shape the car industry of the future—particularly now that other car manufacturers have been researching and deploying driver-assist features in their vehicles. Some early indications of Tesla success included its ability to collect data from over one million intersections every month, even during the 2020 pandemic.[3] Financially, the firm reported a $721 million profit on about $31.5 billion in sales for the full 2020 fiscal year, pushing its valuation to all-time highs (see Figure 9.4) and eclipsing every other automaker. While these numbers are the product of several factors, including the clear move toward electric vehicles and the status of Tesla as a "cult stock," Autopilot is a central element of Tesla's strategy.

Aside from its current success, the question for this chapter is "What would make Tesla's competitive advantage sustainable?" The answer may lay in the data. The hardware and sensors that constitute the building blocks of the Autopilot system are at risk of commoditization. What about the software, then? While, at the date of writing, several companies and start-ups are trying to develop the technology behind autonomous cars, there's a chance for Tesla to play a winner-take-all game because of the way self-driving capabilities are currently made possible. With deep learning instead of coding an impossible set of rules to make a car self-drive, the

---

1 Thompson, C. 2016, October 9. "Elon Musk: Fully autonomous Tesla will drive across the country by the end of 2017." *Business Insider*, retrieved from http://www.businessinsider.com/elon-musk-autonomous-tesla-drive-across-country-by-end-of-2017-2016-10.

2 Musk, E. 2020, August 14. "The FSD improvement will come . . ." (post on Twitter), retrieved from https://twitter.com/elonmusk/status/1294374864657162240.

3 Seeking Alpha. 2020, April 30. "Tesla, Inc. (TSLA) CEO Elon Musk on Q1 2020 results—earnings call transcript." Seeking Alpha, retrieved from https://seekingalpha.com/article/4341298-tesla-inc-tsla-ceo-elon-musk-on-q1-2020-results-earnings-call-transcript.

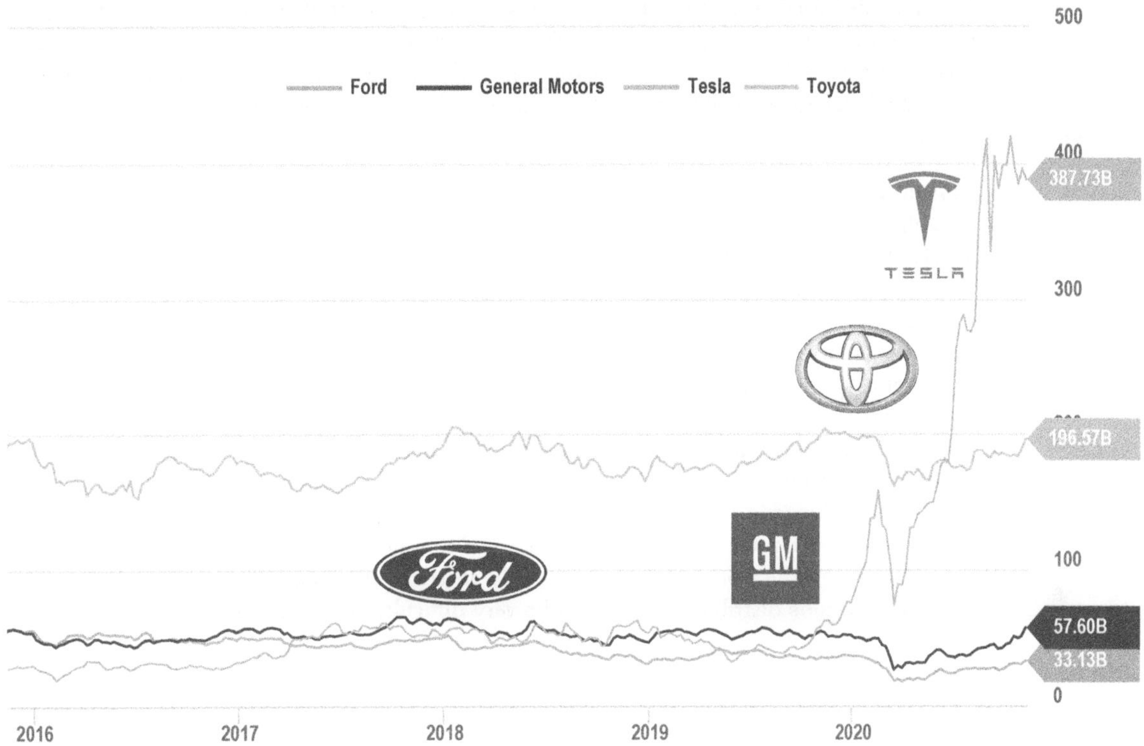

Figure 9.4. The market capitalization of car manufacturers
Source: stockrow.com

software learns from the data; the more data, the better the machine learns. Tesla has been collecting data from all its vehicles since the beginning—not only driving data but the details of the roads and surroundings in which the cars are driven—thus enabling a very precise mapping. In this sense, autonomous driving may be subject to strong network effects, as each car driving on the road is contributing to making all the others "smarter." Tesla may then build its sustainability, leveraging the network effect of the maps and driving data.

If nothing else, the above examples raise the question of whether managers should approach distinct IT-dependent strategic initiatives differently. Put another way, is there a way, a priori, to reduce uncertainty about whether an IT-dependent strategic initiative can help the digital innovator build a moat that will lead to a sustainable advantage? In the remainder of this chapter, we introduce a framework designed to help you in this analysis.

9.2 Appropriating Value over Time: Sustainability Framework

As an attentive reader of this book, you realize that the sustained competitive advantage associated with digital innovation and the strategic use of information systems doesn't just lay in the technology. Housing your new IT infrastructure on Amazon Elastic Compute Cloud or managing your Big Data with MongoDB are choices your competitors can replicate.

By now, you know that information systems are not equivalent to information technology (Chapter 2). Thus creating and appropriating value hinges on successfully deploying a defendable IT-dependent strategic initiative. It follows that the focus on the analysis of sustainability should be the initiative, in all its facets, not just the digital technology at the core. As the opening examples of this chapter show, even when the digital technology elements may be replicable by competitors, it does not follow that the firm's IT-dependent strategic initiative built on that (replicable) technology will be easily copied as well. This realization is the key to understanding digital value appropriation.

Sustainable Competitive Advantage

The ability of a firm to protect its competitive advantage, known as sustainability of the advantage, is often thought of as a binary condition—it is either possible or impossible for competitors to erode the leader's advantage by matching the added value it creates.[1] However, this view is misleading. After all, short of very few **resources**, such as patents or exclusive access to raw materials, almost anything is replicable—at least in theory! Thus it isn't whether the advantage is theoretically replicable that matters in practice; it is the difficulty that competitors face in matching the leader's offer.

Consider the example of Amazon, a firm that is famous for its relentless pursuit of customer service and customer satisfaction. In an effort to improve these important metrics of success, by the year 2020, Amazon had deployed 203 highly automated fulfillment centers—with a staff of robots supporting employees—strategically located throughout the United States and a number more throughout the world (Figure 9.5). While it is true that competitors could theoretically replicate Amazon's distribution, it would be very difficult, time consuming, and expensive for them to do so.

Resource-Based View

In the last couple of decades, much of the thinking in strategic management has coalesced around the resource-based view (RBV) of the firm. Within this approach, a firm is modeled as a bundle of resources (Chapter 8). A key contention of the RBV is that a firm's competitive advantage depends on the characteristics of the resources at its disposal, and when the firm controls resources that are valuable, rare, inimitable, and nonsubstitutable, the advantage will be difficult for competitors to overcome.

- *Valuable.* A resource is valuable when it underpins a value-adding strategy (Chapter 7). In other words, valuable resources enable the firm to offer a value proposition that either is superior to competitors' or, while equivalent, can be offered at a lower cost. For example, when Uber leverages its large network of about one million active drivers to perform its services, the IT-enabled "drivers" resource is valuable.
- *Rare.* A resource is rare when it is idiosyncratically distributed. In other words, it is scarce and not readily available for acquisition by competitors. Consider a firm that has developed a deep understanding of consumer electronics experience design, such as Apple Inc. While there is a vibrant market for design talent, there is no market to which competitors can go to acquire a "design capability." The same argument could not be made for office space, a resource for which (generally) there is a market with adequate supply.
- *Inimitable.* A resource is inimitable when competitors find it impossible, or difficult, to duplicate it. Returning to the example of Apple's seemingly magic touch in designing consumer electronics that have immediate appeal with a loyal customer base,

1 The term *sustainability* is used here in its business strategy connotation. The term has gained much currency lately to refer to environmental sustainability and "green" behaviors. The role of information systems for environmental sustainability will be discussed in Chapter 12.

competitors who seek to imitate such capability find that it is rooted in more than just the hiring and "aggregation" of talent. The difficulty of articulating how this capability comes to be and how it works in practice—a notion known as *causal ambiguity* in the literature—makes it difficult to replicate.

- *Nonsubstitutable.* A resource is nonsubstitutable when competitors are unable to replicate the firm's overall value proposition (e.g., an IT-dependent strategic initiative) using surrogate resources for the ones that are rare, valuable, and inimitable. In other words, if competitors can achieve the same results as the innovator using different resources, the advantage of the leader will be eroded and the inimitability of some of its resources will be rendered irrelevant. Wikipedia was able to produce an equivalent product to an encyclopedia by using crowd-sourced authors instead of distinguished experts in their field as one of the key resources. Experts were successfully substituted by collective intelligence, in this case.[1]

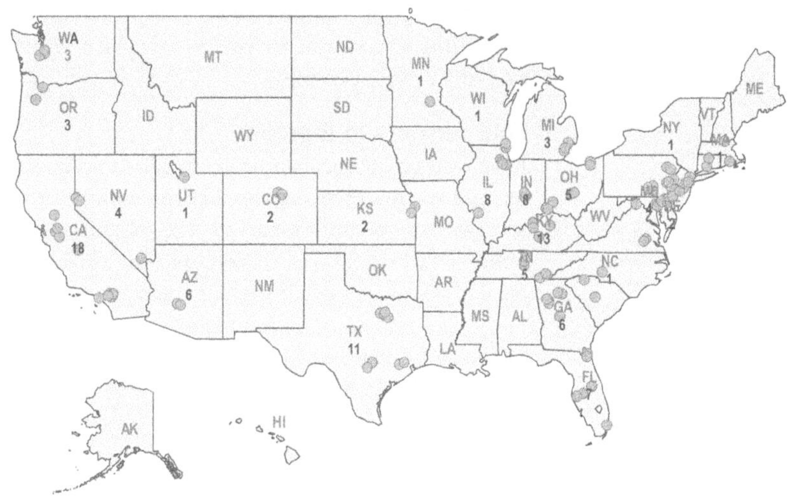

Figure 9.5. Amazon fulfillment centers in the United States
Source: http://www.mwpvl.com/html/amazon_com.html

Response Lag

A practical way to implement the ideas of the RBV framework in the technology arena is to think about sustainability not as a binary condition but rather in terms of how much time and money it would take competitors to erode the advantage that the leading firm has been able to create with its IT-dependent strategic initiative. The higher the "time and money" obstacles—termed here ***barriers to erosion***—the more resilient the firm's advantage.

Competitive retaliation occurs in stages. Once a firm's rivals find themselves at a disadvantage, they search for the sources of the firm's competitive advantage. If they are successful in identifying those sources, the competitors must decide whether they are able and willing to respond and, if they are, what approach they should take. ***Response lag***, the time it takes competitors to respond aggressively enough to erode a firm's competitive advantage, is a measure of the delay in competitive response. The longer the time and the higher the cost of replication, the more resilient the firm's advantage. Thus ***response-lag drivers*** are defined here as the characteristics of the technology, the firm, its competitors, and the value system in which the firm is embedded that combine to make replication of the IT-dependent strategic initiative difficult and costly. Response-lag drivers combine their effects to levy barriers to erosion. Note that this focus on barriers to erosion is critical because, when designing an IT-dependent strategic initiative, you have the opportunity to seed a sustainable competitive advantage—the moat, as Silicon Valley venture capitalists like to call it. The idea is not that, once created, the moat will protect the firm's

1 For more information, see Blanding, M. 2015, January 20. "Wikipedia or Encyclopædia Britannica: Which has more bias?" *Forbes*, retrieved from http://www.forbes.com/sites/hbsworkingknowledge/2015/01/20/wikipedia-or-encyclopaedia-britannica-which-has-more-bias/.

advantage forever. Rather, creating response lag will help make the firm's advantage resilient, giving it time to continue to improve its value proposition and protect it in the (often inevitable) case of strategic missteps.

Four Barriers to Erosion

Response-lag drivers can be grouped into one of the following four barriers to erosion: (1) the technology resources barrier, (2) the complementary resources barrier, (3) the project barrier, and (4) the preemption barrier. The magnitude of each barrier to erosion is determined by the number and strength of its response-lag drivers (Figure 9.6). We briefly describe the barrier to erosion below and detail each of the response-lag drivers associated with them.

Technology Resources Barrier IT-dependent strategic initiatives rely on access to the assets and capabilities necessary to produce and use the digital technology at their core. Two classes of response-lag drivers have historically contributed to the height of the technology resources barrier: IT assets and IT capabilities.

IT assets are tangible and intangible digital technology resources available to the organization, including hardware components and platforms (e.g., a private cloud), software applications and environments (e.g., a proprietary software using custom-developed analytical models), and data repositories. These resources contribute to building response lag directly by simplifying and speeding up the development and introduction of the initiative's digital technology core or indirectly by making it difficult for competitors who have no ready access to the needed technology resources to replicate the leader's initiative. An example of an IT asset is an organization's *digital infrastructure*, the set of digital technology components, both hardware and software, that are interconnected and managed by IS professionals with the objective of providing a set of standard services to the organization.

The legendary examples of the Sabre reservation system, pioneered by American Airlines (AA), and of United Airlines' Apollo are telling. These digital technologies were considered so expensive and risky that it took years for other airlines to invest in similar systems, despite the pioneer's clear success. Far from the big tech spotlight, Domino's Pizza has in recent years built an impressive digital infrastructure. They enabled 13 additional digital ordering channels, including Amazon Alexa, Google Home, Slack, Facebook Messenger, text messaging (yes, you can just send the pizza slice emoji), Twitter, Ford Sync, smart TVs, the voice assistant "Dom," and smart watches. Labeled "Domino's AnyWare," the new infrastructure required the integration of 85,000 data sources and the development of innovative solutions to process, analyze, and distribute the data generated.[1] When the COVID-19 pandemic hit, Domino's digital infrastructure sustained company growth, assuring business continuity.

While cloud services eased the acquisition of the technological elements of the digital infrastructure, its development times are generally still measured in years. Thus the response lag and ensuing barrier to imitation is likely to be substantial. Another classic example of an IT asset is *information repositories*. Information repositories are large data stores containing extensive information about customers, suppliers, products, or operations, organized in a structured form that is accessible and useable for decision-making purposes. Think back about the opening Tesla example. One of the key drivers of the firm's advantage is the camera data generated by the fleet of Tesla cars already on the road. Competitors attempting to replicate the leader's strategic initiative must not only duplicate the digital technology at its core but must also accumulate a comparable information resource—a feat that often takes substantial time.

---

1 Talend. 2018. "Domino's Pizza." Talend, retrieved from https://www.talend.com/customers/dominos-pizza/.

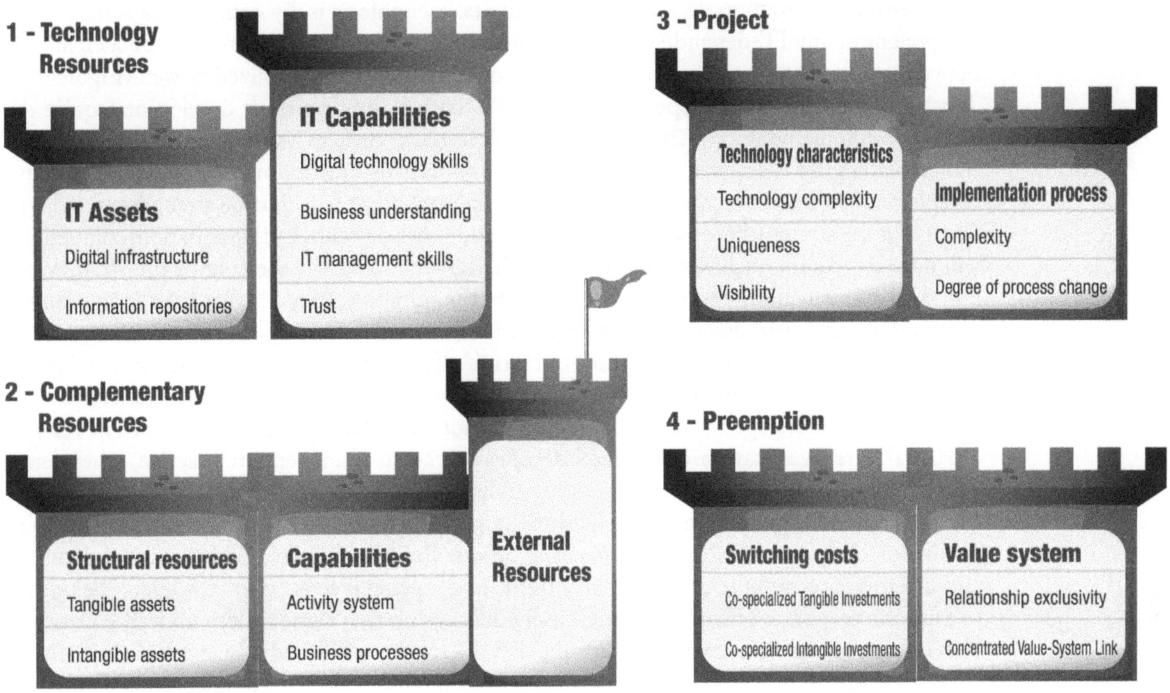

Figure 9.6. Four barriers to the erosion of advantage

IT capabilities are derived from the skills and abilities of the firm's workforce. These capabilities directly influence the response lag associated with the introduction of digital technology because they facilitate the technology's design and development. These capabilities also play a fundamental role in enabling effective and timely implementation, maintenance, and use of the technology. Examples of response-lag drivers in the IT capabilities category include *digital technology skills*, the ability to design and develop effective computer applications, and *business understanding* of the firm's IS professionals, which enables them to envision creative and feasible technical solutions to business problems. A high level of business understanding also contributes to the creation of response lag by mitigating the risks associated with the introduction of the strategic initiative and the relative investments in technology. *IT-management skills* and a *trusting* rapport between the IS function and business managers are other important IT capabilities that develop over time.

The height of the technology barrier increasingly depends on the firm's proficiency with digital resources (Chapter 5). As digital platforms continue to develop into enablers of digital strategy (Chapter 8), organizations rely on their ability to leverage the digital resources that the platforms expose into their strategic initiatives. It follows that some established IT assets and capabilities that organizations may have leveraged in the past to sustain their competitive advantage may lose value. For example, the continuing shift to the cloud makes ownership of large data centers and the skills to manage them less important. Other IT capabilities, such as the ability to effectively integrate digital resources from external partners (e.g., Twilio, Azure, Amazon Web Services [AWS]) into the firm's digital innovations, will become increasingly valuable. As digital technology constantly evolves, however, the importance of leveraging the technology barrier when deploying IT-enabled strategic initiatives will continue to be critical to the ability of your firm to sustain a competitive advantage.

Complementary Resources Barrier While digital technology is, by definition, a fundamental component of any IT-dependent strategic initiative, successful implementation of such an initiative requires that complementary organizational resources be mobilized as well (Figure 9.6). Thus to implement an IT-dependent strategic initiative, the firm must develop or acquire the necessary complementary resources (e.g., physical assets such as warehouses and distribution centers, intangible assets such as a brand). As an initiative becomes more reliant on distinctive complementary resources, the complementary resource barrier to imitation strengthens and replication of the strategy becomes slower, costlier, and more difficult. In this situation, competitors will have to acquire or develop not only the digital technology at the core of the strategy but also the complementary resources that underpin the initiative. (See Sidebar 9.1 for an example of response-lag drivers associated with this barrier.)

Consider again the opening example of Harrah's Entertainment and its business intelligence initiative. While the firm spent a significant amount of money to acquire digital technology resources, it also engaged in a radical reorganization when launching the initiative in the late 1990s. This reorganization challenged decades of casino management practice, where each casino within a chain operated in a highly independent fashion. Through its reorganization, Harrah's asked the general managers of each property to report to divisional presidents, who in turn reported up to Harrah's chief operating officer. The firm also created transfer mechanisms and incentives to support cross-property traffic and a general sense that customers "belonged" to Harrah's corporate office, not to each individual casino that signed them up to the program. This change in organizational structure enabled the success of Harrah's brand-wide initiative.

Fast forward now to the year 2001, when Harrah's was receiving substantial praise and attention for its use of guest data and putting pressure on competitors to imitate. How well positioned were its competitors to replicate Harrah's highly centralized customer data strategy? Not very, since the typical competitor still treated each property as unique and independent, with a distinctive brand and little incentive to share customers and customer data with the other casinos in the chain. Using the terminology of the sustainability framework, we can assert that Harrah's competitors were likely to find it costly and time consuming to successfully imitate Harrah's business intelligence initiative. This is because, at least in part, their organizational structure was not conducive to the strategy, and a change would have been very risky, costly, and time consuming. In other words, Harrah's had access to a unique complementary resource—namely, its idiosyncratic organizational structure—which created substantial response lag and contributed to heightening the complementary resource barrier. By 2015, this strategic initiative collected information on more than 45 million customers and was considered the most valued single asset of the company, worth one billion dollars.

Figure 9.7. A popular conveyor-belt sushi bar

Source: Image created by Chenyun on the Wikipedia project website https://commons.wikimedia.org/wiki/File:Conveyor_belt_sushi_overview.JPG

The analysis of complementary resources is important because, during the design of the initiative, it is often possible to use digital technology to leverage the impact of some idiosyncratic complementary resources. Imagine managing an upscale conveyor-belt sushi restaurant (Figure 9.7). Your store prides itself (and justifies its premium prices) on the superior quality and freshness of its sushi. Using radio-frequency identification (RFID) tags, you can keep constant track of the amount of time each dish has been out on the conveyor belt and thereby ensure that dishes that have been rotating for more than 25 minutes are removed to maintain freshness. With this infrastructure in place, you decide to add a display to each dish in order to show the time each dish has been rotating

SIDEBAR 9.1. Complementary Resources Barrier

Structural Resources

Structural resources comprise non-IT-related tangible and intangible internal assets used by the firm in the enactment of its IT-dependent strategic initiatives.

Tangible Assets

In theory, any tangible resource available to the firm can underpin an IT-dependent strategic initiative. Among these are competitive scope, physical assets, scale of operations and market share, organizational structure, governance, and slack resources.

Intangible Assets

As in the case of tangible resources, nearly any of a firm's intangible resources can support an IT-dependent strategic initiative. Examples of commonly cited intangible resources that can be so applied include corporate culture, top management commitment, and the ability to manage risk. As with tangible IT resources, complementary intangible resources create response lag by making a strategic initiative difficult, costly, and time consuming to imitate.

Capabilities

A firm's capabilities define how the firm carries out its productive activities. These resources specify what activities are performed and what steps or business processes make up those activities. The activities that the firm performs and the manner in which it performs them contribute to response lag and help sustain the competitive advantage created by the initiative.

Activity Systems

A performance-maximizing activity system relies on a set of economic activities that are both interlocking and mutually reinforcing, expressly showing internal consistency (internal fit), and appropriately configured, given the firm's external environment (external fit). Although IT is one of the fundamental components of the strategy, it still must fit within the entire activity system. When a firm has implemented a given configuration of activities and has developed the IT core supporting the linked activities, replication of the technology alone is insufficient for successful imitation. Indeed, narrowly replicating just the IT core leads to further decline of the imitator's position by wasting time, money, and management attention without eroding the leader's competitive advantage. A classic example of a firm that has an idiosyncratic activity system is Southwest Airlines. Because Southwest does not cater flights, does not offer seat assignments, has a standardized fleet of aircrafts, uses less crowded airports, and focuses on point-to-point travel for price-sensitive customers, it is relatively resistant to competitive imitation. Merely imitating one aspect of Southwest's activity system will not suffice, but attempting to duplicate the entire package generates considerable response lag.

Business Processes

We defined a business process as the series of steps that a firm performs in order to complete an economic activity. The notion of business process is related to, but distinct from, that of the economic activities discussed above. Economic activities describe the set of undertakings that the firm performs, while business processes describe the way in which the firm performs them. The contribution that business processes make to response lag and to the height of barriers to imitation depends on their distinctiveness and strategic value. When a firm is able to introduce an IT-dependent strategic initiative built around a business process with characteristics of uniqueness and differentiation, it creates a significant barrier to erosion.

External Resources

External resources are assets (such as brand, reputation, and interorganizational relationships) that do not reside internally with the firm but accumulate with other firms and with consumers. Generally intangible, external resources are usually developed over time.

When a firm's IT-dependent strategic initiative can make use of or contribute to the development of these external resources, response lag increases considerably and barriers to imitation are augmented. Thus the firm forces competitors to develop a comparable level of external resources before producing an effective response.

around. This seems like a simple addition, and technologically, it is not overly complex. However, it is not simple for your lower-rate competitors to replicate, as they would in fact be advertising that their sushi is not nearly as fresh as yours. The lesson is simple. By using digital technology to leverage an idiosyncratic complementary resource, you have made replication of your strategy much more difficult for competitors.

As digital entrepreneurs like Stripe and Twilio continue to innovate and bring new digital resources to the market, savvy managers should realize that once-valuable complementary resources may rapidly become obsolete. Consider electronic commerce (Chapter 5). There are retailers that have invested heavily in their ability to process and fulfill orders. However, as digital platforms like Shopify or digital entrepreneurs like Shippo introduce digital fulfillment capabilities, such investments may lose their strategic value. Competitors may be able to replicate world-class fulfillment services just by integrating with the Shopify Fulfillment Network.[1] Process virtualization theory (Chapter 4) provides guidance for determining which business processes and organizational capabilities are most readily modularized and abstracted into software (i.e., transformed into digital resources). The key is to realize that the potential of response-lag drivers to provide a lasting advantage is constantly evolving and, while not being technological, they are still subject to digital disruption.

Project Barrier IT-dependent strategic initiatives rely on an essential digital technology core. Thus they cannot be implemented until the necessary development and implementation project has been successfully completed. The response-lag drivers of the project barrier depend on the characteristics of the digital technology core and on the characteristics of the implementation process (Figure 9.6).

As you know, digital technologies are not homogeneous, undifferentiated entities. To the contrary—they differ substantially with respect to their intrinsic characteristics, their ability to complement other organizational resources, the context in which they are introduced and used, and the degree of organizational change that needs to occur during the implementation process. Thus sustained advantage for an IT-dependent strategic initiative can stem from the characteristics of the technology core. As software continues to "eat the world" (Chapter 1), you will witness firms in any industry, not just hardware and software companies, insourcing technology development. Tesla, for example, builds and commercializes transportation vehicles (i.e., cars and trucks). This "car company" manufactures its own microchips (Figure 9.8) optimized for machine-learning algorithms in an effort to accelerate the self-driving capabilities it sees as central to its strategic advantage.

Figure 9.8. The Full Self-Driving Chip, the autonomous driving chip designed by Tesla

Source: "FSD Chip—Tesla," 2020, WikiChip, retrieved from https://en.wikichip.org/wiki/tesla_(car_company)/fsd_chip

The response-lag drivers determining the height of the project barrier to erosion include *technology complexity*, *uniqueness*, and *visibility*. The *complexity* of the technology is a function of the bundle of skills and knowledge necessary to effectively design, develop, implement, and use it. Technology complexity raises the project barrier by increasing development lead times for a competitive response. The *uniqueness* response-lag driver ranges from self-contained, off-the-shelf hardware and software applications that need little integration or customization (e.g., an electronic mail system) to custom-developed applications or infrastructure subsystems that are unavailable in the open market (e.g., specialized microchips for self-driving). When the underlying digital technology is not distinctive or is readily available in the form of digital resources (Chapter 5), competitors seeking replication face a lower barrier. Unique digital technology is a stronger response-lag driver that makes replication more difficult since competitors must engage

1 Shopify. 2019, June 19. "Introducing Shopify Fulfillment Network." Shopify, retrieved from https://news.shopify.com/introducing-shopify-fulfillment-network.

in development rather than acquisition of the digital technology core. *Visibility* is the extent to which competitors can observe the enabling technology. We can conceptualize the visibility dimension as a continuum, spanning from custom-developed internal systems, which are virtually invisible to competitors, to immediately visible interorganizational or customer-facing systems that require extensive education of external users or customers (e.g., an online purchasing system). Digital technology that is highly visible and is readily available for inspection by competitors limits the strength of the project barrier.

Another set of response-lag drivers underpinning the project barrier pertains to the implementation process. Since different kinds of information technology are inherently dissimilar, the processes by which they are implemented and become available to the organization also differ. *Implementation-process complexity* is a function of the size and scope of the project, the number of functional units involved, the complexity of user requirements, and possible political issues, among other things. Digital infrastructure projects represent a powerful example of complex systems that have a substantial lead time to completion. While the components may be commodity-like (e.g., sensors, personal computers [PCs], servers, telecommunication equipment), it is difficult to integrate them effectively. This is also true when the infrastructure components are cloud-based rather than owned and operated by the firm.

The other critical response-lag driver in the implementation process is the *degree of process change*. Business processes often need to change to fit a new system (Chapter 2). The challenges escalate when several organizations or operations are involved in the strategic initiative. The more departments that are involved and the more organizational boundaries that are crossed, the harder and riskier the change becomes.

Preemption Barrier You may now be wondering why we term the four forces ensuring sustainability as *barriers to erosion* rather than *barriers to imitation*. We do so because in some cases, even if a competitor is able to replicate an IT-dependent strategic initiative, the response may bear no fruit for the laggard. A discussion of the preemption barrier will clarify this point (Figure 9.6).

In some cases, the IT-dependent strategic initiative pioneered by the first mover creates a *preferential relationship* with customers or other members of the value system and introduces substantial *switching costs*. Under these circumstances, it is not enough for competitors merely to imitate the leader's strategy; they need to either compensate the customer for the cost of switching or provide enough additional value to justify the customer's decision to incur the switching costs. That is, imitators must be "that much better," where "that much" is an amount greater than the current value of all co-specialized investments that the customer has made.[1] (Sidebar 9.2 lists and explains in detail all the response-lag drivers associated with this barrier.)

eBay provided an early example. Aware of its success, two formidable competitors—Amazon and Yahoo!—launched their own online auction sites. Despite having a "digital pedigree," brand recognition, and technical capabilities just as good as eBay's, both firms achieved lackluster results when trying to replicate eBay's IT-dependent strategic initiative. Interestingly, a brief look at both competitors' websites would show that they were remarkably similar to eBay's own. Clearly an imitation strategy didn't pay. Their results were so lackluster, in fact, that both had to close their respective auction sites, unable to viably compete with eBay (at least in the North American market). The reason for eBay's dominance was its ability to harness strong network effects (Chapter 4). By the time Yahoo! and Amazon reacted, both buyers and sellers in the eBay

---

1 The term *co-specialized investment* simply refers to investments made in conjunction with a specific IT-dependent strategic initiative. Because these investments are specific to the initiative, they will lose part or all of their value if those who made the investment switch to a competitor (see Sidebar 9.2).

SIDEBAR 9.2. Preemption Barrier

Switching Costs

Switching costs represent the total costs borne by the parties of an exchange when one of them leaves the exchange. They include not only economic costs but also psychological and physical costs: "Switching costs are the norm, not the exception, in the information economy."[1] IT-dependent strategic initiatives, which rely heavily on the collection, storage, manipulation, and distribution of information, are particularly suited to the creation and exploitation of switching costs.

Co-specialized Tangible Investments

When an IT-dependent strategic initiative is deployed, it may require that the firm's customers to acquire the physical assets necessary to participate in the initiative. The total capital outlay necessary to obtain these assets is termed *co-specialized tangible investments*. These range from computer hardware and telecommunication equipment to software applications and interfaces between the existing customers' systems and the firm's IT. For example, hotel franchisees buy costly interfaces for the franchising brands' reservation system. These interfaces become valueless if the property is rebranded. The extent to which the IT-dependent strategic initiative requires co-specialized tangible investments determines the potential for strong barriers to imitation associated with the initiative.

Co-specialized Intangible Investments

As is true of tangible investments, the deployment of an IT-dependent strategic initiative often necessitates a firm's customers or channel partners to invest time and money to take part in the initiative. An investment of this kind is known as a *co-specialized intangible investment*. For instance, to benefit from customer relationship management (CRM) initiatives (Chapter 3), customers often need to take the time to complete a profile. Co-specialized intangible investments might include "setup" costs as well as ongoing costs (e.g., retraining new travel associates using a reservation system). Data and information repositories represent perhaps the most important class of co-specialized intangible investments in the information age. Considerable switching costs can be built on information accumulated over time. An interesting example is offered by information that is valuable only as long as the customer is using the firm's products or services (e.g., revenue-management models and historical records that are brand specific and become valueless if the hotel is rebranded).[2]

The same situation occurs even when switching costs are not readily apparent. Facebook takes advantage of the privileged and long relationship with customers to reach a position of a "central" platform. Switching from Facebook to any other social platform would probably signify losing contacts' information, pictures, videos, or comments. Each user invested his or her time and effort in Facebook, uploading content and publishing and sharing status updates. Facebook profiles are the result of the continuous interaction of users on the platform. For Facebook, these data are a valuable asset for profiling users and understanding users' behaviors. For most users, the same data represent their social digital identity.

Value System's Structure

A firm does not engage in economic activity in isolation but as a link in a larger value chain or system that includes upstream and downstream members. The structure of this value system can provide opportunities for preemptive strategies and for the exploitation of the response-lag drivers discussed here. The structure of the value system does not directly affect the strength of the preemption barrier to imitation but instead magnifies or diminishes the preemptive effects of switching costs.

Relationship Exclusivity

An exclusive relationship exists when participants in the value system elect to do business with only one firm that provides a particular set of products or services. The firm's counterpart (i.e., customer or supplier) places a premium

1 Shapiro, C., and Varian, H. 1998. *Information rules: A strategic guide to the network economy*. Boston, MA: Harvard Business School Press: 111. Also see this book for an excellent treatment of switching costs in the information age.

2 The software here is neither proprietary nor brand specific, and the data are not acquired over a network or hosted by the brand. Yet the historic data and the models the hotel has developed assume that the hotel has a given brand (e.g., Four Seasons). If the hotel is rebranded while the software, data, and models are retained, its value is much lower because the data and models are specific to the original brand and assume that the hotel sports the related flag (e.g., has access to Four Seasons' brand equity, reservation systems, and loyal customer base).

on dealing with either the firm or one of its competitors, but not both. Relationship exclusivity is the norm with IT-dependent initiatives that provide integration services and that benefit from the accumulation of historical information. When first introduced, the AA Sabre terminal for travel agents created strong incentives for relationship exclusivity, as travel agents did not want to waste valuable office space for competitors' proprietary terminals (e.g., United's Apollo), which were considered essentially duplicates of the Sabre terminal. When a business relationship benefits from exclusivity, the customer faces penalties for multihoming. Multihoming is the sourcing of needed products or services from multiple competitors, as in the example of powerful hardware-based platforms like the Apple iPhone.

Concentrated Value-System Links

At each of the various stages or links in the value system, the degree of concentration in the link is inversely proportional to the number of suitable business entities populating that link—where suitability depends on whether the firm would find the products or services offered by the vendors populating the link acceptable. A highly concentrated link is one where there are relatively few organizations or consumers available for the firm to use or serve. In the case of airline reservation systems, for instance, the total number of travel agents serving the market targeted by the airline sponsoring the system represents the concentrated link.

A market of a given size will support only a finite number of competitors, and achieving a substantial penetration in the concentrated value-system link—by definition, a small market—is necessary to successfully preempt imitation. As the degree of concentration increases, the time necessary to secure a relationship with a substantial proportion of the link decreases—all else being equal. Consequently, the leader has a better chance of capturing a substantial proportion of relationships and being able to use switching costs to "lock out" competitors and maximize its barriers to imitation. Conversely, when a link in the value system comprises a large number of business entities, a firm is unlikely to effectively reach a critical mass of entities and raise substantial barriers to imitation in the same amount of time.

marketplace faced daunting switching costs, and only "wholesale defection" of a large portion of the customer base would have enabled the competitors to catch up. With the language of the sustainability framework, we can assert that eBay, by virtue of being the first company to reach a critical mass of buyers and sellers in a market with strong network effects, was able to erect an insurmountable preemption barrier to erosion.

While the framework for evaluating the sustainability of IT-dependent strategic initiatives breaks the analysis into its component parts—barriers to erosion and response-lag drivers—you should always be mindful that appropriation of value is dependent on these components working together.

IT-dependent strategic initiatives do have the potential to be valuable, rare, inimitable, and nonsubstitutable. However, it is critical that, as a savvy manager, you remember that digital technology does not lead to value creation and appropriation in isolation. Rather, it does so as a component in a sociotechnical system (Chapter 2) that underlies an IT-dependent strategic initiative.

The Dynamics of Sustainability

To further complicate matters, remember that neither digital technology nor competition are static. As high as the barriers to erosion may be, when launching IT-dependent strategic initiatives, a digital innovator may hold the advantage. However, such an advantage is generally short-lived if general and functional managers don't have a plan for continuously remaining ahead of the competition. This means looking for opportunities to reinvigorate and reinforce the barriers to erosion described above. Consider the example of Dell, a firm that maintained its leadership position in PC manufacturing for more than two decades. At the heart of Dell's strategy was its high-velocity, built-to-order production model for direct sales. The firm continually improved the performance of its production system as well as introduced further initiatives that leveraged its core advantage. For example, in the mid-1990s, Dell took its direct sales model to the Internet and began to sell to individual consumers. Later, the firm extended its high-velocity production model to other products, such as high-end servers and consumer electronics—albeit with less positive results.

Netflix offers a more recent example. Their business started in 1997 as the first online DVD rental store—at a time when DVD rental was a flourishing $17 billion market. The Internet was a slow and different place those days. Only after scrapping the idea of a dedicated device and being inspired by YouTube, in 2007, they finally considered the opportunity of streaming feature films. Part of Netflix's success stems from its cloud-based streaming infrastructure. Built primarily on AWS, the streaming infrastructure evolved into a global-scale, omni-device delivery platform on which the firm has increasingly integrated features for tracing errors, analyzing customers' behaviors, and running experiments.[1] Today, Netflix's infrastructure handles more than 500 billion events per day (e.g., viewing statistics, usage statistics, performance monitoring, and troubleshooting data), with peaks of eight million events per second—roughly equivalent to 24 GB of data generated per second.

There are two main dynamics for rejuvenating and strengthening barriers to erosion over time: **capability development** and **asset-stock accumulation**. There is, in fact, a mutually reinforcing dynamic between barriers to erosion and a firm's IT-dependent strategic initiative. Available response-lag drivers offer the digital innovator a "head start" on the competition. The enactment of the strategy allows the leader to engage in the capability development and asset-stock accumulation processes described below, in turn leading to further development of the response-lag drivers and the preservation of barriers to erosion. (See Table 9.1 for a list of response-lag drivers affected by each dynamic.)

Capability Development *Capability development* refers to the process by which an organization improves its performance over time by developing its ability to use resources for maximum effectiveness. When it comes to IT-dependent strategic initiatives, capability development consists of the ability to engage in "learning by using," defined as the process by which a firm becomes more effective over time in utilizing and managing an information system and the digital technology at its core. Note that capability development processes can only be set in motion once the IT-dependent strategic initiative is introduced. In the case of Dell, repeated practice with its high-velocity, built-to-order production model enabled the computer manufacturer to consistently increase inventory turns—thereby strengthening its direct sales initiative over time—and subsequently to leverage its advantage to reach previously unserved consumers and small accounts through the Internet.

Asset-Stock Accumulation While several digital technology elements have been commoditized, most recently by the emergence of cloud computing (Chapter 3) and digital resources (Chapter 5), many of the assets underpinning an IT-dependent strategic initiative cannot be readily acquired, particularly when they are internally developed. For example, internally generated data, specialized databases, and ad hoc forecasting models need to be custom developed; the same goes for a digital infrastructure and for many complementary resources as well.

Asset-stock accumulation represents the process by which a firm accrues or builds up a resource over time. Assets of this kind must accrue as a result of a consistent process of accumulation. Google search-engine data are the classic example of this accumulation process. To provide search results, Google needs to first crawl the web and index the content of each page.

1 Netflix Tech Blog. 2020. Retrieved from https://netflixtechblog.com/.

By 2017, Google indexed more than 55 billion web pages.[1] The larger data asset, coupled with efficient algorithms, provides Google an advantage, increasing the likelihood it will provide user-relevant results. Specifically, every time users click on a link, they are training Google's algorithms by endorsing the quality of the query result. Having a dominant position in searching, with an estimated 89% of queries in North America and 93% in Europe, provides Google with the ability to continuously optimize its results while starving competitors of such critical data streams.[2] Interestingly, as perhaps the best example of an aggregator of demand (Chapter 5), website publishers are helping Google by optimizing and formatting their pages to closely fit the demands of Google's indexing crawlers. The Caesars initiative discussed throughout this chapter is another example of asset-stock accumulation. At the core of the initiative, there is a comprehensive centralized repository of personal and behavioral data about each gambler and a set of predictive models that forecast a player's projected worth. A firm's ability to collect data and develop the predictive models depends on having information systems for data collection, storage, analysis, and distribution. These only became available when Caesars launched its business intelligence strategic initiative. The process of data accumulation required time to complete. Consider a destination customer who visits a Las Vegas property once per quarter to play blackjack—collecting six data points about her (i.e., information on six visits) requires one and a half years.

Table 9.1. Barriers to erosion and response-lag drivers

Barriers to erosion	Response-lag drivers
Technology resources barrier	IT assets • IT infrastructure* • Information repositories* IT capabilities • Technical skills† • IT-management skills† • Relationship assets*
Complementary resources barrier	Complementary resources*†
Project barrier	Technology characteristics • Visibility • Uniqueness • Complexity Implementation process • Complexity • Process change
Preemption barrier	Switching costs • Tangible co-specialized investments* • Intangible co-specialized investments* • Collective switching costs* Value-system structural characteristics • Relationship exclusivity • Concentrated links

* Response-lag drivers subject to asset-stock accumulation processes

† Response-lag drivers subject to capability development processes

The above examples should help you realize that sustainability of a competitive advantage often does not stem from visionary, one-time initiatives. Rather, it rests on evolutionary projects predicated on a commitment to capability building and asset-stock accumulation. On this basis, the firm can develop the strategic initiative, offering a moving target to its competitors by reinforcing its barriers to imitation over time.

Value Appropriation in Digital Platforms and Aggregators

While the digital value appropriation framework described here was developed before the recent rise to prominence of digital platforms and aggregators, its main tenets still apply. Digital platforms that create value by producing and exposing resources for digital innovators can use the framework to analyze the value appropriation potential of their digital resources. Consider Twilio. The firm competes by making digital resources available to those firms that need to incorporate communication into their digital innovations. Thus a customer like the American Red Cross of Chicago and Northern Illinois who seeks to build an efficient volunteer's dispatch system

1 World Wide Web Size. 2020. "The size of the World Wide Web (the Internet)." Worldwidewebsize.com, retrieved from http://www.worldwidewebsize.com.

2 StatCounter Global Stats. 2020. "Search engine market share worldwide." StatCounter Global Stats, retrieved from https://gs.statcounter.com/search-engine-market-share.

will turn to Twilio to enable a broadcast short message service (SMS) able to contact multiple volunteers, provide them with details of the emergency, and enable them to respond to record their departure from the scene.[1] The value of this automation for the Red Cross is in the ability to free time for dispatchers to focus on more complex and important voice interactions (e.g., with the fire department). In this scenario, Twilio is selling a digital capability (SMS dispatch) that competes with similar services traditionally offered by telecom operators and by competing platforms like Nexmo or Plivo. Its ability to withstand competitive retaliation rests on the same four barriers to erosion. Does Twilio have technology or complementary resources underpinning its digital resources that competitors would find difficult to acquire or develop internally? Are there digital technologies or implementation hurdles that will considerably slow down competitors seeking to develop comparable digital resources? Will Twilio be able to preempt imitation via switching costs and/or by leveraging relationship exclusivity with customers it has already acquired, like the Red Cross?

The framework also applies to aggregators that integrate control of a marketplace with their digital platform ownership. However, because of the prominence of network effects in aggregators' strategies (Chapter 8), the preemption barrier is the most relevant for them. In other words, for an aggregator to succeed in digital value creation, it has to be able to ignite network effects in order to aggregate demand and establish its marketplace as the venue were suppliers can reach customers. Having done so, much of its ability to sustain a competitive advantage that enables value appropriation rests on the strength of the collective switching costs associated with the nature of the network effects in the specific industry (Chapter 4). Let's consider again the example of smartphones. When you purchase an iOS or Android device, you are buying not just the hardware but the network where the device and operating system are the base platform on which the application store aggregates developers. Besides the "entry ticket" of the device cost, every time you buy an app, you increase your switching costs, as you cannot migrate your purchases to a different platform (e.g., Android). The same lock-in happens in the video game industry too. In fact, users are increasingly advocating for *cross-play*, the possibility to play their favorite titles over multiple systems, independently from their "home" platform.

9.3 Applying the Framework

When looking to be innovative with information systems and IT, you can easily get wrapped up in wishful thinking about the potential of new ideas and new technologies. Importantly, the sustainability framework is as useful in helping you decide when *not* to pursue an IT-dependent strategic initiative as it is in suggesting when to do it. You can use the framework when evaluating IT-dependent strategic initiatives either as the innovator looking to protect an existing advantage or as the laggard looking for ways to respond. This is done by asking a series of increasingly specific questions.[2]

Prerequisite Questions

Since the focus of the analysis here is on sustainability (i.e., appropriation of value over time), you must assume that the IT-dependent strategic initiative under investigation does indeed create value and is consistent with the firm's priorities. The set of prerequisite questions discussed next can be used as a check.

---

1 Twilio. 2020. "The Red Cross streamlines volunteer dispatch with Twilio." Twilio, retrieved from https://customers.twilio.com/1136/redcross-streamlines-volunteer-dispatch-with-twilio-sms/.

2 We frame the analysis here by referring to a proposed initiative. Thus we take the perspective of the innovator evaluating a new initiative. The same script can be used, with minor adjustments, by followers as well.

Is the Proposed Initiative Aligned with the Firm's Strategy? This crucial question often goes unasked until late in the analysis. (Sometimes it never gets asked!) This question is important because it is necessary for the proponents of the initiative to be able to formulate how the initiative advances the firm's positioning and strategy. If the firm has developed a strategic IS plan (Chapter 6), this question is relatively easy to answer by ensuring that the proposed initiative follows the information system's guidelines.

Is the Proposed Initiative Focused on Reducing the Firm's Cost or Increasing CWP? Rare but particularly coveted initiatives have the potential to accomplish both—decreasing the firm's cost while increasing CWP. As we discussed in Chapter 7, the value of this question is in requiring managers to clearly define the value proposition of the planned initiative.

What Is the IS Design Underpinning the Proposed Initiative? This question is designed to further formalize the analysis begun with the second question. At this stage in the analysis, one needs to achieve clarity with respect to the information processing functionalities of the information system supporting the proposed initiative. Each of the four components—digital technology, people, processes, and organizational structures—also needs to be discussed to evaluate what changes to the current information systems will have to be made and what new resources may be needed. This question is also crucial because it is the first step in evaluating the chances of implementation success of the needed information system (Chapter 2).

Sustainability Questions

While it is impossible to estimate perfectly the magnitude of any particular barrier to erosion, the purpose of this analysis is to refine the design of the IT-dependent strategic initiative, identify areas of potential weakness, and identify areas where changes to the initiative—often small ones at this stage—can substantially strengthen it. Perhaps the most important aspect of this analysis is to identify initiatives that are not sustainable. Because it is important to understand when to avoid investing in expensive projects, the following questions can raise red flags before substantial resources are committed to the initiative.

Which Competitors Are Appropriately Positioned to Replicate the Initiative? Based on a clear understanding of the characteristics of the proposed IT-dependent strategic initiative, the objective of this competitor analysis is to evaluate the strength of the IT-resource and complementary resource barriers to erosion. Competitor analysis allows the innovator to identify sources of asymmetry that can be exploited and amplified through the deployment of the proposed initiative. The objective is to design the initiative so that it takes advantage of the existing sources of asymmetry and provides a basis to reinforce them over time through capability development and asset-stock accumulation.

A powerful opportunity here is to take advantage of competitors' rigidities, which are resources that hamper competitors' abilities to replicate an innovation. A classic example is provided by firms with strong distribution ties (e.g., Compaq computers, Levi's), which could not easily replicate direct sellers' use of the Internet (e.g., Dell, Tesla) because of channel conflict. While Compaq, for example, may have had the ability to sell directly from its website, as Dell does, it wasted precious time early on because it could not risk upsetting its dealers, who were responsible for the bulk of its distribution. As a consequence, it experienced a substantial delay in responding to Dell's move online.

This dynamic is playing out as we speak in the car manufacturing industry, where Tesla has decided to retain control of sales and service instead of creating a network of dealers. Having direct relationships with customers enables Tesla to retain flexibility in the design and implementation of future IT-dependent strategic initiatives.

The result of this analysis is a clearer understanding of which competitors are in a position to respond quickly to the IT-dependent strategic initiative and which ones will instead need to first acquire necessary resources or capabilities. This analysis may also provide guidance as to how hard it would be for competitors to acquire these prerequisite resources. It is clear that when fundamental resources are heterogeneously distributed, substantial response lag can be created. Developing initiatives that amplify and leverage this heterogeneity is a critical step in your analysis.

How Long before Competitors Can Offer the Same Value Proposition? This question is primarily concerned with the response lag associated with the creation, rollout, and infusion of the information systems at the heart of the IT-dependent strategic initiative. This analysis yields an assessment of the strength of the project barrier.

After a visioning stage, where the main characteristics of the initiative are envisioned by managers, the information system at the core needs to be developed and implemented. This process follows a sequential set of stages from inception to full functionality (Chapter 11). It generally includes the following sequential stages: system definition, system build, and system implementation. Upon completion of the process, the cycle often restarts with maintenance and enhancements to the system.[1]

Competitors looking to have the same information processing functionality in place, and thus be able to offer the same value proposition, need to enter a similar development and implementation cycle. The only difference is that a follower will start the process with an awakening phase rather than a visioning phase.

The awakening stage occurs when the competitor realizes that the innovator has an advantage. The timing of the awakening depends on the characteristics of the initiative and can occur when the competitor begins to witness losses (e.g., market share, revenue), when the innovation is first introduced (typically for customer-facing systems), or even before the innovator has launched the initiative. Being able to delay the awakening is critical to build a head start. This is the reason digital entrepreneurs incubate their initiatives in what is called the "stealth" phase. Knowledge of behavior patterns exhibited by competitors may help in gauging the timing of the awakening and of the subsequent stages.

For example, Burger King has traditionally shown a propensity to quickly enter geographical markets pioneered by McDonald's. While not technology related, this type of knowledge of the competition is what allows the innovator to more precisely estimate lead time. In some rare cases, some competitors will find imitation so daunting that they will elect not to follow. The Sabre reservation system, pioneered by AA, and the Apollo reservation system, built by United Airlines, emerged as the dominant airline reservation systems because other airlines elected early on not to follow the lead of these two carriers. The decision not to engage in the design and development of their own reservation systems was based on a consideration of the expenses and risks associated with such projects.

Will Replication Do Competitors Any Good? Armed with an understanding of which competitors will be in a position to respond to the innovator's IT-dependent strategic initiative and a general

---

1 Thoughtful readers will note that the rationale offered by those who suggest that IT is not strategic because it is easily imitable is that the technology can easily be replicated. In other words, this rationale addresses only the build and implementation phases of the IT development cycle. The attentive readers of this book know that there is much more to IT-dependent strategic initiatives than simply technology development!

idea of how long it may take them to have the same functionality in place, the innovating firm must estimate the magnitude of the preemption barrier to erosion. The fact is that being second sometimes means being left behind (remember the eBay example?). Exploiting the characteristics of the innovation and the industry in which they compete, innovators can sometimes preempt any meaningful response by competitors.

Even when outright preemption is not possible, the attentive innovator often has the ability to create substantial obstacles for any prospective imitator by levying switching costs. Preemption is strongest when the firm can identify a link in the value system where few customers or partners (e.g., suppliers) exist and the partners that do exist place a premium on having an exclusive relationship with only one firm. In this scenario, they may eventually sever their relationship with the firm and do business with a competitor, but they won't trade with both at the same time.

Figure 9.9. Buddy Valastro on the set of *Cake Boss*

Source: Photo by Jen Knoedl / CC BY 2.0

Consider, for example, the case of "Buddy" Valastro's Carlo's Bakery.[1] Following the success of his reality series *Cake Boss* (Figure 9.9), orders skyrocketed. His business went from a single shop in Hoboken, New Jersey, to 17 retail stores, a mail-order shop, a cooking school, and a new line of business-to-business sales. To maintain the same high level of customer experience, it was mandatory for Buddy that all stores and the "headquarters" shared information on sales, waiting times, customer details, forecasts, and analyses, such as those offered by CRM solutions like Salesforce. In this case, for the solution to be useful, all of Buddy's business locations must use it. From Salesforce's standpoint, customers (in this case, the bakeshop chain) place a premium on an exclusive relationship. Buddy's business will either use Salesforce for all shops, or it will switch them all to a competitor's solution. In either case, it will not work with two vendors at the same time, as that would defeat the original purpose of sharing customer and sales information.

When such conditions are present or can be created, switching costs have the most power in raising the preemption barrier. When switching costs are high, competitors must indemnify any newly won customers for the cost of switching. As we stated previously, competitors must be that much better than the leader, where "that much" is determined by the magnitude of the switching costs.

The set of three questions offered above should provide the innovator, or any follower who is using this analysis as a diagnostic tool, to study the leader's IT-dependent strategic initiative with an idea of how defendable the initiative is and what options are available to improve its barriers to erosion. No initiative is static, though, and barriers to erosion decay over time as competition runs its course. As a consequence, you should ask one more question to complete the analysis.

What Evolutionary Paths Does the Innovation Create? Sheltered by its lead time, the innovator can and should seek ways to reinforce its barriers to erosion. Based on their understanding of the capability development and asset-stock accumulation processes described previously, the leading firm's managers can chart an evolutionary path for the initiative. While the evolutionary paths thus identified must be revised as the situation changes, the analysis to this point can highlight important response-lag drivers that can strengthen over time. Performing this analysis will also ensure that the evolution of the initiative is intentional rather than haphazard and minimizes the likelihood that opportunities will be missed.

---

1 See Wikipedia. 2020. "*Cake Boss.*" *Wikipedia*, last updated November 13, 2020, retrieved from https://en.wikipedia.org/wiki/Cake_Boss.

Consider, for example, the case of modern hotels. Because of the nature of the lodging service, where guests often volunteer preference and personal information, many hotels have assembled vast databases of guest needs and likes. Yet until recently, the value of guest data for analyses (e.g., customer lifetime-value analysis) was not recognized by managers. (Cynics may suggest that it still largely isn't.) A careful analysis of guest-reward initiatives may have shown that the substantial information repositories that accumulated as a by-product of the initiative are subject to asset-stock accumulation.

More recently, telecom operators have assembled large databases collecting geolocation information from mobile phones connecting to their networks. Companies like Inrix and Foursquare recognized early on the value of this location data to predict traffic flows and create innovative navigation services, challenging incumbents. The value of devices' location information manifested only after an accumulated "critical mass." The diffusion of the mobile platform (Chapter 5) made the location data collected by telecom companies a valuable and sought-after asset for creating navigation services.

9.4 Managerial Takeaways

As we stated in the introduction, business managers historically had a love-hate relationship with digital technology. In large part, such distrust of the strategic potential of digital technology came from the many disappointing investments firms made in initiatives they could not protect. As you learned in the analysis of added value (Chapter 7), you should not engage in digital value creation that does not lead to digital value capture. The main takeaway from this chapter for you as a manager should be the importance of planning for value appropriation. The framework presented in this chapter is designed to help you reduce the inevitable uncertainty surrounding this question. On the basis of the analysis discussed above, you will be in a better position to decide whether to go forward with a proposed initiative or shelve it for future reevaluation. The following are three possible broad outcomes from the analysis.

Develop the IT-Dependent Strategic Initiative Independently

Independent development is warranted if the analysis suggests that strong barriers to erosion exist and the firm foresees the ability to appropriate the value created by the initiative over the long term (i.e., sustainable advantage can be attained). Independent development is also warranted if the leader can reap an acceptable return on its innovation, even though the analysis shows that competitors will eventually be able to overcome the barriers to erosion.

Note, once again, that focus should be on the IT-dependent strategic initiative as a whole. The determination of whether the technology at the core of the initiative should be developed in a proprietary manner will depend on the role that the response lags associated with it play in the sustainability of the advantage (i.e., the project barrier).

Develop the IT-Dependent Strategic Initiative as Part of a Consortium

When the initiative is unlikely to yield sustainable competitive advantage for the innovator but, even after replication by competitors, will improve the overall profitability of the industry, the firm should attempt to create a joint venture with competitors or engage them in a consortium. In this scenario, the leader should strive to minimize costs and risks associated with the initiative and seek to share them with competitors, since all will benefit in the long term.

Shelve the IT-Dependent Strategic Initiative

When the analysis suggests that the initiative will not offer strong barriers to erosion, and retaliation by competitors will degrade the average profitability of the industry (e.g., any value created is driven to customers by competition), the firm should shelve the proposed initiative. If the firm does go ahead with the initiative, the likely outcome is competitors' imitation and the creation of value that will be competed away and appropriated by customers. For these types of initiatives, the firm should refrain from being the innovator and instead plan to follow only when strictly necessary. Because of the fast-declining costs of digital technology and the increasingly flexible option for technology deployment (i.e., cloud computing, digital resources), being a follower with nonsustainable innovations enables the firm at least to replicate the leader's initiative at a much lower cost.

Summary

In this chapter, we demonstrated that quibbling about the strategic potential offered by IT is not a productive use of management time and efforts. A large number of companies, including eBay, Dell, Harrah's Entertainment, and Walmart, provide evidence that IT-dependent strategic initiatives with technology at their core can be a source of sustained advantage for the modern firm.

More important, though, this chapter armed you with the tools to make recommendations about whether a specific firm should pursue a specific IT-dependent strategic initiative or if it is better served by forgoing the financial investment and implementation effort. Specifically, we learned the following:

- When analyzing the potential to defend a competitive advantage created by an IT-dependent strategic initiative, you must estimate the magnitude of the following four barriers to erosion: the technology resources barrier, the complementary resources barrier, the project barrier, and the preemption barrier. The extent to which an IT-dependent strategic initiative can be protected from competitors' retaliation is a function of the presence and significance of the response-lag drivers that underpin it.
- An IT-dependent strategic initiative is defendable when the magnitude, in terms of time and money, of one or more of the barriers to erosion is such to discourage imitation or to render it impossible or impractical.
- Digital technology can be critical to the sustainability of competitive advantage. However, aside from rare occasions, it is not IT itself that ensures sustainability but rather the characteristics of the IT-dependent strategic initiative that technology enables.
- The useful life of an IT-dependent strategic initiative (i.e., the span of time while the firm is able to protect the added value it created) can be extended by rejuvenating the barriers to erosion. Two processes, capability development and asset-stock accumulation, enable the firm to maintain its leadership position.
- Digital platforms that create value by creating and exposing resources for digital innovators can focus on resource creation as their IT-dependent strategic initiative.
- For aggregators, because of the prominence of network effects in their strategies, the preemption barrier is the most relevant.
- The outcome of the analysis is one of three recommendations: (1) pursue the IT-dependent strategic initiative independently when the firm can protect it or reap an acceptable return on investment before competitors can successfully retaliate, (2) pursue the IT-dependent strategic initiative as part of a consortium when the firm cannot protect it but all the firms in the industry will be better off once replication has occurred, or (3) do not pursue the IT-dependent strategic initiative when the firm cannot protect it and industry profitability degrades once replication has occurred.

Study Questions

1. The chief executive officer (CEO) of your company, where you serve as the chief information officer (CIO), recently read the article titled "IT Doesn't Matter."[1] He calls you into his office to "pick your brain" and asks, "Why do we invest money in IT when every one of our competitors can buy the same technology?" It's your reputation at stake! How do you respond?
2. Why is the difference between information systems and information technology so important to the analysis of sustainability?
3. Describe each of the four barriers to erosion.
4. For each barrier to erosion, provide an example of an IT-dependent strategic initiative that, in your opinion, leverages the barrier. Can you identify which response-lag drivers underpin the barriers to erosion in your examples?
5. Review your answers to the questions at the end of the opening minicase. Have they changed? Why or why not?
6. Identify some businesses that currently appear to have a sustainable advantage. How do information systems contribute to this sustainability? Highlight any cases where this sustainability appears to have a non-IS foundation or where there is no complementary IS to support the firm's value proposition.
7. Identify some businesses that currently appear to have no sustainable advantage and as a result are losing market share, are in (or close to) Chapter 11, or have gone out of business in the last year. Did a competitor's IS contribute to this decline, or was there some other fundamental problem?
8. Can IS create a sustainable competitive advantage or only support a firm in achieving a sustainable competitive advantage based on a compelling value proposition? Use evidence to support your argument.
9. Identify an aggregator that, in your opinion, is flourishing because of substantial preemption barriers. Identify and describe which response-lag drivers most affect it, and discuss their sustainability.

Glossary

A glossary of key terms can be found at the book's website. Visit https://www.prospectpressvt.com/textbooks/piccoli-information-systems-for-managers-5-0, then click on "Student Resources" in the horizontal red menu bar.

1 Carr, N. G. 2003. "IT doesn't matter." *Educause Review 38*: 24-38.

PART IV

Getting IT Done

The last part of this book is dedicated to the many issues that surround the management of information systems (IS) and of digital technology in modern organizations—from budgeting and operational planning, to design and development, to ongoing operations. This section of the book concentrates on what general and functional managers need to know to be actively involved in the management of their firm's or function's IS resources.

Your involvement with the decisions discussed in this section is essential. While you can typically avoid worrying about hardware decision making, the same cannot be said of software applications and the issues that surround them. Software applications enable work and constrain how work is done and have a direct impact on organizational operations and climates. Thus as the organizational expert and the person responsible for the success of your business function, you must have a say in the funding and prioritization of projects, you must be intimately involved in the design or selection of new systems, and you must be cognizant of the organizational risks associated with security and privacy failures.

In order to be an asset to the team making the aforementioned decisions, you must understand the processes that surround them, remain up to date on the IS trends that concern these decisions, and be well versed in the vocabulary and issues pertaining to the management of IS resources. Part IV of this book is devoted to these topics:

- *Chapter 10: Managing Information Systems.* This chapter focuses on the decisions and techniques pertaining to the funding and the governance of information systems. It also discusses outsourcing decisions and the outsourcing decision-making process.
- *Chapter 11: Creating Information Systems in the Digital Age.* This chapter describes the process by which IT-enabled information systems come to be. It discusses the three main avenues for new systems creation: systems design and development, systems selection and acquisition, and end-user development.
- *Chapter 12: Information System Trends.* This chapter introduces the enduring IS trends and emerging digital technologies that business managers must be aware of. It then describes and discusses the characteristics and implications of each one.
- *Chapter 13: Cybersecurity, Privacy, and Ethics.* This chapter makes the case for why managers need to be intimately involved in cybersecurity, privacy, and ethics decisions. It then provides the background to partake in the organizational debate of these issues.

Chapter 10

Managing Information Systems

What You Will Learn in This Chapter

We begin Part IV by discussing how modern organizations support, fund, and manage their information systems (IS) efforts. We will explain the relationship between strategic IS planning and the yearly budgeting and prioritization process and discuss the ways in which IS projects are funded and managed.

MINICASE: Budgeting at Performance Boards Inc.

As the chief information officer (CIO) at Performance Boards Inc., you chair the information technology (IT) steering committee. Performance Boards has recently been acquired by Big Sporting Manufacturer Inc. and is currently operating independently as a wholly owned subsidiary. During the yearly IT budgeting process, or the "ultimate fighting championship," as you call it, you are the unwilling center of attention—the arbiter of all disputes. It's that time of year again, as shown by the calls you are receiving from other managers you hardly hear from all year.

Every July, the budgeting process starts with a call for projects. Every functional area responds with a rank-ordered list of initiatives that need funding and their supporting **business cases**. Once the steering committee reviews the preliminary proposals, each executive sponsor presents the case for his or her proposed projects. Armed with this information, the steering committee deliberates and chooses the projects to be presented to the executive team for inclusion in the overall budget. Typically, whatever the steering committee proposes, the executive team approves. The executive team's main concern is overall IT spending. Bjorn Dunkerbeck, the founder and chief executive officer (CEO) of Performance Boards, is adamant that the firm should be in line with the manufacturing industry benchmark of 3.3% of revenue as a yearly IT budget.

This year, the third year of declining revenues for the firm, the ultimate fighting championship is shaping up as an all-time great—not a good thing for you! You had set aside 64% of the budget for the IS function to control, in accord with industry **allocation** benchmarks. Your group needs the money for security, disaster recovery, general maintenance, infrastructure management, and administrative expenses. Yet because of the tightening budgets, for the first time in your tenure as CIO, you are being questioned and required to justify the allocation to the IS function.

At this point, the human resource project and the inventory management projects seem most likely to get green-lighted. The vice president (VP) of human resources has been asking for an upgrade to the benefits package management application for three years now. His business case shows both productivity improvements and higher retention of employees. The chief operating officer (COO) presented the business case for the manufacturing group. He has shown a substantial return on investment (ROI) associated with the proposed supply chain and just-in-time inventory management initiatives.

The VP of accounting and the new director of sales, Robby Naish, are exerting lots of pressure to obtain funding for their projects as well: an upgrade to the accounting management system and a sales force automation (SFA) application, respectively. Naish has just finished reiterating his case to you on the phone. Being new to the firm, he is becoming quite frustrated with Performance Boards' approach to budgeting: "How am I supposed to compete with one project that increases productivity and another that improves efficiencies through automation? They can easily compute an ROI, but my project is not suited to that type of analysis. I can surely come up with some fictitious ROI number. I can pull it out of thin air if this is how you guys do business!"

As you review the current informal ranking of projects, you can't help but think that you need to find a way not to alienate the functional managers and project sponsors. The last thing you need is for the IS function to be perceived as a roadblock to the success of the other functional areas and ultimately of Performance Boards Inc. as a whole.

Discussion Questions

1. What should you do next? What are some of the options at your disposal to ensure that you do not alienate your colleagues?
2. Are there any structural problems with the budgeting process at Performance Boards Inc.? What improvements would you suggest for next year—if any?

10.1 Introduction

As you know, strategic IS planning involves identifying the long-term direction of IS use and management within the organization (Chapter 6). It provides a framework for decision making and project selection. Within this framework, the firm develops yearly operational plans and budgets in order to prioritize IS spending.

As a manager, you need to understand how the budgeting and prioritization processes work so that you can make the most of them. All too often, organizations fund information systems using simple metrics, like percentage of revenue or fixed increments over the previous year's budget. While these metrics have a place in the budgeting process, the yearly budget is an opportunity to formally evaluate competing projects and make the (sometimes tough) comprehensive prioritization decisions necessary to align information systems with the firm's strategy. The firm that fails to do so misses the opportunity to offer guidance and a clear mandate to the IS function. The consequence is that a lack of direction and cohesive effort will degrade service (in many cases, leading to outright failure) and demoralize the firm's IS professionals.

Note that this prioritization role should not be delegated to the IS group but should be made in concert with business managers (i.e., those who need the information systems) and IS professionals (i.e., those who make and manage the information systems). An IT group that does not deliver consistently is often the product of an executive team that fails to provide clear priorities.

10.2 Information Systems Governance

Information systems governance is the set of decisions rights and the guiding accountability framework designed to ensure that digital technology resources are employed appropriately in the organization.[1] IS **governance** in the modern firm has two principal aspects: the management of downside risk and the fostering of upside potential. The first facet, IT risk governance, is concerned with decisions for minimizing threats (e.g., security risks) and failures (e.g., unsuccessful project implementations). The second facet, IT value governance, is concerned with maximizing the return on digital technology investments and the firm's ability to leverage its IS resources.

While to any casual observer it would appear obvious that the board of directors within an organization needs to pay close attention to IS and digital technology decisions, one of the most enduring research findings is that most boards of directors abdicate this duty. For example, the CEO is considered a champion of digital transformation, and the CIO increasingly plays a strategic role, both recognizing the critical role digital technology plays in their firm's success. However, less than 20% of the respondents reported being satisfied with the current level of expertise.[2]

While firm executives need not evaluate every project, they do need to set up the governance framework for decision making about IS resources within their organization.[3] There are a

1 PwC. 2020. "2020 global digital IQ survey." PwC.com, retrieved from https://www.pwc.com/us/en/digital/digital-iq/pwc-2020-global-digital-iq.pdf.

2 Weill, P., and Ross, J. 2004. *IT governance.* Cambridge, MA: Harvard Business School Press.

3 PwC. 2017. "The governance divide—boards and investors in a shifting world." PwC.com, retrieved from https://www.pwc.com/us/en/governance-insights-center/annual-corporate-directors-survey/assets/pwc-2017-annual-corporate--directors--survey.pdf.

Table 10.1. Five categories of risk the board of directors must address

IT competence risk	This risk factor captures the degree of IT-related knowledge of the board of directors. While boards of directors need not all be as knowledgeable as CIOs or IT professionals, it is critical that they have the ability to follow IT discussions and ask relevant questions. Moreover, there should be a leading IT director who maintains an up-to-date competence on IT matters.
Infrastructure risk	A firm's IT infrastructure represents the set of IT components that are interconnected and managed by IT specialists with the objective of providing a set of standard services to the organization. It provides the foundation for the delivery of business applications. The board of directors must be keenly aware of the weaknesses and risks associated with the firm's IT infrastructure.
IT project risk	In Chapter 3, we provided a sample of high-profile IT project failures. IT projects are generally complex and expensive undertakings that, if not properly managed, can put the organization in peril. The board of directors must ensure that the appropriate guiding framework for IT projects is in place. In large organizations, this may require a project office, a certification process for project managers, and a portfolio approach to IT project management (see below).
Business continuity risk	Business continuity refers to the activities a firm performs to ensure that critical business functions remain operational in a crisis and that the organization can withstand unforeseen disasters (Chapter 13). The board of directors must ensure the existence of a business continuity plan and that such a plan is periodically tested and revised.
Information risk	Information risk pertains to the many hazards associated with the collection and use of organizational, partner, and customer data (Chapter 13). The board of directors must craft a governance system that ensures that an officer of the organization has clear responsibility for signing off on and ensuring compliance with established privacy and security policies.

Source: Parent, M., and Reich, B. H., 2009, "Governing information technology risk," ***California Management Review*** ***51***(2): 134–52

number of models and approaches that guide the design of an IT governance framework for an organization. The specifics of these models are beyond the scope of this book. However, research has highlighted five key areas of concern that boards of directors must proactively address (see Table 10.1). In order for a board of directors to take a proactive stance to governance, these five areas of potential risk must be monitored and cyclically evaluated.

Beyond the management of the principal risks associated with IS use in organizations, the governance framework an organization puts in place must also ensure that the use of digital technology resources is aligned with organizational strategy and priorities.

Governance in the Digital Age

Throughout this book, we have shown how recent developments in digital technology, such as cloud computing, have created fertile ground for digital disruption. One of the ways in which such changes impact organizations is by creating flexibility for business units seeking to implement IS initiatives. Until very recently, any effort at introducing digital technology fully depended on the IS function. If the marketing department needed to acquire customer relationship management (CRM) capabilities, it would make the business case to fund a project allowing the firm to purchase a CRM application, and IS professionals would provision servers and install, secure, backup, and manage the hardware and software needed. If the HR department needed a new payroll system, the same type of effort would be required. Even something as simple as upgrading the staff's personal computers could not be executed without the support and direct intervention of the IS function.

The rise to prominence of cloud computing and software as a service (SaaS) architecture (Chapter 3) freed the business units from their technical dependence on IS professionals. SaaS applications require very little infrastructure on the client side, with most running in a web browser. Thus there is no need for any on-premise specialized hardware purchase, software installations, or direct management of the technology. It is the very promise of SaaS companies to provide access to the needed software functionalities without any of the headaches associated with managing the IT infrastructure that supports and enables them. If the marketing group needs

CRM capabilities, they can contract with Salesforce. If the HR department wants workforce management capabilities, it can license Workday.

Cloud-computing revolutionized software provisioning creating unprecedented levels of flexibility for modern organizations. We hope, however, readers of this book already realize that it would be naïve to think that direct provisioning by a firm's business units is an optimal strategy. In fact, executives with a deep understanding of this new environment are being recognized for their expertise—with the 2021 promotion of Andy Jassy to replace Jeff Bezos as the CEO of Amazon being the most famous example.[1] Given your understanding of process and data integration (Chapter 3), the importance of strategic IS planning (Chapter 6), and clear IT governance, you realize that just because it is technically feasible for business units to provision software independently, it does not follow that it is advisable. Moreover, while a lot of software has been servitized, there are many situations where the traditional on-premise model is still a preferred alternative. Therefore, as flexible options for software provisioning and digital technology implementation proliferate, sound IS governance becomes even more important.

Steering Committee

Larger organizations often formalize management involvement in IS decision making by forming a steering committee. The steering committee brings together representatives from the various functional areas, the CEO (or other general management staff), and key IS professionals (e.g., the CIO) who convene regularly to provide guidance to the IS function and share the responsibility for aligning its efforts with the larger business strategy. The steering committee is typically the venue where business cases are presented for approval or as a filter before the budgeting process. It's the steering committee that makes decisions on the investment approach, portfolio, and risk. Furthermore, it is generally responsible for the prioritization of the investments. The steering committee is also the main recipient and evaluator of progress during system development and implementation efforts.

Cloud Center of Excellence

With the growing importance of cloud computing for creating and provisioning the firm's digital infrastructure (infrastructure as a service—IaaS), computing platforms (platform as a service—PaaS), and software applications (SaaS), organizations are increasingly deploying **cloud centers of excellence (CCOE)**.[2] A CCOE brings together employees with a wide range of competencies, including application and domain experts, vendor management specialists, business architects, and security and compliance professionals. A CCOE develops best practices and repeatable processes for managing cloud-computing resources with a life cycle approach (Figure 10.1) based on the recognition that, while some SaaS applications are not critical for the organization (e.g., using WhatsApp to communicate informally with colleagues), failing to holistically govern cloud resources and SaaS applications will result in the kind of haphazard IT infrastructure that the planning process is designed to eliminate.

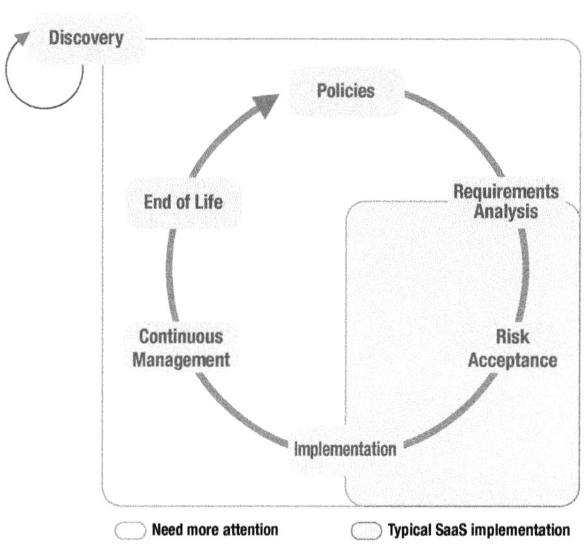

Figure 10.1. SaaS governance life cycle approach

Source: Adapted from Heiser, J., 2020, "How to develop a SaaS governance framework" (no. G00382661, p. 16), Gartner

1 Tilley, A. 2021, February 3. "Andy Jassy's climb to Amazon CEO shows the cloud's rising power." *Wall Street Journal*.

2 Heiser, J. 2020. "How to develop a SaaS governance framework" (no. G00382661, p. 16). Gartner.

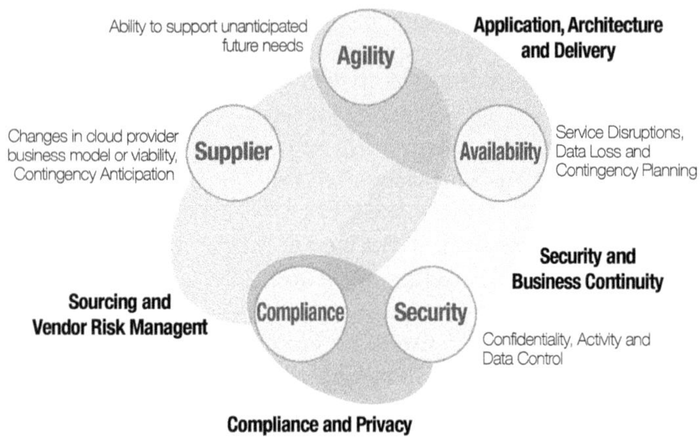

Figure 10.2. SaaS risk domain responsibility

Source: Adapted from Heiser, J., 2020, "How to develop a SaaS governance framework" (no. G00382661, p. 16), Gartner

Because of the flexibility and simplicity of provisioning cloud services, the firm's infrastructure will quickly result in a proliferation of siloed SaaS applications that are difficult to inventory, administer, secure, and integrate. Consider the simple example of ensuring compliance with regulation and cybersecurity policies (Chapter 13). A large organization strives to integrate applications in order to reduce any need for data duplication. IS professionals also work hard to ensure that employees have single sign-on privileges to the applications they need to carry out their work. IS auditors ensure that processes are in place to maintain compliance with regulation, such as securing personal identifiable information for privacy reasons (Chapter 13), and with cybersecurity practices. Imagine a situation now where one business unit uses Dropbox for file synching and exchange, while another uses Box, and a third one uses Google Drive. While this proliferation of document repositories may seem inconsequential, it actually results in a grave security vulnerability and a high likelihood of data replication, redundancy, and inconsistency across departments. With more critical and consequential applications, like human resource systems, the risks proliferate even more (Figure 10.2).

The following list of activities enforced by a well-resourced CCOE with executive support helps ensure the governance of the growing portfolio of SaaS applications and cloud resources used by modern organizations:

- Establish a standard process to evaluate and approve the provisioning of SaaS applications and cloud resources. Such a process formalizes the involvement of the CCOE or, at a minimum, the collaboration of business units seeking the resource and the appropriate IS professionals.
- Formalize ownership of the resource and accountability for its management. Infrastructural and platform cloud resources are typically the purview of the IS function. Conversely, SaaS applications may be owned by the business unit that requested it. In any case, the identification of a responsible party is critical to ensure that the responsibility for policy compliance, user administration, and user support is clearly articulated and accepted.
- Maintain a comprehensive inventory of cloud resources. The firm should require "registration" of any SaaS applications and cloud resources with the IS function that will maintain a list of relevant information (e.g., resource, owner, contract details).

10.3 Funding Information Systems

As with any other organizational asset, the firm must account for and fund IS assets and expenses. Information systems are typically designed, built, implemented, and managed to achieve some business goal (e.g., improve factory floor efficiencies, increase sales effectiveness and customer repurchases). The exception is provided by shared services (e.g., security, planning and administration, business continuity initiatives) and infrastructure investments. There are three main methods used by modern organizations to fund information systems: ***chargeback***, *allocation*, and ***overhead***. Each one offers advantages and disadvantages. The more strategic role played by digital technology and increased diffusion of cloud infrastructures and services have prompted organizations to evolve their funding methods, but the following three approaches remain the general framework for IS resource provision in organizations.

Chargeback

The chargeback approach calls for the direct billing of IS resources and services to the organizational function or department that uses them. It is grounded in the pay-per-use principle. For example, a department may be charged for networking expenses based on the volume of traffic it initiates and for personal productivity software based on the number of concurrent software application licenses it runs.

While several approaches to chargeback have been developed (see Table 10.2), the main advantage of these mechanisms is the perceived fairness and accountability they create for both users and the IS function. Another advantage of such systems is the degree of control they afford managers, who can proactively control their function's IS expenses. However, maintaining such detailed costing mechanisms can generate substantial direct and indirect expenses, including the cost of tracking and those for auditing and dispute resolution of charges.

Chargeback systems typically treat the IS function as a cost center. That is, the units are billed on the basis of actual costs. In some instances, particularly when the IS function has some unique skills, it may become a profit center and compete for service provisions with external vendors (see Table 10.3). In rare cases, it may also sell its services to other firms, not only to internal users, and return the profits to the organization.

Allocation

The allocation approach calls for the direct billing of IS resources and services to the organizational function or department that uses them. However, rather than a pay-per-use metric, it computes allocations based on more stable indicators such as size, revenue, and number of users. For example, a hotel chain may charge individual hotels in the chain on the basis of the number of rooms. A chain of retail outlets may charge each unit a fixed percentage of revenue or square footage.

Table 10.2. Different chargeback methods

Method	Description
Service-based pricing (SBP)	Charges per a specific measured unit of service
Negotiated flat rate (NFR)	Charges based on a negotiated and projected use of a service
Tiered access (TA)	Charges based on service accessibility and level of utilization, expressed in pricing bands
Measured resource usage (MRU)	Charges based on actual measured usage of specific IT resources (e.g., data traffic, storage consumed, energy consumption)
Direct cost (DC)	Charges based on dedicated ownership of the resources (e.g., time and material-based costing)
Low-level allocation (LLA)	Charges based on simple user metrics (e.g., user and server counts)
High-level allocation (HLA)	Charges based on general organizational metrics (e.g., number of employees)

Source: Uptime Institute[1] and Gartner[2]

1. Heslin, K. 2015, September 29. "IT chargeback drives efficiency." ***Uptime Institute***, retrieved from https://journal.uptimeinstitute.com/it-chargeback-drives-efficiency/.

2. Lozada, C., and Subbert, M. 2019. "Key concepts in IT financial management: budgeting, transparency, allocation and funding" (no. G00455007). Gartner, retrieved from https://www.gartner.com/document/3975832.

Table 10.3. IT funding models

Model	Use	Description	Advantages	Challenges
Subsidy	30%	May involve allocation of IT costs but no chargeback	Low administrative overhead. This model is sometimes used as a precursor to the cost center model and can work if the enterprise has strong demand governance.	Constant downward pressure on costs and a high level of scrutiny by the business on any attempts by the IT organization to reinvest in IT
Cost center	30%	Approach to chargeback is asset-based and simplistic—most typically, a straight allocation by head count	The administrative burden is higher in this model than it is with the subsidy model, but there is also a higher level of accountability for IT consumption	Has the same cost pressures of the subsidy model, but it usually exploits chargeback as a mechanism to manage demand
Service center	30%	Involves charging back for IT based on consumption of specific IT services, a necessary step for enterprises that wish to run IT as a business	Can allow for much more strategic funding and business value discussions with business stakeholders	Can be too labor-intensive to take a service-based approach to cost allocation and recovery without a specialized tool
Profit center (or self-funding)	10%	IT services are billed at market pricing, enabling a "profit" to be made for reinvestment into IT	Enables IT to run as a business, selecting internal projects based on return on investment (ROI) like any other business	Real profits may not accrue directly to IT

Source: Gartner, 2017*

* Anderson, J. M., and Lozada, C. L. 2017. "Key concepts in IT financial management: Budgeting, funding, transparency and allocation" (no. G00337779). Gartner.

The allocation method seeks to strike a balance between the pay-per-use fairness and the high cost of the chargeback method. Since rates are typically set once a year, the expenses each unit can expect are also more predictable. Some functional managers prefer the predictability of fixed allocations, while others prefer the higher degree of control offered by the chargeback mechanism.

Overhead

The overhead approach treats information systems as a shared expense to be drawn from the organization's overall budget rather than to be paid for by each unit. This is the simplest approach to funding information systems, since decisions are made once a year during the budget approval process. It also provides the most control to the IS function over spending decisions. As a consequence, the IS function is more likely to experiment with and evaluate new technologies.

The main drawback of the overhead approach is a lack of accountability for both the functional areas and the IS department. On the one hand, since users are not billed directly, they are less likely to proactively manage their usage and filter their requests for service and new projects. Moreover, they remain largely unaware of their impact on the overall IS budget, which often leads to misconceptions about the cost of IT resources and the reasonable expectations for service. On the other hand, since the IS function has little accountability to the individual functional areas, it is more likely to be less responsive and to offer poorer service.

Recognizing their complementarities, some organizations adopt a blend of funding approaches for the IS function. For example, while infrastructure services and maintenance may be billed based on the overhead method, new projects and new systems may be billed based on an allocation approach, while SaaS billing may be covered directly by the organizational unit that provisioned the application.

10.4 The Budgeting and Project Prioritization Process

The yearly budgeting process is a tool that organizations use to communicate plans and enforce control systems. As a planning tool, the budget provides an assessment of what the firm believes future financial flows will be. As a control mechanism, the budget helps encourage and enforce specific behaviors. More subtly, the budget can be used to allocate decision rights and to enforce the firm's governance framework. In 2018, advisory firm Deloitte found in its global CIO survey report that IT spending ranged from 1% to 7.9% of revenue in the industries surveyed (Figure 10.3).

Typically, control over the IT budget is split, with a portion allocated to the IS function for infrastructural expenses and projects that enable the business to operate in a reliable and secure manner (e.g., disaster recovery planning, business continuity initiatives, IS security management). The remainder is controlled by the individual units for funding operations of existing systems and for funding new projects (Sidebar 10.1).

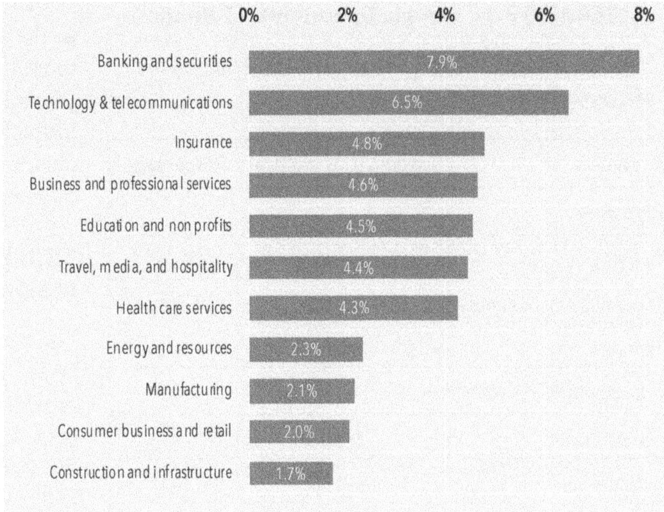

Figure 10.3. IT spending as a percentage of revenue

Source: Kark, K., 2020, ***Reinventing tech finance: The evolution from IT budgets to technology investments***, Deloitte Insights, p. 12

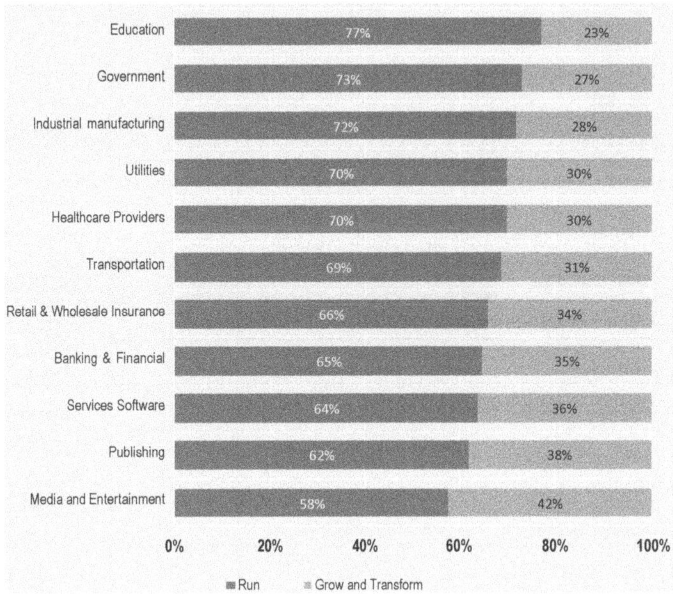

Figure 10.4. IT spending for running and growing by industry

Source: Gartner, 2017, November, "Run, grow and transform the business IT spending: Approaches to categorization and interpretation," retrieved from https://www.gartner.com/document/3357425

Making the Budget

The budgeting process requires trade-offs between diverging interests and the prioritization of projects under resource constraint. These pressures can create a stressful environment within which executives must argue and rally support for their initiatives, all the while attempting to ensure that IS resources are deployed to meet the strategic demands of the business.

The budgeting process varies by organization, but there are typically two decisions to be made: determining the appropriate budget for ongoing operational expenses (e.g., maintenance) and evaluating large capital expenditures (e.g., new systems)—in other words, "keeping the lights on," supporting current business functions on the one hand and evolving business capabilities on the other. Published industry benchmarks can be instrumental in offering some guidance (see Figure 10.4), but it is essential to stress that your firm's budget is dictated by its vision and architectural guidelines (Chapter 6) and the unique contingencies it faces.

For instance, a firm that has been very successful in rationalizing and consolidating its infrastructure can be expected to have below-average ongoing IT expenses. Conversely, if a firm ran a number of legacy applications in need of substantial upkeep, it would be spending above average.

SIDEBAR 10.1. Sample Operational IT Budget

This sample operational IT budget is loosely based on that of a major hotel chain.

Expenses	Amount (U.S. dollars)	Percentage
Personnel		
Payroll	3,140,000	31.4
Travel and entertainment	270,000	2.7
Training	45,000	0.3
Depreciation/amortization		
Depreciation[1]	450,000	4.5
Amortization[2]	1,080,000	10.8
Equipment/hardware		
Purchases[3]	90,000	0.9
Lease/rental		
Maintenance		
Software		
Purchase		
Lease/rental		
Maintenance	680,000	6.8
External services		
Cloud sourcing		6
Software development		
Consulting		
Telecommunications		
Telecommunications expenses	520,000	5.2
Disaster recovery	10,000	0.1
Miscellaneous operating expenses	*180,000*	*1.8*
Corporate IT allocation[4]	*3,530,000*	*35.3*

1. Capitalized equipment is subject to depreciation.
2. Capitalized software is subject to amortization.
3. Equipment below the capitalization threshold.
4. IT allocation from the parent company. For firms without a parent company, this line would spread over the other items in roughly the same percentages.

A sound appreciation for the role of information systems in the firm (i.e., the IS vision discussed in Chapter 6) is even more important when evaluating capital expenditure and new IS projects. This is critical not so much for the evaluation of individual initiatives but for the evaluation of the aggregate degree of IS risk the firm is willing to accept and the business outcomes it expects to achieve.

10.5 Funding Information Systems Projects: Making the Business Case

The business case is a formal document prepared and presented by the general or functional manager sponsoring the project. It provides the rationale for pursuing the opportunity. Its primary objective is to explain and provide evidence to convince the executive team, typically during the budgeting process, that the initiative will pay off and its funding is warranted.

Traditionally, business cases required fact-based investment analyses grounded in financial indicators such as internal rate of return (IRR), return on investment (ROI), net present value (NPV), return on assets (ROA), or payback period. They required an analysis of the timeline of the project and its future cash flow stream and supporting evidence for project benefits and cost estimates. However, in order to provide these analyses, project proponents all too often had to make assumptions and arbitrary judgments.

The business case approach is a standard in modern organizations.[1] However, despite their popularity, business cases have encountered criticism.

Limitations of the Business Case

Skeptics suggest that business cases strictly based on fact will often require so many assumptions and speculations that they will become based on fiction. This problem is less likely with "automation" and efficiency-focused initiatives (i.e., first-order change as defined in Chapter 2), but it is particularly evident for projects that rely on business model innovation (Chapter 5) or emerging digital technology and, as a consequence, are characterized by significant uncertainty as to the final outcome. Financial projections are also difficult to make for projects that have mostly "soft" benefits. Consider as an example a firm that decides to invest in digital initiatives and social media, like the investment in its global pizza-delivery chain Domino's has made aggressively in the last decade (Chapter 8). Today Domino's Pizza has more than 10 million fans on its Facebook page. These fans participate in the chain's social media contests,

1 Weill, P., and Ross, J. 2011, April 25. "Four questions every CEO should ask about IT." *Wall Street Journal.*

polls, and specials—a recent contest asked Instagram users to post photos of restaurants that are still using an outdated Domino's logo and hashtag them with #logoinformants and #sweeps. The initiative was a winner, with more than 4,500 submitted photos. It is quite clear that Domino's has an established presence in social media and has gained significant traction and customer engagement. However, what is the value of such customer engagement? What is the link between such engagement and pizza sales? These remain very difficult questions to answer.

The limitations of a traditional business-case approach are clearly exemplified by an executive of a large financial services institution explaining how his firm justified a recent project consisting in developing a new technology stack comprising more than 15 systems in less than six months:

> People tend to have a microfocus on achieving ROI. . . . But the key is to have a confidence level on the return and purpose of the return. That's the magic. What is the purpose and why are you doing the digital transformation?[1]

Overcoming the Limitations of the Business Case

In order to overcome the limitations of a traditional business case, some observers point to the value of heuristics. A heuristic is a simple rule that is good enough to make decisions, recognizing that adjustments along the way will be necessary. This approach offers another advantage: it systematizes the reevaluation of both the costs and benefits of projects during their development. Another valuable approach consists of relaxing the focus on fact-based business cases and allowing proponents of a project to ground their request on faith (i.e., project rationale based on beliefs about market trends, customer expectations, competition, strategy, etc.) and fear (i.e., need to engage in projects to keep from falling behind the competition or to avert a likely negative outcome). Typically, a well-crafted business case will include all of the above: fact-, faith-, and fear-based arguments.

Recent literature has advanced a more formal approach to modern business cases, one that develops in six sequential steps:[2]

1. *Define business drivers and investment objectives.* Business case writers should begin by establishing the business drivers underlying the need for the project and by clearly spelling out the investment objectives and their relationship with the business drivers. If the firm has engaged in the strategic IS planning process (Chapter 6), much of the work underlying this first stage has already been completed.
2. *Identify benefits, measures, and owners.* Business case writers should then identify all the potential benefits accruing to all stakeholders if the project is successfully implemented. Note that benefits could extend beyond the investment objectives or even be unintended. Once all benefits are identified, the business case writer should explain how benefits will be measured and who will own them (i.e., who will be representing the stakeholder group that the benefit will accrue to).
3. *Structure the benefits.* The value, and the challenge, of business cases revolves around the ability to accurately assess the actual realizable benefits of the project. Structuring the expected benefits using the grid in Figure 10.5 can foster

		Type of Business Change		
		Do New Things	Do Things Better	Stop Doing Things
High ↑	Financial Benefits			
Degree of Explicitness	Quantifiable Benefits			
	Measurable Benefits			
↓ Low	Observable Benefits			

Figure 10.5. Benefits evaluation grid for business cases

1 Ward, J., Daniel, E., and Peppard, J. 2008. "Building better business cases for IT investments." *MIS Quarterly Executive* 7(1): 1–15.

2 Shein, E. 2020, April 6. "Real-world tips for ensuring digital ROI." *CIO Magazine*, retrieved from https://www.cio.com/article/3535849/real-world-tips-for-ensuring-digital-roi.html.

deep, focused discussions that can significantly increase the precision of benefits evaluations.

4. *Identify organizational changes enabling benefits.* Heading each column in Figure 10.5, benefits are categorized as the ones that accrue from doing new things (e.g., creating Facebook-only pizza specials, thus being able to track demographics of interested customers), doing things better (e.g., improving call-center operations by centralizing customer information for easy access by representatives during a call), and ceasing to do things (e.g., stopping the printing and shipping of paper procedure manuals by transferring them online). While this may appear to be a simplistic categorization, it helps to list the type of change explicitly.
5. *Determine the explicit value of each benefit.* Heading each row in Figure 10.5, benefits are categorized by the extent to which they can be made explicit. It is important to specify, during the writing of the business case, the extent to which a benefit is measurable in order to ensure the maximum degree of precision in its estimation. Benefits can be classified as follows:
   a. *Financial benefits.* Computed by applying cost/price metrics or another recognized financial formula to measure the benefit.
   b. *Quantifiable benefits.* Computed by gathering metrics, expressed in number form, that provide evidence of change univocally attributable to the project.
   c. *Measurable benefits.* A measure is available to monitor a given benefit, but changes in such metrics cannot be univocally tied to the project.
   d. *Observable benefits.* There exist agreed-upon criteria, albeit not quantifiable, to evaluate the impact of the project on this class of benefits. Failing to structure this evaluation often results in the business case writer overlooking potential benefits or stopping at a level of precision in their evaluation that is unnecessarily coarse.
6. *Identify costs and risks.* A complete business case concludes with an estimation of the costs the firm will incur to see the project through to completion as well as the degree of uncertainty and risk surrounding successful completion. Project costs will include all technology development expenses, licensing fees, training, and change management initiatives. Beyond the development costs, the firm should include the **total cost of ownership** (TCO). TCO is a financial estimate designed to explicitly recognize the full life cycle costs of IT assets. The costs of information systems and technology typically far exceed the costs of acquisition (e.g., selection, licensing, implementation) and include expenses that occur after the system is up and running but are necessary to maintain it in operation over its life span. These include ongoing trainings of users and support personnel, maintenance, upgrades, security, and business continuity investments. Even the end-of-life costs associated with retiring the asset should be included in TCO calculations. TCO is an imprecise measure that depends on substantial estimation and assumptions about feasible configurations and future events. Nonetheless, it is essential to estimate TCO as comprehensively as possible when developing the business case. Today, there is commercial software that can be used to aid in TCO computations, and major vendors offer TCO tools online (Figure 10.6). Specifically, cloud vendors are providing free TCO calculators in an effort to support business case development for their solutions.

When discussed in the abstract, the business case development process can sound highly theoretical. Conversely, the technique discussed here is very practical and pragmatic. Sidebar 10.2 illustrates it in action.

Individual Project Risk

An important aspect of business case writing is the estimation of project risk. This is an important assessment, both when evaluating the risk of individual initiatives and also particularly when measuring the overall risk of all projects under development in the organization at a given point in time. IS projects are renowned for their high degree of risk and incidence of failure. However, the risks associated with any one particular initiative can vary widely based on the following:

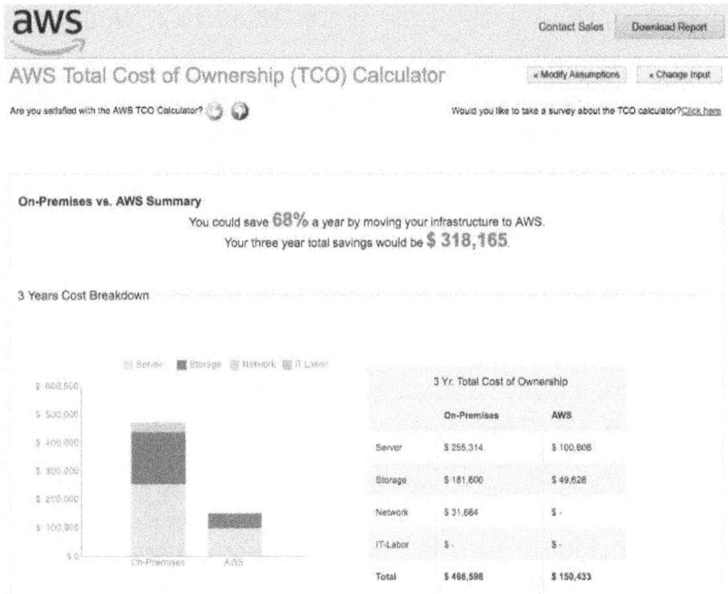

Figure 10.6. TCO computation for a cloud migration project

- *Project size.* Project size, expressed as the estimated monetary investment, is a proxy for project complexity and the potential consequences of failure. Note that you should focus less on the absolute size of the project (e.g., $12 million) than on its size relative to the typical kind of project the information systems group in your firm undertakes. In other words, a $12 million initiative carries different risk for a firm that typically handles $20 million projects than one that is mostly familiar with $150,000 efforts.
- *Experience with technology.* The degree of experience a firm has with the technologies (e.g., hardware, software development, cloud-computing environment) at the heart of the project is a primary determinant of risk. Working with new and unproven technologies carries more risk than using mature ones. However, the relative novelty of the technology to the firm is also an important source of risk.
- *Organizational change.* The degree of organizational change that the project requires is another important determinant of risk. Third-order change (Chapter 2) is much more difficult (and risky) to implement than second- and first-order change.

With respect to the cost/benefits that a system is expected to deliver, different initiatives vary significantly with respect to the size of the investment they call for and the expected useful life over which they will deliver the benefits. A large enterprise system will require a significant up-front investment but will deliver its benefits over decades. Conversely, a redesign of the company website will typically require a fraction of the investment but will also obsolesce much more quickly.

A portfolio approach can be used to optimize the blend of projects under development and ensure a balance of the overall risk as well as proactively manage the cost/benefits of the applications under development.

Portfolio Management

After evaluating the risk of each proposed project, management should take a portfolio approach to IS risk management. The set of ongoing IS projects a firm has underway can be categorized into subclasses based on the objectives of the initiative: Strategic projects seek new value creation and appropriation. Informational projects seek increased control and better information management.

SIDEBAR 10.2. Writing the Business Case[1]

The Business Objective

This company is a major global provider of mobile telephone services to both consumers and businesses. Following an internal restructuring of its service and territorial divisions, the UK consumer division wished to improve the service provided to customers and its ability to promote new network services and features to existing customers. The company believed that excellent customer service was one of the few ways it could differentiate itself from competitors in a very competitive market. It had also invested considerable amounts of money in new network facilities and needed to increase sales of its higher-end services to recoup this investment.

The company identified that service improvement and the promotion of its newer services could be achieved by upgrading its call-center systems. Most service requests from customers came into the call center. If the request was dealt with promptly, the company believed the call-center agent could then discuss newer service offerings with the customer.

In addition to dealing with incoming service requests, agents would also make outbound marketing and promotional calls to customers. A new customer-profiling system would be deployed so that the service being promoted, and the script used, could be tailored to the perceived customer needs. Agents would also collect data from customers during service and promotional calls and record them in the profiling system. This data would be used both to improve future targeting and to develop new service offerings.

Business Drivers

In terms of external drivers, the company perceived competing mobile telephone services as being indistinguishable, so it was difficult to differentiate its offerings on brand alone. In the past, it had tried to compete on price but had found it difficult to sustain this. Hence it saw service as a differentiator.

In terms of internal drivers, the company needed to recoup the high investment it had made in network access and infrastructure by increasing customer take-up of its higher-end services.

Investment Objectives

The company identified two objectives for the investment in upgraded call-center systems:

- Significantly improve the service provided by the call center and reduce service failures.
- Increase the take-up of newer services and collect customer profiling information to better target new services.

Benefits

The main benefits that would be realized by achieving the two objectives are shown in Table 10.4. The total expected financial benefits amount to £1,805,000 ($3,587,400) per annum.

The benefits framework shown in Table 10.4 is typical of many IT investments. It includes a full range of benefit types, from observable to financial. While the senior managers involved were keen to show that the financial benefits provided an acceptable return to the organization, they recognized that the observable benefits were of most interest to the hundreds of call-center agents who were required to use the new systems and adopt new ways of working. The agents' buy-in to the new system was key to making the investment a success.

The benefits in this example also cover the full range of business changes, from discontinuing things the company wished to avoid, such as callbacks to customers due to service call failures, to doing new things, such as promoting new, higher-value services during service calls. The example also demonstrates that it is easier to put a financial value on things the company is already doing and wishes to either stop or do better. It is, however, harder to determine a robust quantity or financial value for benefits resulting from innovation.

Investment Costs and Risks

In this section are detailed investment costs that the mobile phone company incurred in implementing its new call-center systems (Table 10.5) and the risks involved (Table 10.6).

1 Ward, J., Daniel, E., and Peppard, J. 2008. "Building better business cases for IT investments." *MIS Quarterly Executive* 7(1): 1–15.

Table 10.4. Benefits evaluation grid

Objective type	Doing new things	Doing things better	Stop doing things
Financial		*Benefit*: Increased customer retention due to improved service provision *Measure*: Reduction in customer defections; Avoided defections due to service failure = 1,750 per annum; Cost per defection = £500—savings of £875,000 per annum *Benefit owner*: Customer accounts manager *Benefit*: 20% reduction in call servicing costs *Measure*: Cost per service call; Number of calls per annum = 5.6 million; Total servicing costs = £1.2 million—savings of £240,000 per annum *Benefit owner*: Telechannel sales manager	*Benefit*: Stop callbacks to customers after failed service calls *Measure*: Number of callbacks; Number in previous years = 1.5 million; Cost per callback = £0.46—savings of £690,000 per annum *Benefit owner*: Call-center operations manager
Quantifiable			*Benefit*: Eliminate call waiting times more than two minutes for customers *Measure*: Number of calls currently waiting more than two minutes = 1.1 million *Benefit owner*: Call-center operations manager
Measurable	*Benefit*: Call-center staff able to undertake sales calls / promote new services *Measure*: Number of sales calls per staff member or sales per staff member; Current value = 0 (call center currently purely inbound) *Benefit owner*: Telechannel sales manager	*Benefit*: Customers not switching to competitors' products and services *Measure*: Number of defections to competitors; Current number of customers switching = 5,500 per annum *Benefit owner*: Customer accounts manager	
Observable	*Benefit*: Call-center staff motivated by being trained about newer services *Measure*: Increased call-center motivation *Benefit owner*: Call-center staff manager	*Benefit*: Ability to develop future services based on customer data *Measure*: Quantity and quality of customer profile data *Benefit owner*: New service development manager	*Benefit*: Stop customers becoming frustrated/rude because of service failure *Measure*: Call-center staff opinion *Benefit owner*: Call-center staff manager

Table 10.5. Investment costs

Investment costs	
Purchase of new call-center hardware and software	£250,000
Cost of implementing technical consultants	£120,000
Internal systems development costs (for configuration)	£150,000
Infrastructure upgrade costs	£75,000
Business change costs	£270,000
Training costs	£80,000
Total	£945,000
Net increase in annual systems support and license costs	£80,000

Table 10.6. Risks of the project

Risk analysis	
Technical risks	Complexity of the systems functionality Number of system interfaces and systems being replaced
Financial risks	Confidence in some investment costs—especially business change Confidence in the evidence for some of the benefits Business criticality of areas affected by the system
Organizational risks	The extent of changes to call-center processes and practices Limited existing change management capability Call-center staff capability to promote more technical services Customer willingness to share information for profiling purposes

Transactional projects seek efficiency and cost cutting. Infrastructural projects seek integration and streamlining of the firm's IT infrastructure (Figure 10.7).

During the budgeting process, when the firm evaluates the collection of initiatives for the coming year, it should determine the appropriate level of aggregate risks it is willing to accept. The risk profile will change depending on the blend of strategic, informational, transactional, and infrastructure projects the firm chooses to engage in (Figure 10.8). Note that this evaluation is a managerial decision, not an IT decision. While IS professionals must help in evaluating project risk, it is the responsibility of the **steering committee** or the board of directors to decide what overall degree of risk the firm should accept given its overall strategic IS plan (Chapter 6).

A portfolio approach to managing IS risk ensures that the funded initiatives fit the risk profile that the firm has deemed appropriate. For instance, firms in the strategic quadrant (Chapter 6) typically need to take on a much higher aggregate risk than those in the support quadrant. The blend of projects they engage in is also different. This is due to the fact that, in the first case, IS assets and initiatives are instrumental to the success and growth of the firm. The same cannot be said for those organizations that choose a very defensive approach to IS use and management (e.g., support quadrant). In such a case, a high degree of aggregate portfolio risk is a signal that the firm is not managing information systems in accord with its strategic IS plan. It may be that the organization needs to reevaluate and update its IS vision, perhaps moving into the turnaround quadrant. Conversely, it could be that the firm is failing to take advantage of the budgeting and prioritization process to enforce the existing (and appropriate) plan.

10.6 Outsourcing

IS outsourcing is often used as a means of funding IS operations by engaging outside providers. Outsourcing is the general process of acquiring products or services that used to be created internally by the organization from an outside provider. IS outsourcing is the more specific case of contracting with an outside firm to obtain IS services. Such services can range from automation of specific processes (e.g., payroll), to management of specific assets (e.g., data center), to development of new applications, to outright management of the IS function as a whole (i.e., full outsourcing). Outsourcing of IS services is now estimated to be a $330 billion industry, as measured by worldwide spending, and interests all sectors of the economy (Figure 10.9) The growth of cloud services and the digitization of business is impacting the IS outsourcing market, changing its composition. David Tapper, VP of outsourcing and managed cloud services at IDC, suggests,

> The IT outsourcing services market continues to undergo a fundamental shift to a new consumption and delivery model referred to as managed cloud services for which established players are facing increased competition from new

players (e.g., AWS, Google, Azure, and Alibaba) and need to make fundamental changes to their businesses.[1]

Like Jassy at Amazon, Microsoft CEO Satya Nadella ran the Azure Cloud unit before being promoted to the top of the firm—further demonstration of the central role cloud services is carving in the outsourcing and IS provision market.

Drivers of Outsourcing

When outsourcing information systems and digital technology to a specialist, the outsourcing firm is typically driven by one or more of the following considerations:

Reduced Cost Perhaps the primary driver for the outsourcing decision is the intention to capitalize on the provider's ability to create economies of scale in the production of digital services. Large providers can consolidate their infrastructure (e.g., data centers) and enjoy superior bargaining power when dealing with technology vendors (e.g., hardware and software firms). Additionally, IT outsourcing decisions significantly impact non-IT operating costs, such as sales and general and administrative costs.[2] This is an important aspect to consider when building the business case for outsourcing or cloud services, as operating costs are offloaded to the provider.

Access to Superior Talent Many organizations find it difficult to attract top technical talent. The reason is simple: top engineering and computer science graduates want to be challenged and want to remain on the cutting edge of digital technological development. But many organizations cannot (and should not!) make this objective a priority. Conversely, digital service providers are in the business of continually seeking to improve their offerings, evaluate new digital technologies, and attract the best talent. Additionally, new technologies and platforms are emerging on the market, and although they are "open for business," few organizations can leverage their internal experience and talents to profit from them (e.g., Big Data, analytics, robotics).

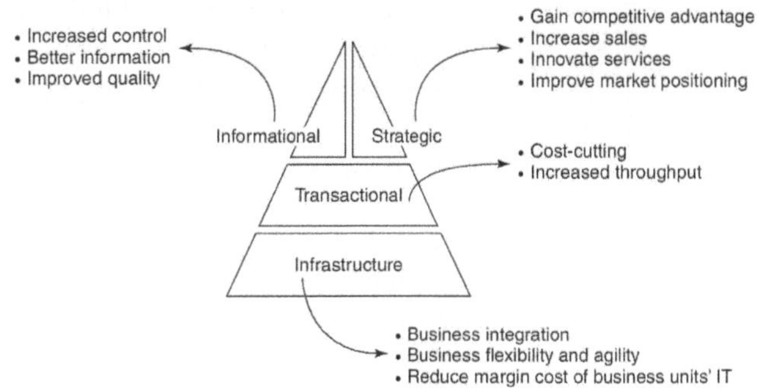

Figure 10.7. IT project portfolio components

Source: Adapted from Weill, P., and Broadbent, M., 1998, ***Leveraging the new infrastructure—how market leaders capitalize on information technology***, Boston, MA: Harvard Business Review Press

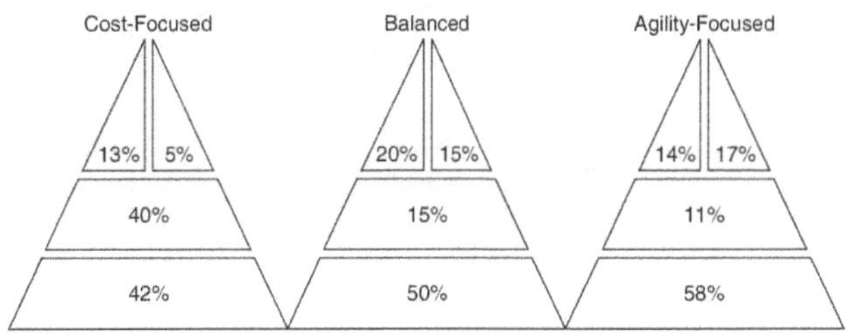

Figure 10.8. IT project portfolio profiles

Source: Adapted from Weill, P., and Broadbent, M., 1998, ***Leveraging the new infrastructure—how market leaders capitalize on information technology***, Boston, MA: Harvard Business School Press

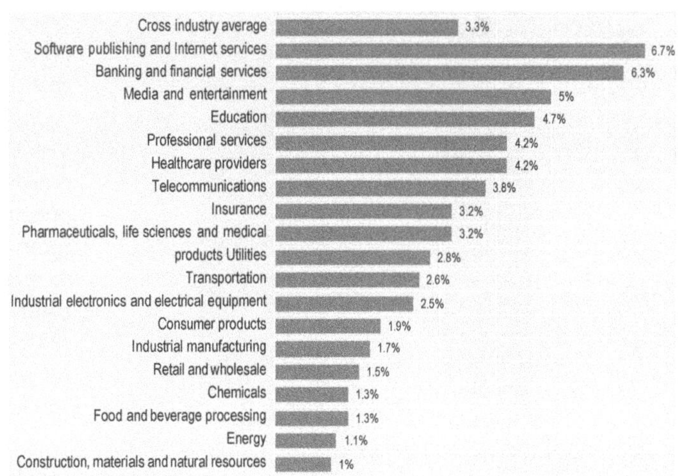

Figure 10.9. IT spending in outsourcing as a percentage of total IT spending by industry

Source: Gartner, 2014, IT spending key metrics data

1 Tapper, D. 2019, April. "Worldwide and U.S. IS outsourcing services 2019–2023 forecast." *IDC Market Analysis.*

2 Han, K., and Mithas, S. 2013. "Information technology outsourcing and non-IT operating costs: An empirical investigation." *MIS Quarterly 37*(1): 315–31.

Because of these structural differences, outsourcing and cloud service contracts offer the opportunity for organizations to access top technical talent and receive an infusion of digital technology and information systems management expertise.

Improved Control In many organizations that resort to outsourcing, particularly full outsourcing, the driving force for the decision was an attempt to reclaim control over the IS function when it was perceived to be inefficient and unable to provide the appropriate level of service to the organization. By engaging in a contractual arrangement with an outside provider, the theory goes, the firm can surface costs, making them explicit, and hold the provider to its service-level agreements.

Improved Strategic Focus For many organizations, IS operations were considered (rightly or wrongly) a nuisance rather than a core strength of the firm. For these organizations, outsourcing held considerable appeal because it enabled the firm to focus on what it considered its strengths and core competencies. However, the accelerating pace of digital disruption over the last decade has forced any organization that seeks to stay competitive to engage in digital transformation and digital innovation (Chapter 5).

Financial Appeal Outsourcing arrangements where the service provider acquired the infrastructure and IT staff of the outsourcing organization had the added advantage of liquidating some of the tangible and intangible assets tied up in the firm's infrastructure. As such, these deals could strengthen the balance sheet, at least in the short term, and held financial appeal. The modern equivalent of this approach is the increasing migration of digital infrastructure services and applications from on-premise operations to the public cloud (Figure 10.10).

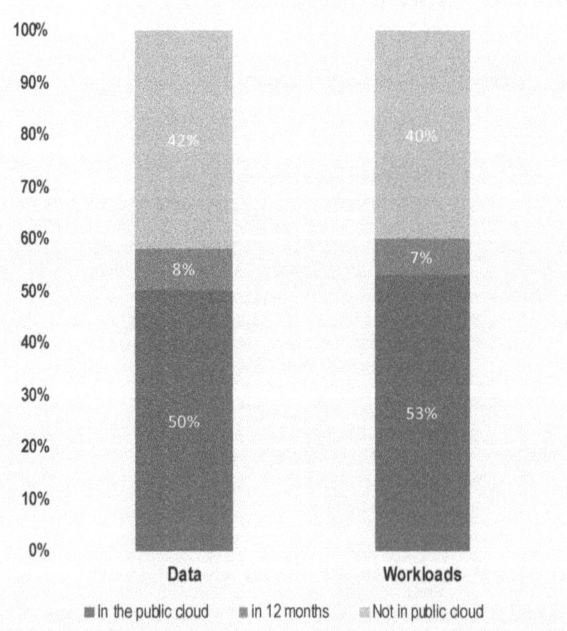

Figure 10.10. Workloads and data are moving into the cloud

Source: Flexera, 2020, "2020 state of the cloud," retrieved from https://resources.flexera.com/web/pdf/report-state-of-the-cloud-2020.pdf

The Risks of Outsourcing

While the drivers of outsourcing described above provide a strong case for the decision, the outsourcing literature and history have provided many examples of outsourcing deals gone bad. The following are the major potential drawbacks that must be carefully evaluated before taking the plunge.

The Outsourcing Paradox Organizations that resort to IS outsourcing often do so out of frustration with their current IT operations. Yet if you have little faith in your ability to manage information systems internally, how can you expect to be able to make sound outsourcing decisions (i.e., draft advantageous contracts) and monitor the performance of the service provider? How can you determine the appropriate amount to pay for the services and the service levels you should expect? How do you know that you are getting what you paid for? (Table 10.7)

The Dark Side of Partnerships The word *partnership*, when used in a business context in particular, can be very misleading. While the outsourcing firm and the service provider are "partners," as they work together to ensure the success of the operation, each has a responsibility to shareholders to maximize its own performance. This conflict of interest sometimes leads to friction and court battles. Imagine, for example, a service provider that is under pressure to cut costs and increase profitability. This result can be achieved by reducing customer service, particularly in areas where service levels are difficult to measure, hard to monitor, have no penalties attached, or have not been clearly specified in the contract.

Table 10.7. Ingress and egress costs per gigabyte (GB) or terabyte (TB) of the main cloud providers, based on monthly usage

Provider	Ingress Price	Egress Price
aws	Free	1 GB–10 TB ~ $0.09 10 TB–50 TB ~ $0.085 50 TB–150 TB ~ $0.07 150 TB–500 TB ~ $0.05 >500 TB on demand
Azure	Free	5 GB–10 TB ~ $0.087 10 TB–50 TB ~ $0.083 50 TB–150 TB ~ $0.07 150 TB–500 TB ~ $0.05 >500 TB on demand
Google Cloud	Free	0–1 TB ~ $0.12 1–10 TB ~ $0.11 >10 TB ~ $0.08

Source: Amazon, https://aws.amazon.com/ec2/pricing/on-demand/#Data_Transfer; Azure, https://azure.microsoft.com/en-in/pricing/details/bandwidth/; Google Cloud, https://cloud.google.com/storage/pricing

Changing Requirements One of the primary dangers in an outsourcing relationship is given by the length of the contracts in relation to the speed of technological and business needs evolution. In the early days of IT outsourcing, the considerable effort and up-front costs associated with the outsourcing decision led to long-term arrangements. However, a 10-year contract will see significant change in organizational requirements before it is up for renewal. As cloud-computing services continue their rise to prominence, servitization and monthly payments have become the norm over long-term contracts. However, as a savvy manager, you should realize that migration to the cloud entails significant switching costs and lock-in potential (Chapter 9). For example, cloud providers such as Amazon, Microsoft, and Google are famous for making it cheap and simple to migrate data to their storage services. But while "getting data in" is easy and cost effective, "getting data out" is not as simple or inexpensive given the sometimes substantial "egress" charges associated with data extraction.

Hidden Coordination Costs One of the biggest surprises awaiting organizations that outsource information systems is the extent of coordination efforts needed to work with the provider. Coordination efforts include communicating requirements, monitoring and measuring the provider's activities and services, handling dispute resolution, and the like. These coordination efforts come at a cost, of course—a cost that must be estimated beforehand and used in the decision-making phase.

The Deceptive Role of Information Systems While many organizations outsourced their digital technology service claiming not to be in the "IT business," many realized after the outsourcing decision the critical role that information systems play as enablers of business success. With the recent trends toward accelerating digital disruption and the growing need for digital transformation, very few organizations would downplay the strategic importance of digital technology to their success and maybe even survival.

10.7 Managerial Takeaway

As with any other complex and far-reaching managerial decision, there are no silver bullet solutions when it comes to outsourcing or the provision of cloud-computing services. The outsourcing decision requires a clear understanding of the characteristics of the organization and of the relationship with the service provider—a debate that general and functional managers must be involved in. This is why it is critical for you to be involved in decisions via traditional approaches, like the steering committee, or newer coordination mechanisms like CCOE. Instruments like the strategic grid (Chapter 6) can provide helpful guidance by mapping the current and future role that information systems are expected to play in the firm. Typically, firms that find themselves in the support and factory quadrant may find it easier to externalize services, given the standardized and well-understood role played by IS resources. More difficult is the decision for firms in the turnaround and strategic quadrants. Here, due to the critical role that IS assets must play in enabling the firm's strategy, outsourcing and cloud provision may be both challenging and risky. In this case, particularly for firms with limited access to new technologies or superior IT talent, outsourcing may be the only viable solution. However, these firms will typically provide IS services in-house.

In most cases, a firm should not resort to full outsourcing, locking itself into one provider, but should instead rely on selective outsourcing arrangements. Selective outsourcing arrangements are those where the firm relies on multiple providers offering different services. In such arrangements, the firm often retains an internal IS group that enables it to compete for the contract against outside service firms.

Perhaps the clearest advice that can be offered to firms considering outsourcing is to maintain a core group of IS specialists and a strong CIO function and invest in a CCOE. First, having the in-house expertise to match the organization's business needs to the appropriate IS services—whether these services are provided in-house or not—is fundamental. Second, it is critical to have a group of internal employees with an allegiance to your firm who understand what the service firms are providing and how best to manage the relationship. Third, it is critical for the in-house IS group to be skilled at negotiation, contract writing, and the ongoing monitoring of service-level agreements.

Even before you rise to the executive suite or are called upon to serve on the steering committee, your involvement is going to be critical. As you understand from the way business cases are developed and their need for deep managerial expertise, business cases cannot be successfully developed if the responsibility to redact them is left with the IS group. Rather it will be you as a functional manager who will need to collect appropriate data, complete comprehensive analyses, and craft the document that can garner approval from the steering committee and, more importantly, deliver value to the organization once implemented.

Summary

This chapter begins our discussion of the techniques and methodologies that modern organizations use to introduce and manage information systems within the framework provided by the strategic IS plan. As a general or functional manager, you must understand this process in order to fund initiatives of interest to your area and partake in the overall budgeting and prioritization process in partnership with other executives and IS professionals.

Specifically, in this chapter, we learned that

- TCO is a financial estimate designed to explicitly recognize the life cycle cost of IT assets. The costs of information systems and technology typically far exceed the cost of acquisition (e.g., selection, licensing, implementation) and include expenses that occur after the system is up and running but are necessary to maintain it in operation over its life span.
- Project sponsors use TCO in the formulation of the business case. The business case is the formal documentation used to garner support and win funding for the initiative. The project sponsor presents the business case to the executive committee or, in larger organizations, to the steering committee. The steering committee, composed of representatives from the various functional areas and IS professionals, provides guidance for the use of IS assets and shares the responsibility for aligning IS efforts with business strategy.
- Modern organizations use one of three approaches for the funding of IS operations and projects: chargeback, allocation, and overhead. The chargeback method, requiring direct billing based on actual usage, gives the most control to users but has the highest administrative costs. The allocation method, requiring direct billing based on measures such as size or revenue, seeks to strike a balance between fair billing and administrative overhead. Finally, the overhead method, drawing

funding directly from the overall organization's budget, is the simplest to administer but reduces accountability of both the IS function and units using the services.

- The yearly budgeting process is the tool organizations use to assess future IS requirements and prioritize funding. The budgeting process enables the firm to encourage and enforce specific behaviors and allocate IS decision rights and control. It can be a fairly stressful and emotionally charged process in which managers compete for funding of their projects.
- During the budgeting process, the firm has an opportunity to evaluate the risk of proposed projects, both individually and as a portfolio. The firm must take this opportunity to evaluate whether the degree of risk associated with its current portfolio of projects matches the risk profile the firm deemed appropriate during strategic IS planning.
- IS outsourcing is the process of contracting with an outside firm to obtain IS services. Modern organizations outsource their complete IS function (i.e., full outsourcing) or some of their IS assets and services (i.e., selective outsourcing), seeking one or more of the following benefits: reduced costs, access to superior IS talent, improved control over IS resources, a freeing up of resources to focus on core competencies, and liquidated IT assets. When evaluating the outsourcing of IS services, you need to consider the following risks: a firm with admittedly poor IS management will have difficulties evaluating providers and negotiating good contracts; outsourcing partners seek to maximize their own performance, which often creates friction; IS requirements evolve rapidly; and IS operations are often more strategic than executives realize.

Study Questions

1. Describe the relationship between strategic IS planning and the yearly budgeting and prioritization processes. What is the objective of each?
2. Why should general and functional managers be involved in decisions about the funding of IS assets and services?
3. What is a business case? What is its purpose? Who should be developing and presenting the business case for a new information system?
4. Define the following terms: *TCO* and *steering committee*.
5. Describe each of the three main IS funding methods and discuss the advantages and disadvantages of each.
6. What is a CCOE? Describe its role and function in harnessing the potentials of SaaS solutions within an organization.
7. What are the key drivers of new IS project risk? Why should a firm evaluate the aggregate risk of its portfolio of projects? What should the organization do if the current level of portfolio risk is not aligned with the degree of risk deemed appropriate according to the strategic IS plan?
8. Define and differentiate the following terms: *outsourcing*, *IS outsourcing*, *full outsourcing*, and *selective outsourcing*. What are the principal drivers and risks associated with IS outsourcing?

Glossary

A glossary of key terms can be found at the book's website. Visit https://www.prospectpressvt.com/textbooks/piccoli-information-systems-for-managers-5-0, then click on "Student Resources" in the horizontal red menu bar.

Chapter 11

Creating Information Systems in the Digital Age

What You Will Learn in This Chapter

This chapter covers a very important subject: the process by which organizational information systems come to be. While as a general or functional manager you may not concern yourself with hardware decisions, you must surely partake in the software design, acquisition, and **implementation** processes. Your involvement is essential because technology professionals rarely can evaluate the cost/benefit trade-off and impact of new information systems (IS) on the organization and its business success drivers. But in order for you to be a productive partner for the IS professionals who will carry out the technical work for new IS creation, you need to understand the IS development, acquisition, and implementation processes. As you will see, there are some techniques and practices that have been in use since the early days of computerized information systems as well as new emerging practices ushered in by the digital age.

MINICASE: Project Management Blues

"What am I going to do now?" you find yourself asking out loud while staring at the ceiling in your office. "Should I de-escalate this project or press on?" It felt like you were in one of those management case studies—except that it was real and it was you!

You replayed the events leading up to this dilemma. It all started when you were appointed the lead of the project team in charge of creating the new human resources (HR) benefits package management system. You had made a very successful business case presentation and received public praise from the executive team. "Finally! Someone who does not speak techno-mumbo-jumbo but can present an IT project in business terms!" L. J. Lalli, the chief executive officer (CEO), had exclaimed. It had been your ability to interface with both the developers and the business stakeholders that had landed you the project manager position. You were the first project manager in your firm to come from a functional area (HR) instead of the IS function.

The project had proceeded very well, with great support from the user community—your former colleagues, of course. This was due in large part to your knowledge of HR and your stakeholder-friendly approach. You had made a conscious choice to seek user feedback and to honor as many requests for enhancements as possible. "You have to freeze the requirements," Erik Khan, the lead **system analyst**, had objected. "Otherwise it's going to be anarchy." But you had dismissed his complaints as "development team grumblings." Those guys were never happy with a little uncertainty. Having been on "the other side" as a stakeholder in a number of system development projects, you knew full well that unhappy users were the fastest route to system failure.

Now you were beginning to second-guess your decision. The original schedule called for releasing the beta version of the application for user testing later this week. Instead, you had only 40% of the approved functionality coded. Moreover, your team was looking at a list of 22 enhancements, 2 of which would require a change in the database structure. Projected completion, without the proposed enhancements, entailed seven more months (a 45% increase on the original timeline).

It was now apparent that the original project had also been underfunded. The current estimate for finishing the project with the approved set of requirements called for a 62% budget increase (over the original). It was unclear how much more it would cost to exceed the requirements since the 22 proposed enhancements had yet to be evaluated by the **system architect**.

You were due to present a progress report to Ms. Lalli tomorrow afternoon, but you were still unsure about what course to take. The only certainty at this point was that you had to make your pitch for a project extension and ask for further funding at the meeting. Your plan was to report on the current state of affairs, paint a picture of the final product, and seek support. But what was the final product going to be?

Discussion Questions

1. What should your agenda for tomorrow's meeting be? Should you press on with your strategy, or is a change of course in order?
2. What would you do differently, if anything, given the chance to start this project all over again?

11.1 Introduction

Once a firm has developed a strategic plan for the use of IS resources (Chapter 6) and has gone through the budgeting and prioritization processes (Chapter 10) to identify what specific information systems it needs, it is time to act. Whether the information systems rely on custom-developed digital solutions or commercial off-the-shelf (COTS) software, it is critical that you as a general or functional manager understand how information systems come to be. Armed with this knowledge, you can proactively participate in the process.

While managers need not be concerned with hardware decisions, they must take part in the software design, acquisition, and implementation processes. Aside from the significant portion of your budget devoted to IS management and development, managers' involvement in IS funding and design is essential because never before has a firm's success depended so much on the use of the right **software applications.** Deciding what the characteristics of the "right" applications are is a business decision. It is a decision based more on the business case and the understanding of the business processes the software will enable (or constrain) than on any technical consideration.

How Hard Can InoIT Be?

Consider the following three examples. On a separate sheet of paper, answer the question before reading on to find out what really happened:

- The U.S. subsidiary of one of the major food producers in the world inked a deal to implement SAP (the leading enterprise system application) in an effort to centralize and rationalize operations across its nine divisions. The project required streamlining processes, standardizing software applications, and implementing the same organizational structure across units. How much time and how much money would you budget for this project?
- A large hospitality company with more than 2,000 branded hotels developed a customer information system to enable its customer relationship management (CRM) strategy. The custom-developed functionalities of the software application at the heart of the information system included a property-management system, the loyalty and CRM applications, and the reporting modules. How much time and how much money would you budget for this project?
- A major telecommunication carrier scheduled an upgrade of its customer service systems from version 6 to version 7 of a leading **off-the-shelf application.** The newer, more powerful version, exchanging information with 15 other systems (e.g., billing), would simplify customer service representatives' access to customer data and would increase the amount of customer information available for sales and service

transactions. How much time and how much money would you budget for this project?

There are few technologies and products that have evolved as far and as fast as digital technology has. However, the astounding successes of information technology (IT) can be misleading, tricking you into severely underestimating what it takes to build and implement a stable, robust, secure system that will work under a wide array of organizational conditions. Should you check your answers one last time before reading on? OK, here's what happened:

- The implementation of SAP by the major food service company took more than six years and cost more than $200 million. It was mired by setbacks and dead ends, with high-profile casualties, including the project leader, who was reassigned midway through the implementation.
- The large hospitality firm invested more than $50 million in the design and development of the application and in integrating it with the other applications in the firm's infrastructure. The project took about two years, and by its conclusion, it had cost about $120 million. The resulting system, the firm's largest investment in recent history, was considered a success.
- The upgrade at the telecommunication company was a complete failure. The new system was unstable, crashing for days at a time, and the old system was no longer usable. The customer service difficulties translated into an estimated $100 million in lost revenue during the three months it took to complete the upgrade. A rival acquired the firm, which was mired in difficulties, for half its original valuation.

The critical insight to be gained from this simple exercise is that organizational information systems usher in a wealth of complexities that go far beyond those associated with the personal computing environment that is most familiar to the typical end user (e.g., purchasing and installing Microsoft Office; Figure 11.1). Unfortunately, managers are surrounded by the misleading rhetoric of statements like "IT is easy; the hard part is people," or "Today, firms can easily develop

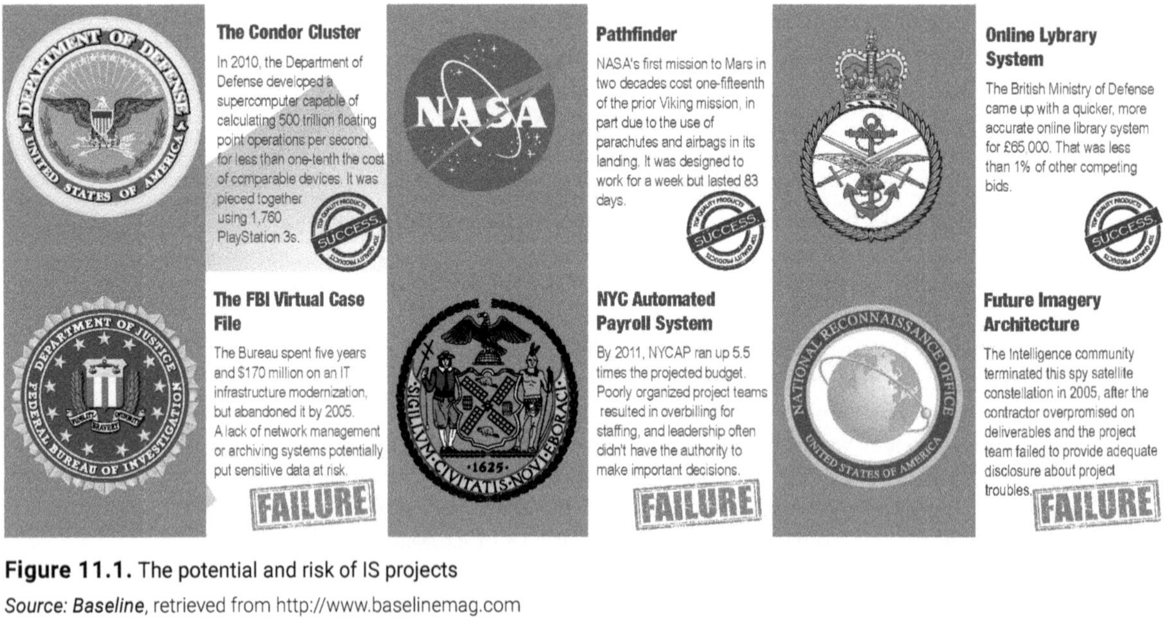

Figure 11.1. The potential and risk of IS projects

*Source: **Baseline***, retrieved from http://www.baselinemag.com

or purchase digital technology to obtain the capabilities to rapidly match their competitors," or "Digital technology is a commodity."

These views are gross oversimplifications of reality. When they are held by those who have never been involved in large-scale IS development efforts, they dangerously hide the truth: organizational IS development efforts are very complex and risky endeavors. The latest CHAOS report by the Standish Group that reviewed 50,000 IT projects showed that only 35% of projects were deemed successful by being completed on time, on budget, and with satisfactory results (Figure 11.2).[1]

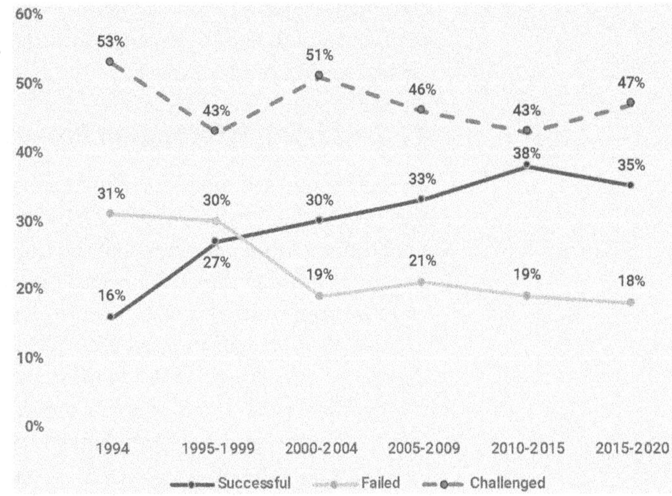

Figure 11.2. 25 years of IT project success

Source: The Standish Group, "CHAOS 2020: Beyond infinity," retrieved from https://www.standishgroup.com/

But more interesting, only 14% of these projects delivered high customer satisfaction or high return on value to the organization. Interestingly, despite much attention and efforts, the number of failed projects remained practically constant over the last two decades.[2] The current focus on digital transformation initiatives is increasing the complexity and ambitions of IS projects, impacting, often negatively, their chances of success.[3] While organizational issues, overoptimism, and poor objectives are the typical culprits, a lack of technology skills, poor governance, and sheer technical and implementation complexity represented important contributing causes to these failures.

IS projects are complex and risky precisely because they involve both technical and social challenges—and the intersection of the two. Consider the following explanation of why IS projects will likely never be as disciplined and predictable as other engineering processes (e.g., the building of a factory):

> In classic IT terms, important "requirements" are often not discernible in advance. If this statement sounds wrong to you, try on the alternative—that it's always possible to discern all the important requirements in advance, regardless of the size and complexity of the system and the rate of technological and business change. . . . Indeed, what makes a system great in the end, usually, is not just that it satisfies requirements that were known in advance. The difference between a great, value-adding IT system and a clunky dog that everyone hates is often in the details that are discovered along the way, as the system is implemented and users begin to have a more tangible sense of how it will work.[4]

The interplay of many different actors (often with divergent agendas), the sheer size of modern organizational systems, the myriad of expected and unforeseen organizational conditions the system must support, and the constantly evolving nature of the modern firm make these projects

1 Johnson, J. 2020. *CHAOS 2020: Beyond Infinity*. The Standish Group, retrieved from https://www.standishgroup.com/.
2 For more details on the "CHAOS 2020: Beyond infinity" report, you may consider watching the presentation of the results by Standish's European Research Director, Hans Mulder: https://www.youtube.com/watch?v=W4tmgk2QZBw.
3 KPMG, and Nash, H. 2017. "CIO survey 2017." KPMG.com, retrieved from https://home.kpmg.com/content/dam/kpmg/kz/pdf/cio-survey-harvey-nash-2017-us.pdf.
4 Austin, R. 2005, July 1. "No crystal ball for IT." *CIO Magazine*, retrieved from http://www.cio.com/article/2448685/project-management/no-crystal-ball-for-i-t-.html.

extremely challenging. IS projects require technical expertise. They also call for a big dose of managerial skill and informed involvement by general and functional managers who contribute critical subject matter expertise.

11.2 Fulfilling Information Processing Needs

The primary reason modern organizations introduce information systems is to fulfill their information processing needs (Chapter 2). Information systems leverage digital technology at their core to optimize the manner in which the firm captures, processes, stores, and distributes information. Thus they become the enablers and pillars upon which a firm executes its digital strategy.

How does the firm go about introducing the information processing functionalities needed to fulfill its information processing needs? How do firms develop the system that enables their digital transformation? How do information systems come to be in today's organizations? At the most general level of analysis, this process has two main elements—digital technology development and information system deployment:

- *Digital technology development.* Modern information systems are built on a digital technology core. Whether the technology is acquired and integrated into the existing firm's infrastructure or it is custom built by (or for) the organization, generating the digital technology core is a prerequisite to delivering the needed information processing functionalities.
- *Information system development.* Creating the needed digital technology core is not sufficient to fulfill the information processing needs of the firm (Chapter 2). The organization must successfully integrate the technology with the other components of the organization (i.e., people, processes, structure) to develop a working information system. This is the implementation process.

The technology development and implementation processes are intertwined, not sequential. Because of systemic effects (Chapter 2), the components of an information system must interact with one another without friction. Thus the design of a new software program (i.e., technology development) must take into account how the technology will be employed (i.e., processes), by whom (i.e., people), and under what organizational enablers and constraints (i.e., structure).

Three Approaches

There are three general approaches to the acquisition of information processing functionalities and the introduction of information systems built upon a digital technology core. Note that each of these approaches encompasses both the technology development and implementation processes. However, the critical differences among them pertain to how the digital technology components—and more specifically, the software that defines the capabilities of the system—are designed and developed:

1. *Build: custom design and development.* With this approach, the organization implements a software application that is expressly made, whether internally or through outsourced development, for the unique needs of the firm.
2. *Buy:* **system selection and acquisition**. With this approach, the organization implements a software application developed by a vendor and available on the market.
3. **End-user development**. With this approach, the organization uses a software application created ad hoc by its end users rather than the firm's IS professionals. While still relatively minor as compared to the *build* or the *buy* approaches, end-user

development is increasingly popular with tech-savvy employees who have access to ever more powerful digital technologies.

We describe below the advantages and risks associated with each approach. We also introduce the most prevalent methodologies used to articulate the IS design and development process in each case. Our objective here is not for you to become an expert in systems design and development—there are whole courses devoted to that objective. Rather, it is to help you clearly understand the process so that you can successfully take part in it.

Build versus Buy

In some cases, custom-developing the software at the heart of a new information system is not an option for your firm; it is a necessity. This is the case when the system must enable a new initiative and no market for such a product already exists. For example, when Amazon first introduced its personal recommendation system, electronic commerce was largely uncharted territory, and the firm was indeed shaping the online retailing industry. Waiting for an off-the-shelf product to be developed was not an option.

Typically, though, the firm will have to weigh the choice between custom development and purchase of the needed technology. Each approach offers some advantages and disadvantages.

Advantages of Custom Development While prepackaged software is available in the marketplace, in many cases, the firm will still engage in custom design and development to capitalize on the advantages of this process. Such advantages include the following.

Unique Tailoring The defining characteristic of custom-developed software applications is that they are molded to fit the unique features of the firm that commissions them. COTS software, like clothes you buy at a department store, will "fit" well in some areas of the firm but may create problems in others and require some adjustment from the organization.[1] Conversely, custom-made software, like a tailor-made suit, can be designed to fit perfectly with the organization's characteristics and needs.

Note that while every organization is unique, not all its processes are. For example, while Lands' End and Eddie Bauer are two different organizations, they both provide email for their employees and do so in a very similar fashion. Standard mail server software will likely serve the needs of both firms quite well. Conversely, if the business processes that the software is designed to enable are unique and value adding (i.e., a source of competitive advantage), COTS software may undermine their uniqueness and be detrimental to the firm.

Flexibility and Control Custom-developed software applications offer the highest degree of flexibility and control to the organization. Because the project team builds the system from scratch, the software can be molded into any form the stakeholders (e.g., management, end users) would like it to take, and the firm owes no licensing fees to software vendors. Moreover, since the firm retains control over the code, the system can be evolved, at any time and with the proper resources, in any direction the firm chooses.

This level of control is not achievable with COTS software purchased from vendors, since software houses need to develop applications that serve the needs of a large number of buyers. Moreover, the software house has to prioritize the features that will be coded into the upgrades.

1 As we will see, this adjustment of the organization typically takes place during implementation, when, taking the software as given, a working information system is developed.

Typically, they are the ones that have the broadest appeal rather than niche requests from individual clients.

Advantages of Purchasing As the software industry has evolved and grown dramatically over the last 30 years, the off-the-shelf offer has become comprehensive. Perhaps the greatest impetus behind the use of COTS applications has come in recent years from the viability of reliable cloud-computing solutions—specifically, the rise to prominence of software-as-a-service (SaaS) solutions (Chapter 3). Purchasing software from a vendor, whether on the cloud or to be installed in the firm's own data center, yields a number of advantages to the purchasing organization.

Faster Roll-Out An organization that purchases new software is typically interested in the information processing functionalities it enables. Thus how quickly the firm can be "up and running" with the new information system is an important concern. Purchased software dramatically reduces the time it takes to obtain the software and begin the implementation process. Rather than engaging in the lengthy custom-development process, the firm researches and evaluates existing applications before selecting one. Upon purchasing the selected application, the implementation phase is ready to start. This lag between the go-ahead decision and a working information system is even more compressed when the software is delivered in the cloud (i.e., SaaS model). The firm needs not develop the software, as with any COTS solution. But in this case, there is no hardware to provision for the firm's data center, no installation of software to perform, and no security, backup, and recovery procedures to implement since these services are all included in the SaaS model (Chapter 3).

Knowledge Infusion Another advantage offered by off-the-shelf applications is access to the expertise coded in the software. Because software programs can enable and constrain the manner in which users complete a task or execute a business process (Chapter 2), an organization that purchases prepackaged software also acquires a "way of doing business." Consider the example of a call-center operator who takes orders from catalog shoppers. The design of the application will determine the order in which the interaction takes place (e.g., greeting, items to be shipped, verification of address, payment) and what data are necessary to complete the transaction (e.g., no order can be completed without a valid phone number).

This notion of knowledge infusion is now an important design and marketing tool for software vendors, particularly SaaS providers, who proactively seek out best practices in order to code them into their applications. Returning to the call-center example, an often-mentioned best practice in call-center operations is to enable personalized interactions with each customer. Thus a call-center software vendor may code a feature set in its application that automatically brings up the customer's order history or prompts for a targeted offer so that the representative may engage in an informed—and more profitable—conversation with the customer.

Economic Attractiveness While it is always difficult to generalize when it comes to system design, maintenance, and development costs, purchasing off-the-shelf applications typically allows the firm to capitalize on the economies of scale achieved by the vendor. As with the example of mass-produced and tailor-made clothing, when a vendor can produce many units of the same software application, it enjoys declining fixed costs that, in turn, lower the unit cost. This is particularly true with software, a classic information good (Chapter 4) characterized by very high costs of producing the first copy but negligible reproduction and distribution costs. When the software is servitized, an added advantage is the ability of the firm to pay over time (e.g., monthly) using a subscription or pay-per-use model. Trading capital expenditures (CAPEX) for operating expenditures (OPEX) has the additional advantage of aligning expenses with the trajectory of business growth.

High Quality A great deal of debate surrounds the issue of software quality, with the skeptics pointing to the many examples of prepackaged applications that have significant bugs. Yet large software houses with mature products will point to their sizable testing budgets and large installed base of users as evidence that their applications have been put through the paces and thus all major problems have surfaced. When software runs in the cloud, there is the added advantage that the provider can continually update its products, yielding software that does not need to be patched or upgraded when bugs are discovered or new features are added.

Build and Buy

The build versus buy decision is typically treated as a dichotomous one (i.e., the firm must choose one approach or the other).[1] In practice, the general rule of thumb is that software acquisition should be preferred to custom development when the off-the-shelf solution meets 80% of the required functionality. But what happens if that 20% is strategic for your business? Organizations may adopt a blended approach, first acquiring a COTS solution and then modifying it extensively. This is another area where the emergence of cloud computing is creating more options for organizations. Cloud software vendors innovate and release updates and new functionalities faster than ever (Chapter 3)—in some cases, weekly! Such speed of deployment clearly reduced the need for internal customization. In fact, heavily modified or custom applications may limit the capacity of the organization to modernize their infrastructure or to migrate to the cloud.[2]

Keeping with the current integration and technological trends, organizations are expected to develop custom applications where strategically needed and leverage the configurability—the inhering possibility to customize the application—of COTS software in the other cases.

11.3 Systems Design and Development

Until the rise to prominence of the software industry, the acquisition of prepackaged software was the exception, rather than the norm, for most organizations. Its long tradition notwithstanding, designing and developing organizational software applications and information systems has always been a complex, failure-prone undertaking. Viewed by many as more akin to alchemy than to a reliable science, systems design and development continues to frighten non-IT managers, who perceive it as a minefield of technical, behavioral, and managerial challenges (Figure 11.3). **Custom-designed software** programs are increasingly developed by software houses that "fill in" for the firm's IS professionals. These arrangements, typically called ***software development outsourcing***, vary greatly, with some firms only outsourcing the programming and testing stages, while others resort to an external provider to see them through the entire **systems development life cycle (SDLC)**.

Whether designed internally by the firm that will use it or developed in outsourcing, the creation of **custom software** and the introduction of new information systems carry significant risks. In order to manage such risk and complexity, IS specialists, academics, and consultants have contributed to the creation of a number of system designs and development methodologies. In this section, we first describe the general process underpinning them, the SDLC, and we

1 To simplify the discussion, we discuss each approach separately as well.

2 Dawson, P. 2017. "Custom-built applications restrict infrastructure modernization" (no. G00345370). Gartner, retrieved from https://www.gartner.com/document/3826268?ref=solrAll&refval=195729113&qid=49c631a43f31bc7f97ac3d75276f934d.

SIDEBAR 11.1. Build versus Die

As software continues to "eat the world" (Chapter 1), organizations find themselves increasingly having to manage digital relationships with their customers (Chapter 5). In this environment, Jeff Lawson, CEO of Twilio, proposed, only half-jokingly, that the new paradigm should be called "Build versus Die":

> It used to be that the big decision in companies was "Build versus Buy," the classic IT decision. It was either like, "Well, we're going to go just buy some solution that gets forklifted in by vendor," or "We're going to go build, we're going to go build something from scratch. It's going to be very hard." So, you had vendors regularly coming in and telling you smart sounding things like, "Well, you don't want to reinvent the wheel, we've already built the thing" or "Is that value added? Is that your core competency?" and those things all sounded pretty good so most companies would just buy most of the software they needed to run the business. But as it turns out, as the world has turned digital, the interface that we have with most of the companies that we do business with is now a digital one. So, our view of our bank is now a screen and same for almost all the companies that we do business with. In that world, you can't just buy the same software that every company in your industry, all your competitors have, and expect to be differentiated. You need to build. Oftentimes, it's not entirely apparent in an industry until some startup arrives. It's a bunch of young people, they don't know any better and they are a bunch of software engineers and they just say, "Well, we're just going to build something," and they challenged the status quo. . . . Then the incumbents start doing and they start building. Pretty soon, it's not a matter of "Build versus buy," it's "Build versus die," because you have this Darwinian evolution going on of the companies that adapt quickly, and the pace of innovation accelerates in that industry. The builders compete on who's got a better take on what the customer needs and who's building a higher quality experience. The ones who are sitting there with, "Well, I bought a thing ten years ago," well, they just start losing their relevance in the eyes of customers. So, every industry seems to undergo this build versus die evolution where the people, whether they're startups or incumbents, are able to essentially become **agile** and listen to customers and build.[1]

Lawson is perhaps biased in his argument, since his company is one of those selling the digital resources (Chapter 5) that organizations need when they switch from "buy" to "build." But the attentive reader of this book will recognize quickly that Lawson's argument is a direct extension of many of the concepts we already discussed. As infrastructure (Chapter 3) and servitized digital resources (Chapter 5) have become engines of digital transformation along with enduring digital technology trends (Chapter 1), incumbents in almost every industry find themselves having to engage in digital innovation and, at times, outright transformation, under pressure from digital entrepreneurs (Chapter 5). But while Lawson's argument is appealing, you should realize that the type of "building" he advocates is out of reach for many organizations—particularly those who, due to size or financial constraints, are unable to marshal the capital necessary and attract the employees with the sophisticated development skills needed to successfully build. For these organizations that are likely to compete within digital platforms rather than build their own (Chapter 8), acquiring the digital technology core needed to enable their information systems is a viable and appropriate strategy.

---

1 Thompson, B. 2020, October 15. "Twilio acquires Segment, what is Segment, an interview with Twilio CEO Jeff Lawson." Stratechery, retrieved from https://stratechery.com/2020/twilio-acquires-segment-what-is-segment-an-interview-with-twilio-ceo-jeff-lawson/.

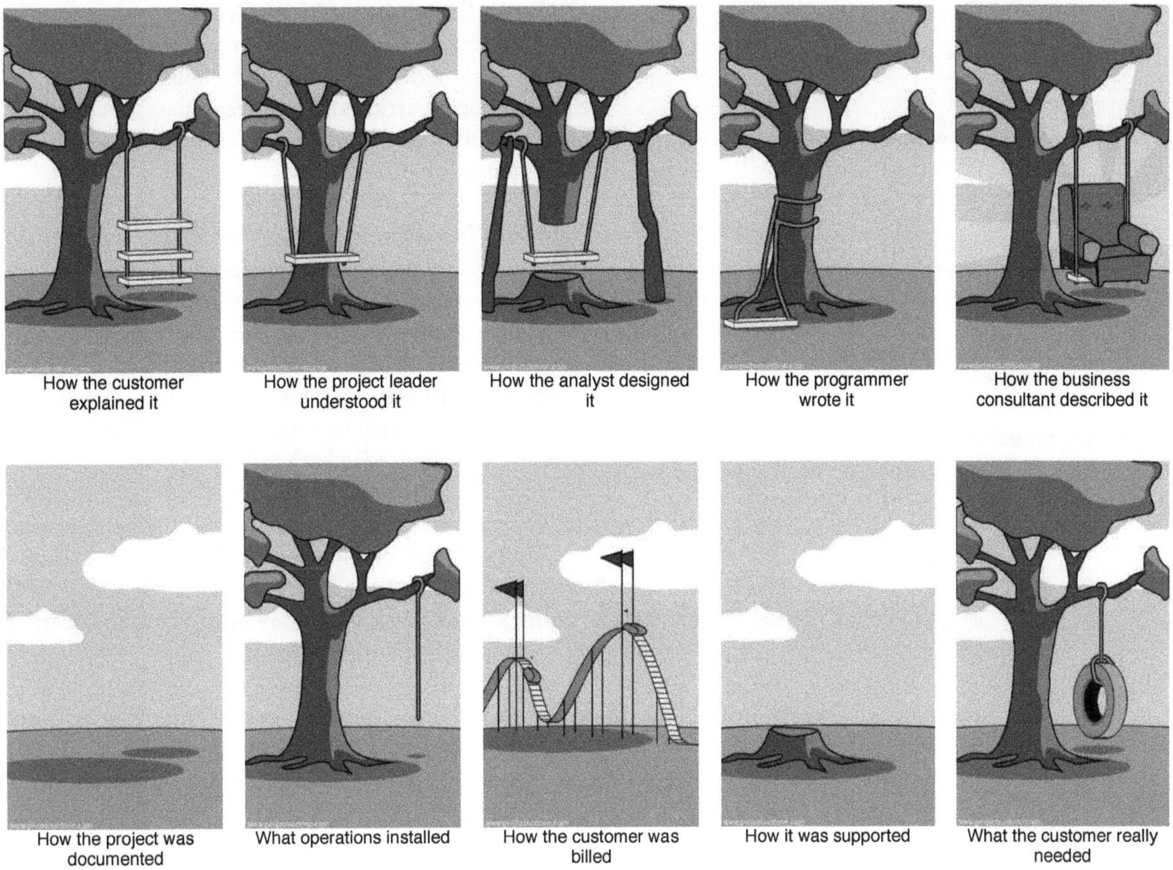

Figure 11.3. The SDLC as perceived by many managers

Source: ProjectCartoon.com

then compare two commonly used methodologies: the **waterfall** implementation of the SDLC and agile development. While more organizations fully adopt agile methodologies (47%), particularly younger firms in the high-tech sector, the waterfall approach to the SDLC is still quite popular in a number of contexts (41%; Figure 11.4). While we describe the two approaches as neatly separate to ensure clarity of exposition, they often coexist in the same organization (Figure 11.5). You should also note that, as digital technologies have taken center stage, the influence of agile principles has gone well beyond the four walls of IT departments. Agile practices are now common in digital innovation, product development, marketing, and HR. In the near future, you will likely serve on an agile team or you will need to adapt your work routine to agile practices. Understanding the agile philosophy and its methodology, benefits, and challenges is a critical skill of managers.

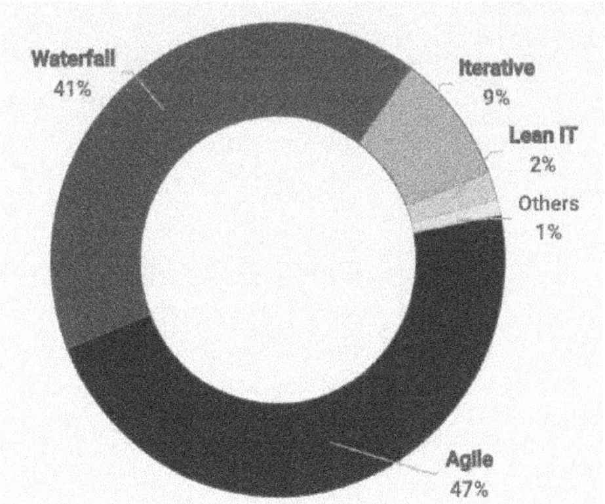

Figure 11.4. Main development methodologies

Source: Holz, B., and West, M., 2019, "Results summary: Agile in the enterprise 2019," Gartner, retrieved from https://circle.gartner.com/Portals/2/Resources/pdf/Agile%20in%20the%20Enterprise%202019%20-%20Results%20Summary%20(updated).pdf

Figure 11.5. Agile in application development

Source: Holz, B., and West, M., 2019, "Results summary: Agile in the enterprise 2019," Gartner, retrieved from https://circle.gartner.com/Portals/2/Resources/pdf/Agile%20in%20the%20Enterprise%202019%20-%20Results%20Summary%20(updated).pdf

Systems Development Life Cycle

The SDLC describes a general engineering approach to system design, development, and implementation. The SDLC process articulates in three phases—**definition**, **build**, and implementation—each one further divided into three steps (Table 11.1).[1]

Definition The definition phase of the SDLC is concerned with clearly identifying the features of the proposed information system. The critical actors in this phase are the prospective end users and the general or functional managers who represent the main stakeholders.

From the IS staff, systems and business analysts get involved. Systems analysts are highly skilled IS professionals who are well versed in both technology issues and communication. Their role is to help users identify and articulate the system requirements and serve as a liaison with the technical staff (i.e., developers). Business analysts are individuals with expertise in business process redesign as well as technology. They help ensure that the business processes and software programs at the heart of an information system are jointly optimized and work smoothly together.

Investigation During investigation, proponents of the new system must identify what business issues the system will pertain to. Managers who envision new ways of operating are the driving force at this stage, as they formulate the main goals, scope, and value proposition of the new system. This stage is typically very informal. The next stage brings a greater discipline to the analysis.

Table 11.1. Principal phases of the SDLC
Definition
Investigation Feasibility analysis System analysis
Build
System design Programming Testing
Implementation
Installation Operations Maintenance

Feasibility Analysis In order to ensure that scarce organizational resources are put to best use, the project team must heavily scrutinize the proposed project prior to giving the formal go-ahead. The business case provides the basis for this analysis (Chapter 10). Specifically, the team must evaluate the technical, operational, and economic feasibility of the project.[2]

Technical feasibility is the evaluation of whether the proposed system is viable from an IT standpoint. The team must ask whether the state of the art in hardware, software, and telecommunication equipment is such that the proposed system will work as intended (e.g., it will have enough storage capacity and an acceptable response time). The history of new systems development abounds, with examples of technology implementations that predated their time, thus undermining system success.

Operational feasibility, sometimes called behavioral feasibility, is the evaluation of whether the information system as a whole, not just the technology component, is viable. This analysis requires an evaluation of the other three components to make sure that employees have the skills necessary to utilize the new technology and that they will accept (or can be given incentives to

1 Note that the specific labels of the SDLC phases may vary, but the value of this discussion for you as a manager is in understanding the overall process.

2 Note that the feasibility analysis for a new system is informed by its business case (Chapter 10), but there are differences. The business case mainly focuses on justifying the pursuit of the project (i.e., cost/benefit analysis), while the feasibility analysis looks comprehensively at all the factors that can hamper system success—technical, operational, and economic.

accept) the new work system. During this phase, the project team must envision how business processes will be redesigned and must foresee possible drivers of user resistance and rejection.

Economic feasibility is the evaluation of the financial viability of the proposed system. A number of techniques have been developed over time to justify the proposed investment, including return on investment (ROI), payback, and net present value computations. Ultimately, evaluating financial feasibility consists of performing a cost/benefit analysis in order to ensure that the money to be spent on the system design and development project meets the firm's financial hurdles for investment.

The outcome of the feasibility analysis is a document culminating in a "go" or "no-go" recommendation. At this point, the firm has invested a small amount of resources relative to the full project cost; thus if the project is to be called off, this is an appropriate time to do so.

System Analysis Once a decision has been made that the system is worth pursuing, the project team needs to identify and articulate the system requirements. Systems analysts and the stakeholders (i.e., end users, management) take center stage at this point. If the system is not simply automating existing tasks but is instead enabling redesigned business processes, business analysts will join the team.

In large systems implementations, it is impossible to involve all users at this stage; rather, a subset of the user population joins the team, sometimes full time. Note that it is critical to choose users who are representative of the broader population—that is, the team should include not only the users who are highest performing or most well versed with technology (so-called superusers) but also underperforming and, most importantly, dissenting users (those individuals who may indeed resist rather than support the new system).

As the outcome of this stage, the project team produces the systems requirements document (Figure 11.6). This document details what inputs the system will accept, what outputs it will produce, which users will have access to what information, and so on. The document typically includes mock-up screens and scenarios (Figure 11.7) and is sent to the stakeholders for review and approval.

Reservation functions	❖ Clear all fields by clicking a Reset button ❖ Retrieve guest profile info for the selected main guest ❖ Modify guest profile info for selected main guest ❖ Retrieve previous reservations for modification ❖ Select a main guest by typing his/her name in full ❖ Select a main guest by typing the first few letters, then selecting from a suggested list of 5 names ❖ Add multiple rooms or a single room for multiple days to a reservation ❖ View the calculated tax for a reservation ❖ View the calculated total cost (before and after tax) ❖ Modify rates for individual room nights on the fly ❖ Record prepaid amount and method of payment ❖ Add notes specific to a reservation

Figure 11.6. User requirements

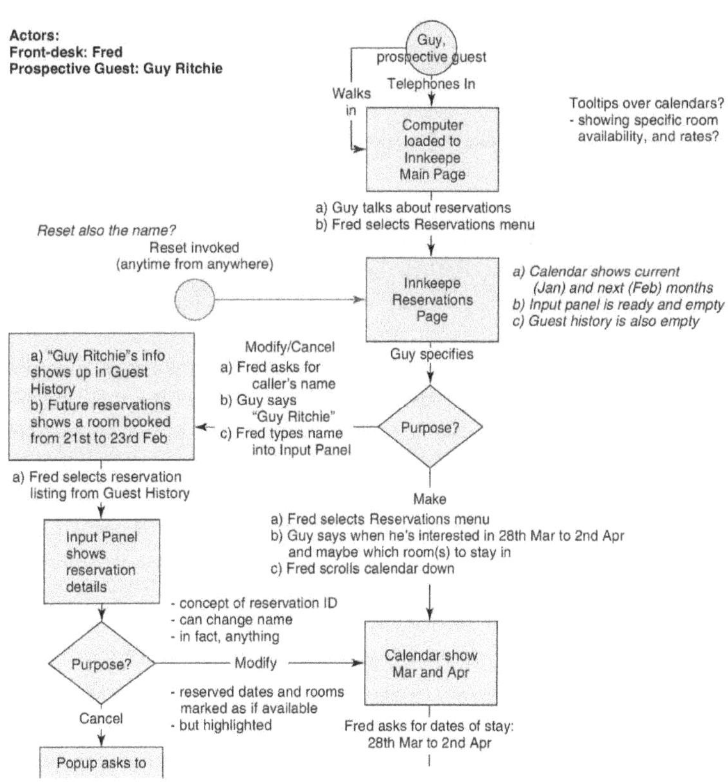

Build The build phase of the SDLC is the most technical and the one that most people picture when they imagine how software is designed and developed. This phase is the primary domain of technical staff: systems architects and developers (Chapter 1). The objective is to take the system requirements document and produce a robust, secure, and efficient application.

Figure 11.7. Sample scenario for the hotel reservation process

System Design The build phase begins with the system design stage. Taking the results of the definition phase (i.e., what the application should do), architects create the structure of the system (i.e., how the application will perform its tasks). At this stage, the team makes critical technical decisions, such as what hardware will be used, what languages will be adopted, and what data structures are needed. The output is a precise set of documents that developers use to write code.

Programming This is the process of translating the abstract software design into a set of commands or instructions that can be executed by the hardware. If the application requires the creation of new databases, their structure is also developed at this stage (Figure 11.8).

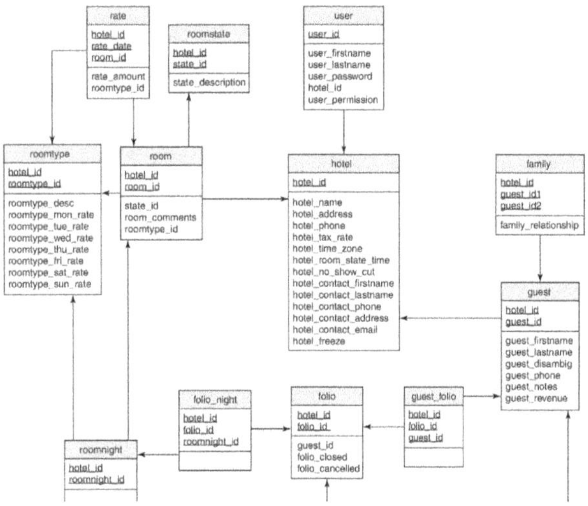

Figure 11.8. Database design underlying a hotel property management system

An important element of the programming stage, but one that developers often detest, is the documentation. Thorough and clear documentation is essential in organizational software programs because they are large, complex, and expected to last for a number of years. Without adequate documentation, such systems become impossible to support and maintain, let alone upgrade and evolve over time.

Testing While system testing is a process that **programmers** are constantly engaged in as they develop, a formalized assessment of components and subsequently of the complete application is an essential stage in the SDLC. While most individuals who are not IS professionals rarely think about testing, this stage can take as much time and resources as the programming stage. The testing phase is articulated in alpha testing, carried out by developers themselves, and beta testing, carried out by releasing the "beta version" to a limited set of actual users who are asked to report any problems they identify while using it.

Note that the objective of the testing stage is not to identify and correct all the possible bugs plaguing the system, as this is uneconomical and rarely needed. Rather, the testing phase is designed to stress the system, to make sure that it can perform under production circumstances, and to correct the most important errors.

Implementation Once the software has been developed and tested, the project team needs to ensure that it is properly integrated with the other components of the information system. This is the implementation phase, an extremely delicate time when project management skills and executives' involvement are essential.

Installation During the installation stage, the system is loaded on the production hardware and the databases are populated. Installation typically takes place during slow periods for the organization and, if at all possible, while the system is not needed (e.g., over a weekend, at night). If an existing system is being replaced, the firm migrates from the old one to the new one following one of four approaches (Figure 11.9):

- *Parallel.* The old and new systems are run for a time together. This approach is the most conservative, as it offers insurance against failure of the new application. It is also the costliest, as it requires a significant redundancy of efforts. In some cases, this approach is the only option (e.g., systems that must operate 24/7/365).

- *Direct.* The old system is suddenly discontinued, and the firm cuts over to the new one. This is the most radical approach but one that sometimes cannot be avoided (e.g., the old system becomes unusable).
- *Phased.* The new system progressively replaces the functionalities of the old one. This approach is best suited to modular or componentized applications that can be rolled out in stages.
- *Pilot.* Well suited for multiunit operations (e.g., hotels, chain retailers), this approach allows the firm to run the new system in one business unit or in one of the firm's departments before rolling it out completely.

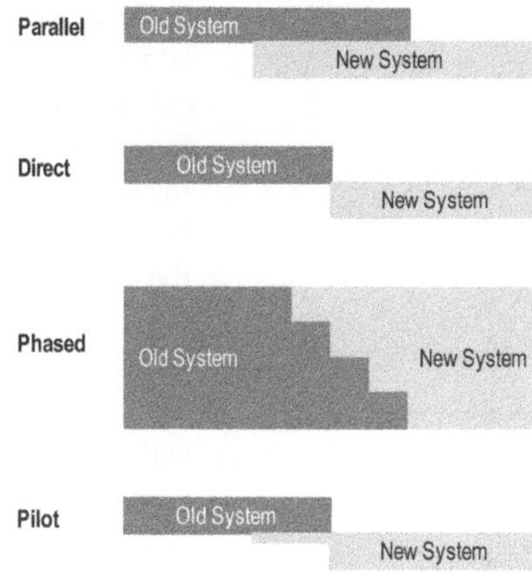

Figure 11.9. Migration approaches

Beyond the technical aspects of the installation phase, there are two critical processes that take place at this time: end-user training and change management. End-user training typically occurs in formal settings, such as classrooms or makeshift computer labs. Change management is the process of smoothing the transition from the way the various stakeholders interacted with the previous system and carried out their work to the new work practices. User resistance and inertia are the biggest dangers at this point. To the extent that stakeholders had been actively involved in the early stages of the SDLC and the design of the system, this phase will be less traumatic, thus minimizing risks of rejection.

Operations At this stage, the system is up and running, and the organization begins to use it. The project team is disbanded, and the new system becomes a permanent asset of the firm to be maintained and managed.

Maintenance Once the system is in place and is up and running, both errors that had escaped the testing phase and enhancements that had escaped the requirements definition phase begin to emerge. Maintenance is the process of compiling this information, prioritizing requests, and implementing both fixes and improvements. Note that as comprehensive and well designed as a new system may be, the organization is in continuous evolution. As a consequence, it is normal for a gap to emerge over time between the current system's functionalities and the firm's needs.

The functionality gap is closed on an ongoing basis by way of upgrades and additions until such ongoing maintenance becomes economically unfeasible and management makes the case for the development of a new information system.

Waterfall Methodology

A popular approach to the SDLC in information systems implementations is the so-called waterfall methodology. In the waterfall model, the focus is on limiting iteration between phases, as in a waterfall where water only flows downstream. The approach was dominant in the early days of computing because, given the cost and complexity of IS design and development, requiring detailed justification and up-front planning helped minimize risk and allocate responsibility for changes. In a strict application of the waterfall methodology, once the stakeholders approve the systems requirements, they are "frozen," and the cost of future changes, if any are requested, becomes the responsibility of the stakeholders. This step is necessary to minimize the impact of scope creep—the phenomenon by which stakeholders add or change requirements during the build phase of the SDLC, thus significantly increasing cost and considerably delaying development.

While the evolution of digital technology has enabled more agile methodologies (see below), there are many situations where the stage-gated approach to system design and development is warranted—for instance, when there are stringent regulatory concerns or when the system is being deployed in high security environments, or when the outcome is well established (e.g., time, budget) and defined. This is typically the case when the project team has knowledge of the problem and can reliably specify the requirements and overall software design. In all these cases, spending considerable time up front, the project team improves the chances of solving the right business problem with the right information system design while maintaining the needed level of control over the process. For this reason, in the waterfall methodology, the outputs of one SDLC stage become the inputs of the next, and the project team strives to keep changes to a minimum after the project has started.

Advantages of the Waterfall Methodology The waterfall implementation of the SDLC is a highly structured methodology that provides a systematic approach to reducing the uncertainty and risk associated with systems design and development projects. It clearly identifies roles and expectations for the members of the project team, and it offers a blueprint for how these individuals should interact. By demanding a thorough justification and requirements definition, it is particularly well suited for large-scale projects where changes that occur during development or implementation can be very costly and for projects where requirements are particularly sensitive. In these cases, a structured and formalized vehicle for communication and negotiation between the project team and the many project stakeholders ensures clarity of purpose and responsibilities. It does so by requiring the evaluation and approval of deliverables for every phase, thereby stimulating discussion, facilitating the identification of priorities, and surfacing hidden trade-offs.

Limitations By its very nature, the waterfall methodology is a highly structured approach. Thus its critics point out that it creates substantial overhead in terms of time and cost and that it does not enable the project team to properly address the inevitable changes that occur during the life of complex projects.

Agile Development

Recognizing the limitations inherent in the waterfall implementation of the SDLC, a number of new approaches were introduced over time. They all share the belief that it is impossible to clearly estimate and plan in detail such complex endeavors as IS design and development projects. Agile development, an umbrella term for a variety of approaches characterized by rapid development and deployment, has evolved substantially since first being promoted in 2001. Today, agile methods are pervasively adopted, with over 90% of firms reporting some level of use in their development efforts.[1]

The proponents of the agile software development philosophy captured it in a document called the "agile manifesto." The opening statement in that document reads,[2]

> We are uncovering better ways of developing software by doing it and helping others do it. Through this work we have come to value:
>
> - *Individuals and interactions* over processes and tools
> - *Working software* over comprehensive documentation
> - *Customer collaboration* over contract negotiation
> - *Responding to change* over following a plan
>
> That is, while there is value in the items on the right, we value the items on the left more.

1 DigitalAI. 2020. "14th annual state of agile report." Versionone, retrieved from https://stateofagile.com/.

2 Agile Manifesto. 2001. "Manifesto for agile software development." Agilemanifesto.org, retrieved from http://agilemanifesto.org.

The agile manifesto and the twelve supporting principles[1] define the agile approach, specifying the priorities and key characteristics of agile methodologies. First and foremost are adaptability and speed (Table 11.2). Agile methods do not call for the substantial amount of planning that characterizes traditional methodologies and are then more apt to situations lacking clarity about requirements and objectives. Agile proponents believe that planning for all requirements and contingencies is impossible in all but trivial system development projects. Significant planning will in fact backfire and limit the ability of the development team to adjust to new information. Agile teams, therefore, focus on developing applications quickly and releasing the system often. It is the rapid iteration of development and release—typically in two- to three-week sprints—that surfaces accurate requirements and enables the development team to converge to a system that meets customers' expectations and delivers value.

Table 11.2. Waterfall versus Agile

Waterfall	Agile
Fixed statement of requirements	Frequently changing requirements
Top-down assignment of tasks	Highly collaborative
Strict change control	Accepting of frequent changes
Large but infrequent releases	Short sprints measured in weeks
Disciplined governance	Self-organized governance
Phase gates	Review scope between sprints

Source: Barton, N., van der Heiden, G., and Wilson, N., 2018, "How to contract for agile development services" (no. G00325642), Gartner, retrieved from https://www.gartner.com/en

A second defining characteristic of agile development methodologies is teamwork in open-space offices that facilitate communication. Agile proponents advocate the use of small, cross-functional teams with a customer representative and daily face-to-face meetings. In order to organize the work and maintain the aggressive schedules that characterize agile projects, developers "chunk" the work into manageable yet self-standing components. The team then iterates through all the phases of development—from requirements elicitation through testing and customer acceptance. An interesting characteristic of agile methodologies consists of setting fixed time and resources for delivering the expected functionalities.

The agile approach has far-reaching managerial implications demanding a radical shift in project management culture away from the control enforced by the waterfall methodology. As each team manages the entire development cycle from requirements to operations, it self-determines how it will achieve the outcomes (e.g., selection of platforms, languages, frameworks, tools).

Benefits Agile methods accelerate software development and reduce project risk. Agile projects are increasingly proving more successful than traditional ones, particularly for larger undertakings—for which failure rate is traditionally high.[2] While you should not consider agile methods the "silver bullet" for project success, firms that transitioned to this development method generally experienced accelerated product delivery and greater adaptability to changing priorities (Figure 11.10). Agile methods are best suited to projects where problems are complex,

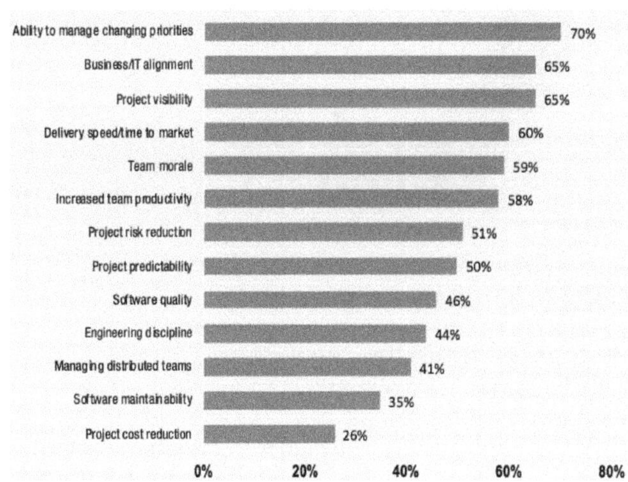

Figure 11.10. Agile method benefits

Source: DigitalAI, 2020, "14th annual state of agile report," Versionone, retrieved from https://stateofagile.com/

1 Agile Manifesto. 2001. "Principles behind the Agile Manifesto." Agilemanifesto.org, retrieved from https://agilemanifesto.org/principles.html.

2 The Standish Group. 2019. "CHAOS 2018: Decision latency theory." The Standish Group, retrieved from https://www.standishgroup.com/news/37.

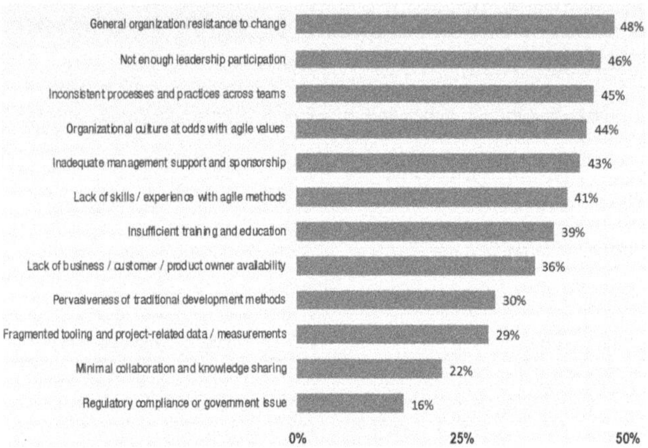

Figure 11.11. Agile method challenges

Source: DigitalAI, 2020, "14th annual state of agile report," Versionone, retrieved from https://stateofagile.com/

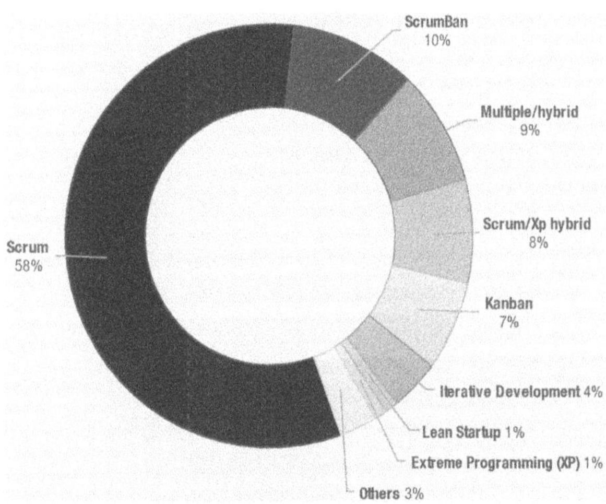

Figure 11.12. Agile method frameworks

Source: DigitalAI, 2020, "14th annual state of agile report," Versionone, retrieved from https://stateofagile.com/

objectives or requirements may vary, solutions are uncertain, the work can be modularized, and both customer involvement and cross-collaboration are feasible.[1] By relying on cross-functional teams, agile methods also facilitate the alignment between the IT staff and business leaders' objectives, tackling, at least partially, the disconnect that historically characterized their relationship.

Limitations Agile methods engender challenges (Figure 11.11). For many organizations, the new practices represent a major change in their way of working, affecting not only personal behavior but the entire organizational culture. It is then not surprising that firms experience strong resistance to change when their organizational context is not favorable to an agile culture (e.g., established traditional development methods, hierarchical decision making, risk-averse culture). Furthermore, the limited planning and specification of requirements and objectives typical of agile methods increases the chances of an uncontrolled growth of project scope and related requirements, engendering delays and raising overall costs. Rapidly changing specifications, additional requirements, and short deadlines are all favorable conditions for losing control of a project scope.

Agile Frameworks While *agile* identifies the overall development philosophy, *agile frameworks* are specific prescriptive approaches to software development. Frameworks are useful to organize and structure the development process, and each one presents its own strengths and weaknesses. While, in practice, teams often adopt frameworks resulting from the combination of different approaches, including elements of legacy practices, **Scrum** has become the leading framework for agile teams (Figure 11.12).

Scrum Started by Jeff Sutherland in 1993, it is the most used among agile frameworks due to its ability to deal with rapidly changing user requirements. Its strength lies in the simplicity of the approach and the rapid feedback cycles it enforces. Scrum software development progresses via a series of iterations called sprints, usually lasting two to four weeks. Sprints begin with a brief planning phase and end with a review and retrospective. For the sprint's duration, the team meets every day. The meeting is called the daily scrum. During the sprint, the daily scrum consists of a short, 15-minute meeting to check the results, impediments, and goals to advance the development. The *product backlog* is composed of the entire set of new features, changes, and fixes the

1 Rigby, D. K., Sutherland, J., and Takeuchi, H. 2016, May 1. "Embracing agile." *Harvard Business Review*, retrieved from https://hbr.org/2016/05/embracing-agile.

team should deliver to complete the final product. The part of product backlog completed during a sprint is called *product increment* (Figure 11.13).

The idea behind the sprint is that at the end of each sprint, the software is "done," meaning ready for delivery—at least in principle.

The Scrum team is in charge of the delivery of the product increments and consists of the following:

- A *product owner*, an internal or external customer who drives the product backlog. The backlog is a prioritized list of all the features and work that go into making the system.
- The *development team*, a group of three to nine members. It is capable of delivering the product increment. Thus team members are assembled by considering the complementary skills required to move the development forward.
- The *Scrum Master*, the individual who facilitates and guides the development team, playing both an internal and external role. Internally, she assures goals are well understood, suggests proper practices and techniques for their achievement, and coaches team members. Externally, he facilitates coordination and information exchanges between the development team and other organizational units or experts that could provide knowledge or skills the team needs.

You have probably noticed that the Scrum framework well suits the team level of a development effort. However, the same approach can be "scaled" to larger projects where multiple agile

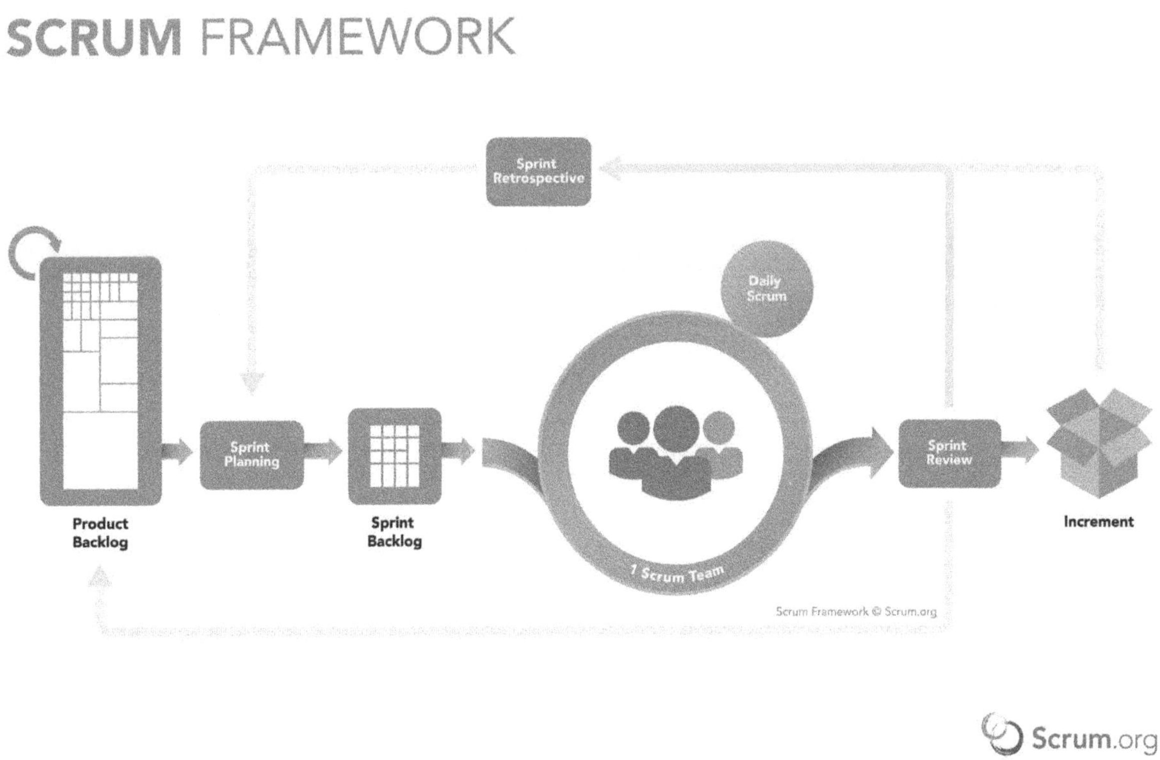

Figure 11.13. The Scrum framework

Source: "What is Scrum?," Scrum.org, retrieved from https://www.scrum.org/resources/what-is-scrum

teams need to coordinate for an enterprise-wide effort. Among the multiple frameworks and approaches to support scaling, Scrum of Scrums (SoS) and SAFe (Scalable Agile Framework) are the most popular. SoS fosters the collaboration and communication among the development teams extending the practice of the daily meetings to an all-teams event, the Scrum of Scrums. Each team delegates the Scrum Master or another member to report, there, the completed tasks, next steps, and issues. The Scrum of Scrums team follows nearly identical practices as the individual development teams but focuses on integration and delivery of teams' work and the solution to the impediments to the sprint's goals.

SAFe is a more structured approach than Scrum, aiming at scaling an overall agile approach to the whole organization. While Scrum mainly focuses on the teams' role, SAFe provides four different configurations—essential, large, portfolio, and full. These configurations provide a set of options depending on the organization's size, its experience, and the extent of the targeted agile scaling. By being a broader, enterprise-wide framework, it can accommodate the other agile approaches such as Scrum, Kanban, or XP. Compared to competing frameworks, SAFe is more structured and prescriptive, but at the same time, it can engender a lack of flexibility.

We emphasize that despite the popularity of these approaches, they serve diverse needs—therefore, the characteristics of the project and organization should always come first when considering their use.

Sourcing *Sourcing* is a term referring to commissioning a software development project. When sourcing agile developments, it is critical to ensure that contracts and charging methods fit project goals and the agile framework characteristics. A key challenge is that agile methods assume a lack of goal clarity, flexibility, adaptation, and cross-functional collaboration. How can such a project be contracted for up front with certainty and clarity (as it is done with more traditional implementations of the SDLC process)?

While traditional contracts define deliverables and times over multiple months through rigid planning, sourcing agile developments requires a flexible arrangement and appropriate charging mechanisms (see Table 11.3). A client could then decide to contract only a certain number of

Table 11.3. Agile charging mechanism

Charging mechanism	Use today	Characteristics
Time and materials	69%	The vendor is paid based on the time spent performing the work and the related materials specific to the project. It is the most widely used model for application development. The customer assumes project risks (e.g., over time and over budget).
Managed capacity	47%	In a managed capacity model, the vendor provides a certain number of agile teams that fit project requirements (e.g., skills, experience). The customer sets goals and priorities for the team and assigns its personnel to oversee teams' outcomes.
Unit-based	3%	The vendor and the customer agree on a unit measure upon which to base price. Several practices exist to assess these metrics, including the number of functions implemented or their difficulty (e.g., function and story points).
Fixed price per sprint	13%	In this case, each sprint is treated as a mini fixed-price project. This embraces agile principles of flexibility, limiting the risks derived from uncertain objectives and changing requirements.
Business outcomes	9%	An emerging charging mechanism in which the customer, instead of agreeing with the vendor on the cost for delivering the service, defines and purchases the expected business outcomes (e.g., room occupancy in hospitality, patient nights per nurse in health care).

Source: Adapted from Gupta, G., van der Heiden, G., and Barton, N., 2019, "5 Steps to effectively select and contract with an agile development service provider" (no. G00432872), Gartner, retrieved from https://www.gartner.com/en

sprints and stop if the project fails to progress or meet the established goals, similar to a rapid **prototyping** approach.[1]

While agile development methodologies are still relatively new and evolving, they have gained broad acceptance within the software development community. They are now applied in projects of varying complexity and contexts. As such, you may soon find yourself partaking in agile development as a customer representative on a cross-functional agile team.

DevOps

The agile paradigm has undeniably impacted development practices and has more recently extended its influence to IT operations. Historically, there was a clear separation between IS development and operations. You can see those divisions in the SDLC and the separation of the development and implementation processes. In the worst stereotypical cases, operation (Ops) team members saw developers (Devs) as disconnected from the real world of production environments, individuals unconcerned with the pragmatic issues of business operations who wanted every new feature they designed and built to be immediately moved to production environments. Devs, on the other hand, saw the Ops team as a bunch of slow, risk-averse individuals set in their ways and reluctant to change. But in practice, the two groups are more related than one would imagine. For example, delays in the deployment of new software affect the development process because feedback from customers is delayed, and unproductive downtime is injected. To mitigate the disconnect between development and operations, an innovative approach, called **DevOps**, is rapidly taking hold.

DevOps, as the term implies, integrates development and operations, taking inspiration from the agile movement. The goal is to build mutual trust and understanding between these traditionally siloed teams. More tangibly, the DevOps approach seeks to reduce the amount of software rework and communication overhead stemming from the lack of coordination between the Devs and Ops teams (see Table 11.4). In many cases, the development and operations teams are merged into a single group responsible for an application's entire development life cycle. Consequently, DevOps demands a significant change to organizational structure and culture, as traditionally separate groups are fused. The new group takes responsibility and ownership of the applications or digital services they manage. Automation and software tools feature prominently in DevOps as a way to maximize Devs' productivity and Ops' reliability, streamlining many of the traditionally manual elements of the deployment process. The key practices are continuous integration and continuous delivery.

Continuous integration (CI): Traditionally, developers worked in isolation for extended periods of time. CI requires the regular merging of code changes into a central repository where automated builds and tests run to provide ready access to the system's most updated version. A critical element of CI is automated testing, which is triggered when code changes are committed.

Continuous delivery (CD): Deployment is the process of moving software code into production. CD builds on CI to create workflows where software updated with changes is automatically built, tested, and prepared for production,

Table 11.4. DevOps principles and implications

Inspiring principles	Implications
Trust	Get all process stakeholders to work effectively together through disciplined feedback loops
Performance	Automate where possible to reduce delays and overheads and engage in shared metrics
Communication and collaboration	Develop tools to support automated activity monitoring and tracking and shared repositories
Continuous development and deployment	Provide rapid development (thus apt to accommodate agile methods) without scarifying quality and innovation

1 Barton, N., van der Heiden, G., and Wilson, N. 2018. "How to contract for agile development services" (no. G00325642). Gartner, retrieved from https://www.gartner.com/en.

thus quickly releasing frequent updates. As part of the CD pipeline, the team performs automated tests to evaluate the system's production readiness (e.g., UI testing, load testing), and the process stops if the tests fail. CD's added benefit is that deployments can be rolled back to a previous reliable version if new code introduces critical bugs or exhibits other problems.

Embedding a logic of continuous improvement, DevOps jointly improves the software, the environment, the infrastructure, and the overall delivery process, enabling the firms that adopt it to focus on rapid innovation. As Kurt Bittner, principal analyst at Forrester Research, put it, "If agile was the opening act, **continuous delivery** is the headliner."[1] You should appreciate how different DevOps is from a strict implementation of the SDLC, where deployment cannot even be planned until the developers complete the build stage and formally turn over responsibility to the Ops team (see Sidebar 11.2 for an example of DevOps in action). Conversely, under the DevOps paradigm, code is released almost continuously by applying the CI and CD processes. Consider Duetto, the revenue management start-up (Chapter 1). Duetto prides itself on releasing software to clients two to three times per week: "Our engineers, in partnership with the product team, maintain an enviable pace of two to three releases per week of feature-rich and stable code, and our customers feel the difference."[2]

Implementing a DevOps approach requires significant investments in terms of cultural change, process redesign, new technical skills acquisition, and automation. In fact, while the benefits of a CI/CD strategy are significant, they come at high costs for the organization adopting it. Following are some important considerations:

- *Industry competitiveness.* A DevOps approach is more appropriate in hypercompetitive industries, where organizations need to rapidly improve customer experience or test new capabilities. Many large business-to-consumer (B2C) software products, such as Netflix or Facebook, fall under this category. Conversely, an upscale restaurant chain (e.g., Morton Steakhouse) may not. However, as digital technology becomes pervasive, the average level of competitiveness is rising across the board. Consider, for example, Tesla. While in traditional cars, the infotainment system would remain the same for the life of the automobile, Tesla has mastered the art of updating the software of their entire fleet regularly through over-the-air updates.
- *Business criticality.* Given their critical role inside a business or society and because of practical or legal reasons, some systems might not be suited to an agile DevOps approach. Banks' core systems are a prototypical example of systems that cannot be changed rapidly or often. In other cases, software is legally regulated, as potential bugs can have significant societal effects. Consider, for example, the software governing aircrafts, medical equipment, or power plants. Updates to business-critical systems often require manual auditing, thus reducing the utility of a DevOps approach. Yet as digital disruptors enter these industries, the DevOps approach will increasingly take root.
- *Test coverage.* A CI/CD pipeline is predicated on the availably of a comprehensive set of tests during the software development and deployment phases of an application. Test coverage is the technique responsible for determining whether sufficient test cases have been formalized in order to guarantee that the application is working as expected. Therefore, before testing can be automated, each test case needs to

1 Bittner, K. 2013, July 22. "Continuous delivery is reshaping the future of ALM." *Forrester*, retrieved from https://www.forrester.com/report/Continuous+Delivery+Is+Reshaping+The+Future+Of+ALM.

2 Bosworth, P. 2013, March 15. "A year in, and Duetto keeps evolving." Duettocloud.com, retrieved from http://duettocloud.com/a-year-in-and-duetto-keeps-evolving/.

SIDEBAR 11.2. DevOps at ReimageIT

Founded in 2019, ReimageIT commercializes a software solution that can automate the grading of Excel assignments and exams. ReimageIT's solution is used by thousands of students practicing their accounting, finance, marketing, or general Excel skills. Students can access practice instructions from a web application and receive automated feedback on their work after submitting Excel files with their solutions to exercises. ReimageIT's strategy is predicated on developing a superior product compared to large, established textbook publishers. Thus constant improvement and deployment of the ReimageIT web application and grading engine in response to user feedback is a critical success factor. Like many other tech start-ups, all ReimageIT's systems are cloud-based, using Amazon Web Services (AWS). The following is the technology stack and automated DevOps workflows that ReimageIT has implemented with AWS tools:

- The ReimageIT development team uses Visual Studio Code, an integrated development environment (IDE), to write and debug the code. Visual Studio Code is configured to work with GitHub, the leading online source code repository, to maintain a consistent and centralized system to track and control each code change.
- ReimageIT implements a continuous delivery approach to automatically deploy code changes to a testing and production environment after the build stage. The entire CI/CD process, from the change commit to the application release to customers, is enabled by the AWS CodePipeline service.

The following is the DevOps process at ReimageIT:

- The first step is sourcing and testing of the latest codebase committed to the central repository. Once the code is sourced from GitHub, CodeBuild programmatically performs unit tests to verify that the codebase does not present technical bugs in its subroutines and overall functions.
- The second step is building, which starts only if the unit tests are successful. CodeBuild creates the application by automatically provisioning servers and any other AWS cloud resources or services the application needs. For example, the grading system requires building more than 30 AWS Lambda functions. The web application requires provisioning AWS EC2 computing instances across different regions.
- The third step is publishing. CodeBuild makes available the full application in a test environment that the ReimageIT team uses to ensure that the application is working as expected and that all code changes appear as planned (e.g., new UI elements are present and work as planned). At this point, further automated testing occurs for the application as a whole. ReimageIT manually codified the expected result from hundreds of test Excel files, thus enabling automated evaluation of the grading engine's performance. Only after each file is evaluated correctly according to the standardized tests, CodePipeline starts the migration of the application to the production environment.
- The final step is release. CodePipeline triggers the deployment process in production by leveraging AWS CodeDeploy. CodeDeploy is a tool that automatically creates a production environment with the same configurations as the one tested in step three. Because ReimageIT implements a continuous delivery and not a continuous deployment process, an administrator is required to manually approve the release of the production build to all customers. This is necessary to ensure that deployment into production occurs in the most effective manner, following considerations like the ones discussed in the installation phase of the SDLC (see earlier discussion of the four migration approaches). The application is available to thousands of students around the world as soon as the DevOps team grants approval.

be formalized and codified. Formalizing tests is an insurmountable task for many organizations where knowledge is tacit and difficult to verbalize. Furthermore, considering the complexity of some applications, a complete test coverage might require years to complete, thus limiting the impact for DevOps.

11.4 Buying Off-the-Shelf Applications

The SDLC provides the basis for the system selection and purchasing process that organizations use to design and develop information systems based on off-the-shelf software programs (Table 11.5). The systems selection process often starts when managers learn about the capabilities of a new application that is being advertised, described in the press, or promoted by consulting firms. Following the systems selection process is important because it enables a systematic investigation of these applications as well as competing products—thus ensuring that all issues are considered and that the firm chooses the best solution for its current needs. A systematic approach to the selection and procurement of software applications is even more imperative in contemporary organizations because, with the continuing migration toward cloud computing (Chapter 3), business units are increasingly able to procure SaaS solutions without the need to involve IS professionals (Chapter 10).

Table 11.5. Phases of the systems selection process

Definition
Investigation Feasibility analysis System analysis Formulation of evaluation criteria Compiling short list of vendors Compiling and distributing request for proposal (RFP) Evaluating alternatives Contract negotiation
Build
System design (customizations) Programming (customizations) Testing
Implementation
Installation Operations Maintenance

Definition

Both the investigation and feasibility analysis stages are qualitatively similar to those in the SDLC. At this time, the proponents of the system articulate a vision for the proposed information system and evaluate its technical, operational, and economic viability. It is in the remaining stages of the definition phase that the major idiosyncrasies of the systems selection process occur.

System Analysis During the system analysis stage, the selection committee focuses on eliciting the specific functionalities required of the proposed system. As with the SDLC, this phase entails the interplay of system analysts and stakeholders. However, the degree of precision and detail sought by the selection committee is less than that needed for custom development. The objective here is to have enough of an understanding of the system requirements to formulate evaluation criteria.

Formulation of Evaluation Criteria Systems selection is the structured attempt to evaluate all commercially available software solutions that can enable the proposed information system. In order to do so, it is necessary to develop a set of metrics that can be uniformly applied to all packages under investigation. The criteria must also be amenable to communication to software vendors by way of the request for proposal (RFP) process.

A common approach is to use the system requirements document to identify the features that appropriate applications should have and group them into three categories:

- *Essential features.* Those capabilities that the system must have. Systems that miss any one of these features are automatically discarded.
- *Value-adding features.* Those capabilities that, while not essential, offer significant advantages for which the firm would be willing to pay a premium.
- *Nonessential features.* Those capabilities that are "nice to have" but produce small tangible advantages for which the firm is not willing to pay a premium.

Compiling Short List of Vendors Armed with the evaluation criteria, the selection committee seeks information about existing solutions. Websites, trade press, vendor brochures, and trade expos are all viable sources of information. The information gathered is used to identify a preliminary, short list of vendors.

This stage is important for two reasons. First, creating targeted RFPs that yield high-quality responses, and then evaluating those responses, is time consuming. Second, products that fail to meet necessary requirements can be identified fairly quickly, and vendors will appreciate not being asked to respond to RFPs that they have no chance of fulfilling.

Compiling and Distributing the RFP The RFP is the formal communication document used to elicit substantial, detailed information from the short-listed vendors. Most organizations have a template for such documents. The RFP should explain what the selection committee has identified as the critical system requirements, the environment in which the system will be used, and any required performance metrics and expectations.

Upon its distribution to the short-listed vendors, those interested will respond to the RFP. The selection team should ask vendors to adhere to a template, making cross-comparisons of the applications simple, and should ask for pricing information. Vendors should also provide such pricing information according to the firm's template, ensuring that the pricing mechanisms for the applications, any customization, and ongoing maintenance and upgrades are clear and comparable across vendors. Finally, the selection committee should specify a deadline by which vendors must respond in order to be considered.

Evaluating Alternatives Once all interested vendors have responded to the RFP, the competing solutions are evaluated using the criteria developed earlier. The selection committee compiles the list of top vendors and seeks any further information needed to make a final decision. This includes on-site demonstrations, evaluation of reference sites, and the like. The outcome of this stage in the selection process is a rank-ordered list of the acceptable candidates.

Contract Negotiation Negotiations can be relatively quick and simple, or they can be a very involved process requiring input from professionals and legal counsel. The objective of the selection committee is to draft and sign a contract that provides the firm with the needed solution and insulates it from future risks. Common elements of negotiation are the components and magnitude of costs (e.g., installation, training, customization, maintenance, upgrades), the eventual liabilities and service-level agreements, control over the intellectual property, and the extent to which modifications are allowed.

Build

The build phase in the system selection process mirrors that of the SDLC but is much narrower in scope. When the software program is to be installed without configuration or customization, as is the case with simple applications, the firm can move directly to formal testing and implementation. However, as we mentioned above, it is becoming increasingly common for organizations that purchase off-the-shelf applications to configure and customize them extensively.[1]

To the extent that customization is necessary, the firm will engage in system design and programming of the required enhancements. This is where a tightly written contract becomes

1 *Configuration* is the process of tailoring a software program using built-in parameters. *Customization* is the process of changing the functionality of the software program beyond available options. Customization requires that the new functionalities be coded using a programming language.

important, as the customization process can add significant time and cost to the project. The contract should specify who is responsible for customizing the application—the vendor, the firm, or a third party (e.g., independent consulting firm, integrator)—and the conditions of the customization effort (e.g., schedule). Whether customized or not, the firm should test the system. In the case of off-the-shelf applications, the testing stage is mostly concerned with system performance rather than with the identification and correction of bugs.

The build phase is even quicker when the firm is procuring SaaS applications, since there is no need for any installation. Moreover, SaaS applications typically are not customized specifically for one customer. Rather, they offer different versions, or tiers, of varying functionality. Thus the firm needs to match the available options to its needs and select the appropriate tier.

Implementation

When implementation is necessary, for software that is installed on-premise or in a private cloud, the implementation phase is quite similar to the one described earlier regarding the SDLC. Interestingly, even within the same class of software applications (e.g., ERP), a firm may choose different approaches to move from the old system to the new one. Note as well that the degree of process change and training required to get buy-in from users is typically greater when implementing off-the-shelf applications. This is because a prepackaged program is not designed with the idiosyncrasies of your organization in mind. Rather, the software house builds the program to appeal to the broadest market possible.

Asking for stakeholders' input during the selection and evaluation of competing solutions is one way to enroll them in the process and reduce rejection risks. However, you should still plan to invest considerable resources during the implementation phase to set up the application, train employees, and engage in change management—particularly when the application is larger in scope and forces a change in traditional work practices.

11.5 Open Source

Open source software is no longer an emerging trend. Today, open source solutions are a viable option for managers to consider when planning new system design and implementations. Gartner estimates that open source code is present in the portfolio of mission-critical software (not just any application!) of 90% of mainstream IT organizations worldwide, composing the foundational elements of almost all digital platforms and aggregators and substantially contributing to digital innovation.[1]

The trend toward open source software use gained significant momentum with the emergence of the Internet, but it remains today an important and evolving digital technology opportunity for modern organizations. Linux, the operating system that runs in most data centers, still represents the best example of open source software development. But open source solutions are now mainstream with household names like Android, Apache Web Server, Apache Hadoop, MySQL RDBMS, Eclipse IDE, and MongoDB. These open source projects matured into systems and applications that are often industry-leading products in their respective categories. Today, much of the code that runs cutting-edge machine-learning applications (Chapter 12), from computer vision to speech recognition, is available for download on GitHub and freely reusable thanks to open sourcing.

1 Driver, M. 2017. "What every CIO must know about open-source software" (no. G00301608). Gartner, retrieved from https://www.gartner.com/document/3660617; Batchu, A., Klinect, T., and Thomas, A. 2020. "Hype cycle for open-source software, 2020" (no. G00450395). Gartner, retrieved from https://www.gartner.com/document/3987447.

Open Source Definition

Open source software is often confused with free software—as in *free of charge*. In fact, licenses for open source software may or may not be offered at no cost (free-of-charge software is called freeware). Rather, the term *open source* is used to differentiate it from closed source, or proprietary, programs that prevent users from accessing and modifying the source code. Software programs are created by software engineers, who design the algorithm, and programmers, who code it using a specific programming language. The code generated by the programmers, which can be understood by anyone who is well versed in the programming language used, is called the *source code* (see Figure 11.14).

```
dBDate = CDate(BirthDate)
dRelDate = CDate (RelativeTo)
iAns = Year(dRelDate) - Year (dBDate)
if Month(dBDate) <> Month(dRelDate) Then
    bSubtractOne = Month(dBDate) > Month(dRelDate)
Else
    bSubtractOne = Day(dBdate) > Day(dRelDate)
EndIf
```

Figure 11.14. Sample source code

In order for the program to work, the source code has to be transformed (i.e., interpreted or compiled) into a format that a computer can execute, called the *object code*. Typically, when you buy a software license from a software company (e.g., Microsoft Office, Oracle Database 12c), you are given the object code and the right to run it on your computers, but you are not provided with the source code. In fact, any effort to reverse engineer the object code to gain access to the source code is considered a violation of the intellectual property of the software house and will land you a well-founded lawsuit.

Unlike proprietary software, open source programs are distributed with the express intent of enabling users to gain access, modify, and improve the source code. An open source license typically exhibits the following characteristics:

- *Free redistribution.* The software can be freely given away or sold.
- *Available source code.* The source code is published and freely obtainable.
- *Derived works.* Licensees can modify the software and redistribute it under the same license terms as the original.
- *No discrimination.* The license is available to any entity, including for-profit organizations and commercial users.
- *Technology neutrality.* The license and all of its provisions must be free of restrictions tied to the use of any technologies or type of interface.

Analysis of the open source licensing characteristics shows that the open source movement encourages, rather than opposes, commercial applications and commercial redistribution. This friendliness toward business applications has been the catalyst for widespread acceptance and growth of open-source software as a viable alternative to proprietary programs. Indeed, many organizations consider open source a practical alternative to the traditional make-or-buy software development decision discussed above. Others see open source as a way to foster innovation and development. For example, Google has released more than 2,000 open source applications and recently open sourced TensorFlow, its machine-learning library.

Advantages and Disadvantages of Open Source Software

As a future manager, you will without a doubt be part of a system selection committee. Increasingly, such committees have the option of adopting open source software rather than purchasing proprietary programs. While decisions of this kind are very context specific, below we identify the main benefits and drawbacks of open source.

Advantages The primary advantages of open source touted by its proponents are a function of the ability of open source projects to leverage a large community of developers, programmers, testers, and customers. These advantages include the following:

- *Robustness*. Proponents of open source software claim that mature projects (e.g., Linux) are more robust, more reliable, and generally higher quality than comparable proprietary applications (e.g., Microsoft Windows).
- *Creativity*. Open source software harnesses the creativity of thousands of developers around the world. As such, it is more likely to generate new breakthrough solutions (e.g., Firefox tabbed browsing, now a standard in this class of applications) than traditional products created by a small community within one software house. Many edge and innovative markets (e.g., Big Data, artificial intelligence, cloud) are indeed based on open source developments.
- *Limited lock-in*. Open source software is not without switching costs, but supporters claim that such costs are much lower than those associated with proprietary software. For example, customers of open source software can make their own modifications to the source code rather than having to rely on the software vendor to do so.
- *Simplified licensing*. Because of the structure of an open source license, customers need not worry about complex legal constraints (e.g., number of concurrent users). They simply install as many copies of the program as they need.
- *Free licensing*. While not regarded as one of the chief benefits of open source by the open source movement, total cost of ownership is still an important factor to those firms that adopt open source applications. Because open source generally can be licensed for free, costs are lower than those associated with proprietary applications.

Disadvantages Software is by no means simple (or cheap) to install and operate, and open source software is no exception. Thus skeptics respond by raising the following concerns:

- *Unpredictable costs*. Skeptics like to say that free software is like a free puppy—yes, you get it for nothing, but then you will encounter many (often unplanned) costs along the way. Recent studies suggest that while in 80% of cases, open source software is deployed to reduce costs, savings can be proved in only 50% of them.[1] Thus you need to carefully evaluate an open source installation based on total cost of ownership (Chapter 10).
- *Support varies widely*. Depending on the product and where your firm acquired it, support can range from high quality to nonexistent, even in relation to the maturity—and widespread adoption—of the software.
- *Security*. Skeptics claim that publishing source code gives an advantage to those who want to break its security. Proponents of open source respond that a large community of developers will identify and close more weaknesses than a small team of company developers.
- *Compatibility*. Standardization of products using one or a few vendors simplifies compatibility and integration. There is no guarantee that open source solutions will be compatible with one another and/or with proprietary software.
- *The legal landscape*. Open source software requires that no portion of the code is protected by copyright. Recent court challenges have raised the specter that there is no way to ensure that copyrighted code will not make it into open source solutions, thus opening customers to liability. In response, some of the firms that support open source software (e.g., JBoss, HP, Red Hat) have adopted indemnification clauses in their licensing agreements.

1 Driver, M. 2019. "Start with these three best practices to maximize open-source software value" (no. G00392337). Gartner, retrieved from https://www.gartner.com/document/3956330.

11.6 End-User Development

The ease of use of information technology has steadily increased, while its cost has declined dramatically over the years (Chapter 1). These two forces have conspired to bring the power of software development to the masses in the form of end-user development. *End-user development* is an umbrella term capturing the many ways in which knowledge workers, not IS professionals, create software.

End-user-developed systems range from spreadsheet models (e.g., an ROI calculator written in Microsoft Excel), to personal or departmental databases, to full-fledged software programs built with user-friendly computer languages (e.g., Visual Basic for Applications) or development tools such as fourth-generation languages. End-user development has received renewed impetus with the surge in availability of SaaS platforms, analytics, and data science projects. Aided by the growing ecosystems around powerful programming languages such as R and Python, you may soon find yourself scraping your own data, building your own analytics models, and even creating dashboards for your coworkers. The tools and data are there; all you need are the skills! As exciting as these "**shadow systems**" are, their growing presence in modern organizations should be carefully managed.[1] For you as a manager, understanding the benefits and limitations of end-user development is a start.

The Benefits of End-User Development

The chief benefits of end-user development stem from user empowerment and the fact that some of the burden on typically overworked IS departments is lifted. The benefits include the following:

- *Increased speed of development.* The user community typically must direct requests for new systems, and improvements to current ones, to the IS function. In turn, the IS function must prioritize the deployment of its scarce resources. As a consequence, those projects that end users can complete independently will be completed faster by virtue of not entering the queue.
- *End-user satisfaction.* One of the main problems with new systems is users' dissatisfaction or outright rejection. When users create their own applications, they are more likely to be satisfied with the result; they either have created the functionalities they wanted or have themselves decided which features to forgo.
- *Reduced pressure on the IS function.* End-user development can limit the number of requests the IS function receives, enabling them to be more focused on the projects that, because of their scope and complexity, really require their attention.

The Risks of End-User Development

Unfortunately, end-user development presents a number of difficult-to-manage risks that limit its value to the organization:

- *Unreliable quality standards.* There is a reason software development is a lengthy process. Quality software requires a number of activities that may not be readily apparent but are necessary—such as testing, documentation, security, integration, and the like. Because of the limited skill set and knowledge of most end users, the quality of their work varies dramatically.
- *High incidence of errors.* Audits of spreadsheets used in organizations show that a sizable percentage, between 20% and 40% (sometimes 90%), contain errors. The focus on outcomes (i.e., what the program does) and rapid development typically conspires to increase the likelihood of errors in end-user-developed applications.

---

1 Mingay, S., and Topham, D. 2020. "Integrate 'shadow IT' and business-led IT into the I&T operating model to enable enterprise agility" (no. G00365598). Gartner, retrieved from https://www.gartner.com/document/code/365598.

- *Continuity risks.* Because end-user development often does not comply with traditional system development methodologies, it may be difficult for anyone but the individual who wrote the program to understand it, enhance it, and support it. Lack of documentation compounds this problem. A common scenario involves people like you who develop great applications during internships only to see them fade into company oblivion once the developer leaves the firm. Furthermore, errors or mismanagement may expose the firm to security breaches (Chapter 13), noncompliance with legal or regulatory mandates, and ultimately the loss of intellectual property.
- *Increased pressure on the IS function.* While end-user development can relieve some of the development demands on the IS function, it often creates more requests for assistance during the development process and, over time, more requests for help managing the applications after release.

Summary

This chapter continued our discussion of the techniques and methodologies modern organizations use to introduce and manage information systems within the framework provided by the strategic IS plan.

Specifically, in this chapter, we focused on the three approaches used to introduce new organizational information systems: custom design and development, system selection and acquisition, and end-user development. We learned the following:

- The astounding progress that has characterized IT over the last 40 years often misleads general and functional managers. Being mostly familiar with personal computing, they underestimate how much time and money it takes to build a stable, robust, and secure system that will work under a wide array of organizational conditions. In order to avoid these misconceptions, managers must become familiar with the process by which IT-based information systems come to be in modern organizations.
- Introducing an organizational information system is a two-step process requiring technology development and the implementation process. These two processes, while often described separately, are complementary and intertwined. More recently, DevOps has emerged as a practice in the context of agile software development to attain shorter response times when it comes to the delivery of features or bug fixes utilizing **continuous integration** and continuous deployment.
- Modern firms introduce new information systems using one of the following approaches: custom design and development, system selection and acquisition, or end-user development. The critical difference among them is the manner in which the software applications at the core of the information system are developed. In the first approach, IT professionals within the organization or those who are contracted develop uniquely tailored software for the firm's needs. In the second approach, the selection committee chooses an off-the-shelf application. In the third approach, it is the firm's end users, rather than the IT professionals, who create the software.
- The main methodology for custom system development is the SDLC. The SDLC, predicated on the notion that detailed up-front planning is the vehicle to reduce risk and uncertainty in systems design and development efforts, is best suited for the development of large, complex software applications. The SDLC is articulated over three main phases—definition, build, and implementation—and nine stages. The primary limitation of the SDLC is the creation of substantial overhead and rigidity that limit the project team's ability to address the inevitable changes.
- Agile methods represent an innovative and alternate development method that places emphasis on the development team and user involvement. Each iteration introduces features or changes into the product, and these are reviewed by the development team and the users. Agile methodologies provide the ability to change development direction later in the process. While agile methods may bring tangible benefits to the firm, they challenge current organizational culture and practices.
- Agile methods best suit software developments where problems are complex, objectives or requirements may vary, solutions are uncertain, the work can be modularized, and both customer involvement and cross-collaboration are feasible.

- When sourcing agile developments, it is critical to assess that contracts and charging methods fit project goals and agile framework characteristics. A client could then decide to contract only a certain number of sprints and retire if the project fails to progress or meet the established goals.
- Open source software programs, which enable the adopting firm to receive and modify the source code, are increasingly becoming a viable option for organizations. When weighing the decision to adopt open source instead of a proprietary software program, you need to evaluate the following advantages and disadvantages of open source projects. The pros include robustness, creativity, limited lock-in, simplified licensing, and free licenses. The cons include unpredictable costs, varying degrees of quality support, security concerns, compatibility concerns, and a potentially complex legal landscape.
- The software industry has grown to a point where almost any application a firm needs is available off the shelf. When building information systems around prepackaged software applications, the firm must engage in a formal systems selection and acquisition process. Doing so ensures that the selection team evaluates all possible solutions and acquires the one that is best suited to the firm's needs. The selection and acquisition process mirrors the SDLC, with some important variations during the definition and build phases.
- The advent of powerful and easy-to-use computer languages, software development tools, and cloud services and infrastructure has enabled an unprecedented degree of software development by end users (i.e., non-IT professionals). The benefits of end-user development include increased speed and end-user satisfaction and a reduced pressure on the IS function to develop new applications. The risks of end-user development include unreliable quality standards, high incidence of errors in the applications, continuity risks, and increased pressure on the IS function to support development and management of end-user applications.

Study Questions

1. Describe the reasons general and functional managers often fail to understand the complexities of organizational IS development. Can you provide an example from your experience?
2. What is the difference between technology development and IS development? What is the relationship between these two processes?
3. How do the three IS development approaches in use today in modern organizations differ? Can you provide an example of each?
4. Provide arguments in support of both the make and the buy approaches. What are the principal advantages of each decision? Increasingly, firms approach IS development as a "buy and make" process. What do we mean by "buy and make"? Why is this approach gaining increasing popularity today?
5. Describe the SDLC methodology in the context of a "real" example. In other words, think about (or imagine) a situation where you proposed the need for a new information system. For this system development effort, describe what happened (or should happen) during the definition, build, and implementation phases.
6. Repeat question 5, this time using Scrum.
7. Repeat question 5, this time using the systems selection and acquisition methodology.
8. Articulate the advantages and disadvantages of agile methods and DevOps.
9. What is open source software? What are the main advantages and disadvantages of open source software? When would you consider an open source software implementation in your organization? When would you not?
10. Articulate the advantages and disadvantages of end-user development.

Glossary

A glossary of key terms can be found at the book's website. Visit https://www.prospectpressvt.com/textbooks/piccoli-information-systems-for-managers-5-0, then click on "Student Resources" in the horizontal red menu bar.

Chapter 12

Information System Trends

What You Will Learn in This Chapter

In this chapter, we introduce emerging and enduring trends in information systems (IS) and digital innovation. Understanding these trends, the digital technologies underlying them, the associated vocabulary, and the benefits and risks they engender for modern organizations is critical for you as a general or functional manager. You will hear much of this vocabulary at conferences, from consultants, and in the media, and you must learn to navigate it successfully. More important, as a manager, you will be (or should be!) called on to participate in the debate about whether your firm should embark on the type of digital transformation initiatives resulting from the introduction of the digital innovations described in this chapter.

MINICASE: Improving Environmental Sustainability through Platforms

You never believed this could happen! After five years, working 90 hours per week for a major consulting firm, you were overdue for a change! You liked consulting—you liked the pace, the challenge every new project offered, and the constant learning it required. But finding meaning in your work was increasingly difficult, and you felt the need for more social engagement. It was during a dinner with your friend Eric that you had the chance to make the turn. The former chief financial officer (CFO) at a major telecom firm you had consulted for, Eric had decided to run for mayor of Grenoble, the town in the French Alps you both called home. The information was still confidential—as he was in the process of assembling the team—but he was thinking you should be deputy mayor for digital transformation. "You know, we need disruptive, impactful new ideas. We need to transform the city; we need to realize its environmental transformation. We can't do so without a digital transformation first. You'd be perfect for the position!" Eric stated with infectious excitement in his voice.

He was looking straight in your eyes; he knew he had hit the right buttons and that this was the opportunity you had been craving to find more meaning in your work. "I need you to help me revolutionize mobility for our citizens. This is the most direct way to reduce pollution and the use of fossil fuels. How do we do it?" The project was ambitious, but you couldn't expect less, considering that the city was recently named the 2020 European Green Capital. Voters' and stakeholders' expectations were equally high. Despite being caught off-guard by Eric, an idea was already forming in your mind. The Grenoble Urban Area (Région Urbaine Grenobloise) had a population of almost 500,000 inhabitants who made over 2.7 million daily journeys. Its three main junctions saw over 50,000 vehicles per day go by. Also, while 59% of residents still used cars, they were increasingly shifting toward public transport and cycling. With the right incentives and a functioning infrastructure, that number could grow dramatically.

Besides buses, trains, taxis, railway networks, and the unique cable car to visit the Bastille fortress, the city already welcomed innovative sustainable transportation solutions. Cityz was offering a self-service, car-sharing fleet of over 100 vehicles. Metrovélo provided nearly 8,500 bicycles for hire in two central agencies. TIER Mobility, the European micromobility solutions provider, had recently extended its network in Grenoble, providing more than 500 e-scooters. Interestingly, while the sustainable transport network was developing fast, it lacked the necessary integration. "You know, Eric, we need to integrate all these mobility options. We may build a unique platform to converge air pollution data, traffic, mobility recommendations, vehicle booking, tickets, and payments. Based on traffic, air pollution, weather, and other information, we can go beyond a mobility pass. We can leverage the platform to recommend and influence citizens' behaviors when they move. We can run simulations and scenario analyses, and then we can open all this up and see what the private sector and universities can do with the data. We can be disruptive and change the city forever!" With a twinkle in his eyes, Eric looked at you and said, "That's the spirit! I knew you would jump at the opportunity. The devil is in details, we need a feasibility study before we announce my candidacy." This was your chance to have the meaningful social impact you had been craving. Digital technology would have to feature prominently in your design, no question, but what else? With a full-time job still in consulting, you knew there would be many nights of work ahead . . . but this was work you'd be happy to do!

Discussion Questions

1. What will be the focus of your proposal to Eric?
2. What do you believe is the optimal design for your proposed platform? Who are the key partners you need to sign up to ensure its success?
3. What are the main challenges that will derail your ambitious project if you don't manage them appropriately?

12.1 Introduction

In this chapter, we discuss the most relevant and influential emerging trends in information systems and digital innovation. We focus on those emerging and enduring trends that are capturing media attention and that consulting companies are promoting today. These are the digital innovations that you will need to confront as you join the workforce in the immediate future.

Even the visionary Steve Jobs, the pioneer who led the mobile revolution, missed the mark completely in 2003. In an interview with the prominent *Rolling Stone* magazine, he boldly proclaimed, commenting on the future of the music industry,

> People don't want to buy their music as a subscription. They bought 45s, then they bought LPs, they bought cassettes, they bought 8-tracks, then they bought CDs. They're going to want to buy downloads. The subscription model of buying music is bankrupt. I think you could make available the Second Coming in a subscription model, and it might not be successful.[1]

While 2003 was a different time and the Internet was nowhere near the same network it is today, this example shows how the future evolution of today's trends is hard to predict reliably. However, as a manager, you nonetheless need to study them and formulate your own ideas and plans when new technology developments emerge.

We organize this chapter around the digital technologies at the core of each trend. This is a conscious approach that mirrors how decisions are made in today's organizations. Typically, an organization will become aware of an emerging digital trend through publications, consulting companies, or conferences and events. For example, the consulting firm Gartner developed the Hype Cycle methodology to classify emergent technologies based on their expected potential and to help managers discern drivers from vendors' hype.[2] Typically, these trends are defined by the functionalities of the technology at the core or the features and characteristics of a new class of software applications being introduced—from which the trend typically takes its name and its impetus. It is critical for you to remember that, no matter how sophisticated a digital technology

1 Goodell, J. 2011, October 6. "Steve Jobs: Rolling Stone's 2003 Interview." *Rolling Stone*, retrieved from https://www.rollingstone.com/culture/culture-news/steve-jobs-rolling-stones-2003-interview-243284/.

2 Gartner. 2021. "Hype Cycle Research Methodology." Gartner, retrieved from https://www.gartner.com/en/research/methodologies/gartner-hype-cycle.

may be, in order for it to have a positive impact on the organization, you must be able to design an information system around it (Chapter 2). Thus for each of the trends discussed in this chapter, we analyze the technological capabilities as a departure point to understand the organizational impacts that they engender.

12.2 Digital Data Genesis

Two established trends we discussed in this book, ubiquitous computing (Chapter 1) and the Internet of things (IoT; Chapter 5) have moved us toward a world where events, transactions, and processes are all digitized in real time as they occur. As described by process virtualization theory (Chapter 4), a process (e.g., purchasing a book) is virtualized when it is carried out through a digital device (e.g., purchasing a book online). Conversely, when an entity is digitized (e.g., a radio-frequency identification [RFID]-enabled gaming chip), we have the ability to access an informational representation of that entity (e.g., the position of the chip on the gaming table) and thus generate relevant data from it (e.g., how many times the chip has been played). Interestingly, such data are natively generated in digital form—a process we have termed *digital data genesis* (Chapter 1).[1] Recall the example of the multibillion-dollar Wynn Las Vegas casino. Historically, casinos have not placed significant value on customer data. A large casino's standard operating procedure was to value customers on the basis of judgments made by hosts and pit bosses. This approach tended to give undue weight to the contribution of a few big gamblers—the so-called whales—while undervaluing the multitude of smaller but often more valuable players. With the advent of state-of-the-art digital slot machines, it became feasible and cost effective to build comprehensive profiles of avid players, without disrupting their experience, in order to craft a targeted rewards strategy.

Casinos are now expanding their use of technology to capture valuable customer data by embedding RFID transceivers in the chips used at table games. Embedding RFID transceivers in chips means that table games' data are generated in real time and with the utmost precision. Then they are recorded in an easily storable and retrievable format without interfering with the customer's enjoyment. Today, as gaming chips become digitized, monitoring, guessing, and data entry are unnecessary behaviors because data are "born digital" and can be automatically collected and stored in a computer. Examples of digital data genesis are all around us (Figure 12.1). When we type search terms into Google, we are generating data in a digital form (e.g., data about what is interesting to us); when we place calls with our mobile phones, we are generating data in a digital form (e.g., data about our location, the person we called, and the like); and when a smart electrical outlet is recording power flow, it is generating data in a digital form (e.g., data about the electricity usage profile of the device that's plugged in).

Digital Data Streaming

With digital data genesis (DDG), we refer to a paradigm shift in data generation. Data are "born digital," and thus they can be automatically collected and stored in a digital computing device. Upon digital data generation, the data can be streamed. ***Digital data streams* (DDSs)** are the continuous digital encoding and transmission of data describing a related class of events.[2] They represent the flow of data originating from DDG activities and can be intercepted and used by organizations or individuals. The transmission or flow of these digital representations of events

---

1 Piccoli, G., and Watson, R. T. 2008. "Profit from customer data by identifying strategic opportunities and adopting the 'born digital' approach." *MIS Quarterly Executive* 7(3).

2 Pigni, F., Piccoli, G., and Watson, R. 2016. "Digital data streams: Creating value from the real-time flow of Big Data." *California Management Review* 58(3): 5–25.

may be human generated (e.g., a tweet, an Instagram post) or machine generated (e.g., a CO_2 reading, a GPS location). Consider Uber as an example. It is the world's largest "taxi" company, but it owns no vehicles! It harnesses Android or iOS platform application program interfaces (APIs) to generate real-time DDSs of its drivers' cars and matches them with the real-time demand for rides.

At the highest level of abstraction and generality, a DDS can capture and thus represent up to six basic elements describing an event (Table 12.1). These elements are "primitives," meaning they cannot be described in terms of other elements or inferred from them. These primitives derive from what are commonly known as the 5W+H of narrative (who, what, when, where, why, and how).

We now turn our attention to the DDS's "life cycle" and its two main aspects: DDS creation and DDS exploitation (Figure 12.2).

DDS becomes available to organizations through three consecutive stages:

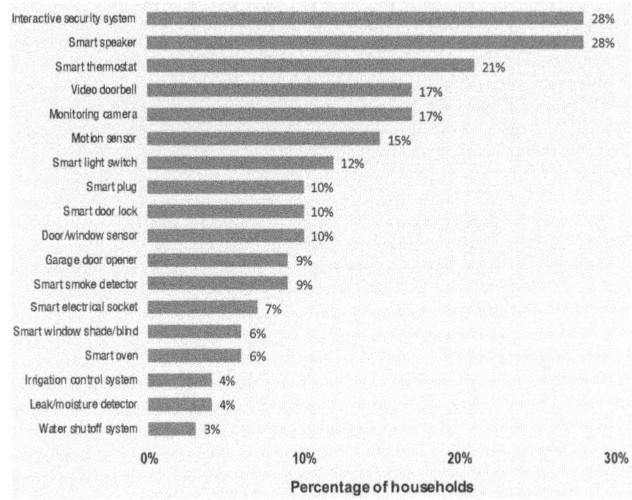

Figure 12.1. US households' adopted connected technologies (Q4 2019)

Source: "Strategy analytics at CES: Most homes are now smart homes," 2020, January 7, retrieved from https://news.strategyanalytics.com/press-releases/press-release-details/2020/Strategy-Analytics-at-CES-Most-Homes-Are-Now-Smart-Homes/default.aspx

1. *Generate.* This is the stage at which DDG happens and the digital information of the event occurs (i.e., a tweet, a Google search, the GPS position of an object). When such a DDG event is not isolated but rather is part of a series (or a stream) of DDG events, there is an opportunity for DDS generation. For example, a single click on a hyperlink is the digital representation of a person's decision (a DDG event). Activity on a website is a stream of personal decisions and is aptly called the clickstream, which is a well-known example of a DDS. Sensor readings, smart metering, or a person's activity on Pinterest, Twitter, or Uber are all activities that lead to DDS generation.
2. *Stream.* When data are available, channeled, and transmitted as a continuous flow, we refer to them as a DDS. The streaming phase concerns the manner and format in which the data become available. The stream is characterized by (1) the type of technologies used to create the channel (e.g., APIs), (2) the nature of the content

Table 12.1. Elements of a DDS segment

Element	Description	Example
When	The time when the data segment was created	A time stamp with date, time, and time zone
Where	The location of the entity when the segment was created	Latitude, longitude, elevation
Who	The unique identifier of the entity that caused the data segment to be created	Person's customer number, RFID of a pallet, URL of a website
What	The activity that caused the segment to be created	The identifier of an item in a sales transaction, the arrival of a ship in a port
How	The means by which the event was initiated, authorized, or completed	Credit card number for payment, status of arriving flight (e.g., safe landing)
Why	Motivation for the action related to data segment creation	Birthday gift, planned destination

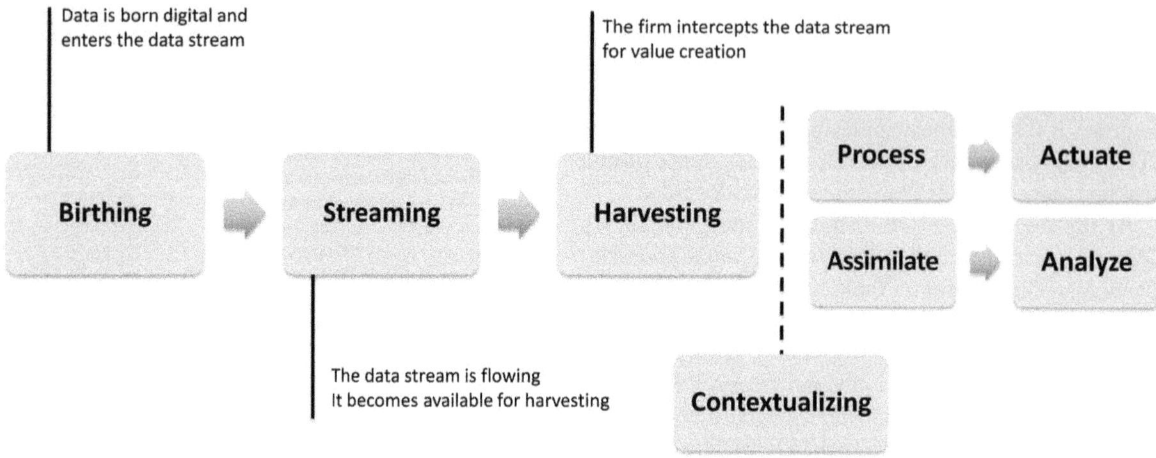

Figure 12.2. The main stages of DDS creation and exploitation

(e.g., video, blog post, review, sensor reading, including considerations of accuracy), (3) the source (e.g., public, business, individual, or community), and (4) the legal status of the data contained in the stream or derived from it (e.g., rights reserved and sensitivity).

3. *Harvest.* At this stage, an organization taps into the DDS and extracts some or all the data being streamed. The harvest stage is described in terms of the technologies adopted to perform the data harvesting. The technology used are complementary to those used for streaming. So for example, if the stream is accessible through an API that returns a JSON-formatted response, a proper method to post the request and JSON parsers should be used.

Upon successfully harvesting the DDS, an organization can put it to use. While considerations about value creation and appropriation are discussed in Chapter 7, it is important at this point to discuss the mechanics of DDS utilization. Organizations exploit DDS via two general classes of activities: process-to-actuate and assimilate-to-analyze (Figure 12.2).

- *Process-to-actuate* occurs when a firm takes action based on real-time DDS processing. An insurance company monitoring a weather forecast data stream and sending text messages to its customers in the area where hail is expected in the next 30 minutes illustrates the immediacy of process-to-actuate. The firm combines events that are currently streaming in a DDS (i.e., real-time, location-specific, short-term weather forecasts) and the results of a static database query and other contextual data in order to alert its potentially affected customers in a timely manner. The result is superior customer service and fewer insurance claims because customers have been able to garage their vehicles at the right time.
- *Assimilate-to-analyze* occurs when a firm merges multiple data streams and static databases in an effort to analyze the data. The focus is on extraction of insights rather than immediate action—as in the process-to-actuate approach. To avoid the financial risks associated with planning errors, some firms have integrated external DDSs into their demand forecasting systems. For instance, Tesco and other retailers merge and analyze data from multiple DDSs to generate forecasts to estimate demand. Predictions are based on information generated from store location, product

characteristics, recent weather history, and weather forecasts. Note how the result of the analysis is not immediate automatic action, as in process-to-actuate, but rather the presentation of superior insight that enables better decision making.

Naturally, firms can combine these two approaches so that a prescribed action is the result of the analysis of multiple data streams and database records. In some cases, the process-to-actuate approach requires the combination of multiple DDSs.

Digital Twins

By leveraging DDSs, firms can observe customers before, during, and after a purchase because a person's social media postings can be monitored in real time and matched with their specific website behavior. Because DDSs typically contain spatial-temporal data, new DDS-based digital services include a wide variety of messages based on time and location (e.g., weather alerts and coffee shop offers). As DDSs become wider (more sources) and deeper (more data), they enable the creation and calibration of highly detailed representations of reality and entities in the physical world: ***digital twins* (DTs)**. A DT is a digital representation of a unique physical object implemented as a software program capable of mirroring, in the virtual space, the object's characteristics and properties. A DT is associated, or tethered, through DDSs to the physical object so that, for example, sensor readings can be used to know the state of the object, make predictions on activities and performance, and react to changes in the object or the environment (Figure 12.3). A DT is based on a detailed digital model of an object, and it is capable of representing its relevant features. While DTs may appear as enhanced 3D models, they have distinct properties. First of all, a DT is unique for each physical object it is associated with, and it represents an instance of the general model used to describe the class of objects. Second, the model is detailed to a level where the properties of the object allow the model to be part of a simulation, where the DT properties can be evaluated in context. Thus unlike traditional 3D models or other digital representations of physical objects, DT are live software objects connected to their physical counterparts through DDSs (e.g., sensor readings, contextual information, manufacturing data, and other data sources—see Figure 12.3).[1]

For example, the Italian car manufacturer Maserati used DT for designing and manufacturing the Ghibli. The DT of the car, as was increasingly specified during the design process, was used to optimize the assembly line. Both the car and the assembly line were digitally mirrored "100 percent true to the original, down to the last screw."[2] The DT of the Ghibli enabled the firm to simulate competing configurations of the plant and identify the optimal production solutions. The simulation environment was capable of not only evaluating the process time but anticipating the physical constraints of the different equipment, robots, and human staff even before the car was ultimately completed.[3]

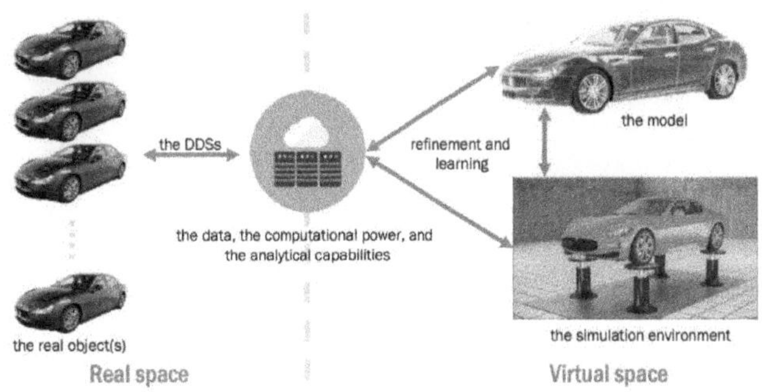

Figure 12.3. DT connecting the physical and digital worlds

1 For a general introduction to DT, see GE Digital. 2016, November 18. "Minds + machines: Meet a digital twin." YouTube, retrieved from https://www.youtube.com/watch?v=2dCz3oL2rTw.

2 Siemens. 2015, October 1. "Getting to market quickly." Siemens, retrieved from https://new.siemens.com/global/en/company/stories/industry/getting-to-market-quickly.html.

3 Siemens. 2015, April 13. "Maserati and the future of manufacturing." YouTube, retrieved from https://www.youtube.com/watch?v=Hk__pjSe3a8.

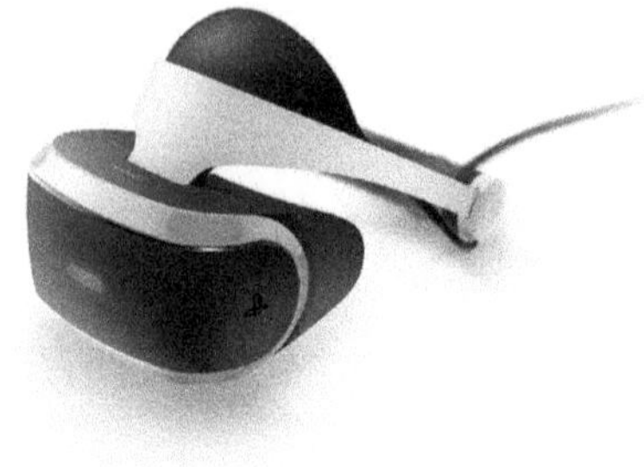

Figure 12.4. The Sony PlayStation VR headset

Another important characteristic of DTs is that they can leverage **artificial intelligence (AI)** and machine-learning algorithms to dynamically evolve, adjust, and learn. For example, the data generated by the physical twin may be employed to improve the software models inside the DT. DDSs may lead to more accurate and realistic models, which in turn will feed back to the physical twin to improve its performance in the physical world.

12.3 Virtual and Augmented Reality

Virtual reality (VR) is a class of technologies that immerses users in digitally rendered artificial environments. VR users, typically wearing a head-mounted display or headset (Figure 12.4), experience multiple sensory stimuli, including visual, audio, and tactile signals. Unlike the experience at a 3D movie, however, the VR environment reacts to the actions of the user—just like in the real world.

Figure 12.5. No masks, no gel, and a customized avatar for Grenoble's 2020 back-to-school event.

Source: Saviotti, M., 2020, September 15, "Virtual reality 2020 back-to-school challenge," Grenoble Ecole de Management, retrieved from https://en.grenoble-em.com/news-virtual-reality-2020-back-school-challenge

While VR has been researched for decades, it has yet to gain mainstream adoption due to the computational power intensity of VR representations. Many software companies, Google and Facebook above all, have invested heavily in VR projects. Yet computer gaming represents the only industry where applications are being used regularly by a subset of the consumer base. The COVID-19 pandemic provided a great acceleration in the use of virtual environments. Several schools adopted VR campus tours during these difficult times when in-person visits were limited (Figure 12.5).

Despite intriguing scenarios and use cases proposed for areas such as employee training, travel and tourism, or education, VR still awaits the "killer application" that will take it mainstream (Figure 12.6).

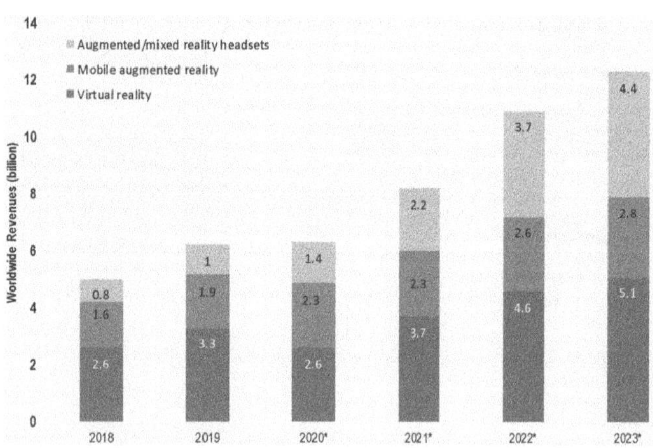

Figure 12.6. Forecasted revenues (billion) of augmented and virtual reality headsets from 2018 to 2023

Source: "SuperData XR quarterly update," 2020, October, SuperData, a Nielsen Company, retrieved from https://www.superdataresearch.com/blog/superdata-xr-update

Conversely, **augmented reality (AR)** appears to be on the brink of widespread acceptance thanks to the pervasiveness of mobile devices. In its simplest form, AR consists of superimposing an information layer on a real image. Current applications use the smartphone's camera, GPS receiver, compass, and accelerometer to identify a scene or objects and then, using a database of entries downloaded to the device, superimpose contextual information on recognized objects.

An intriguing early example was offered by Italian gaming company Illusion Networks. The firm created in 2010 an impressive AR application for the iPhone 3GS called Voyager Xdrive (Figure 12.7). It allows travelers who are walking around the Roman Forum to see a three-dimensional reconstruction of the Forum the way it was during Constantine's rule

(320 AD) and to see and hear a description of the buildings and monuments currently in view.

Another early example was the interactive content enhancement IKEA experimented with for its 2014 catalog app, allowing customers to experience digital furniture over real space by taking pictures of the augmented environment. To use the application, it was sufficient to point the smartphone at selected catalog pages and then place the catalog where the furniture was supposed to be placed to see it virtually appear.

Figure 12.7. Voyager Xdrive in action

The development of AR received further impetus in 2017 with the release of software development kits (SDK) by both Apple and Google. The first one, announced as part of iOS 11 under the moniker ARKit (Figure 12.8), and today at its fourth installment, enabled leveraging the iPhone 12 Pro's LiDAR scanner through the Depth API. LiDAR is an acronym for "light detection and ranging" and consists of emitting a laser and a sensor capable of measuring light return times and wavelengths to build 3D representations of scanned areas. Its use enhances the positioning and blending of 3D virtual objects in the real space. Besides being useful for AR applications and 3D scanning, it additionally enhances focus accuracy and depth of field information, thus making photos three dimensional. ARKit is a framework that developers can use to build reality experiences for blending digital objects and information with the environment around the user (Figure 12.9).[1] Not to be undone, Google released ARCore for high-end Android phones.

Figure 12.8. Tesla Model 3 design studio created with ARKit

Source: https://youtu.be/xCdjlDnCtps

AR is not confined to the mobile platform. Car manufacturers, for example, are beginning to use the windshield of a car as the surface on which to superimpose information layers. But on the mobile platform, AR may be disruptive, mainly because it will further reduce information barriers and information asymmetry and will do so "on the go" rather than forcing users to be sitting at a computer (Figure 12.10). Like the commercial Internet two decades ago, AR has the potential to arm users with an unprecedented quantity of easily accessible and usable information. The difference is that such information will be available to them seamlessly anytime and anywhere. While examples of AR applications delivering customer value are still relatively few, they are becoming more common, and as Tim Cook recently put it, "You rarely have a new technology where business and consumer both see it as key to them. So, I think the answer is that's the reason that I think it's going to pervade your life."[2]

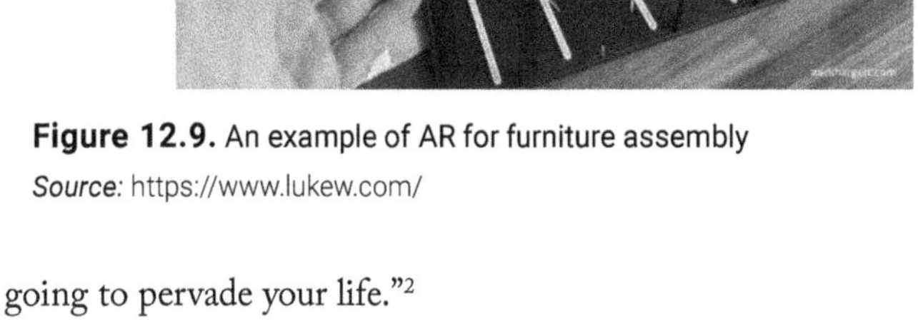

Figure 12.9. An example of AR for furniture assembly

Source: https://www.lukew.com/

1 Apple Inc. 2020. "ARKit." Apple Developer, retrieved from https://developer.apple.com/arkit/.

2 Eadicicco, L. 2020, January 2. "Tim Cook keeps talking about why this one technology will 'pervade' our lives in the future." *Business Insider France*, retrieved from https://www.businessinsider.fr/us/apple-q1-earnings-call-tim-cook-augmented-reality-ar-2020-1.

Figure 12.10. AR for indoor navigation

Source: https://www.lukew.com/

12.4 Digital Manufacturing

AR represents an example of digital objects seamlessly inhabiting the real world as we perceive it and see it. The development of 3D printing is taking this trend a step further. Digital objects can materialize in the physical world. Interestingly, the process is exactly the opposite of traditional manufacturing techniques like lathing or milling. 3D printing is an additive process where layers of material are laid down following a fully digital blueprint.

In a few years, the range of applications for 3D printing expanded enormously from prototyping to the manufacture of fully functional components. The rapid technological developments made 3D printers cheaper while increasing their precision, resolution, speed, and printable materials, making them viable alternatives to current manufacturing processes. From a managerial perspective, the core aspect to balance is the trade-off of flexibility and unitary cost. While 3D printing can't match traditional production methods' economies of scale yet, it provides unmatched levels of flexibility. 3D printing offers an endless set of shapes, colors, sizes, and other customizations, with minimal increase of manufacturing cost. This is the result of a reduced need for assembling product components, as the whole object can be printed from the inside. For example, GE aviation moved to 3D printing for a series of fuel nozzles for their engines that otherwise needed assembly for their components. Even on a volume of 45,000 units per year, they were able to cut costs by 75%.[1]

Managers should understand and anticipate the transformations brought about by the digitization of manufacturing:[2]

- *Offerings.* Accelerated product development cycles are the benefits of 3D printing. Prototypes can be built and delivered to customers as soon as their digital design is available. The diffusion of 3D printers could deeply impact the product offering. You may consider just selling or "loaning" the digital design, allowing the customers to self-manufacture the product. However, once the digital design is available, protecting it from copying or redistribution will require special attention, as with all information goods. If the competitive advantage is not found in the manufacturing process anymore, the design and engineering skills become the resources on which your competitive advantage is built.
- *Operations.* As more products are manufactured through 3D printing, operations and supply chain management will need to change radically. Operational efficiency will emerge directly from the reconfiguration of the phases of the manufacturing and delivery of products. A scenario where your Amazon order is not only shipped but manufactured and shipped in 24 hours is not farfetched anymore. These are the promises of "instantaneous inventory management," where stock will no longer exist.
- *Capabilities.* Transitioning from the physical world to the digital world and back again is the challenge organizations face with 3D printing. Organizations will need to refocus their skills to differentiate through their own design and product

1 D'Aveni, R. 2015, May. "The 3-D printing revolution." *Harvard Business Review*: 40–48.

2 D'Aveni, R. A. 2018, July 1. "The 3-D printing playbook." *Harvard Business Review*, retrieved from https://hbr.org/2018/07/the-3-d-printing-playbook.

Table 12.2. Possible digital manufacturing business models

Business model	Description	Example
Mass customization	Creating one-off products that are precisely adjusted to the needs or whims of individual buyers—adjustments that can be carried out by simply uploading each customer's digital file into a 3D printer. Thanks to the efficiency and precision of digital technology, these products cost less than conventionally manufactured items but fit individuals' specifications more exactly.	Hearing aids, orthodontic braces, prostheses, sunglasses, car and motorcycle accessories, and Christmas tree decorations
Mass variety	Creating one-off products, targeting customers who have strong and varying preferences but don't need products adjusted to their personal specifications. Manufacturers can skip the process of collecting personal information and offer a wide variety of options at affordable prices.	Jewelry with customized basic designs
Mass segmentation	Creating a limited variety of products for which a specific version targets an addressable market segment. This is a model that can be suited to serve seasonal, cyclical, or short-term fad markets. Viability of the model depends on design and switchover costs.	Components for popular business-to-business (B2B) or business-to-consumer (B2C) products (e.g., bicycle components, smartphone handlers), limited-run productions, specialized parts, customized sneakers
Mass modularization	Selling a 3D-printed body with interchangeable modules for insertion. The model is challenged by the decision on what to embed in the base unit and what to leave modular.	Military modular platforms, modular consumer electronics
Mass complexity	This model exploits the digital manufacturing ability to make products with intricate designs that conventional manufacturing can't achieve and to produce unusual shapes and embed sensors and other elements.	Generative design-based products, new designs
Mass standardization	This model leverages the greater efficiency of additive manufacturing for reducing waste and direct costs or by limiting supply chain extension, inventory, elaborate assembly, or investments in specialized equipment.	OLED screens, mascara brushes

Source: Adapted from D'Aveni, R. A., 2018, July 1, "The 3-D printing playbook," ***Harvard Business Review***, July–August 2018, retrieved from https://hbr.org/2018/07/the-3-d-printing-playbook

customizations in a fiercely competitive digital space. At the operational level, organizations have traditionally designed for manufacturing, but designing for printing requires a different set of skills.

- *New competitors and ecosystems.* Initially, systems with lower entry barriers will see the birth of new business for quickly manufacturing niche products and customer designs. The impacts of the digitization of manufacturing may involve increased competition from new intermediaries. Paralleling the experience of the music industry, specialized intermediaries could emerge as platforms for the distribution of the digital designs. Makers, too, could challenge your products with improved designs, released under creative commons licensing and put in the public domain. As we will see in the next sections, open source designs may emerge from the collaboration of makers and users, competing directly with commercial products. This discussion should remind you of the unique role of digital technologies in enabling new ways of doing business (see Chapter 5). The unique characteristics of **digital manufacturing** enable new business models (Table 12.2).[1]

1 D'Aveni, R. A. 2018, July 1. "The 3-D printing playbook." *Harvard Business Review*, retrieved from https://hbr.org/2018/07/the-3-d-printing-playbook.

Digital manufacturing is the next stop on the progression from digital to physical and "has the potential to revolutionize the way we make just about everything."[1] Interestingly, digital manufacturing offers a reversal of 60 years of digital revolution, when physical objects and occurrences received a digital representation. With 3D printing, it is the digital representations that take physical shape.

12.5 Machine Learning

Let computers learn and adapt. While just a few years ago, this would sound like science fiction, machine learning is today a mainstream topic of discussion—not only in research universities but also in business and government. As a discipline, machine learning sits at the intersection of computer science and statistics. If data mining represents the broad category of activities for knowledge extraction performed on the data, machine learning is the computer science approach to accomplishing the same goal. Knowingly or not, you have already come in contact with machine-learning applications. Search engine web page classification, Facebook's "People You Might Know" feature, Airbnb Search Ranking, the CAPTCHA,[2] some video games' player-matching scenarios, product or movie recommendations, and face and object detection are all areas where machine learning is intensively being used.

At a more conceptual level, the idea of machine learning is to develop algorithms that let digital computers identify or discriminate patterns (e.g., the pattern of pixels in an image, the pattern of terms in a tweet) without needing the programmers to code an explicit procedure (i.e., rules) but letting the machine learn the patterns from the data themselves. Machine learning is particularly useful for solving problems that are difficult to create rules for. Consider face recognition. While humans intuitively perform this activity, could you explain *how* you do it? Could you codify the set of "rules" your brain applies to every face to distinguish known from unknown ones? What about the procedure to put a name to the face? When Facebook suggests a tag, the application is asking confirmation for associating the name and profile of a person to the pattern of facial features it has detected. In that case, you are training Facebook machine-learning algorithms to recognize that person.

You can see why machine learning has become so popular as of late. The trends we discussed throughout this book, such as the growing amount and variety of data being continuously produced by humans and machines alike, make it viable to train and useful (almost necessary!) to adopt machine-learning algorithms. There are two general categories of machine-learning algorithms: supervised and unsupervised (Table 12.3).

Supervised Machine Learning This generates a predictive model from a known set of training examples known as the gold standard. In other words, the terms—or features—of the problem are known, and it is up to the machine to learn how to correctly answer given a new input (Figure 12.11). The data fed to the algorithm is "labeled," meaning that it has already been manually classified and is then more difficult and expensive to develop. Supervised machine learning can perform extremely well in contexts where there is a

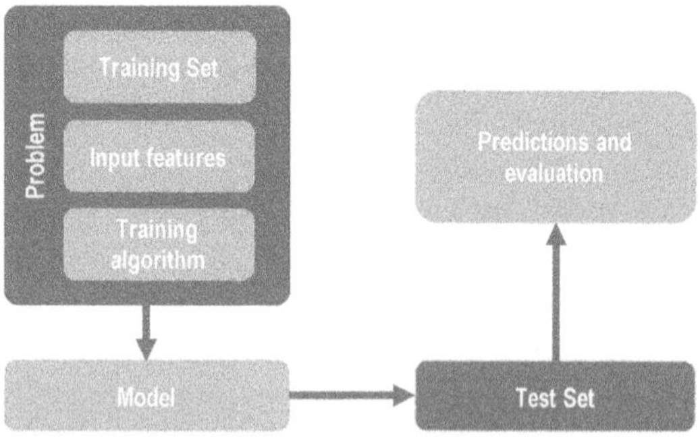

Figure 12.11. A representation of the supervised machine-learning process

1 Gross, D. 2013, February. "Obama's speech highlights rise of 3-D printing." CNN, retrieved from http://edition.cnn.com/2013/02/13/tech/innovation/obama-3d-printing/.

2 CAPTCHA is an acronym for "Completely Automated Public Turing test to tell Computers and Humans Apart." It is a challenge-response system to verify that the user is actually a human. On the other hand, CAPTCHAs are used to create training sets for machine-learning algorithms . . . so every time you are proving that you are not a robot, you are indeed aiding in the training of better robots!

Table 12.3. Supervised and unsupervised machine learning

Type	Supervised	Unsupervised
Input data	• Data are labeled or classified • Gold standard	• Data are not labeled or classified
Class of problems	• Using a model to learn a pattern association between input examples and a target variable in the presence of an "instructor" supervising the process • Classify data predicting a class (classification) or numerical label (regression)	• Using a model to describe or extract relationships in data; it operates without an output or target variable. • Find groups in data (clustering) or summarizes their distribution (density estimation)
Computational intensity	• Low	• High
Uses	• Identifying risk factors for diseases and planning preventive measures • Classifying whether an email is spam • Predicting house prices • Predicting customer churn • Classifying words as nouns and verbs • Predicting rainfall and weather conditions • Predicting whether people will vote for a given candidate • Sales forecasting and risk assessment analysis in health insurance companies • Finding out whether a loan applicant is low risk or high risk • Predicting the failure of mechanical parts in automobile engines • Predicting social media share scores and performance scores	• Grouping customers by purchase behavior • Identifying associations in customer data (for instance, people who buy a certain style of shoe may also be interested in a certain style of bag) • Segmenting data by purchase history • Classifying people based on different interests • Grouping inventories by manufacturing and sales metrics
Requirements	• Obtain or create labeled data and a gold standard set to optimize the machine-learning process for classification and prediction problems • Make sure the algorithm has the right amount of data; of equal importance is ensuring that the algorithm has relevant data for the specific problem in hand • Divide the data in a training dataset and a validation dataset—each of which should be large enough and diverse enough to represent the variety of learning cases and reduce possible bias • Make sure output values are labeled and annotated and bias is minimized before feeding data to a supervised learning algorithm; data labeling is critical for the success of supervised learning • Perform a detailed assessment on the quality of key data attributes, enabling focused improvements that deliver better business value	• Apply unsupervised learning to use cases involving clustering (i.e., discovering the inherent groupings in the data) or association (i.e., discovering rules that describe large portions of your data) • Check for sufficient representation and diversity of data; although unsupervised learning does not require labeled data, it still requires statistical validation to reduce bias and draw reliable conclusions • Work with data scientists to try to select the most effective algorithm for your particular data and goal • Unsupervised learning is also used to prepare data for subsequent supervised learning; this is done by identifying patterns or features that can be used to categorize, compress, and reduce the dimensionality of data

Source: Adapted from Alaybeyi, S., and Linden, A., 2020, "3 Types of machine learning for the enterprise" (no. G00465941, p. 10), Gartner reliable gold standard (i.e., a training set), or it can be developed, and a number of success stories in machine learning leverage established supervised algorithms.[1] Consider the example of detecting fake and fraudulent online reviews. A team of scientists at Cornell University used supervised machine learning to achieve 89% accuracy when classifying TripAdvisor reviews. To do so, they commissioned 400 known fake reviews through Amazon Mechanical Turk and built a gold standard to train their classifier.[2] Handwriting recognition provides a similar example, as does image recognition when algorithms are trained by humans who classified a small set of images (e.g., tagging your friends' faces on Facebook, or identifying an object in a series of pictures when challenged by the CAPTCHA test).

Unsupervised Machine Learning This is helpful in situations where a gold standard does not exist or the domain is constantly evolving and a gold standard would not be relevant over time. The unsupervised algorithm is then used to explore that data to detect possible patterns and extract the relevant features for classifying data or distill relevant information. The main advantage of unsupervised algorithms is that they do not need user intervention to "label" or categorize the data, making the solution scalable and applicable to a broad range of situations.

A typical example is that of machines being able to successfully play games. The first high-profile example of its kind was the famous IBM Deep Blue—the computer program that in 1997 beat reigning World Chess Champion Garry Kasparov under tournament conditions. Unsupervised machine learning has since improved dramatically, and the consensus is now that an algorithm that is allowed to play a game with a clear set of rules and objective function (i.e., a goal) will always beat unaided humans, no matter how complex the game. The defining moment for this recognition was AlphaGo's win against 18-time world Go champion Lee Sedol in March 2016.[3] Go is an ancient Chinese strategy board game, similar to chess, long considered the ultimate frontier for machine learning due to the staggering number of possible board combinations that can occur during a match.

Deep Learning

Much of the current progress of machine learning is in an area called **deep learning**. Deep learning is an approach to machine learning that, using neural networks, mimics the way the brain works. Deep learning models seek to parameterize (i.e., learn) a discriminant hierarchical structure of features directly from input data. In perspective, this should make learning algorithms better and more autonomous. More specifically, the aim is to reduce the efforts in determining the features necessary for successful discrimination—today still a relatively labor-intensive process. In other words, with deep learning, algorithms' features are classified through different hierarchical layers, each one focusing on a different interpretation of the data. Deep learning gained early popularity in 2011 from the success of the Google Brain project, where a deep-learning algorithm was able to recognize high-level concepts like faces, human bodies, and cats after being trained with only a random sample of 200 × 200 pixel images from YouTube.[4]

---

1 Beyer, D. 2015, May 3. "On the evolution of machine learning." O'Reilly, retrieved from http://radar.oreilly.com/2015/05/on-the-evolution-of-machine-learning.html.

2 Streitfeld, D. 2011. "In a race to out-rave, 5-star web reviews go for $5." *New York Times*, retrieved from http://www.nytimes.com/2011/08/20/technology/finding-fake-reviews-online.html?_r=0.

3 Wikipedia. n.d. "AlphaGo versus Lee Sedol." *Wikipedia*, last updated April 4, 2018, retrieved from https://en.wikipedia.org/wiki/AlphaGo_versus_Lee_Sedol.

4 For the science behind the experiment, see Le, Q. V. 2013. "Building high-level features using large scale unsupervised learning" (paper presented at the IEEE International Conference on Acoustics, Speech and Signal Processing [ICASSP],

The technical breakthrough in deep learning, and machine learning in general, is the development of programmable graphical processing units (GPUs). GPUs are specialized microchips with an instruction set optimized for image processing. They excel at performing the same operation on many data elements very quickly. This enables them to carry out functions such as shading an area of the screen, moving a window or an icon from one part of the screen to another, or decoding compressed video frames very quickly. While programmable GPUs were originally designed to speed up graphics processing, the type of fundamental computations required by machine learning have similar characteristics. More recently, specialized hardware has been introduced that dramatically increases the efficiency and speed of deep-learning and machine-learning algorithms. In Chapter 9, we presented Tesla's Full Self-Driving Chip—a chip specifically optimized for machine-learning algorithms that the firm built in an effort to accelerate the self-driving capabilities of their cars. Similarly, smartphones have increasingly leveraged deep-learning and machine-learning digital capabilities to enable new applications and features and providing an overall better user experience. It is not surprising, then, to find an increased number of these specialized chips at the heart of IoT devices. Starting with the iPhone 11, Apple introduced a dedicated neural processing unit (NPU), the Apple Neural Engine, to accelerate and support machine-learning applications.

Despite all the recent success and the press hype, it is important for you as a manager to realize the remaining current limitations of machine learning. Machine-learning algorithms perform best in scenarios where there are millions of reliably labeled data (e.g., cat pictures versus images where no cats appear). Alternatively, as in deep-learning algorithms that can play games, it is necessary that the "game" has clear rules and that it is possible to run millions of simulations or training experiments. Despite high-profile research examples, there are still limited real-world scenarios where these conditions are met in business life. Murray Campbell, one of the original creators of IBM Deep Blue, put it best: "While poker is a step more complex than perfect information games, it's still a long way to go to get to the messiness of the real world."

A Note about Artificial Intelligence

Attentive readers will have noted that we use exclusively the term *machine learning* in this section, and unlike much of the business press, we do not treat the term as synonymous with AI. AI is a computer science concept that is almost as old as the discipline. The general test for artificial intelligence systems is the Turing test, named after the British mathematician Alan Turing, who proposed it in 1950. As normally interpreted, the test suggests that a machine can be said to exhibit "thinking" ability if it can fool a human, making him or her unable to tell the machine answers apart from those of a real person. However, the term *AI* is confusing because it engenders visions of machines that become sentient and similar to humans. It should, however, be fairly clear from the above discussion of machine learning and deep learning that computers are not (as of yet, at least) intelligent. Rather, under specific conditions, algorithms such as neural networks are able to parameterize thousands of mathematical functions based on available training data in order to engage in reliable classification and prediction. In other words, "learning" for a computer is about performing mathematical operations. Thus machine learning is very different from human learning, and framing it in terms of artificial "intelligence" is more confusing than helpful. Some experts even argue that the term *AI* should be banned. François Chollet, one of the foremost deep-learning experts at Google and the author of the Keras framework, explained it best:

IEEE): 8595–98. doi: 10.1109/ICASSP.2013.6639343. For the story, you may check Clark, L. 2012, June 26. "Google's artificial brain learns to find cat videos." *Wired*, retrieved from https://www.wired.com/2012/06/google-x-neural-network/.

> A lot of well-funded, large-scale gradient-descent projects get carried out as a way to generate bombastic press articles that misleadingly suggest that human-level AI is perhaps a few years away, and many people have staked a lot on this illusion. But it's still an illusion.[1]

In conclusion, while fears of the near-term development of "artificial super intelligence" or the "rise of the machines" may be overblown, there is no doubt that machine-learning algorithms will continue to revolutionize various aspects of our lives. The most productive approach for you as a manager will be to think about machine learning as a foundational technology that will become increasingly embedded in information systems and applied to a wide array of problems. Consider computer vision and face recognition, a problem that has largely been "solved" by deep learning. Benedict Evans, a former partner with the venture capital firm Andreessen Horowitz, made this point (Figure 12.12):

> Eric Raymond proposed that a computer should never ask the user for any information that it can autodetect, copy, or deduce; computer vision changes what the computer has to ask. So it's not, really, a camera, taking photos—it's more like an eye, that can see.[2]

Tangible applications of what Evans means are already available. On the one hand, computer vision is automating and systematizing existing processes, like the scanning and detection of known persons of interest at airport checkpoints or other high-traffic venues (e.g., stadiums). As with any technology, however, computer vision will start by automating existing processes but very rapidly thereafter will begin to change the way we perform work and create opportunities for novel activities. In 2017, Google launched Google Clips (Figure 12.13), somehow epitomizing the success of computational photography, where machine-learning algorithms are used to detect the objects in a scene, light, and shooting conditions (e.g., a night scene) to optimize the shooting mode and the camera parameters to obtain the best possible photos. While the increasing processing power of NPU is making similar applications possible in real time on the smartphone by leveraging the cloud—and users' labeled photos—these algorithms will improve over time.[3]

Figure 12.12. Benedict Evans's tweet

Figure 12.13. The Google Clip smart camera

Source: https://images.idgesg.net/images/article/2017/10/google_clips-100738128-large.jpg

Explainable AI One of the primary challenges of AI, and many machine-learning algorithms in general, is their unobservability. Unlike traditional rule-based algorithms, where a person knowledgeable in the programming language can predict outputs based on observed inputs, it is impossible to do the same with machine learning and AI. Explainable AI is critical to ensure trust in these algorithms and their predictions, recommendations, and actions with tools capable of exposing the operations of

1 Ray, T. 2019, November 26. "Keras inventor Chollet charts a new direction for AI: A Q&A." ZDNet, retrieved from https://www.zdnet.com/article/keras-creator-chollets-new-direction-for-ai-a-q-a/.

2 Evans, B. 2016, August 15. "Imaging, Snapchat and mobile." *Benedict Evans* (blog), retrieved from https://www.ben-evans.com/benedictevans/2016/8/15/imaging-snapchat-and-mobile.

3 Byford, S. 2019, January 31. "How AI is changing photography." The Verge, retrieved from https://www.theverge.com/2019/1/31/18203363/ai-artificial-intelligence-photography-google-photos-apple-huawei.

"black-box" machine-learning algorithms. For example, Google's Explainable AI framework consists of a set of tools developers may use to determine possible bias in the data that may affect the results or to identify for every result the factors that weighed more in the decision.[1]

12.6 Blockchain

As happens often with successful products, you may be more used to **Bitcoin** than to **blockchain**, the underlying technology that made it possible. Bitcoin is a decentralized cryptocurrency system conceived by a mysterious person (or group of people) using the pseudonym "Satoshi Nakamoto." The interest in Bitcoin and cryptocurrencies in general stems from the promises that they may revolutionize the way we think about money and the transfer of value in general.[2] After a quiet launch in 2009, Bitcoin has evolved into a multibillion-dollar industry, and it has inspired the creation of hundreds of similar cryptocurrency systems. While cryptocurrencies may be in the spotlight—probably caused by the rampant financial speculation around them (Figure 12.14)—it is much more relevant for managers to understand how the blockchain works. The reason is that blockchain technology holds the potential to revolutionize record keeping, contract registration, and transaction management in a way that parallels the introduction of Structured Query Language (SQL) and relational database management systems (see Chapter 3) in the 1970s.

The first element to consider concerns the nature of contracts and the associated counterparty risks. Prior to the introduction of blockchain, to avoid trading parties failing to fulfill their contractual obligations, a trusted third party was typically involved. Banks, credit card issuers, notaries, and lawyers acted as the "guarantor" of the terms of the transaction. They attested to the availability of funds, parties' identities, the effective ownership or rights on a property, the condition and existence of an asset, the agreement of the parties, the lack of coercion, and any other required elements of the transaction. However, involving a trusted third party has a cost and yields control to such an authority. With blockchain, the opportunity lays in enabling transactions directly between parties, without anyone else involved. By some accounts, blockchains are creating a paradigmatic shift from an Internet of information and an Internet of things to an "Internet of value."

With this understanding of the blockchain as a reliable system of record, an immutable accounting log or ledger, we can use the famous example of Bitcoin as an illustration. The blockchain is engaged when a user wants to initiate a transaction, which in the case of Bitcoin is a financial transaction (e.g., the sale of an asset). The user digitally signs the transaction with his or her private key, a message referencing a previous transaction like the previous sale of the asset, or in the case of Bitcoin, the unspent transaction outputs (UTXO) from a previous transaction.[3] The user then indicates the recipient's public address and the amount of Bitcoin or the asset that will be transferred. The combination of public and private keys acts as a unique identifier of the users

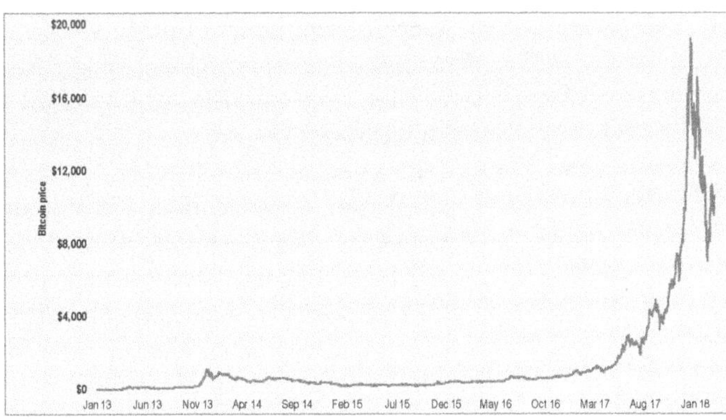

Figure 12.14. Bitcoin prices since their introduction
Source: https://charts.bitcoin.com/chart/price

1 "Technology can't fix this." 2020. *Nature Machine Intelligence* 2(7): 363. doi: 10.1038/s42256-020-0210-5.
2 Nakamoto, S. 2008. "Bitcoin: A peer-to-peer electronic cash system." Bitcoin.org, retrieved from https://bitcoin.org/bitcoin.pdf.
3 As transactions may use just fractions of the cryptocurrency unit, the unspent value is then made available to complete other transactions.

and removes the need for a central authority to assign accounts and identities. With the users authenticated and the transaction validated, the transaction is then broadcasted and propagated via the peer-to-peer blockchain network to all full nodes. All full nodes keep a complete copy of the global ledger so that they can independently verify that the asset belongs to the entity claiming ownership. This shared governance is established in lieu of a central authority or trusted third party that ensures the accuracy and legitimacy of the transactions. The full nodes, in fact, have a complete history of all transactions, or, in other words, the full ledger. For this reason, the blockchain is a form of "distributed ledger," fully available at all full nodes and on which all nodes agree. It is this architecture that enables the distribution of control previously concentrated in one guarantor. However, to secure the blockchain, it is necessary to discourage fraud and secure the "blocks," the lines of the ledger that record the transactions. For a new transaction to be recorded on the blockchain, a second type of node, the *miners*, takes available unconfirmed transactions and groups them together in a candidate block. The miners compete to validate the candidate block by engaging in a computational race that consists of solving a cryptographic challenge to find a special number called *nonce*. The miner who manages to solve the puzzle first immediately submits the block and a proof-of-work (PoW) to the rest of the network, which accepts the block as valid if it meets all requirements. All miners then begin searching for a new block that will reference this newly recorded valid block. It is the continuous chain of all these cryptographically linked blocks—the blockchain, as it is called—that provides the instrument for record keeping. All full nodes on the network thus maintain a shared state of the global ledger that everybody agrees on by independently recomputing the whole history of transactions starting from the genesis block, the first block mined to launch the network. Ironically, the first Bitcoin block, block 0, contains a reference to an article in the *London Times* on January 3, 2009, titled "Chancellor on brink of second bailout for banks." This is a clear reference to the potential disruption—or fix, depending on the point of view—that the blockchain could imply for the overall banking system.

The blockchain system works because miners who provide costly resources, mostly electricity and computing power, do so in search of a monetary reward, a fee plus some amount of new currency (e.g., Bitcoin) that is released for each new valid block added to the blockchain. The core innovation of Bitcoin was not the blockchain itself but rather the so-called *Nakamoto Consensus*—an approach ensuring that miners behave honestly by making it more profitable for them to support the integrity of the system rather than undermine it by allowing unverified transactions.

In practical terms, blockchains represent, then, an innovative system to secure and store information permanently in a distributed ledger, thus enabling fault tolerance, resilience, and availability. Our description of the blockchain is not designed to make you an expert on the technology, Bitcoin, or cryptocurrencies in general. There are a lot of freely available resources for those who want to wade deeper into the technical aspects of the blockchain. But even with this limited understanding, you can begin to appreciate the business appeal of this new technology. While we focused our description on public, open source blockchains, private blockchains are also finding an application in several industries. In public blockchains, any stakeholder can join the network and participate as a user, miner, developer, or simple community member. Private blockchains are generally more centralized, managed by a single actor, and their access is limited only to the authorized parties. For example, Facebook attempted, first with Libra and then with Diem, to introduce a new digital private currency controlled by the Diem Association. In summary, you can think of a blockchain as a database with very desirable properties:

- *Distributed ledger*. All transactions in blockchain applications are recorded in a permanent ledger of all valid transactions that have occurred. The nodes in the network replicate and agree on the ledger to assure that no party tampers with the records. Inconsistent nodes are then ignored by the network.

- *Immutability*. Validated transaction records are added to the distributed ledger and are stored permanently in a "block." Unless the entire network agrees to roll back the ledger—for example, in case of an error—it cannot be altered.
- *Shared governance*. Transaction records, collected in validated blocks, can be stored by any entity interested in doing so. The distributed architecture adopted to maintain the ledger grants that no single party can control the network and that parties can directly transact with each other. In such a system, there's no need for a central authority or institution to regulate or certify the transactions. Furthermore, changes or amendments to the network require the agreement and coordination of all the stakeholders. The lack of political governance in blockchains may produce "forks," splits in the network that create new blockchains and rely on different communities and stakeholders.
- *Heightened security*. No individual entity represents a concentrated point of failure for the overall record-keeping system. The distributed nature of the ledger makes blockchains fault tolerant, resilient, and highly available.
- *Built-in validation*. Because of the requirements for block validation, the blockchain ensures that no one individual entity can tamper with the records. Old transactions are preserved forever, and new additions are irreversible.
- *Transparency*. Anyone who joins the blockchain can check the ledger and reconstruct the full history of transactions that have occurred since the system's inception.

Incentivized by the success of Bitcoin and by the promise of blockchain technology, hundreds of start-ups have entered the space. Most notable of all is Ethereum, whose website does not mention currency but rather reads, "Ethereum is a global, open-source platform for decentralized applications."[1] In the music industry, where major labels traditionally acted as gatekeepers, the Open Music Initiative is actively working on developing a shared API based on blockchain to simplify the way music creators and rights owners are identified and compensated.[2] While Spotify is using blockchain technology to handle compensations, music creators are advocating for a widespread adoption of the technology, potentially disrupting the industry.[3] As with machine learning, we believe that the power of the blockchain is in its promise to dramatically change the way institutions and organizations work.

12.7 Emerging Computing Architectures

In this chapter, we introduced some fundamental innovations (one of which is beyond digital!) that will shape the future of information systems. These innovations represent enabling technologies with the potential to generate paradigmatic shifts in the way we consume, produce, and distribute applications.

5G Networks

5G is the fifth-generation wireless network standard. While much of media attention remains focused on faster connectivity, 5G represents a major milestone in wireless communication beyond speed. A multitude of digital devices and diverse network technologies (e.g., satellite, cellular,

---

1 See the home page at https://www.ethereum.org/.

2 Heap, I. 2017, June 5. "Blockchain could help musicians make money again." *Harvard Business Review*, retrieved from https://hbr.org/2017/06/blockchain-could-help-musicians-make-money-again.

3 Hissong, S. 2021, February 4. "A field guide to music's potential crypto boom." *Rolling Stone*, retrieved from https://www.rollingstone.com/pro/features/music-crypto-blockchain-nfts-guide-1116327/.

cable, fiber, phone) support applications requiring different speeds and response delays, often enabled by different providers. 5G is a single framework, harmonizing a plethora of different protocols and private solutions, facilitating interoperability, standardization, and digital innovation. 5G represents a leap forward for the mobile platform as we know it today, as it harnesses its potential to cover a broader set of possible applications. The engineering challenge of 5G is that it had to meet three divergent types of connectivity expectations to support the digital economy:

- *Massive machine type communication (mMTC).* This requirement originates from the necessity to support the billions of devices and smart objects constituting the IoT (Chapter 5). Current support is fragmented, and several not interoperable solutions risk becoming dead ends lacking support and scalability. With 5G, the possibility of supporting industrial automation and control to coordinate large fleets or devices may encourage the development of applications including AI-enhanced actuation (e.g., car fleet updates, self-driving on less congested roads), massive data streaming, and analysis.
- *Ultrareliable and low latency communications (URLLC).* While in the consumer space latency is rarely an issue, business applications may greatly benefit from defined and low latency communication. Latency in a network represents the time a message takes to reach its destination and depends on distance, the characteristics of the network, and the physical layer used for the communication (e.g., wireless, fiber, satellite, cable) in which the propagation of the signal varies. For example, if you are an online gamer, low network latency is critical to avoid being "fragged" because the game lags. While on a different scale, it's the same problem of receiving images from satellites around the moon or Mars. For example, in the 2021 landing of the Perseverance rover on Mars, it took 11 minutes for the signal to reach the people at NASA's mission operations center. Networks exhibit different latency depending on their configurations, and until now, this was not a design requirement for public networks. 5G has now the capability to support critical services that rely on real-time operations, including public safety, remote surgery, or industrial applications. Able to guarantee a reliable connection, the wireless network becomes a likely substitute for traditional "wired" solutions.
- *Enhanced mobile broadband (eMBB).* The speed of 5G is the most advertised feature in the consumer space. The objective is to provide, in mobility, enough bandwidth to support high-quality video streaming and immersive services such as AR and VR, thus facilitating their use for applications such as maintenance, safety, design, and collaboration.

Unlike the previous generation of communication networks, 5G is designed to enable the following:

- *Standardization.* 5G will be offered under a single standard that will facilitate interoperability between different applications and will guarantee the scalability of the applications including IoT-based services.
- *The "virtualization" of the telecommunication infrastructure.* If "software is eating the world," then 5G software is eating the infrastructure. The major difference compared to previous generations of networks is that with 5G, the network infrastructure is controlled by software and can be configured dynamically based on the service's requirements. This will allow running virtual networks optimized for each business case on top of the same physical communication network.[1]

---

1 GSMA. 2017. "An introduction to network slicing." Global System for Mobile Communications Association, retrieved from https://www.gsma.com/futurenetworks/wp-content/uploads/2017/11/GSMA-An-Introduction-to-Network-Slicing.pdf: 20.

- *Ubiquity via multiconnectivity*. 5G can potentially achieve complete geographical coverage, without white zones or dropping connections. By supporting IoT devices and satellites to everything in between, 5G can coordinate all the different technologies within its framework to assure a stable connection. 5G devices can "hand over" the communication from one type of connectivity to another, according to the service level required. Multiconnectivity has the additional advantage of allowing new levels of localization and tracking. Covering both indoor and outdoor connectivity, 5G enhances the traceability of objects and assets.

5G, and its evolution to 6G, will become the backbone on the Internet of the future composed, as we have already discussed in Chapter 5, of billions of individuals and connected devices. Instead of relying on centralized servers, the future Internet will leverage 5G characteristics to create "edge nodes" where computing capabilities are moved closer to users where data are generated.[1] This solution provides faster response times, reduces network congestion to central cloud systems, and processes data faster. The enhanced digital capability will support the growing demand for machine-learning and AI applications, including self-driving capabilities (e.g., for the coordination of different vehicles) and industrial IoT (e.g., for controlling equipment in real time).

Serverless Computing

Serverless computing refers to an emerging software development architecture that assumes the existence of cloud-computing resources (Chapter 3) and enables developers to build applications without having to provision servers or programming environments. Our discussion of digital resources and the examples of Twilio and Instacart (Chapter 5) provided you with a preview of the way cloud-native applications are developed. Applications are increasingly modular, orchestrating multiple digital resources available as services through the cloud-computing infrastructure. In fact, you can conceptualize serverless computing as the progression of cloud-computing architectures (Chapter 3). Software as a service (SaaS) provides servitized use of full-fledged applications, platform as a service (PaaS) provides servitized control of a hardware/software platform, and infrastructure as a service (IaaS) provides servitized access to virtualized hardware. An even lower level of access is servitized access to individual computing resources, such as storage or processing. Thus the serverless computing paradigm assumes that developers are best served by accessing ready-made digital resources, such as Amazon Web Services (AWS) S3 storage or Stripe payment, or being given the tools to create their own. The major cloud providers offer function as a service (FaaS) capabilities (e.g., AWS Lambda, Microsoft Azure Functions, Google Cloud Functions).

In the serverless architecture, a software application or other software program is the result of a dynamic orchestration of FaaS resources that are called upon as needed at run time (i.e., when the software program is running). Consider ReimageIT, the firm that can grade your Excel homework immediately after you drop the file into the application (Chapter 10). What happens behind the scenes is that the file is stored in an AWS S3 bucket (a storage service). The writing of the file triggers an AWS Lambda Function (a compute service) that unpacks the file and writes the source code into another S3 bucket. Such writing triggers the reading of the exercise solution (a key file) and then triggers another AWS Lambda Function that compares the two and writes the results into MongoDB (a storage service). The application's results page reads results from

1 Arkenberg, C., and Essing, N. H. 2020, November 20. "The intelligent edge is here." *WSJ CIO Journal*, retrieved from https://www2.deloitte.com/us/en/insights/industry/technology/intelligent-edge-network.html.

the MongoDB service and, as soon as those have been written, displays them to the user. The whole process, barring any errors, takes a few seconds. The key is to realize that this application is not written as a complete software program but rather as a network of digital resources that are activated as needed.

The main advantage of the serverless computing architecture is its flexibility and efficiency. Unlike any previous cloud service (e.g., IaaS), in serverless computing, there is no need for developers to manage any aspects of the infrastructure. This creates efficiencies, as they are free to spend all their time on coding the functionality of their software application instead. More specifically, there is no need for provisioning, configuring, scaling, and maintaining servers to run the computing and storage services. It is the cloud provider who takes care of all those requirements and charges a pay-per-use fee (Chapter 5) for its service. Conceptually, from the developer's standpoint, there are no servers—hence the label *serverless computing*. Servers are, of course, still running, but their management is fully transparent to the client, who is free to work as if they did not exist.

Consider the ReimageIT example one last time. When the grading of a student's file is completed, the AWS Lambda Function becomes dormant, waiting for the next trigger. Moreover, being highly scalable and based on a pay-per-use model, as the deadline for the homework assignment approaches and all students drop in files, the FaaS provider automatically scales the service by triggering as many AWS Lambda Functions as there are files, allowing uninterrupted service irrespective of the load (i.e., the number of students submitting files at the same time).

As a manager, you immediately realize that this shift in computing architecture requires a parallel shift in the skills and competencies of the IS professionals who design and develop those applications. From a strategic standpoint, you also realize that the serverless architecture may increase the dependency from the cloud provider even more than traditional cloud-native applications. Once multiple functionalities are implemented and benefit from the native integration functionalities made available by the cloud provider, migration to other vendors proves increasingly difficult or requires an architectural overhaul.

Quantum Computing

The interesting part of **quantum computing** is that, by definition, it's not a digital innovation! Quantum computing uses properties of matter at the quantum level (superposition and entanglement) to solve specific classes of problems more efficiently than the current digital technology (Figure 12.15).[1]

Relying on the qubit (quantum bit) instead of the bit, quantum computers can explore many solutions of an algorithm all at once. While a "traditional" bit can assume only two states to encode information—0 or 1 (Chapter 1)—qubit may maintain a state of superposition where both states are simultaneously possible. This particular characteristic allows quantum computers to represent all possible input variable values at once, thus solving problems in a single step. And the more qubits are available, the more input variables and complex problems can be solved in the quantum environment. At the time of writing, the most powerful quantum computers use around 64 qubits, with relatively high error rates and short exploitation times ready for computation. This is still a bit off if you consider that quantum computers should not be ready for business before they can use 1,000 qubits. But the technology is progressing fast, and by some calculations, processing

1 Here "more efficiently" means orders of magnitude faster, to levels that would nullify, for example, the effectiveness of current encryption algorithms that cannot be cracked by brute force (Chapter 13).

power is growing "doubly exponentially" every year (2^{2^1}, 2^{2^2}, 2^{2^3} . . . 2^{2^n}).[1] For comparison, consider that Moore's law grows with a single exponential 2^n (Chapter 1). At this rate, in five years, quantum computers should find an application in several domains. Among them are the following:[2]

Figure 12.15. Google's Sycamore processor

Source: Photo by Ceselin, R., retrieved from https://phys.org/news/2020-08-google-largest-chemical-simulation-quantum.html

1. *Security.* Quantum properties make data inherently secure as safely coping, or eavesdropping, requires violating the laws of physics. While current systems rely on the complexity of the computation to prevent hackers from wreaking havoc, quantum encryption is based on the laws of physics.
2. *Simulation.* Quantum computers provide better results in predicting molecular systems and transition states, facilitating drug developments. Both qubits and molecules operate on the same "hardware"—the laws of physics—resulting in more efficient, and faster, computations.
3. *Optimization.* The probabilistic nature of quantum computers makes them best suited to solve traditionally computationally intensive problems where multiple combinations of variables are used to identify the best solutions. For example, for solving problems of the "traveling salesman" class, for which complexity grows exponentially, digital computers reach their limit with just 25 cities (with 25 cities, the number of possible routes to explore is 3.1×10^{23}).[3] Not so with quantum computers because they can leverage superposition and algorithms requiring fewer operations to solve the problem than digital computers.[4]

While extremely promising, developments are still in their infancy, and modeling problems for quantum solutions is challenging for most organizations. Traditional infrastructures, practices, and application development tools cannot be applied in the quantum paradigm. Furthermore, handling the errors to which current quantum computers are prone remains unsolved, and multiple, competitive solutions are under study. Yet the proliferation of "quantum supremacy" claims—where quantum computers outperform digital supercomputers—fuel continued optimism. Consider, however, that within the span of your career, quantum computing will not replace digital computers or current architectures—rather, they will complement them in specific applications.

---

1 Lucero, E. 2019, June 18. "Does Neven's law describe quantum computing's rise?" *Quanta Magazine*, retrieved from https://www.quantamagazine.org/does-nevens-law-describe-quantum-computings-rise-20190618/.

2 Ghose, S. 2020, September 17. "Are you ready for the quantum computing revolution?" *Harvard Business Review*, retrieved from https://hbr.org/2020/09/are-you-ready-for-the-quantum-computing-revolution.

3 The number of cities is calculated as $(n-1)!/2$, where n is the number of cities to visit. Stęchły, M. 2019, March 13. "Solving the traveling salesman problem using quantum computer." Musty Thoughts, retrieved from https://www.mustythoughts.com/solving_tsp_with_qc.html.

4 Aspuru-Guzik, A. 2018, September 14. "Algorithms for quantum computers." *Scientific American*, retrieved from https://www.scientificamerican.com/article/algorithms-for-quantum-computers/.

Summary

In this chapter, we introduced emerging and enduring trends in IS and technology management. Understanding these trends and technologies, the associated vocabulary, and the benefits and risks they engender for modern organizations is critical for you as a general or functional manager, as you will be called upon to participate in the debate about whether your firm should embark on initiatives that leverage the technologies and trends discussed in this chapter.

Specifically, in this chapter, we learned the following:

- The widespread adoption of information technology and the increasing computer mediation of organizational and social processes has created the possibility of utilizing data that are born digital. The DDG trend creates this opportunity. The sensors introduced in objects like the smartphone are an illustrative example of this trend. The continuous digital encoding and transmission of data related to the captured events generate DDSs.
- DTs are digital representations of a unique physical object implemented as a software program capable of mirroring, in the virtual space, an object's useful characteristics and properties. By leveraging DDS and simulations, DTs allow firms to acquire greater visibility, situational awareness, enhanced predictions, and learning. When models and simulation environments are sufficiently detailed to allows precise predictions, DTs may substitute for current forms of prototyping and tests.
- Virtual and augmented realities are increasingly representative classes of technologies capable of immersing the user in digital environments. These technologies are bridging the separation between the artificial and the real world with the aim of augmenting the sensorial experience, thus enabling new interaction models.
- With digital manufacturing, it is possible to directly print objects from their digital representation and design. The flexibility of the process and 3D printers' increasing capabilities generate the opportunity of a new breed of products and services, disrupting current product manufacturing practices.
- Advanced analytics move the interest of data analysis and gathering to external sources, providing the opportunity for greater insight. The heterogeneity of the data, their massive volume, and greater speed challenge established analysis practices, skills, and technologies.
- Machine learning, and deep learning in particular, allows algorithms to identify occurrences and unknown patterns in data. These algorithms can be trained to look for the same occurrences in new data sets. It's like having an analyst capable of looking for trends within amounts of data that would be otherwise unbearable for humans. This opens a new category of applications that can derive reliable predictions from data. Instead of causality, machine learning derives empirical models fitting the available data.
- The renewed interest in AI, fueled by the success of machine-learning-based applications, is revamping the debate on artificial cognition. While still far away from sentient machines capable of fooling humans by passing the Turing test, machine-learning technologies are getting increasingly embedded in information systems and applied to a growing array of problems. Their success has raised ethical concerns, and it is increasingly requested that AI systems be explainable, meaning the decision path an AI takes for a decision needs to be interpretable by humans (e.g., which factors weighed more in the decision).
- Blockchain-distributed ledgers gave birth to a flourishing ecosystem of applications leveraging characteristics of distributed ownership, built-in validation, and transparency. Bitcoin is the most notable example of cryptocurrencies based on blockchain technology, in which the absence of centralized authority is challenging the role of the banking system as the trustee for monetary exchanges.
- 5G networks are based on the fifth-generation wireless network standard comprising a single framework for all the usage that today is covered by a plethora of different protocols and private solutions. 5G deployment will facilitate interoperability, standardization, and digital innovation thanks to its capacity to support faster, more reliable, and ubiquitous connectivity.
- Serverless computing is a cloud-based model where application functions are run in the cloud and the cloud provider dynamically manages the allocation of digital resources. Pricing is determined by the effective execution time of the application function, reducing capacity waste.
- Quantum computing represents a new computing paradigm based on totally different technologies and assumptions compared to digital computing. While quantum computers will not replace current systems,

their development proceeds faster than Moore's law. Quantum computers are already apt at solving complex optimization problems faster than traditional architecture. Quantum encryption is the sought-after solution for data security, as "breaking the encryption" will equal breaking the laws of physics.

Study Questions

1. Identify two DDSs that you believe are relevant in a sport and suggest an IT-dependent strategic initiative based on both of them.
2. What is a DT? Which are the fundamental components of a DT?
3. What is VR? How is VR different from AR?
4. What is digital manufacturing? How is digital manufacturing set to transform current offerings, operations, and competition?
5. What is machine learning? What is deep learning? Discuss how they are different from traditional statistics methods.
6. What is AI? How is machine learning different from human learning?
7. Should all AIs be explainable? Define the concept of an explainable AI and justify your answer.
8. Which machine-learning approach would you use if you had to build a system to predict consumers' credit risk and you had past knowledge of the transactions and the delinquency rate on loans?
9. What is the difference between data mining and machine learning?
10. What is a blockchain? What are the main properties of blockchains?
11. What are the connectivity requirements at the base of 5G networks?
12. Define serverless computing and discuss its main characteristics and challenges.
13. Will quantum computers become a substitute for their current digital counterparts? Justify your answer.

Glossary

A glossary of key terms can be found at the book's website. Visit https://www.prospectpressvt.com/textbooks/piccoli-information-systems-for-managers-5-0, then click on "Student Resources" in the horizontal red menu bar.

Chapter 13

Cybersecurity, Privacy, and Ethics

What You Will Learn in This Chapter

This chapter discusses some important topics of managerial interest that are often delegated to information technology (IT) specialists: security and **digital risk management**, **privacy**, and information systems (IS) ethics. Our first objective is to convince you that, as future general and functional managers, you will have to be involved in these decisions. The second objective is to help you gain an understanding of the circumstances in which choices and trade-offs are made so that you can actively participate in decision making.

MINICASE: Reinventravel.com Comes under Fire

As you watch the sun setting over the San Francisco skyline from your hotel room window, you can't avoid feeling that you really dropped the ball this time. You can still hear Clive Sturling, your chief information officer (CIO), as he tells you, "Don't worry about security. That's techie stuff; I'll take care of it. Just grow the business. That's what you are good at." You had not asked about security again after that conversation, perfectly happy to leave the "techie stuff" to him, and that was before you launched the company over two years ago!

Well, it was him on the phone a minute ago, ruining what had been a perfectly good day. In a daze, you replay the conversation in your mind: "We have been attacked," Clive had said. "It was a distributed denial-of-service (DDoS) attack—not much we could do with our current security infrastructure. The site was unavailable for about 70 minutes; it wasn't defaced or otherwise ruined, just down. I don't think many people noticed. The attack ended about an hour ago. I didn't want to call you before checking if they had compromised any files or stolen customers' data. It doesn't look like it."

Not much we could do? Isn't he the one who said not to worry about security? The site was down for "only 70 minutes." Does he know that in that amount of time, Reinventravel.com typically processed 19,000 transactions? Granted, evenings were a bit slower, but there must have been at least 4,500 customers who noted the outage. Your emotions kept mixing at a dizzying pace. You were angry at Clive; you trusted him, and he let you down. However, you felt sympathetic to his position as well. You had been the one who told him to "run IT on a shoestring" to help you speed the path to profitability as much as possible.

Oddly enough, as you begin to recover from the shock of the news, your college days flash into your mind, bringing a smile to your face. You had started in this field only three and a half years before, when you learned in one of your classes about the opportunity to revolutionize how people seek and purchase travel products. That day in your information systems class seemed like decades ago; now you were the chief executive officer (CEO) of a growing company with 52 employees, over 70,000 active customers and members, and revenues approaching eight million dollars. Clive had built the search engine in just eight months alone! He was a wizard with that kind of stuff. Half the time, you had no idea what he was doing, but as for the user interface, you certainly appreciated and understood that part of his work; everyone did! It was so far superior to anything that had been seen before . . . it was that fabulous demo that got you your first round of venture capital financing.

Financing . . . that word snapped you back to reality! You had to get ready for dinner. The meeting with your venture capitalists was in less than an hour, and you had yet to take a shower. With the first round of financing beginning to run out and minimal profits, a second round was a must. You had hoped to spend the evening discussing your plan for growing the customer base and beginning to monetize your membership, seeking their guidance and help with regard to the three potential partners you were evaluating. "Well, that ain't going to happen," you mumble.

What should you do? Should you tell your venture capitalists about the **denial-of-service attack**? It may not be your choice; these guys liked to do their homework, and the odds were good that they were poking around the site when the outage happened. No time to call your legal counsel; you had to go it alone on this one.

Clive had been very unclear about whether an **intrusion** had occurred along with the denial-of-service attack. At this point, you had little faith with regard to his staff's ability to find out; it seems that security and monitoring had not been ranking very high on their priority list! Reinventravel.com stored quite a bit of personal information about customers, including identifying information and credit card data. Should you communicate to the customers that an attack had occurred? Should you issue a press release? There was no evidence that security had been compromised and even less that **personal data** had been stolen. A denial-of-service attack only made a website unavailable for some time, did it not? "No way, Clive and his staff would know if data had been stolen," you told yourself.

This was increasingly looking like a situation you were ill prepared to address. But as your father always said, "You wanted the bicycle—now you have to pedal." As you begin to feel the adrenaline pumping again, you exclaim, "Here we go!" and jump up from your chair. You had 55 minutes to develop your plan before dinner.

Discussion Questions

1. Do you agree with the assessment that you had dropped the ball? Or are you being unduly harsh on yourself?
2. Who do you think should be making security calls at Reinventravel.com? Shouldn't this be the CIO's job?
3. What should you do tonight? Should you approach the topic at dinner or wait and see if anyone else raises the issue?
4. What should you do in the next few days? Should you issue a press release? Should you contact your customers directly? Should you focus on overhauling your security safeguards to prevent future similar problems and forget today's incident?

13.1 Introduction

This chapter focuses on three topics: cybersecurity and digital risk management, privacy, and IS ethics. These areas, while distinct, are connected by a common thread. Cybersecurity, privacy, and ethical concerns were born along with the introduction of computer systems and IT in organizations. However, digital transformation (Chapter 5), the widespread adoption of cloud solutions (Chapter 3), and the proliferation of digital business have dramatically amplified these threats. The digital security industry, for example, is estimated to be already in the $100 billion range, with Gartner research valuing worldwide security spending at about $123 billion in 2020. In 2020, cybersecurity was the main priority among a majority of organizations.[1]

Security breaches have significant business impact. PwC reports that 53% of securities exchanges have experienced cyberattacks. During the 2020 COVID-19 pandemic, cyberattacks multiplied, exploiting a context where **hackers** could leverage "existing attack tactics, techniques and procedures to exploit new opportunities created by a massive increase of employees working from home, children using home computers for schooling, as well as the human factor and emotions caused by the pandemic."[2] The economic and political relevance of digital security is well illustrated by the words of FBI director Christopher Wray, who declared in a recent speech,

1 Scavo, F., Wagner, D., Dunlap, T., Newton, B., Scavo, J., Agrawal, N., and Gupta, A. 2017. "IT spending and staffing benchmarks 2020/2021." *Computer Economics*, retrieved from https://www.computereconomics.com/page.cfm?name=it-spending-and-staffing-study.
2 ECHO. 2020. "The COVID-19 hackers mind-set" (no. 1; ECHO White Paper, p. 8). Retrieved from https://echonetwork.eu/wp-content/uploads/2020/04/20200408-ECHO-WhitePaper-Hackers-Mindset-FINAL.pdf.

"We've now reached the point where the FBI is opening a new China-related counterintelligence case about every 10 hours."[1]

A failure in security, privacy, or ethics can have dramatic repercussions on an organization, because of both its potentially damaging direct effects (e.g., computer outages, disruptions to operations) and its increasingly negative indirect effects (e.g., legal recourse, image damage). Consider the following three examples:

The Largest Ransomware Attack On May 12, 2017, WannaCry **ransomware** took the world by surprise, infecting hundreds of thousands of computers in banks, hospitals, and other organizations. The attack took place in just a matter of days and caused damages estimated to be in the billions of dollars.[2] In the United Kingdom, hospitals had to turn away noncritical patients, 19,000 appointments were canceled, and staff had to revert back to paper and pencils as the main information technology.[3] The car manufacturer Renault saw its production disrupted when at least five factories were compromised by WannaCry (Figure 13.1).

The WannaCry ransomware attack made use of "EternalBlue," a national security agency (NSA)-developed exploit that leveraged a vulnerability in the Windows 7 file-sharing system. That is, a 2017 attack exploited outdated technology that was still in use in critical systems in organizations handling very sensitive data, such as hospitals. The exploit was leaked by hackers who had compromised some of NSA's **malware** toolset just two months before this incident. The WannaCry ransomware encrypted victim's system data and requested a monetary ransom in Bitcoin (Chapter 12) to unlock them. Interestingly, perpetrators made relatively little money from the attack (estimated at less than $200,000). But the WannaCry attack was the latest of many wake-up calls to both private businesses and policy makers.

Figure 13.1. The WannaCry ransom request

The Largest Identity Theft On September 7, 2017, Equifax, one of the three major U.S. credit agencies, publicly disclosed that its database of customer data had been compromised months before, in mid-May.[4] The firm discovered the breach on July 29, but it took six weeks to report the incident. Stolen Equifax data included names, birth dates, addresses, and social security numbers of 147 million people—almost half the population of the United States. Additionally, perpetrators were able to take credit card and driver's license numbers of hundreds of thousands

1 "The Threat Posed by the Chinese Government and the Chinese Communist Party to the Economic and National Security of the United States." 2020, July 7. Speech. Federal Bureau of Investigation, retrieved from https://www.fbi.gov/news/speeches/the-threat-posed-by-the-chinese-government-and-the-chinese-communist-party-to-the-economic-and-national-security-of-the-united-states.

2 Berr, J. 2017, May 16. "'WannaCry' ransomware attack losses could reach $4 billion." CBS News, retrieved from https://www.cbsnews.com/news/wannacry-ransomware-attacks-wannacry-virus-losses/.

3 Neville, S. 2017, October 26. "NHS cyber attack far more extensive than thought, says report." *Financial Times*, retrieved from https://www.ft.com/content/4110069a-ba3d-11e7-8c12-5661783e5589.

4 If it weren't so tragic in its outcomes and the countless lives upended by the data breach, the long list of tragicomic errors in the Equifax debacle would be funny. In fact, there is a very funny John Oliver segment on the topic (note, it is late-night cable TV with language to match. Don't watch if you are easily offended by foul language): https://youtu.be/mPjgRKW_Jmk.

of people. Hackers accessed Equifax databases through an unpatched software vulnerability the firm was aware of in March, two months before the actual breach. Both human negligence and a problem with the security audit software prevented the firm from fixing the vulnerability before hackers first accessed sensitive information on May 13. This allowed perpetrators to exfiltrate terabytes of data through July, when the renewal of an expired security certificate revealed the hack. In August, and ahead of the breach's public announcement in September, four Equifax executives sold $1.8 million of the firm's stock—an action that resulted in a charge for insider trading.[1]

Beside the expected share drop—14% the day after the breach was disclosed—the breach directly impacted Equifax's bottom line for the last quarter of 2017. Service renewals and subscriptions dropped, affecting the company's total revenues by 2%. At the same time, Equifax had to face legal, cybersecurity, and related services costs of $87.5 million, and the final settlement included $425 million to support the people affected by the data breach.[2]

The First Carjacking If you believed that at least your car you would be safe from digital security threats, you should think again. In July 2015, hackers Chris Valasek and Charlie Miller caught the automotive industry by surprise when they wirelessly carjacked a 2014 Jeep Cherokee on the highway from 10 miles away.[3] This pressed Chrysler to issue the first "cybersecurity recall" for 1.4 million vehicles, as the vulnerabilities affected several other models, and to issue a software patch (Figure 13.2). It also led Sprint—the provider of the cars' connectivity—to block the vulnerable ports. Interestingly, the case showed thc complete absence of precedents, pushing both the industry and the government to action.[4] Since then, the reports of hacking activity targeting car manufacturers multiplied. A white-hat hacker took control of the entire Tesla fleet in 2017, allowing him to arbitrarily "summon" any car.[5] In 2020, other hacks involved the car's screen, the keyless entry system, and the autopilot, fooling the car into a sudden stop by flashing a stop signal from a digital billboard.[6]

Figure 13.2. Chris Valasek's tweet after the Jeep Cherokee was patched

A New Attack Vector—Digital Resources In a world where systems are increasingly built by leveraging servitized digital resources (Chapter 5), firms are exposed to the risk of incorporating third-party vulnerabilities. The recent SolarWinds case provides an early example. SolarWinds is a Texas-based IT firm whose Orion system was used by more than 33,000 customers to monitor and

1 Moyer, L. 2018, March 14. "Former Equifax executive charged with insider trading for dumping nearly $1 million in stock ahead of data breach." CNBC, retrieved from https://www.cnbc.com/2018/03/14/former-equifax-executive-charged-with-insider-trading-ahead-of-data-breach.html.

2 "Equifax data breach settlement." 2019, July 11. Federal Trade Commission, retrieved from https://www.ftc.gov/enforcement/cases-proceedings/refunds/equifax-data-breach-settlement.

3 Greenberg, A. 2015, July 21. "Hackers remotely kill a Jeep on the highway—with me in it." *Wired*, retrieved from http://www.wired.com/2015/07/hackers-remotely-kill-jeep-highway/.

4 Greenberg, A. 2015, July 21. "Senate bill seeks standards for cars' defenses from hackers." *Wired*, retrieved from http://www.wired.com/2015/07/senate-bill-seeks-standards-cars-defenses-hackers/.

5 The label "white-hat" refers to computer hackers who focus their skills on uncovering vulnerabilities in software to communicate them to the software developers to fix. In many cases, when doing so, the hackers collect a bounty (in this case, $5,000 from Tesla).

6 Lambert, F. 2020, August 20. "The big Tesla hack: A hacker gained control over the entire fleet, but fortunately he's a good guy." Electrek, retrieved from https://electrek.co/2020/08/27/tesla-hack-control-over-entire-fleet/; Greenberg, A. 2020, November 23. "This Bluetooth attack can steal a Tesla Model X in minutes." *Wired*, retrieved from https://www.wired.com/story/tesla-model-x-hack-bluetooth/.

manage IT infrastructure. Attackers breached SolarWinds' systems, compromising their software update suite and modifying the code to install a **backdoor**—obviously termed *Sunburst*. Once SolarWinds' customers, in compliance with their security policies, installed the updates to Orion, they introduced the backdoor in their networks. The breach remained undetected for months, potentially from March 2020, leaving perpetrators plenty of time to access and exfiltrate data from the compromised client networks. Up to 18,000 customers, including high-profile business and governmental agencies, were exposed. Around 100 private sector firms suffered from the breach. Among them were tech giants Microsoft, Cisco, Intel, and Nvidia, consulting firm Deloitte, cloud-computing software maker VMware, and routers manufacturer Belkin. U.S. government agencies were the clear objective of this attack, likely sponsored by the Russian military. It directly hit the Pentagon, the Department of Homeland Security, the State Department, the Department of Energy, the National Nuclear Security Administration, the Internal Revenue Service, and the Treasury.[1] To use the words of Anne Neuberger, deputy national security advisor for cyber and emerging technology, "The techniques that were used lead us to believe that any files of emails on a compromised network were likely to be compromised."[2] Even more alarming, "many of the private sector compromises are technology companies, including networks of companies whose products could be used to launch additional intrusions." While at the time of writing, the story is still unfolding, SolarWinds showed the critical role of digital resources and private and public networks, painfully reminding us that software "supply chains" are interdependent and cybersecurity in an interconnected world becomes an interorganizational issue.

Cybersecurity, privacy, and ethics are areas where, as managers, you cannot abdicate your responsibility. Yet in order to actively participate in decision making on these three fronts, you must be able to understand under what circumstances choices and trade-offs are made and what the principal threats and responses are.

13.2 Digital Risk Management and Cybersecurity

IS security, or *cybersecurity*, as it is generally called, refers to the set of defenses an organization puts in place to mitigate threats to its technology infrastructure and digital assets. Digital risk management is the process by which the firm attempts to identify and measure digital security risks and devise the optimal mitigation strategy.

Cybersecurity is an area that has increased in importance along with the widespread adoption of digital technology and even more so with the increasing connectivity. As software continues to "eat the world" (Chapter 1), software vulnerabilities become increasingly important. The development of the mobile and Internet of things (IoT) ecosystems are bringing new challenges at both the individual level (e.g., personal account hijacking) and the organizational level (e.g., hacks of executives' smartphones or smart devices). Cybersecurity and digital risk management have come to the forefront of managerial attention because of the increasing threat of cybercrime, cyberterrorism, and cyberwarfare. As digital systems are increasingly underpinning the infrastructure of developed economies and digital assets and capabilities become essential factors of production, they become primary targets for criminals, terrorists, and state and industry spies.

---

1 Jibilian, I., and Canales, K. 2021, February 25. "Here's a simple explanation of how the massive SolarWinds hack happened and why it's such a big deal." *Business Insider France*, retrieved from https://www.businessinsider.fr/us/solarwinds-hack-explained-government-agencies-cyber-security-2020-12.

2 "Press briefing by Press Secretary Jen Psaki and Deputy National Security Advisor for Cyber and Emerging Technology Anne Neuberger, February 17, 2021." 2021, February 17. White House, retrieved from https://www.whitehouse.gov/briefing-room/press-briefings/2021/02/17/press-briefing-by-press-secretary-jen-psaki-and-deputy-national-security-advisor-for-cyber-and-emerging-technology-anne-neuberger-february-17-2021/.

Why Is Cybersecurity *Not* Just an IT Problem?

The pervasiveness and possible cost of cybersecurity threats should suffice to convince managers that cybersecurity is a matter of strategic interest, not something that "the IT people should worry about." A 2020 Ernst & Young survey found that while 92% of company boards of directors are involved in cybersecurity, in 59% of the surveyed organizations, the relationship between the cybersecurity team and business areas are at best neutral—when not outright based in mistrust.[1]

Cybersecurity should be on managers' radar not only because of its cost but also because of its peculiar characteristics that run the risk of leaving it underfunded. The game of chess offers a great metaphor for the information security management and digital risk management processes. In the game of chess, the objective of the players is to circumvent the defenses of the opponent in order to checkmate him or her. Cybersecurity is a constantly evolving game of chess—one where current defenses, and their limitations, are the basis for future attacks. But what do you get if you win the security chess game? Nothing tangible or immediate. Very few companies have been able to parlay their cybersecurity prowess into financial results, and certainly not quickly. For most firms, being successful in cybersecurity simply means . . . nothing happening. As in the opening minicase, all the "excitement" occurs when your security has been breached. We are not suggesting that "nothing happening" is a bad outcome. Much to the contrary, what we are showing you is why cybersecurity so often goes underfunded.

More specifically, cybersecurity is a negative deliverable. All the money spent on managing digital risk and the firm's digital infrastructure and assets produces no revenue and creates no efficiencies. It does not yield a positive return on investment (ROI). As a consequence, historically, it was difficult to ensure funding for cybersecurity efforts, a tendency nicely captured by Gene Spafford, a professor of computer science and cybersecurity expert: "People in general are not interested in paying extra for increased safety. At the beginning, seat belts cost $200 and nobody bought them."[2] This trend is reversing these days. The seemingly endless stream of cybersecurity attacks and breaches is finally convincing business and government leaders that serious cybersecurity threats are indeed serious, almost existential.

Another reason security is a negative deliverable is that it is difficult to take credit for doing a great job when all you have to show for your efforts is that nothing bad has happened. This is particularly true when "lucky" firms around you also have not suffered an attack (or have yet to notice that one has taken place!) and skeptics in your organization can point to them as "proof" that you are overinvesting in cybersecurity. The paradox is that while 48% of firms believe that their business will be more than moderately affected by data breaches, for chief information and security officers (CISOs), the greatest challenge remains that of proving the value of their activities and securing an adequate budget.[3]

Because security is the type of investment that is difficult to gain funding for, particularly when competing for limited resources with projects that promise big results (efficiency improvements, revenue enhancements, and the like), it is all the more critical that it is not left to the cybersecurity team to make the case. Managers must get involved in the cybersecurity discussion, understand the threats, and assess the degree of risk that the firm should be allowed to take. It is your responsibility as a manager to weigh and balance the decisions concerning purchasing more

---

1 Ernst & Young. 2020. "EY global information security survey." Ernst & Young, retrieved from https://assets.ey.com/content/dam/ey-sites/ey-com/en_gl/topics/advisory/ey-global-information-security-survey-2020-single-pages.pdf: 26.
2 Sullivan, B. 1999, March 30. "Is Microsoft to blame for Melissa?" ZDNet, retrieved from http://www.zdnet.com/article/is-microsoft-to-blame-for-melissa/.
3 Ernst & Young. 2020. "EY global information security survey." Ernst & Young, retrieved from https://assets.ey.com/content/dam/ey-sites/ey-com/en_gl/topics/advisory/ey-global-information-security-survey-2020-single-pages.pdf: 26.

cybersecurity or accepting higher risks. For some organizations, cybersecurity and privacy offer opportunities to innovate and differentiate their offerings. Consider Apple. The firm emphasizes its ability to protect customer data from hackers and governments. For Apple, the security features of their product and services represents a distinctive value-adding feature.

As future managers, you should participate in the cybersecurity assessments and decisions in your organizations. While when dealing with risk management, you need to understand the basic threats and fundamental trade-offs engendered by cybersecurity, you should be ready to consider the opportunities where security and privacy can also directly contribute to value creation (Chapter 8). This does not mean that you must develop an extraordinary amount of technical knowledge or became an expert in cybersecurity, as you will not be called on to personally implement the security measures. Instead, you must understand the managerial process of risk management.

Risk Assessment

The **risk assessment** process consists of auditing the current resources, technological as well as human, in an effort to map the state of the art of IS security in the organization. An understanding of the current resources will provide an idea of the current set of vulnerabilities the firm is facing.

For instance, Amazon's website is not only the "face" of the company; it is also its main source of revenue. If customers cannot access it, Amazon loses hundreds of thousand dollars by the minute. Thus for Amazon, the risks associated with a denial-of-service attack that brings the website down is a very tangible one. The same could not be said for the Boatyard Grill, a restaurant in Ithaca, New York, and its website (http://www.boatyardgrill.com).

The risk audit is useful because it provides the basis for a **risk analysis**. Risk analysis is the process by which the firm attempts to quantify the hazards identified in the audit. We use the word *attempt* to stress that precisely quantifying the monetary consequences of some of these risks is impossible. What do you think is the value of the loss of customer confidence in Equifax after the hack? What is the value of the loss of confidence your customers may feel if they visit your website or Twitter account only to see it defaced with questionable pictures or comments (Figure 13.3)?

The impact of some cybersecurity risks is harder to measure than others. However, the exercise is useful insomuch as rational decision making suggests that the amount you invest in cybersecurity safeguards should be proportional to the extent of the threat and its potential negative effects. This is a critical point. Because cybersecurity risks are really business risks—nobody would argue that the loss of customer confidence is "an IT issue"—managers must be deeply involved in digital security prioritization decisions. The SolarWinds case provides an additional example in support of this point. As firms integrate external servitized digital resources in their services, they increasingly rely on the "digital supply chain" to create and maintain digital relationships (Chapter 5). While they use external code to create those relationships, they remain ultimately responsible toward their customers. The direct consequence is a growing size of the exposure surface to cybersecurity threats, with decreasing direct control over execution. As you can see, these trade-offs are not trivial, and they can't be fully offloaded to IS professionals but rather require the knowledgeable input of savvy general and functional managers.

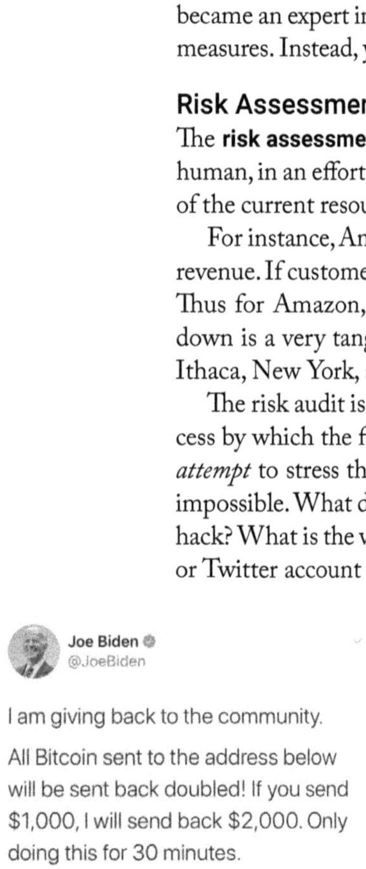

Figure 13.3. The now U.S. president Joe Biden's Twitter account compromised on July 15, 2020

Risk Mitigation

Risk mitigation is the process of matching the appropriate response to the cybersecurity threats your firm has identified. As Bruce Schneier, the noted computer security and cryptography expert, aptly put it, "There are two types of **encryption**: one that will prevent your sister from reading your diary and one that will prevent your government."[1]

Risk mitigation allows your organization to devise the optimal strategy given the set of cybersecurity risks it faces. Such optimal strategies are ones that yield the best trade-off between the degree of security the firm attains and the total investment in countermeasures necessary to achieve it (Figure 13.4). The total cost for cybersecurity is a combination of anticipation costs, those expenditures designed to anticipate and mitigate the threats (e.g., purchasing and installing vulnerability detection software), and failure costs, the negative financial fallout ensuing from a breach of cybersecurity (e.g., loss of revenue during a website outage).

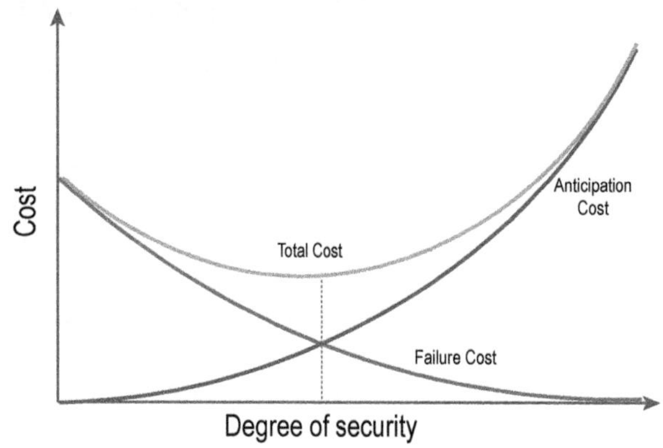

Figure 13.4. Cost/security trade-offs

While we don't dig deeper into this matter, you should remember that risk can be estimated and, at least roughly, quantified.[2] Intuitively, you may calculate the ROI of cybersecurity expenditures through standard risk formulas: for example, the expected ROI of a cybersecurity initiative as a function of the value of the asset, the exposure factor as a percentage of the loss of an asset, the cost of the security safeguard, the estimated cost per incident, and the number of incidents expected per period. You may consider probabilistic methods, where continuous and measurable mitigation efforts reduce the probability of negative events. When faced with a cybersecurity threat, the firm has three mitigation strategies available. Note than none of the three strategies described below is superior or inferior to the other in the absolute. The typical organization uses, consciously or unconsciously, a blend of all three.

1. *Risk acceptance.* This strategy consists of not investing in countermeasures and not reducing the security risk. The more an organization gravitates toward this strategy, the higher the potential failure cost it faces while minimizing anticipation costs.
2. *Risk reduction.* This strategy consists of actively investing in the safeguards designed to mitigate security threats. The more an organization gravitates toward this strategy, the higher the anticipation cost it faces while actively reducing failure costs.
3. *Risk transference.* This strategy consists of passing a portion (or all) of the risks associated with cybersecurity to a third party (e.g., by outsourcing security or buying insurance).

As the firm seeks to identify the optimal risk management strategy, it will endeavor to identify the optimal blend of the three mitigation strategies. Excessive security can negatively impact a firm's competitive advantage. If security processes and systems are too rigid, they will reduce the firm's flexibility and ability to adapt quickly, choke the execution of regular business activities, and negatively influence innovation. The goal is to achieve "security as a business enabler" rather than security that will get in the way of business progress. The ideal portfolio of cybersecurity and risk management measures is based on the specific security threats the organization faces as well as management's willingness to accept these risks. The major threats confronting the modern organization, and the safeguards available to respond to these threats, are discussed below.

1 Schneier, B. 1996. *Applied cryptography: Protocols, algorithms, and source code in C.* Somerset, NJ: John Wiley & Sons.
2 Hubbard, D. W., and Seiersen, R. 2016. *How to measure anything in cybersecurity risk.* Hoboken, NJ: Wiley: 280.

13.3 The Cybersecurity Threats

The security of digital resources and infrastructures is challenged both internally and externally (Table 13.1).

Internal Cybersecurity Threats These are threats posed by individuals who have direct, on-premises access to the firm's technology infrastructure or those who have legitimate reasons to be using the firm's assets. Internal cybersecurity threats are important because the firm that is able to secure its assets against improper internal use not only has mitigated an important risk but is well on its way to mitigating the outside threat. In their annual survey, the telecom operator Verizon found that for 30% of companies, employees were the sources of cybersecurity incidents.[1] When addressing internal security threats, we can separate them into two broad categories: intentional malicious behavior and careless behavior. Before the advent of the Internet and widespread connectivity, the importance of internal security threats far outweighed the danger posed by hackers and other outsiders. Mitigating the outside threat prior to pervasive networking simply amounted to physically securing the firm's digital assets.

External Cybersecurity Threats These threats today represent a major concern with geopolitical consequences. **Viruses, Trojan horses, worms**, time bombs, **spyware**, keyloggers, ransomware, spoofing, snooping, sniffers—these are just some of the most popular examples of malicious codes and techniques that digital organizations find themselves fighting off. Couple this seemingly unabated tide of new releases and new forms of harmful software with human threats like crackers, thieves, social engineers, and industrial espionage contractors and you realize why cybersecurity is continually ranked as one of the top executives' worries.

Those individuals who attack an organization's digital infrastructure are typically called hackers. Some consider this a misnomer, contending that the term *hacker* simply means someone who possesses superior computer skills. While ethical hacking may be perpetrated by "white-hat" hackers to determine the vulnerabilities of a computer system, the term has come to be associated in the media and general terminology with more or less maliciously intentioned individuals who attempt to subvert computer security defenses—so-called black-hat hackers.

In the table below, we address some of the main internal and external cybersecurity threats confronting current organizations.

Table 13.1. Internal and external security threats

Threat	Description
Internal threats	
Intentional malicious behavior	A threat typically posed by disgruntled or ill-willed employees that may exfiltrate data or compromise a firm's systems. These threats are difficult to detect because they are potentially coming from authorized users who can access the data and, at the same time, have a reason for leaking or tampering with them. Cybercriminals have resorted to extortion and bribing to recruit internal employees, even as part of nation-sponsored cyberattack programs.
Careless behavior	This type of threat is typically associated with individuals' ignorance of, or disinterest in, cybersecurity policies. Several different behaviors fall into this category, including failing to modify default passwords, breaking the organization's policy on Internet and web usage, not following guidelines about saving data on personal or portable devices, or failing to destroy sensitive data according to planned schedules.

(continued)

1 "2020 data breach investigation report." 2020. Verizon, retrieved from https://enterprise.verizon.com/resources/reports/2020-data-breach-investigations-report.pdf: 119.

Table 13.1. Internal and external security threats (*continued*)	
Threat	**Description**
External threats	
Intrusion	Threats consisting of the unauthorized access to an organization's digital infrastructure and resources.
Social engineering	The practice of obtaining restricted or private information by somehow convincing legitimate users, or other people who have the information, to share it. This is typically done online or by phone, arousing emotions (e.g., fear, panic, urgency) and leading the victim to reveal sensitive information promptly.
Phishing	A social engineering technique consisting of sending an official-sounding message (i.e., unwanted email) from known institutions (e.g., Mastercard) to a potential victim. The message indicates that the institution needs the recipient to confirm or provide some private data and contains a link to a fake web page, mimicking the original (e.g., logo and descriptions), with fields for providing the "missing" information, luring the victims into sharing access credentials, paying methods, and personal data.
Backdoors and exploits	Undocumented features or code weaknesses that can be leveraged by attackers for gaining control or compromising an organization's systems. Backdoors refer to default settings or undocumented features that can bypass the security controls. Exploits are errors, bugs, or weaknesses in a system's software that could be exploited by an adversary to access and control it.
Malicious code	Software especially created with the objective to compromise a system.
Virus	A type of malicious code that spreads by attaching itself to other legitimate, executable software programs. Once the legitimate software program runs, the virus runs with it, replicating itself and releasing its payload. The payload is the harmful set of actions that the virus is designed to perform. They may range from simply annoying the user to wreaking havoc and bringing significant damage to the user. Some viruses deliver their payload immediately after infection, while others, known as time bombs, deliver it at a specific point in time or when a certain action is performed.
Trojan horses	Malicious code that claims to partially deliver some useful functionality or that appears as a known software or app but that executes a payload similar to a virus. Instead of replicating like the former, trojan horses spread by the diffusion of the software of which they imitate the functionalities.
Worms	Malicious code that exploits cybersecurity holes in network software to replicate itself up to the organization's digital resources depletion. Strictly speaking, a worm does not deliver a payload like a virus. A worm simply replicates over and over and continues to scan the network for machines to infect.
Spyware	Spyware is software that runs without the user's awareness that data and information are collected or exfiltrated. Broadly speaking, spyware is software that, unbeknown to the owner of the computer, monitors behavior, collects information, and either transfers this information to a third party via the Internet or performs unwanted operations.
Ransomware	Ransomware represents a broad category of software that allows an attacker to limit users' access to digital resources and systems, demanding a ransom in exchange. Generally, it spreads as an innocuous attachment to email messages or through social engineering. When executed, the payload begins to encrypt users' files (e.g., Office documents, pictures, shared files) and then displays a warning requesting a ransom payment, generally in cryptocurrency, or else the decryption key will be destroyed.
Denial-of-service	A digital assault carried out over a computer network to overwhelm an online service so as to force it offline. To generate all these requests, attackers may exploit a *botnet*, an extensive network of compromised systems, and realize a DDoS. The multiplication of connected devices (e.g., routers, IP cameras, and digital video recorders) and scalable cloud-based digital resources is providing hackers with the opportunity to heighten the DDoS threat.
Online fake news	The targeted diffusion of false information that gains traction among the target audience (Figure 13.5). The ubiquitous access to social media has reduced the entry barrier to the business of disinformation and the speed with which fake news can spread. ***Troll factories*** are new organizations, either private or state-sponsored, specializing in the writing and posting of targeted fake news articles, online reviews, and comments. These messages seek to either discredit or promote a product, firm, or other aspects of an organization's reputation.

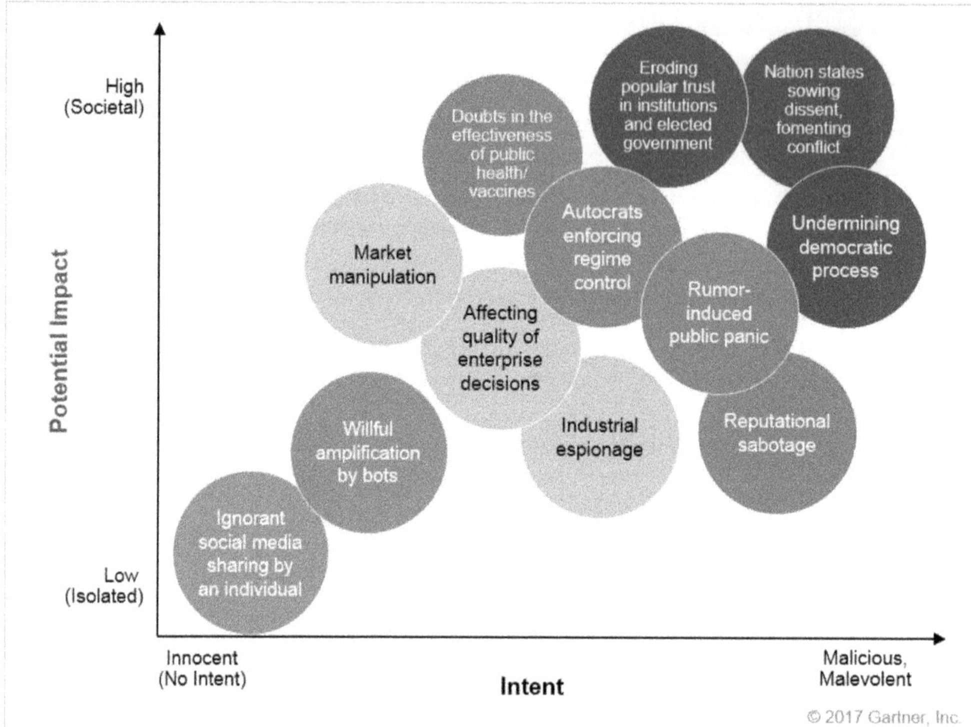

Figure 13.5. The different types of fake news attacks

Source: Fabre, S., and Holgate, R., 2017, "Building corporate resilience and restoring faith in facts in the fake news era" (no. G00334787), Gartner, retrieved from https://www.gartner.com/document/3808165

Mobile and IoT Cybersecurity Threats

Smartphones and IoT devices are full-fledged computers and are therefore subjected to the same cybersecurity threats that affect other digital technologies. Routers, smart watches, video game consoles, smart speakers, connected cars, thermostats, electric scooters, lightbulbs, toys, and doorbells are just some representatives from this growing class of devices.

Smartphones and IoT devices are now central in supporting firms' new digital initiatives, increasing the risks of being exposed to cybersecurity threats. This is a dangerous trend when considering the results of a recent study from cybersecurity firm Palo Alto Networks stating that 57% of IoT devices are vulnerable to attacks.[1] Most users do not seem to realize the severity of the cyberthreats involving IoT devices and smartphones. Because of ignorance and careless behavior, users of IoT and mobile devices are exposed to even higher risks compared to traditional desktop users. Mobile users store personal and sensitive information, such as photos, email, PIN codes for bank cards, passwords, all their contacts, and other sensitive information. Paradoxically, the device that contains your most personal and intimate data is likely to be the one you are worrying least about securing!

As these devices become more sophisticated, hackers are finding creative ways to break into them. Creativity is required, as mobile platforms and cloud services represent a paradigm shift

1 "The connected enterprise: IoT security report 2020." 2020. Palo Alto Networks, retrieved from https://www.paloaltonetworks.com/apps/pan/public/downloadResource?pagePath=/content/pan/en_US/resources/research/connected-enterprise-iot-security-report-2020: 6.

of software release and installation. Consider mobile users again—as you are surely well aware, they install software (i.e., apps) mainly through application stores where software is reviewed and approved by the platform owner before making it available for download. The review process should guarantee that the available software on the store does not contain malware or any violation of the terms and conditions. However, hackers proved several times that malicious code can be hidden from reviewers. This proved particularly true for the Android platform, as hundreds of applications in the Google Play Store turned out to be spyware collecting data about users' conversations, app information, photos, and locations. For example, in summer 2019, more than 1,300 Android apps were found to harvest personal data (i.e., location, email addresses, phone and IMEI numbers) without requiring user consent.[1]

Responding to Cybersecurity Threats

The management of IS security is a continuous effort. The principal objective is to identify the different threats and develop safeguards that match them and limit their incidence of success. This objective may be achieved in different ways, and different approaches to cybersecurity have emerged. **Industry and regulatory standards**, company guidelines, and organizational policies are all means for protecting digital assets. For example, Lockheed Martin advanced the Cyber Kill Chain framework based on the military-inspired "kill chain." The inspiring principle is to model how hackers operate when they perpetrate an attack to elaborate a strategy to prevent the attack. The life cycle of the attack is modeled around seven phases hackers must complete to achieve their objectives (see Figure 13.6). Interrupting a hacker action along the phases can serve to stop the attack. However, this approach requires organizations to commit a large amount of resources to effectively implement it. Each type of attack needs to be modeled, and specific countermeasures need to be implemented.

Even more worrisome, rarely do attackers precisely follow the playbook! They may focus on certain phases and neglect others and eventually step back. As a security expert recently put it, "The kill chain as invented by Lockheed Martin is malware focused, and that makes certain attacks invisible."[2]

The framework proposed by standardization bodies like the U.S. National Institute of Standards and Technology (NIST) and the International Organization for Standardization (ISO) go beyond attack prevention and represent advanced methodologies for assessing and mitigating the overall organizational cybersecurity risks. As we discussed before, managers need to make informed decisions on the risks to which their digital assets are exposed, and then size and prioritize the relevant initiatives to manage them. It is critical for you to understand these initiatives and the extent of their implications.

Cybersecurity Frameworks

The NIST Cybersecurity Framework and ISO 27001 are reference methodologies based on risk management that companies may adopt to manage their IS security. The two approaches present several similarities. Both are technology neutral and can be applied to any type of technology or organization. Additionally, they aim at delivering business benefits while complying with legal and regulatory requirements. The main differences are that the ISO 27001 is internationally recognized and can be certified. In other words, following an audit, an organization can prove

---

1 Wagenseil, P. 2019, July 8. "More than 1,000 Android apps steal your data without permission." Tom's Guide, retrieved from https://www.tomsguide.com/news/more-than-1000-android-apps-steal-your-data-without-permission.

2 Korolov, M., and Myers, L. 2017, November 7. "What is the cyber kill chain? Why it's not always the right approach to cyber attacks." CSO, retrieved from https://www.csoonline.com/article/2134037/cyber-attacks-espionage/strategic-planning-erm-the-practicality-of-the-cyber-kill-chain-approach-to-security.html.

Reconnaissance

Research, identification and selection of targets. The adversaries are in the planning phase of their operation. They conduct research to understand which targets will enable them to meet their objectives. Detecting reconnaissance as it happens can be very difficult, but when defenders discover recon – even well after the fact – it can reveal the intent of the adversaries.

Weaponization

The adversaries are in the preparation and staging phase of their operation creating a deliverable payload by pairing remote access malware with exploits. Malware generation is likely done through automated tools. This is an essential phase for defenders to understand. Though they cannot detect weaponization as it happens, they can infer by analyzing malware artifacts. Detections against weaponizer artifacts are often the most durable and resilient defenses.

Delivery

The adversaries convey the malware to the target pairing remote access malware with exploits into a deliverable payload. This is the first and most important opportunity for defenders to block the operation. A key measure of effectiveness is the fraction of intrusion attempts that are blocked at delivery stage.

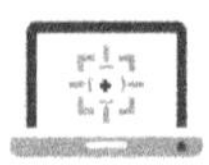

Exploitation

Once delivered, the weapon's code is triggered, exploiting vulnerable applications or systems. The so-called "hardening" may not suffice dealing with new and "zero-day" exploits.

Installation

Typically, the adversaries install a persistent backdoor or implant in the victim environment to maintain access for an extended period of time. Defenders should detect and log installation activity. Analyzing the installation phase during malware monitoring may help create new endpoint mitigations.

Command & Control

Malware opens a command channel to enable the adversary to remotely manipulate the target's network. This is defender's last best chance to block the operation. If adversaries can't issue commands, defenders can prevent impact.

Actions and Objectives

With hands-on keyboard access, intruders accomplish their goals. What happens next depends on who is on the keyboard. The longer an adversary has access, the greater the impact. Defenders must detect this stage as quickly as possible by using forensic evidence.

Figure 13.6. The Cyber Kill Chain framework

Source: Adapted from "Cyber Kill Chain®," Lockheed Martin, https://www.lockheedmartin.com/en-us/capabilities/cyber/cyber-kill-chain.html

to its customers, partners, and other stakeholders that it complies with the standard, thus being able to guarantee a certain level of security. The NIST framework is commonly adopted in the United States and was originally developed for organizations and governmental agencies in charge of critical national infrastructure (e.g., electricity, transportation, security services). The NIST framework provides an exemplar architecture for organizing cybersecurity management. It considers the core dimensions, the implementation level, and the target profile of an organization's cybersecurity management efforts (Figure 13.7).

The core dimension contains an array of activities, outcomes, and references for five core functions (identify, protect, detect, respond, recover). Functions provide a high-level, strategic view of the life cycle of cybersecurity risk management. The activities can be considered as organization capabilities for facing cybersecurity events along the overall cybersecurity life cycle, from prevention to recovery after a hack. The framework core then identifies categories and subcategories of outcomes and associated detailed approaches to cybersecurity. We review the main core functions.

Identify. The overall objective of this function is to develop an understanding of the context in which the organization operates to manage cybersecurity risk to systems, assets, data, and capabilities—in other words, to understand the organization and the relevant threats. Because every organization is unique, the cybersecurity effort needs to be tailored to the distinct requirements of each firm. For example, a hospital may recognize that a cybersecurity breach could compromise its operations by rendering inoperable several devices (e.g., blood gas analyzers, X-ray equipment), resulting in financial and reputation damages—if not loss of life.

Protect. This function consists of the development and implementation of the appropriate safeguards to ensure business continuity, limiting or containing the impact of cybersecurity events—for example, restricting the access to physical (e.g., computers) and digital assets (e.g., databases) using a biometric control system.

Detect. The objective is to develop and implement ways to identify cybersecurity breaches and events—for example, using cybersecurity monitoring tools to enable the timely discovery of intrusions.

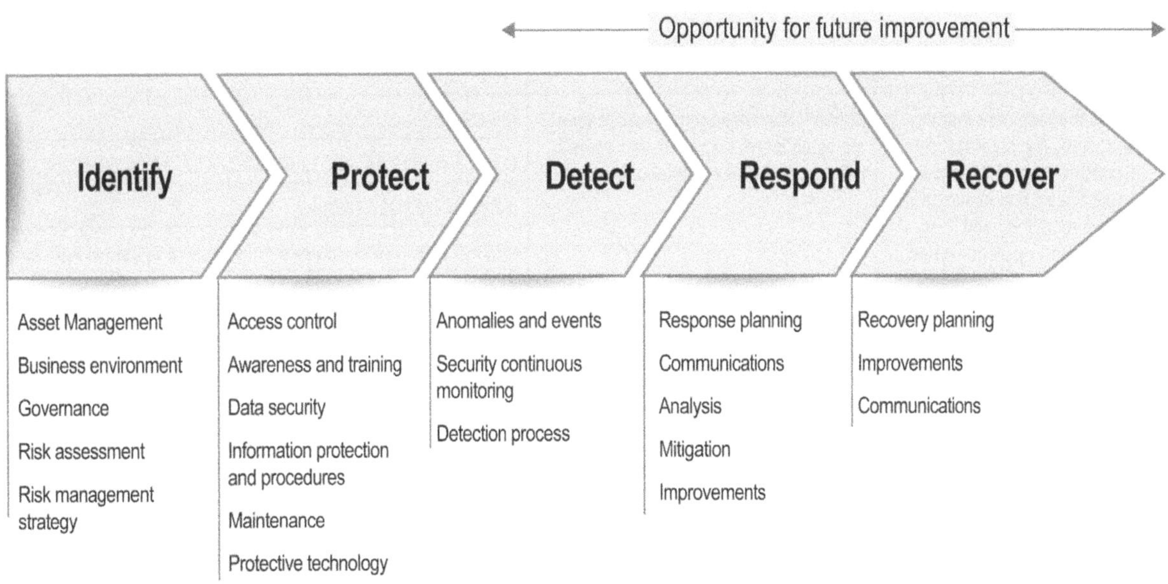

Figure 13.7. The five core functions and respective categories of the NIST Cybersecurity Framework

Respond. This function seeks to develop the capability of the organization to take action on a detected incident—for example, by defining a response and communication plan to comply with regulations and notify stakeholders or by analyzing and containing the impact of the cybersecurity incident.

Recover. An organization needs to prepare for recovering from a hack by defining the appropriate measures to restore the compromised services or capabilities. The objective is a timely return to normal operations to reduce the impact of a cybersecurity incident—for example, defining a recovery plan including stakeholders and external partners other than restoring the organizational data.

To support the cybersecurity effort, for each function, category, and subcategory, the NIST framework makes additional references to standards or frameworks that could be further adopted or are already in place (Table 13.2).

The implementation tiers identify four levels (partial, informed, repeatable, adaptive) representing the extent and sophistication of an organization's approach to cybersecurity. Tiers characterize an organization based on the characteristics defined in the framework over an increasingly sophisticated approach (Table 13.3). Each tier depends on the assessment of the level of the risk management process, the integrated management program, and the external participation that an organization has put into place. While resembling a maturity model (Chapter 10), tiers define the level of integration of the cybersecurity management practice and should be used to align the cybersecurity effort with the firm's objectives, regulatory compliance, and risk profile. Thus the progression between tiers will depend on the assessment of the trade-off between the reduced cybersecurity risk and its cost effectiveness.

The profile is determined on the basis of the core category and subcategory outcomes of the NIST framework. An organization identifies and prioritizes the opportunities for improvement comparing its current profile (what the organization currently does) and the desirable risk level (the target profile). Interestingly, profiles can be used for setting the minimum requirements and communicating with both internal and external stakeholders (e.g., suppliers).

Table 13.2. An example of function, category, subcategory, and reference standard

Function	Category	Subcategory	Reference standard
Detect	Risk management strategy (ID.RM): The organization's priorities, constraints, risk tolerances, and assumptions are established and used to support operational risk decisions.	ID.RM-1: Risk management processes are established, managed, and agreed to by organizational stakeholders.	CIS CSC 4
			COBIT 5 APO12.04, APO12.05, APO13.02, BAI02.03, BAI04.02
			ISA 62443-2-1:2009 4.3.4.2
			ISO/IEC 27001:2013 Clause 6.1.3, Clause 8.3, Clause 9.3
			NIST SP 800-53 Rev. 4 PM-9
		ID.RM-2: Organizational risk tolerance is determined and clearly expressed.	COBIT 5 APO12.06
			ISA 62443-2-1:2009 4.3.2.6.5
			ISO/IEC 27001:2013 Clause 6.1.3, Clause 8.3
			NIST SP 800-53 Rev. 4 PM-9
		ID.RM-3: The organization's determination of risk tolerance is informed by its role in critical infrastructure and sector-specific risk analysis.	COBIT 5 APO12.02
			ISO/IEC 27001:2013 Clause 6.1.3, Clause 8.3
			NIST SP 800-53 Rev. 4 SA-14, PM-8, PM-9, PM-11

Table 13.3. The implementation tiers

	Risk management process	Integrated management process	External participation
Partial	Not formalized Ad hoc Reactive manner	Limited cybersecurity awareness Irregular implementation Cybersecurity information is not shared	No external collaboration
Informed	Approved practices Not Established organization-wide	Risk awareness Cybersecurity information is shared informally Risk assessment of organizational and external assets occurs	Not formalized Cybersupply chain awareness
Repeatable	Practices approved as policies established organization-wide	Organization-wide approach Risk informed policies, processes, and procedures are established and reviewed Consistent risk monitoring Cybersecurity managerial roles	Collaboration Receive information Acts formally on the cybersupply chain
Adaptive	Continuous improvement	Organization-wide approach Alignment between risks and organizational objectives Continuous awareness Cybersecurity organizational culture	Actively shared Formally and informally proactive on the cybersupply chain

Cybersecurity frameworks are critical management tools because they enable productive conversations between management and IT professionals and the identification of major gaps in current practices.

Cybersecurity Responses

With the overall cybersecurity framework in place, we discuss a number of specific activities and safeguards that organizations can enact in response to cybersecurity threats.

Internal Cybersecurity Threats Prevention of internal threats is no simple feat, since cybersecurity products and technologies can only partially help. Prevention of internal threats requires the development and enforcement of cybersecurity policies and auditing standards designed to ensure that such policies are understood and respected by those within the organization. Most organizations tend to rely on technology to respond to insider threats, but relying on organizational measures, such as developing a cybersecurity-aware culture, is equally important. The first step for raising the awareness of the risk of internal threats consists of performing a comprehensive assessment to understand the effects of potential scenarios, their likelihood, and their potential harm.

Security Policies The most easily preventable cybersecurity risks are those caused by ignorance of sound cybersecurity practices. A security policy spells out what the organization believes are the behaviors that individual employees and groups within the firm should follow in order to minimize security risks. They include what computing services will be made available and what computing services will not be made available within the firm. They specify what password standards the firm should follow (e.g., length, characters to be used, renewal schedules) and what rights different types of users will have. They specify the level of care that employees need to use with their passwords (e.g., do not share passwords with anyone, do not send passwords over clear channels such as unencrypted mail messages), what computing resources and what data should be accessible within the organization, and what data can be downloaded to personal devices and what needs to remain within the company. The policy should address legitimate uses of portable devices, what data can be downloaded to them, and how such devices should be secured by those who own them. A cybersecurity policy may even address the level of care that employees should exercise when they leave the premises (e.g., not reviewing sensitive data on laptops while on airplanes).

Beyond having comprehensive cybersecurity policies in place, the organization must audit them to ensure compliance. For example, accounts of terminated employees need to be swiftly deleted or made unavailable to prevent access from former employees who at best will see material they should not and at worst will be able to damage company resources (e.g., delete data).

Monitoring Employees are a primary threat to an organization's cybersecurity. The classical approach is to monitor and limit the access to dangerous websites or to prevent the installation and execution of unauthorized software. Indirectly, employees will tend to behave more cautiously when "feeling" the firm actively monitoring and controlling their activities. Advanced analytics and machine learning are increasingly used to profile employees and detect suspicious activities and anomalies. While these systems challenge the delicate balance between organizational security and employees' privacy, they present the actual advantage of detecting unknown threats while scaling easily to large numbers of users.

External Cybersecurity Threats In addition, the prevention of external threats is also no simple feat.

Intrusion A number of techniques and technologies to prevent intrusion have been developed over the years. The cornerstone of securing against intrusion is the use of passwords. Passwords ensure that resources are only made available to those who have the appropriate authentication levels. Thus a password can be used to block unauthorized external users as well as indicate to whom resources should be available among the legitimate users.

Organizations typically enforce standards to ensure that passwords are reasonably difficult to guess. For example, many cybersecurity policies require that passwords have a minimum length; they use letters, numbers, and special characters; they don't use dictionary words; and so on. However, there is an inherent trade-off between the complexity of a password and human ability. In other words, if passwords are too difficult to remember, people will write them down or reuse them, creating a new security risk! For this reason, the computer security industry is hard at work devising more robust identification schemes, such as **biometrics**—the use of physical traits (e.g., fingerprints, iris scans) as a means to uniquely identify users (Figure 13.8).

A **firewall** is a software tool designed to screen and manage traffic in and out of a computer network. Thus a firewall is used to secure the perimeter of the organization's computing resources, employing a number of technologies and techniques. Firewalls can also be used to enforce cybersecurity policies (e.g., blocking traffic from websites deemed risky and blocking the download of some file types).

Figure 13.8. Fingerprint scanner

Source: U.S. Customs and Border Protection, https://en.wikipedia.org/wiki/File:US-VISIT_(CBP).jpg

Firewalls are a very important cybersecurity tool, but you need to remember that perimeter protection is only as strong as its weakest link—much like the perimeter protection of a castle. For example, no matter how powerful your firewall is, if there are unsecured or compromised endpoints (e.g., employees' laptops, servers) that an attacker can control, the network is not secure. Once an attacker gains access to the organization's resources, the strongest perimeter security is made useless—the intruder is working on the inside now.

Consider as well that any resource that is brought outside of the perimeter is not secured by the firewall—hence the inherent danger associated with the proliferation of mobile and IoT devices.

For instance, if an employee copies sensitive data to her laptop and then takes the machine on the road, the firm's firewall is useless in protecting such data. With data becoming increasingly prevalent, portable devices achieving widespread adoption, and small devices capable of storing large amounts of data, perimeter security is increasingly insufficient.

Another technique to safeguard against the intrusion threat is encryption. Through the encryption process, content is scrambled in such a way that it is rendered unreadable to all recipients, except those who hold the key to decrypt it. Encryption ensures that if the wrong individuals gain access to the data, they will be unable to make out its meaning. Encryption techniques are used to secure transmissions as well as to secure stored data. Consider once again the example of the lost laptop discussed above. If the data it contained had been encrypted, the loss to the firm would simply amount to the cost of the asset—less than a thousand dollars, in most cases.

Cybersecurity policies and audits will help a firm ensure that no backdoors are left open after the installation of a new software program. With respect to cybersecurity holes and weaknesses in software due to bugs, the IT staff in the organization must monitor bug reports and install patches—pieces of add-on code published by the software house that wrote the program designed to eliminate weaknesses that surface after the release of the software program. The problem is that many organizations lack the staff to constantly monitor this information and may fall behind in patch installations, thus opening the firm to unnecessary risks. To obviate this problem, many of the large software houses (e.g., Microsoft) have now developed patch management software that automatically alerts users to the availability of newly released patches—and, if configured to do so, downloads and installs them automatically. As both the Equifax and SolarWinds breaches well illustrate, patch management is critical to fix digital infrastructure and capability vulnerabilities. With new threats emerging daily, ensuring the protection of digital assets becomes a business priority. Systems affected by known vulnerabilities are the ideal target for attackers.

Malware Safeguarding against malware requires that the firm's IT professionals install the appropriate detection software (e.g., antivirus, spyware sweepers). With the large number of new viruses being released, antivirus and other detection software is only as good as its most recent update. For this reason, organizations that manage their own networks are increasingly attempting to centralize these applications and push updates to individual users so as to ensure that safeguards against malware are up to date.

Training and policies can also be very helpful in mitigating the malware threat. Simple behaviors, such as not opening email attachments from accounts you don't recognize (e.g., intriguing @sexylips.com) or limiting downloads from the Internet to trusted websites, go a long way in preventing infection.

Denial-of-Service Attack Preventing a denial-of-service attack is very difficult. This is because in a well-orchestrated denial-of-service attack, the requests for the service are not issued from the same few locations, which would make it easy to recognize and block. Instead, in what's called a DDoS attack, the attacker will hijack or spoof multiple machines and initiate the attack from these multiple locations.

Managing Cybersecurity: Overall Guidelines

As manager, you are not likely to be involved in the technical details of the procedures and safeguards chosen to mitigate the specific cybersecurity threats identified in risk assessment. However, you should expect to be involved in setting the agenda for how the overall set of risks is to be addressed. In this role, it is paramount that you recognize one important characteristic of cybersecurity investments: cybersecurity is a negative deliverable, one that produces no upside but helps in limiting damage ensuing from an uncertain negative event. For these types of investments, it is difficult to obtain appropriate funding.

Moreover, when it comes to cybersecurity, it is impossible to ensure success. In other words, it is impossible to claim that the organization is absolutely "secure." Rather, it is possible to find out that the organization was not secure after a breach had occurred. Amid all these difficulties, managers play a critical role.

Have a Plan and Specify Responsibilities You would be surprised to find out how many organizations do not devise formal plans to be enacted during an attack or fail to assign formal responsibility for cybersecurity design and enforcement. This is particularly true for the design of new applications, as requirements requested by the business sometimes weaken the security of the applications. In this case, the overall responsibility for security choices and trade-offs should reside with a business owner or other appropriate senior person, not with IT. When outside contractors are engaged in development, cybersecurity requirements should be spelled out in the contract.

A crisis management plan should specify who needs to be contacted in an emergency and what their roles should be. The plan must address questions such as what the first reaction measures should be—Should the systems under attack be shut down or left operational? When and how should authorities, such as the FBI, be contacted? What will the firm disclose about the attack, if anything, and who should be in charge of press releases or of customer communication? What recovery plans need to be enacted under the various scenarios? In short, the plan should script all those decisions that are difficult to make when the firm is actually under attack.

Revisit Often Cybersecurity is a constantly evolving area for a number of reasons—first and foremost, the breathtaking pace of technical evolution. Every new technology and software program your firm adopts ushers in a unique set of cybersecurity and risk management challenges that should be proactively addressed—whether that means taking specific steps to manage them or consciously accepting the risk. Thus you must ensure that cybersecurity audits and a reassessment of security plans are done periodically.

Develop a Mitigation Plan A well-architected cybersecurity infrastructure and plan can go a long way toward tempering the many security threats contemporary firms face. But no matter how good your cybersecurity is, there is always the chance that your firm will be successfully attacked and that your defenses will be breached. What you do in this case can be critical, particularly when you become vulnerable to an intrusion.

The first reaction to an attack is often to shut everything down. This is a mistake, since diagnosing where the attack is coming from, its severity, and its reach is much easier to do if the system is maintained as operational and the attacker is kept unaware of the fact that you spotted the cybersecurity breach.

The first order of business at this point should be to determine how the attack took place in order to eliminate its chance of occurring again. The next step requires an assessment of the damage, particularly as it pertains to the loss of sensitive data. A series of laws have been recently passed across the globe requiring firms to immediately notify those parties whose data may have been compromised. Whether this is a necessity or not in your jurisdiction, it is a wise move to immediately communicate the problem to those affected. As much as you would like to keep the matter private to avoid the negative publicity, people understand that cybersecurity breaches do sometimes occur, but they will be much less forgiving if they discover an attempt to cover up the problem.

13.4 Privacy

Privacy concerns emerge in the relationship between individuals and organizations because, when dealing with business firms or nonprofit and governmental organizations, customers (and employees) often provide personal information under the assumption that the organization will take "good care" of it. We as customers and/or employees provide personal information in order to receive the benefits of the transaction (e.g., employment, value proposition). Yet a prerequisite to the transaction is that we hold some degree of trust in the organization that we are dealing with. We trust that our information is reasonably secure, despite many data breaches over the years (Figure 13.9), and that the organization has taken steps to prevent unauthorized use of it by those who have no legitimate reason to access it.

In other words, we must have trust in the firm's IS security. Even more important, we must trust that the organization will be a steward of our personal information and that it will refrain from employing it in ways that will be harmful to us or our interests. For instance, even if we have great faith in the firm's cybersecurity, we still must believe that the firm is ethical enough not to collect our data and immediately turn around and sell it to spammers in order to make a "quick buck."

Consider the following incident:

> In 2003, the public learned that JetBlue Airways had turned over records on more than a million of its passengers to a government contractor. The contractor sought the information to test a security application designed to identify suspected terrorists. However, the sharing violated JetBlue's privacy policy, which promised that the company would not provide personal information to third parties. The CEO issued a public apology and indicated that he had no

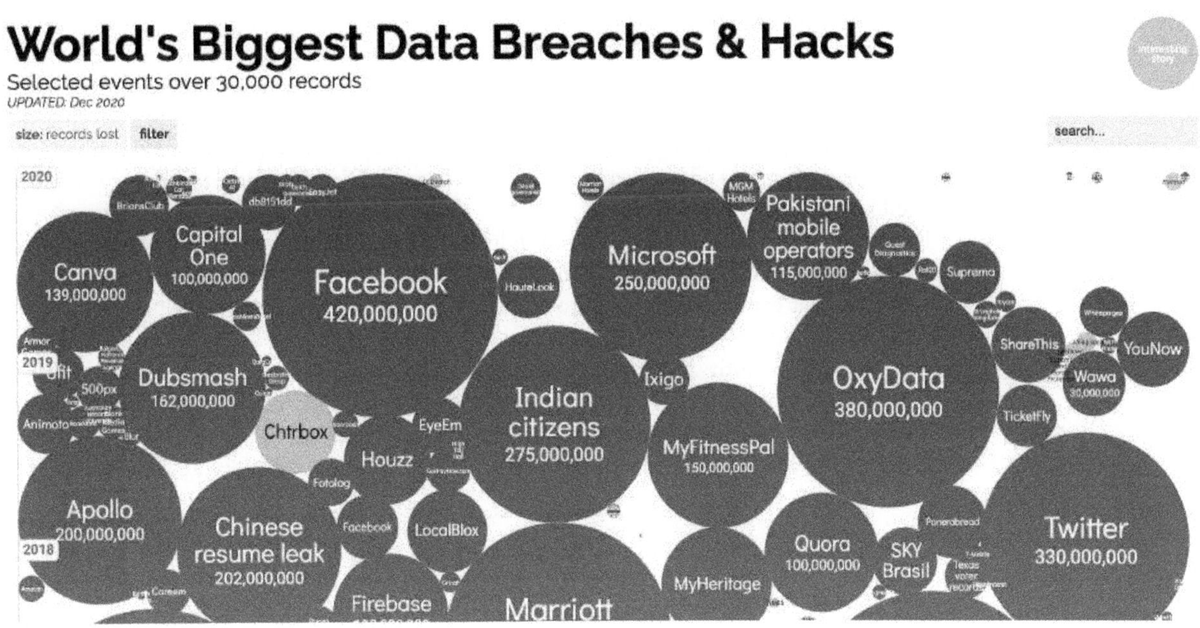

Figure 13.9. The greatest data breaches of all time

Source: "World's biggest data breaches & hacks," 2020, December 13, Information Is Beautiful, retrieved from http://www.informationisbeautiful.net/visualizations/worlds-biggest-data-breaches-hacks/

> knowledge of the data transfer at the time it was made. Nonetheless, JetBlue had to confront a lawsuit from privacy groups and passengers.[1]

What was the problem? After all, JetBlue was only trying to aid in efforts to prevent terrorist attacks and ultimately improve the safety of its own passengers! The problem is that JetBlue had not developed an appropriate process for dealing with privacy concerns in the face of the proliferation of customer data it stewards and the mounting number of potential (perhaps legitimate) uses of those data.

More recently, Tim Cook—with a clear interest in reinforcing the positive perception of Apple's strategy concerning data privacy—stated,

> Some of the most prominent and successful companies [in Silicon Valley] have built their businesses by lulling their customers into complacency about their personal information. They're gobbling up everything they can learn about you and trying to monetize it. We think that's wrong. And it's not the kind of company that Apple wants to be.[2]

Cook's words well depict how the offer of personalized services based on technologies such as Big Data analytics often comes together with unsolved privacy concerns. In fact, Big Data analytics are presented as the technology that will support the future of the knowledge economy.[3] This technology undeniably bears new promises for product and service development for all sectors of activity. Traditional sectors such as the banking industry as illustrated in the example above could use Big Data to improve the provision of their services. New sectors such as smart sensing applications in the field of sports and well-being could benefit from Big Data technologies and provide new services tailored to users' specific needs.

Privacy Defined

There is no doubt that privacy is not an "IT issue." Given the ethical concerns and the potential for liability associated with privacy, general and functional managers are typically front and center in identifying and responding to the privacy concerns of the organization. But what is privacy exactly? First of all, it is important to understand that, from a legal perspective, there are two distinct approaches to privacy regulation: an American concept, based on precautionary principle reasoning, and a European concept, based on the idea of private information as a form of intellectual property.

Privacy can be defined as the ability of individuals to control the terms and conditions under which their personal information is collected, managed, and utilized. Private information is the information that can be traced back to the individual—for instance, a person's name, picture, address, tax identification number, or medical history. It is evident from the above definition that privacy subsumes security. A firm that is unable to secure customer or employee data will not be able to ensure privacy. More specifically, privacy is about informed consent and permission to collect and use identifying information, while cybersecurity is about safekeeping the collected data. Privacy concerns are not only tied to perceptions or attitudes at an individual level

---

1 Culnan, M. J. 2006. "Privacy in search of governance." *Cutter Benchmark Review* 6(1): 5.

2 Panzarino, M. 2015, June 2. "Apple's Tim Cook delivers blistering speech on encryption, privacy." Tech Crunch, retrieved from http://techcrunch.com/2015/06/02/apples-tim-cook-delivers-blistering-speech-on-encryption-privacy/.

3 See, for example, EU Council. 2013, October 25. "Conclusions." Consilium, retrieved from https://www.consilium.europa.eu/uedocs/cms_data/docs/pressdata/en/ec/139197.pdf; OECD. 2013, June 18. "Exploring data-driven innovation as a new source of growth: Mapping the policy issues raised by 'Big Data.'" OECD, retrieved from http://www.oecd-ilibrary.org/science-and-technology/exploring-data-driven-innovation-as-a-new-source-of-growth_5k47zw3fcp43-en.

but connected to organizational and institutional elements like privacy policies and reputation. Indeed, privacy concerns arise because of the existence of a perceived value of privacy by users and therefore from an appreciation for the protection of private information.[1]

Privacy Risks

Privacy risks are a by-product of the success that firms have been enjoying with their use of digital technology. In a world where competition is global and it is not possible to have a personal relationship with the thousands or millions of customers your firm is trying to reach, information systems have created the ability to "know" individuals we do not interact with directly. In many cases, these developments were welcomed by those involved (e.g., customers).[2] Many of us enjoy the personal recommendations produced by online platforms and aggregators like Amazon or Netflix (Chapter 4) or the fact that we don't have to repeat our preferences every time we book a reservation with the same hotel chain or airline. But advances in technology that allow organizations to better tailor products and services to customer needs also create the potential for highly damaging privacy violations. Organizations increasingly collect and process the data of employees, customers, patients, consumers, and citizens—data that if leaked may reduce loyalty, sales, or trust. Privacy concerns have a number of causes:

Function Creep Function creep occurs when data collected for a stated or implied purpose are then reused for other, unrelated, objectives. When function creep occurs, individuals lose control over their personal data. Regulators fined the Italian mobile operator TIM €27.8 million for having used both customers' and prospects' personal data collected for technical purposes to push forward an aggressive marketing campaign involving around 13 million users.[3]

Proliferating Data Sources As the digital revolution continues unabated, we are surrounded by digital devices generating data about us. We are very confident in claiming that, as you read this paragraph, there are multiple apps on your smartphone that are aware of your location. And we are confident in predicting that you are not aware of many of them doing so. Moreover, today's consumers seem to revel in providing more information on a voluntary basis—from Instagram to TikTok to Twitch.

Improving Data Management Technologies Not only is personal data easier to generate than ever before, but it is increasingly simple and cost effective for organizations to merge multiple data repositories and handle enormous quantities of data (Chapter 3). The most recent development in this area includes data exchanges. Start-ups like Snowflake and established players like Amazon are creating platforms (Chapter 4) for simplifying cross-firm data exchange and merging. Data sharing across organizations, unless expressly agreed to by customers, is a sure example of function creep.

The Legal Landscape Digital technology evolution outpaces legal development. Only recently, particularly in Singapore, Australia, and Europe, governments have started comprehensively legislating digital privacy (see below). Yet given the complexity of the challenge, such legislation

---

1 Xu, H., et al. 2011. "Information privacy concerns: Linking individual perceptions with institutional privacy assurances." *Journal of the Association for Information Systems 12*(12): 798–824; Martin, K. D., and Murphy, P. E. 2017. "The role of data privacy in marketing." *Journal of the Academy of Marketing Science 45*(2): 135–55.

2 While the arguments here can be extended to other entities, such as employees or suppliers, we use the example of customers throughout for simplicity.

3 "5 biggest GDPR fines so far." 2020, December 7. Data Privacy Manager, retrieved from https://dataprivacymanager.net/5-biggest-gdpr-fines-so-far-2020/.

is moving slowly, and at times, it generates unintended consequences. If you are reading this book from Europe, you are probably used to clicking the **General Data Protection Regulation (GDPR)** banners about cookie tracking (Figure 13.10)—banners that almost nobody reads.

Safeguarding Privacy

Fair information practices have been proposed as a basis for privacy governance. Fair information practices are based on the five principles of notice, choice, access, security, and enforcement.

- *Notice* refers to the right of individuals to be informed when their personal data are being collected and to be informed about how they are or will be used.
- *Choice* calls for the ability of individuals to be informed of, and object to, function creep, whether within one firm or across firms who share information.
- *Access* refers to the right of individuals to be able to access their information and correct any errors that may have occurred in their records.
- *Security* calls for organizations that house individuals' private information to ensure its safekeeping and to protect it from unauthorized access.
- *Enforcement* calls for organizations that collect and use private information to develop enforceable procedures to ensure that the above principles are upheld.

While firms tried the road of self- and industry-regulation, privacy legislation is being advanced in many jurisdictions. Europe's GDPR and the **California Consumer Privacy Act (CCPA)** established a core set of rules and practices for privacy protection and management. These two regulations empower consumers' control of their data and attempt to raise awareness concerning the collection, treatment, management, and transfer of personal information (Figure 13.10). While established with similar intentions, the two regulations differ significantly beyond the scope of application.

General Data Protection Regulation The GDPR defines *personal data* as any information relating to an individual who could be identified through an attribute such as a name, an identification number, location data, or an online identifier or by one or more factors specific to the person's physical, physiological, genetic, mental, economic, cultural, or social identity. The directive establishes seven main principles that implement the spirit of the fair information practices introduced above:[1]

1. *Lawfulness, fairness, and transparency*. Data processing should be both legal and understandable. It should occur in a way that could have unjustified adverse effects on the individual.
2. *Purpose limitation*: Data should only be collected for a defined scope. Harvesting data as an end in itself is not an acceptable purpose.
3. *Data minimization*. Personal data collected and processed must be strictly necessary to fulfill that purpose.
4. *Accuracy*. Personal data must be accurate and up-to-date with respect to the data collection purpose. When it is not, it should be corrected.
5. *Storage limitations*. Data can be archived for no longer than necessary for the purposes for which it was collected.
6. *Integrity and confidentiality*. Personal data must be processed in ways apt to ensure appropriate technical and organizational safeguards, including protection against unauthorized or unlawful processing and against accidental loss, destruction, or damage.

1 Art. 5., Regulation (EU) 2016/679 of the European Parliament and of the Council of 27 April 2016 on the protection of natural persons with regard to the processing of personal data and on the free movement of such data, and repealing Directive 95/46/EC (General Data Protection Regulation).

7. *Accountability*. The firm is responsible and must be able to demonstrate compliance.

The firm or the organization that collects and processes personal data is requested to respect these principles.

GDPR brings legal clarity to individual rights and firms' responsibilities in data collection and processing.[1] At the same time, GDPR imposes a relevant compliance burden on organizations, and as a consequence, it has been accused of stifling innovation. Furthermore, the right of the individual to request and obtain the total deletion of any personal data challenges most firms' data governance. Few organizations were equipped to trace an individual's data across the different systems, cloud platforms, and applications when GDPR was first enacted. Recognizing the limits of GDPR, the EU is advancing the ePrivacy Regulation, which will clarify and enhance the GDPR framework in the context of digital communications. More specifically, the ePrivacy Regulation will focus on limiting unsolicited marketing communications, safeguarding communication confidentiality, and specifying cookie use.[2]

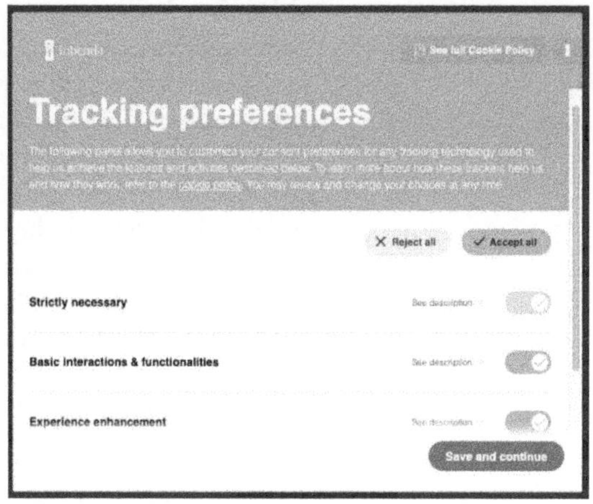

Figure 13.10. GDPR- and CCPA-compliant control tool for opting in and out

California Consumer Privacy Act The CCPA represents an important step forward in privacy regulation considering the United States still lacks a comprehensive federal law. While, ideally, the CCPA may facilitate the harmonization between US and EU regulations, their different approach to similar principles is an obstacle to harmonization. The CCPA grants consumers both rights and control over their personal data, including the right to know, the right to delete, and the right to opt out from the sale of information or its sharing beyond the original collector. Under the CCPA, organizations must leave to consumers the choice to share their data and should not discriminate against individuals opting out. However, the regulation allows the firm to offer financial incentives to consumers for opting in as long as they can then withdraw their consent. Thus a firm can apply a different pricing scheme depending on an individual's privacy choice.

Another important measure of the CCPA concerns customers' right to know both what data the firm collected and with whom the data was shared within the last 12 months.

In the changing landscape of privacy regulation and customer awareness, many organizations opt to educate users and explain to them the reasons for data collection and the implications of an opt-out request. Transparency is clearly an important achievement, but privacy is a fast-moving area, and you can expect the legal and regulatory landscape to continue to evolve.

13.5 Ethics

Webster's Dictionary defines *ethics* as "the discipline dealing with what is good and bad and with moral duty and obligation." Ethics is the branch of philosophy that concerns itself with morality by studying right and wrong and attempting to draw a distinction between good and evil.

Examples of unethical behavior in business abound. The corporate scandals that have led to the demise of one-time stock market darlings like Enron, Global Crossing, Lehman Brothers, or Bernie

---

1 Poritskiy, N., Oliveira, F., and Almeida, F. 2019. "The benefits and challenges of general data protection regulation for the information technology sector." *Digital Policy, Regulation and Governance 21*(5): 510–24. doi: 10.1108/DPRG-05-2019-0039.

2 The proposal of the ePrivacy Regulation: https://eur-lex.europa.eu/legal-content/EN/TXT/HTML/?uri=CELEX:52017PC0010&from=EN.

Madoff or Volkswagen cheating in its reporting on pollution emissions have clearly shown that we cannot assume that all of today's managers are equipped to make appropriate ethical choices.

While many react with outrage to unethical behavior, arguing that "everyone knows what's right and what's wrong," this stance oversimplifies reality. Because of the intense technical training that most business and management schools focus on, the fact that managers will make ethical decisions is often mistakenly taken for granted. Most managers are ill equipped to make ethical decisions because they lack formal training in the area and because their attention is usually on the objective they are trying to reach. A former CIO at Metro-Goldwyn-Mayer Studios aptly captured this dilemma: "When your job is building the best-performing database you can, you don't always think about the ethical implications of how that data will be used."[1]

Moreover, aside from those spilling over into illegal behavior, ethical choices are rarely straightforward. In fact, ethical dilemmas are typically rooted in the choice between multiple suboptimal courses of action that force well-intentioned individuals to make difficult trade-offs.

Information Systems Ethics

Information systems and digital technologies, with their penchant for enabling new ways of doing business, constantly introduce the potential for ethical dilemmas. Moreover, because of the rapid pace of technology evolution and the slow pace at which legislation is passed, formal explicit rules lag behind the possibilities offered by new technologies. Ethical guidelines should fill the void, providing direction in the absence of explicit laws. Consider the following scenarios:

- As the digital innovation director for your organization, you have some leeway with the priorities you assign to various projects. You recently reallocated resources and delayed the cloud migration requested by marketing to speed up the smartphone app roll-out that is already running considerably behind schedule. You have reason to believe that Jack, one of your project managers, forwarded your email about the shift of resources to the vice president (VP) of marketing.

  As the digital innovation director, you are well aware that all company emails are backed up, and you possess the credentials that enables access to every account. As you recover a lost email late one evening, you tell yourself that it is critical that a general be able to fully trust his troops—you must find out whether you can trust Jack.
- It was your college-days' dream: running your own company! Reinventravel.com had been it for the last four-and-a-half years. An industry magazine called it "the intermediary that reshaped how people buy travel." But now your dream had turned into a nightmare. After filing for Chapter 11 bankruptcy protection and trying to restructure, it was clear that Reinventravel.com would not make it.

  The decision was tough. You had been offered \$7.2 million by your largest competitor—MightyTravel—for the preferences and historical transaction data of your customers. While you never liked the folks over at MightyTravel, they assured you that they would use the data to offer a more targeted and personal travel experience to your former customers. Your privacy policy never explicitly addressed what you would do with customer data in the case of bankruptcy. The \$7.2 million will allow you to honor salary and pension commitments to your employees. As you sign the contract, you reassure yourself, thinking that your customers will appreciate receiving more targeted offers from MightyTravel. Everyone prefers targeted offers . . . right?

---

1 Wilder, C., and Soat, J. 2001, May 14. "The ethics of data." *InformationWeek*, retrieved March 7, 2011, from http://www.informationweek.com/837/dataethics/htm.

What is common to these scenarios is the fact that each of the ethical dilemmas they capture would not have been possible just a few years ago. New technologies brought them about as a by-product of their enabling new ways to collect, process, store, and distribute information. More important, both scenarios paint an accurate picture of typical ethical dilemmas: the appropriate course of action is far from clear, and no choice is without a negative impact on other individuals.

Ensuring Ethical Uses of Information Systems

There are no silver bullets to ensure ethical behavior in the context of organizational information systems. Developing a culture of ethical decision making is critical. Such a culture should create the preconditions for and reward behavior that strives for harm minimization, respect, and consistency. Professional communities, such as the Association for Computing Machinery (ACM), the Association of Information Technology Professionals (AITP), and the Institute of Electrical and Electronics Engineers (IEEE) Computer Society, have attempted to create an ethical culture that goes beyond individual organizations by promulgating codes of ethics. For example, the ACM first introduced its code of ethics in 1992 with the following preamble:

> Commitment to ethical professional conduct is expected of every member (voting members, associate members, and student members) of the Association for Computing Machinery (ACM).
>
> This Code, consisting of 24 imperatives formulated as statements of personal responsibility, identifies the elements of such a commitment. It contains many, but not all, issues professionals are likely to face. . . .
>
> The Code shall be supplemented by a set of Guidelines, which provide explanation to assist members in dealing with the various issues contained in the Code. It is expected that the Guidelines will be changed more frequently than the Code.
>
> The Code and its supplemented Guidelines are intended to serve as a basis for ethical decision making in the conduct of professional work. Secondarily, they may serve as a basis for judging the merit of a formal complaint pertaining to violation of professional ethical standards.[1]

Ethical dilemmas typically pit the interest of one person or group (e.g., shareholders) against that of another (e.g., customers). Applying the principle of harm minimization, one needs to weigh the relative impact that the decision will have on all individuals affected and strive to moderate damage to any one individual or group. The principle of respect requires that IS decisions be made in an effort to treat each of the affected parties with the utmost consideration. Finally, the principle of consistency provides a test for evaluating decisions. It requires a person confronted with an ethical dilemma to consider whether he would approve if everyone else made the same choice he is considering.

A practical and pragmatic approach to foster a culture of ethical decision making is to establish an IS ethics code of conduct. A code of conduct, typically used by professional associations, offers two advantages. On the one hand, it communicates to all parties the organization's principles of ethical IS use. Thus it can be used as an educational mechanism to point employees in the right direction. On the other hand, it identifies the firm's formal stance, thus enabling detection of, and distancing from, unethical choices made by any member of the organization.

---

1 ACM. n.d. "ACM code of ethics and professional conduct." ACM.org, retrieved from http://www.acm.org/about/code-of-ethics/#sect2.

Summary

In this chapter, we focused on three topics of interest to general and functional managers: IS security and digital risk management, privacy, and IS ethics. A failure in cybersecurity, privacy, or ethics can have dramatic repercussions on the organization, both because of potentially damaging direct effects (e.g., computer outages, disruptions to operations) and increasingly negative indirect effects (e.g., legal recourse, image damage).

In this chapter, we sought to convince you that, as future general and functional managers, you will have to be involved in these decisions. We also helped you gain an understanding of the circumstances in which choices and trade-offs are made so that you can actively participate in decision making.

- Digital infrastructure and resources must be secured against both internal and external threats. The internal threat is due to either ill-willed or careless members of the organization and is mitigated by way of security policies and training. The external threat comes from skilled individuals, referred to as *hackers*. The external threat takes the form of malware, intrusion attempts, and denial-of-service attacks. Each of the threats is matched by the appropriate safeguard.
- Digital security and risk management are not "IT issues." Because of the impact of the security breaches on the current and future viability of the organization, it is critical that general and functional managers take an active role in security decision making. This is done by participating in risk assessment, the process designed to evaluate the potential impact of threats confronting the firm, and risk mitigation, the process of identifying the appropriate response to these security threats. This involvement is necessary in order to make the appropriate trade-off decision among risk acceptance, risk reduction, and risk transference.
- Various framework and industry practices have been developed for helping organizations that are facing increased cybersecurity risks. All frameworks clearly rely on collaboration with stakeholders, either internal or external to the organization. In particular, communication efforts may prove fundamental for reducing the negative fallout after a cybersecurity incident.
- Privacy concerns, like security threats, need general and functional managers' full attention. This is because privacy, like security, becomes visible only as a negative deliverable—that is, investments in privacy help the organization avoid a possible negative occurrence (e.g., lawsuit, loss of customer trust, negative impact on the firm's image) rather than generate benefits, such as improved efficiency or increased revenues.
- In order for the firm to safeguard the privacy of its employees and customers, it must subscribe to fair information practices. Fair information practices are based on the five principles of notice, choice, access, security, and enforcement. Moreover, the firm should produce a codified set of security policies, monitor and enforce compliance with them, and document both the policies and the processes it has developed to ensure privacy.
- Europe's GDPR and the CCPA established a core set of rules and descending practices for privacy protection and management. These two regulations empower consumers' control of their data and raise awareness concerning their collection, treatment, management, and transfer.
- The recent flurry of corporate scandals has ignited interest in business ethics. When it comes to information systems, ethics becomes a crucial guiding light for management behavior, as legislation often lags behind technology improvements. Thus developing a culture of ethical decision making is essential for modern organizations.

Study Questions

1. Imagine that you have just been hired by a retail financial institution. How would you explain to your CEO that she needs to get involved in IS decisions?
2. What are the three costs associated with IS security? What is the relationship among them?
3. Define what is meant by internal and external threats. How do the two differ? How are they related?
4. Define and provide an example of each of the different types of intrusion threats. Describe the appropriate countermeasure for each of your examples.
5. Define and provide an example of each of the different types of malicious code threats. Describe the appropriate countermeasure.

6. What is a denial-of-service attack? Why are these attacks particularly dangerous?
7. What is the “kill chain”? Can you describe the inspiring principles of Lockheed’s Cyber Kill Chain framework?
8. How is privacy defined? What are the principal privacy risks? Can you provide examples of each one?
9. What is ethics? What are the principal challenges associated with IS ethics?

Glossary

A glossary of key terms can be found at the book’s website. Visit https://www.prospectpressvt.com/textbooks/piccoli-information-systems-for-managers-5-0, then click on “Student Resources” in the horizontal red menu bar.

PART V

Cases

Case Study for Chapter 2

Troubleshooting Information Systems at the Royal Hotel

Introduction[1]

In early May 2019, after his first year in the master's program at the Very Famous University (VFU), Blake Cantera landed a summer internship with Fancy Consultants & Company (FC). Upon receiving FC's call, Blake was thrilled. FC was a highly regarded local IT consulting firm serving the needs of clients ranging from independent midsized hotels to large multinational grocery chains.

While small and nimble, FC afforded the opportunity to work with large clients on cutting-edge projects. It also offered significant potential for personal growth, and with its flat organizational structure, FC prided itself on picking independent and self-reliant young consultants who'd work immediately on projects rather than toil in the "analysts limbo" for years. This was the most appealing aspect of FC for Blake, who knew that he would be able to do some real work over the summer.

After a brief introduction to FC systems and culture and a two-week orientation discussing the FC approach to consulting, Blake was assigned to his first account. As expected, this was a relatively small account. On the bright side, Blake was sent alone to manage the whole project, from start to finish. He was thrilled; July had not even started and he was already doing some real work!

The Royal Hotel

The Royal Hotel in New York City was a luxury all-suite hotel primarily serving an executive clientele visiting Manhattan on business. Typically, these business guests stayed for three to six days, during which time they used their suite as a temporary office. Thus, Royal Hotel's management had positioned the property to cater to the many needs of this busy and demanding audience. Amenities included in-suite plain paper fax, printer, and copier; three two-line telephones with voice mail and remote message alert; high-speed Internet access; and plasma TVs and entertainment centers in each of the 482 guest suites. The Royal Hotel also provided three restaurants and a coffee shop on the premises, a 24-hour business center, a fitness center, suite dining, laundry service, complimentary shoe shine, and dedicated high-speed elevators. While business fluctuated in relation to the economic cycle, the Royal welcomed more than 150,000 guests per year, bringing in total revenues upward of $30M.

This made for a fairly complex operation that needed to run smoothly and consistently. Given the high percentage of repeat guests, it was important that guest rooms be spotless and consistently in working order.

The Task

As he arrived at the property for a one-week assignment, all expenses paid, Blake thought to himself with a smile, "I can get used to this . . ." But, with just enough time to take a shower, he had to get ready for a dinner meeting with the general manager (GM).

The Royal Hotel's GM was a no-nonsense, old-school hotelier with a distinctive German accent. He quickly zeroed in on the task, in response to Blake's comment about the "very good" quality of service provided by the Royal Hotel:

> Our level of service is unacceptable! We are very good by most hotels' standard, but we are not "most hotels." Our guests are extremely discerning; it is completely unacceptable to have a light bulb out in the bathroom when the guest checks in, particularly if she is a returning guest. And that's not as bad as a stain on the carpet or a clogged toilet. I had one of my best customers call down to report a clogged toilet last week; can you imagine? Unacceptable! I need you to make sure this never happens again.

As he sat listening to the GM, Blake briefly questioned the wisdom of taking on so much responsibility so quickly; he had not even finished his master's! But this was a brief moment of doubt, and he remembered one of his father's famous sayings: "Did you want the bicycle? Now you have to pedal!" Blake silently chuckled to himself and tuned back into the GM's tirades with a confident smile. He already had the answer to the problem.

The Solution

After examining the property and interviewing a number of people, including the directors of housekeeping, maintenance,

---

1 While based on a real hotel and public sources, this is a fictitious case developed by the textbook authors with the exclusive intent of supporting class discussion.

and IT, Blake recommended that the Royal Hotel adopt M-Tech's Cappuccino! Rapid Response Solution (see Case Appendix for a description of the product). In his presentation to the executive team, highlighting the main advantages of the proposed information system, he mentioned the following:

- *Rapid response*: The Cappuccino! application enabled the use of a phone interface, allowing housekeepers to report problems with the room (e.g., light bulb out) as soon as the problem was identified rather than having to wait until the housekeeper ended the shift and verbally communicated the problem to the maintenance department.
- *Quality control*: Since the new information system allowed immediate reporting of problems, it reduced the chance of "slippage through the cracks" occurring when housekeepers at the end of the shift forgot to communicate the problem. It also eliminated the risk that maintenance would forget or claim it did not receive the request.
- *Preventive maintenance*: The maintenance department would be able to identify recurrent problems and stop them before they occurred again.
- *Reporting*: Management would be able to extract a number of extremely valuable reports from the system (see Case Appendix for details). This would allow managers to reward best performers and motivate employees.

Upon receiving the go-ahead from the executive team, Blake negotiated with the vendor for the application license, configuration and startup costs, support, and a week of onsite training. But as he was preparing for the upcoming roll-out and implementation, he was called to a new account. This unexpected call was bittersweet. Yes, he would not be able to see his very first project through, but the partner at FC must have noticed his performance since he was being reassigned to a project with a regional credit union in Cortland, New York. Not quite New York City, but the project was larger and more high profile. This was a good move for a summer intern!

As Blake handed the Royal Hotel project to his replacement and classmate at VFU, Jack Scarso, he was a bit nervous. Having been on a couple of teams with Jack back at school, Blake did not hold him in the highest esteem. Yet, telling himself that Jack deserved a fair shake, Blake turned over all the paperwork and his draft information system design, adding a word of caution:

> Jack, the GM is very impatient about this project. Make sure you don't let his anxiety for an operational system rush you into a half-baked design. This is a complex operation, there is a heck-of-a-lot going on here. Good luck!

Sic Transit Gloria Mundi[1]

A month and a half had gone by since Blake left the Royal Hotel. While he heard from Jack a couple of times regarding minor questions, he assumed everything had gone well. He felt good about the quality of the material he had left with Jack as well as the quality of the Cappuccino! application and the contract he had negotiated.

He had missed staying at the Royal Hotel, having traded down to a Ramada Inn across the street from the bank headquarters. But he felt good about the project as he wrapped up the documentation. A full-time offer was a sure bet!

"Here it comes," Blake smiled as he recognized the cell phone ring tone associated with his boss's personal cell phone. As he picked up, Blake quickly realized he was in for a surprise. Blake's boss sounded quite unhappy as he said,

> What happened at the Royal Hotel? I just got a call from the GM over there. He said that they did what you and Jack proposed and they wasted a bunch of money on a system nobody is using! I had my doubts about Jack, but I thought you'd have no problem with this project. You don't start school for another two weeks, right? My assistant just booked you on a flight back to NYC; you should have confirmation in your inbox.

Blake realized that this was not the time to voice his own doubts about Jack. Rather, he simply took ownership of solving the problem and began modifying his plans on the fly. Out were the pre-class barbeques and trading summer internship stories with classmates. Two weeks was probably just enough to attempt to straighten out the mess made by Jack. Blake's attempts to get in touch with Jack were futile. Jack's internship had ended, and he was backpacking through the woods of Utah to, as he put it, relieve stress and recharge his batteries before school started again.

Upon returning to the Royal Hotel, Blake found that the machine dedicated to Cappuccino! was sitting in a corner collecting dust. It looked like it was abandoned soon after roll-out, a suspicion confirmed by the director of IT, who mentioned that the installation and training session had been smooth sailing. Employees had been very eager to learn about the system but seemed to lose that interest rapidly afterward.

The director of housekeeping and the director of maintenance did not have much to add, simply noting that employees found the old manual system to work much better for their needs. The GM, on the other hand, had much to say, his German accent more pronounced than ever. The words were still ringing in Blake's ears as he left the meeting: "I invested a lot

1 A phrase meaning "thus passes away the glory of the world," used to remind us that nothing is permanent and that we must stay humble. As Blake found out, so fleeting can be the fortunes of a summer internship.

of money in this software. You better deliver all the results you promised in your presentation."

As Blake prepared to troubleshoot this failed information system and to devise a solution to make it work, he remembered the words of his information systems professor at VFU: "Focus on the information system design, not the technology investment!" "Therein lay the solution," Blake thought with a tinge of hope.

Discussion Questions

1. Despite having relatively little specific information about why the system failed, what do you think are the main reasons for such failure?
2. How could you fix these problems?
3. Reflecting on this experience, what do you think were the main mistakes, if any, that Blake made in handling the engagement?

Appendix: Brochure of Cappuccino!

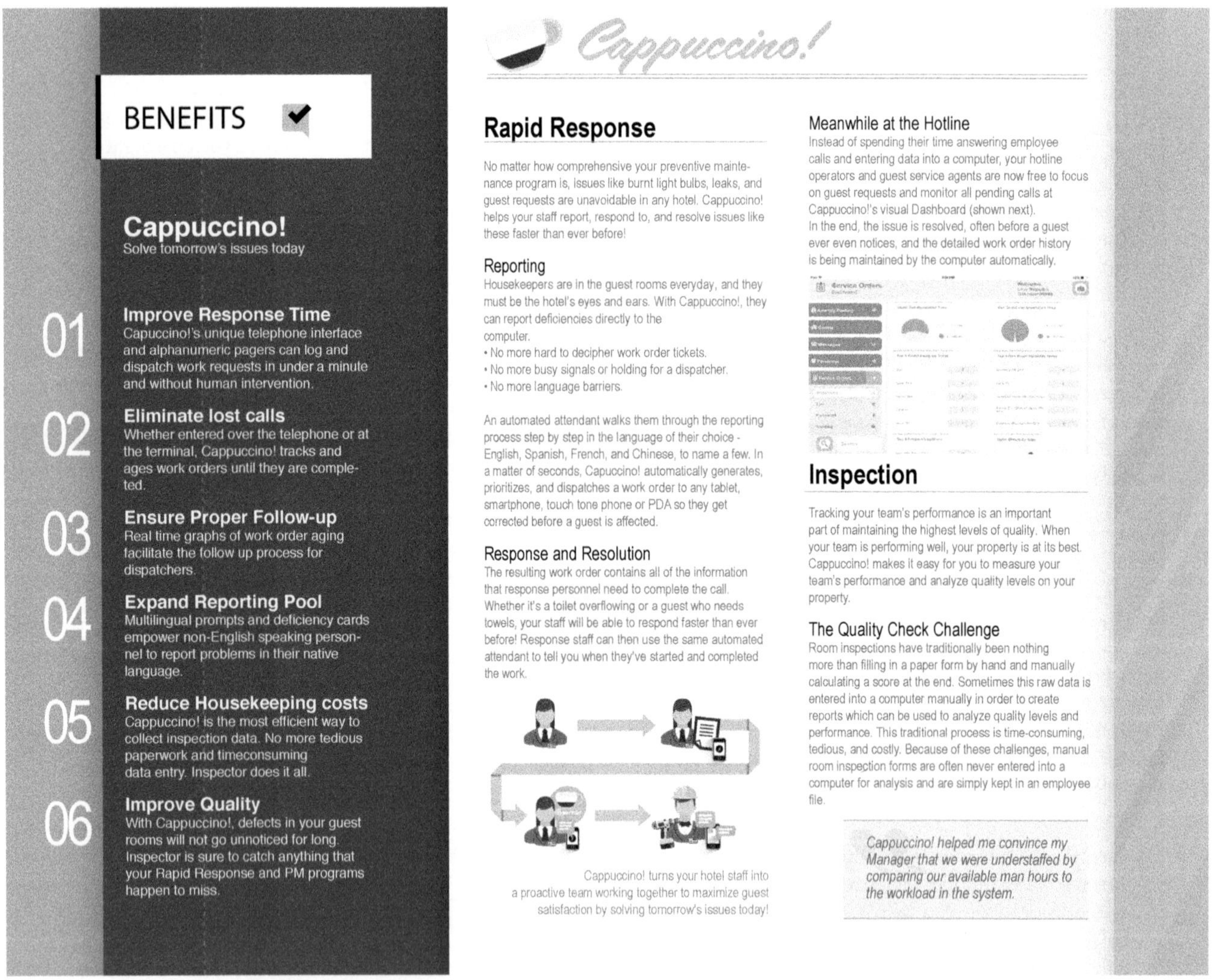

BENEFITS

Cappuccino!
Solve tomorrow's issues today

01 **Improve Response Time**
Capucinol's unique telephone interface and alphanumeric pagers can log and dispatch work requests in under a minute and without human intervention.

02 **Eliminate lost calls**
Whether entered over the telephone or at the terminal, Cappuccino! tracks and ages work orders until they are completed.

03 **Ensure Proper Follow-up**
Real time graphs of work order aging facilitate the follow up process for dispatchers.

04 **Expand Reporting Pool**
Multilingual prompts and deficiency cards empower non-English speaking personnel to report problems in their native language.

05 **Reduce Housekeeping costs**
Cappuccino! is the most efficient way to collect inspection data. No more tedious paperwork and timeconsuming data entry. Inspector does it all.

06 **Improve Quality**
With Cappuccino!, defects in your guest rooms will not go unnoticed for long. Inspector is sure to catch anything that your Rapid Response and PM programs happen to miss.

Cappuccino!

Rapid Response

No matter how comprehensive your preventive maintenance program is, issues like burnt light bulbs, leaks, and guest requests are unavoidable in any hotel. Cappuccino! helps your staff report, respond to, and resolve issues like these faster than ever before!

Reporting
Housekeepers are in the guest rooms everyday, and they must be the hotel's eyes and ears. With Cappuccino!, they can report deficiencies directly to the computer.
• No more hard to decipher work order tickets.
• No more busy signals or holding for a dispatcher.
• No more language barriers.

An automated attendant walks them through the reporting process step by step in the language of their choice - English, Spanish, French, and Chinese, to name a few. In a matter of seconds, Capuccino! automatically generates, prioritizes, and dispatches a work order to any tablet, smartphone, touch tone phone or PDA so they get corrected before a guest is affected.

Response and Resolution
The resulting work order contains all of the information that response personnel need to complete the call. Whether it's a toilet overflowing or a guest who needs towels, your staff will be able to respond faster than ever before! Response staff can then use the same automated attendant to tell you when they've started and completed the work.

Cappuccino! turns your hotel staff into a proactive team working together to maximize guest satisfaction by solving tomorrow's issues today!

Meanwhile at the Hotline
Instead of spending their time answering employee calls and entering data into a computer, your hotline operators and guest service agents are now free to focus on guest requests and monitor all pending calls at Cappuccino!'s visual Dashboard (shown next).
In the end, the issue is resolved, often before a guest ever even notices, and the detailed work order history is being maintained by the computer automatically.

Inspection

Tracking your team's performance is an important part of maintaining the highest levels of quality. When your team is performing well, your property is at its best. Cappuccino! makes it easy for you to measure your team's performance and analyze quality levels on your property.

The Quality Check Challenge
Room inspections have traditionally been nothing more than filling in a paper form by hand and manually calculating a score at the end. Sometimes this raw data is entered into a computer manually in order to create reports which can be used to analyze quality levels and performance. This traditional process is time-consuming, tedious, and costly. Because of these challenges, manual room inspection forms are often never entered into a computer for analysis and are simply kept in an employee file.

Cappuccino! helped me convince my Manager that we were understaffed by comparing our available man hours to the workload in the system.

REPORTS

Cappuccino!

see what you ever wanted to know, but you naver dared to ask!

01 **Most frequently occurring recurring issues**
Cappuccino! is the most effective way to collect recurring issues data. No more paperwork and timeconsuming data entry.

02 **Top reporting and/or completing performers**
With Cappuccino!, you finally measure your employees, easily identifying your best performers!

03 **Response time reports**
How quickly did you solve a customer's issues? With Cappuccino! get perfect visibility on guest request fulfillment and easily track repeat guests and recurring issues.

04 **Productivity reports**
Productivity will virtually leap out of Cappuccino!'s report generator to show you how well your team is performing. Once armed with this kind of data, you'll be able to target training programs and increase productivity like never before.

Cappuccino!

Mobile Capabilities

Cappuccino! provides features that were previously available only on a PC so users can access information via tablets and smartphones. Utilizing a team-based approach, service orders are automatically directed into group "buckets." Staff members can be assigned to single or multiple teams, thereby receiving notice for only those requests that pertain to their skill set. This better utilizes the strengths and abilities of individual employees, which reduces costs and increases productivity.

Smart Scheduling

Cappuccino! is designed to handle one of the most common issues guest room attendants encounter: trying to clean a room only to find that a guest is still in it. Cappuccino! provides guest room attendants with a way to set a time to return later and then re-issues the room to the most appropriate guest room attendant when the time has elapsed. This ensures that no room is forgotten and staff members are utilized to maximum effectiveness.

When cleaning a room, guest room attendants are provided with information about the guest and their preferences. In addition to the guest name, special instructions can be provided such as their desire for extra pillows.

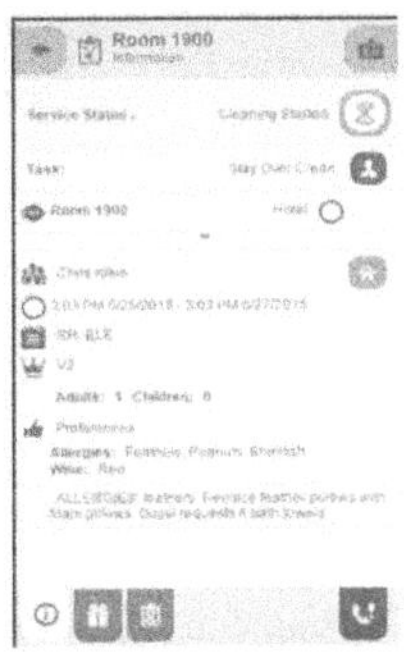

Empower Staff Members

Increasingly the industry is moving away from traditional and simple guidebooks and guidelines to empower their employees. Cappuccino! is an innovative solution to enable your staff to better relate with individual guests and coordinate among them.
Gone are the days of anonymous guests, and wrong communication between room attendants and maintenance staff.

Key Features

- Requests are received and handled by staff members in the field
- Guest related issues include guest name for recognition
- Every step is tracked including start, complete, response time, issue and more
- Guest initiated requests are flagged as incident or request
- Upon completion, issue is flagged for follow-up by guest agent
- Escalation rules ensure prompt attention

> *Cappuccino! provides a simple but thorough system, and we couldn't be more pleased with the solution...*

Case Study for Chapter 3

eLoanDocs: Riding the Tide of Technology without Wiping Out

Introduction[1]

On a warm summer evening in Northeast Ohio, Albert Michaels, the Chief Technology Officer (CTO) of local software company eLoanDocs, was enjoying his evening drive home. Though his eyes were on the road in front of him, his mind was stuck on the topic of the day behind him: the "cloud." How could eLoanDocs take advantage of this emerging and exciting new technology platform? Cloud computing held the promise of greatly reduced costs and nearly unlimited scalability for a company like his and seemed like it might be the wave of the future for hosted software providers. But the barriers to his customers' adopting the cloud were potentially high. And if those barriers were overcome, the competitive landscape in which eLoanDocs operated might shift in unfavorable ways. As a technology professional, adopting the cloud seemed to him to be a forgone conclusion. But his years of experience had shown him that it's rarely easy to be one of the early adopters.

Inefficiencies in the Mortgage Industry

The home mortgage closing process in the early 1990s was slow, paper intensive, and ripe for innovation. Realtors, mortgage lenders, title companies, and borrowers met and collaborated in primarily local marketplaces. The myriad documents required to support the mortgage approval process were exchanged through a combination of fax, mail, courier, and in-person reviews. Realtors, mortgage brokers, and escrow officers worked together to ensure that all of the necessary documents were generated, supporting services such as appraisals were ordered and performed, and required documents were signed by the borrower. The average time between a consumer application for a mortgage loan and the final closing was about 90 days. Closings were often delayed or rescheduled when late-breaking changes in the loan terms or associated costs required the lender to generate new documents. The majority of documents required for the mortgage closing were generated by the mortgage lender, but these documents were traditionally reviewed and signed by the borrower at the place of settlement (closing), generally at the title company. Mortgage lenders sent documents to the title company and to the borrower through mail, overnight express delivery, or courier. A successful closing required that the mortgage lender generate final documents and send them to the title company at least one day before the scheduled closing.

Technology to the Rescue

In 1994, a Cleveland-based title and settlement services company, Premium Title, was determined to reduce their costs and differentiate their service to the market by adding technology to the mortgage-closing process. Premium Title's owners created a separate company, eLoanDocs, to connect the various parties involved in the process by using technology. eLoanDocs' founders wanted to improve the speed and accuracy of the mortgage-closing process while increasing market share for Premium Title and other connected business partners. The founders believed that they could create a company that would grow quickly and that would generate significant return for their investors.

In industries where larger companies with dedicated IT staff existed, standard protocols had been developed to exchange information electronically. For example, in the automotive industry, Electronic Data Interchange (EDI) had been used for years to exchange purchasing and billing information between manufacturers and their suppliers.[2] There were no standards for electronic communication between business partners in the mortgage industry, and since the Internet was not being used broadly for commercial purposes, intercompany data exchange was dependent on proprietary communication networks.

eLoanDocs launched a proprietary electronic interchange in 1995 that connected Premium Title with several mortgage lenders in the Cleveland area along with a few local service providers such as appraisal vendors and surveyors. Proprietary

---

1 Excerpted from Cain, J., Levorchick, M., Matuszak, A., Pohlman, A., and Havelka, D. 2015. "eLoanDocs: Riding the tide of technology without wiping out." *Communications of the Association of Information Systems 36*. Article 38. Used with permission from Association for Information Systems, Atlanta, GA; 404-413-7445; www.aisnet.org. All rights reserved.

2 Hart, P. J., and Saunders, C. S. 1998. "Emerging electronic partnerships: Antecedents and dimensions of EDI use from the supplier's perspective." *Journal of Management Information Systems 14*(4): 87–111.

Roos, D. n.d. "The history of e-commerce." HowStuffWorks.com. Retrieved from http://www.howstuffworks.com/history-e-commerce1.htm.

data formats were defined for title insurance and appraisal orders, and mortgage documents were delivered electronically using the common HP Printer Command Language (PCL) print stream data format. The PCL[1] is a page description language (PDL) that allows a document's appearance to be described at a high level. This allowed Premium Title, using equipment commonly available at the time, to define the documents needed in their industry and share them with the necessary business partners.

eLoanDocs purchased off-the-shelf communications software and customized it to their needs; they also purchased computer servers, network equipment, and modems to run their electronic interchange. The computer equipment and telephone lines were hosted in their modest office space in Cleveland, Ohio. The small network of participating companies each installed modems, standard communication software, and eLoanDocs' proprietary software application to exchange documents that represented orders for services and the delivered real estate products such as appraisals, flood search certificates, and surveys. The electronically delivered documents replaced slower, lower-quality, or less-reliable courier and fax deliveries. eLoanDocs was successful in building a network of local mortgage service providers but struggled to extend the technology and business model outside of Northeast Ohio.

Right Technology, Right Place, Right Time

In the late 1990s, eLoanDocs realized that the emergence of the Internet as a driver of commerce would present both a threat to their network and an opportunity to extend their mortgage data interchange to more parties across the country at a lower cost. In 2000, eLoanDocs re-launched their mortgage industry electronic collaboration network on the Internet with the debut of their new software product, Document Posting Service (DPS). DPS used standard communication protocols such as HTTPS and SFTP over the Internet, which eliminated the need for modems and proprietary communications software. DPS also featured HTML web user interfaces for settlement agents to avoid the need for software to be installed at each customer location. DPS was a multitenant application (Figure 3.1) that provided software as a service (SaaS) to the mortgage industry.

SaaS allows customers to use software that is owned, delivered, and managed remotely by one (or more) providers.[2] This model allows the provider to maintain one set of code and data for many different customers. In essence, SaaS allows customers

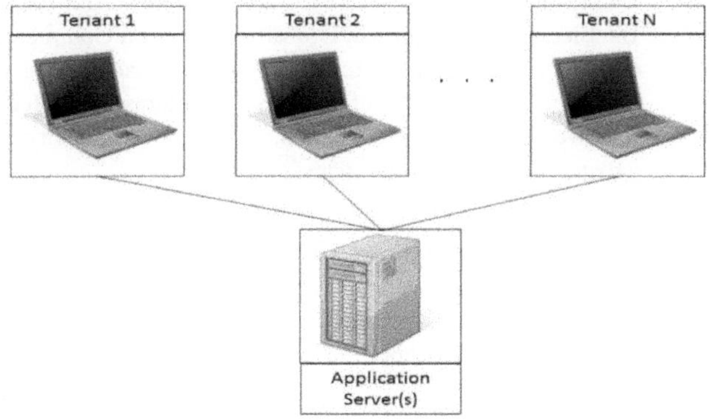

Figure 3.1. Multitenant Architecture

to rent software rather than buy it. The advantages of SaaS for customers include cost savings, scalability, accessibility, upgrades without disruption, and resilience. Some disadvantages also exist, the primary one being security.[3]

Market acceptance for DPS was tremendous, with several major mortgage lenders signing contracts to deliver all of their closing documents to settlement agents using eLoanDocs. As a small company facing growth challenges brought on in part by a boom-or-bust mortgage industry, eLoanDocs took a pragmatic approach to new product development. Product development investments were guided by immediate opportunities with existing customers that would lead to short-term revenue and possible broader market appeal. Following this model, eLoanDocs extended their product line beyond closing-document delivery to include borrower-disclosure delivery and electronic-signature capability.

Supportive Regulatory Changes

Federal and state legislation in 1999 opened the market for electronic signatures in the real estate industry, and eLoanDocs developed services to take advantage of this legislation. The Uniform Electronic Transaction Act (UETA) was first adopted by California and Pennsylvania in 1999.[4] At the time of writing in 2015, 47 of the 50 U.S. states have adopted this act. The remaining three states (New York, Illinois, and Washington) have not adopted the act but have statutes pertaining to electronic transactions. The UETA's purpose is to bring into line the differing state laws over such areas as retention of paper records and the validity of electronic signatures to support the

1 Hewlett Packard. 1992. "PCL 5 printer language technical reference manual." Retrieved from http://www.hp.com/ctg/Manual/bpl13210.pdf.

Wikipedia n.d. "Printer command language." Retrieved from http://en.wikipedia.org/wiki/Printer_Command_Language#See_also.

2 Singleton, D. 2011. "What is SaaS? 10 frequently asked questions about software as a service." Software Advice. Retrieved from http://blog.softwareadvice.com/articles/enterprise/saas-faqs-1072811/.

3 McLellan, C. 2013. "SaaS: Pros, cons and leading vendors." ZDNet. Retrieved from http://www.zdnet.com/article/saas-pros-cons-and-leading-vendors/.

4 Whitaker, D. R. 1999. "Rules under the uniform electronic transactions act for an electronic equivalent to a negotiable promissory note." *The Business Lawyer* 55(1): 437–53.

validity of electronic contracts as a viable medium of agreement. The Electronic Signatures in Global and National Commerce Act (ESIGN) is similar to the UETA with the exception that it pertains to the validity of electronic signatures on the federal level instead of the state level.[1] It also brings validity to signatures for foreign commerce.

The Changing Business Tides

eLoanDocs' business was growing fast, but the computers that hosted eLoanDocs' services were still run out of a small office computer room. On a hot summer day in late July 2000, Cleveland faced scattered power outages due to heavy draw on the power grid for air conditioning. Power was lost in eLoanDocs' office for over eight hours, well beyond the two-hour battery backup that was in place to support the computer systems. Dave Griffith, data center manager for eLoanDocs, said, "We tried to find portable generators for rent but there was nothing available big enough and we couldn't even get the generators close enough to our computer room to run extension cords" (personal communication). eLoanDocs' electronic services were unavailable to customers for most of the day. Customers suffered costly business delays due to this extended system outage on one of the busiest days of the month for mortgage closings.

It was clear that eLoanDocs needed to improve their computer hosting infrastructure in order to maintain a leadership position as a provider of electronic services to the mortgage industry. Up until this time, eLoanDocs did not have the necessary financial strength or the technical management experience to bring their computer infrastructure up to the needed levels of scalability and reliability. With major new customers ready to sign contracts and the memory of the 2000 power outage fresh in their minds, eLoanDocs management raised the needed capital and engaged a technology consulting firm to prepare for the next level of capability. In early 2001, eLoanDocs moved their computer servers to a private cage in a dedicated third-party co-location data center in Chicago, Secure Hosting. The Secure Hosting facility in Chicago featured redundant power feeds, on-site generators, multiple Internet providers, and state-of-the-art physical and network security. Secure Hosting quickly became eLoanDocs' most important and most expensive vendor.

Security Considerations

By 2004, eLoanDocs was doing business with seven of the top 10 mortgage lenders in the U.S., and documents and data for over 50% of the mortgages in the country flowed through eLoanDocs' systems. eLoanDocs had become a critical part of the mortgage industry, but with fewer than 50 employees and under $15 million in annual revenue, the company was hundreds of times smaller than most of its giant financial institution customers.

Given the sensitive nature of the information that eLoanDocs was handling, the attention given to cybersecurity breaches at well-known companies (like Apple, JP Morgan Chase, Target, and the Home Depot), and the consequences of these breaches,[2] many of eLoanDocs' largest customers began to demand that it demonstrate the reliability and security of their computer hosting facility through extensive load testing, system failure testing, and third-party security audits. Some customers sent their own security teams to the eLoanDocs office in Cleveland and to the Secure Hosting data center in Chicago to review eLoanDocs' policies, procedures, and capabilities. Paul Hunter, eLoanDocs CEO, was excited to show off Secure Hosting to the top mortgage companies:

> The first time the National Mortgage security team visited the Secure Hosting facility they were thrilled to see the biometric security, diesel generators with 3 days of fuel on-site, and our private cage that was secured on all sides. eLoanDocs finally looks like the big player that we are. (personal communication)

By 2007, demands for additional capacity in the network and customer requirements to maintain an active disaster recovery data center drove eLoanDocs to make several significant investments. First, eLoanDocs acquired a competing mortgage technology company based in Seattle, WA, FastForms. eLoanDocs then moved their primary data center from Secure Hosting in Chicago to FastForms's co-location provider in Seattle, SunGuard. Finally, eLoanDocs built an identical redundant hosting facility in Cleveland using another co-location provider. In late 2009, eLoanDocs completed implementing a highly scalable and virtualized computer hosting infrastructure in Seattle with real-time replication of all customer documents and data to the backup site in Cleveland. The Cleveland facility could automatically take over all of eLoanDocs' services in the event of an extended outage in the Seattle data center (Figure 3.2). The time and expense required to build and maintain their services in secure and redundant data centers gave eLoanDocs a significant advantage in the market because few technology providers could make the necessary investments in infrastructure and software required to compete. In addition, eLoanDocs implemented best practices for disaster recovery (DR) planning including risk assessment and business impact analysis and training for and testing of the DR plan.[3]

---

1 Stern, J. E. 2001. "The electronic signature in global and national commerce act." *Berkeley Technology Law Journal 16*(1): 391–415.

2 Snyder, B. 2014. "5 huge cybersecurity breaches at companies you know." *Fortune*. Retrieved from http://fortune.com/2014/10/03/5-huge-cybersecurity-breaches-at-big-companies/.

3 ComputerWeekly.com. 2011. "IT disaster recovery plan best practices: Fundamentals in DR Planning." Retrieved from http://www

With the new infrastructure in place, eLoanDocs met customer service–level agreements (SLAs) for 99.9% uptime of services in 2011 and 2012. eLoanDocs had developed a mature set of policies and procedures around information security and had published results of a third-party SSAE 16 Type II compliance audit twice a year to customers. The organization had six full-time staff dedicated to data center operations and a full-time information security officer. Their internal staff had accumulated significant expertise in data center operations, but the company experienced, on average, a 20% annual turnover rate due to an active job market for their staff members' highly sought-after skills. One eLoanDocs employee was recruited to manage networks for Microsoft's hosting facilities in Washington. Michelle Fletcher, eLoanDocs' Director of Technical Operations, complained, "I'm having a hard time keeping my best people working here at eLoanDocs. We just don't have enough scale to keep these people challenged and there is no way that eLoanDocs can match the pay of the big guys" (personal communication).

eLoanDocs' annual vendor expenses for data center hosting, data networks, computer hardware maintenance, and software support subscriptions were nearly USD2 million per year. Employee costs and third-party audit expenses brought the overall cost of eLoanDocs' data center hosting, security, and compliance to about USD3.5 million annually.

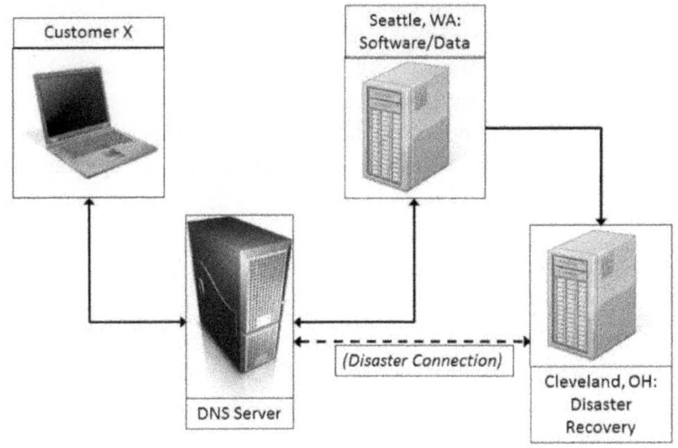

Figure 3.2. Existing Infrastructure

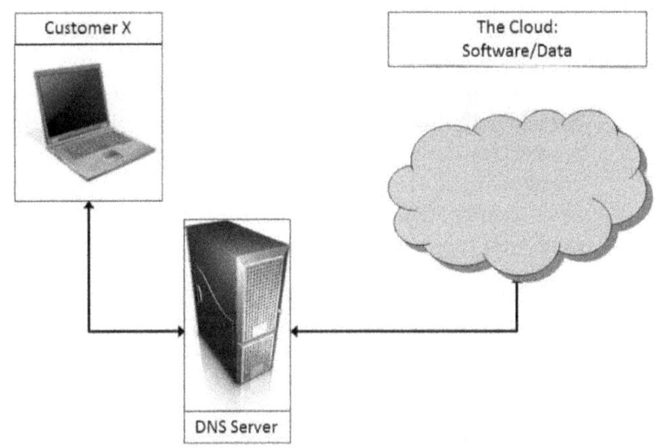

Figure 3.3. Cloud Infrastructure

Clouds Ahead?

Just as the rise of the Internet enabled eLoanDocs' explosive growth in the 2000s, technological changes beginning in 2010 led to new opportunities and competitive challenges for the company. Giant technology vendors such as Amazon began to offer comprehensive computer hosting services with a new model: cloud computing (see Figure 3.3). Amazon, Microsoft, Google, and other companies built data centers at massive scale that were designed to allow them to sell computing capacity to the market at prices significantly below what companies could achieve on their own.[1] Cloud computing vendors offered a model where a company could simply purchase the needed amount of processing power, memory, disk storage, and Internet bandwidth on a monthly subscription model. Customers could increase or decrease their usage on demand. Public cloud providers also offered high availability, multisite data replication, and full disaster recovery capabilities as optional or standard services.[2]

Using a cloud hosting service such as Amazon Web Services (AWS), a small software vendor could launch a new service with twice the computing capacity of eLoanDocs in a matter of days. Matt Pittman, VP of Sales for eLoanDocs, said,

.computerweekly.com/report/IT-disaster-recovery-plan-best-practices-Fundamentals-in-DR-planning.

McBeth, C. B. 2014. "Business continuity and disaster recovery best practices from the availability trenches." *Forbes*. Retrieved from http://www.forbes.com/sites/sungardas/2014/11/19/business-continuity-and-disaster-recovery-best-practices-from-the-availability-trenches/.

1 Cloud Computing tutorials: http://www.tutorialspoint.com/cloud_computing.

2 Beal, V. 2015. "Cloud computing." Retrieved from http://www.webopedia.com/TERM/C/cloud_computing.html.

IBM. n.d. "What is cloud?" Retrieved from http://www.ibm.com/cloud-computing/us/en/what-is-cloud-computing.html.

Strickland, J. n.d. "How cloud computing works." HowStuffWorks.com. Retrieved from http://computer.howstuffworks.com/cloud-computing/cloud-computing.htm.

> I can't compete on price with mortgage technology competitors like EchoSign that launched their products hosted at Amazon. Their costs are so low that they are giving away basic services with a freemium model. I just hope that companies like this don't start cutting into our core client base. (personal communication)

Services like AWS were examples of the "public cloud"—inexpensive computing capacity that can be purchased on demand, with many different customers' workloads and information intermingled on the same computer servers and storage devices. Cost advantages available to customers of public cloud services were enhanced by aggressive competition in the industry, which sparked an ongoing price war between providers. Amazon reduced its prices a total of 41 times between 2008 and late 2013. CFO Marty Buckley had calculated that, by switching to a public cloud provider for all of their data center needs, eLoanDocs' annual technology costs (including expected staff reductions) would be $750,000 less than current spending levels.

Public cloud providers also maintained strict security policies and published third-party security audit results, but large financial institutions were not ready to trust their most critical information and systems to the public cloud as of 2013. eLoanDocs' security officer Randy Wallace had his doubts about the viability of cloud hosting for eLoanDocs:

> I just finished another grueling vendor audit from a giant mortgage lender's security team. These guys want visibility into all of our processes and they want to make sure that eLoanDocs has control over every aspect of our systems. I just don't know how we could ever convince them that a cloud service is secure. (personal communication)

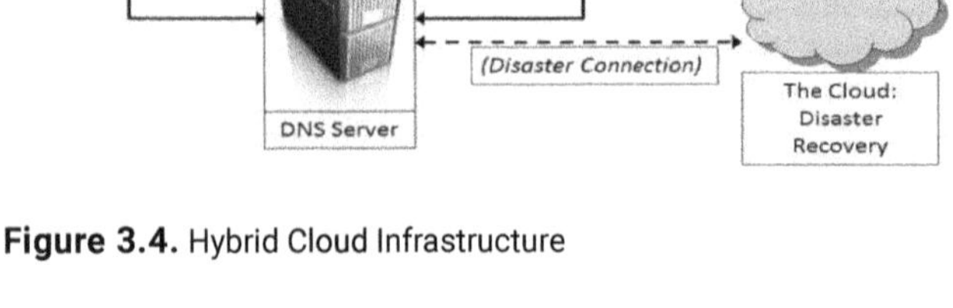

Figure 3.4. Hybrid Cloud Infrastructure

Concerns regarding the security of data stored in the cloud continue to be an ongoing challenge for many IT executives.[1]

Recognizing the need for more secure and more flexible cloud computing options, computer hosting vendors such as Rackspace began to offer private cloud solutions to the market. Rackspace provisioned and supported a set of dedicated hardware to any customer that wanted to keep its applications and information segregated from other customers. The private cloud offerings used the same technologies as public cloud providers and still provided cost advantages due to economies of scale. Buckley had calculated a USD350,000 annual savings should eLoanDocs move to a private cloud solution. However, Albert Michaels was concerned about service availability and uptime with a third-party private cloud solution:

> With our services hosted in our data centers I know 100% for certain that my team can find the source of any problem and fix it within minutes, helping us to meet our customer SLAs. How do I know that a cloud provider will have the same ability and motivation to get things back up and running when there's a problem? (personal communication)

A third cloud hosting model appeared called hybrid cloud (Figure 3.4). This model allowed customers to take a measured approach to moving some of their computing to outsourced cloud providers. With a hybrid cloud offering such as VMware's vCloud, a software company could easily host some of its applications and data on its internal servers while moving its development, test, or disaster recovery systems to the cloud. Hybrid cloud solutions offered many of the security benefits of internally hosted systems while also providing scalability on demand. Arlene Christianson, eLoanDocs' VP of operations, felt that hybrid cloud was not a good fit for eLoanDocs because "if we go with a hybrid cloud solution, we will need two separate security and compliance audits and sets of controls" (personal communication). Buckley estimated that moving to a hybrid cloud solution would reduce eLoanDocs' technology costs by about USD200,000 per year.

A Necessary Decision

By the middle of 2014, the computer systems that eLoanDocs had installed in 2009 were nearing the end of their useful life and had no more capacity for expansion. As eLoanDocs prepares for their next generation of data center hosting architecture for 2015 and beyond, the choices they face are complex and will have significant implications for the future of the company.

1 Corbin, K. 2015. "Security concerns cloud federal data center overhaul." CIO. Retrieved from http://www.cio.com/article/2878657/data-center/security-concerns-cloud-federal-data-center-overhaul.html.

- Should eLoanDocs continue with their current model of designing, building, and managing their own computer hosting infrastructure using their co-location partners?
- Would eLoanDocs' financial institution customers accept a move of eLoanDocs' services to a public or private cloud provider?
- How should eLoanDocs go about choosing a cloud hosting provider?

As Albert Michaels considered his options, his mind was roiled by a number of questions:

- Will eLoanDocs' customers—with their focus on data security—accept a cloud-based solution? If so, to what degree? And how many customers would accept some form of cloud-based solution?
- If customers do accept one of the cost-saving cloud services solutions, what barriers exist to prevent new competitors from rapidly entering the market and eroding eLoanDocs' market share?
- What is the value that customers believe they are receiving from eLoanDocs?
- Could it be that customers actually value the secure environment that they can visit and audit in person? If so, might convincing those customers to adopt a cloud-based solution to their document delivery problems actually be a damaging move to eLoanDocs in the long term?
- Speaking of security, which solution actually provides better protection of customers' data? Though ownership of the hosting hardware enables eLoanDocs to literally pull the plug if a breach is detected, how does that compare to the security benefits associated with outsourcing to a cloud provider? Is one solution more likely than the other to be targeted for attack? Is either solution better able to detect and prevent intrusions?
- Assuming that the system will be attacked at some point, what ability will eLoanDocs have to identify the compromised data? How might that ability change if hosting services are outsourced to a cloud provider?
- How robust is the existing disaster recovery strategy? Which solution best fits the redundancy needs of eLoanDocs?
- How might the eLoanDocs employees react to adoption of a cloud-based hosting solution?

As Albert considered these and other questions, the only answer he felt sure about was that it was an exciting time to be alive and working in the technology industry.

Glossary

Cloud computing: Using network resources to perform computations without the need, or often the ability, to determine the exact resources used at any time.

Co-location facility: Physical location that provides reliable power, secure physical facilities, and networking services to clients for a monthly operating fee. Clients provide their own hardware to run in the co-location facility.

Disaster recovery: Alternative to normal system operations intended to be used in case of catastrophic events (e.g., widespread power outages, local natural disasters).

High availability: High rate of system uptime, typically in excess of 99%. Also refers to technologies required to achieve a high rate of system uptime, such as redundant hardware components.

Hybrid cloud: System configuration in which some combination of public cloud, private cloud, and dedicated server solutions are mixed and used together to form the complete system.

Infrastructure as a service (IaaS): A type of cloud computing where the cloud provider provisions and maintains the computer hardware, storage, and networking for their clients, while the client is responsible for maintaining the operating systems and software.

Multitenant: Software configuration in which a single instance of the system serves multiple clients. Clients typically have no visibility or awareness of the data (or even the existence) of other clients.

Private cloud: Ownership and management of cloud computing resources within an organization's firewall or optionally dedicated equipment managed by a cloud hosting provider on behalf of a customer.

Public cloud: Computing resources that are hypothetically available to any user connected to the same cloud service provider.

Recovery point objective (RPO): Amount of time for which data may be lost due to catastrophic events.

Recovery time objective (RTO): Amount of time that a system may be unavailable due to unexpected circumstances (e.g., a catastrophic event that prevents the function of the system).

Replication: Act of making exact copies of systems. Disaster recovery plans often use replication in order to minimize the RPO of a running system by using identical hardware located at geographically remote sites and synchronizing the data storage in real time.

Software as a service (SaaS): Licensing software solutions such that the hardware and the software are typically remote to the licensees and administered and maintained by the licensors.

SSAE 16 Type II: Statement on Standards for Attestation Engagements (SSAE) 16 is the professional standard used for issuing reports in accordance with the American Institute of Certified Public Accountants' Service Organization Control (SOC) reporting framework, which consists of SOC 1 (SSAE 16) along with SOC 2 and SOC 3 (AT 101) reporting. Additionally, the SSAE 16 standard effectively replaced the aging and antiquated SAS 70 auditing standard that had been in use for approximately 20 years.

Virtualization: Creating a logical instance of a real system in such a way that it appears to an end user as a real system. A virtual machine—configurations of powerful servers so that multiple operating systems can be run with their own disk storage partitions—is a common example of virtualization.

Case Study for Chapter 4

Online Education

Thirty years from now the big University campuses will be relicts. Universities won't survive. It is as large a change as when we first got the printed book.

—Peter Drucker[1]

Introduction[2]

Almost 25 years after Peter Drucker had made his famous prediction, the digital disruption of countless industries was in full swing. Yet, the education industry had been a difficult nut to crack for technology companies and Silicon Valley investors—despite multiple tries. This was ironic, as lectures were classic information goods, and the process of acquiring knowledge seemed perfectly virtualizable. But on the eerily quiet campus of Ivey University in the midst of the COVID-19 pandemic, Drucker's words sounded like a warning echoing from the past to Federico Pigni. The dean of the School of Information Management recalled the quote while trying to organize lectures on the main campus amid ever-changing governmental restrictions, student demands, and professors' struggles to adapt to online teaching at the main campus and the many affiliated programs around the world. Could this pandemic be the "acceleration event" that would finally prove Drucker right?

The seemingly relentless announcements of education ventures, one after another, powered by the Internet and the digital innovations that leveraged it, were making administrators like Dean Pigni question themselves about the future of the institution and programs they had been entrusted to lead. Just yesterday, Pigni had seen that Techcrunch had published a story titled "Professor Scott Galloway just raised $30 million for an online school that upskills managers fast."[3] Galloway, an adjunct professor at NYU Stern School of Business, shot to prominence for a viral email directed to a student with the bold exhortation: "get your s@#! together."[4] Prof. Galloway was a frequent commentator on branding and marketing strategy who claimed to have successfully enrolled a number of high profile academics to his venture, Section 4, "dedicated to making an elite business education more affordable and accessible."[5] And a minute ago, one of Pigni's colleagues had forwarded the announcement that Google was expanding its Professional Certificate Training program, claiming the following: "No experience necessary," "Learn at your own pace," "Stand out to employers," "A path to in-demand jobs."[6] With paths to jobs like IT support, data analytics, project management, and UX design, Google was encroaching right into Ivey University's territory.

Whereas these new online players seemed to have had minimal incremental effect on prestigious research universities such as Ivey, the future appeared far more uncertain for "middle-tier" institutions that had suffered mightily from the forced move to online education the COVID-19 crisis had sprung on them out of nowhere.[7] Enrollment in undergraduate programs at big-name colleges was more selective than ever due to rising demand. Campuses were teeming with construction workers developing new buildings, adding to existing ones, remodeling teaching and office space, equipping ever-more-sophisticated labs. Ivey had just finished building a multimillion-dollar robotic experience lab where students could come to experiment with new digital technologies in perfectly recreated environments, including a sector of downtown, a supermarket, and a connected shop. Then the pandemic hit.

1 Lenzner, R., and Johnson, S. S. 1997. "Seeing things as they really are." *Forbes*, retrieved from https://www.forbes.com/forbes/1997/0310/5905122a.html.

2 While based on a real educational institution and public sources, this is a fictitious case developed by the textbook authors with the exclusive intent of supporting class discussion.

3 Loizos, C. 2021, March 11. "Professor Scott Galloway just raised $30 million for an online school that upskills managers fast." *TechCrunch*, retrieved from https://social.techcrunch.com/2021/03/11/professor-scott-galloway-just-raised-30-million-for-an-online-school-that-upskills-managers-fast/.

4 Daulerio, A. J. 2010, February 22. "NYU business school professor has mastered the art of email flaming." *Deadspin*, retrieved from https://deadspin.com/nyu-business-school-professor-has-mastered-the-art-of-e-5477230.

5 "Business Education for All." 2021. *Section4*, retrieved from https://www.section4.com.

6 "Professional Certificate Training Programs—Grow with Google." 2021. *Grow with Google*, retrieved from https://grow.google/certificates/#?modal_active=none.

7 Hubler, S. 2020, October 26. "Colleges slash budgets in the pandemic, with 'nothing off-limits.'" *New York Times*, retrieved from https://www.nytimes.com/2020/10/26/us/colleges-coronavirus-budget-cuts.html.

Nietzel, M. T. 2020, October 7. "Liberal arts colleges face the full force of the pandemic." *Forbes*, retrieved from https://www.forbes.com/sites/michaeltnietzel/2020/10/07/liberal-arts-colleges-face-the-full-force-of-the-pandemic/.

Executive education programs had been a different story, with numbers declining steadily over the last decade. Some blamed growing competition and a strong job market, but it was hard to ignore the challenge from established and new online learning outfits like Coursera, Udemy, and a whole host of nondegree executive education.[1] However, as Dean Pigni pondered the future, he recalled a passage from an article in the *Economist* that he had seen a few years before. The memory brought back some of Dean Pigni's own uneasiness: "The innate conservatism of the academic profession does not help. The modern university was born in a very different world from the current one, a world where only a tiny minority of the population went into higher education, yet many academics have been reluctant to make any allowances for massification."[2]

Was everyone missing the forest for the trees? Was the Internet finally a mature enough technology to sustain the disruptive innovation that would finally digitally transform education as it had done with the news and entertainment industries? Was the pandemic crisis that made Zoom a household name and worth over $100 billion a passing storm or a secular change of behaviors and expectations that would soon blindside slow-to-react incumbents?

Education and Research at Ivey

Like its peers, Ivey University had a complex mission and a large community of stakeholders, ranging from students and faculty to alumni and the local and global community. At the highest level of analysis, Ivey performed two main activities: the creation of new knowledge (i.e., research) and the dissemination of knowledge (i.e., education).

As a prestigious research institution, Ivey spent a considerable amount of resources supporting the development of new knowledge by hiring some of the brightest young faculty members and accomplished researchers it could find. Rating always at the top of all rankings, Ivey University boasted three main world-class research centers.

While the research mission was pursued in experimental labs and offices throughout campus, the most evident manifestation of Ivey's contribution to society was its teaching mission. A large school like the School of Information Management at Ivey University had truly global reach.

Its largest population was about 1,200 undergraduate students. The school also trained master's students, who left the workforce for one or two years (a substantial opportunity cost on top of the direct costs of going back to school) to gain an advanced degree and the skills to accelerate their careers. Ivey had a medium-sized, but very selective, master's program with about 100 students enrolled. Finally, the school educated the next generation of faculty and researchers by way of its PhD program.

A very recognized brand in the business world, Ivey also offered a number of executive education and professional education programs. These were typically highly condensed courses, held on Ivey's own campus or satellite locations, designed to serve the needs of corporations seeking to update the skills of their workforce or to offer working students a chance to access the wealth of knowledge that the school's faculty had to offer without having to resign from their job.

Global Expansions

Because of its brand recognition around the world, the School of Information Management and a number of other schools at Ivey had been focused on global expansion through partnerships and the opening of satellite campuses. The school had more than 500 international partners, enabling it to develop new, complementary teaching methods and content, and offering students the opportunity to create and participate in enriching transcontinental educational experiences. Partners were mainly located in Asia, Africa, and Europe and the school was currently evaluating whether to enter the South American market with a new campus.

The reason for global expansion was simple: with the skyrocketing demand for high-quality education in emerging markets around the globe, there was great opportunity to extend the Ivey brand. Expansion was not without challenges, with revenue models being at times challenged and a myriad of logistics and quality assurance hurdles to be overcome. However, with almost every other recognized education brand entering the new markets, a wait-and-see attitude could be extremely risky.

Online Players: A Real Threat or a Nuisance?

Since Peter Drucker's prediction, there had been a significant amount of development in online educational offerings. University of Phoenix, the largest for-profit institution in the USA, had about 100,000 enrolled students now, with a peak of 470,000 in 2010. While quality concerns lingered, not just on prestigious university campuses, online universities seemed to be gaining traction.

Perhaps even more interesting, and threatening, were open-source content creation and delivery entities. The best example was offered by the for-profit education platform Coursera—the brainchild of Daphne Koller and Andrew Ng, who offered their computer science courses online, drawing the attention of more than 100,000 students. Coursera, now backed by more than $400 million in funding and readying its IPO, had come a long way, making 3,100 courses available to interested students from all over the world and keeping a running count of more than 50 million learners, 2 million course completions, and 136 university and 50 industry partners. Coursera was focused on academic courses in topics ranging from math and science

1 *Financial Times*. 2015, May 17. "New market entrants create competition for business schools." *Financial Times*, retrieved from http://on.ft.com/1JrEyeH.

2 *Economist*. 2014, June 28. "The digital degree." *Economist*, retrieved from http://www.economist.com/news/briefing/21605899-staid-higher-education-business-about-experience-welcome-earthquake-digital.

to history and the humanities, and it was originally geared toward college students. As Dr. Koller put it, "Online education is a more effective way to teach basic facts and skills in part because students can learn at their own pace. Taking classes online gives people the opportunity to pause and reflect and grapple with the material."[1] Another innovator, Lambda School, had pioneered the a revenue model that challenged tradition head-on. The school focused on coding skills and promised, "Train remotely to become a web developer or a data scientist and pay no tuition until you're hired."[2] With the cost of education skyrocketing, particularly in North America, and the increasing focus on tangible skills rather than degrees for technical hires, the Lambda approach could be very disruptive for established players.

Not all online educational offerings were by upstarts. Indeed, traditional universities had their own offerings. Ivey itself had launched its own online education effort only six years ago, where distance learning was a necessity for students. As an example, Ivey's online program for top athletes was explicitly designed to allow students to attend classes even during their intense training and in the months preceding the Olympic Games.

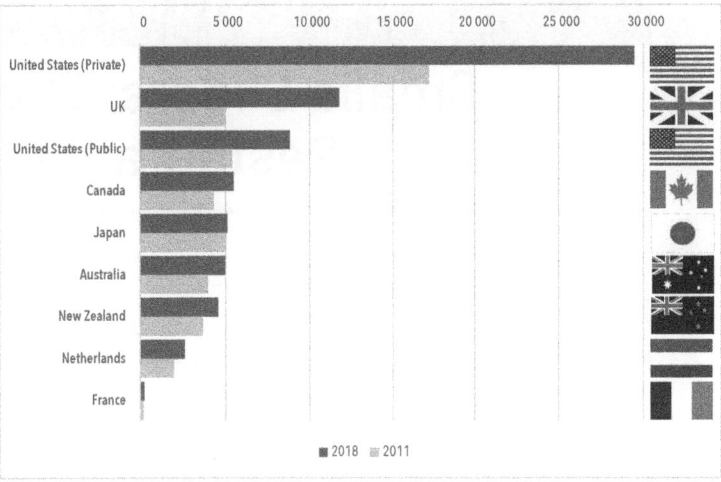

Figure 4.1. Bachelor's tuition fees compared

Source: OECD, "Education at a Glance," 2013, 2018

Note: Estimated annual average tuition fees in equivalent USD converted using purchasing power parities.

The Future

As Dr. Pigni watched the snow drop a fresh dusting of white powder on the city skyline and the nearby mountains, he pondered some of the words of the article he read:

> Other industries next in line for disruption like education and health care would be wise to pay attention. Most of what they do depends on the control of information that will soon no longer be scarce. Education reformers have long predicted a world where top professors spread their knowledge across the globe through electronic tools. But the knowledge students need is not only located in those few professors' minds. Once we digitize not just the distribution of knowledge but the production of it, the existing university system loses its *raison d'etre*. Why would people come to a single physical location at higher and higher costs when the knowledge it houses is no longer scarce?[3]

The words of Dr. Koller in this recent *Wall Street Journal* article seemed to echo Drucker's words: "School experience will be like turning the tap—and great education comes out for anybody. We'll have data from hundreds of thousands, millions of people at a level that's unprecedented. I think we're at the cusp of a revolution of treating human learning as science."[4]

Would this really happen? And how would it affect a top university like Ivey? As the dean of the School of Information Management, Dean Pigni was not only entrusted with the future of the school he led, but he also felt a responsibility to help the university community at large thrive in the network economy. Would Ivey miss the wave of the future? "Not on my watch!" Dean Pigni told himself while getting ready for the first of the day's many meetings.

Discussion Questions

1. Do you agree with Peter Drucker's opening quote?
2. Is the Internet fostering disruptive innovation in the education industry in general? What about for Ivey's School of Information Management in particular?

---

1 Wolfe, A. 2015. "Daphne Koller on the future of online education." *Wall Street Journal*, retrieved from http://www.wsj.com/articles/daphne-koller-on-the-future-of-online-education-1433532321.

2 "Our concept." 2021. *Lambda School*, retrieved from https://www.raisingcanes.com/our-concept.

3 McQuivey, J. 2011. "Why the end of scarcity will change the economics of everything." *Mashable*, retrieved from http://mashable.com/2011/04/28/scarcity-economics/?utm_source=feedburner&utm_medium=feed&utm_campaign=Feed%3A+Mashable+%28Mashable%29&utm_content=Google+Feedfetcher.

4 Wolfe, A. 2015. "Daphne Koller on the future of online education." *Wall Street Journal*, retrieved from http://www.wsj.com/articles/daphne-koller-on-the-future-of-online-education-1433532321.

Case Study for Chapter 5

Smalls Sliders: The Composable Restaurant Franchise

We like to keep things pretty simple. At Smalls, we have one mission: simplify everyday life with moments of delicious food and true gratitude. We want to create the ideal food experience for our customers: simple, fun, delicious.

—Smalls Sliders[1]

Introduction[2]

Recalling when he had been invited to the warm opening of Smalls Sliders,[3] Joaquin Rodriguez could not stop thinking about how he would manage the growth of the chain if he had been in charge. Joaquin, a PhD candidate at the E. J. Ourso College of Business at Louisiana State University, was writing his dissertation on the strategic impact of digital technology and had been focusing on online food delivery as the context of his empirical work.

During the warm opening, when he had toured the back-of-the-house of the restaurant, he formulated the idea that Smalls, as the founders and employees liked to call the concept, was perfectly positioned to take advantage of two colliding trends: the rise to prominence of digital channels in the restaurant industry and the availability of digital resources in an increasingly infrastructural, combinatorial and servitized competitive environment. And that was months before the COVID-19 pandemic crisis ravaged the restaurant industry. Even as the US was seemingly coming out of the crisis, many customer behaviors had been changed forever.[4] All this work he was doing, he thought, was not just "academic" stuff; it could really guide decision-making for real businesses.

The Making of Smalls

Smalls Sliders was the brainchild of Brandon Landry, successful restaurateur and cofounder of the Walk-On's franchise Sport Bistraux, based in Baton Rouge, Louisiana. "Smalls was created a lot like Walk-On's, 'Hey, I have an idea!'" Landry said. "I told Drew about the concept, and he was all in."[5] Drew is Drew Breeze, NFL superstar and partner in the Walk-On's restaurant concept. The duo launched the concept with cofounders Jacob Dugas and Scott Fargason. "We couldn't be more excited to launch this unique concept in our hometown," Dugas was quoted as saying in a press release, adding that the restaurant was focused on helping guests "celebrate the small things."[6] As the name suggested, the menu was focused exclusively on slider burgers as the main dish (Figure 5.1).

Local monthly lifestyle periodical *225 Magazine* described the concept:

> Smalls Sliders is not your average fast food joint. Located [near Louisiana State University], the bright orange burger restaurant designed from repurposed shipping containers makes simplicity look cool. . . . With this new concept, less is more. The menu has four main slider combos (with one, two, three or four sliders depending on your appetite) that come with fries and a drink. You can also order vanilla and chocolate milkshakes, children's options and party packs.[7]

The concept was unique in that the whole restaurant fit in a shipping container (Figure 5.2). Dugas explained on the radio show *Foodie Friday*, "The whole building is modular, we build everything off-site, all of the equipment is put in there,

---

1 Smalls Sliders. n.d. Homepage. *Smalls Sliders*, retrieved from https://www.smallssliders.com/.

2 While based on a real restaurant concept and public sources, this is a fictitious case developed by the textbook authors with the exclusive intent of supporting class discussion. All product names, logos, and brands are property of their respective owners. All company, product and service names used in this note are for identification purposes only. Use of these names, logos, and brands does not imply endorsement.

3 A "warm opening" of a restaurant is a pre-opening phase where friends and family are invited to dine for free at a restaurant while the operation works out the inevitable imperfections in food preparation and service operations.

4 Watson, R., Ives, B., and Piccoli, G. 2020. "Guest editorial: Practice-oriented research contributions in the COVID-19 forged new normal." *MIS Quarterly Executive 19*(2). Article 2. Retrieved from https://aisel.aisnet.org/misqe/vol19/iss2/2.

5 Fastcasual.com. 2019, March 15. "Smalls Sliders will make debut in Baton Rouge." *Fastcasual.com*, retrieved from https://www.fastcasual.com/news/smalls-sliders-will-make-debut-in-baton-rouge/.

6 Ibid.

7 Corfah, C. 2019, October 3. "Fast break: The new Smalls Sliders on Nicholson Drive brings small bites with big flavor." *225 Baton Rouge*, retrieved from https://www.225batonrouge.com/food-drink/fast-break-new-smalls-sliders-nicholson-drive-brings-small-bites-big-flavor.

Figure 5.1. Smalls Sliders menu

Figure 5.2. Architectural rendering of the Smalls Sliders concept

the flooring is in, it comes in and its plug and play. In 60 days we have a restaurant."[1] Locations had a small outside area, reminiscent of 1950s roadside burger joints, where food could be consumed. However, the whole concept was geared toward take out, with the drive-through lane being the primary source of traffic. A year after the warm opening in September 2019, Smalls had delivered it's 500,000th slider. Dugas commented, "With this whole pandemic [it] was kind of an up-and-down, it was a rollercoaster. . . . With LSU [students] being sent home, your whole customer base is gone. But we did some grassroots marketing, we took the food truck out, people started hearing about us. . . . End of March, mid-April [2020] we definitely took a hit, but after that it really took off."[2]

The company took great pride in the quality of the ingredients, with food quality and perfect execution of the recipes being the critical success factor in the founders' minds. Thus, the staff prepared the 2-ounce (56g) cooked-to-order burger patties daily, almost 1,200 a day on average, and ordered the buttery buns from a local provider.[3] But food was just the beginning in the creation of a successful operations. "We know that it takes a lot more than a product to make a quick-service restaurant," Dugas said in an early interview. "In particular, the culture of a brand. Who you are as a company."[4] At Smalls, this was evident in every aspect of the branding, from the building, to the colors to the service. From inception, it appeared that the

1 Talk 107.3. 2020, August 7. "Foodie Friday featuring Smalls Sliders." *Talk 107.3*, retrieved from https://talk1073.com/2020/08/07/foodie-friday-featuring-smalls-sliders/.
2 Ibid.
3 Ibid.
4 Klein, D. 2019, March 25. "With help from Walk-On's, Smalls Sliders cooks up big plan." *QSR*, retrieved from https://www.qsrmagazine.com/exclusives/help-walk-ons-smalls-sliders-cooks-big-plan.

founders were laser-focused on making a highly recognizable, highly scalable, and easily franchisable restaurant concept—not unlike Raising Cane's, another restaurant with a limited menu started right outside the gates of LSU:

> Our concept is simple and unique . . . we only have ONE LOVE—quality chicken finger meals! At Raising Cane's® you get an exceptionally high quality product served quickly and conveniently. We can do this because we offer a limited menu. The specialized systems developed by Raising Cane's® allow us to maintain a level of quality unmatched in the industry. Our commitment to this concept will not allow us to compromise our quality, cut corners or clutter our menu with new products that do not fit our core menu offering.[1]

Launched in 1996, Raising Cane's had grown dramatically in recent years. Forbes wrote,

> Raising Cane's is on a growth trajectory unlike most concepts in the crowded restaurant space. Consider the chain's sales in 2016 were about $576 million with about 250 units. In 2019, they were about $1.18 billion with about 400 units. Raising Cane's just surpassed $1.5 billion in annual sales, tripling in size in just four years. The company is about to open its 500th restaurant. Even more impressive, average unit volumes are approaching $3.6 million. For context, McDonald's pulls in about $2.8 million in AUVs.[2]

For Raising Cane's, a franchisee typically required an upfront fee of about $45,000. But the biggest hurdle to getting started was the investment in real estate, equipment, and startup costs, estimated around $768,100 to $1,937,500. Ongoing operation required anywhere from $90,000 to $250,000 in working capital and a 5% royalty fee.[3] These numbers were fairly standard for the industry. For example, recent data about Checkers & Rally's franchise agreement listed a franchise fee at $30,000 with a 4% of net sales royalty fee and a required advertising contribution of 4.5% of net sales for units delivering about $1.2 million in AUV (average unit volume).[4] Expansion was key to a franchisor's success because the nature of what the franchisor offered was information—a brand, information in the customers' mind, and operational know-how—that would be at the disposal of the franchisees. Like any "information good," investments in support of the franchise network had high fixed cost but scaled very well. As long as a franchisor was able to ensure the quality of the product offered by the franchisees under its brand, incremental units delivered high margins.

Pandemic Shock

Just a few months after Joaquin had been at the Smalls' warm opening event, on March 11, 2020, the World Health Organization (WHO) declared the novel coronavirus (COVID-19) outbreak a global pandemic. The crisis hit restaurants very hard, with the closure of dining halls and later restrictions on the number of guests a location could host. More importantly, the pandemic had a profound impact on customer habits. As people were increasingly confined in their houses, digital ordering, contactless pickup, and delivery became preferred options to obtain restaurant food.

Restaurants with a sizeable take-out and delivery operation, after an initial hit, were able to navigate the pressures posed by the crisis. For example, two months into the pandemic lockdown restrictions, Todd Graves, founder of Raising Cane's stated, "Through the drive-thru we got quicker, we got more proficient, and went from 30% [sales drop] to 25% to 20%, and now we're actually even with our original forecasted sales. Other restaurants that don't have the drive-thru, my heart goes out to them, but thank goodness we haven't laid off any crew."[5] Domino's Pizza, the chain exclusively focused on delivery and take out, actually benefitted from the pandemic. A digital pioneer in the restaurant industry (Figure 5.3), Domino's same-stores sales grew 11.2% in the last quarter of 2020, up from 3.4% in the year before. Furthermore, in 2020, the chain saw a 5% increase in digital sales, which by the end of 2020 accounted for 75% of total sales.[6] Domino's financial results were impressive, especially since most restaurants experienced reductions in sales during 2020. Despite its remarkable results, Domino's leadership believed the firm had to continue investing in both customer facing (e.g., car-side delivery) and operational (e.g., AI-enabled demand forecasting) innovations if it was to maintain its leadership.

---

1 Raising Cane's. 2013, April 4. "Our concept." *Raising Cane's*, retrieved from https://www.raisingcanes.com/our-concept.

2 Kelso, A. 2020, March 13. "Raising Cane's now offering 'industry-leading' compensation, including the chance to make $1 million." *Forbes*, retrieved from https://www.forbes.com/sites/aliciakelso/2020/03/13/raising-canes-is-now-offering-industry-leading-compensation-including-the-chance-to-make-1m/.

3 Franchise Grade. 2021. "Raising Cane's franchise review." *Franchise Grade*, retrieved from https://www.franchisegrade.com/franchises/raising-cane-s.

4 "Checkers franchise information." 2021. *1851 Franchise*, retrieved from https://1851franchise.com/checkers/info.

5 Chung, H. 2020, April 30. "Raising Cane's CEO: We're not going to hurry back and reopen stores." *Yahoo! Finance*, retrieved from https://finance.yahoo.com/news/raising-canes-ceo-were-not-going-to-hurry-back-and-reopen-stores-182318462.html.

6 Motley Fool Transcribers. 2021, February 25. "Domino's Pizza Inc (DPZ) Q4 2020 earnings call transcript." *MSN*, retrieved from https://www.msn.com/en-us/money/companies/dominos-pizza-inc-dpz-q4-2020-earnings-call-transcript/ar-BB1e1y4B.

While the restaurant industry remained insulated for nearly three decades from the digital disruptions pervading other hospitality sectors (e.g., air travel and accommodation), digital disruption was happening rapidly, with the pandemic impressing that acceleration. Multiple restaurants—such as Shake Shack, Chick-fil-A, McDonald's, Five Guys, KFC, and Burger King—were investing heavily in their digital channels (e.g., mobile and web apps), like the airline industry had done in the late 1990s. Furthermore, the rise to prominence of aggregators of demand that promised incremental demand and delivery services with no up-front investment (e.g., UberEats, Deliveroo, DoorDash, and GrubHub) was reminiscent of the rising power of online travel agencies (e.g., Expedia, Booking) in the lodging sector during the early 2000s economic recession. While requiring no fixed investment, the typical restaurant aggregator charged substantial commissions (e.g., 30%) in an industry with razor-thin margins. More importantly, aggregators disintermediated the relationship with customers, preventing restaurants from create differentiated value propositions.

As the "digital disruption" of the restaurant industry accelerated, many restaurants were creating new concepts that leveraged these opportunities. Wendy's was experimenting with drive-through-only restaurants.[1] Chipotle was redesigning its physical stores to incentivize the use of its digital channels and pick-up and delivery services.[2] McDonald's leveraged menu optimizations and AI-powered digital menu boards to improve its service speed and increase product upselling.[3] While the crisis hit the industry very hard, those restaurants that could take advantage of digital technology and new post-pandemic customer habits seemed poised for growth.

New entrants to the industry were "digital first." Companies like Virtual Dining Concepts, the brainchild of long-time industry player and Planet Hollywood–founder Robert Earl, promised restaurants reeling from the pandemic crisis, "You're already cooking, Virtual Dining Concepts (VDC) is the secret sauce to increase your profits. VDC offers restaurant owners an easy solution to generate more revenue out of your existing kitchens with minimal impact to current operations. Attract new customers with delivery-only menus using food delivery services."[4] VDC enabled restaurants to prepare food for virtual brands it owned and operated. As the firm put it, "You focus on food. We do all the rest." The firm aggressively partnered with celebrities to theme their virtual brands and had an in-house culinary team that created the menu. One of the high-profile VDC brands was MrBeast Burgers, developed in partnership with YouTube creator Jimmy Donaldson (aka MrBeast). Thanks to Donaldson's popularity and his more than 55 million YouTube subscribers, the MrBeast ordering app "quickly surged to the top spot on the Apple Store among food and beverage apps,"[5] and the launch video uploaded on December 19, 2020, to MrBeast's YouTube channel quickly surpassed 50 million views. Within three months, the MrBeast "chain" counted 450 outlets and was on its way to a target 2,000 units.[6] Fans could order the food via the proprietary app or website, as well as leading food delivery aggregators like DoorDash. While food consistency remained a challenge, the concept's ability to generate demand was not in question.[7]

Figure 5.3. Domino's Anyware initiatives are at the forefront of digital ordering and customer's experience

1 Luna, N. 2020, November 4. "Wendy's breakfast gains steam as quick-service brand looks at building drive-thru-only restaurants." *Nation's Restaurant News*, retrieved from https://www.nrn.com/quick-service/wendy-s-breakfast-gaining-steam-brand-looks-building-drive-thru-only-restaurants.

2 Luna, N. 2020, February 14. "Chipotle Mexican Grill takes a page from quick-service playbook to drive digital sales." *Nation's Restaurant News*, retrieved from https://www.nrn.com/fast-casual/chipotle-mexican-grill-takes-page-quick-service-playbook-drive-digital-sales.

3 Luna, N. 2020, October 2. "McDonald's drive-thru lanes are nearly 30 seconds faster thanks to limited menu, 11,000 Dynamic Yield menu boards." *Nation's Restaurant News*, retrieved from https://www.nrn.com/quick-service/mcdonald-s-drive-thru-lanes-are-nearly-30-seconds-faster-thanks-limited-menu-11000.

4 Virtual Dining Concepts. n.d. Homepage. https://www.virtualdiningconcepts.com/.

5 Maze, J. 2021, February 12. "Why MrBeast Burger is the most important restaurant concept in the U.S. right now," *Restaurant Business*, retrieved from https://www.restaurantbusinessonline.com/financing/why-mrbeast-burger-most-important-restaurant-concept-us-right-now.

6 Lessin, S. 2021, March 24. "'MrBeast' explains his plans to help YouTube creators raise equity finance," *Information*, retrieved from https://www.theinformation.com/articles/mrbeast-explains-his-plans-to-help-youtube-creators-raise-equity-finance.

7 Watson, A. 2021, January 7. "The untold truth of MrBeast Burger." *Mashed.com*, retrieved from https://www.mashed.com/308763/the-untold-truth-of-mrbeast-burger/.

Post-pandemic Growth

Coming to work one morning, Joaquin saw the announcement of a second location being planned for Smalls Sliders in Baton Rouge. The article mentioned that the firm had applied for a permit to open its second restaurant in a local shopping center: "Plans are to install an 800 square foot [74 square meters] modular building. . . . There will be a drive through, a pick up window and a covered dining area, similar to the setup at the first Smalls."[1] This was welcomed news, and it coincided with the hope that was beginning to spread in the US as the vaccination plans were kicking into high gear. The article made Joaquin think back on the prepandemic growth plans he had heard from Smalls. With experienced cofounders like Landry and a strong financial backing, the concept could be feasibly taken nationwide through an aggressive franchising model. Before the first restaurant even opened, *QSR* magazine wrote, "Smalls Sliders has national growth potential. One indicator of what's to come: it's a 950-square-foot [88 square meters] shipping container building. Similar to one of Checkers & Rally's top growth vehicles, the construction is built off-site, at a controlled cost, and shipped to location. It makes for a quicker and more operator-friendly approach to start-up fees."[2] The article quoted Dugas as saying, "You could be doing your site work while the building is being built. . . . It's a really cool process. And you can ship it anywhere in the country." At the time, Checkers & Rally's branded about 900 locations[3] in the United States, while another competitor, Five Guys Burgers and Fries, operated and franchised about 1,700 units globally, with another 1,500 under development.[4]

The world had dramatically changed with the pandemic. But, Joaquin thought, "perhaps those changes are a blessing for a concept like Smalls." The article got him thinking about how he would advise the firm based on his research and his knowledge of digital disruption. Was the postpandemic world setting up the conditions for growth, or was it a challenge to a concept like Smalls Sliders? What would be the growth model, and what would be the role of digital technology?

Discussion Questions

1. Do you believe that the Smalls Slider concept is well suited for rapid growth in a postpandemic world?
2. What is the role of digital technology in the current version of the Smalls Sliders approach?
3. What should Joaquin suggest to Smalls executives concerning the use of digital resources to support the customer experience? Is the approach you advocate an example of digital transformation or digital innovation?
4. What do you see as the role of food delivery aggregators? Are they a threat or an asset to a concept like Smalls?

1 Boone, T. 2021, March 1. "Smalls Sliders plans second location," *Advocate*, retrieved from https://www.theadvocate.com/baton_rouge/news/business/article_57f926d0-7ae1-11eb-908d-5bd0b7e0adb2.html.

2 Klein, D. 2019, March 25. "With help from Walk-On's, Smalls Sliders cooks up big plan." *QSR*, retrieved from https://www.qsrmagazine.com/exclusives/help-walk-ons-smalls-sliders-cooks-big-plan.

3 Silverman, Ellie. 2019, January 10. "Checkers and Rally's looks to add dozens of stores in Philadelphia." *Philadelphia Inquirer*, retrieved from https://www.inquirer.com/business/retail/checkers-rallys-fast-food-restaurant-philadelphia-20190110.html.

4 "Five Guys story." 2021. *Five Guys*. https://www.fiveguys.com/Fans/The-Five-Guys-Story.

Case Study for Chapter 6

Outrigger Hotels and Resorts

I am involved with every decision that senior management takes. They look to me for an IS slant to it—whether an IT solution can capitalize on opportunities or eliminate threats. They also expect my team to independently develop an IS strategy that will further the business.

—Joe Durocher, SVP & CIO

Every manager must have an IT strategy. You can't delegate to technologists and only worry about your allocated cost or what training your employees need. You must understand how to be master of your own destiny and make IT work best for you. Too many managers still don't get that.

—Rob Solomon, SVP Sales & Marketing

Outrigger History[1]

On Black Friday, September 13, 1929, Roy C. Kelley arrived in Hawaii with his wife, Estelle. An architect by training, Mr. Kelley joined the firm of C. W. Dickey and was responsible for designing many of Honolulu's landmark buildings, including the main building of the old Halekulani Hotel and the Waikiki Theater on Kalakaua Avenue.

Nine years later Kelley set out on his own and opened his architecture firm, building numerous homes, apartment buildings, and hotels on the island of Oahu. In 1963, Kelley took over the land occupied by the old Outrigger Canoe Club, and Outrigger Hotels became a reality with the mission of bringing the dream of a vacation in paradise within the reach of the middle-class traveler. Included in the agreement were leases on three Waikiki lots that later became the Outrigger East, Outrigger West, and Coral Reef hotels. The Outrigger Waikiki Hotel was built on the site of the old canoe club, arguably the prime spot on Waikiki beach, in 1967. Throughout the next two decades, Outrigger Hotels Hawaii, as the company was named, continued its expansion in Waikiki. When in the seventies the zoning authority put a cap on new construction in Waikiki, Outrigger began to expand through acquisition rather than construction, ultimately becoming the largest chain in the State of Hawaii with over 7,000 rooms and a total of 15 properties concentrated in Waikiki (see Exhibit 6.1). Thanks to its clustered configuration, Outrigger Hotels Hawaii was able to maintain a centralized management structure fitting Mr. Kelley's "management by walking around" style.

In 1989, Outrigger Hotels Hawaii, now under the leadership of Roy Kelley's son, Dr. Richard Kelley, took over management of The Royal Waikoloan Hotel on the Big Island of Hawaii. When Hurricane Iniki, heading for Waikiki in 1992, barely missed Honolulu and ravaged the island of Kauai, it provided further impetus for Outrigger's geographical diversification strategy to and beyond neighboring islands. The firm, now expanding into management agreements with third-party owners, added properties on Maui and Kauai and ultimately grew to a total of 26 locations in the Hawaiian Islands (see Exhibit 6.2).

In 1996 the firm made its first international foray, opening the Outrigger Marshall Island Resort on Majuro Atoll in the Republic of the Marshall Islands. Through partnerships, joint ventures, acquisitions, and new developments, the firm continued to grow internationally, adding properties in Guam, Fiji, Tahiti, Australia, and New Zealand (see Exhibit 6.3).

While growing geographically, Outrigger Hotels Hawaii also began to diversify its product portfolio with the addition of condominium resorts beginning in 1990. Because of its geographical and product diversification, in 1995 Outrigger Hotels Hawaii changed its name to Outrigger Hotels and Resorts, and in 1999 it re-branded fifteen of its hotels in Waikiki to launch a new hotel brand called OHANA Hotels of Hawaii. Reflecting on the decision, President and CEO David Carey commented:

> We had an identity crisis because the market moved up, we upgraded the on-beach properties where we had higher demand and bought some nice properties in neighboring islands. But we had huge variation in the portfolio—if you stayed at a budget property vs. a beach front property, you'd be very confused as to what an Outrigger was.

In an effort to bank on the name awareness that the Outrigger brand had developed with consumers, the on-beach properties became upscale full-service hotels under the Outrigger brand. The condos, also typically on-beach upscale locations, maintained the Outrigger brand. Conversely, the OHANA brand was positioned to cater to the budget traveler looking for value on off-beach properties. Perry Sorenson, COO, explained the OHANA value proposition:

> OHANA hotels are something between a Holiday Inn and a Hampton Inn. No expectation of restaurants,

1 Excerpted from Piccoli, G. 2015. "Outrigger hotels and resorts: A case study." *Communications of the Association of Information Systems 15*. Article 5. Used with permission from Association for Information Systems, Atlanta, GA; 404-413-7445; www.aisnet.org. All rights reserved.

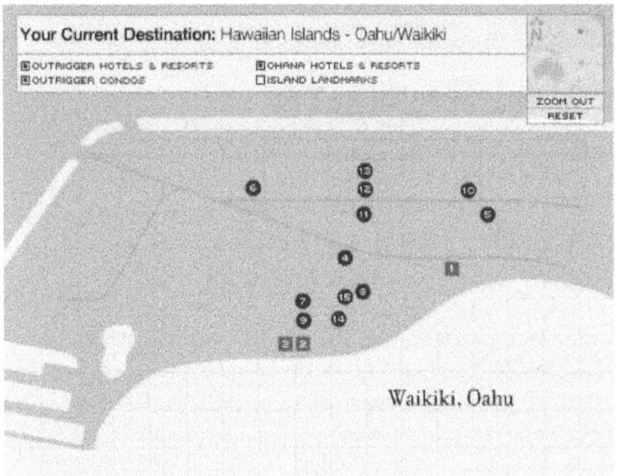

Exhibit 6.1: Outrigger properties in Waikiki

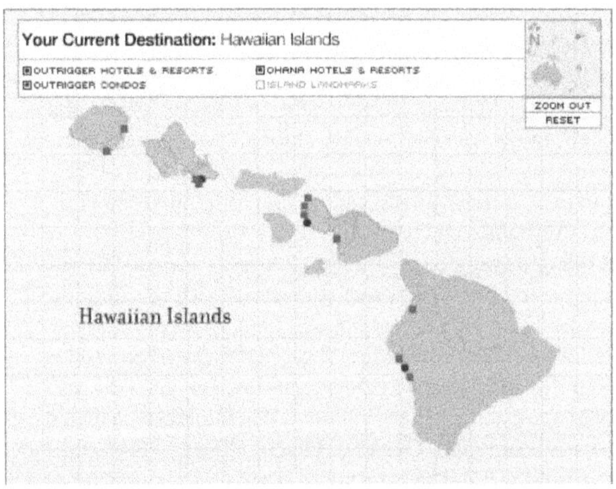

Exhibit 6.2: Outrigger properties in the Hawaiian islands

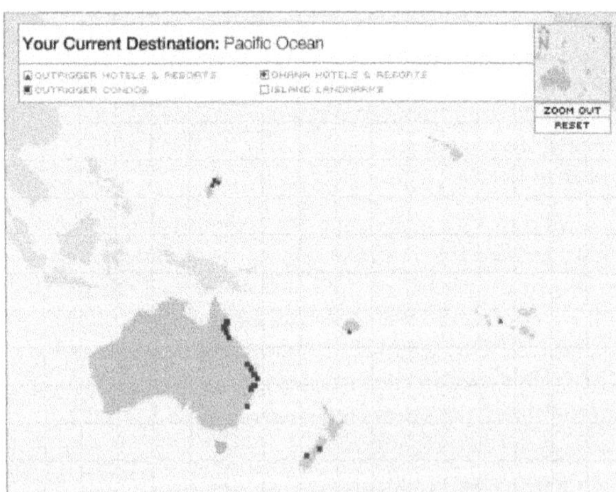

Exhibit 6.3: Properties managed by Outriggers Hotels and Resorts (International)

> but expectations that you have a friendly staff, that the room is going to be clean, and you will be taken care of. Not a lot of extras, but good value.

Condominiums represented an increasingly important share of the total portfolio of properties (see Exhibit 6.4), even though the firm had sort of stumbled upon the opportunity condominiums offered. Condominiums appealed to the independent traveler who would do much research and planning on his own. Condominiums were also very complex, non-standard products that travel agents and wholesalers found hard to sell. As Sorenson explained:

> The addition of condominium properties was a customer driven initiative. We kept receiving inquiries about condominium vacations and had to direct customers to competitors who also ran hotels. That did not make any sense.

As the firm learned over time, condominiums were very different than traditional hotel and resort operations. While management agreements with condominiums varied substantially, unit owners typically had the option to join a pool of units that Outrigger was responsible for marketing and managing. Owners typically received 55% of the gross income the units generated and Outrigger funded its operations with the remainder. Beyond labor costs, the primary expenses included the costs of marketing and selling the properties, front desk and housekeeping operations, and in-unit maintenance. Maintenance of the common areas, defined as anything from the unit's inside wall paint outward, was the responsibility of the AOAO (the owners association) and was funded through annual dues. This state of affairs was simpler in the Australian condominiums—referred to as strata title properties. There, the management company had to buy and control the lobby area, and the contracts were generally 25 years in length and required standardization of revenue splits. This approach created simplicity and clarity that made it more efficient for the management company to operate.

Because condos were rarely built as business ventures but rather were designed as primary or vacation homes for the tenants, they offered little office or staging space for management companies to operate in. They also lacked many of the typical hotel services and departments such as food and beverage, room service, laundry, and daily maid service. Working with a relatively unsophisticated and widespread ownership base, with some condominiums having almost one owner (i.e., one contract) per unit, presented significant challenges. Jim Hill, Regional Director—Maui, summarized the challenge:

> The thing that is hardest to do in condos is to change anything. You'll sit in a board meeting with the association and they'll say no, no, no, when the next property over offers a more appealing layout and better amenities. But that same person will ask you in another meeting why isn't the revenue higher?

Exhibit 6.4: Rooms breakdown in each of the three products		
Outrigger Resorts		
Outrigger Waikiki on the Beach	Oahu	525 Rooms
Outrigger Reef on the Beach	Oahu	858 Rooms
Outrigger Guam Resort on Tumon Bay	Guam	600 Rooms
Outrigger Reef Fiji	Fiji	207 Rooms and 47 bures
Te Tiare Beach, An Outrigger Resort	Tahiti	41 Bungalows
OHANA Hotels		
OHANA East	Oahu	445 Rooms
OHANA Islander Waikiki	Oahu	283 Rooms
OHANA Reef Lanai	Oahu	110 Rooms
OHANA Waikiki Tower	Oahu	439 Rooms
OHANA Waikiki Village	Oahu	442 Rooms
OHANA Waikiki West	Oahu	663 Rooms
OHANA Reef Towers	Oahu	480 Rooms
OHANA Waikiki Malia	Oahu	327 Rooms
OHANA Waikiki Surf East	Oahu	102 Rooms
OHANA Royal Islander	Oahu	101 Rooms
OHANA Waikiki Surf	Oahu	302 Rooms
OHANA Maile Sky Court	Oahu	596 Rooms
Best Western The Plaza Hotel	Oahu	274 Rooms
Honolulu Airport	Oahu	307 Rooms
OHANA Maui Islander	Maui	360 Rooms
OHANA Keauhou Beach Resort	Hawaii (Big Island)	309 Rooms
OHANA Oceanview Guam	Guam	191 Rooms
OHANA Bayview Guam	Guam	148 Rooms
Outrigger Condominiums		
Outrigger Waikiki Shore	Oahu	25 Apartments
Outrigger Luana Waikiki	Oahu	N/A
Outrigger Palms at Wailea	Maui	89 Apartments
Outrigger Maui Eldorado	Maui	100 Apartments
Outrigger Royal Kahana	Maui	191 Apartments
Outrigger Napili Shores	Maui	101 Apartments
Outrigger Kiahuna Plantation	Kauai	190 Apartments
Outrigger at Lae Nani	Kauai	60 Apartments

(continued)

Exhibit 6.4: Rooms breakdown in each of the three products (*continued*)		
Outrigger Condominiums		
Outrigger Kanaloa at Kona	Hawaii (Big Island)	83 Apartments
Outrigger Royal Sea Cliff	Hawaii (Big Island)	61 Apartments
Outrigger Fairway Villas	Hawaii (Big Island)	78 Apartments
Outrigger at the Beacon, Queenstown	New Zealand	14 Rooms, 23 Apartments
Outrigger at Clearwater Resort	New Zealand	N/A
Outrigger on the Beach at Salt Kingscliff	New South Wales	N/A
Outrigger Ettalong Beach Resort	New South Wales	N/A
Outrigger Heritage Port Douglas	Australia	15 Rooms, 42 Apartments
Outrigger in the Village Port Douglas	Australia	8 Rooms, 13 Apartments
Outrigger on the Inlet Port Douglas	Australia	10 Rooms, 21 Apartments
Outrigger Beach Club & Spa Palm Cove	Australia	104 Rooms, 195 Suites
Cairns Resort by Outrigger	Australia	127 Rooms
Outrigger 1770 at Agnes Water	Australia	N/A
Outrigger Hervey Bay	Australia	27 Rooms, 139 Suites
Outrigger Mooloolaba International Beach Resort	Australia	201 Apartments
Outrigger Sun City Resort	Australia	266 Apartments
Outrigger Coolangatta Beach Resort	Australia	90 Apartments

These difficulties notwithstanding, Outrigger found the condo business appealing when it made its first foray into it in the early 1990s, because it provided a means for expansion through management contracts without the need to acquire expensive properties. Condo products varied widely, ranging from studios to two bedroom apartments, and did not have all the services typically associated with a hotel, like room service, on-property restaurants, and retail shops.

Outrigger had grown to a sizable firm, encompassing about 3,600 employees (of which about 230 were at corporate), a portfolio of properties exceeding U.S. $1.4 billion,[1] and approximate revenues of U.S. $45 million.

The Hotels and Resorts Industry

As the new millennium dawned, the global lodging industry was estimated to exceed $295 billion in sales, about 11% of the world's economic output, and employed more than 250 million workers (see Table 6.1 for performance indicators).[2] The leisure travel segment accounted for about 45% of total volume.[3]

With respect to the Hawaiian market, which was Outrigger's traditional stronghold, recent figures showed performance levels above the average of the global industry (see Table 6.2). Being quite isolated from any large population pool, Hawaii was a classic destination market with an exclusive fly-in customer base. The major feeders were U.S. westbound traffic and Japanese eastbound traffic. These markets were thought to yield very high return rates—estimated by some to be around 50% westbound and over 65% eastbound. This trend made for a very location-savvy customer base. Peculiar to this market was also the trend of multi-island stays, with guests visiting more than one destination during the same trip.

Because the Hawaii and Pacific Rim markets were exclusive destination markets, the use of packages—including air and accommodations—was pervasive. Historically, packages were

1 "Outrigger's president and CEO David Carey named hotel person of the year by Travel Agent magazine." (2003, January). Hotel Online Special Report. Retrieved June 1, 2004, from http://www.hotel-online.com/News/PR2003_1st/Jan03_DCarey.html.

2 "Hotels and other lodging places." *Encyclopedia of Global Industries*, 3rd ed., 2003. Business and Company Resource Center. Infotrac. http://infotrac.galegroup.com/itweb/nysl_sc_cornl?db=bcrc.

3 Worldwide Hotel Industry Study, 2002, Horwarth International.

Table 6.1. Performance of global hotel industry	
Occupancy	63.5%
Avg. Number of rooms	706
ADR	$91.62
RevPAR	$58.18

Table 6.2. Performance of Hawaii hotel market	
Occupancy	72.1%
Avg. Number of rooms	706
ADR	$198.41
Revenue*	$78,488

* Amounts per available room

assembled and sold by wholesalers and tour operators who purchased both air and hotel rooms in bulk and re-marketed them to the traveling public. With the widespread adoption of the Internet, a new type of package was emerging under the leadership of large online travel agencies: dynamic packages. A dynamic package was one that enabled the guest to choose air, hotel, car rental, and even activities, ticket them independently, and then price them out as a bundle. Dynamic packages were appealing to suppliers because the price of each item was not disclosed, making price comparison difficult and alleviating commoditization fears. They were appealing to perspective travelers because they increased choice and fostered flexibility. Finally, they appealed to online travel agents because they built upon their value proposition—customer choice—and had the potential to improve their margins.

As a mature destination, Hawaii had been entered by many of the larger branded hospitality and resort companies. The largest hospitality firms, such as Marriott International, Hilton Hotels and Resorts, and Starwood, had a significant presence with eight, five, and eleven properties respectively. But the largest operators in Hawaii were geographically- and leisure-focused players such as Outrigger, ASTON Hotels & Resorts Hawaii (with twenty-eight properties), and Marc Resorts Hawaii (with eleven properties).

Outriggers Organization

Outrigger Hotels and Resorts was a management company wholly owned by a holding corporation called Outrigger Enterprises. Reflecting its real estate development roots, Outrigger Enterprises also owned a real estate ownership company called Outrigger Properties (Exhibit 6.5).

Outrigger Properties wrote and managed real estate contracts with third-party owners and supervised the owned assets (accounting for about a third of all properties in the Outrigger portfolio), as well as the development of new properties. The firm also monitored the real estate market for optimal times to invest in available properties or sell assets in the portfolio and raise needed capital. Outrigger Properties managed leasing contracts with the many independent retailers occupying food and beverage outlets—rarely run internally by Outrigger—and shops within the hotels and resorts in its portfolio. Sorenson explained the tradeoffs associated with this decision:

> We are the third largest retail landlord in Hawaii, with about 300,000 square feet of retail space, so we have access to the best restaurant operators. Leasing restaurants allows us to focus our energies on hospitality and [the] profitability of the rest of the hotel, but of course you lose some control when you outsource. It takes a hotel mentality to do room service very well for example.

Outrigger Hotels and Resorts, the operating arm of Outrigger Enterprises, was responsible for the writing of new management contracts as well as overseeing property renovations and operations of the managed hotels, resorts, and condos. Outrigger Properties generally negotiated a base and a percentage of revenue with tenants; revenues from leased space were assigned to the hosting property's own profit and loss (P&L). Room revenue made up the bulk of each property's revenue, with rental income as low as 5% in hotels with little retail space and as high as 20% in some of the most appealing locations. Other more marginal revenue lines were parking, in-room entertainment, telecommunications, and kids' clubs operations.

Outrigger Hotels and Resorts had historically maintained a highly centralized organizational structure. As the firm grew in size and geographical distribution, a more traditional structure emerged, but, reflecting its roots, Outrigger Hotels and Resorts remained consolidated where possible. For example, the two beach-front Outrigger Hotels on Waikiki beach were managed as one. As Chuck Shishido, OHANA Hotels VP of operations and a 33-year veteran of the company, explained:

> We have centralized services—accounting, IT, finance, engineering, purchasing, and special projects—that support all the properties on Oahu, as well as indirectly the neighboring islands. There is also one executive housekeeper in charge of all properties. We run the OHANA Hotels like a 4,200 room distributed hotel. It is very efficient.

Since each property in the Outrigger family had its own P&L, these shared services were charged back to them based on room count, revenue, or usage, depending on the service. As the firm expanded internationally, it became more decentralized, with resorts in the Pacific Rim working much more like independent operations and organized like traditional resorts. Recognizing the significant advantages offered by its

Exhibit 6.5: Organization chart and bios of key personnel

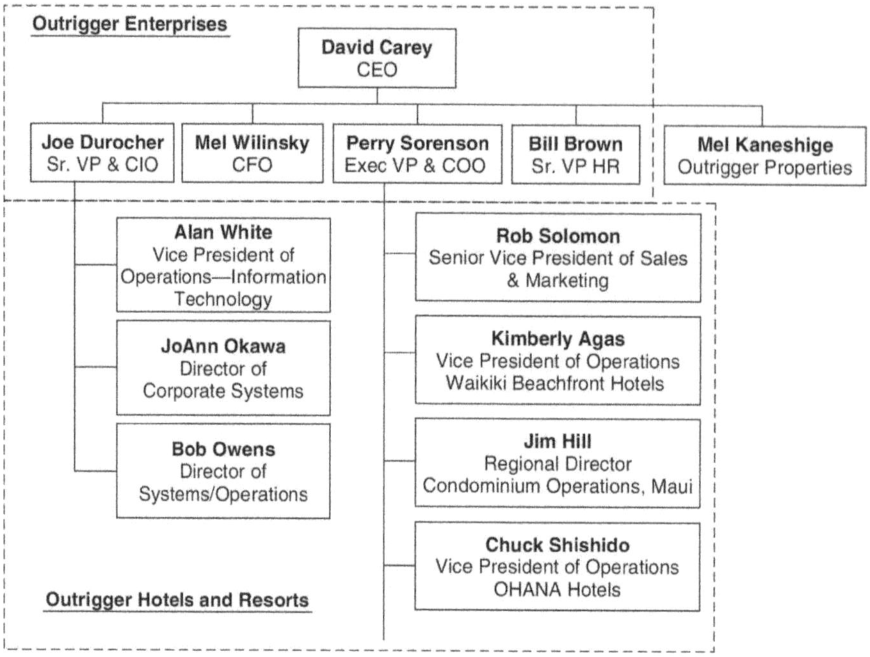

W. David P. Carey III, President and Chief Executive Officer

David Carey joined Outrigger Enterprises, Inc. as executive vice president and general counsel in 1986 and was named president of the company in 1988 and chief executive officer in 1994. After graduation in 1982, Carey moved to Honolulu and was an attorney specializing in corporate and real estate law at Carlsmith Wichman Case Mukai and Ichiki, where Outrigger Hotels was one of his major clients. Carey is a member of numerous business and community organizations, including the Hawaii Tourism Authority, Hawaii Hotel Association, and others. Carey has a BS in electrical engineering from Stanford University; a JD, cum laude; and an MBA, with distinction, from the Santa Clara University. He was a member of the Beta Gamma Sigma Honor Society.

Joe Durocher, Senior Vice President and Chief Information Officer

Joe Durocher first joined Outrigger in 1986 as vice president of information systems. During his tenure, he was instrumental in the installation and maintenance of the company's Stellex reservations and front desk computer system. After 10 years with Outrigger, Durocher left the company to join Hilton Hotels Corporation as SVP (senior vice president) and CIO, where he was responsible for data processing and related strategies for all of Hilton's non-gaming hotels, amounting to well over 300 properties with over 110,000 rooms worldwide. While with Hilton, Durocher was instrumental in the replacement of Hilton's Hilton central reservation system with Hilstar, a new state-of-the-art central reservations system. Durocher rejoined Outrigger Enterprises in 2000. Born and raised in Hawaii, Durocher received his BS in Electrical Engineering and MBA from the University of Hawaii. He is a Certified Systems Professional (CSP) and Certified Data Processing Professional (CDP). Durocher is also a member of the Beta Gamma Sigma Honor Society.

Perry Sorenson, Chief Operating Officer

In his position as a chief operating officer, Perry Sorenson is responsible for all aspects of hotel operations as well as Outrigger's current expansion across the Pacific. He joined Outrigger as executive vice president in 1991. Sorenson's career spans over 20 years in the hospitality industry. He was previously EVP and COO for Embassy Suites, Inc., the world's largest all-suite hotel chain. Sorenson directed the rapid growth of Embassy Suites operations as it expanded from five to 105 hotels in five years. Prior to joining Embassy Suites, Sorenson was vice president of operations for Holiday Inns Inc., where he was the recipient of the Holiday Inn Corporation Chairman's Award for Service Excellence. He has also held management positions with Radisson Hotel Corp. and Rockresorts, Inc. He is currently a member of the Native Hawaiian Tourism & Hospitality Association and sits on the board of various other community and industry organizations. Sorenson received his BA in psychology and MBA from the University of Utah.

Robert L. Solomon, Senior Vice President, Sales & Marketing

Before joining Outrigger, Solomon spent 15 years with Dollar Rent A Car as senior vice president, with additional responsibilities as vice president, Pacific sales. Solomon's professional career has also included senior management positions in California with various corporations as well as government agencies in California. Solomon is well known within Hawaii's visitor industry, having served on numerous boards and committees, including the research and marketing committees of the Hawaii Visitors & Conventions Bureau and the Oahu Visitors Bureau. He is the immediate past chair of the Marketing Committee for the Oahu Visitors Bureau and presently chairs the Internet Subcommittee for the Hawaii Visitors and Convention Bureau Advisory Committee. Solomon holds master's degrees from Princeton and Yale universities in the fields of public and international affairs and Southeast Asia studies.

Alan White, Vice President of Operations—Information Technology

White joined Outrigger Hotels & Resorts in 2001 and is responsible for all data processing for hotel operations. In his time with Outrigger, he has produced numerous new interfaces to the company's proprietary Central Reservations System (CRS) and Property Management System (PMS). White also created electronic interfaces to more than 13 different suppliers, who account for nearly 30% of the total reservations for Outrigger and OHANA Hotels.

White has extensive operation experience in hotels ranging from the food and beverage (F&B) area to the front desk to general management. Prior to joining Outrigger, White worked at Pegasus Solutions, Inc. for over a decade. Born in Paris, France, White was educated at Providence College in Rhode Island, the Universitat Freiburg in Switzerland, and George Washington University in Washington, D.C. He did his advanced graduate studies at the Sino-Soviet Institute and the Osteuropa-Institut.

Kimberly Agas, VP Operations, Waikiki Beachfront Division

Agas joined Outrigger in 1984 while attending college and has worked at eight different properties throughout her years with the company. Prior to her tenure at the Outrigger Waikiki, she was hotel manager for the OHANA Village and OHANA Coral Seas hotels. Recently, she was promoted to the newly created position of Vice President-Operations, Waikiki Beachfront Division, and she will assume the duties of general manager of the Outrigger Reef on the Beach.

Agas is a graduate of the liberal arts program at Leeward Community College and the University of Hawaii at Manoa, where she studied economics. Since joining the company, she has been awarded the company's prestigious President's Award six times.

Chuck Shishido, VP Operations, OHANA Hotels & Resorts

Chuck Shishido, certified hotel administrator (CHA), joined the company in 1972 as a part-time worker in the housekeeping department of the then Outrigger Reef Towers while pursuing an accounting degree at the University of Hawaii. In 1975, Shishido was promoted to assistant hotel manager at the Outrigger Reef. A year later, he transferred to the Outrigger Reef Towers as hotel manager. For the next thirteen years, Shishido moved between six different Outrigger properties as hotel manager. He was promoted to group manager in 1990 and was responsible for four properties in the Lewers Street area. In 1994, he was appointed general manager of the Outrigger Waikiki on the Beach and remained there until he was called upon to assist in the launch of OHANA Hotels & Resorts.

Shishido holds a bachelor of arts degree in accounting from the University of Hawaii and is a member of the Hawaii Hotel & Lodging Association.

Jim Hill, Regional Director, Condominium Operations, Maui

Jim Hill is responsible for the daily operation of the Outrigger Royal Kahana, in capacity as general manager, as well as overseeing operations for Outrigger's three other condominium resorts on Maui—the Outrigger Napili Shores, Outrigger Maui Eldorado, and the Outrigger Palms at Wailea. Hill, who has been with Outrigger for over 17 years, previously served as general manager for the Outrigger Kiahuna Plantation located on the island of Kauai.

---

centralized structure, Outrigger was looking at the possibility of better integrating its international resorts. However, distance presented new challenges—1,800 miles separated its southernmost Australian property from its northernmost property alone. Sorenson explained:

> We need a reservation solution for Australia, a real-time coordination with a central reservation service. They are operated as individual hotels; the central 800 number today is just switched to the correct hotel. A centralized system would offer tremendous value because we get drive-in business and substantial potential cross-property traffic.

The Outrigger Customers and Competition

Outrigger's original mission was to bring the opportunity for a vacation in paradise within the reach of middle-class families. As the firm began to diversify its portfolio, the profile of its customers and the competition changed as well. The typical Outrigger guest was often a multigenerational customer with a sense of loyalty to the Outrigger family (about 25% of guests were returning to Outrigger) and an annual income exceeding $75k. Outrigger guests were almost exclusively leisure travelers, with some mixing business and travel (e.g., attending a conference and extending the stay for some leisure time with the family afterwards). This customer base created seasonality, with winter and summer being the high seasons when properties like the Outrigger Waikiki on the Beach reached an average daily rate (ADR) of $260 and an overall occupancy around 90%. Group business was limited, with some overflow from the conference center looking for meeting space for break-out events. Solomon profiled Outrigger's customer base:

> Our customers are independent-minded and look for an experience that is more regional and attuned to the destination, but still within their comfort zone. They may stay with big brands in their road warrior capacity, but that's not what they are looking for in a tropical destination.

Competing for these customers, Outrigger went head-to-head with such major brands as Marriott International, Hilton Hotels and Resorts, and Starwood Hotels and Resorts—the latter having a big Sheraton presence in Waikiki. These brands enjoyed name recognition, significant brand awareness among the traveling public, a flow of customers redeeming points, available capital, and availability of programs for employees such as discounted travel beyond Hawaii and the Pacific region. In response, Outrigger leveraged some of the premier locations in the markets it competed in, like the Outrigger Waikiki on the Beach, strong name recognition and long-term relationships with the travel distribution network, a strategic focus on vacation destinations, a deep local knowledge and community ties, and good employee relations. Kimberly Agas, VP of operations for Outrigger's Waikiki Beachfront Hotels and a 20-year veteran with the company, explained:

> Our employees are trusted to help the guests have a wonderful stay, and have the flexibility to act on their initiative. The teams of employees include our partners in the retail, restaurants and activities. We are concerned with the holistic experience, an all encompassing experience. In much of our unionized competition everyone has a narrow job [description] and outside of that they will refer you to a colleague, out of respect, because of contract restrictions, or because that's how they look at it: "this is all I have to do."

The typical OHANA guest was a value-minded and Hawaii-savvy leisure traveler with income below $100k a year. Typically, OHANA guests had visited Hawaii multiple times, stayed longer than average, and visited more often. Business travel was mainly composed of military and corporations with operations on multiple islands. Groups accounted for less than 10% of OHANA's overall traffic. Shishido explained:

> We have about 50% return guests. Your first trip you want a beach front hotel, the atmosphere, the ambiance—you want the full Hawaii experience. When you come more often, you still want the experience, but you look for more value and instead of spending $250-$300 a night for a beachfront you can stay longer off-beach for $70-$80 a night.

With seasonality similar to that of the full-service Outrigger Hotels, OHANA Hotels typically achieved an ADR around $66 and approximate occupancy levels of 75% over the year. A number of small regional chains (such as Marc Resorts and Castle Resorts) and many off-beach independent hotels existed in the Waikiki market. But Outrigger's senior management thought that OHANA hotels had no direct competition. Solomon explained:

> There is no quality branded competitor for OHANA. Because of the real estate costs and lack of efficiencies, competitors like Holiday Inn, Cendant, and Choice can't build and operate their product at their price point here. There are many independent no-name products.

Pricing for off-beach properties was much harder to manage because of the commodity nature of the hotels not enjoying a premium location, even though wholesalers concerned about their own brand and customer satisfaction were more willing to carry the OHANA brand over independents because Outrigger backed OHANA's quality promise. OHANA was the largest operator in Waikiki and the largest Hawaii owned operator.

Two types of customers were typically staying at the condominiums. On the lower side of the $90k to $160k income brackets were families visiting during school breaks and looking to control expenses and control their vacation experience. They valued the full kitchen—a standard in every unit—and the two bedrooms and two baths. This was substantiated by the fact that condos had four times as many reservations coming from the Internet direct and tended to recover faster after a soft economy. On the upper side of the spectrum were "newlyweds" and "nearly dead" couples who liked the privacy and space afforded by a condo. As Hill explained:

> On the upper end of the scale people like the convenience to have a full size refrigerator and kitchen amenities but they may never cook. If they want to cook the kitchen is available, but chances are they'll never use it. That's why, against conventional wisdom, the new trend is to put restaurants with more resort services in condominium properties.

While high degrees of variability existed between properties, returning guests to the same property ranged typically between 20% to 40%. With its expanding portfolio, Outrigger believed it enjoyed significant cross-property traffic as well, but it had little hard data on this.

Condominiums enjoyed almost no branded competition. Instead, because of the need to convince individual owners to join the pool of Outrigger managed units, the firm competed with small local management companies and individual owners' beliefs that they could do a better job alone. This idiosyncrasy of condominium operations amounted to having two customers—the unit owners and the guests—who, unaware of the workings of condo operations, were looking for the same level of service they would receive at a resort. On average, a condominium with mostly two bedroom units would achieve ADRs around $175, while properties with mostly studio and one bedroom units would settle around $140.

Outrigger operated a Central Reservation Office (CRO) in Denver, Colorado, with anywhere from 40 to 70 reservationists (FTEs), mainly depending on the volume of business. A corporate marketing staff of 12 people, allocated about 6% of revenue, was responsible for managing the brand and for going to market. An additional 2% of revenue was used to fund reservation and other distribution costs. Reservations were centralized for all properties in Hawaii. Outrigger Hotels had a staff of two or three people at each property to follow through (e.g., to reconcile inconsistencies, create rooming lists, and identify VIPs). For the OHANA hotels, this was done as a shared service. Beyond Hawaii, reservations were only taken at each property.

Outrigger executives believed that distribution was a cornerstone of its success, with about 50% of the business coming from wholesalers (classified as anything that is on a contract basis) who often sold Outrigger products as part of a package.

Consumer direct (via voice or the web), travel agents, government and military, and corporate clients made up the difference. For international properties the source of business percentage from wholesalers was close to 80%, and almost all reservations were faxed to the property. But the lines were blurring with the increasing prominence of online travel agents (e.g., Expedia) and the advent of dynamic packages. Solomon explained:

> We strive to make distribution as broad as possible, and for each pipeline (voice, Web direct, GDS, fax) we want to make it as efficient and user friendly as possible. The customer is in control; more than half of those who transact in the wholesale channel can pick the hotel.

The firm felt that it had been able to capitalize on the use of technology to increase distribution efficiencies in the face of ever rising labor costs. Conversion rates at the CRO had improved from 20% to 45% to 50% with widespread consumer adoption of the Internet. The firm estimated that as much as 60% of callers had already researched the Outrigger website and had made a purchase decision but, as Solomon put it, "had one more question." In an effort to provide support right on the website, the firm introduced live chat functionalities and also offered e-mail confirmation for significant savings in labor and postage costs.

Outrigger Strategy

At the heart of Outrigger Hotels and Resorts' strategy was a drive to position its properties in places where people could enjoy a vacation experience leveraging Outriggers' own core competencies. The firm was very careful not to create a cookie cutter approach but instead to deliver an experience that was respectful of the culture and the special characteristics of the localities in which it operated—a "sense of place" as the firm called it. As Carey put it:

> Our business is really about being a "window" to an experience, not the experience itself. We are the enabler through which people can engage in the leisure experience they desire. We don't try to export Hawaii when we go elsewhere, but we do honor the same values in the places we operate hotels and resorts.

The firm was embarking in a $315M renovation of the heart of Waikiki that required the leveling of five existing OHANA hotels—almost 2,000 rooms—that required significant investment in asset maintenance. In this area the firm planned to create about 500 rooms—a substantial reduction in room count—with a sizeable retail component. The firm's real estate ownership in the area totaled 7.9 contiguous acres and, with its biggest real estate investment being in Waikiki, a renewal of the area had benefits beyond the creation of new hotels and retail space. This bold project limited the firm's ability to expand in the short term, but Outrigger remained committed to growth and expansion once the renovation was completed. Carey explained the rationale for the firm's growth strategy:

> Given our key competencies, expanding to Guam, the Pacific, and Australia was a source-customer or distribution-driven growth strategy. It leveraged both markets where the customers knew us because they had experienced our hotels before and [markets] where we had relationships with global distribution channels.

Outrigger's senior management felt that its key competencies resided in providing hospitality to guests visiting their properties and successfully marketing those properties through leisure distribution channels, which before the widespread adoption of the Internet by consumers had accounted for over 80% of travel to Hawaii and other fly-in leisure destinations. To complement these basic competencies, Outrigger felt it had developed a superior capability to manage in a multicultural environment, including multicultural and multilingual employees and guests.

Outrigger had its roots in the economy segment of the market, but the firm's executives believed that it was not feasible to compete on price alone and had begun to focus on service delivery as well to build customer preference. Aided by a turnover rate in the single digits in the tight Hawaii labor market, and an average of 25 years of employee tenure with the company, it implemented a value-based management system that called for upward evaluation of managers and a focus on helping employees understand the principles behind Outrigger's service delivery strategy. As a testament to the firm's ability to fulfill its employees' needs, Outrigger had managed to be a mostly non-union shop in the heavily unionized Hawaii labor market. Carey summarized the firm's positioning:

> We operate properties that have good locations, we have a strong travel distribution network, and our employees really provide hospitality from the heart. That creates a differentiated product making price less important.

Beyond maintenance of the product under capital constraints, at the operational level three dimensions were deemed fundamental for success: providing guests with a rewarding experience and a sense of place, enabling employees to reach their potential, and being an integral part of the community. These dimensions were reflected in the firm's Management Incentive Plan (MIP), structured along three dimensions: Operating Cash Flow, guest satisfaction surveys (using reports produced by the independent research firm Leisure Trends), and employee satisfaction.

Beyond these critical success factors within the firm's control, Outrigger was wedded to the success of its destination markets given the proliferation of competing choices for consumers' entertainment budgets. Moreover, the firm was dependent on airlines. As leisure decisions became more and more impulse-driven, it became more difficult for travelers to find available seats at suitable times and prices. Carey summarized these challenges:

> If Hawaii does well, so do we. I spend a lot of time working with local tourism authorities to improve the appeal of the destinations we operate in. But airlines can be a bottleneck. We may not have available lift at times when we need it. If the airlines are full or they have decided in their yield model that they are going to only sell their top fares, there is nothing we can do. From purely the hotels' perspective, the best thing for us is an airline price war to Hawaii.

The major carriers, those driving the most traffic into Hawaii and the Pacific region, were under constant financial pressures. The events of September 11, 2001, and the recession that had hit the United States had depressed airline travel and had negatively impacted Outrigger's own financial performance. While the firm had been able to recover from these setbacks and was seeing high occupancies, terrorist threats and alerts remained a significant concern.

Outrigger IT Infrastructure

Joe Durocher, the CIO of Outrigger Enterprises, was hired by David Carey in 1986. Durocher recalled his early days with the firm:

> Mr. Roy Kelly was a hands-on manager. He once told me he hated two things: computers and vice presidents. As the VP of IT, I had two strikes against me. Yet, in 1986 I was brought in to overhaul Outrigger's IT infrastructure and we built Stellex—our integrated CRS/PMS. At the time all our properties were in Waikiki, within one square mile of each other.

Stellex, introduced in 1987, was a COBOL application running on a Tandem NonStop platform and a proprietary Enscribe database management system that guaranteed complete redundancy and 24 x 365 uptime. In 1992 Outrigger introduced Stellex 2.0, its first major update to Stellex, which ran on a Sun Microsystems UNIX platform and provided revenue management functionality and reservation center support. Because of its unique need for substantial wholesale interaction, Outrigger engaged Opus to build their revenue management module for Stellex 2.0. Outrigger retained control of the source code and over the years made substantial enhancements, mainly to manage wholesale relationships. The firm felt that its centralized IT infrastructure was a source of competitive advantage. Durocher discussed the trade-offs associated with centralized IT:

> Decentralizing IT would decrease our capabilities while increasing overall costs. But centralized IT creates friction at times. When a hotel is sold for example, the IT allocation may increase for other properties.

Stellex provided the anchor to which all other operational systems, including telephone switches, call accounting, and in-room entertainment, connected. All of the properties in the Hawaiian Islands had access to Outrigger's centralized IT systems, served from the Honolulu-based data center, through the firm's proprietary Wide Area Network (Exhibit 6.6).

Stellex, for example, was accessed using an ASP model by all the properties in the Hawaiian Islands, the firm's Denver-based CRO, and the Portland-based web servers, greatly simplifying the achievement of single image inventory, disaster recovery, and overall IT management. This enabled the properties to operate with PCs (as few as 12 in a typical 500-room property) and networking equipment. The Point of Sales (POS) systems were not centralized, since Outrigger leased retail and restaurant space. This state of affairs generated some friction at times, as Alan White, VP of property technology, explained:

> We offer to interface their POS to Stellex and pay for interfaces to automate room charges. But many of those POS are old and can't interface, they must be upgraded first. Restaurants have to write a manual charge voucher and walk it to the front desk for input. It's not a popular or efficient way to do it.

Due to the need for local support, the high telecommunication costs, and the reliability of international networks deemed unacceptable, Outrigger had yet to extend this centralized model to its operations in Australia and the Pacific. The properties in Australia and New Zealand, all condominiums, used a highly specialized PMS particularly well suited for strata title properties and their special tax code requirements. Durocher explained:

> None of the properties in Hawaii has a server on property. In the outer regions we have standalone PMS's and on-property reservations. We don't even try to keep Stellex in sync, they just open and close. If a date is getting full, they issue a stop-sell. Reservations that are taken centrally are automatically emailed.

Outrigger's IT function comprised a staff of twenty-six full-time employees, including four data entry operators and three developers housed in a separate limited liability company designed to help Outrigger take advantage of tax incentives offered by the state of Hawaii. One corporate IT professional supported the Australian properties' application needs. Hardware support was contracted out to local vendors. The function was organized along user needs rather than traditional departmental lines (e.g., data entry, application development, support). Alan White, VP of property technology, led the group in charge of creating and supporting IT solutions for the hotels. JoAnn Okawa, director of corporate systems, led the group in charge of creating and supporting IT solutions for the firm's back-office needs (e.g., general accounting, HR, payroll, purchasing). Bob Owens, director of system operations, and his group managed the data center and supported the other two groups. They also performed advisory work for the international properties that had local MIS managers in charge of procuring and managing technology solutions locally. This organization enabled operations personnel to unequivocally ask the Property Technology group for support, while administrative personnel referred to the Corporate Information Service group.

The IT function at Outrigger was designated as a cost center. Its operations were funded through allocations to the business units and to each property using four different methods. A charge, based on room count, was assessed for use of property technology. The same mechanism was used to account for use of administrative systems. Group sales software (i.e., Newmarket's Delphi) was charged based on each property's meeting space. Finally, any ad-hoc solution (e.g., the writing of a specialized report) was charged directly to the requesting unit. Traditional metrics to measure success (e.g., on-time and on-budget project delivery) were used, and the IT function had recently introduced service-level agreements. Durocher explained the rationale for the decision:

> Service-level agreements enable the management of expectations, increase accountability, and offer choice to user-managers. If you feel you are paying too much, you can reduce your allocation accepting less service. Or you can request more service and we'll adjust your charge. Of course, we still get some of the "I want more service but I don't want to pay for it."

Beyond maintaining and upgrading Stellex, Outrigger's IT professionals engaged in minimal application development—mainly writing customized reports, and configuring and interfacing off-the-shelf applications. Outrigger had implemented JD Edwards ERP as the cornerstone of its back-office operations in 1990, years before the ERP craze swept the business world. JD Edwards ran on an IBM AS 400—a very mature and stable platform. The use of outsourcing was limited to the website, developed and hosted by a third party in Portland, Oregon. Yet, in order to maintain the integration of direct channels, Stellex served as the website's booking engine through an XML interface that Outrigger's IT group used as the proof of concept for the interfaces with wholesalers—a key initiative for Outrigger. Durocher explained:

> With many wholesalers we have real-time electronic interfaces—they can check availability and we get their reservations instantaneously. Without the interface, if they create a reservation six or three months out, we don't see it until reporting time, ten days out, when we receive a fax and manually input it. It is virtually impossible to revenue manage like that. Many big brands have great revenue management systems, but don't have real-time wholesaler data. Moreover, we can write wholesale contracts brand-wide.

Exhibit 6.6: Property-level IT infrastructure (Hawaii properties)

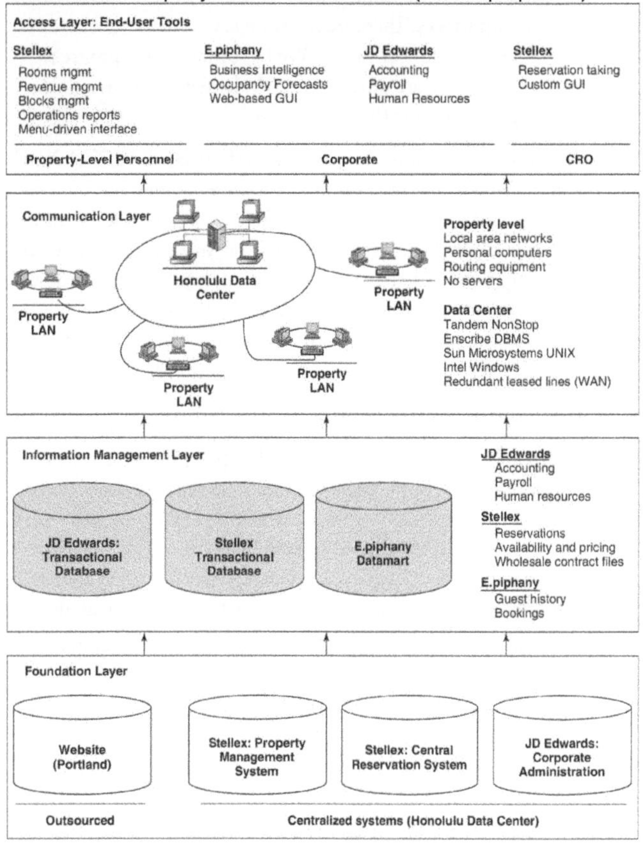

Outrigger felt that its electronic interfaces afforded it a competitive advantage and preferential treatment from interface-enabled wholesalers, a relationship that proved particularly important during slow periods or a soft economy. Electronic interfaces generated substantial efficiencies, including automatic billing and invoicing without human handling, lowering estimated costs to $0.75 from an estimated $10 for manually handled ones. But not all wholesalers were able or interested in automating reservation processing. This was particularly true for small operations or those for whom Hawaii and the Pacific represented a small percentage of business. Solomon summarized the challenge:

> The industry is a mess from a connectivity standpoint. We are fortunate that we have the in-house expertise and the recognition from senior management of how important this is. Even the big companies often don't understand the conditions for success. The dirty little secret of the travel industry is that the fax machine still rules.

White added:

> I spend 30–40 hours a week working with wholesalers on interfaces. There are many legacy systems out there; the fax is state of the art. We have made great progress with the more advance[d] wholesalers or those that upgraded recently.

Outrigger found the Open Travel Alliance (OTA) XML standards, specifying common message format and common content, of great help. But being able to pick the right partner, and avoid costly failures, remained the major challenge. While Outrigger felt it had been successful to date, with an estimated 33% of total reservations received electronically through the various channels, it still handled more than half a million faxes a year—about eight hundred a day from its largest wholesaler alone before its recent migration to the electronic interface.

The firm had recently acquired business intelligence software, a data mart, and analytical tools from E.piphany running on a Windows 2000 platform. The data mart held detailed data for three years, enabling analysis down to the individual guest folio. Data were consolidated afterwards, enabling only aggregate analyses. While E.piphany was a recent purchase, Outrigger had been disciplined in collecting data for some time. White explained:

> We had 10 years of high quality data from Stellex; we are very rigid about data capture standardization like room category, naming conventions, request codes, [and] what goes where. For example, postal and country codes are mandatory fields. Our employees' long tenure helps, and peer pressure is a great asset—nobody wants to be the one that ruins the value of these reports for all.

The data collected by Stellex, including source of business, stay information, and consumption, were extracted every night by load programs that scrubbed (i.e., cleaned) them, and transferred them to JD Edwards for accounting and E.piphany for analysis. Feeding historical data and forward-looking availability and reservation activity, Outrigger had learned to harness the analytical power of E.piphany to do forecasts and generate business intelligence both at the source of business and at guest levels. White elaborated:

> We want the marketing data. It is stupid to have a treasure trove like that and not use it. We mine it. We send thank you letters to recurring guests, we can give you history on who visited, how they got here, what in-flight magazine we should hit. We sold a resort once and they figured they would have to hire 3 people to achieve manually what our reports gave them automatically. They even set their rates based on E.piphany forecasts.

The IT group served as custodian of the data, but any user with security clearance had access to E.piphany data though a web interface; the data was used for marketing and operational analysis (e.g., analysis of call patterns to evaluate the appeal of Voice over IP solutions). More challenging was to incorporate the information into daily operations. Outrigger found it hard to justify a frequent guest program—with an average repurchase cycle for returning guests of three years, a once a year purchase was considered very high in resort operations. Speaking about recognition programs, Sorenson explained:

> Individual properties have their own customer database and a strong informal recognition system. We haven't been able to justify the investment technologically to do it brand wide. It would be a natural extension of the recognition we give our return guests, but it must be cost-effective.

Agas added:

> If a guest did not tell us he is returning when making the reservation, our current system does not have a database with guest history. Many times we recognize our frequent return guests at the door, or during check in at the front desk, but without any guest history in on current system a new employee may not acknowledge the return guest or their special occasion. We have special programs (e.g., for honeymooners, wedding anniversaries), but we need to know their history to appropriately acknowledge them.

IS Assessment

Outrigger's senior executives found technology to be a great asset to enable communication, as Outrigger's operations spanned 11 time zones, and felt confident that the IT function was enabling the firm to compete effectively. Carey indicated:

> We think that our IT capability in the leisure travel space exceeds the major chains and we have an ability to implement things very quickly. [That's] the advantage of being small.

The IT function was thought to be able to operate more efficiently than the competition, often offering the same level of service with one or no property-level IS professionals when the competition needed three to six. Outrigger also felt that its size enabled it to move faster than the competition. Bob Owns, director of system operations, explained:

> We don't do anything slow here. Major systems in other firms take a year to plan, a year in committees that assign responsibilities, and two or three years to build. A year is a really along time here to develop and implement anything. But we are not a huge company, and capital is a constraint, so we are always challenged to get way ahead of the curve, speculate, and build with a forward thinking mentality. You don't get bored here.

As the firm was expanding aggressively, and had yet to find an integrated solution for its international properties, some questioned the viability of reinvesting in Stellex. Its rapid geographical and product growth notwithstanding, the IS group felt that its legacy technology—specifically its mature ERP, integrated PMS/CRS, and electronic interfaces with distribution partners—was serving the firm well. White explained:

> Stellex is 18 years old. So three years ago we developed the business case for PMS and CRS functionalities. We could not find anything better, with one exception—Stellex is a green screen application that needs a windows GUI.

The firm was prompted to re-evaluate the role of Stellex after a failed attempt to migrate to a more modern platform thought to simplify connectivity with the other off-the-shelf computer systems in the portfolio. After testing in two properties over an eight month period the project was aborted, principally blaming the difficulty in effectively managing wholesale relationships and billing manually with the new PMS.

Outrigger engaged in limited formal technology training and relied mainly on on-the-job training when it came to software applications. While this created difficulties for people who were hired from outside the firm, Sorenson explained:

> Our people have been working with Stellex so long that they have effective workarounds when necessary, and we have very low employee turnover. If someone new comes in we have many experienced employees to help them; this makes training easier.

As guests became used to ever increasing technology choices and availability both at home and on the road, even resorts focused on the leisure traveler felt the pressure to provide it to guests—whether they used it or not. But for a mid-size company like Outrigger, chasing the technology curve could be dangerous. Agas articulated the challenge:

> Our guests say: "I do wireless at home, why can't I do it here?" As a company we use our buying power to do what's best for the company. But as two beachfront properties with guests paying the highest ADR and expecting more, sometimes we are held back when it gets to technology as we explore what is best for all.

The Future

Outrigger's senior management felt that the firm could leverage its hospitality and marketing expertise, as well as big brand name recognition, by entering into management agreements with third-party owners and large brands. While it remained committed to growing and strengthening the Outrigger family of brands, it also had plans to engage in this type of partnership.

Another important trend affecting Outrigger's future strategy was the rapidly changing hospitality distribution landscape and the role of the retail travel agent. Travel agents had historically provided significant amounts of information, counseling, and reassurance to leisure travelers, but more and more consumers were now turning to the Internet for this information. This presented Outrigger with the challenge of populating the new electronic world. The emergence of powerful online agencies (e.g., Expedia, Orbitz) was creating significant opportunities and threats. Carey captured them:

> We have grown up with wholesalers; we know how to yield manage the merchant model. The major chains are not yet embracing the capabilities of the internet. They look at Internet bookings through third party providers as a threat. We see it as just another wholesaler; we know how to revenue manage the merchant model. We all must recognize the consumer's desire to shop before they buy. The single website solution will not work in my opinion.

This was particularly true with wholesalers using electronic interfaces. With these partners, Outrigger was able to open and close rates dynamically. Yet, questions remained as to the long-term effects that powerful online intermediaries were having on customer loyalty and brand preference. As some senior managers put it: "Whose customer is it, Expedia's or ours?" For a company with relatively small scale and a niche positioning, the commoditization threat could be quite dangerous. Durocher summarized the challenge:

> In the days of Mr. Kelley and Dr. Kelley, Waikiki was running at 98% occupancy annually. Get the reservations in accurately was the main concern. That world has changed, now we compete in mature destinations.

With the increasing competition in its key markets, Outrigger felt that strengthening electronic relationships with distributors, improving its trademark hospitality and customer service, better managing inventory yield, and better integrating its international properties were crucial steppingstones to the firm's continued success. The right information systems strategy was crucial to enabling these goals.

Discussion Questions

1. What is Outrigger Hotels and Resorts' strategic position? What are its strengths and weaknesses? What are the firm's Critical Success Factors (CSF)?
2. How well are current IS resources serving the needs of Outrigger Hotels and Resorts?
3. What should be, in your opinion, the role of the IS function at Outrigger Hotels and Resorts?
4. Can you articulate both the IS vision and the IS guidelines for the firm?
5. Based on your proposed IS plan for Outrigger Hotels and Resorts, what strategic initiatives would you propose?

Case Study for Chapter 7

TRIPBAM: Leveraging Digital Data Streams to Unleash Savings (A)

Introduction[1]

> We have a solution unlike anything that is on the market today or that has been on the market before. . . . Our features provide real savings for business travelers, companies, and leisure travelers that puts us well ahead of the rest of the travel market. (TRIPBAM, 2013)

It was April 2013 when Steve Reynolds found himself taking a new and unexpected journey. After 25 years in the travel industry, he had successfully funded two travel companies in the corporate space and had held positions as executive and consultant. Now he was incorporating an innovative consumer service: TRIPBAM. TRIPBAM's value proposition consisted of discovering cheaper hotel rates for travelers who had already booked a trip.

This turn in Reynolds's career came about rather serendipitously. First, at a board meeting, a colleague explained how he managed to always get the best rates by having his secretary shop every day until the day of travel. Then he noticed a corporate travel agency offering a similar manual rate-checking service, called the platinum desk, to their best clients. The catalyst was an observation at a four-way intersection on the way to the Dallas airport. Four comparable hotels stood on each corner. Reynolds recalled thinking, "What if we automated rate monitoring and provide it as a service? Could we significantly lower clients' hotel cost while preserving flexibility and quality?" From this intuition, Reynolds grew TRIPBAM into both a consumer-oriented and business-to-business (B2B) solution. In 2014, the firm won the Business Travel Supplier Innovation Award at Business Travel News' Innovate Conference and, in 2015, the Phocuswright Battleground Contest. It also won recognition from corporate travel publications in the United Kingdom and the United States (TRIPBAM, 2013).

With the product in place and seed funding secured from Thayer Ventures, the leading hospitality venture capital firm, it was critical for Reynolds to grow TRIPBAM from a clever idea into a viable business. The success or failure of the company depended on growing the user base, beating an increasing number of competitors, avoiding industry retaliation, and building a sustainable business model.

Industry Background

More than 50 years had passed since the first reservation system had been introduced in the airline industry. In 1963, Sabre, a joint effort of American Airlines and IBM, demonstrated the capability of IT to efficiently manage the inventory of airline seats. It was just a matter of time for hospitality firms to implement a similar solution to manage the daunting problem of pricing hotel rooms. Two years later, Holiday Inn's HOLIDEX was a reality, and the hospitality industry changed forever. In the late 1970s and early 1980s, the introduction of central reservation systems (CRSs) showcased the power of distribution networks, which created the efficiencies and distribution services needed for the franchising model to work. Franchising ushered in an era of consolidation during which large multinational chains replaced family-run hotels. By the 1990s, the U.S. lodging industry was dominated by franchisors and management companies with strong brands in each market segment.

The widespread adoption of the Internet in the mid-1990s subverted the established order. Hospitality operators lagged behind, and new entrants with technology backgrounds and the strong entrepreneurial spirit of the dot-com era entered and grew quickly. These new ventures quickly captured a substantial amount of the value created in the industry by intermediating reservations online.

The dominant players included Expedia, originally a business unit of technology giant Microsoft, and Priceline, the brainchild of technology entrepreneur Jay Walker.

In early 2015, the market continued to consolidate. Expedia purchased Travelocity for $280 million on January 23, Orbitz on February 12 for $1.34 billion, and HomeAway on November 4 for $3.9 billion. After its acquisition spree, Expedia became the world's largest online travel agency (OTA) by total bookings and, with Priceline, established a de facto duopoly in the industry (see Exhibit 7.1). Large OTAs drove revenue through commissions and placement costs.[2]

1 Excerpted from Piccoli, G., and Pigni, F. 2016. "TRIPBAM: Leveraging digital data streams to unleash savings." *Communications of the Association of Information Systems 39*. Article 25. Used with permission from Association for Information Systems, Atlanta, GA; 404-413-7445; www.aisnet.org. All rights reserved.

2 Commissions are paid to OTAs as a percentage of the total booking value. Placement costs consist of additional fees OTAs may request

However, unlike the use of fixed commission as in the "pre-Internet days," current arrangements created costs of acquisition through intermediaries ranging anywhere from 15% to 35%. Furthermore, the OTAs' share of guest-paid room revenue increased along all hotel segments. For economy and midscale hotels, it increased from 7.8% in 2011 to 15.8% in 2014; for upper-midscale and upscale hotels, it increased from 7.7% to 11.7% during the same period (Mayock, 2015); and for upper-upscale and luxury hotels, the OTAs' share increased from 6.6% to 8.8%. In 2014, while lodging operators were still receiving the largest share of their bookings from their websites and through their call centers, the OTAs' share of the online hotel market increased to 48% (up from 46% in 2012; Walsh 2014). This trend was generally thought to reflect the propensity of millennials, who represented 33% of U.S. leisure hotel guests (Harteveldt, 2013), to interact and book directly online through apps or social media. As Douglas Quinby, a principal analyst with the travel research firm Phocuswright, put it, since about 2005, there had been "pretty amazing growth of both the funding that's gone into online travel and travel technology startups as well as the number of companies that have . . . made their way into the arena" (Robinson-Jacobs, 2013).

OTAs were not, however, the only players in the distribution space. Metasearch sites like Kayak (a Priceline property), Skyscanner, and Hipmunk were often the place where savvy travelers would shop first (Figure 7.1).

CRSs, travel agencies and travel management companies (TMCs) were other important players in the lodging distribution chain. While travel agents serviced a niche market of travelers with complex itinerary requests in the unmanaged consumer market, business travelers relied on internal travel managers or TMCs like Concur or American Express for their travel needs. Travel agents' revenues were primarily composed of service fees (e.g., ticketing, booking, expense management) and transaction fees (e.g., credit card discount rate, processing costs, global distribution system [GDS] fees). In some contracts, they could also garner commissions, but these financial flows represented a fraction of their total revenue.

Business travel management was a complex process encompassing corporate travel policies (e.g., maximum allowed rates, allowed categories, and daily expense limits), itinerary planning and booking, expense management, and reporting. It also included yearly rate negotiations with suppliers. Up 4% and totaling $112 billion in 2015, U.S. corporate travel represented one-third of the total travel market (Figure 7.2). Despite hoteliers' efforts to attract direct bookings, more than three-quarters ($87 billion, 78%) of corporate travel was still managed by an intermediary (Fadnis 2016).

to hotels for gaining better positioning in hotel and price comparison searches.

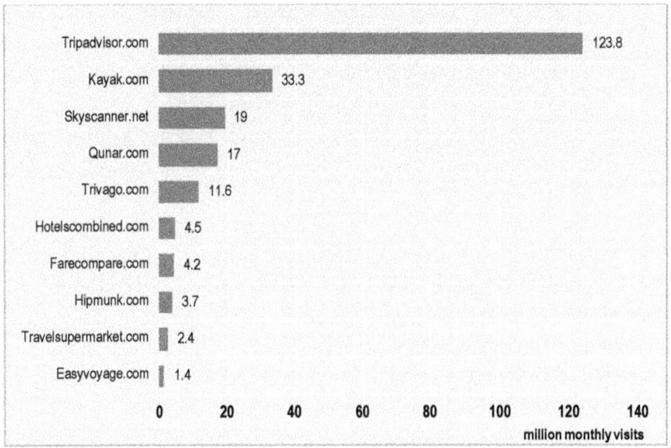

Figure 7.1. Key global metasearch engines in travel
Source: Data from SimilarWeb, February 1, 2016

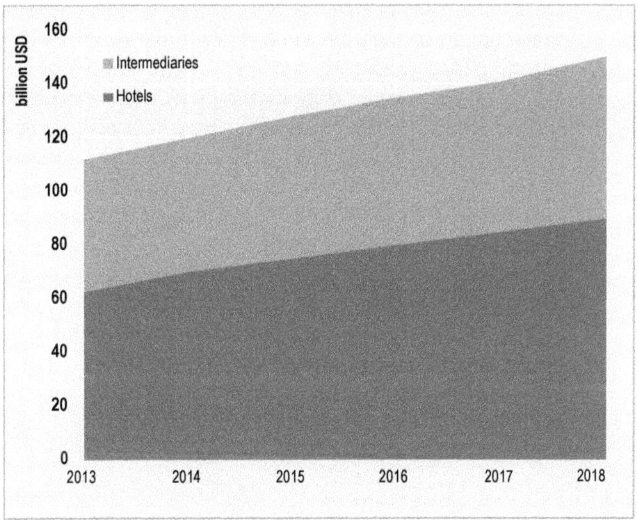

Figure 7.2. Worldwide digital travel
Source: eMarketer (2015)

The rising power of OTAs on the consumer side imposed high customer acquisition costs on hotels and an increasing dependency upon intermediaries in the distribution channels, particularly for independent operators. In addition, given the strength of their cash flow, OTAs were able to invest in marketing and communication to attract consumers. They also continually innovated on their online offerings by introducing faster and better search and comparison tools. In 2015, ease of use was still the main reason travelers favored OTAs and other booking sites, even ahead of price (Gasdia, Liu, and Blutstein, 2015). Additionally, OTAs introduced their own loyalty programs in an effort to offset the value proposition of the major chains' proprietary frequent traveler incentives. Sixteen percent of consumer travelers booked rooms four or more times annually, and many of them (67%) belonged to hotel loyalty programs (Gasdia, Liu, and Blutstein, 2015).

While observers did not consider lodging a pioneering industry when it came to IT, the travel value chain was controlled by software in the form of GDSs, CRSs, and property management systems. By 2012, four major GDSs dominated the hotel reservation landscape: Amadeus, Sabre, and Travelport's Worldspan and Galileo. GDSs represented the backbone of the travel industry, where hotels published information such as room availabilities and rates. Other than hotels, the access to GDSs was granted only to travel agents or authorized travel resellers. Smaller brands accessed GDSs indirectly through intermediate distribution systems called switches. Pegasus, Genares, Dhisco, and InnLink were representative players.

Internally, hotels used revenue management systems to optimize their occupancy and profits, acting directly on room availabilities and rates. During periods of expected high occupancy, the hotel raised rates to extract greater margins while tolerating lower rates in times of low projected occupancy. The Internet increased rate visibility for guests and competitors alike, enabling revenue management systems to take market conditions into account when creating forecasts. The ultimate outcome of these trends was a drastic increase in volatility of room rates and the complexity of rate structures that customers had to understand and work with.

Interestingly, the rise of OTAs forced reservation and distribution systems to evolve. The OTA model, enabling consumers to shop directly for travel, utterly increased the number of hits to the GDS. These hits were called availability requests or "looks." But the number of actual confirmed reservations sold, the so-called books, did not increase accordingly. The net result was a significant growth in the look-to-book (L2B) ratio (Figure 7.3). Such growth required that technology intermediaries like the switch (e.g., Pegasus), the GDSs, and CRSs continuously maintain and update their technology infrastructure. They were paid, however, for confirmed reservations. Therefore, they had to insulate themselves against the risk of intermediaries who would shop (create cost) without booking (create revenue). Intermediaries then charged escalating fees for the access to their connectivity services, fees tied to L2B ratios.

It is in this landscape that TRIPBAM was born, with the objective of helping travelers navigate industry complexity and take advantage of rate competition. Industry figures looked promising: $40 billion of hotel bookings were placed through intermediaries (Figure 7.4). Of all travelers, 92% averaged two bookings per year (Gasdia, Liu, and Blutstein, 2015), and the average daily room rate had been steadily rising. Reynolds had reason to believe that TRIPBAM was set for a profitable journey.

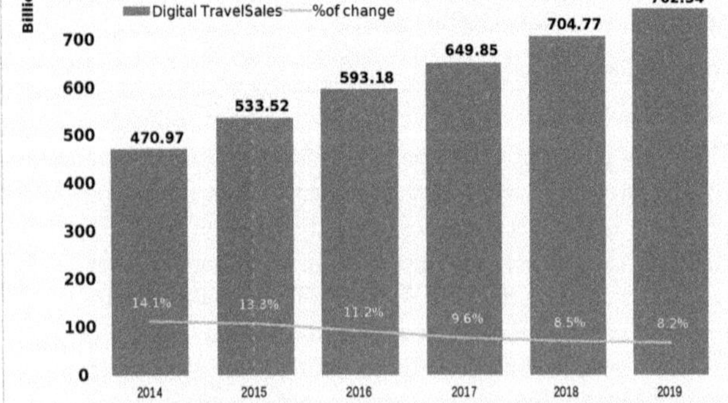

Figure 7.3. The historical evolution of L2B ratios for GDS and alternate distribution systems

Source: Peacock (2010)

Building TRIPBAM

Steve Reynolds was a founding member of two technology companies in the travel vertical. The first, Houston-based CTI, provided travel systems and services to Fortune 1000 companies in the late 1980s and early 1990s and was purchased by American Express. The second, Travel Technologies Group (TTG), funded in June 1992 and based in Dallas, Texas, grew to $25M in annual revenues before merging with two other companies in 2000 to become TRX. TRX employed 1,600 people in 14 countries and processed more than 80% of the corporate travel transactions in the United States. After leaving his executive position at TRX in 2003, Reynolds held multiple managing and consulting positions and in 2015 was named one of the top 25 most influential executives within corporate travel. Inspiration struck in 2012 when he wanted to prove the viability of TRIPBAM's automated hotel rate shopping. Reynolds explained these early tests:

> We went out and bought historical hotel rates for 2012 to 2013 and we built a little analytic engine to go back and look at how often, within a cluster of hotels, we observed rate fluctuations. We wanted to know if we could deliver savings. The key was to beat OTA's non-refundable rates by tracking rates across a set of hotels on a daily basis. Results came out better than we expected. We found savings 50–70% of the times and the savings averaged around $60 per night. That's when I knew we were on to something.

TRIPBAM found that in the majority of cases, rates granting free cancelation (i.e., refundable rates) were found twice per booking and were beating OTAs' rates when approaching the day of the stay. The last piece of the puzzle was to ascertain whether the commissions earned on bookings offset the costs for the daily shopping of room rates. Access to GDS and other distribution systems varied but generally required the payment of a one-time setup fee and a monthly fee. Such fees could range substantially depending on the size of the intermediary and the volume of reservations being generated, but they usually hovered around $5,000 and $2,500, respectively. OTAs like Expedia provided free API[1] access to their affiliates as soon as they could comply with an L2B ratio of 500 to 1 for room rates and availability requests. As Reynolds commented, "Shopping costs have come down compared to the mainframe era because hardware and infrastructure costs have decreased dramatically. Today, it's micro pennies per hit."

In November 2011, once he became convinced of the opportunity available to TRIPBAM, Reynolds reached out Caleb Blanton, a former colleague at TRX. After leaving TRX in 2007, Blanton had created his own IT consulting company, Creativa Consulting. Blanton recalled,

> Steve told me the story of how he came up with the idea. We had our share of experience with booking tools and we knew that the travel space was really congested, but I immediately realized this wasn't a booking tool. It was a shopping service and in an industry this congested it's pretty unique to find an emerging niche.

But between an idea and an operating startup, there was significant work to be done. Blanton explained, "I took Steve's coarse-grained business requirements and developed finer grained software solutions. It was challenging, but I am particularly good at bridging the gap between the business and the technology."

Reynolds financed the initial development of TRIPBAM's prototype. It was a proof of concept limited to core functionalities and based only on static data. Reynolds recounted, "I needed a minimum viable product to pitch investors, to demonstrate our initial success. Investors were critical partners to fuel our growth."

TRIPBAM was incorporated in Dallas in April 2013, and it was the first service that provided clustered-rate shopping to consumers. This unique feature of TRIPBAM enabled users to shop rates from a set of similar hotels and within a geographical area.

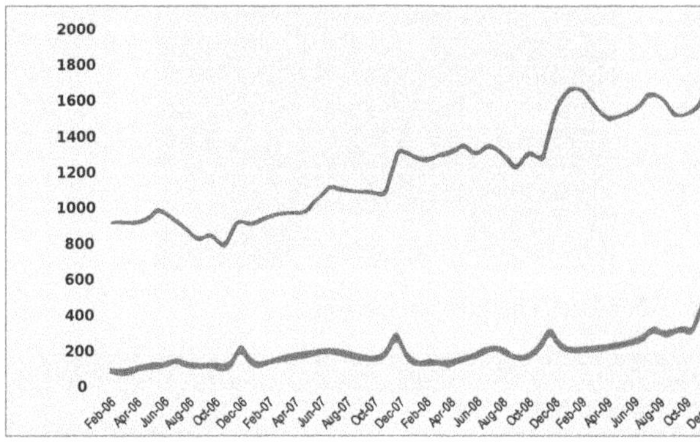

Figure 7.4. Hotel global online sales by channel, projections for 2014 and onward

Source: Rossini (2014), 11

Initially, TRIPBAM development was outsourced to Blanton's software company, but in 2015 his whole team joined TRIPBAM full time, raising the employee count to 10. In August 2013, Thayer Ventures injected $1M in seed funding. A second wave of funding followed in January 2015 for an additional $525K of equity to fund TRIPBAM's development and market penetration in the business segment. To secure TRIPBAM's intellectual property in September 2013, the firm applied for a utility patent covering the clustered rate shopping concept, receiving it in March 2014 (Reynolds, 2014).

The Product

At its core, TRIPBAM was an automated cloud-based service enabling customers to monitor and rebook hotel reservations prior to traveling. On average, stays were booked between one week and two months in advance, and in that period, rates may drop once or twice (Gasdia, Liu, and Blutstein, 2015). In the simplest use case, a customer would create a set of specifications for a reservation the customer already held: date of arrival and departure, type of room, discounts the customer was qualified for (e.g., AAA), rate booked, and of course hotel name and location. TRIPBAM referred to these specifications as the "search." Once the search had been confirmed, TRIPBAM would monitor the hotel for rate fluctuations and alert the customer when rebooking would yield savings (Figure 7.5).

A more powerful approach to searching occurred when a customer created a cluster. A cluster was a group of hotels the traveler deemed comparable to the one he or she had originally booked and was willing to stay at for the current trip. By dragging hotels within a geographical radius into a list, the customer constructed the cluster (Figure 7.6). For customers creating clusters in unfamiliar locations, TRIPBAM produced customized suggestions based on inclusion criteria like distance,

1 An application programming interface (API) is a standardized method to allow a set of data and functions of a computer program to be shared with other applications. For example, both TripAdvisor and Yelp use Google Maps API to show the location and provide directions to users' desired venues. More specifically, Expedia's API provided affiliates the access to several core functions for managing the booking process (e.g., listing hotels, room availability, room images, payment types, geolocation). See https://developer.expediapartnersolutions.com/getting-started/ for more details.

Figure 7.5. Search confirmation message

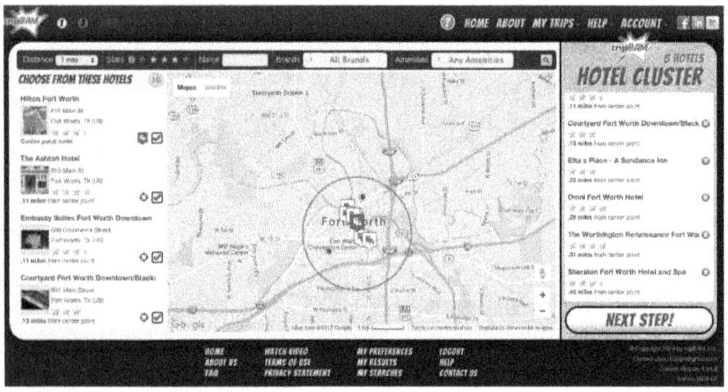

Figure 7.6. Creating the cluster on TRIPBAM

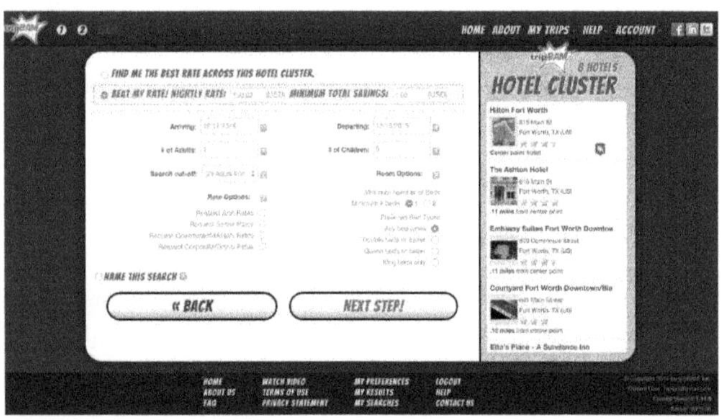

Figure 7.7. Clustered search design page

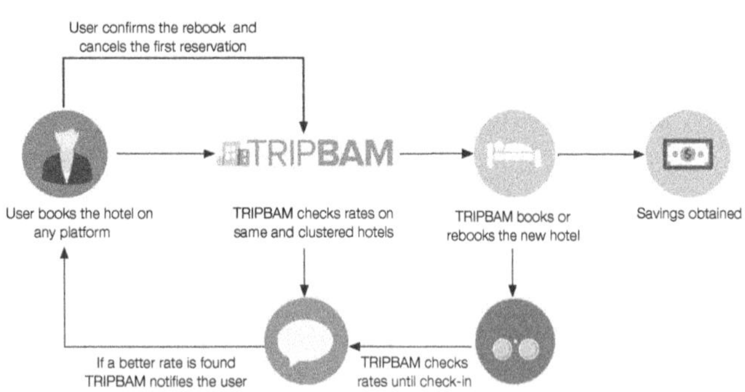

Figure 7.8. TRIPBAM overall clustered rate shopping process

Source: Case writers

brand, and amenities. The firm touted the benefits of clustered search on its website:

> Only looking at rates at an individual hotel severely limits your options. Expanding your search to a group of preferred hotels will yield better results. Create a hotel cluster within a defined geographic area with an equivalent quality level to increase your odds of finding a great rate.

Once a cluster had been established, the traveler designed and stored a search as if monitoring a single hotel. For those individuals without a reservation, TRIPBAM would select the best available rate within the cluster. For those who had already booked, TRIPBAM sought to beat their current rate (Figure 7.7). Setting up a clustered search also enabled customers to take advantage of all the different ways in which TRIPBAM algorithms could uncover advantageous rates. TRIPBAM stated on its website,

> Fine tuning the amenity mix is a great way to reveal hidden rates. For example, you can save a lot of money by switching from a king-size bed to a queen. Look at all types of combinations to further lower your costs.

While current search design allowed limited selections, mostly focusing on bed types, the firm had plans to expand the list. The strategy for what rates to book also created opportunities. The firm website explained,

> In the early stages of a search, it could be risky to book a non-refundable rate. However, a day before check-in it might be worth considering. We include penalty rates when it makes sense to get you rates at rock-bottom prices.

Once TRIPBAM found a lower rate at the same hotel the customer had reserved or one in the acceptable cluster, it sent an automatic e-mail asking the customer to accept or decline (Figure 7.9). If the traveler accepted, then TRIPBAM would rebook the stay at the hotel and send a new confirmation e-mail. Otherwise, no action was taken, and TRIPBAM used the information to improve its knowledge of the customer's preferences (see Figure 7.8 for the overall process). While customers were maximizing savings with clustered shopping, their participation in loyalty programs could dampen their interest in clusters. However, hotels had invested in loyalty programs for retaining and attracting new customers with uncertain results (Lee et al., 2014).

As customers became well versed in the use of TRIPBAM and more confident with the service, they could store their payment information with TRIPBAM and set searches to automatically rebook (Figure 7.9).

3.1 Core Technology

The TRIPBAM system had to determine the cluster starting from the booking, extrapolating geographic and hotel information (e.g., stars, brand affiliation). The firm harvested and consolidated hotel rates from multiple sources, but its principal connection was to the GDS via the Dhisco switch.

In 2016, the core application resided on a private cloud in a datacenter in Miami. TRIPBAM was developed as a client-server web application. Server side, TRIPBAM followed a three-tier architecture, where the presentation, business logic, and data access tiers were separated (Figure 7.10). This separation increased the maintainability and scalability of the application, as tiers could be updated independently.

Blanton took particular care in designing the data providers' access. He wanted the application to easily integrate new data sources without affecting the way the logic layer dealt with that information. By doing so, developers could concentrate on data use and manipulation instead of investing precious time and development dollars on interface and component integration.

The business logic layer, including the rules engine for rate comparison and clustering, was based on the Microsoft .NET framework. The database management system was Microsoft SQL Server.

The user interface was built on DotNetNuke, an open source content management system and application framework for delivering the web user experience. The user interface tier implemented additional components based on non-Microsoft technologies like AngularJS (an open source web application framework). Blanton intentionally decided on a best-of-breed strategy for TRIPBAM as he wanted to use "the right technology for the job" at each step.

Interestingly, TRIPBAM's architecture was mainly transactional. Static hotel data and a subset of booking information for analysis and operations were stored in the database. Data providers were polled for rates, availabilities, and bookings. Customers' personal data were not stored or processed outside of booking fulfillment. The whole system architecture was conceived to support an application platform internally. TRIPBAM planned to seamlessly expose internally and externally its system capabilities through an API. Or as Blanton put it, "It's always good to drink your own champagne."

TRIPBAM collected rates and availability information for the hotels in the monitored lists at a predetermined time. Timely notification to users was an important part of the value proposition as special rates were available for limited time or quantities.

Blanton's team built the rules engine on a proprietary framework and determined when a monitored rate fit a customer's preferences, the cluster characteristics, and the saving threshold he or she identified, thus triggering the alert notification (Figure 7.11).

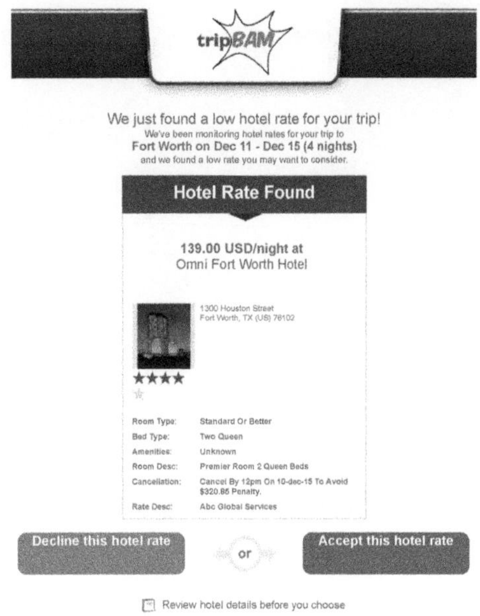

Figure 7.9. New rate alert

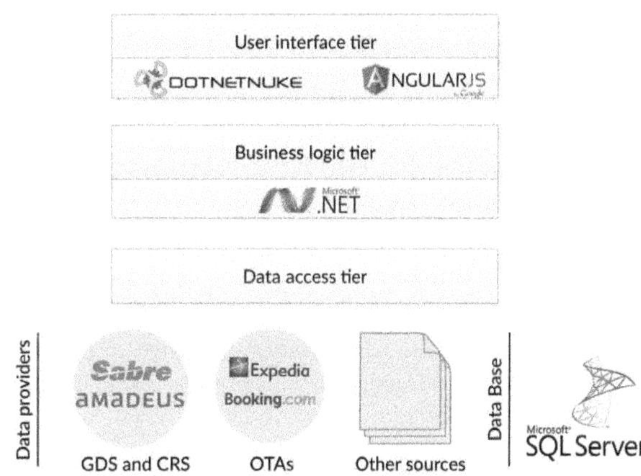

Figure 7.10. Overall logical architecture

Source: Case writers based on company data

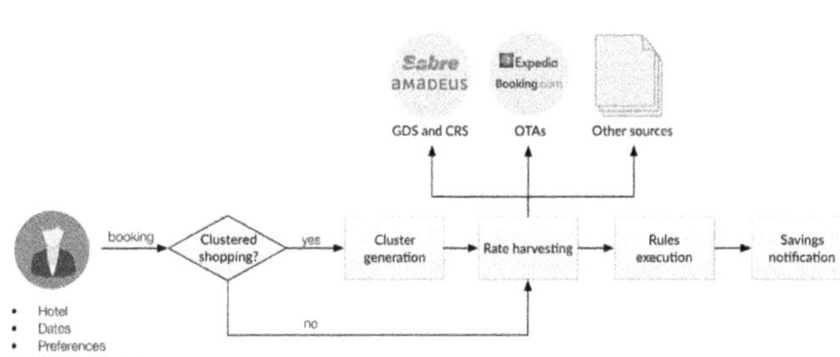

Figure 7.11. TRIPBAM workflow

Source: Case writers based on company data

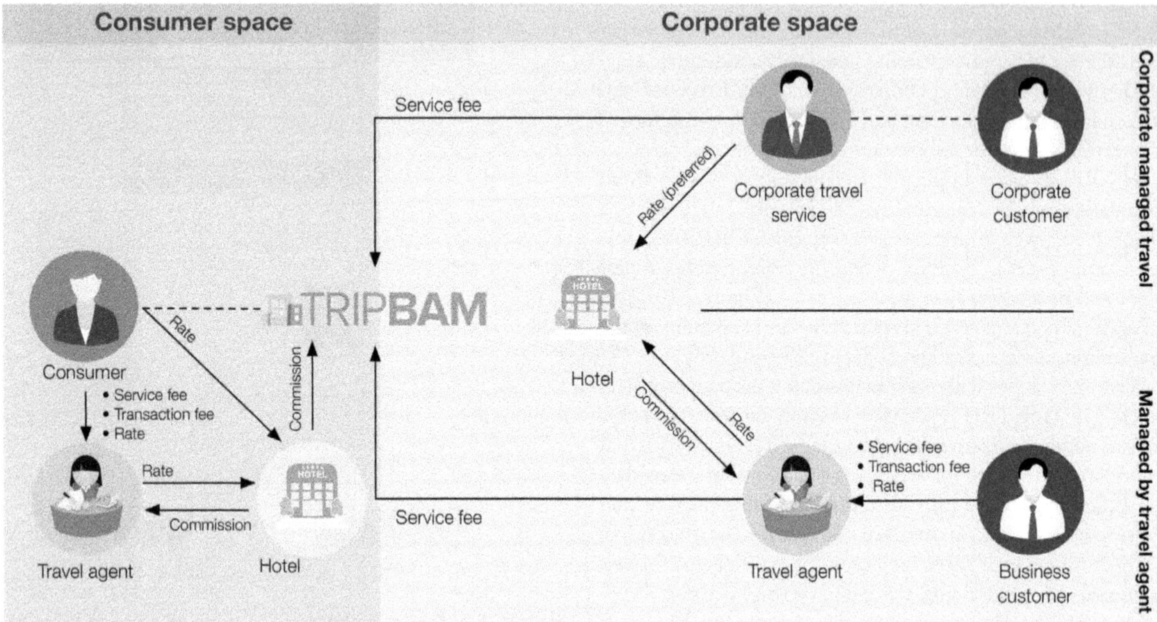

Figure 7.12. TRIPBAM consumer and corporate revenue models

Source: Case writers based on company data

3.2 The Corporate Opportunity

Just two months after TRIPBAM's launch, something unexpected happened. Reynolds recalled,

> Large corporations called up and said they wanted to use TRIPBAM for all of their corporate bookings. We were surprised, but we started to work with them. Based on those relationships, we started to develop more corporate-oriented functionalities like a customizable rules engine, reporting and analytics.

Large corporations easily exceeded 15,000 bookings per year (Hermes Management Consulting, 2013). The team had to work hard to refine the service for corporate customers. In particular, compliance was a critical element, because in corporate travel, companies had specific requirements on preferred property programs, negotiated rates, preferred brands, and agreements with hotel brands. While most corporations had automated policy compliance tools, 8 of 10 companies manually monitored and controlled travel expenses (Amadeus, 2015). The use of TRIPBAM was substantially different compared to the consumer market. In the corporate space, bookings were initiated and managed by the customer on reservation systems like Concur, by travel agents, or directly on brand.com sites. Thus TRIPBAM had to pivot its value proposition in order to market itself as a service for generating savings and enforcing travel policies' compliance. The service agreement between TRIPBAM and a business customer defined the engine shopping parameters (e.g., same hotel, same room, preferred cluster, nonpreferred cluster, etc.). In general, customers provided the bookings they wanted to monitor through TRIPBAM's API, and rates were shopped accordingly. The customer was then notified in the case of a lower rate.

While different from what TRIPBAM planned for the consumer market, the B2B opportunity appeared to be real. Reynolds recalled, "Thayer Venture's Jeff Jackson, a real expert in the B2B space, pushed us to pursue B2B since he saw it as the best opportunity."

3.3 Revenue Model

In the consumer space, TRIPBAM sought commissions for brokering the reservation like travel agents did (Figure 7.12). Commissions were directly collected when TRIPBAM was the travel agent confirming the booking that the customer actually used—the last booking. Reynolds explained the inherent challenge:

> Today many consumers use TRIPBAM for shopping. We have to figure out how to increase conversions and make these shoppers book with us. We planned to limit the ability to shop for free, providing one or two rebookings before asking a fee.

On the corporate side, the revenue model was different because bookings were initiated and managed independently from TRIPBAM. Reynolds commented, "TRIPBAM can be used to improve compliance as much as cost savings. One of our clients has improved preferred property usage by more than 20 percent."

On the corporate side, TRIPBAM was marketed as a subscription service with a very minimal monthly fee. While the minimum subscription discouraged smaller customers, it helped TRIPBAM contain interface development costs.

3.4 Competition

While TRIPBAM had been a pioneer, there was increasing competition. OTAs were offering price guarantees only on the day of purchase and not aggressively moving into TRIPBAM's territory. Reynolds explained, "While we are about cost savings, OTAs interest lays in revenues improvement. They are so dependent upon hotel commission revenues that copying us would actually mean diminishing their revenues."

Reynolds worried more about the number of other startups or existing companies that would look to copy them in the event that their service became widely accepted. In 2012, Tingo, owned by TripAdvisor, introduced the concept of automatic rebooking and refund. The application leveraged the Expedia Affiliate Network to book and monitor hotel rates. In case of price drops or free upgrades, the service would automatically rebook and refund the price difference directly on customers' credit cards. In 2013, the European startup TripRebel launched on Tingo's basic principles. The system featured the same automatic room rebooking when upgrades, better rates, or additional amenities were made available at the same hotel. Additionally, it enabled the user to rebook at better rates within a predefined "favorite" list of properties.

In 2014, WorldMate, a traveler's mobile application, received the 2014 Phocuswright Most Innovative Established Company Award for its Hotel Price Alerts and counter offers features, which notified users of lower rates available at the same or in a comparable property (Figure 7.13). WorldMate disclosed that these features were generating an average savings per trip of $88 for their 11 million users and that the rebooking process took less than one minute and required eight clicks (WorldMate, 2014). As WorldMate CEO Amir Kirshenboim put it, "Price Alerts and Counter Offers are the culmination of over 100 man-years of engineering and are only possible because of the amazing efforts of our talented and dedicated team" ("WorldMate Wins the PhoCusWright," 2014).

Owned by Carlson Wagonlit Travel (the fourth-largest travel agency in the world with $23.4 billion in revenues from 58 million transactions in 2015) and strengthened by the partnership with Expedia, WorldMate showed that rate shopping was not only successfully replicable by competitors but also exploitable by established travel agents. WorldMate vice president Ian Berman further stated that he could envision a managed version

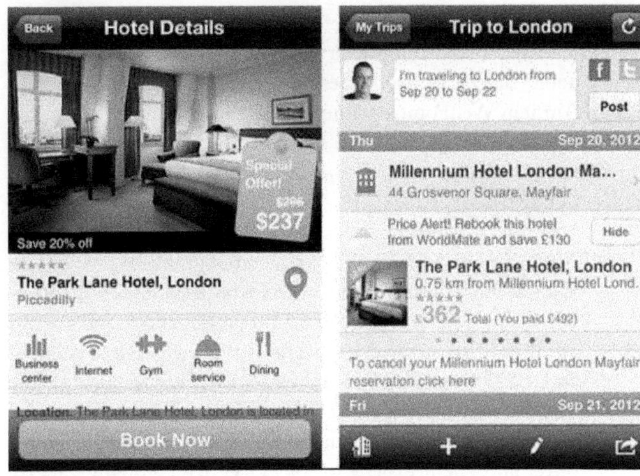

Figure 7.13. WorldMate's app screenshots

Source: Belic (2012)

Figure 7.14. Yapta RoomIQ service

Source: "RoomIQ Hotel Price" (2016)

incorporating policies and preferred suppliers for corporate customers.

In early 2015, Yapta introduced the RoomIQ feature that, coupled with FareIQ, offered a service similar to TRIPBAM on both flight and hotel reservations (Figure 7.14).

While Yapta traditionally targeted the managed corporate travel for airfare refunds, it started to challenge TRIPBAM on hotels using GDS data and same-hotel-rate shopping. It was backed by Amadeus and Concur and had secured more than $20 million in funding.

Early Results

TRIPBAM consumer base growth was slow. However, TRIPBAM caught the attention of the mainstream press, with stories in the *New York Times* and the *Wall Street Journal* that generated several thousand users, even before any marketing effort. On the corporate side, initial testing of TRIPBAM returned savings averaging $44 per room night on the same hotel bookings where TRIPBAM found a better rate. Tests with prospective clients entailed taking a set of active reservations and applying TRIPBAM's proprietary shopping algorithms to compute potential savings. Cindy Heston, director of travel and events at Anthem WellPoint, commented on these results: "This initial test, even on such a small scale, shows there are new, incremental savings to be found in traditional travel categories" (TRIPBAM, 2014).

While adopted by only 20% of corporate customers, the clustering approach proved to generate significant savings (see Exhibit 7.2). For those customers who used clusters, TRIPBAM shopped an average of 6.5 properties and found savings on 39% of monitored reservations with average savings of about $174 per stay (Campbell, 2015). Beside savings, the transparency gained on hotel rates had additional value for corporate customers. Michelle De Costa, director of global travel and client experience of IT service company Liberty Mutual, observed, "This could be a game-changer, because I'll know how my negotiated rates are stacking up out there" (Baker, 2014).

Agencies and corporate customers gained access to data supporting contract negotiation. By 2016, TRIPBAM was processing about 170,000 bookings monthly, identifying more than $3 million in potential savings per month (see Exhibit 7.3). Potential savings represented the difference between the initial bookings and the final rate TRIPBAM identified before the stay. Interestingly, firms were still conservative concerning rate shopping and rebooked for lower rates in only about one in two bookings. TRIPBAM's business was gaining traction in the corporate space, having closed deals with 15 of the 20 top travel management companies, enabling the firm to serve almost 1,000 business customers (i.e., companies). Combined, these corporate agreements represented a total of $81 million in monitored hotel bookings. In the same year, TRIPBAM closed deals with two major European agencies. It also inked an agreement with a consortium of travel agents that would make TRIPBAM's service available to its network of 1,000 associates. Corporate bookings represented more than 90% of TRIPBAM monitored reservations.

The hotel industry was less enthusiastic about TRIPBAM's innovation. Maxine Taylor, executive vice president of asset management for Chartres Lodging Group, commented, "I think it's a great model from the customer's perspective. From our perspective, obviously it's horrid. We're commoditized already. Now it's even more so" (Mayock, 2014).

Reminiscent of the lodging industry's reaction to the emergence of the Internet in the early 1990s, some hoteliers resisted TRIPBAM's innovation. Others understood that technological change was relentless, and they embraced the possibilities introduced by the firm. One major chain was evaluating the opportunity to use TRIPBAM to offer an enhanced-rate insurance service to their loyalty program members by rate shopping on a cluster of other preferred properties nearby. Other forward-looking operators were exploring viable options for using TRIPBAM's monitoring service to their advantage.

Charting the Future

As Reynolds focused on turning TRIPBAM into a resilient, viable going concern, many questions remained. Was the consumer space the one that had the greatest potential, or should the firm focus exclusively on the corporate market? How could TRIPBAM protect its early advantage from the inevitable imitators? Both the consumer and business opportunities looked wide open, despite each side presenting distinct challenges to growth. To achieve profitability, TRIPBAM needed to not just grow the user base but also seek a unique positioning in the market to defend against the increasing number of startups entering its space. One issue that was front and center was the need for the technological infrastructure to evolve in order to accommodate the growth TRIPBAM was already experiencing. Blanton explained,

> One of the agencies we are working with has the potential to generate millions of bookings per month, alone. Even if we are mainly transactional, every hotel rate we see, at a certain point, needs to be stored. And this can impact analytics.

While confident about the future of TRIPBAM, Reynolds reflected on the road ahead:

> Our biggest challenge is focusing on an initial market. We've gotten so much attention from multiple market segments that it's been difficult to stay focused. Our service has value for consumers, travel agencies, independent contractor agents, executive assistants, corporations, convention management companies, and more. (Vivion, 2013)

References

Amadeus. (2015). *2015 European Business Travel and Expense Analysis*. Amadeus. https://amadeus.com/documents/en/corporations/research-report/2015-european-business-travel-and-expense-analysis.pdf?crt=DownloadRequest.

Baker, M. B. (2014). BCD Travel Tests TripBam with Client Hotel Programs. *The Beat*. http://www.thebeat.travel/post/2014/07/21/BCD-Travel-Tests-TripBam-With-Client-Hotel-Programs.aspx.

Belic, D. (2012). WorldMate's iOS App Gets Hotel Price Alerts. *IntoMobile*. http://www.intomobile.com/2012/08/24/worldmates-ios-app-gets-hotel-price-alerts/.

Campbell, J. (2015). TripBam, Yapta Offer Savings in Hotel Rate Assurance. *The Company Dime*. http://www.thecompanydime.com/yapta-tripbam/.

eMarketer. (2015). *Worldwide Digital Travel Sales Will Total More Than $533 Billion in 2015*. http://www.emarketer.com/Article/Worldwide-Digital-Travel-Sales-Will-Total-More-than-533-Billion-2015/1013392.

Fadnis, C. (2016). *U.S. Corporate Travel: Market Sizing and Trends*. Phocuswright.

Gasdia, M., Liu, B., and Blutstein, M. (2015). *U.S. Consumer Travel Report Seventh Edition*. Phocuswright.

Harteveldt, H. H. (2013). *Who's Sleeping with You?* Adara and Hudson Crossing. http://adara.com/images/uploads/white_papers/Detailed-Look-Into-The-US-Online-Hotel-Guest-2013-2014.pdf.

Hermes Management Consulting. (2013). *Corporate Travel Management in Western Europe: Challenges and Opportunities*. Amadeus. https://amadeus.com/en/insights/research-report/corporate-travel-management-in-western-europe.

Lee, J. J., Capella, M. L., Taylor, C. R., Luo, M. (Meg), and Gabler, C. B. (2014). The Financial Impact of Loyalty Programs in the Hotel Industry: A Social Exchange Theory Perspective. *Journal of Business Research 67*(10): 2139–2146.

Markets Data. (2016). *Financial Times*. http://markets.ft.com/research/Markets/.

Mayock, P. (2014). In the Hot Seat: HotelTonight, TripBam, Nor1. *Hotel News Now*. http://www.hotelnewsnow.com/articles/24597/In-the-hot-seat-HotelTonight-TripBam.

———. (2015). Study: OTAs Continue to Steal Market Share. *Hotel News Now*. http://www.hotelnewsnow.com/Articles/27250/Study-OTAs-continue-to-steal-market-share.

Peacock, M. (2010). *Attacking High Look-to-Book Ratios*. http://www.slideshare.net/peacock.ma/attacking-high-looktobook-ratios.

Reynolds, S. D. (2014). Rate Oscillation Monitoring Hotel Reservation System. *Google Patents*. http://www.google.com/patents/US20140081678.

Robinson-Jacobs, K. (2013). Dallas Travel Startup Hopes to Tap into Increasing Funding. *Dallas Morning News*. https://www.dallasnews.com/business/2013/09/17/dallas-travel-startup-hopes-to-tap-into-increasing-funding/.

RoomIQ Hotel Price Tracking for Corporate Travel. (2016). *Yapta*. http://yapta.com/roomiq/.

Rossini, A. (2014). Online Travel Intermediaries: A Fast Changing Competitive Landscape. *Euromonitor*.

TRIPBAM. (2013). Patents Pending for tripBAM Hotel Booking Technology. *PRNewswire*. http://www.prnewswire.com/news-releases/patents-pending-for-tripbam-hotel-booking-technology-227811731.html.

———. (2014). BCD Travel Pilots tripBAM, Driving Hotel Savings for Corporate Clients. *BusinessWire*. http://www.businesswire.com/news/home/20140721005239/en/BCD-Travel-Pilots-tripBAM-Driving-Hotel-Savings.

Vivion, N. (2013). TripBAM Promises Cash Savings on Hotels with Minimal Effort. *Tnooz*. https://www.tnooz.com/article/tripbam-promises-cash-savings-on-hotels-with-minimal-effort/.

Walsh, C. S. (2014). OTA Hotel Play Pays Off: Double-Digit Gains Boost Outlook for U.S. Intermediaries. *Phocuswright*. http://goo.gl/DNtvco.

WorldMate. (2014). WorldMate PhoCusWright Presentation. https://www.youtube.com/watch?v=r2O2nJkFUE8.

WorldMate Wins the PhoCusWright Travel Innovation Summit Award. (2014). *Carlson Wagonlit Travel*. https://www.prnewswire.com/news-releases/worldmate-wins-the-phocuswright-travel-innovation-summit-award-282599591.html.

Appendix A: Exhibits

Exhibit 7.1. Key statistics: OTAs and hotels

	Market capitalization	Revenues	Revenues growth	ROA	ROE	Operating income
OTAs						
The Priceline Group	$60.3B	$8.44B	24.27%	15.7%	29%	$3.07B
Amadeus	€17.6B	€3.42B	10.12%	10.53%	30%	€956M
Expedia	$15.5B	$5.76B	20.80%	7.26%	39%	$518M
Sabre	$7.5B	$2.63B	4.27%	5.66%	106%	$388M
Hotels						
Hilton Worldwide	$19.8B	$10.50B	7.88%	2.90%	14.95%	$1.71B
Marriott International	$16.4B	$13.80B	7.92%	13.14%	NM	$1.16B
Starwood Hotels & Resorts	€11.1B	$5.98B	-2.16%	6.84%	33.42%	$882M
Accor	€8.6B	€5.45B	0.53%	3.21%	6.14%	€536M
Wyndham Hotel Group	$8.1B	$5.28B	5.43%	5.73%	45.54%	$941M
InterContinental Hotels Group	$6.3B	$1.86B	-4.23%	16.6%	NM	$680M
Hyatt Hotels Corp.	$5.95B	$4.42B	5.52%	3.42%	6.10%	$249M
Extended Stay America	$3.1B	$1.21B	7.12%	3.94%	6.74%	$338M
Choice Hotels International	$2.6B	$758M	4.60%	18.0%	NM	$215M
La Quinta Inns & Suites	$1.6B	$977M	11.79%	0.46%	1.78%	$135M
Meliá Hotels International	€2.3B	€1.46B	6.99%	NM	NM	€132M
China Lodging Group	$1.9B	¥4.96B	19.10%	6.21%	12.36%	¥389M
Home Inns	$1.63B	¥6.27B	5.21%	4.35%	7.85%	¥641M

Source: "Markets Data" (2016)

Note: Market capitalization is as of January 7, 2016. Revenues, operating income, ROA, and ROE are calculated based on data as of December 31, 2014.

Exhibit 7.2. TRIPBAM estimated average savings

Shopping method	Probability of saving	Average saving per stay
Like for like (same hotel, same room type)	5%	$60
Same hotel (same hotel, different room type)	15%	$80
Selective cluster (preferred hotels, different room type)	40%	$100
Large cluster (different hotels, different room type)	50%	$120

Source: Corporate documents

Exhibit 7.3. Selected TRIPBAM metrics

Year	Month	B2B metrics			
		Number of corporate clients	*Number of reservations monitored*	*Savings identified*	*Savings realized*
2015	January	350	20,000	$249,000	$136,950
	February	378	30,000	$373,500	$205,425
	March	408	40,000	$498,000	$273,900
	April	441	50,000	$622,500	$342,375
	May	476	60,000	$747,000	$410,850
	June	514	70,000	$871,500	$479,325
	July	555	80,000	$996,000	$547,800
	August	600	90,000	$1,120,500	$616,275
	September	648	100,000	$1,245,000	$684,750
	October	700	110,000	$1,369,500	$753,225
	November	756	120,000	$1,494,000	$821,700
	December	787	130,000	$1,618,500	$890,175
2016	January	826	140,000	$1,743,000	$958,650
	February	868	154,000	$1,917,300	$1,054,515
	March	911	169,400	$2,109,030	$1,159,967

Source: Corporate documents

Case Study for Chapter 8

TRIPBAM: Creating Digital Value at the Time of the COVID-19 Pandemic (B)

1 Introduction[1]

TRIPBAM, founded in late October 2012 by Steve Reynolds had grown into the leading hotel rates shopping and compliance engine for travel managers.[2]

The original intuition behind TRIPBAM came from Steve Reynolds' observation that, in most cases, rates granting free cancelation (i.e., refundable rates) were found twice per booking and were beating the best rates that online travel agents (OTAs) could propose when approaching the day of the stay.

At its core, TRIPBAM was an automated cloud-based service enabling customers to monitor rates changes and re-book hotel reservations prior to travelling. While stays were on average booked between 1 week and 2 months in advance, prices may have drop once or twice (Gasdia et al., 2015), resulting in customers' savings. In the simplest use case, on TRIPBAM a customer could create a set of specifications for a reservation she already held: the date of arrival and departure, the type of room, discounts the customer was qualified for (e.g., AAA), the rate booked and of course the hotel name and location. TRIPBAM referred to these specifications as the "search." Once the search had been confirmed, TRIPBAM would monitor the hotel for rate fluctuations and alert the customer when rebooking would yield savings (Figure 8.1).

TRIPBAM also offered a more advanced service that monitored rates within a defined hotel cluster. A cluster was a group of hotels the traveler deemed comparable to the one she had originally booked and was willing to stay at for the current trip. The customer could form the cluster by dragging hotels within a geographical radius into a list. For customers creating clusters in unfamiliar locations, TRIPBAM produced customized suggestions based on inclusion criteria like distance, brand, and amenities.

While initially Reynolds believed that TRIPBAM could flourish in the consumer space, corporate customers started to show their interest, attracted by the potential savings they could achieve. The addressable B2B market was huge as, on average, a corporate customer could easily manage over 15,000 bookings per year. In particular, compliance was a critical element in the corporate space. Each company had specific requirements on preferred property program, negotiated rates, preferred brands, and agreements with hotel brands. While most corporations had automated policy compliance tools, most companies still manually monitored and controlled travel expenses!

By 2018 it was clear that TRIPBAM had consolidated its position as the leader in hotel rate shopping for travel managers in the corporate space. The firm had evolved from a solution offering little more than clustered rate shopping, into a full-fledged hotel reservation and compliance management system. There was no doubt that TRIPBAM created value for their large corporate customers that needed to monitor and enforce their travel policies. Compliance and the prospective savings that TRIPBAM was unleashing had fueled the firm's growth. However, OTAs were still reluctant to use TRIPBAM or propose similar features themselves as they were hesitant about its impacts on their bottom line. Who could ever want to offer customers a service that results in lower revenues?

However, leading travel agencies and management firms saw the opportunity to provide their customers with additional valuable services. What could go wrong?

1 Excerpted from Piccoli, B., Pigni, F., Wixom, B., and Rodriguez, J. 2019. "TRIPBAM B: Creating digital value at the time of the COVID-19 pandemic." *Communications of the Association for Information Systems 44*. Used with permission from Association for Information Systems, Atlanta, GA; 404-413-7445; www.aisnet.org. All rights reserved.

2 This case is designed to be used in conjunction with case study for chapter 7: TRIPBAM: Leveraging Digital Data Streams to Unleash Savings (A). Much of the background information on the travel industry, the underlying technology and systems and the history of TRIPBAM can be found there.

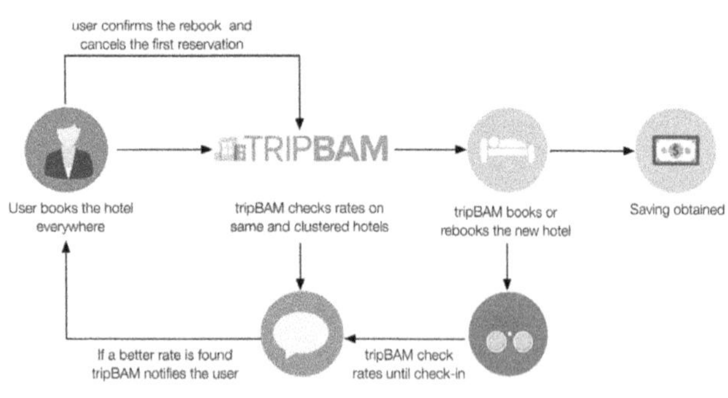

Figure 8.1. TRIPBAM overall clustered rate shopping process

2 COVID-19, the Unexpected Guest

In early 2020, TRIPBAM had found its stride, hitting record levels for number of customers, transactions and revenues in a robust business travel market (Figure 8.2).

The firm had achieved metrics that made it a very desirable acquisition target. Then, as the United States went into progressive lockdown due to the COVID-19 pandemic, travel stopped, and with it, hotel reservations ground to a virtual halt (Figure 8.3).

With everyone stuck at home, TRIPBAM leadership appreciated the good fortune of being a software company. After the initial shock, work could continue and with the right moves the firm may even emerge in a better competitive position. TRIPBAM had been architected to run a lean shop, employing just over 30 people. With an annual run rate exceeding $10 million,[1] the firm monitored almost $10 billion in annual travel spend on behalf of customers. As the pandemic hit the travel industry, Reynolds recalibrated staff and rallied employees to focus efforts on optimally positioning for the inevitable recession. As the old adage goes, it was no time to waste a good crisis. Reynolds explained:

> This crisis is helping us take step back and say: hey, this initiative we've been trying has not really worked out. In a normal environment you just let it chug along, but now we are making some hard choices. Pulling the plug on some underperforming initiatives and making some bold moves.

Reynolds described an example that could benefit the firm while helping the industry:

> We are creating a referral network, and we are supporting laid off and furloughed travel industry executives to promote our product. For example, if one of their contacts watches a sales presentation, we will pay for the referral. The executives can supplement their income in these hard times, and we reach new potential customers.

Other initiatives included partnering with consultants who typically support negotiations between corporations and suppliers. Since all deals had been put on hold, consultants had the time to become certified TRIPBAM administrators. After certification, they realized earnings by helping TRIPBAM's clients action the savings opportunities found by the system. Having become expert users of the TRIPBAM's product, industry consultants could then evangelize it with their own clients and become account executives for any new client they procured.

But one of the greatest "gifts" of the crisis was the time and incentive to think creatively beyond the transactional model that had historically driven the firm's growth. One such new initiative was the hotel closures report. Because of its visibility into

1 Run rate is calculated on the base of 2019's annual data.

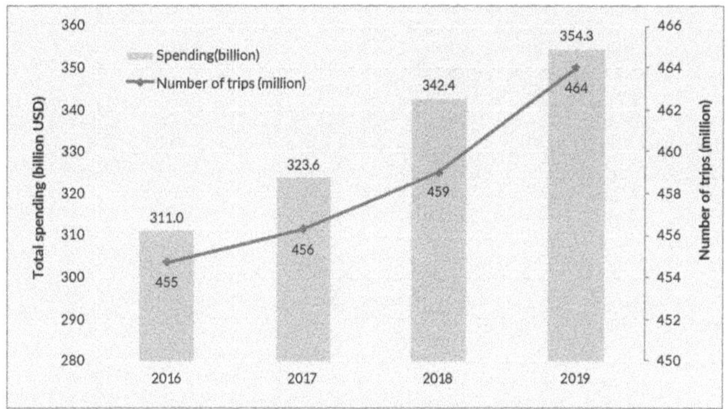

Figure 8.2. US domestic business trips and spending (data on total spending from WTTC, 2019 and on number of trips from the U.S. Travel Association, 2019)*

* Person-trip is defined as one person on a trip away from home overnight in paid accommodations or on a day or overnight trip to places 50 miles or more [one-way] away from home.

availability of hotel rates, TRIPBAM could determine whether a hotel was operating, or it had shuttered its door. The firm published the weekly report for free, to provide valuable information to the industry and policy makers dealing with the crisis. This type of report was a simple demonstration of the value that could be extracted from the multiple digital data streams that TRIPBAM captured. What other opportunities existed for monetizing TRIPBAM's data? How could the firm take advantage of new opportunities without losing focus on the main product?

3 Growing TRIPBAM

While it was originally envisioned as a consumer play, TRIPBAM quickly focused on B2B. Since signing up its first mega agency, BCD Travel in 2013, the firm had inked deals with 12 of the largest agencies[2] and integrated with 19 out of the 20 larger Travel Management Companies (Figure 8.4).[3] After signing its first corporate client in late 2014, TRIPBAM had grown to over 2,000 corporations under contract, with about half of the Fortune 100 and a sizable proportion of the Fortune 500 as direct customers. John Berkley, TRIPBAM's head

2 Mega agencies were categorized as the top earning global travel agencies, including powerhouses such as American Express and BCD Travel—companies with tens of billions in sales and thousands of employees.

3 A TMC, also known as a business travel agency or corporate travel provider, is travel agent that fully manages the business travel requirements delegated by an individual, company, or organization. Compared to retail travel agencies, the TMC offered services like rebooking when travel was disrupted or plans changed, keeping control of a travel policy, and allowing the client to spend less time on time-consuming travel arrangements.

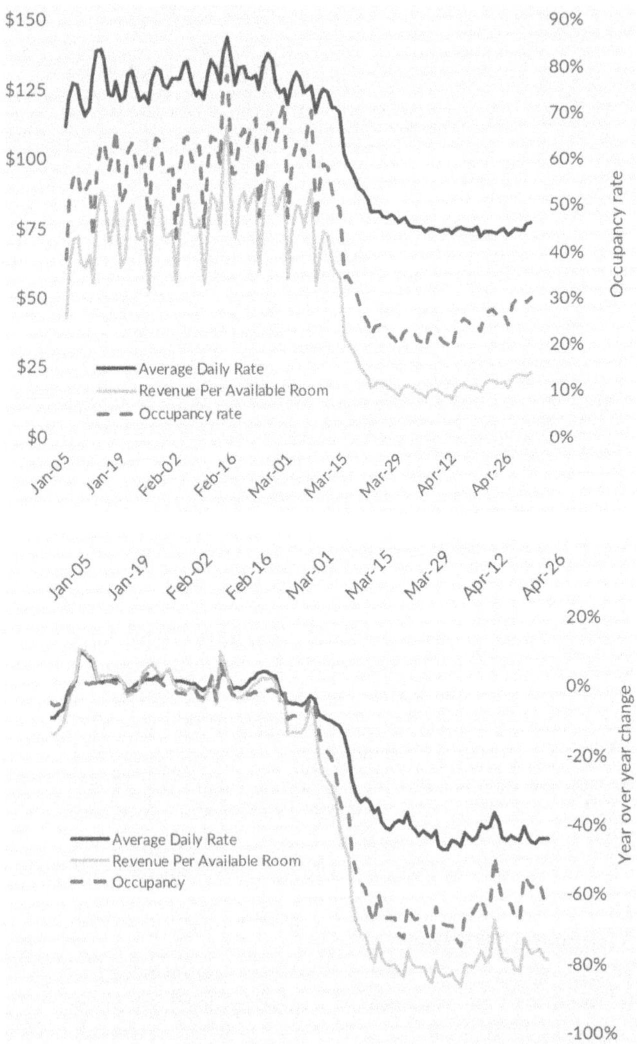

Figure 8.3. Average daily rate (ADR), revenue per available room (RevPAR), and Occupancy rate in the US and Year over year change (data from STR Global)

of business development, explained how the firm viewed the market opportunity:

> Large buyers have more than $10+M in yearly hotel spend. There may be 5–700 of those in the US. The mid-tier spends $5–10M and there are probably 1,000 of those. Then you have a very varied landscape with many viable candidates, as long as they actively manage travel and direct travelers to booking channels we can monitor.

TRIPBAM had achieved a major milestone in October 2019: 20 million reservations processed, and in the months preceding the US lockdown of nonessential travel, the firm was monitoring almost four hundred thousand daily shops. The "daily shops" metric referred to reservations currently being monitored by TRIPBAM's system on any given day.

Beyond rate shopping, TRIPBAM's value proposition included compliance, and the firm had continued to grow its portfolio of services. Major reports and aids to travel buyers now included negotiated rate auditing, daily rate assurance, sourcing solutions, automatic alerting to hotels out of compliance (Table 8.1).

4 The Business Model

Since its inception, TRIPBAM's most obvious value proposition was helping customers uncover travel savings by shopping hotel rates after a reservation had been booked by the traveler. Lodging was a sizable portion of corporate travel budgets (Figure 8.5). By keeping tab on constantly changing prices at the booked hotel, as well as a cluster of comparable hotels nearby, TRIPBAM could produce substantial savings for their corporate customers. Reynolds explained:

> We try to get 2–3% in overall saving on total travel spend for our customers, at least. Right now, with the rate volatility in the market that number is way up to 5–6% and even 10% in some cases.

Large customers acquired the service by paying a monthly subscription fee, ranging from the low thousands to more than sixty thousand dollars, depending on the average number of reservations. By monitoring savings in relation to monthly subscription costs, TRIPBAM's leadership could compute a clear ROI for each client and could ensure the service was delivering appropriate value. As a testament to the value TRIPBAM created with its shopping engine, it had virtually no customer churn. As Reynolds put it: *We make sure each customer has at least a 4x ROI, why would they churn?* Berkley added:

> When you get to the large corporations, the economics and financials are so crystal clear, it's hard not to buy TRIPBAM. Even if you set it up at the most conservative level, with no traveler friction or employee resistance, you already save 1% on pure rate arbitrage—and then it grows from there.

Some smaller clients, unable to reliably predict the number of monitored reservations they would need, opted for a gainsharing model. The gainsharing model entitled TRIPBAM to receive 25% of the realized savings for any reservation that TRIPBAM rebooked on the client's behalf. Those customers had the option to flip over to a subscription model if volumes did grow to meet thresholds that made the conversion advantageous.

Finally, TRIPBAM had cleverly devised a way to move the gainsharing from the corporate level to the individual traveler's budget. After finding a lower rate, when the reservation was rebooked, the traveler's credit card would be refunded the

Table 8.1. Partial list of TRIPBAM offerings (company's website)

1	Monitoring room rate fluctuations
2	Clustered rate monitoring and shopping on preferred properties
3	Amenities adjustments
4	Automatic and Assisted rebooking
5	Monitoring and reporting of Last Room Availability
6	Monitoring and reporting of Lowest Qualified Rate
7	Comparison of Last Room Availability and Lowest Qualified Rate
8	Corporate program evaluation
9	Identification of negotiated rate opportunities

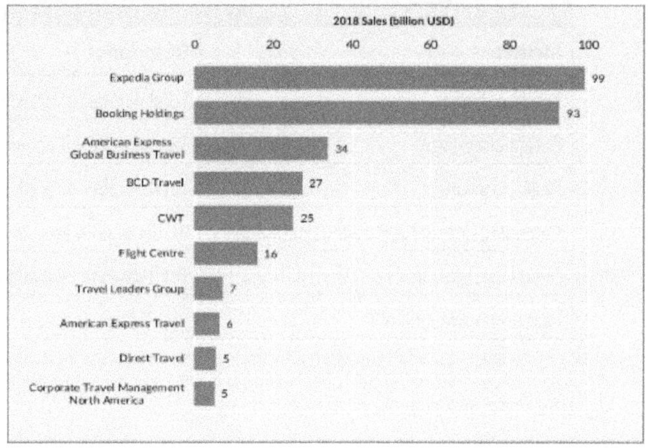

Figure 8.4. Ranking of the largest travel companies by sales (Travel Weekly, 2019)

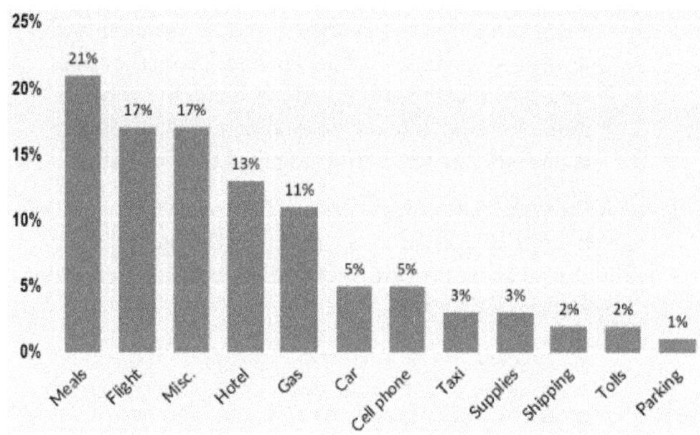

Figure 8.5. Average business travel spending (Certify, 2019)

original rate, charged the new lower rate, and debited a separate gainsharing fee. This approach helped to attract customers in the few cases where the corporate travel office did not have the budget to purchase the TRIPBAM solution. Because it shifted the cost of the service to the travel budget of the ultimate beneficiary, the corporate traveler's own departments, TRIPBAM enabled the corporate travel office to obtain its service and overcome resource constraints.

The subscription and the gainsharing models were mostly designed for larger corporations and accounted for about 50% of TRIPBAM's revenue. The remaining 50% came from bookings associated with smaller corporations thru travel agencies. TRIPBAM had tried to crack a difficult problem: how to move down market and reach smaller corporations with limited travel spend. Reynolds explained the challenge:

> It's hard to figure out a solution that has seamless onboarding, whereby a customer can self-provision our service similar to a SaaS[1] model. Go to the website, click a few buttons and start saving. To capture their bookings, you still need to involve a TMC.

Having to involve the travel agencies required TRIPBAM to be able to provide appropriate incentives. A reseller model was the option the firm settled on, because it provided a win-win for all involved. The TMC would onboard its own clients onto the service using TRIPBAM's enterprise product, which allowed the travel agents to easily configure accounts using standard features. Sales, onboarding and account management was done by the TMC who could be paid for the service and share the savings uncovered by TRIPBAM with their clients.

TRIPBAM in turn was paid a small percentage of savings. In this model, every client had a standard configuration that made it easy for travel agents to onboard and manage the account, while forgoing some ad-hoc features typically needed by large corporations (e.g., international versions, VIP traveler exceptions). Berkley explained:

> The regional TMCs that are being successful are moving to a consultative model and bundling TRIPBAM with their other services. By doing so they can charge a monthly fee for service to their corporate clients.

More subtle, but no less important, was the value of insight TRIPBAM could uncover on behalf of travel buyers. This included employee compliance with the firm's travel policies and suppliers' compliance with complex preferred hotel agreements. Corporate clients negotiated such agreements through a Request For Proposal (RFP) process. Berkley explained:

1 Software as a Service (SaaS) is a software delivery approach in which a provider hosts the application in its data centers and the customer accesses the needed functionalities over a computer network. Gmail and Dropbox are two examples in the consumer space.

Table 8.2. Primary hotel level metrics

Metrics	Description
Total bookings	Total number of bookings monitored for a hotel
Potential savings	Total number of bookings where a lower rate has been uncovered by TRIPBAM
Realized savings	Total number of bookings where the client accepted the lower rate uncovered by TRIPBAM
Percentage of realized savings captured	Percentage of total bookings where the client accepted a lower rate uncovered by TRIPBAM
Overall net savings	Total savings divided by total travel spend
Last Room Availability*	Percentage of time a negotiated discount was correctly applied

* LRA measures then the number of times the hotel correctly applied the negotiated fee. In the context of negotiated rates, customers have the right to book even the last room—typically pricing higher—at the negotiated rate. However, in case of high occupancy, and corresponding high room rates, hotels have less incentive to apply a discount.

> The RFP process is a very laborious manual process that does not take advantage of any digital data stream. Large corporations hire their TMC or consultants to analyze a year's worth of data in Excel, then they have a back and forth with sometimes hundreds of hotels. And of course, the hotel uses different data so there is no single source of truth to guide the discussion.

While the relationship with suppliers could be contentious, given TRIPBAM's strict allegiance to the buyers, the firms could play an important intermediary role in negotiations. Reynolds explained:

> Chains are trying to convince corporate travel buyers to move to a dynamic discount—a percentage off BAR[1]—rather than a fixed corporate rate. That benefits everyone because it takes into account seasonality for hotels and it can reduce cost for travel managers.

Blanton added:

> Dynamic deals are ideal, but you need auditing and rate limits. We offer ongoing monitoring of BAR, LQR and the actual discounts being provided. This gives full visibility and auditing to corporate buyers. We are the only ones who do it like this. When this auditing is coupled with automated sourcing, rate caps and rate targets by market, you no longer need annual RFPs. You renegotiate when needed, and our automation handles the rest. We call this the "TRIPBAM Way" and we feel like while the world has essentially pushed pause on travel, now is the perfect time to disrupt this unreasonably cumbersome, legacy RFP process and bring a superior hotel procurement process to our clients.

Dynamic discounts held great appeal for both buyers and suppliers who could benefit financially, and they dramatically simplified the contractual process. However, a discount off of a dynamic rate was much more difficult for travel buyers to evaluate than a fixed corporate rate. In other words, how could a corporate travel manager who had negotiated a 20% discount off of BAR be sure that the discount was applied to the BAR and not some higher rate? Having harvested and cleansed reservation data and shopping results, TRIPBAM could compute a staggering array of daily, weekly and monthly metrics (Table 8.2). Each of these metrics was available to TRIPBAM's customers and their account executives. Further, TRIPBAM was in a position to audit transactions to ensure that discounts were applied correctly. As Blanton put it:

> You learn a lot when you dig deep and you know what you are looking for. We are able to grade hotels based on their compliance with the specifics of their contract with our clients—are they providing LRA as negotiated? Are the rates good compared to the market? Are they rate squatting? We can even identify hotel squatters.[2]

Once reliable auditing was in place buyers and suppliers could do away with the yearly RFP process, negotiate a recurring discount and let TRIPBAM monitor the accuracy and fairness of the contract application. While the entire industry was facing the negative effects of the pandemic, TRIPBAM was offering real cost cuttings and efficiencies, and a new form

1 Best Available Rate represents a pricing mechanism based on the lowest available rate for a room not requiring pre-payment and not imposing particular change or cancellation charges. In this sense, the price may vary and be different each day of the stay.

2 Hotel squatting occurred when a hotel that did not have, or no longer had, a negotiated rate with a corporation would publish rates that signal otherwise in the rate descriptions.

of smart sourcing that could change the industry permanently. TRIPBAM survival was clearly not at stake.

5 Competition

TRIPBAM's competition was constantly evolving. In the early days, when the firms entered the consumer space, several startups had similar models: Tingo, TripRebel, Waylo, Dreamcheaper, Rebookey.com, RoomsNinja, and Pruvo. Customer demand never truly materialized. Tingo, a TripAdvisor company, ended its operations in 2019. Both TripRebel brand and website were acquired by Ashton Travel, but the shopping service was discontinued. Waylo tried to differentiate replicating the formula Hopper used for airline fares. By analyzing the daily prices Waylo focused on forecasting a price drop, offering users to rebook at the predicted lower rate. RoomsNinja and Rebookey.com never seemed to go past the seed investment funding. Pruvo, an Israeli startup, aimed at targeting the B2B space, and was able to secure a $1.1M seed funding round as the pandemic broke out. It remained unclear how resilient the company could be through the crisis (Lerner, 2020).

Yapta, one of the early entrants in the corporate travel market, remained the only main competitor. However, in January 2020, the firm was purchased by Coupa Software, a business spend management platform similar to Concur:

> Coupa Software has acquired business travel price assurance technology Yapta, adding it to the company's travel and expense management technology portfolio. Yapta will now be marketed as Coupa TravelSaver for Airlines and Hotels, respectively, according to James Filsinger, who served as president and CEO of Yapta prior to the acquisition. (Perrotta, 2020)

Before the acquisition, Yapta claimed a portfolio of 30 TMC partners distributing or using Yapta products and technology, resulting in 8,500 corporate clients and half of the Fortune 100, including Amazon, the National Basketball Association, and Shell Oil (Perrotta, 2020). However, these customers appeared to mostly utilize Yapta's airfare shopping service and TRIPBAM rarely found itself in direct competition. As a consequence of the acquisition, it was expected that the firm would no longer serve agencies or corporations that did not use Coupa Software for travel management. Reflecting on competition Reynolds concluded:

> I don't feel like there is anyone doing what we do for the transient corporate travel market. Most travel agencies who want to offer a service like ours try on their own, they lose patience after a year or two and then they sign on with us when their large customers demand it.

TRIPBAM was setting up to come out of the crisis stronger and more differentiated. It had a number of efforts focused on increasing transactions, and also going down-market—something that had proven difficult before.

6 Digital Infrastructure to Fuel Hypergrowth

As a technology company, TRIPBAM's core strength was in its proprietary software. The IT group was headed by Caleb Blanton, who originally created the proof of concept for the idea. With the basic business model and technology proven, the TRIPBAM leadership was focused on growth. With the number of transactions rising steadily, it became clear that moving to the public cloud would be advantageous. Blanton explained:

> It was fortuitous timing for us, but we had planned to migrate to the cloud before the crisis hit. We decided to migrate to achieve hyperscale agility. We use Microsoft Azure and it allows us to quickly scale up (or down) and scale out. You also get security out of the box and a lot of DevOps tools and support.

The cloud infrastructure enabled cost control and flexibility, which came in handy during the crisis when corporate travel dropped by an unprecedented percentage. TRIPBAM pursued two major technology development growth vectors: rate shopping and insight actioning for customers. Data capture could occur via direct connections to the TMC's own system, but for the bulk of reservations TRIPBAM harvested bookings from the GDS (Figure 8.6).[1] Blanton explained:

> Most of our data capture is done on the Global Distribution Systems. A travel agent needs to have its customers

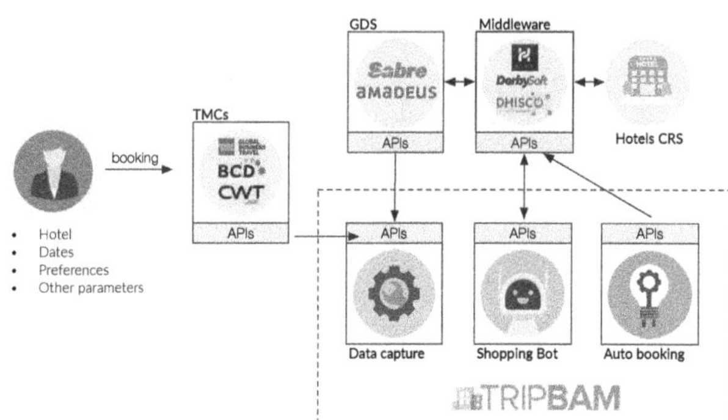

Figure 8.6. Simplified ecosystem diagram (from company documents)

1 The Global Distribution Systems (GDS) represented the backbone of the travel industry. Hotels published room availabilities and rates in these systems to make them visible to travel agents.

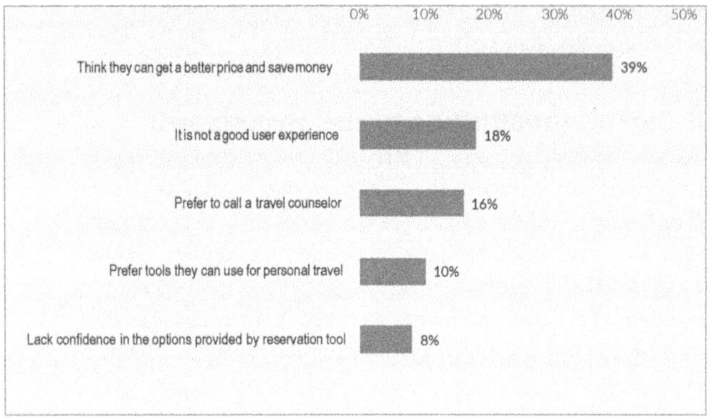

Figure 8.7. Reasons for noncompliance (ACTE Research, 2019)

> sign a DRA[1] and we have a solution that has been vetted and it works reliably to capture the reservations.

However, capturing reservations reliably and accurately was no simple feat. A typical reservation contained structured data (e.g., date of check-in and check-out), but most of the required information was in free form text fields. Blanton explained the associated challenges:

> Booking systems, like the GDS, were designed to be used by humans. Room types, available amenities, the specifics of cancellation policies, often even the nightly rate being charged is stored in blobs of text. But you have to extract it precisely and reliably to make sure you shop a comparable product.

Constantly on the lookout to improve the scope and range of data sources, TRIPBAM was in talks to create direct data capture interfaces with the major lodging chains. As hotels were incentivizing corporations to book directly and bypass costly intermediaries, the travel buyers had leverage to require the suppliers to share direct reservations with TRIPBAM. *We are discussing a direct interface with several chain central reservation systems*, Reynolds remarked. *It is mostly valuable for the corporation, but it can also help the chain lower its cost structure and improve its brand control.*

The value of TRIPBAM services grew proportionally with the coverage of the travel spend of the client. In other words, the type of insight a corporate travel manager could glean from TRIPBAM reports and the type of actions they could implement once armed with the insight (e.g., new sourcing deals), depended on having a sizeable proportion of the bookings tracked. Blanton framed the challenge: *The main issue is travelers following hotel reward points, but sometimes travelers can find better deals online.* Many corporations were unable or unwilling to force employees to adhere to travel policies and use the corporate booking systems exclusively. As a consequence, it wasn't uncommon for corporate travelers to book preferred hotels[2] directly or even to reserve rooms at non-preferred hotels. In many cases it was just convenience that spurred the traveler to book out of channel (Figure 8.7).[3] But such behavior created a major challenge for TRIPBAM. Blanton explained:

> It's a big challenge to identify out of channel reservations where travelers go to Hyatt.com and book their own reservation because they want frequent guest reward points or perhaps are shopping for better deals. This is the hotel attachment problem. Capturing those bookings reliably is incredibly hard, and sometimes they account for 40% of a corporation's reservations.

With respect to actioning insights and reports, the challenge stemmed from the fact that travel was rarely seen as a strategic unit in modern firms, despite being a sizeable expense for many organizations (Table 8.3). The typical corporate travel office had to fight hard for budget to fund its operations. Yet, given the complexity and fragmentation of the travel landscape, there was a lot of opportunity to create efficiencies once armed with the right data. As the travel buyer's advocate, and as the company that would share in the success of the corporate travel managers, TRIPBAM was constantly looking for ways to help corporate travel offices do more with less. Reynolds explained:

> In the early days we would show reports to our customers and they would say: This is great insight, but I don't have any staff to action it. That's why we are constantly building and improving alerts and automations. For example, we can fire off an email to a hotel that goes outside of bounds to say, hey, you need to fix this, or you get kicked out of the program. Which is a big deal to some of these hotels—it's a lot of volume for them.

One automation that proved fundamental was AutoRebooking, the ability to confidently cancel a previous reservation and move the traveler into a new room that was comparable, or better, at a lower rate. In a typical transaction, when a lower rate was uncovered, the travel agent was alerted and asked to manually approve and make the change. The key was to do so without requiring a travel agent to first verify that the new

1 A Data Release Authorization (DRA) is a contract provision that allows a travel agent to receive, process, and/or transfer personal travel data from a client's travelers.

2 The term preferred hotels refers to properties that the corporation has a contractual relationship and negotiated rates with.

3 Out of channel reservations are those performed by employees outside of the mandated corporate reservation system or TMC, generally via consumer online travel agencies (e.g., Expedia) or direct with large chains (e.g., Hilton.com)

Table 8.3. Top corporations by business travel spending in the US (from author research and BTN, 2019)

Rank	Company	Travel and Expenses*		Booked Air Volume*	TRIPBAM Client	Expense Supplier	TMC
		USA	*Global*	*USA*			
1	Deloitte	1,800		559		SAP	BCD
2	IBM			417	Yes	Concur	Amex
3	PwC			324	Yes	Concur	Amex
4	Apple			315	Yes		BCD
5	EY	932	2,100	311	Yes	Proprietary	Amex
6	McKinsey & Co.			285	Yes		Amex
7	Accenture	572	1,300	260		MyTE Internal System	CWT
8	Microsoft	450		250	Yes	Dynamics	Amex
9	Boeing		567	244		Concur	
10	ExxonMobil			240	Yes	Chrome River	CWT
11	Google			239	Yes		
12	Amazon			229			CWT
13	Lockheed Martin	392	464	200		Concur	BCD
14	GE			195		Concur	CWT
15	The World Bank		475	195		SAP	Amex
16	KPMG LLP	584		187	Yes	Concur	Amex
17	Bank of America			183	Yes		Amex
18	Facebook			175		Concur	CWT
19	JPMorgan Chase			164			CWT
20	BCG			150			Amex
21	Oracle			145	Yes		CWT

* In million USD; [I] empty cells represent unavailable data

room was indeed comparable or better on all dimensions (e.g., bed type, amenities, cancellation policies). Reynolds explained:

> When you are 100% certain that you are meeting the company's criteria and you have a lower rate, you can enable automatic rebooking. We are the only ones on the planet who can do that. We are touchless end-to-end.

7 Charting an Uncertain Future

Despite the uncertainty brought about by the crisis, early investors in TRIPBAM remained excited about the opportunity. Chris Hemmeter, the Managing Director of early TRIPBAM investor Thayer Ventures, remarked:

> TRIPBAM is well positioned for a mid-market private equity exit. All the components are there: topline growth, EBITDA margin, defensible market position and massive addressable field of play. The only one of these key measures effected by the COVID-19 crisis is topline growth and we have every reason to expect that to return in force.

As with any technology firm positioned at the nexus of multiple data streams, TRIPBAM had the ability to extract insight from the data it captured. Such insight could be monetized to impress an acceleration to topline growth during the unavoidable recession. Blanton explained:

> We are starting to see clients becoming more dependent on our data to manage their hotel programs. For example, we have a dashboard for large clients who can see for every hotel they have a discounted rate, how often these preferred hotels are producing the promised discounts and how strong those discounts are compared to the market and the benchmarks from other clients. We rate hotels from A to F to give buyers a sense of the strength of their supplier partnerships.

This level of insight and knowledge enabled the TRIPBAM account executives to play a consulting role to their accounts. Individual customer level metrics (Table 8.2) could be aggregated into anonymized benchmarks yielding comparative reports. A recently launched feature enabled travel managers to evaluate hotels, brands, and chains' performance based on the discounts and the LRA percentage obtained by aggregating the data by city (Airoldi, 2019). In an industry characterized by long distribution chains with many and varied intermediaries, reliable and auditable information was at a premium. Reynolds offered an example:

> We have a series of reports, called Hotel Intelligence, which shows our clients what their metrics look like compared to their peer group. So, we can tell a client that they are getting an average discount which is lower than that obtained by companies doing similar volume in the same destinations. They cannot get this type of insight anywhere else.

Basic analytics about the performance of the company's program were typically bundled into a premium subscription that large corporations purchased. Smaller corporations did not receive in-depth reports, but they were provided to their TMC so that they could be consultative and offer greater value. Specialized reports such as Hotel Intelligence and sourcing reports[1] were sold separately. TRIPBAM was constantly looking for ways to better serve its customers by leveraging its data assets. Ideas were forming around the opportunity to train customers on how they could maximize the value of their own data. As an information good, data was not consumed by use[2] and Reynolds wondered:

> How can we monetize this data by packaging it up in such a way that creates a lot of value for other stakeholders? While of course being acceptable to our primary customer. We are and we will remain the buyers' advocate.

With his leadership team homebound due to lockdown restrictions, this was the time to get creative and evaluate all possible options, no matter how crazy they seemed.

References

ACTE Research. (2019). *Booking Tools and Technologies* (p. 12). Association of Corporate Travel Executives. https://3rxg9qea18zhtl6s2u8jammft-wpengine.netdna-ssl.com/wp-content/uploads/2019/06/ACTE-OBT-Whitepaper-04Jun19-1.pdf.

Airoldi, D. M. (2019, August 4). *Tripbam Updates Hotel Intelligence Solution*. Business Travel News. https://www.businesstravelnews.com/Procurement/Tripbam-Updates-Hotel-Intelligence-Solution.

BTN. (2019). *Corporate Travel 100: Business Travel News*. Business Travel News. https://www.businesstravelnews.com/Corporate-Travel-100/2019.

Certify. (2019). *The Average Cost of Business Travel*. Certify.com. https://www.certify.com/Infographic-TheAverageCostOfBusinessTravel.aspx.

Lerner, S. (2020, March 20). *Hospitality startup Pruvo closes $1.1M seed round amid coronavirus outbreak*. PhocusWire. https://www.phocuswire.com/pruvo-1-1m-seed-round.

Perrotta, A. (2020, January 8). *Coupa Software Acquires Yapta: Business Travel*. Business Travel News. https://www.businesstravelnews.com/Technology/Coupa-Software-Acquires-Yapta.

Travel Weekly. (2019). *Travel Weekly Power List 2019: Travel Weekly*. https://www.travelweekly.com/Power-List-2019.

U.S. Travel Association. (2019). *Travel Forecast—Fall 2019*. U.S. Travel Association. https://www.ustravel.org/system/files/media_root/document/Research_Travel-Forecast_Summary-Table.pdf.

WTTC. (2019). *WTTC Data Tool*. World Travel & Tourism Council. https://tool.wttc.org/.

1 The sourcing report enabled a customer to gain visibility into hotels they had no preferred relationship with. If such hotels were providing favorable discounts to others in the market they could be flagged for a customer as potential leads they should inquire with.

2 Information goods are those products or services the customer acquires strictly for their information content. A classic example is software, books, and music. A defining characteristic of information goods is that they are not consumed by use, unlike physical goods, and can therefore be reused and resold multiple times.

Case Study for Chapter 9

Seller Labs: Democratizing E-Merchants Competitiveness

1 Introduction[1]

In April 2013, Brandon Checketts and Paul Johnson decided to leave their business as Amazon and eBay sellers and create a software business: Seller Labs. The founders seized the opportunity to develop software tools that solved common problems faced by the rapidly growing number of Amazon's third-party sellers. Despite the presence of Amazon's proprietary software solution (Seller Central), the founders recognized there was a market for advanced software tools for data analytics and work automation that facilitated the pursuit of diversified strategies and improved the competitiveness of Amazon's Marketplace third-party sellers.

The first Seller Labs' product was a messaging system (i.e., Feedback Genius), that enabled targeted email communication between customers and sellers. On the impetus of this early success, Seller Labs developed a range of software products for sellers' advertising, market research, inventory management, and financial reporting. Five years after the original launch, Seller Labs was one of the major complementors[2] on the Amazon platform, with thousands of customers, and an invitation to the Amazon Marketplace developer council.

As the Amazon Marketplace grew and became more competitive (see Table 9.1), numerous complementors developed comparable software products supporting third-party sellers. For the first time in Seller Labs' history mounting competition required a reevaluation of the firm's strategic positioning. Hank Harris had been thinking about a way forward since taking over the CEO reigns in November 2017:

> I had seen the same dynamics playing out similarly on the Google Advertisement platform, where after years of intense competition, only two or three companies were successful. When looking at Seller Labs, I knew that we were in the middle of that same competitive dynamic, as keeping up with the continuous innovations of the Amazon Marketplace Web Services and Advertisement platforms was becoming increasingly challenging. Our most successful products, such as buyer-seller messaging, could be copied by other start-ups, or Amazon itself. We needed a differentiation strategy that uniquely positioned Seller Labs in the market.

Table 9.1. Sellers in US Marketplace (Marketplace Pulse 2020)

Year	Worldwide	Based in China
2009	900,000	
2010	1,300,000	
2011	1,800,000	
2012	2,300,000	
2013	2,800,000	
2014	3,400,000	
2015	4,000,000	
2016	4,700,000	1,090,400
2017	5,700,000	1,943,700
2018	6,800,000	2,652,000
2019	8,000,000	3,732,000

After evaluating several options, Seller Labs acquired X-Cart in October 2019, an established e-commerce platform[3] with tens of employees. X-Cart, founded in 2001, offered the opportunity of implementing a differentiating strategy that would enable Seller Labs to expand its offerings outside the Amazon's Marketplace. With the acquisition of X-Cart, Seller Labs could establish a more comprehensive suite of products and services in the broad e-commerce space. However, the acquisition put Seller Labs at a crossroad. Should the firm continue to stay true

1 Excerpted from Rodriguez, J., Watson, R. T., and Piccoli, G. 2019. "Seller Labs: Democratizing e-merchants competitiveness." *Communications of the Association for Information Systems 44*. Used with permission from Association for Information Systems, Atlanta, GA; 404-413-7445; www.aisnet.org. All rights reserved.

2 A complementor is a firm that leverages platform resources (e.g., Amazon MWS API) to create products and services that address the needs of a large and heterogeneous group of suppliers (e.g., third-party sellers).

3 A platform is an evolving socio-technical system that exposes digital resources supporting suppliers and complementors' digital innovations.

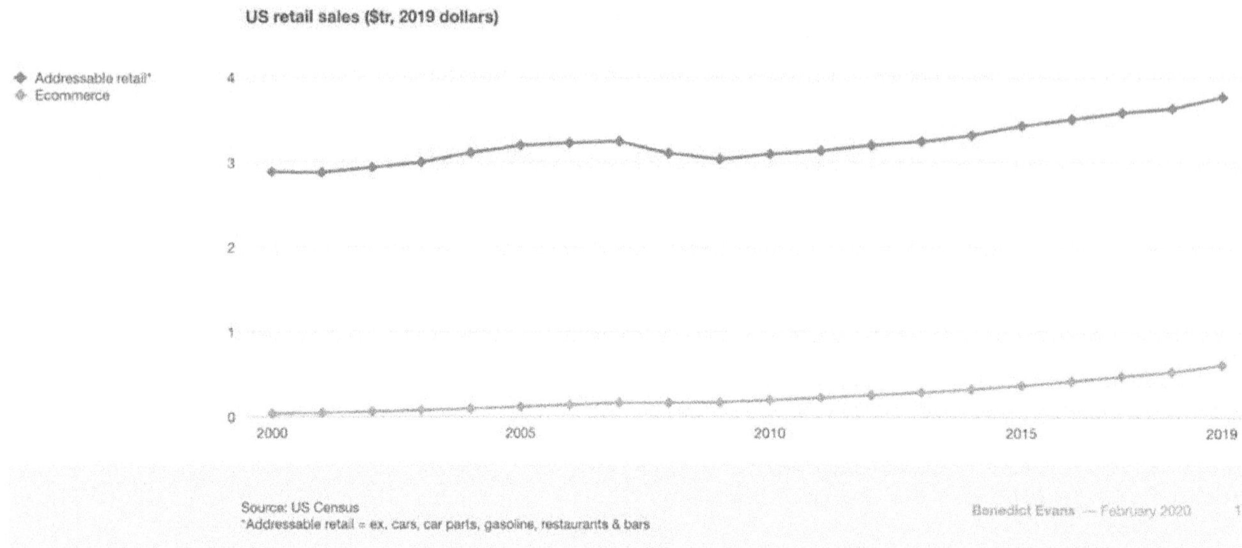

Figure 9.1. US E-commerce Share (Evans 2020)

to the original vision and primarily focus on the development of solutions targeted to Amazon's sellers? Or should it use the firm's limited resources to pivot outside of the Amazon Marketplace and developed an integrated e-commerce platform?

2 Industry Background

The widespread availability of computers and the Internet has driven a constant year-over-year increase in online sales since the early 2000s. Online sales in the US alone reached more than $600 billion in 2019, up from around $50 billion in 2000, and $200 billion in 2010. Although the total online GMV[1] (Gross Merchandise Volume) in 2019 was staggering compared to 20 years ago, online sales still only represented 16% of the US addressable retail market[2] (i.e., $3.5 trillion) (Figure 9.1). Thus, online commerce was still at a relatively infant stage, and many opportunities still existed for firms that wanted to capitalize on this market as more products and services moved online.

The increase in online commerce saw the emergence of online native companies that tailored their strategies and capabilities around this emerging opportunity. In the US, Amazon positioned itself as the clear maker leader, experiencing a 30% year-over-year revenue increase for almost two decades. In 2018, Amazon's GMV reached nearly $280 billion, around 45% of the US online market. Amazon's success was apparent when comparing it to the other e-commerce companies. eBay and Walmart, who ranked second and third by online GMV, captured "only" 6% and 4.5% respectively of the US online market (eMarketer 2019). At the beginning of 2020, it was estimated that around 50% of customers start their product searches on Amazon.com. Although barely noticeable to most customers, the driver of Amazon's success was its effective transformation into an e-commerce platform, where third-party sellers could leverage the end-to-end customer experience and the digital resources provided by Amazon for selling their products.

After experimenting with an ultimately failed auction model called zShops, Amazon launched the Amazon Marketplace in 2010. The Marketplace offered third-party sellers, from small resellers to large manufacturers, the possibility to list their products on Amazon's digital "shelves." Amazon's transition into a platform for third-party sellers required the exposure of digital resources that previously were available only internally to the organization. Most importantly, the creation of a platform required the design and implementation of a governance model through which Amazon determined the "rules of the competition" inside the marketplace, which allowed the standardized management of thousands, and then millions of third-party sellers. Furthermore, by controlling the demand side of the market (i.e., end-customers), Amazon could almost unilaterally determine the roadmap of the software tools that they made available to third-party sellers for enabling competition inside the marketplace. Much of the

1 GMV is used in online retailing to indicate the total sales value of merchandise sold over a certain time frame through a particular marketplace.

2 The addressable market represents the revenue opportunity available for a product or service.

"marketplace governance" was implemented algorithmically, and sellers could join the platform, transact with customers, and compete with other sellers without ever having to interact with an Amazon employee.

As a result of an intense competition between sellers, Amazon's customers benefitted from improved customer service, a vast product selection, and lower prices. Amazon's sales growth, especially in the 2014–18 period, was driven by its Amazon Marketplace initiative. Third-party sales on Amazon grew from $70 billion in GMV in 2014, to $160 billion in 2018, representing nearly 60% of all Amazon sales. In his 2019 letter to shareholders, Jeff Bezos synthesized this trend in a paragraph titled, "*Third-party sellers are kicking our first party butt. Badly.*" Bezos explained the reason for the Marketplace success: "*We helped independent sellers compete against our first-party business by investing in and offering them the very best-selling tools we could imagine and build*" (Amazon 2019). Those tools would of course be all software tools and algorithms (i.e., digital resources) that sellers could self-provision and use without any interaction or consultation with Amazon employees.

Several large established organizations joined the Amazon Marketplace over the years (e.g., Apple, Nike). However, the vast majority of third-party sellers' sales originated from small and medium resellers, who joined dynamically in response to market opportunities. In the US, there were 2.1 million sellers on Amazon, although only 1.1 million were active (Table 9.2).[1] Yet a small proportion of retailers captured most of the sales (Table 9.3).

As stated by Bezos, the large number of third-party sellers present stemmed from Amazon's success in creating services and software tools that sellers could leverage. Amazon Seller Central, an intuitive web application where sellers could manage and monitor their offerings and sales, was a critical element of Amazon's value proposition. Seller Central functionalities included the management of product listings, tracking of product inventories, and monitoring of sales. For example, sellers could leverage Seller Central for uploading new product images or changing prices. Amazon services included Fulfillment by Amazon (FBA), which allowed third-party sellers to outsource storage, shipment, and returns management to Amazon. Despite all of these advantages, operating as a seller on Amazon was challenging. The sheer number of sellers on the marketplace resulted in hypercompetition and, at times, anticompetitive actions by sellers (e.g., "black hat tactics").[2] It was not uncommon to have tens of sellers competing with

Table 9.2. Number of Active Amazon Sellers by Marketplace (Marketplace Pulse 2020)

Marketplace	Active Sellers 2019	Growth Since 2018	
USA	1,114,388	54,388	(+5.1%)
UK	281,257	43,727	(+18.4%)
Germany	244,425	36,425	(+17.5%)
Italy	216,610	51,110	(+30.9%)
France	211,859	47,859	(+29.2%)
India	205,884	36,184	(+21.3%)
Spain	203,413	52,063	(+34.4%)
Japan	173,483	28,303	(+19.5%)
Canada	163,595	35,555	(+27.8%)
Mexico	51,087	17,357	(+51.5%)
Australia	24,227	5,587	(+30.0%)
UAE	20,842		
Brazil	15,605	8,335	(+114.6%)
Turkey	5,987	1,967	(+48.9%)
Singapore	1,721		
Total	**2,934,383**	**441,423**	**(+17.7%)**

Table 9.3. Active Sellers in US Marketplace (Marketplace Pulse2020)

Marketplace	Active Sellers 2019
Professional sellers	514,000
Individual sellers	384,000
Sellers with at least one sale	898,000
Less than $100,000 in sales	672,000
With $100,000 in sales	168,000
With $500,000 in sales	40,000
With $1,000,000 in sales	18,000

1 An active seller performed at least one sale in the previous year.

2 Black hat tactics consist of a seller taking advantage of a vulnerability of the Amazon technology or governance rules in such a way that it can gain a benefit or detrimentally affect a competitor. For example, sellers could force the suspension of a competitor's account by paying a third party to create fake complaints.

the exact same product, as well as with Amazon itself, who shrewdly entered the most profitable product categories. Amazon's behavior was at the center of sellers' distrust and also triggered a series of antitrust inquiries (Satariano 2019).

Because Amazon controlled the main storefront and platform, sellers were required to monitor and quickly adapt to their policy and algorithmic changes. For example, in 2016, Amazon changed its policy and determined that it was against its terms of services to provide free or discounted products in exchange for products reviews, which was a widespread practice among sellers. Similarly, in 2019, Amazon changed its policy regarding Personally Identifiable Information (PII), as a result it ceased the sharing of information that many sellers depended upon for their marketing campaigns. During the Covid-19 pandemic, Amazon prioritized the restocking of household staples, medical supplies, and other high-demand products going into its fulfillment centers. As a result, many sellers saw their operations disrupted as Amazon's unilaterally decided not to accept their products in their fulfillment centers and to reduce their speed of delivery. From the standpoint of visibility on the Marketplace, Amazon had been known to make changes to the algorithm that ranks listings, resulting in third-party sellers' rankings changing—sometimes overnight (Mattioli 2019). Because Amazon regularly changed its algorithms for "improving the customer experience," sellers could see their competitive position in the marketplace harmed, without any prior notice from Amazon.

2.1 Edging Bets

Despite the dominant position Amazon had achieved and the substantial tailwind the Amazon platform could provide to independent sellers, with services such as Prime and FBA, many were weary of being excessively dependent on Amazon. Third-party sellers started exploring differentiating strategies that could improve their market positioning while also decreasing their dependency on Amazon. Several technology experts backed up sellers' concerns and predicted crucial disruptions in the e-commerce space. In fact, after years of growth, some argued that Amazon's market share would decline. In fact, while Amazon's ability to aggregate millions of sellers and customers under a homogeneous and standardized experience was considered at the core of its competitive advantage, it was also a limiting factor to its future growth. As an increasing share of the economy moved online, e-commerce was expected to expand beyond the products categories sold by Amazon. Furthermore, many opportunities existed for rethinking business models and online experiences for those products sold on Amazon.

As a testament to this shift, the market for direct sales e-commerce observed a significant increase and was expected to exceed $20 billion in GMV by 2021 (eMarketer 2020). Sellers who wanted to create a direct online presence were not limited to Amazon. They could leverage a wide range of services and software tools that required little IS knowledge. Shopify, BigCommerce, Skubana, Magento, and many others facilitated the creation of an online presence through which sellers could maintain a direct relationship with customers. As further evidence of the widespread interest and success of the direct sales model, Shopify exceeded a $90 billion market capitalization in early 2020 only five years after its IPO (Figure 9.2), and Magento, another longtime e-commerce platform, was acquired by Adobe in 2018 for $1.68 billion.

3 Starting Seller Labs

In 2010, Checketts was buying pallets of books from the Mail Recovery Center[1] in Atlanta to resell online. Three years before, he had started BookScouter.com, a website that helped people find the online marketplace where they could sell their used books at the highest price. He soon became the largest user of BookScouter.com, as he methodically determined the online marketplace where he could resell for most profit the thousands of used books he was buying. His software determined that Amazon was the marketplace where he could obtain the most profit for around 80% of the books.

Driven by early success, and now with the help of Paul Johnson, Checketts expanded his inventory and started buying pallets of CDs and DVDs. Although for most transactions he received high review scores, several customers returned the products and wrote negative reviews. After analyzing the problem, Checketts approached the problem with a technical solution:

> When we began selling CDs and DVDs, we noticed that our seller rating dropped from the 98–100% level to the range of 95–96%. When I read through the negative feedback, the common theme was that customers hadn't read our detailed description and were disappointed when the product arrived in a generic case without the original artwork.
>
> After a couple of weeks developing some software, I created a prototype that would send a message to customers explaining the CDs and DVDs condition. This message was sent as soon as the order was placed, allowing customers to cancel the order if they wanted. For other products that we sold, the software would cross-reference data available from shipping carriers like UPS and the US Postal Service to send the buyer a message shortly after their product was delivered.

As the initial version of Feedback Genius was starting to have a positive impact on his Amazon business, Checketts started gauging the potential for establishing a software company:

1 The Mail Recovery Center (MRC) is the official lost and found department of the U.S. Postal Service. The MRC auctions the lost mail after a 90-day period during which the legitimate owner could state a claim.

Figure 9.2. Shopify Stock Quote (Yahoo Finance 2020)

> The software worked very well, and within a month or two, our feedback rating returned to the 98–100% level that we were used to. As the software business launched, it became very apparent where I wanted to spend my time. In the used book and media business, growth meant bigger warehouses and more overhead, for a relatively slim margin and an increasing amount of capital tied up in inventory. The economics of software are quite different. It requires significant up-front investment, but if done successfully yields higher margins and much more potential for scale and growth.

Seller Labs was incorporated in Athens, Georgia in April 2013. The timing could not have been better, as the company leveraged its early market entrance capitalizing on the quickly growing Amazon third-party sellers base (Table 9.4). The company's early success allowed it to become cash flow positive within the first two years from its incorporation. Unlike many technology startups, Seller Labs had the opportunity to pursue organic growth without raising capital from investors (Table 9.A.1). By year 2019, Seller Labs was established as one of the most recognized software tools for Amazon sellers, with thousands of customers and a sound market share.

Table 9.4. Active Sellers in US Marketplace (Marketplace Pulse 2020)

Marketplace	Active Sellers 2019
Professional sellers	514,000
Individual sellers	384,000
Sellers with at least one sale	898,000
Less than $100,000 in sales	672,000
With $100,000 in sales	168,000
With $500,000 in sales	40,000
With $1,000,000 in sales	18,000
Sellers with at least one sale	**898,000**

4 Seller Labs Products and Technology

4.1 Seller Labs Product Suite

Shortly after incorporation, Feedback Genius became popular among Amazon sellers. Leveraging the initial success, Seller Labs launched other products that addressed the most critical needs of its growing customer base. The result was a comprehensive software suite tailored at improving sellers' efficiency for the most critical activities they perform when managing their Amazon business. More importantly, Seller Labs' products had the potential of significantly improving a seller's competitiveness in the marketplace. In the words of Jeff Cohen, the VP of Marketing:

> Success on Amazon depends on three basic marketing components: product, price, and placement. Great products have to continuously evolve as the result of feedback loops with customers. They need adequate pricing given the value they create and the level of competition they face. And they need the right advertisement push. Our mission is to provide the best software tools for monitoring and improving offerings on these three dimensions.

At the beginning of 2020, Seller Labs' suite of products could automate many day-to-day activities, facilitate recurrent tasks, and provide analytics and actionable recommendations, among many other functionalities. Seller Labs primarily targeted small and medium sellers that were in need of software tools that enhanced their productivity and enabled the implementation of diversified strategies on the Amazon Marketplace. The sheer complexity of the Amazon marketplace was daunting, especially for small and medium businesses, as they do not have the resources, capabilities, or IS infrastructure, for competing with established sellers. The Seller Labs Pro suite addressed the four most important business cases that an Amazon seller had to get right for improving its chances of success: manage online reputation; perform market research; monitor performance and profitability; secure placement via advertisement campaigns (see Table 9.5 for a full product description). Katie Reilly (VP of Product) captures Seller Labs' core value proposition:

> Admittedly there is no single solution that does it all. You could give all the tools they need to win on Amazon, but that becomes unbearable and risks paralyzing sellers as they are overflown by information. Our goal is to address the most important areas, simplify the customer experience, and make recommendations tailored to seller's goals in a way that is digestible and actionable. For example, our solutions can look at a seller's performance and compare it to benchmarks in their category, or similar types of business, and recommend ways they could tune their advertising, improve their listing, or increase conversion.

Another important element of Seller Labs success was the Managed Services group, which acted on a consultant basis for sellers that needed help with the management of their business. Seller Labs' experts combined their deep expertise of the Amazon Marketplace with their knowledge of Seller Labs suite of products. The group provided a white-glove service in the areas of listing optimization, advertisement management, and messaging optimization. In addition to contributing to the overall profitability, Managed Services helped reduce the churn of those sellers that required higher levels of customization and support.

Over the years, Seller Labs' role of facilitating seller's success extended beyond providing products and professional services. The company significantly contributed to the creation of articles, guides, and webinars that informed sellers about the latest news, trends, and best practices they needed to consider for establishing a successful business on Amazon. In addition, Seller Labs was active in all major conferences, while also hosting its conference, Resonate, since 2016. Resonate's objective was to provide experienced sellers a venue to connect and learn advanced selling strategies from each other and from Seller Labs Managed Services' specialists.

4.2 Seller Labs Technology

Seller Labs understood from the beginning the need for an agile IS infrastructure that allowed rapid development and release of new products and features. From day one, Seller Labs decided that it would leverage Amazon Web Services (AWS) cloud computing[1] platform and that all products would be offered with a Software as a Service (SaaS)[2] model. The decision to leverage AWS enabled Seller Labs to implement a flexible cost structure and the possibility of focusing on the development of software functionalities rather than the management of a proprietary and complex IS infrastructure. Furthermore, the SaaS model was critical for Seller Labs' success, as it meant that sellers were not required to install any software or buy specific hardware when using its products. Any seller with an Amazon professional sellers' account,[3] a PC, and an Internet connection could take full advantage of all products without needing any specific hardware, software, or capital investment. Furthermore, by leveraging AWS, Seller Labs had access to

1 Cloud computing is the on-demand availability of computer system resources (e.g., data storage and computing power) over the Internet and without direct active management by the user.

2 Software as a service (SaaS) is a software licensing and delivery model in which software is licensed on a subscription basis and is centrally hosted. Generally, SaaS applications are accessed via "thin" clients, such as web browsers.

3 Only third-party sellers with professional accounts are granted access to the Amazon Marketplace Service and Advertisement platforms. Professional accounts require the payment of a monthly fee to Amazon, contrarily to the free Individual accounts.

Table 9.5. Seller Labs Suite of Products

Product Name	Product Description
Feedback Genius	Manage buyer-seller email messaging to share product information or solicit a product review. By sending targeted emails to buyers, sellers can improve their online reputation and increase their brand awareness. Feedback Genius offers email analytics and the possibility of performing A/B testing to compare the effectiveness of different messages.
Scope	Find product opportunities and perform market research on competitors' products. Scope helps understand the potential profits for any product available on Amazon.com by analyzing keywords, search volume, average Cost-per-Click (CPC) bid, products fees, and estimated sales volumes.
Ignite	Manage advertisement campaign optimization on Amazon. Ignite provides personalized suggestions based on a seller's products, advertisement budget, and goals. Furthermore, Ignite enables campaign automations that can improve advertisement efficiency (e.g., avoid adverting during non-peak times) and drive Average Cost of Sales (ACoS) down.
Quantify	Monitor the inventory and financial performance. Quantify provides detailed profitability analysis for each product by automatically calculating gross and net margins. Furthermore, it offers customizable inventory management metrics for managing and optimizing storage fees and stock levels.

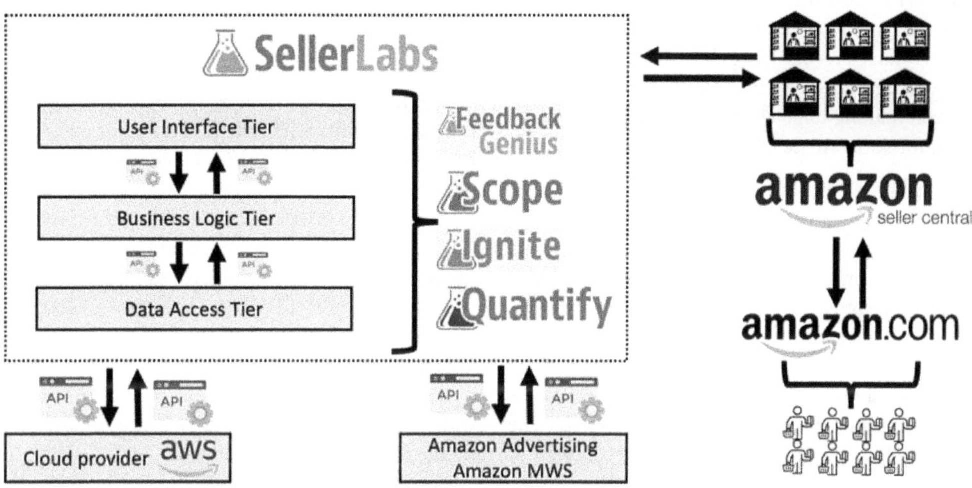

Figure 9.3. Seller Labs IT Infrastructure

advanced IT services (e.g., computing and storage resources), without needing to own an IT infrastructure (e.g., servers) or having to make long-term financial commitments. Offering a SaaS solution enabled Seller Labs full control over its application. For example, the company could release new versions of its products without requiring sellers to make a software update. This was essential as sellers often depended on Seller Labs for quickly introducing or removing features as a consequence of Amazon's changes to its platform and marketplace policies. Also, as sellers' data were not stored on their premises, they had an incentive in maintaining a recurrent relationship with Seller Labs to preserve transaction history.

Seller Labs adopted a three-tier architecture (see Figure 9.3), which separated the presentation, business logic, and data access level, and the whole architecture used internal APIs[1] to expose data, and systems capabilities. Its modular architecture enabled fast product iterations, and the ability to quickly launch completely new products (e.g., Ignite), or dismiss old features no longer supported.

At the core of Seller Labs' offering was the Amazon Marketplace Web Service (MWS) and Amazon Advertisement platforms.[2] Both platforms exposed resources, in the form of

1 An Application Programming Interface (API) is set of functions and procedures allowing the creation of applications that access the features or data of an operating system, application, or other service.

2 We refer to the combination of both the Amazon MWS and Amazon Advertisement APIs as: Marketplace APIs.

web service APIs, that enabled third-party sellers to programmatically exchange data with Amazon. When launched in 2009, Amazon MWS allowed only the manual upload of sellers' inventory through an XML and flat file feed. After receiving interest from sellers and complementors, over the years Amazon released new APIs allowing the programmatic exchange of product listings, orders, payments, and reports with Amazon's servers. Checketts recalls the integration efforts when first leveraging the MWS platform:

> The original Marketplace APIs that Amazon exposed were not really intended for third-party developers such as ourselves. They seem to have been originally intended for sophisticated sellers that had their own IT department so they could integrate with Amazon to get their own data out of Amazon to put into their own systems. For some time, the team at Amazon did not promote the use of the APIs and did not make major efforts to improve them. However, over the last five years Amazon has started recognizing the value of having third-party integrations and providers like ourselves and has been making huge strides in improving the experience for software developers.

Amazon Advertisement APIs enabled the programmatic creation of advertising campaigns for maximizing the visibility of sellers' listings in Amazon.com. The launch of the Advertisement platform in 2016 was a critical in enabling Ignite's product conceptualization and development. In fact, from a technology infrastructure perspective, Seller Labs products were dependent on the availability of the Marketplace APIs functionalities and data transfers. The Marketplace APIs provide the building blocks (i.e., APIs) that enable the creation of Seller Labs products through the exchange of data (e.g., inventory levels) and software instructions (e.g., product price change) (see Figure 9.3). For example, the financial reports available in Quantify were generated from raw data sourced from the Amazon MWS Reports API. Similarly, most of the product functionalities (e.g., change of product's price or description) were enabled by programmatically exchanging data with the Marketplace APIs. Checketts summarized the challenges of working with the Amazon's Marketplace APIs:

> One of the challenges we now face is that the pace of innovation at Amazon, especially in Advertising, is quite rapid. Every quarter new APIs and important functionality is released, requiring updates to our software to keep up. For some features, competitors have been able to adopt new functionality before us, keeping us in our toes.

5 X-Cart Acquisition

X-Cart was a Russian-based startup that developed the first downloadable PHP/MySQL shopping cart platform[1] in 2001. When first released, it was one of the leading solutions that enabled the creation of an e-commerce store by sellers. Despite its early entrance into the market, the company never became a market leader. Over the years, X-Cart lost its early advantage as other companies challenged its position (e.g., Magento), and its business model remained anchored to a one-time license fee. For almost two decades, X-Cart earned a significant portion of its revenue by providing customization and maintenance services for those sellers that wanted to create an e-commerce store based on the X-Cart platform. X-Cart economics and organizational incentives were strongly influenced by the service group within the company. As a consequence, their product roadmap could not evolve as the product advancements of its competitors.

Despite the many challenges faced by X-Cart, it had significant growth potential. Its e-commerce platform was broad and all-inclusive, as it comprised all the functionalities needed for creating an e-commerce storefront: store builder, product listings management, designs optimized for search engines, and the compatibility with over 120 different payment gateways.[2] By adopting X-Cart, a seller could create a full-fledge independent e-commerce presence. More importantly, X-Cart was architected as a platform, thus enabling the creation of add-ons (i.e., software extensions) that could increase functionalities to the core product. For example, the Google Ads add-on enabled the advertisement of seller products on Google Shopping, YouTube, and Gmail. Similarly, DHL integration enabled the inclusion of real-time shipping quotes and the comparison of different delivery methods. X-Cart had tens of add-ons for a wide range of purposes, such as order fulfillment, shipment, marketing, and tax management.

The X-Cart platform was adopted by more than 35,000 customers worldwide over the years. However, due to the one-time license model, a significant number of these customers were not recurrent, and did not interact with the company on an ongoing basis. In fact, many of X-Cart customers were small or medium business that entirely hosted and managed their e-commerce stores on-premise without ever interacting with X-Cart. As the company prepared for

1 Shopping carts are software products that enable visitors to an e-commerce site to browse, select, add to a shopping cart, and purchase a list of items. Full-fledged shopping carts, as in the case of X-Cart, have three main components: storefront, administration, and a payment gateway.

2 A payment gateway is a merchant service provided by an e-commerce application service provider that authorizes credit card or direct payments processing for e-businesses, online retailers, bricks and clicks, or traditional brick and mortar.

releasing a cloud-based version of X-Cart with a SaaS model, Harris reflected on the challenges and opportunities of the acquisition:

> The acquisition presented many challenges in terms of reorganizing and integrating our different business models, technological stacks, and organizational culture. However, I do think that X-Cart is a critical piece of the puzzle to become a diversified and sustainable company. The release of X-Cart Cloud is a first step in the journey for achieving that diversification. In many ways, X-Cart is simply another marketplace that we can provide to our customers and with the release of the first cloud-based version of it we can now facilitate its adoption.

6 Competition

The competitive environment of software tools for Amazon sellers was mounting. At the same time, the acquisition of X-Cart expanded Seller Labs' competitive territory to the entire e-commerce space. As a consequence, Seller Labs faced increasing competition from multiple fronts.

The number of startups offering software tools for Amazon sellers was substantial. The market was crowded with small and medium-sized firms providing products and services comparable to those of Seller Labs. While some firms only focused on specific products (e.g., keyword optimization), or particular regions or countries (e.g., Europe, China), each competitor presented a potential threat to Seller Labs. However, the biggest threat for Seller Labs was three existing companies that could potentially imitate their entire product suite and steal market share (see Table 9.A.2, and Table 9.A.4 for a comparison). Furthermore, as some of its competitors had access to venture capital funding, they could make significant investments in improving their products and market penetration. Sellics, a major competitor based in the Bay Area, raised Series A funding of $10 million in September 2019, signaling major plans to expand its current offerings. The increased competition created an arms race for developing new features and products, or rapidly imitating the ones introduced by direct competitors. Furthermore, some competitors were also starting to pursue diversification strategies by expanding their offerings to other marketplaces, such as Walmart, eBay, and Target.

In the broad e-commerce space, Seller Labs competed with several platforms explicitly developed to enable the creation of online stores by retailers. Shopify had emerged as the market leader in the space, with more than 1,000,000 customers globally and a GMV of its business exceeding $40 billion. Furthermore, with a big enough market share and access to capital, Shopify was starting to offer many platform resources previously only available on Amazon. For example, through Shopify's proprietary fulfillment network, sellers could reliably store and deliver their products, while keeping absolute control over their customer's experience and data.

The multiple technological stacks and the variety of seller needs created significant opportunities for a vast number of companies beyond Shopify. Magento, WooCommerce, BigCommerce, Feedonomics, were all successfully established e-commerce platforms, with substantial access to capital and market share. As a result of the head-to-head competition between the different e-commerce platform, sellers had several first-rate alternatives for easily creating and managing their online presence. Furthermore, many e-commerce platforms were starting to integrate horizontally into different channels, including Amazon and Walmart. Despite the existence of many established players, Hank Harris remained positive about the future:

> Just because there are dominant players in the industry with large market shares and access to capital does not mean the industry cannot be disrupted. We have seen multiple times in the technology industry how dominant positions can be quickly eroded by small entrants. Consider for example what happened to Skype and WebEx, which were the dominant videoconference software a few years ago. Now they are only minor players, in a market dominated by Zoom.

7 Charting the Future

Despite still being a startup, Seller Labs had already achieved many successes. The company grew from just two employees to more than a hundred. The company has also been praised multiple times by business press, ranked two years in a row among the top fastest growing private US companies, and was a founding member of the Amazon Marketplace Developer Council.

However, Seller Labs could not stand still contemplating its previous successes. The competitive environment was becoming increasingly hostile as competition was escalating rapidly from multiple angles. Numerous direct competitors were engaging in an arms race for developing increasingly advanced features for Amazon sellers and rapidly expanding internationally. Furthermore, the COVID-19 pandemic anticipated a decade of expected online commerce growth in just a few months, creating significant opportunities and challenges. Because of social distancing and shutdowns, a significant number of customers had online commerce as their only option for buying products during the pandemic, and several companies capitalized on this opportunity. Google made freely available to all sellers its Google Shopping service, independently of whether sellers advertised on Google. Furthermore, Google partnered with Shopify and WooCommerce to make it easy for small retailers to access Google's results. Facebook launched Facebook Shops, an online store where sellers could create their catalog, customize their offerings, and promote them within Facebook's applications (i.e., Instagram Stories or Facebook Ads). This meant that any seller could create an online business and connect with the millions of Facebook users. At the same time, Amazon was introducing a new Selling Partner API (SP-API) platform, which offered all the functionality already available in Amazon MWS platform plus a new set of functionalities. While the Amazon AWS platform was still available it would eventually be deprecated in favor of the more technologically advanced SP-API.

With the acquisition of X-Cart and the commercialization of X-Cart Cloud, Seller Labs had made its first move on its journey of becoming a differentiated sustainable business. Yet, the company was entering uncharted territories and without a clear strategy going forward it could put at risk the earlier successes. Despite the many challenges the company was facing, Harris remained positive:

> We have been investing so much in infrastructure for integrating the two companies over the last months. The significance of these investments, both in terms of human and capital resources, made us fall behind some of our competitors. We are finally back in a position where we can start releasing features and products again. I am incredibly exited that our customers will finally see what we have been working on for the past few months and I am looking forward to the opportunities ahead.

References

Amazon. 2019. "Amazon's Annual Shareholder Letter," *Security Exchange Commission.* (https://www.sec.gov/Archives/edgar/data/1018724/000119312519103013/d727605dex991.htm, accessed September 2, 2020).

eMarketer. 2019. "Digital Investments Pay Off for Walmart in Ecommerce Race," *EMarketer*, February 14. (https://www.emarketer.com/content/digital-investments-pay-off-for-walmart-in-ecommerce-race, accessed September 2, 2020).

eMarketer. 2020. "US Direct-to-Consumer Ecommerce Sales Will Rise to Nearly $18 Billion in 2020," *EMarketer*, April 1. (https://www.emarketer.com/content/us-direct-to-consumer-ecommerce-sales-will-rise-to-nearly-18-billion-in-2020, accessed September 2, 2020).

Evans, B. 2020. "The Shoulders of Giants." (https://www.ben-evans.com/presentations, accessed August 10, 2020).

Marketplace Pulse. 2019. "Marketplaces Year in Review 2019," *Marketplace Pulse*, December 16. (https://www.marketplacepulse.com/marketplaces-year-in-review-2019, accessed October 1, 2020).

Mattioli, D. 2019. "Amazon Changed Search Algorithm in Ways That Boost Its Own Products," *Wall Street Journal*, September 16. (https://www.wsj.com/articles/amazon-changed-search-algorithm-in-ways-that-boost-its-own-products-11568645345, accessed August 20, 2020).

Satariano, A. 2019. "Amazon Faces E.U. Inquiry Over Data From Independent Sellers," *The New York Times.* (https://www.nytimes.com/2019/07/17/technology/amazon-eu.html).

Yahoo Finance. 2020. "Yahoo Finance—Stock Market Live, Quotes, Business & Finance News." (https://uk.finance.yahoo.com/, accessed August 10, 2020).

Appendix A: Exhibits

Table 9.A.1. Seller Labs Financial Data (Casewriters)

Marketplace	2014	2015	2016	2017	2018	2019
Products	162,100	2,170,000	3,090,000	4,485,000	5,928,000	6,230,000
Services	66,211	930,000	1,321,624	2,015,000	1,672,000	2,101,429
Revenue	**228,311**	**3,100,000**	**4,411,624**	**6,500,000**	**7,600,000**	**8,331,429**
Total Customers	810	3,200	4,300	5,900	8,400	9,200

Table 9.A.2. Average Website Visits During February-April (SimilarWeb 2020)

Company name	Monthly Visits	Unique Visits*
Seller Labs	19,900	8,870
Sellics	106,660	16,930
Helium10	1,111,000	166,798
Junglescout	423,138	103,810

* Unique visits are the average number of unique visitors (not duplicated by device) to a website for a time period.

Table 9.A.5. Churn of Top 10,000 Amazon Sellers (US Marketplace) (Marketplace Pulse 2020)

Year	2015	2016	2017	2018	2019
2015	100%				
2016	95%	100%			
2017	88%	95%	100%		
2018	80%	89%	94%	100%	
2019	70%	80%	87%	95%	100%

Table 9.A.3. Website Geographic Access during February-April (SimilarWeb 2020)

Country	Seller Labs	Helium10	Sellics	Junglescout
United States	65.59%	29.37%	25.10%	33.61%
United Kingdom	7.15%	7.16%	10.98%	7.23%
China	4.40%	13.83%		24.42%
Taiwan	3.93%		3.26%	
India	2.63%		3.15%	
Germany	2.19%	7.02%	30.22%	3.33%
Spain	2.10%			3.79%
Canada	2.01%	4.22%	3.49%	4.78%
Pakistan		2.90%		1.78%
Sweden			5.02%	
Netherlands			3.16%	
Philippines		2.20%	2.74%	
Israel			2.44%	
Ukraine	1.11%	2.35%		
Australia	1.15%	2.07%		1.99%
Others	7.74%	28.88%	10.44%	19.07%
Total	**100%**	**100%**	**100%**	**100%**

Table 9.A.4. Competitors Comparison (Casewriters)

Feature	Seller Labs	Helium10	Sellics	Junglescout
Buyer-seller messaging	X	X		X
Marketing campaign automation	X	X	X	
Listing optimization automation		X	X	X
Amazon keywords analytics	X	X	X	X
Financial and inventory reporting	X	X	X	X

Case Study for Chapter 10

IT Planning at ModMeters[1]

Brian Smith, CIO of ModMeters, groaned inwardly as he listened to CEO John Johnson wrapping up his remarks.[2] "So our executive team thinks there are real business opportunities for us in developing these two new strategic thrusts. But before I go to the board for final approval next month, I need to know that our IT, marketing, and sales plans will support us all the way," Johnson concluded.

Brian mentally calculated the impact these new initiatives would have on his organization. He had heard rumors from his boss, the COO, that something big was coming down. He had even been asked his opinion about whether these strategies were technically doable, *theoretically*. But *both* at once? Resources—people, time, and money—were tight, as usual. ModMeters was making a reasonable profit, but the CFO, Stan Abrams, had always kept the lid screwed down tightly on IT spending. Brian had to fight for every dime. How he was going to find the wherewithal to support not one but *two* new strategic initiatives, he didn't know.

The other VPs at this strategy presentation were smiling. Taking ModMeters global from a North American operation seemed to be a logical next step for the company. Its products, metering components of all types, were highly specialized and in great demand from such diverse customers as utility companies, manufacturers, and a host of other industries. Originally founded as Modern Meters, the firm had grown steadily as demand for its metering expertise and components had grown over the past century or so. Today ModMeters was the largest producer of metering components in the world with a full range of both mechanical and, now, digital products. Expanding into meter assembly with plants in Asia and Eastern Europe was a good plan, thought Brian, but he wasn't exactly sure how he was going to get the infrastructure in place to support it. "Many of these countries simply don't have the telecommunications and equipment we are going to need, and the training and new systems we have to put in place are going to be substantial," he said.

But it was the second strategic thrust that was going to give him nightmares, he predicted. How on earth did they expect him to put direct-to-customer sales in place so they could sell "green" electric meters to individual users? His attention was jerked back to the present by a flashy new logo on an easel that the CEO had just unveiled.

"In keeping with our updated strategy, may I present our new name—MM!" Johnson announced portentously.

"Oh, this is just great," thought Brian. "Now I have to go into every single application and every single document this company produces and change our name!"

Because of its age and scientific orientation, ModMeters (as he still preferred to call it) had been in the IT business a long time. Starting back in the early 1960s, the company had gradually automated almost every aspect of its business from finance and accounting to supply-chain management. About the only thing it didn't have was a fancy website for consumers, although even *that* was about to change. ModMeters currently had systems reflecting just about every era of computers from punch cards to PCs. Unfortunately, the company never seemed to have the resources to invest in reengineering its existing systems. It just layered more systems on top of the others. A diagram of all the interactions among systems looked like a plate of spaghetti. There was *no way* they were going to be able to support two new strategic thrusts with their current budget levels, he thought as he applauded the new design along with the others. "Next week's IT budget meeting is going to be a doozy!"

Sure enough, the following week found them all, except for the CEO, back in the same meeting room, ready to do battle. Holding his fire, Brian waited until all the VPs had presented their essential IT initiatives. In addition to what needed to be done to support the new business strategies, each division had a full laundry list of essentials for maintaining the *current* business of the firm. Even Abrams had gotten into the act this year because of new legislation that gave the firm's outside auditors immense scope to peer into the inner workings of every financial and governance process the organization had.

After listening carefully to each speaker in turn, Brian stood up. "As many of you know, we have always been cautious about how we spend our IT budget. We have been given a budget that is equal to 2% of revenues, which seriously limits what we in IT have been able to do for the company. Every year we spend a lot of time paring our project list down to bare bones, and every year we make do with a patchwork of infrastructure investments. We are now at the point where 80% of our budget in IT is fixed. Here's how we spend our money." Brian clicked on a PowerPoint presentation showing a multicolored pie chart.

1 Reproduced by permission of the authors Heather A. Smith and James D. McKeen.

2 This case was originally published as Smith, H. A., and McKeen, J. D. (2005, September). "IT planning at ModMeters." #1-L05-1-008, Queen's School of Business. Reproduced by permission of Smith School of Business, Queen's University.

"This large chunk in blue is just about half our budget," he stated. "This is simply the cost of keeping the lights on—running our systems and replacing a bare minimum of equipment. The red chunk is about 30% of the pie. This is the stuff we *have* to do—fixing errors, dealing with changes mandated by government and our own industry, and providing essential services like the help desk. How we divide up the remainder of the pie is what this meeting is all about."

Brian clicked to a second slide showing a second pie chart. "As you know, we have typically divided up the remaining IT budget proportionately, according to who has the biggest overall operating budget. This large pink chunk is you, Fred." Brian gestured at Fred Tompkins, head of manufacturing and the most powerful executive in the room. It was his division that made the firm's profit. The pink chunk easily took up more than half of the pie. Tompkins smiled. Brian went on, pointing out the slice that each part of the firm had been allotted in the previous year. "Finally, we come to Harriet and Brenda," he said with a smile. Harriet Simpson and Brenda Barnes were the VPs of human resources and marketing, respectively. Their tiny slivers were barely visible—just a few percent of the total budget.

"This approach to divvying up our IT budget may have served us well over the years"—Brian didn't think it had, but he wasn't going to fight past battles—"however, we all heard what John said last week, and this approach to budgeting doesn't give us *any* room to develop our new strategies *or* cover our new infrastructure or staffing needs. Although we might get a little more money to obtain some new applications and buy some more computers"—Abrams nodded slightly—"it won't get us where we need to go in the future."

A third graph went up on the screen, showing the next five years. "If we don't do something *now* to address our IT challenges, within five years our entire IT budget will be eaten up by just operations and maintenance. In the past we have paid minimal attention to our infrastructure or our information and technology architecture or to reengineering our existing systems and processes." A diagram of the "spaghetti" flashed on. "This is what you're asking me to manage in a cost-effective manner. It isn't pretty. We need a better plan for making our systems more robust and flexible. If we are going to be moving in new directions with this firm, the foundation just isn't there. Stan, you *should* be worried that we won't be able to give our auditors what they ask for. But you should also be worried about our risk exposure if one of these systems fails and about how we are going to integrate two new business ventures into this mess."

Tompkins looked up from his papers. It was clear he wasn't pleased with where this presentation was headed. "Well, I, for one, *need* everything I've asked for on my list," he stated flatly. "You can't expect me to be the cash cow of the organization and not enable me to make the money we need to invest elsewhere."

Brian was conciliatory. "I'm not saying that you don't, Fred. I'm just saying that we've been given a new strategic direction from the top and that some things are going to have to change to enable IT to support the whole enterprise better. For example, until now, we have always prioritized divisional IT projects on the basis of ROI. How should we prioritize these new strategic initiatives? Furthermore, these new ventures will require a *lot* of additional infrastructure, so we need to figure out a way to afford this. And right now our systems don't 'talk' to the ones running in other divisions because they don't use the same terminology. But in the future, if we're going to have systems that won't cost increasing amounts of our budget, we are going to have to simplify and integrate them better."

Tompkins clearly hadn't considered the enterprise's needs at all. He scowled but said nothing. Brian continued, "We are being asked to do some new things in the company. Obviously, John hopes there's going to be a payback, but it may take a while. New strategies don't always bear fruit right away." Now looking at Abrams, he said pointedly, "There's more to IT value than short-term profit. Part of our business strategy is to *make* new markets for our company. That requires investment, not only in equipment and product but also in the underlying processes and information we need to manage and monitor that investment."

Harriet Simpson spoke for the first time. "It's like when we hire someone new in R&D. We hire for quality because we want their ideas and innovation, not just a warm body. I think we need to better understand how we are going to translate our five key corporate objectives into IT projects. Yes, we need to make a profit, but Stan needs to satisfy regulators and Brenda's going to be on the hot seat when we start marketing to individuals. And we haven't even spoken about Ted's needs." As the VP of R&D, Ted Kwok was tasked with keeping one or more steps ahead of the competition. New types of products and customer needs would mean expansion in his area as well.

Abrams cleared his throat. "*All* of you are right. As I see it, we are going to have to keep the cash flowing from Fred's area while we expand. But Brian's got a point. We may be being penny wise and pound foolish if we don't think things through more carefully. We've put a lot of effort into developing this new strategy, and there *will* be some extra money for IT but not enough to do that plus everything all of you want. We need to retrench and regroup *and* move forward at the same time."

There was silence in the room. Abrams had an annoying way of stating the obvious without really helping to move the ball forward. Brian spoke again. "The way I see it, we have to understand two things before we can really make a new budget. First, we need to figure out how each of the IT projects we've got on the table contributes to one of our key corporate objectives. Second, we need to figure out a way to determine the *value* of each to ModMeters so that we can prioritize it. Then I need to incorporate a reasonable amount of IT regeneration so that we can continue to do new projects at all."

Everyone was nodding now. Brian breathed a small sigh of relief. That was step one accomplished. But step two was going to be harder. "We have a month to get back to the board with our assurances that the IT plan can incorporate the new strategies and what we're going to need in terms of extra funds to do this. As I said earlier, this is *not* just a matter of throwing money at the problem. What we need is a *process* for IT planning and budgeting that will serve us well over the next few years. This process will need to accomplish a number of things:

- It will need to take an *enterprise* perspective on IT. We're all in these new strategies together.
- It will have to incorporate all types of IT initiatives—our new strategies, the needs of Fred and others for the new IT to operate and improve our existing business, Stan's new auditing needs, and our operations and maintenance needs.
- In addition, we *must* find some way of allocating some of the budget to fixing the mess we have in IT right now.
- It must provide a better way to connect new IT work with our corporate objectives.
- It must help us prioritize projects with different types of value.
- Finally, it must ensure we have the business *and* IT resources in place to deliver that value."

Looking at each of his colleagues in turn, he asked, "Now, how are we going to do this?"

Discussion Question

1. Develop an IT planning process for ModMeters to accomplish the demands as set out above.
2. What do you believe would be the optimal governance structure for the IS function at ModMeters. Why?

Case Study for Chapter 11

Pearson's SuccessMaker: Putting the Customer First in Transforming Product Development Processes

"Complacency Kills" was circled and underlined in Greg Adams-Woodford's notes. It was the image of those words that had stuck with him over the weekend. Now it was Monday and, as the vice-president of product management, it was time to determine the product development road map for the next five years. It was hard for Adams-Woodford to believe it had only been a year since he first joined Pearson. He was proud of the positive impact his work had on improving the market position of the newest version of the SuccessMaker product in such a short span of time.

When Adams-Woodford first joined Pearson, the entire digital learning division was in the midst of the first major upgrade of its SuccessMaker product. SuccessMaker was Pearson's premier software product for helping elementary and middle school students to improve their math and reading skills, so this upgrade was a strategically important project for the firm. The upgrade involved significant changes to the foundational technology of the product—from an older client-server model to a new Internet delivered model—and a complete revamp all of the instructional content in the product. Despite the best efforts of those involved, the new versions of SuccessMaker were failing to meet Pearson's internal goals and the needs of its most important customers. The difficulty of the situation was painfully illustrated in an unsolicited letter from a strategic customer delivered early in Adams-Woodford's tenure. In this letter, the customer bitterly complained of broken promises and missed expectations since work on the new version had begun 18 months earlier. This letter ultimately became the rallying cry for a complete reengineering of the SuccessMaker product development process. Adams-Woodford and his team undertook a complete flash-cut[1] from a more traditional and established development process (waterfall) to a new (and unproven within Pearson) development methodology (Agile) (see Exhibits 1 and 2). Agile's focus on short development iterations and communication over documentation was fundamentally different from the waterfall approach they had previously used, which emphasized complete and detailed documentation before development began. Now, a year later, many of the previous challenges were behind them. Both Adams-Woodford's key customers and internal groups agreed SuccessMaker had gone through a completely positive transformation. Adams-Woodford was pleased that his clear vision for the future

1 "Flash-cut" is a standard term for the immediate implementation of a new methodology or system.

IVEY | Publishing

T. S. Raghu and Collin Sellman wrote this case solely to provide material for class discussion. The authors do not intend to illustrate either effective or ineffective handling of a managerial situation. The authors may have disguised certain names and other identifying information to protect confidentiality.

Richard Ivey School of Business Foundation prohibits any form of reproduction, storage or transmission without its written permission. Reproduction of this material is not covered under authorization by any reproduction rights organization. To order copies or request permission to reproduce materials, contact Ivey Publishing, Richard Ivey School of Business Foundation, The University of Western Ontario, London, Ontario, Canada, N6A 3K7; phone (519) 661-3208; fax (519) 661-3882; email cases@ivey.uwo.ca.

Copyright © 2012, Richard Ivey School of Business Foundation

One time permission to reproduce granted by Richard Ivey School of Business Foundation on April 12, 2021.

had overcome many of the most significant issues with the development of SuccessMaker.

The new path forward was not as clear, however. The development team began to experience growing pains with the specific Agile methodology they chose to implement—Scrum. Some influential members of the team proposed Scrum should be abandoned in favor of a Kanban system. Kanban was developed in the manufacturing industry and was more focused on individual task management whereas Scrum focused short development cycles called sprints that included a whole set of tasks (see Exhibit 3). These trusted team members felt the Scrum methodology did not fit with the current lifecycle phase of the product. The data seemed to support this conclusion as the team's initial increases in productivity were beginning to trend back down. While Kanban was clearly successful in manufacturing firms, Adams-Woodford had no personal experience using it as a software development methodology. His initial reaction was to stick with what he knew. Adams-Woodford had advocated continually for the Scrum methodology over the past year; so the thought of throwing those processes away and implementing a new system did not sit well with him.

There was also the question of whether to extend Agile into other teams involved in developing SuccessMaker. In his strategic planning meetings the previous week, several of Adams-Woodford's peers suggested he take a broader role inside Pearson, assisting other development groups with implementing the Agile system. While the new responsibility was appealing, Adams-Woodford worried these new challenges might jeopardize his focus during a critical time for the broader SuccessMaker team as they evaluated transitioning to a new Agile system beyond software development alone.

Pearson and Pearson Digital Learning

Listed on both the London and New York stock exchange (LSE: PSON; NYSE: PSO), Pearson had three primary business units. The Pearson Education unit was focused on the education market and was one of the leading companies that developed learning products for teachers and students. The Financial Times group provided business and financial news publication and online services. The Penguin group published books in a wide range of genres under the Penguin brand.

Pearson reported US $8.7 billion in revenues in 2010 (see Exhibit 4). Pearson Education group was the largest of the three divisions and accounted for 80 per cent of the firm's profits. The North American education market was the largest business segment for Pearson and accounted for 46 per cent of the firm's worldwide revenues.[1] The business sustained its growth even during the deep recession of 2007–8. Pearson's myLab—a digital learning, homework and assessment service—grew to over 7.3 million registered students in 2010. A recently introduced suite of learning management technologies aimed at the higher education market—eCollege and Fronter—also had online student enrollment of over 8.3 million. Pearson Digital Learning focused on digital supplemental products, viewed as a market segment with strong potential for growth.

Despite an overall strong growth record, Pearson Education faced some headwinds as it moved into 2011. Continuing slow economic growth put considerable strain on education budgets in the United States. The budget crisis had a direct effect on education budgets at all levels. Pearson attempted to address these market difficulties with a focus on providing premium products and services. Additionally, they sought to take advantage of schools shifting to online and digital curriculum, instead of relying exclusively on the demand for traditional textbooks. Pearson also made some key acquisitions to keep offering a broad range of services to the education market. The firm recognized technological changes as the biggest risk to all its product lines and services. Due to these changes, transforming Pearson Education's products and services for digital channels was a key element of the firm's growth strategy.

SuccessMaker

SuccessMaker was one of Pearson Digital Learning's (PDL) products focused on helping elementary and middle school students learn reading and math at their own pace. SuccessMaker was an all-digital product and was highly popular with the students who used it. The software also provided teachers with valuable assessment data, enabling targeted interventions and coaching. SuccessMaker was customized to individual markets to meet state mandated standards at each grade level.

The SuccessMaker product was derived from one of the earliest works of research on computer-aided learning. Researchers Patrick Suppes and Richard Atkinson lead the development of a decision theory based instructional program in the 1960s that eventually was marketed by the Computer Curriculum Corporation. In 1990, Simon & Schuster acquired Computer Curriculum Corporation. Pearson made a successful bid for Simon & Schuster's education operations (including the SuccessMaker product) and created the Pearson digital business unit in 1998.

The underlying approach to the design of SuccessMaker had remained unchanged since its creation. Students were given specific assignments to complete by the teacher in the SuccessMaker program (see Exhibit 5). Based on how students responded to these assignments, the software used a sophisticated, predictive algorithm to move the students across multiple skill levels of curriculum. Students who quickly mastered skills could move forward more quickly than students requiring additional assistance. The software was intelligent enough to know not only where to initially place students in a given set of curriculum, but also

1 http://ar2010.pearson.com/files/pdf/Pearson_AR10.pdf accessed on January 12, 2012.

the optimal way to pace them through it. Through daily use of SuccessMaker, students were often able to achieve significantly better results on standardized tests compared to students using other instructional methodologies or products.

While SuccessMaker had undergone many improvements over the years, the most recent version was part of a major technology and content upgrade that began in 2005. This new version, codenamed *SuccessMaker: Next Generation* was intended to offer customers a more robust and scalable platform as well as a more modern curriculum. This new curriculum used a variety of new instructional tools including 3-D animation and an interactive avatar to guide the student through the program.

While the intent of the new release was to address critical market issues, the reality was that it was missing the mark. When Greg joined Pearson in 2007, he told the senior vice president of product management, "We have spent $30 million on the new SuccessMaker development and even six to eight months after launching the first version, it has not been well adopted by the market. The few early adopters complained it did not meet any of their requirements."

One of Greg's first insights into the root cause of SuccessMaker's problems was that the development team was driving all the decisions instead of product management. As a result, there was little direct communication between the development team and customers causing many well-intended features to completely miss the mark when they were released. The SuccessMaker team used the waterfall development methodology (see Exhibit 1), emphasizing freezing product requirements at the beginning of a development cycle through formal documentation. With a major portion of software being developed offshore in India, miscommunication between the development teams in India and the United States was a common occurrence. Due to contractual constraints, changes to product features midway during the development cycle were either costly or impossible.

The waterfall process also required the organization to commit to a multi-year road map of features that did not allow for any significant changes in scope. These features were individually budgeted and committed in the business unit's strategic plan. Any changes to the road map were generally viewed negatively. The team referred to such changes using the pejorative industry term "scope creep." The waterfall process did make internal planning easier as the scope was essentially frozen for multiple years. It also meant little effort was put toward confirming committed features would meet customer needs or exploring alternative, unanticipated features once the road map was set.

The Push for Agile Methodology

Adams-Woodford successfully used Agile software methodologies at two early-stage firms prior to joining Pearson. Agile software development methodology advocated an iterative and evolving approach to software as opposed to the formal stage model of the waterfall approach. Agile methodology enforced a customer-driven approach to development with short development cycles—typically two week "sprints" delivering a small subset of product features using very little formal documentation. Short, "user stories" written on a single index card replaced the 100-plus page requirements documents of the past. There was no commitment to future sprints until the previous one was complete. This deferred commitment approach enabled teams to adapt, design and prioritize product features through ongoing feedback from product managers and customers.

A key assumption of the methodology was that product management *owned* the product, not the developers. Adams-Woodford lectured the development team on the role of a product manager, "Customers don't pay you to create documents; they want you to create a product. Product management must understand the market, distinctive competence, listen to its customers and prioritize features that create the highest value for its customers."

Adams-Woodford's first task was to change the basic structure and culture of the organization, starting with the software development team, including the developers, quality assurance (QA) engineers and product managers. The overall product development team was composed of a diverse group of individuals with expertise in different areas inside Pearson—product managers, software developers, QA experts, business analysts, curriculum/content experts, instructional designers, animators and artists. Before the formal shift to Agile, Adams-Woodford transitioned all offshore software development to in-house by hiring several new software developers in Arizona. Agile methodology required close cooperation between team members on almost a daily basis. Communication with the off-shore team had been a constant point of contention both with the Pearson team and their vendor. Adams-Woodford realized he had to create a more productive dialog focused on building a better product.

The Move to Agile

The transition was executed as a flash-cut from the previous waterfall process to the new Agile methodology. On a Friday, the software development teams had been following the waterfall process and when they returned the following Monday, they were 100 per cent Agile. The move was timed right after the second major version of the product had been released to manufacturing (burning to DVD/CDs). This dramatic change met with skepticism from a vocal subset of team members. Adams-Woodford stressed the transition was as much a cultural shift as it was a change in processes. A few members quit as the changes were instated.

As soon as the cut was made, training was delivered to all the software developers, QA engineers and product managers, detailing the new methodology and everyone's role within it.

Adams-Woodford brought in an independent Agile coach to teach the team about the new development methodology. All the teams spent two full days with the coach and the training was conducted in three phases. The first phase focused on explaining the underlying rationale of an Agile approach to ensure everyone understood the core Agile principles. The second phase addressed working within an Agile development framework, discussing specific work practices required. The final phase focused on the SuccessMaker team, asking members to discuss why Agile would not work in their environment. Proactively addressing these concerns was suggested by the Agile coach, who had previously observed lingering concerns about Agile often manifested in team behavior weeks or even months after making the shift from more traditional processes.

Since all teams and their top management were present in the training sessions, it resulted in the teams jointly exploring solutions to most of the problems that were surfaced. They spent a considerable amount of time discussing how some groups that were not going to be in Agile mode would engage with developers and product management groups given they were still integral to the larger development effort. Another issue that was discussed was that the development and QA teams had worked as separate and largely autonomous groups previously. It was agreed that the two teams had to be combined in an Agile environment, with extensive cross-training in their respective specialties. Under this new system, the most important change involved how product management would engage with development. Product management would now drive all development efforts and prioritization of new features. Product management would need to prioritize user stories based on market needs on an on-going basis—an uncomfortable change for both the development and product groups. Adams-Woodford and Eric Wagner the newly hired vice-president of the development teams—tried to help the SuccessMaker team with this transition by ensuring they both were consistent and unwavering in their application of the Agile process.

The development teams were reorganized for SuccessMaker Release 3 as they finished their development-related work for Release 2. After some initial experimentation with different team sizes, the development team was split into four groups. Two teams each were responsible for development work involving the Java programming language. One team was responsible for handling coding of reports and curriculum logic. Another team created all the Adobe Flash-based animations required for the product. In addition to these teams, a *scrum master* was designated in each team to assist the team in removing any roadblocks.

Growing Pains for Release 3

Adams-Woodford made a point to walk the halls and interact with as many of the team members as possible to keep in touch with their feelings and attitudes about the ongoing changes. After the initial novelty of the new methodology began to fade, Adams-Woodford observed development teams trying to reintroduce some of their previous waterfall processes. Specifically, he noticed the QA team was not comfortable with their new roles. As he chatted informally with members of this team, it became clear many of them were not sure how to go about their jobs without being able to rely on detailed requirements documents as they had in the past. This impacted the team's relationships with the developers, who felt individuals on QA team were not committed to the fundamental tenets of Agile methodology (see Exhibit 2) and were actively trying to return to previous practices.

The main difficulty facing the QA team was its joint responsibility with the developers for the quality of the release. This meant previously clear lines between what a developer did and what a QA engineer did had blurred. Previously, the QA engineers had no interaction with the developers and simply waited for the code to be completed so they could check it against the test cases developed independently from the original and very detailed requirements documents. If they discovered software defects, the engineers' only responsibility was to identify the issues and pass them back to the development team. In the new model, there were no requirements documents and the entire team—developers and QA engineers—was responsible for delivering working code at the end of the defined sprint. This team approach meant everyone needed to contribute with little regard for formal functional roles. Individuals on the QA team struggled with not only this ambiguity, but also with what they perceived to be a lack of accountability and trackability for features they were developing.

Adams-Woodford also noticed managers of both the developers and QA engineers were increasingly involved with task management on the sprint teams. The scrum process allowed managers to attend the daily stand-up meeting, but also dictated no one other than the members should be involved in team-level decision-making during these meetings. The meetings were called *stand-ups* given all attendees stand during the meeting in order to encourage the team to be concise and keep the meeting to no more than 15 minutes. Adams-Woodford attended all the stand-ups and found that he increasingly had to remind managers the teams were to be self-managing in relation to the daily plan. Adams-Woodford believed that unless the managers were completely committed to their employees managing the daily work, the benefits of Agile would be significantly reduced.

Of all the people he spoke to during his hallway conversations, the product managers and owners seemed to adjust most quickly and experienced the most dramatic change in their outlook for the future. In the scrum process, the product managers and product owners were the business representatives responsible for creating and prioritizing the user stories for each development sprint. Adams-Woodford recalled his early days at Pearson when these individuals were by far the most jaded and ambivalent of any of the cross-functional teams. They all seemed resigned to having little control over what the development teams produced. Most of their time was spent performing damage control when new releases failed to meet expectations. Now, however, these managers and owners were committed to the new methodology, becoming outspoken advocates of the Agile mindset across the business unit. Several took it upon themselves to take classes and join discussion groups about the methodology, while many established good relationships with both customers and the sales team.

When Adams-Woodford was not out talking to the teams, he was in his office reviewing key performance metrics for the development teams. Since the switch to Agile, the code base for SuccessMaker had been cut by half even while the number of features had increased by up to 75 per cent. The new features had been very favorably received by clients. The development team size had also shrunk by at least a third, mainly through eliminating off-shore development teams. Despite these improvements, the first few sprints of Release 3 showed significant performance degradation in the velocity of the developers. Velocity was calculated by converting each user story into a standard measure of effort as story points and determining how many points were completed in each sprint. Adams-Woodford did not think it was a coincidence when the velocity began to slow as the teams began regressing into their previous waterfall processes. This development prompted Adams-Woodford's key developers to suggest a change from Scrum to Kanban in order to jumpstart the team's overall velocity.

The Path Forward

While the move to Agile development improved SuccessMaker's position with both customers and the sales team, a number of challenges remained outside the software development team. Out of several different cross-functional teams, the software development team was the only functional group to transition to Agile. Other teams involved with the product such as the content team still used a waterfall approach. Team leaders on the software team also recognized a number of issues requiring immediate attention beyond the QA engineers' concerns about their roles and software testing. End-to-end integration testing and complete automation of the testing procedures were still works-in-progress. There was a tendency to follow a *mini waterfall* approach when this testing was required. The teams wanted to avoid this tendency in future.

As far as planning for the final release, which was made up of many sprints, the product owners were pleased to note they were able to plan up to two sprints for each development team accurately. As all the teams achieved higher levels of maturity in Agile methodology, they hoped to plan out to three to four iterations. This required greater stability and predictability in achieving the iteration goals. Adams Woodford realized this weakness was an important hurdle to overcome, stating, "Teams don't consistently understand what their velocity is. They still cannot estimate accurately how many story points they can develop in a planning period. This makes it hard to commit to releases with customers." Adams-Woodford also worried teams were becoming too internally focused on estimating a static backlog of features. Unless they were also continuously validating that the requirements in the backlog were the most important to the market, these teams could begin to slip back into developing features unable to meet the market need.

Beyond the planning of the sprints, the activities within an individual sprint had room for improvement. David Foster, the chief architect for the division, noted the product development teams engaged in *context switching*. In other words, development teams often switched from one feature to another distinct feature within the same development iteration. This resulted in reduced velocity since the developers had to review and familiarize themselves with technical aspects of the system components each time they switched from one feature to another. To eliminate context switching, Foster noted the need for the product and development teams to recognize the mapping between features and systems components. Foster also observed, "We haven't become very efficient at maintenance releases. With Scrum, it shouldn't matter, but we are still not comfortable with them. That is still a rough spot."

Adams-Woodford often insisted the developers adopt a startup mentality, stating, "We are diversifying our product lines rapidly. We are going to have to be creative in converting something that some people might think is really boring into something very compelling and focus on our distinctive competence." For this to happen he noted that he needed to keep his development teams fresh and energized. Agile development frameworks can make software development a "grind" and monotonous. Thus, it was important for him to keep his developers engaged with the products and customers. To keep them inspired and motivated, Adams-Woodford encouraged cross-team deployment of development team members across the product family.

Adams-Woodford also saw challenges in the product's strategic position in the market, stating,

> "When you are using SuccessMaker, it still feels like a dated product. We have made great strides in the platform technology, but our content road map has remained essentially the same for the past three years. The product is also expensive to maintain compared to other web-based, next-generation products. We have to provide different options—hosting, subscription-based service etc. There are only so many features you can add to the product. The strategy is to take the good brand, build

ancillary products, reach into the home and connect with the parents and students. We have to make the product entertaining for kids to play on their own at home, while it reports back to the teachers and parents. We must use the data, make it interoperable at the district level. Inside Pearson, we must make sure that the data is portable across products—this is really important."

All of these issues swirled around Adams-Woodford's mind as he considered how to approach the new product road map and where he would focus his efforts in the future. Although Adams-Woodford had made great progress inside Pearson by implementing Agile with the development teams, the challenges he faced now were much more complex and potentially more important in the long-run for the prospects of his product portfolio.

"Well," Adams-Woodford thought to himself. "If Agile works for development, maybe it will work for me." Adams-Woodford went to the developers' planning area and brought back a pack of index cards to his office. Adams-Woodford began to outline user stories for how he would implement the product road map, his place within it, and the broader Agile effort at Pearson.

Exhibit 11.1

A Summary of Waterfall Software Development Approach

Software teams have traditionally followed a phased approach when building systems. The term "waterfall" refers to the inherent rigidity in development lifecycle where one phase is presumed to begin only after the completion of a previous phase. However, it is common to see software teams iterate between phases. The main phases in the software development lifecycle (SDLC) include—(1) requirements analysis, (2) design, (3) development, (4) testing, (5) deployment, and (6) maintenance. Managers emphasize the need to extensively analyze and document the user requirements since they consider changes made to requirements and design during the latter stages of the development lifecycle to be highly expensive and disruptive. As a result, phased development models are documentation intensive.

A key assumption of a phased approach is that the dependency between phases is linear and uni-directional. However, in most software projects, dependencies between components are complex and bi-directional. To address this complexity, managers attempt to split the projects into relatively independent modules and integrate the modules towards the end of the project lifecycle.

Despite adaptations to the basic concept of a "waterfall" model, this development philosophy has many drawbacks. In most cases, it is difficult for users to commit to requirements since they have difficulties in visualizing the final product. Moreover, business requirements may change during the course of development and cause the project to backtrack to earlier phases thus leading to wasted efforts.

Unrealized dependencies between systems, components or modules may lead to long delays during systems integration. As a result, many software development projects have embraced iterative development approaches such as Spiral, Rapid Application Development or Agile models.

Source: Richard E. Fairley, ***Managing and Leading Software Projects***, John Wiley, Hoboken, New Jersey, 2009, pp. 55–58

Exhibit 11.2

Agile Software Development Methodology

To understand the philosophy of Agile development methodologies, it is helpful to note the twelve principles as proposed by the Agile Alliance:

- Our highest priority is to satisfy the customer through early and continuous delivery of valuable software.
- Welcome changing requirements, even late in development. Agile processes harness change for the customer's competitive advantage.
- Deliver working software frequently, from a couple of weeks to a couple of months, with a preference to the shorter timescale.
- Business people and developers must work together daily throughout the project.
- Build projects around motivated individuals. Give them the environment and support they need, and trust them to get the job done.
- The most efficient and effective method of conveying information to and within a development team is face-to-face conversation.
- Working software is the primary measure of progress.
- Agile processes promote sustainable development. The sponsors, developers, and users should be able to maintain a constant pace indefinitely.
- Continuous attention to technical excellence and good design enhances agility.
- Simplicity—the art of maximizing the amount of work not done—is essential.
- The best architectures, requirements and designs emerge from self-organizing teams.
- At regular intervals, the team reflects on how to become more effective, then tunes and adjusts its behavior accordingly.

Agile methodology takes an adaptive approach to development. At the end of an iteration (known as *sprint*), the team delivers a set of specific features that are tested and integrated into the product. Teams have the opportunity to review progress and revisit user requirements at the beginning of each iteration. As a result, all stakeholders have the opportunity to react to changing product needs and provide inputs on evolving product features.

One of the more popular variants of an Agile process is the Scrum approach. In a scrum team, the scrum master represents the developers and manages the development process. A product owner prioritizes features based on stakeholder needs; and the development teams directly engage the product owners and other relevant stakeholders continuously throughout the development process. Agreement on what features to develop in a sprint cycle is made during a planning meeting. In a sprint cycle, the development team strives to meet the feature commitments made during the planning meeting. Teams and stakeholders typically meet at the end of the sprint cycle to review progress and discover opportunities for continuous improvement.

Source: https://www.agilealliance.org/agile101/12-principles-behind-the-agile-manifesto/ accessed January 12, 2012

Exhibit 11.3

Kanban

Kanban is a Japanese term (看板) which literally translates to "signboard." Kanban as a production process has its roots in Lean manufacturing and the Toyota Production System (TPS). Kanban focuses on optimizing the flow of the work through the entire production process by managing the requests for additional components/work from upstream functions. These requests are made via kanban cards that downstream functions pass to upstream functions when more components are required. The cards are visible throughout the production process, so it is possible to literally see the flow of work by examining all of the cards and their queues. Kanban is one tool that is commonly used in Just in Time (JIT) manufacturing given additional inventory is only requested as it is needed. Kanban is also known as a "pull" system given downstream processes pull additional work from upstream functions through the use of these cards. Beyond the use of cards, a Kanban system includes the analysis and optimization of the production flow in order to decrease the waste (known as muda or 無駄) by imposing limits to the amount of work that is in process within the system and individual functions. This optimization seeks to improve the overall throughput of the production process.

Kanban as a software development methodology is a more recent adaptation of the traditional manufacturing application. Kanban in this context is often associated with Lean software development processes. In software, Kanban embraces the philosophy of the manufacturing implementation and applies it to various steps of developing a product or application. These steps often include user story creation, development, test and deployment. Unlike more traditional Agile processes like Scrum, Kanban does not utilize a *time box* for each development sprint. Instead, Kanban imposes work in progress limits for each development step. The overall process can be optimized by imposing and adjusting these limits. Team members focus on system bottlenecks to reduce the overall feature throughput time by eliminating waste (muda) throughout the process. Unlike other Agile processes which use velocity as their key metric, Kanban uses throughput for individual features as its key metric. Kanban has some similarities with more traditional Agile processes including allowing business owners to put off feature commitments until the feature enters development and seeking to allow teams to work together more collaboratively by making the workflow, which can be physical or components of a software application, visible on the Kanban board.

Source: D. Anderson, *Kanban: Successful Evolutionary Change for Your Technology Business*, Blue Hole Press, Sequim, 2010, pp. 11–16

Exhibit 11.4

Pearson Five Year Financial Performance from 2006–10 (in $ Millions except per Share)					
	2010	**2009**	**2008**	**2007**	**2006**
Sales	8,716.49	9,092.32	7,033.20	8,258.66	8,102.73
Cost of Goods Sold	3,715.63	3,847.75	2,991.05	3,625.32	3,619.49
Selling, General, and Administrative Expense	3,544.78	3,697.39	2,803.92	3,266.16	3,221.90
Operating Income Before Depreciation	1,456.08	1,547.18	1,238.23	1,367.18	1,261.34
Depreciation and Amortization	358.63	375.08	286.53	273.83	250.70
Interest Expense	126.21	148.74	154.96	226.21	229.16
Nonoperating Income (Expense)	76.96	43.65	58.48	61.51	131.23
Pretax Income	1,031.26	1,067.02	855.21	928.65	912.71
Income Taxes—Total	224.72	320.11	251.45	259.94	21.55
Income Before Extraordinary Items	814.24	687.10	558.45	617.12	846.12
Extraordinary Items and Discontinued Operations	1,182.11	0.00	(131.57)	(53.58)	27.42
Net Income (Loss)	1,996.34	687.10	426.88	563.54	873.54
Earnings Per Share (Primary)—Excluding Extraordinary Items	1.02	0.86	0.70	0.77	1.06
Earnings Per Share (Primary)—Including Extraordinary Items	2.49	0.86	0.54	0.71	1.10
Common Shares Used to Calculate Primary EPS	801.20	799.30	797.00	796.80	798.40

Source: Standard & Poor's Compustat Data, accessed November 12, 2011

Exhibit 11.5

SuccessMaker

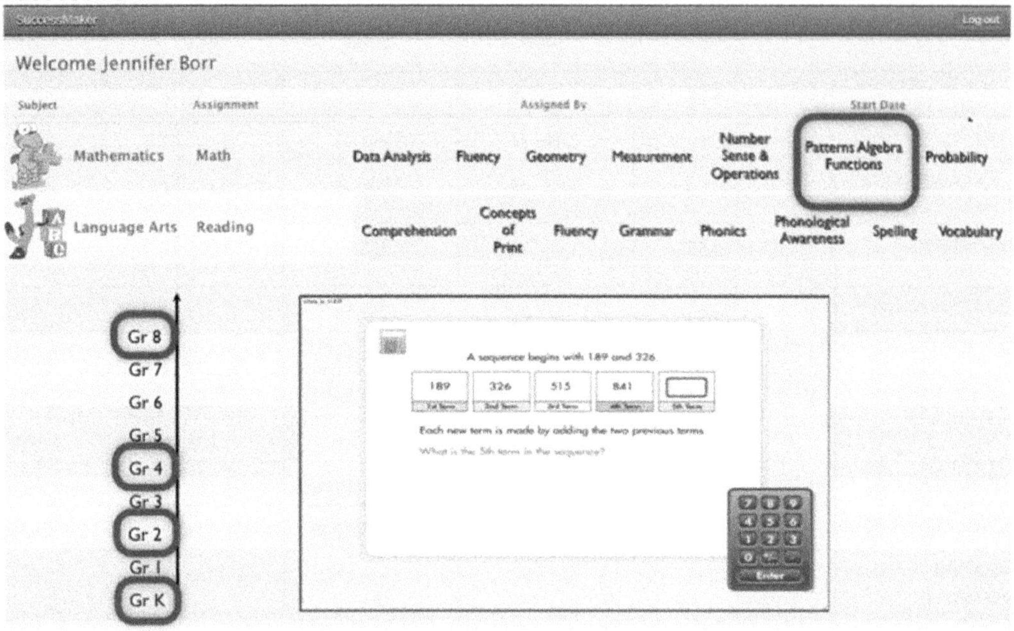

Grade 4		
Strand 1: Number and Operations		
Concept 1: Number Sense		
Understand and apply numbers, ways of representing numbers, and the relationships among numbers and different number systems.		
PO 1. Express whole numbers, fractions, decimals, and percents using and connecting multiple representations.	88%	8
PO 3. Express fractions as fair sharing, parts of a whole, parts of a set, and locations on a real number line.	100%	1
PO 5. Use simple ratios to describe problems in context.	100%	2
Concept 2: Numerical Operations		
Understand and apply numerical operations and their relationship to one another.		
PO 1. Add and subtract decimals through hundredths including money to $1000.00 and fractions with like denominators.	62%	69
PO 2. Use multiple strategies to multiply whole numbers		
two-digit by two-digit and	72%	25
multi-digit by one-digit.	63%	43
PO 3. Demonstrate fluency of multiplication and division facts through 12.	58%	134
PO 4. Use multiple strategies to divide whole numbers.	49%	102
Concept 3: Estimation		
Use estimation strategies reasonably and fluently while integrating content from each of the other strands.		
PO 1. Use benchmarks as meaningful points of comparison for whole numbers, decimals, and fractions.	82%	49
PO 2. Make estimates appropriate to a given situation or computation with whole numbers and fractions.	100%	4
Strand 2: Data Analysis, Probability, and Discrete Mathematics		
Concept 1: Data Analysis (Statistics)		

Source: Pearson Digital Learning

Case Study for Chapter 12

Fake News, Technology and Ethics: Can AI and Blockchains Restore Integrity?

Abstract

This teaching case explores the advantages and disadvantages of battling fake news with advanced information technologies, such as artificial intelligence (AI) and blockchains. Students will explore the purposes of, proliferation of, susceptibility to, and consequences of fake news and assess the efficacy of new interventions that rely on emerging technologies. Key questions students will explore: How can we properly balance freedom of speech and the prevention of fake news? What ethical guidelines should be applied to the use of AI and blockchains to ensure they do more good than harm? Will technology be enough to stop fake news?

Keywords

Blockchains, artificial intelligence and neuroscience, fake news, social media, human biases, ICTs and society

The Fake News Problem

Fake news is the deliberate spread of verifiably false information under the disguise of being an authentic news story. There are many types of fake news, including news satires and parodies intended to entertain. For example, Orson Welles' 1938 radio adaptation of H.G. Wells' *The War of the Worlds* was designed as entertainment in the format of a serious radio broadcast. The broadcaster (CBS), writers, and actors were surprised to learn that more than one million listeners (out of six million) thought it was an actual news bulletin (Hadley et al., 1940).

Beyond entertainment, this teaching case focuses on fake news fabrication, where there is no implicit understanding between the author and the audience that the story is concocted. We examine fake news designed to purposely misinform. The most common fake news fabrications are *political*, aiming to spread false information about politicians and public policies; *anti-scientific*, aiming to spread false information on such subjects as aliens, vaccines, diseases, cures, and nutrition; and *economic*, aiming to manipulate markets, stock prices, and to sell fraudulent products (Lazer et al., 2018).

The practice of creating and disseminating fake news for personal gain is not new. Consider these three stories from the past.

Back in the first century, Octavian became emperor of the Roman Empire, in part because he created fake news about his rival. Octavian portrayed Marcus Antonius (Mark Anthony) as Cleopatra's puppet and published a false document that claimed Anthony's last will requested that he be buried with the Ptolemaic pharaohs (MacDonald, 2017). In Ancient Rome, the dissemination of fake news would require handmade copies of documents and the use of 'praecones' (town criers). Many students might be familiar with the praeco from the HBO series *Rome*. The citizens of Rome faced the same challenge we face today: How could they know whether the news was legitimate?

As an example of scientific fake news, a reporter for the *New York Sun* in 1835 posted several stories that astronomers discovered a civilisation on the moon, including humanoids with bat wings. Many people believed the stories, and sales of the newspaper increased by 500%, which was the presumed aim of the story.[1]

Moving to an example of fake news designed to manipulate to the value of a stock, in 1814, Lord Cochrane, a Radical member of UK Parliament and a naval officer, along with other conspirators, bought £1.1 million in UK government-based stocks. Cochrane then proceeded to spread false news that Napoleon was dead in an attempt to raise stock values. (Napoleon lived until 1821.) The fake news was spread to the British Admiralty via telegraph and by word of mouth at many English inns. It worked. The stock price soared until the hoax was revealed. Lord Cochrane was imprisoned, fined, and stripped of his naval rank (Johnson, 2002).

Fast-forward to the 21st century, in which digital technologies have amplified the creation and distribution of fake news. Today, humans quickly propagate false information via Internet searches, comments on posted content, and by clicking 'like' and/or 'share'.

---

1 The original story can be found at the Internet Archives. Available at: www.archive.org/stream/moonhoax00Lock/moonhoax00Lock_djvu.txt (accessed 6 April 2021).

Lacity, M. Fake news, technology and ethics: Can AI and blockchains restore integrity? ***Journal of Information Technology Teaching Cases*** OnlineFirst, pp. 1–14. Copyright © 2021 by Association for Information Technology Trust. Reprinted by permission of SAGE Publications Ltd.

Figure 12.1. Can you spot the fake news? Many high school students in a Stanford University study could not.

Source: Wineburg et al. (2016). Taken from https://imgur.com/gallery/BZWWx. Originating from a Twitter post by @san_kaido.

Consider the massive dissemination of fake news stories during the 2016 United States (US) presidential election. (It will take us a while to track down the extent of fake news for the 2020 election.) The preponderance of these bogus stories were pro-Donald Trump and anti-Hillary Clinton (Silverman, 2016). In one study by economic professors at New York University and Stanford University, 115 pro-Trump fake stories were shared on Facebook 30 million times, and 41 pro-Clinton fake stories were shared 7.6 million times (Allcott and Gentzkow, 2017). Among the top fake stories were claims that Pope Frances endorsed Trump, which was shared or commented on 960,000 times and that WikiLeaks confirmed that Hillary Clinton sold weapons to ICIS, with 789,000 shares (Silverman, 2016). Both stories have since been removed. According to BuzzFeed news, there were more fake news story engagements (8.7 million) on Facebook than legitimate news story engagements (7.3 million) in the months leading to the US election (Silverman, 2016). Another academic study found that 25% of the 171 million tweets on Twitter in the 5 months prior to the 2016 election were fake or seriously erroneous news (Bovet and Makse, 2019). Although much of the misinformation was attributed to the British political consulting firm, Cambridge Analytica (Zuboff, 2019), Russia successfully inserted fake news on Facebook and Twitter during the 2016 US election according to testimony presented to a US Senate Judiciary Committee (2017). Academic scholars have done post-election surveys to determine whether voters were swayed by fake news, but a direct causal link from fake news exposure to voting behaviour is difficult to ascertain.

Of course, fake news is not limited to the United States; it is a worldwide phenomenon. In the United Kingdom (UK), the *Independent* reported in 2020 that Brittany Kaiser, prior business development director of Cambridge Analytica, claimed that the elections in at least 68 countries included many of the same disinformation tactics seen during the US presidential election and the UK's Brexit vote in 2016 (Wood, 2020).

Fake news stories have serious societal consequences. Fake news undermines the public's confidence in the free press and, moreover, when believed, results in a misinformed citizenry (Jang and Kim, 2018). A 2019 global survey of more than 25,000 Internet users found that 44 percent said their trust in media has declined as a consequence of fake news (Ipsos Public Affairs and the Centre for International Governance Innovation, 2019). It is not just the trust of the press that is at stake; fake news inflames social conflict; results in poor health outcomes (like swallowing bleach to avoid COVID19); gives rise to radicalism and terrorism; undermines the integrity of elections; manipulates markets; and so much more (Asian News International, 2020; Margetts, 2019; Muqsith, 2019). In short, fake news threatens the social trust we have in our institutions and in each other.

Why Do People Believe Fake News?

Academic studies have rigorously sought to uncover the scope of the problem and why some people believe fake news to be true. Stanford University researchers conducted field experiments and surveys with 7804 US students from middle school, high school, and universities. The study found that, across the board, most students were not able to differentiate between fake news and real news. Most high schoolers, for example, simply accepted the picture in Figure 12.1 as fact. More than 80% of middle schoolers believed that advertising content was a real news story (Wineburg et al., 2016). According to a 2019 global survey of more than 25,000 Internet users, 86% admitted to believing at least one fake news story (Ipsos Public Affairs and the Centre for International Governance Innovation, 2019).

Why do so many people believe what they see and read? While our coverage of the research here is far from exhaus tive, it highlights that humans are generally poor at identifying false information and that some personality traits and demographic factors affect this propensity.

Human Biases

More than 50 years of research on human decision-making shows that human behaviour systematically deviates from what rational choice models predict (Santos and Rosati, 2015). Instead, humans are highly influenced by a number of cognitive biases. According to an article in *Science*, '*research demonstrates that people prefer information that confirms their pre-existing attitudes (selective exposure), view information consistent with their pre-existing beliefs as more persuasive than dissonant information (confirmation bias), and are inclined to accept information that pleases them (desirability bias)*' (Lazer et al., 2018). With so many news sites, such biases may prompt individuals to select 'echo chambers'—that is, news sites that merely echo their ideologies.

One interesting twist on human biases is that many people recognise these partialities in others but underestimate their influences on themselves. An academic survey of 1299 people in the United States investigated individuals' perceptions about the effect of fake news on others. As predicted, individuals in this study thought that fake news had greater effects on out-group members (those from a different political party) than in-group members (Jang and Kim, 2018).

Repetition

Another academic study shows that individuals are more likely to accept familiar information as true, suggesting that the more a person is exposed to a fake news story, the more he or she will believe its contents (Swire et al., 2017). This makes the proliferation of fake news even more worrisome.

Personality Traits

While common vernacular would suggest that some people are more *gullible* than others, some academics call this trait 'reflexive open-mindedness'. Reflexive open-mindedness is defined as 'a tendency to unthinkingly accept incoming information as being valid and true' (Pennycook and Rand, 2020). In three studies with 1606 participants, professors from the University of Regina and MIT examined the associations of an individual's psychological profile with their perceptions of the accuracy of fake news. Across the studies, individuals with a general tendency of reflexive open-mindedness tended to believe fake news regardless of the familiarity of the news source or the presence/absence of the headline's source (Pennycook and Rand, 2020).

Demographic Factors

These studies illuminate the demographic factors of people who are better or worse at identifying fake news. For example, a study by professors at New York University and Stanford presented both real and fake news stories about the 2016 US presidential election to people after the election. Respondents were asked to judge whether the stories were true or false. Respondents who spent more time consuming news, those with higher education, and older people were better at discerning real and fake news (Allcott and Gentzkow, 2017).

Traditional Means to Combat Fake News

The commonly used solutions to combat fake news today are deployed after the fact—they aim to detect or punish fakes after the stories have already been disseminated. These include the following:

- *Fact-checking websites.* According to Duke University, there are more than 100 nonpartisan groups worldwide that provide fact-checking services (Duke Reporters' Lab, n.d.). These include professional sites like FactCheck.org, Snopes.com, and BBC's Reality Check.
- *Legal liability.* The United Nations (1948) Universal Declaration of Human Rights places freedom of the press as a human right: *'Everyone has the right to freedom of opinion and expression; this right includes freedom to hold opinions without interference, and to seek, receive, and impart information and ideas through any media regardless of frontiers'*. Many countries protect the freedom of the press in their constitutions. For example, the First Amendment to the US Constitution guarantees freedom of the press in America.[1] Because a free press is a hallmark of any democracy, government regulations to prevent fake news are not generally welcome as a solution. Today, a defamation lawsuit by someone hurt by the story is the main legal recourse against fake news. Defamation liability also extends to anyone who republishes it (Haskins, 2019). It is quite clear, however, that the threat of litigation is an inadequate deterrent.
- *Education.* As suggested by the demographic factors that contribute to a person's propensity to believe untruths, an educated citizenry is a powerful tool to combat fake news. Numerous examples of lesson plans for people of all ages are available online. Readers can take action by fact checking stories themselves, unsubscribing to news from specific outlets, changing the way they use social media, and flagging a story they suspect is made up.

Can Advanced Information Technologies Combat Fake News?

Here, we explore two advanced information technologies to combat fake news—namely artificial intelligence and blockchain technologies. For readers unfamiliar with these technologies, Appendix A provides an overview of AI, and Appendix B provides an overview of blockchain technologies.

AI Solutions

Social media platform providers are relying on AI tools to identify and prevent the spread of fake news. AI solutions are designed to ingest and categorise large amounts of data, a task that is too expensive to do by humans on social media platforms with millions of posts per day. Many AI solutions are designed to detect and flag fake news by searching for sensational words, by checking the geolocation of the source, and by assigning

1 In the United States, the First Amendment to the Constitution guarantees freedom of the press. 'Congress shall make no law respecting an establishment of religion, or prohibiting the free exercise thereof; or abridging the freedom of speech, or of the press; or the right of the people peaceably to assemble, and to petition the Government for a redress of grievances'.

reputation scores to particular sites based on the accuracy of facts (Maruti Techlabs, 2020).

Mark Zuckerberg, Facebook's founder and CEO, for example, mentioned AI as a solution to fake news more than 30 times during his testimony to Congress after the 2016 presidential election (Woolley, 2020). Jump ahead to 2020, and Facebook was using AI to block fake news about COVID-19. By April 2020, Facebook had deleted 50 million posts full of falsehoods about the coronavirus (Some, 2020).

Twitter reported in 2019 that use of AI was identifying and removing 50% of abusive tweets (Newcomb, 2019). However, Twitter does not always delete fake news posts, but sometimes simply flags them. For example, Twitter flagged hundreds of Donald Trump's posts as 'disputed' after the 2020 presidential election.

Following the insurrection at the US Capitol on January 6, 2020, by Trump supporters, Twitter, Facebook, Snapchat, Instagram, YouTube, and Shopify restricted/blocked Trump's social media accounts. Such actions by social media platforms beg a deeper debate about freedom of speech versus the prevention of illegal actions spawned by fake news (Hutchinson, 2021). Where should the line be drawn?

We also note that AI contributes to the problem of fake news: Deepfakes are content manipulated by artificial intelligence, such as a video altered to make the speaker appear to say something she did not say. Microsoft launched a deepfake detector for videos and images in September 2020 (Burt and Horvitz, 2020). This Video Authenticator uses AI to analyse a still photo or video to provide a percentage chance, or confidence score, that the media is artificially manipulated. It works by detecting elements imperceptible to humans. The DeepTrust Alliance (a 501[c]) is also tackling deepfakes. Kathryn Harrison, former Director of Global Product Management for IBM Blockchain Platform, started DeepTrust because she believed a coalition of industry, civil society, and services is needed to 'create the standards that verify the origins of content to combat digital fakes' (DeepTrust Alliance, n.d.). So, here we have examples of good AI fighting bad AI.

But, the battle is young. So far, the AI-driven solutions, such as Facebook's Deeptext and Google's Perspective software, are better at tagging hate speech than detecting fake news (Woolley, 2020).

More recently, some news media outlets have turned to blockchain technologies.

Blockchain Solutions

Gartner, a Global IT research and advisory company, estimates by 2023, 30% of world news (including videos) will rely on blockchain technologies for authentication (Gartner, 2019). There will likely be multiple blockchain-enabled solutions that provide services such as establishing content authenticity, tracking provenance of content over time, blacklisting imposters, spotting deepfakes, and tying digital content to the physical world—for example, by tagging the GPS location of a photo (Mearian, 2020).

Many of these solutions aim to prevent *imposters*, a type of fake news story that appears to be from a legitimate news agency (Wardle Claire, 2017). According to an article in *Science*, imposters are 'particularly pernicious in that it is parasitic on standard news outlets, simultaneously benefiting from and undermining their credibility' (Lazer et al., 2018).

ANSAcheck is one such example of using blockchain technology to prevent imposters. It was developed by Agenzia Nazionale Stampa Associata (ANSA), Italy's top news wire service, along with Ernst & Young (EY), one of the world's largest professional services firms. Like most legitimate news outlets, ANSA has suffered from several imposter stories. In March 2020, for example, there were at least three imposter stories related to COVID-19. One falsely reported Italy's first death as that of a 24-year-old female. Another announced false homemade COVID-19 cures (Sephton, 2020). And a third falsely reported on COVID-19 policies by the Italian government. The fake stories were distributed using the ANSA brand, format, and signature (De Alessandri and Perrone, 2020).

Stefano De Alessandri, ANSA's CEO and managing director, said,

> Fake news is one of the biggest challenges facing traditional media organisations and social media platforms, as it undermines the trust they have built with the public and advertisers, undermining their strategic asset that is their reputation. If we lose trust, we lose everything. (De Alessandri and Perrone, 2020)

Instances like these prompted ANSA to launch the ANSAcheck project. The goal was to create an innovative solution to provide a guarantee for the origin of an ANSA-created story and for that story to be traced through its entire history of updates and reposts. EY knew that blockchain technologies could provide the tamper-resistant authenticity of the story, traceability of the story over time, and would allow readers or publishers to verify the story at any point in time.

In Phase I, completed in March 2020, ANSAcheck was deployed on ANSA's website. The system works by assigning a unique hash ID to an ANSA-created news story.[1] If even one letter in the story is changed, the system will detect that it is not identical to the original story. Story IDs are batched and posted every 15 min to half-hour to Ethereum, a public blockchain network, thereby creating a permanent record of

---

1 A hash is an algorithm for transforming one input into a different output. Given a specific input, the identical output will always be reproduced. A good hash algorithm makes it practically impossible to determine the input value based on the output value, which is why hashes are called one-way functions. ANSAcheck uses the MD5 hashing algorithm, which always produces a 128-bit output.

Figure 12.2. ANSAcheck sticker This sticker appears on all ANSA stories posted on ANSA's website.

Source: www.ansa.it*.

*ANSAcheck is language sensitive, so viewers who translate the website from Italian to another language will not see the specific stickers adhered to each news story; English-translation users will just see an explanation of the ANSAcheck solution as depicted in Figure 12.4.

the story on the public blockchain. The blockchain creates a tamper-resistant record that includes the story ID, transaction ID, timestamp, and the block number on Ethereum's ledger where the transaction is stored.

EY selected Ethereum because they believe it is the safest and most mature platform, with 49 million addresses, 227,000 smart contracts, and 10,000 network nodes by 2020. Ethereum was launched as an open network in 2015. It is supported by Ethereum Foundation, a non-profit organisation based in Switzerland, as well as a vast global network of volunteer developers, node operators, and users.

Each ANSA story that is secured on the blockchain is accompanied with an ANSAcheck sticker (see Figures 12.2 and 12.3) to signal its authenticity to readers. Users can click on the ANSACheck sticker to query Ethereum about the source of the story. If ANSA updates the story, another entry is recorded on Ethereum and linked back to the original entry to form a chain of provenance. The solution is registering about 3000 news stories per day.

Mr. De Alessandri said of Phase I,

> It's not the definitive solution on fake news, but rather the first pass. ANSA is a serious company that employs professional journalists and checks everything that we publish. This first pass does not guarantee the complete accuracy of the contents; rather, it guarantees that this piece of news comes from ANSA, not from someone else. (De Alessandri and Perrone, 2020)

The solution will be enhanced with new features and launched on a platform as learning is incorporated from Phase I.

For Phase II, EY and ANSA are inviting other publishers to join the platform. As of September 2020, EY was working with three other Italian publishers who repost ANSA stories. The ANSAcheck solution is expanding its reach not only throughout Italy, but also to other publishers and social media platforms in the United States and Europe. Many enterprises, including Facebook, have seen demonstrations of the solution.

EY plans to add other services to the ANSAcheck solution. Giuseppe Perrone, EY's blockchain HUB MED leader

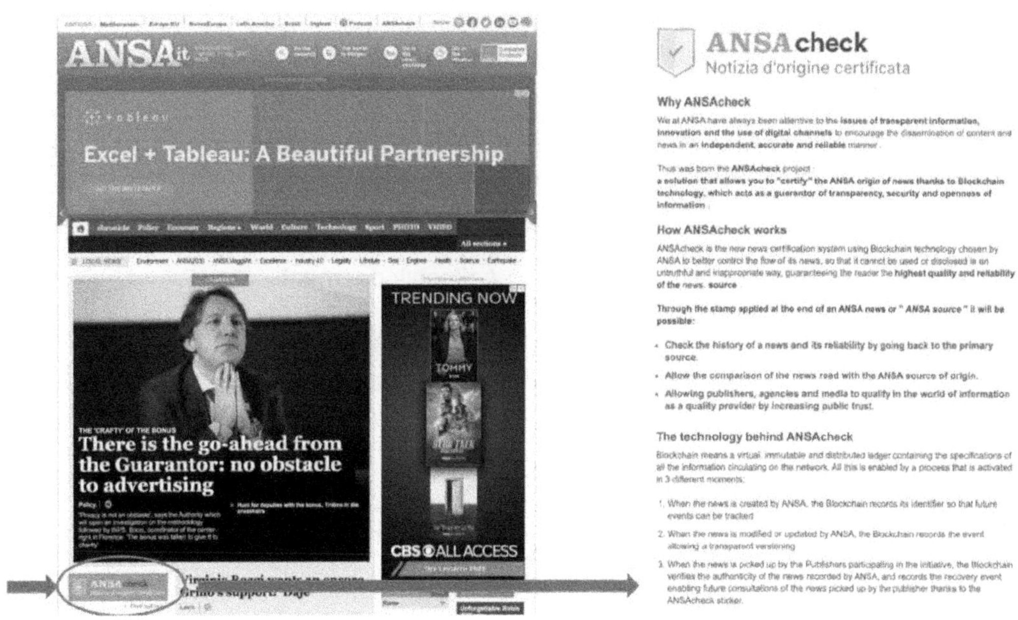

Figure 12.3. ANSA Homepage with pointer to ANSAcheck description (August 11, 2020).

Source: www.ansa.it (Google translation to English).

said, 'The solution will become more sophisticated in terms of functionality and components, such as fact-checking functions, semantic language analysis, and picture data protection'. In the first generation of the solution, a story must be an identical copy of the original, as a change to even a single letter will signal that the source is not the original. In the next generation, the solution will provide a measure of overlap, similar to the plagiarism software familiar to many students. This way, publishers who repost a story can add their own commentary while still verifying the original news source.

Other blockchain solutions include Poex.io and Po.et. Each provides a simple 'proof of existence' service on the Bitcoin blockchain.[1] Creators of content can hash the content and post the hash on the Bitcoin blockchain to demonstrate when the content first existed.

Other solutions are tackling the provenance of visual content, such as photos that accompany news stories. For example, *The New York Times* is working with IBM as its technical partner on the News Provenance Project (NPP). This project aims to protect the authenticity of visual content. In 2019, the NPP conducted interviews to determine how readers decide whether a photo is credible. This research helped to identify several design principles, such as displaying multiple perspectives with their associated source.[2]

Of course, blockchain technologies also have the potential to exacerbate the problem. People can transact on public blockchains anonymously, which is why criminals and terrorists use Bitcoin and other cryptocurrencies to finance illegal operations. Silk Road is perhaps the earliest and most visible example of a nefarious use of blockchains. Silk Road was an anonymous marketplace founded by Ross William Ulbricht in 2011, when he was just 27 years old (Popper, 2015). The marketplace combined the anonymity of Bitcoin with the obscurity of Tor, the network protocol that masks the identity of servers. (One can spot a Tor website by the website address's suffix '.onion'). One can hardly believe the audacity of Silk Road's openness (see Figure 12.4 for a screenshot showing illegal drugs for sale). After an intense search that comprised cyber and behavioural investigative methods—such as posing as drug buyers—the US Federal Bureau of Investigation (FBI) arrested Ulbricht in 2013. He was sentenced to life in prison without the possibility of parole (Weiser, 2015). This story is nuanced in a way that will not be covered here, but the documentary film *Deep Web*, directed by Alex Winter, provides thorough coverage of the rise and fall of Silk Road.

In summary, while AI and blockchain solutions are advancing the ability to protect real news and minimise fake news, they may also contribute to its spread. A combination of ethically guided human and technical interventions are ultimately needed to protect confidence in a free and independent press.

Case Discussion Questions

1. One of the key findings from human bias research is that many of us think other people are more susceptible to fake news than we are ourselves. Can you think of a fake news story that you initially believed to be true but later discovered it was verifiably false? Can you detect the computer-generated faces from real photographs on the 'Which Face is Real' game at www.whichfaceisreal.com/index.php?
2. To what degree can AI and blockchains prevent fake news today?
3. How might AI and blockchains actually contribute to fake news and polarisation of opinions?
4. How can we get the balance right between freedom of speech and the prevention of fake news? What ethical guidelines (see Appendix C) should be applied to the use of AI and blockchains to ensure they do more good than harm?
5. No single solution will eradicate fake news. All parties in the ecosystem, including governments, journalists, publishers, social media platforms, and citizens, have roles to play. What ideas, technical and nontechnical, do you have for preventing the creation and dissemination of fake news?
6. What is the best way to approach a family member or friend who believes a fake news story? Is it possible to have a productive conversation, or is it best to avoid the topic?

Acknowledgements

This teaching case was based upon, with significant extensions, the following sources:

- Lacity M and Conway D (2020). *Authenticating Real News with ANSAcheck, a Blockchain-Enabled Solution Developed by ANSA and EY* (Research Paper Series). Fayetteville, AR: Blockchain Centre of Excellence (BCoE), University of Arkansas.
- Appendix A: Lacity M and Willcocks L (2018). *Robotic Process and Cognitive Automation*. Stratford-upon-Avon: SB Publishing.
- Appendix B & C: Lacity M (2020) *Blockchain Foundations for the Internet of Value*. Fayetteville, AR: Epic Books/ University of Arkansas Press.

---

1 Sources: https://poex.io/; https://www.po.et/

2 Sources: https://www.newsprovenanceproject.com/about-npp; https://www.youtube.com/watch?v=GfZ1xGc3n5Q; https://www.ansa.it/sito/static/ansa_check.html; https://journalistsresource.org/studies/society/Internet/fake-news-conspiracy-theories-journalism-research/; https://papers.ssrn.com/sol3/papers.cfm?abstract_id=3023545; https://onlinelibrary.wiley.com/doi/epdf/10.1111/jopy.12476; https://pubs.aeaweb.org/doi/pdfplus/10.1257/jep.31.2.211

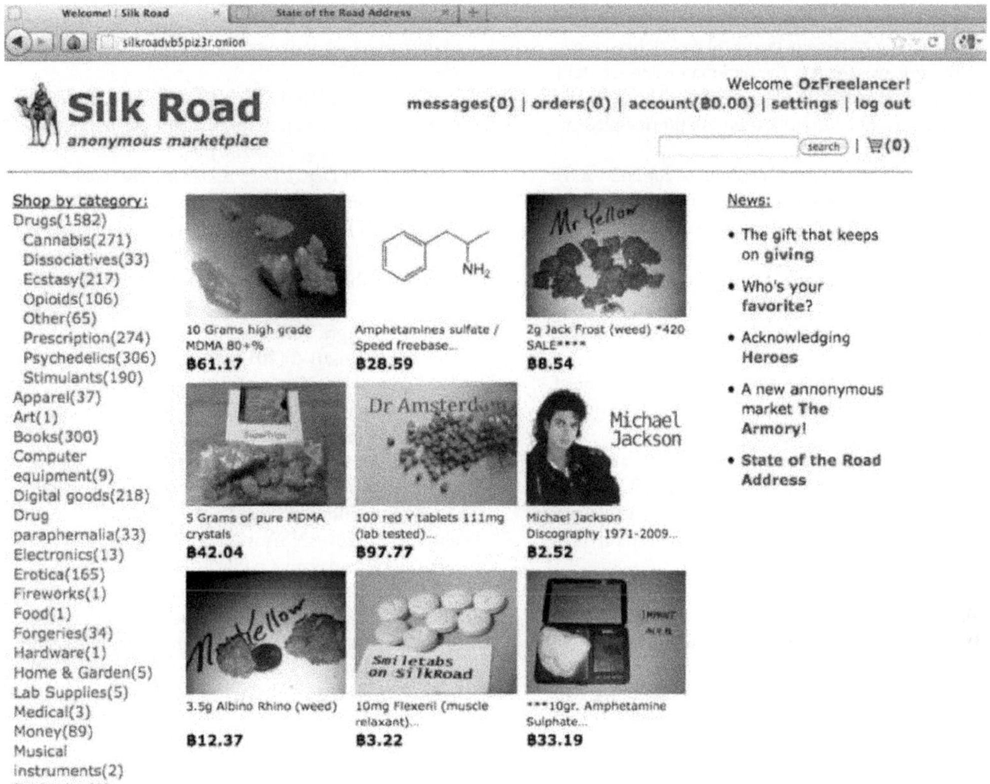

Figure 12.4. Silk Road Website screenshot.

Source: Techrepublic (Silk Road, n.d.; Reese, 2017).

Declaration of Conflicting Interests

The author(s) declared no potential conflicts of interest with respect to the research, authorship and/or publication of this article.

Funding

The author(s) received no financial support for the research, authorship, and/or publication of this article.

ORCID ID

Mary C Lacity https://orcid.org/0000-0003-3363-7086

References

Allcott H and Gentzkow M (2017) Social media and fake news in the 2016 election. *Journal of Economic Perspectives 31*(2): 211–236.

Asian News International (2020) Fake social media accounts root cause of riots; plea in SC seeks to weed them out. *Asian News International*, 4 March. Available at: https://www.business-standard.com/article/news-ani/fake-social-media-accounts-root-cause-of-riots-plea-in-sc-seeks-to-weed-them-out-120030400805_1.html (accessed 6 April 2021).

Bovet A and Makse H (2019) Influence of fake news in Twitter during the 2016 US presidential election. *Nature Communications 10*: 7. Available at: www.nature.com/articles/s41467-018-07761-2 (accessed 6 April 2021).

Burt T and Horvitz E (2020) New steps to combat disinformation. *Microsoft Press Release*, 1 September. Available at: https://blogs.microsoft.com/on-the-issues/2020/09/01/disinformation-deepfakes-newsguard-video-authenticator/ (accessed 6 April 2021).

Daniels A (2018) The rise of private permissionless blockchains. *Medium*, 18 October. Available at: www.medium.com/ltonetwork/the-rise-of-private-permissionless-blockchains-part-1-4c39bea2e2be (accessed 6 April 2021).

De Alessandri S and Perrone G (2020) Going public: EY Global Blockchain Summit 2020. Available at: https://assets.ey.com/content/dam/ey-sites/ey-com/en_au/pdfs/going-public-how-public-blockchains-will-create-exponential-growth.pdf (accessed 19 June 2021).

DeepTrust Alliance (n.d.). Available at: www.deeptrustalliance.org/about (accessed 6 April 2021).

Duke Reporters' Lab (n.d.). Available at: www.reporterslab.org/fact-checking/ (accessed 6 April 2021).

EY Blockchain/Nightfall (n.d.). Available at: www.github.com/EYBlockchain/nightfall (accessed 6 April 2021).

Fjeld J, Achten N, Hilligoss H, et al. (2020) Principled artificial intelligence: Mapping consensus in ethical and rights-based approaches to principles for AI. The Berkman Klein Center for Internet & Society. Research Publication No. 2020–1. Available at: https://cyber.harvard.edu/publication/2020/principled-ai (accessed 6 April 2021).

Gartner (2019) Gartner top strategic predictions for 2020 and beyond. Available at: www.gartner.com/smarterwithgartner/gartner-top-strategic-predictions-for-2020-and-beyond/ (accessed 6 April 2021).

Hadley C, Gaudet H and Herzog H (1940) *The Invasion from Mars: A Study in the Psychology of Panic with the Complete Script of the Famous Orson Welles Broadcast*. Princeton, NJ: Princeton University Press.

Haskins J (2019) Fake news: What laws are designed to protect. Available at: www.legalzoom.com/articles/fake-news-what-laws-are-designed-to-protect (accessed 6 April 2021).

Hutchinson A (2021) President Trump banned from various social networks after US capitol riots. *SocialMediaToday*, 7 January. Available at: www.socialmediatoday.com/news/president-trump-banned-from-various-social-networks-after-us-capitol-riots/593007/ (accessed 6 April 2021).

Ipsos Public Affairs and the Centre for International Governance Innovation (2019) CIGI-Ipsos global survey: Internet security & trust. Available at: www.cigionline.org/internet-survey-2019 (accessed 6 April 2021).

Jang S and Kim J (2018) Third-person effects of fake news: Fake news regulation and media literacy interventions. *Computers in Human Behavior 80*(1): 295–302.

Johnson P (2002) *Civilising Mammon: Fraud and Profit in Nineteenth-Century*. London: London School of Economics and Political Science. Available at: www.web.archive.org/web/20110223134926/http://www.fathom.com/feature/121984/ (accessed 6 April 2021).

Lazer D, Baum M, Benkler Y, et al. (2018). The science of fake news. *Science 359*(6380): 1094–1096.

MacDonald E (2017) The fake news that sealed the fate of Antony and Cleopatra. *The Conversation*, 13 January. Available at: https://theconversation.com/the-fate-of-antony-and-cleopatra-71287 (accessed 19 June 2021).

Margetts H (2019) Rethinking democracy with social media. *The Political Quarterly 90*(1): 107–123.

Maruti Techlabs (2020) Is artificial intelligence the key to combat fake news? Available at: www.marutitech.com/artificial-intelligence-fake-news/ (accessed 6 April 2021).

Mearian L (2020) How blockchain could help block fake news. *ComputerWorld*, 17 February. Available at: www.computerworld.com/article/3526427/how-blockchain-could-help-block-fake-news.html (accessed 6 April 2021).

Muqsith M (2019) Effect fake news for democracy. *Jurnal cita hukum 7*(3): 307–318.

Newcomb A (2019) Twitter says A.I. is now removing over half of its abusive tweets before they're flagged. *Fortune Magazine*, 24 October. Available at: www.fortune.com/2019/10/24/twitter-abuse-tweets/ (accessed 6 April 2021).

PBS(2012) Smartestmachineon Earth. *NOVADocumentary*, 2 May. Available at: www.youtube.com/watch?v=X0JZMHhupTs; www.pbs.org/wgbh/nova/tech/smartest-machine-on-earth.html (accessed 6 April 2021).

Pedersen A, Risius M and Beck R (2019) A ten-step decision path to determine when to use blockchain technologies. *MIS Quarterly Executive 18*(2): 3.

Pennycook G and Rand D (2020) Who falls for fake news? The roles of bullshit receptivity, overclaiming, familiarity and analytic thinking. *Journal of Personality 88*: 185–200.

Popper N (2015) *Digital Gold: Bitcoin and the Inside History of the Misfits and Millionaires Trying to Reinvent Money*. New York: Harper.

Reese H (2017) How the founder of the Silk Road made millions on his illegal startup on the Dark Web. *TechRepublic*, 10 May. Available at: www.techrepublic.com/article/how-online-marketplace-silk-road-became-the-craigslist-for-illegal-drugs/ (accessed 6 April 2021).

Santos L and Rosati A (2015) The evolutionary rots of human decision making. *Annual Review of Psychology 66*: 321–347.

Senate Judiciary Committee (2017) Extremist content and Russian disinformation online: Working with tech to find solutions. Available at: www.judiciary.senate.gov/meetings/extremist-content-and-russian-disinformation-online-working-with-tech-to-find-solutions (accessed 6 April 2021).

Sephton C (2020) ANSA will put the sources and history of 1000 news stories a day on a publicly available blockchain, but bigger challenges lie ahead. *Modern Consensus*, 8 April. Available at: www.modernconsensus.com/people/media/its-brand-used-to-spread-fake-news-italian-media-outlet-turns-to-blockchain/ (accessed 6 April 2021).

Silk Road (n.d.). Available at: https://tr2.cbsistatic.com/hub/i/r/2017/05/10/709e488c-6c51-407f-ae13-5115b14d86c4/resize/770x/ac1af617d15f84b986b574770d9a67de/screen-shot-2012-04-24-at-2-02-25-am.png (accessed 6 April 2021).

Silverman C (2016) This analysis shows how fake election news stories outperformed real news on Facebook. *BuzzFeed News*, 16 November. Available at: https://www.buzzfeednews.com/article/craigsilverman/viral-fake-election-news-outperformed-real-news-on-facebook (accessed 19 June 2021).

Sims G (2015) What is machine learning? *Android Authority*, 7 July. Available at: www.androidauthority.com/what-is-machine-learning-621659/ (accessed 6 April 2021).

Some K (2020) Facebook uses AI to fight coronavirus misinformation and fake news. Available at: www.analyticsinsight.net/facebook-uses-ai-fight-coronavirus-misinformation-fake-news/ (accessed 6 April 2021).

Swire B, Ecker UKH and Lewandowsky S (2017) The role of familiarity in correcting inaccurate information. *Journal of Experimental Psychology: Learning, Memory and Cognition 43*(12): 1948–1961.

United Nations (1948) Universal declaration of human rights. Available at: www.un.org/en/about-us/universal-declaration-of-human-rights (accessed 6 April 2021).

Wardle Claire (2017) Fake news. It's complicated. *First Draft*, 16 February. Available at: www.firstdraftnews.org/latest/fake-news-complicated/ (accessed 6 April 2021).

Weiser B (2015) Ross Ulbricht, creator of Silk Road Website, is sentenced to life in prison. *The New York Times*, 29 May. Available at: www.nytimes.com/2015/05/30/nyregion/ross-ulbricht-creator-of-silk-road-website-is-sentenced-to-life-in-prison.html (accessed 6 April 2021).

Wineburg S, McGrew S, Breakstone J, et al. (2016) Evaluating information: The cornerstone of civic online reasoning. *Stanford*

Digital Repository. Available at: http://purl.stanford.edu/fv751yt5934 (accessed 6 April 2021).
Wood V (2020) Fake news worse during election campaign than Brexit referendum, whistleblower says. *The Independent*, 27 January. Available at: www.independent.co.uk/news/uk/politics/brexit-fake-news-2019-election-facebook-cambridge-analytica-brittany-kaisar-eu-referendum-a9304821.html (accessed 6 April 2021).
Woolley S (2020) We're fighting fake news AI bots by using more AI. That's a mistake. *MIT Technology Review*, 8 January. Available at: www.technologyreview.com/2020/01/08/130983/were-fighting-fake-news-ai-bots-by-using-more-ai-thats-a-mistake/ (accessed 6 April 2021).
Zuboff S (2019) *The Age of Surveillance Capitalism*. New York: Public Affairs.

Author Biography

Mary C Lacity is Walton Professor of Information Systems and Director of the Blockchain Centre of Excellence in the Sam M. Walton College of Business at The University of Arkansas. She was previously Curators' Distinguished Professor at the University of Missouri-St. Louis. She has held visiting positions at MIT, the London School of Economics, Washington University, and Oxford University. She is a Certified Outsourcing Professional ® and Senior Editor for *MIS Quarterly Executive*. Her recent research focuses on improving business services using Robotic Process Automation (RPA), Cognitive Automation (CA), and Blockchain technologies. She has conducted case studies and surveys of hundreds of organisations on their adoption journeys. She has given keynote speeches and executive seminars worldwide and has served as an expert witness for the U.S. Congress. She was inducted into the IAOP's Outsourcing Hall of Fame in 2014, one of only three academics to ever be inducted. She has published 30 books, most recently, Blockchain Foundations for the Internet of Value (2020) and Becoming Strategic with Robotic Process Automation (2019). Her publications have appeared in the *Harvard Business Review, Sloan Management Review, MIS Quarterly, MIS Quarterly Executive, IEEE Computer, Communications of the ACM*, and many other academic and practitioner outlets. According to Google Scholar, her work has been cited over 19,000 times, with an h-index of 58.

Appendix A

Overview of AI

Fans of science fiction are familiar with the notion of 'artificial intelligence'—the idea that machines are capable of human-level (and even greater) intelligence. Popular AI figures include C3PO and R2D2 of *Star Wars*; the character Data of *Star Trek Next Generation*; and the 'Terminator' from the *Terminator* movie franchise. These characters from science fiction display 'strong' AI characteristics, meaning they have general intelligence that enables them to function competently across many knowledge domains and adapt quickly to new environments. In the real world, AI technologies are still quite far away from producing the capabilities we see in science fiction. And when one understands how they work, AI technologies can be quite disappointing compared to our science fiction characters.

To understand why AI is different, we must first know how traditional computer programming works. Traditional programming applies logic rules to process structured data to produce a single correct answer. Logic rules are deterministic, in the form of IF-THEN-ELSE statements, as well as statements to repeat the logic steps on a set of similar data (called a logic loop). For example, a customer might be asked to press the number '2' on a phone keypad if they lost their password. The computer system would be programmed with IF-THEN logic such as, 'If a customer presses '2', then the customer needs a password reset'.

AI technologies rely on a new way of programming computers called machine learning. Machine learning is a way to programme computers so that they perform tasks competently based on prior examples, not just traditional logic rules like IF-THEN-ELSE. With machine learning, the computer is fed data, often in an unstructured form like natural language text or voice, and inference-based algorithms search for patterns in the data to produce probabilistic outcomes, such as, 'There is a 95 percent probability that this customer forgot their password'.

Machine learning algorithms can become quite competent, but only in narrowly defined knowledge domains, such as playing a game, processing customer calls to a help desk, or categorising images. For example, IBM's Watson was programmed to play *Jeopardy!*, a US game show where humans compete to ask trivia questions after they are presented with the clues. iPsoft's virtual assistant Amelia aims to automate customer service interactions, just like a human agent would do, but with higher speed, efficiency, and consistency. Amelia can read 300 pages in 30 seconds and manage thousands of conversations simultaneously. Such examples are called 'weak' or 'narrow' AI because they only function capably in limited contexts (see Figure 12.5).

While there are many types of machine learning concepts, here we focus on supervised and unsupervised machine learning.

Supervised Machine Learning

With supervised machine learning, the inputs to the computer include data labels for each example. When the computer is input with new data it has not encountered, the algorithm[1] instructs the computer to match the new input to the closest pattern.[2] Supervised machine learning makes it easier for the computer to perform well, but it may be prohibitively expensive to label all examples.

---

1 Many supervised machine-learning algorithms use decision trees and Bayesian statistics, where the probability of a hypothesis being true is updated as more data are introduced.

2 In actuality, there are many other types of machine learning algorithms; Wikipedia lists nearly 200 common algorithms. Available at: https://en.wikipedia.org/wiki/Outline_of_machine_learning (accessed 6 April 2021).

IBM's Watson playing *Jeopardy!* in 2011

iPsoft's Virtual Assistant 'Amelia'

Figure 12.5. Two examples of 'weak' AI programming.

Sources: IBM's Watson playing ***Jeopardy!***—https://en.wikipedia.org/wiki/Watson_(computer). iPsoft's Virtual Assistant 'Amelia'—supplied by Amelia, an iPsoft Company (https://amelia.com).

IBM's Watson is an example of supervised machine learning. In 2004, Charles Lickel, IBM's research manager, pitched the idea of building a computer that could play *Jeopardy!* The game is complex; it requires fast (under 1 s) processing of unstructured, natural language. IBM spent years on the challenge using traditional computer programming, but it did not make significant progress until they applied supervised machine learning. Watson was loaded with every previous *Jeopardy!* clue/answer pair, which numbered more than 10,000. Watson hunted through the pairs to extract patterns about correct clues-answers, as well as evidence to support the correct answers. Watson used statistics to weight the importance of evidence in a clue. Take, for example, the *Jeopardy!* clue, 'Keanu Reeves had a Nokia phone, but it took a landline to slip in and out of this, the title of a 1999 sci-fi flick'. Watson recognized–based on patterns of previous questions—to heavily weight the evidence, '1999 science fiction movies' and 'Keanu Reeves' and to lightly weight the evidence about the phones. Watson correctly answered, 'What is The Matrix?' After supervised machine learning, Watson went on to become a *Jeopardy!* champion by beating the two human champions, Brad Rutter and Ken Jennings, in February 2011.[1]

Unsupervised Machine Learning

With unsupervised machine learning, there are no data labels. Instead, a machine-learning algorithm extracts patterns based only on the data. Unsupervised machine-learning algorithms are commonly used to manage large amounts of data, such as categorising and labelling images on the Internet, for example, 'This is likely an image of a man'. Unsupervised machine-learning algorithms typically need massive amounts of examples, or data, to perform competently (Sims, 2015).[2]

Overall, many of the underlying statistical algorithms used in machine learning have been around for decades including Bayesian, regression, and clustering statistics—but we only recently advanced computational power to execute them. Computing techniques like Hadoop[3] and specialised hardware like Google's Tensor Processing Unit (TPU)[4] provide the horsepower we lacked a decade ago.

Appendix B

Overview of Blockchain Technologies

Bitcoin, the first blockchain application released in January 2009, and other blockchain innovations that followed are moving us from an 'Internet of information' to an 'Internet of value', where people transact value—money and other assets—directly with each other. This pivot requires innovative solutions to some very old problems—like counter-party risks, the risks each trading party bears that the other party will not fulfil its contractual obligations—as well as to newer problems like

1 See PBS (2012). Access to either of these sources depends upon geographical location.

2 Unsupervised learning algorithms commonly use cluster analysis, anomaly detection, principle components analysis or generative topographic maps.

3 Hadoop is a programming framework that distributes data across multiple machines for faster processing and replicates it multiple times to prevent failures. Hadoop-enabled queries can get answers in seconds compared to an hour or more in a pre-Hadoop world.

4 Google's TPU units were used in the AlphaGo programme that beat Lee Se-dol in the game Go in 2016.

cybercrime. Before blockchains, banks, credit card companies, money transmitters, notaries, lawyers, and other trusted third parties (TTPs) mitigated counter-party risks. TTPs provide independent truth attestations, such as making sure accounts have sufficient funds before transferring value; notarising signatures; verifying identity; verifying ownership; authenticating assets; and attesting that agreements have been properly executed.

After blockchains, we rely on (1) technologies like distributed ledgers, smart contracts, cryptography, and digital assets and (2) an incentivised community to solve these problems. The value proposition is that TTPs, which are expensive, opaque, and slow, are no longer needed to mitigate counterparty risks. The key features of blockchain technologies are:

- *Distributed ledger upon which all parties agree*: As a component of a blockchain application, a distributed ledger is a time-stamped, permanent record of all valid transactions that have occurred within a given blockchain application. Each node of the blockchain network has an identical copy; no node is in charge. The nodes in the network constantly chatter with each other to make sure no party tampers with the records after the fact. If anyone cheats, the other parties' nodes automatically ignore the node.
- *Automation*: Pieces of software called 'smart contracts' store and execute predefined agreements among trading partners. These contracts specify the rules for exchanging 'digital assets' that represent either (1) a fungible (nonunique) asset, such as a cryptocurrency, loyalty rewards, and airline frequent flyer miles in which tokens are interchangeable; or (2) a nonfungible (unique) asset where the token represents a particular asset in the real world, such as a plot of land, a work of art, or a news story.
- *Shared governance*: No single party controls the blockchain network. The community shares decision-making rights.
- *Immutability*: Once a transaction is approved and added to the distributed ledger, it cannot be changed (unless the community agrees to roll back the ledger in the rare instance of a severe error).
- *Heightened security*: Blockchain applications still function properly, even if a high percentage of nodes are faulty—even malicious—enabling fault tolerance, resiliency, and 100% availability. In theory, the only way to break a blockchain application is to commandeer more than 50% of the nodes.

While Bitcoin was the first blockchain application, there are now hundreds of blockchain networks. These can be categorised into four types, based on rights of participation and rights of validation. Rights of participation define who is allowed to submit transactions to the blockchain network. Rights of validation define who is allowed to run validator nodes in the blockchain network.

At a very high level, rights of participation are either open to the public, like Bitcoin and Ethereum, or restricted to authorised participants. Private network examples include MediLedger (tracks pharmaceuticals from manufacturer to retail pharmacy), the IBM Food Trust (traces food from farm to fork), and TradeLens (tracks shipping containers from origin to destination).

Rights of validation are either permissionless (anyone may operate a validator node) or permissioned (an individual or institution needs permission or must be selected/voted upon to run a validator node). Bitcoin, Ethereum, Monero, and EOS are examples of permissionless networks—anyone may run a validator node by downloading the source code from Github. Permissionless blockchains need strong incentives to attract independent, 'well-behaving' validator nodes. Bitcoin, Ethereum, and Monero incentivise good behaviour by awarding cryptocurrencies to miners.

For an example of a permissioned network, when institutions join the Ripple network, they select which nodes they want to perform validation checks. Many permissioned blockchains have authorised participants take turns validating and adding the next set of transactions to the digital ledger.

However, there are nuances. EOS, for example, distinguishes between node validators, which anyone may run; and block producers, which must be voted upon by the community.

Plotting these two dimensions—rights of participation and rights of validation—yields four types of blockchain networks (see Figure 12.6).

Private-permissionless blockchains long were deemed to be either theoretical or nonsensical.[1] Theoretically, people described that a private-permissionless blockchain could exist, say, by deploying a smart contract on a permissionless network that restricts access and use to specific public keys (Daniels, 2018). In 2019, Ernst & Young (EY) took a major step towards making it a reality by launching Nightfall on Github (EY Blockchain/Nightfall, n.d.). Then in 2020, EY, ConsenSys, and Microsoft released the Baseline protocol as open-source software on Github. Both protocols use zero knowledge proofs so that companies can make completely confidential transactions on public Ethereum. In essence, EY's idea is a 'virtual private blockchain' similar to a virtual private network (VPN) that is connected to the public Internet, but data remains private from anyone not authorised to see the transaction.

1 Some authors only consider three types of blockchains, for example, Pedersen et al. (2019).

		Who can operate a validator node?	
		Permissionless (Anyone)	**Permissioned** (Requires permission, selection, or election)
Who can submit transactions?	**Public** (Anyone)	**Public-permissionless** • Bitcoin • Ethereum • Monero • EOS (node validators)	**Public-permissioned** • Ripple • EOS (block producers)
	Private (requires keys to access)	**Private-Permissionless** • Baseline protocol	**Private-permissioned** • MediLedger • IBM Food Trust • TradeLens

Figure 12.6. Four types of Blockchain networks.

Appendix C

Overview of Ethical Technology Principles

Information technologies solutions should be rooted in ethical principles. Ethical design principles are particularly relevant to new and advanced technologies, where developers need to think through the implications of their designs. It helps to have an overarching ethical framework, so here we rely on a study by the Berkman Klein Centre for Internet and Society at Harvard University to identify ethical design themes. Researchers at the centre, Fjeld et al. (2020), published a research paper on the themes of ethical artificial intelligence (AI). Their ethical AI themes are broad-reaching and apply to any technology solution.

Based on a review of 36 ethics documents published by civil societies, governments, intergovernment organisations, and the private sector, the authors extracted eight themes of ethical and rights-based approaches to design. These are privacy; accountability; safety and security; transparency and explainability; fairness and nondiscrimination; human control of technology; professional responsibility; and the promotion of human values (Fjeld et al., 2020). Within these eight themes, there are nearly 50 specific ethical principles, some of which are highlighted in Table 1.

Table 1. Ethical design principles.		
Ethical themes	**Examples of specific principles**	
1. Privacy	• Privacy by design • Rights of rectification • Rights to erasure	• Consent • Control over the use of data • Ability to restrict processes
2. Accountability	• Verifiability and replicability • Auditing • Impact assessments • Environmental responsibility	• Creation of a monitoring body • Ability to appeal • Liability and legal responsibility
3. Safety and security	• Safe by design • Secure by design • Predictable	
4. Transparency and explainability	• Transparent by design • Explainable by design • Open source • Regular reporting	• Notification when technology makes a decision about an individual
5. Fairness and nondiscrimination	• Inclusive by design • Nondiscrimination and prevention of bias • Representative data	• High-quality data • Fairness • Equality
6. Human control of technology	• Human review of automated decisions • Human control of the technology • Ability to opt out	
7. Professional responsibility	• Responsible by design • Accuracy of algorithms • Consideration of long-term effects	• Multi-stakeholder collaboration • Scientific integrity
8. Promotion of human values	• Leveraged to benefit society • Human values and human flourishing	

Source: Extracted from Fjeld et al. (2020).

1. *Privacy*. As an ethical design theme, the right to privacy is viewed as an inalienable human right. If data pertains to individuals' personal information, ethical privacy design calls for consent. Moreover, individuals should have the right to control access to and processing of data collected about them. They also should have the right to have data rectified in cases of errors, and the right to erasure.
2. *Accountability*. Who is accountable for decisions made by a technology? Fjeld et al. (2020) cover the mechanisms for accountability, including the verifiability and replicability of results; reliable audit trails; assessments of impacts, including environmental impacts; creation of a neutral monitoring body; and legal liability for damages caused by the technology.
3. *Safety and security*. Fjeld et al. (2020) express requirements for system safety, performance as intended, and resistance to being compromised by unauthorised parties.
4. *Transparency and explainability*. These ethical principles address the right to understand how technologies collect, process, store, and distribute information.
5. *Fairness and nondiscrimination*. Fjeld et al. (2020)'s fairness and nondiscrimination principles call for systems to be designed and used to maximise fairness, promote inclusivity, and avoid bias.
6. *Human control of technology*. Fjeld et al. (2020) argue that any important decisions made by technologies should be subject to human review.
7. *Professional responsibility*. Fjeld et al. (2020) recognise that humans design technologies and call on their professionalism and integrity to ensure that the appropriate stakeholders are consulted, and long-term effects are considered. As IT developers, we have a responsibility to make sure that solutions are accurate, multiple parties are invited to collaborate, and scientific methods are used to assess results.
8. *Promotion of human values*. Fjeld et al. (2020) discuss the principles of promoting human values by using technologies to help humans flourish; by providing access to many; and by leveraging technologies to better society.

Case Study for Chapter 13

Targeting Target with a 100 Million Dollar Data Breach

Abstract[1]

In January 2014, the CEO of the renowned U.S. discount retailer Target wrote an open letter to its customers apologizing for the massive data breach the company experienced during the 2013 holiday season. Attackers were able to steal credit card data of 40 million customers, and more were probably at risk. Share prices, profits, but above all reputation were all now at stake. How did it happen? What was really stolen? What happened to the data? How could Target win consumer confidence back?

While the company managed the consequences of the attack, and operations went slowly back to normal, in the aftermath the data breach cost hundreds of million dollars. Customers, banks, and all the major payment card companies took legal action against Target. Some of these litigations remained unsettled three years later. The importance of the breach lies in its far broader consequences, rippling through the U.S. Congress and raising consumer and industry awareness on cyber security.

The case provides substantial data and information, allowing students to step into the shoes of Target executives as they seek answers to the above questions.

Keywords: Teaching case, Cyber security, Hacking, Data Breach, Target, Information Systems

Introduction

On January 13 and 14, 2014, Greg Steinhafel, chairman, president, and CEO of Target, published an open letter to customers (Steinhafel, 2014) in the *New York Times*, the *Wall Street Journal, USA Today*, and the *Washington Post*, as well as in local papers of the firm's 50 largest markets. In the letter he apologized for the massive data breach his company experienced during the 2013 holiday season.

> Target learned in mid-December that criminals forced their way into our systems, gaining access to guest credit and debit card information. As a part of the ongoing forensic investigation, it was determined last week that certain guest information, including names, mailing addresses, phone numbers or e-mail addresses, was also taken.
>
> I know this breach has had a real impact on you, creating a great deal of confusion and frustration. I share those feelings. You expect more from us and deserve better. We want to earn back your trust and confidence and ensure that we deliver the Target experience you know and love.

The breach, announced to the public six days before Christmas, included credit card data from 40 million customers. It was later discovered that data for another 70 million customers were also at risk.

Target Inc.

Target's chain of discount stores sold low-cost clothing, items for the home, and—in some stores—groceries. Major competitors in the U.S. included Walmart, Kmart, Costco, Kohl's, J. C. Penney, and in Target's still small but growing online segment, Amazon. The first Target store, a low-cost subsidiary of the department store chain Dayton Hudson, opened in 1962; by December of 2014, Target's 366,000 employees staffed a network of nearly 2,000 stores located in the U.S. (1,801) and Canada (133). Target's stores also included larger SuperTarget stores, smaller CityTarget stores, and still smaller Target Express stores. In 2014, Target reported revenues of USD 73 billion.

Headquartered in Minneapolis, Target differentiated itself from low-cost competitors by offering Target brands, exclusive deals with other brands, and quality and trendy goods, as well as fashion items from well-known designers—all at modest prices; *Fortune* magazine characterized Target's merchandising focus as "Cheap and Chic" (Wahba, 2014).

The Breach

Target announced the data breach (see Exhibit 13.1) one day after an independent reporter and investigator of Internet security, Brian Krebs, broke the story on his blog:

> . . . Target is investigating a data breach potentially involving millions of customer credit and debit card

---

1 Pigni, F., et al. 2017. "Targeting Target with a 100 million dollar data breach." *Journal of Information Technology Teaching Cases 8*(1), 9–23. Copyright by Association for Information Technology Trust. Reprinted by permission of SAGE Publications Ltd.

> records . . . According to sources at two different top 10 credit card issuers, the breach extends to nearly all Target locations nationwide, and involves the theft of data stored on the magnetic stripe of cards used at the stores (Krebs, 2013).

For several days prior to Kreb's posting, banks had witnessed an uptick in illegal card activity, with a disproportionate number of those transactions traceable to card numbers recently used by Target customers. The banks notified the Federal Bureau of Investigation (FBI). The U.S. Department of Justice (DOJ) alerted Target on the evening of December 12th. The following day, DOJ and U.S. Secret Service personnel met with Target executives. By December 15th, outside experts, hired by Target, helped to discover and remove malware in Target's point-of-sale (POS) terminals and on several of the company's servers. On December 16th, Target notified banks and payment processors (e.g., Visa) that it had been breached.

From November 27th onwards, debit and credit transactions from Target's U.S. store's point-of-sale checkout terminals had been compromised and customer data stolen. By December 15, the hemorrhaging had slowed to a trickle, and by December 18, it was stopped. By then the data contained on magnetic stripes of 40 million debit and credit cards had been copied and, through a circuitous route, transmitted to a server in Russia. Almost immediately, customer credit card data surfaced on the black market at Internet "card shops."

On December 27th, Target announced that encrypted personal identification number (PIN) data from some cards had also been scraped. Then, on January 10, 2014, Target reported that non-financial data from as many as 70 million additional customers had also been stolen from Target servers; included were names, addresses, phone numbers, and e-mail addresses. Because of duplicates between the two sets of data, the total number of customers affected was approximately 100 million.

Data Breaches

The Identity Theft Resource Center (ITRC) defines a data breach as (ITRC, 2015, p. 2):

> An incident in which an individual name plus a Social Security number, driver's license number, medical record or financial record (credit/debit cards included) is potentially put at risk because of exposure.

Data breaches were classified in several ways. Breaches could be criminal or accidental, carried out by insiders or outsiders, computer-based or manual. The external, computer-based, criminal variety often involved changes to, or tapping into, the network, computer, or terminal hardware (called skimming). For instance, fake ATM fronts or card readers were surreptitiously attached to ATM machines; or for as little as USD1,000, an ATM could be acquired and set up as a honeypot for capturing unencrypted data from legitimate cards (Satanovsky, 2011). An alternative approach, called RAM or Memory Scraping (Zetter, 2014), required the use of software tools, either malware or legitimate software employed in an illegitimate manner on customer-facing devices, including ATMs, POS, or even consumers' own computers or phones. Scraping, unlike skimming, required no physical access; it could be carried out from anywhere in the world, thus lowering the risk to the perpetrator, while presenting still greater exposure to the victims.

The Target data breach was but one of an increasingly common phenomenon. One compilation (ITRC, 2015) identified 781 breaches in the U.S. that exposed 169 million records in 2015, a significant increase from 498 reported breaches and 22 million records reported six years earlier (Figure 13.1). In ten years, the ITRC had identified more than 6,000 breaches exposing more than 850 million records—a fourfold increase in a decade, affecting financial services, business, education, government, and healthcare sectors. As many breaches went unreported, these were conservative numbers.

Seven U.S. firms reported having had more than a million records exposed in the year following the Target breach; among them were three retailers: Home Depot, Michael's Stores, and Neiman Marcus. In each case the perpetrators appeared to have employed tools, and taken advantage of organizational lapses, in ways similar to Target's breach. Among notable other victims of data breaches in 2014 were AliExpress (owned by Alibaba.com), American Express, Korean Credit Bureau, J. P. Morgan, The U.S. Postal Service, the U.S. Internal Revenue Service, Rumbler .ru, and perhaps most notoriously, SONY Pictures.

In 2016 data breaches were still increasing 15% year on year, and the number of stolen records was growing at twice that pace (31%), with an average of 3 million records stolen per day. North America (see Figure 13.2) was experiencing the largest number of data breaches, accounting for almost 80% of the world total (Breach Level Index, 2016). The United States led the world in data breaches with over 400 million compromised records (70% of the total). Europe, the next highest, accounted for 10% of the total breaches with close to 50 million stolen records. The Asia and Pacific region was close behind in breaches (8%) but far outstripped Europe with 110 million compromised records (20%). U.S. security breach notification laws and European directives and regulations (e.g., the General Data Protection Regulation 2016/679) required organizations to disclose and to inform promptly customers, authorities, and other parties when personal data were stolen or compromised; an obligation not all countries were under. These regulations had the double objective of encouraging firms to improve their practices and consequently reduce consumers' risk.

Healthcare, government, financial, retail, education, and technology were the main target sectors for data breaches. In the U.S., 2016 saw an increase in breaches to POS systems at several hotel chains and retailers (see Figure 13.3).

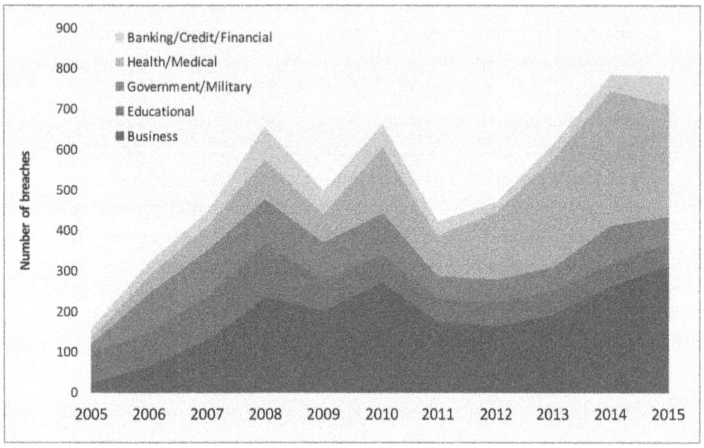

Figure 13.1. Evolution of data breaches in the U.S. (ITRC, 2016)

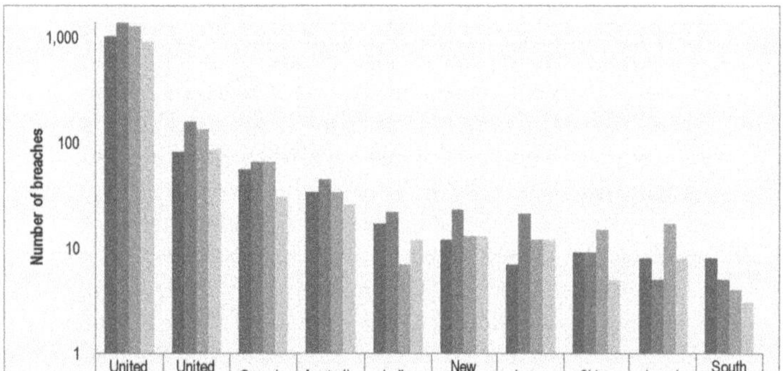

	United States	United Kingdom	Canada	Australia	India	New Zealand	Japan	China	Israel	South Africa
2016	1008	82	55	34	17	12	7	9	8	8
2015	1370	158	65	45	22	23	21	9	5	5
2014	1259	135	65	34	7	13	12	15	17	4
2013	911	86	30	26	12	13	12	5	8	3

Figure 13.2. Data breaches by country—logarithmic scale (authors on Gemalto's data, October 2016—http://www.breachlevelindex.com/data-breach-database)

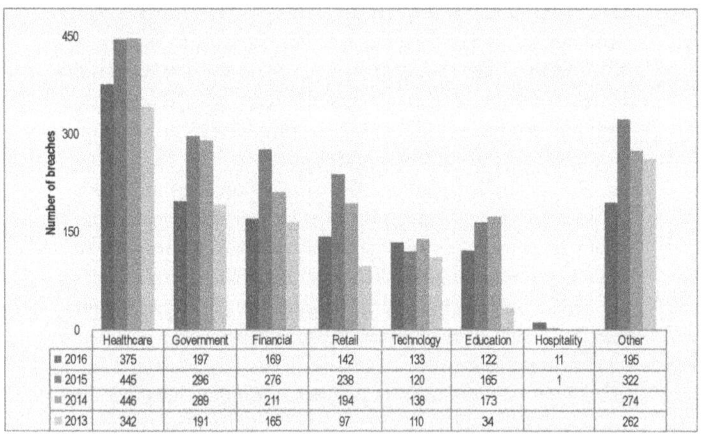

	Healthcare	Government	Financial	Retail	Technology	Education	Hospitality	Other
2016	375	197	169	142	133	122	11	195
2015	445	296	276	238	120	165	1	322
2014	446	289	211	194	138	173		274
2013	342	191	165	97	110	34		262

Figure 13.3. Data breaches by industry (authors on Gemalto's data, October 2016—http://www.breachlevelindex.com/data-breach-database)

Senior management's rising concern regarding computer and network security were on display in the results of the 2016 PwC Annual Global CEO Survey, where 61% of the executives interviewed described cyber threats and lack of data security as a threat to both national and commercial interests (PwC, 2016). Moreover, an even higher proportion (78%) of them considered cyber security technologies to be strategically important for their firms.

While security became a top priority in CEOs' agendas and a prominent topic in boardroom discussions, the data showed that corporations were losing ground in responding to the threat.

Payment Systems and Fraud

The U.S. Federal Reserve Bank reported (Federal Reserve Board, 2014, p. 41) in 2012 that credit cards made up 21% of the total number of non-cash transactions in the U.S. and 1.4% of the non-cash value; the corresponding numbers for debit cards were 38% and 1% and for checks, 15% and 14.8%. For Automated Clearing House (ACH) transactions, such as online bill pay and wire transfers, commonly used for large, non-retail transactions, the transaction and value numbers were 18% and 83%. Cash, an essentially anonymous payment system, was still the most common payment method, constituting 40% of transactions in the U.S. (Bennett et al., 2014, p. 3). An average consumer in the month of October 2012 used cash for 23 of 59 payments (Bennett et al., 2014, p. 2). Cash, however, was primarily used for small dollar value purchases, constituting only 14% of purchases at retail and averaging USD21 per transaction (Bennett et al., 2014, p. 3). At brick and mortar stores such as Target, a high, and increasing, proportion of purchases were made with credit or debit cards.

Payment cards, particularly credit and non-pin protected debit cards and prepaid cash cards, presented tempting, and still relatively risk-free, opportunities for criminals. The ability to tap into U.S. payment systems from other countries, particularly those with weak enforcement or no extradition treaties with the U.S., further lowered the risk. In 2012, the Federal Reserve reported over 31 million fraudulent payment transactions with a value of over USD 6 billion; twenty-six million of these transactions, and over USD 4 billion of value, were from credit, signature-only debit or prepaid cash cards. Pin protected debit cards were far more secure, experiencing only 20% of the fraud rates of signature debit cards (Federal Reserve Board, 2014).

The biggest vulnerability in card payment systems in the U.S. was the card's magnetic stripe. The data written on the "magstripe" included the primary account number, the account holder's name, the expiration date, a service code indicating the types of charges that could be accepted, and discretionary data, such as a PIN code. Once compromised, either by scraping or skimming, this data could be used to make online purchases or to legitimate counterfeit cards, which could then be used in physical stores. While in-store use might seem risky, it did not require a mailing address to collect the ordered merchandise. Moreover, the stolen merchandise, mostly electronics or gift cards, could often be immediately resold.

"Big Box" and discount retailers were particularly vulnerable to payment card fraud and data breaches due to the size of their customer population, their high daily transaction volumes, the liquidity of some of their merchandise, and their customers' desire for fast and convenient checkout. Moreover, huge past investments in point-of-sale checkout devices, as well as the typical customer's comfort with mag-stripe credit and debit cards, had retarded retailers' transition to more secure technologies (Geuss, 2015).

The complexity of the payment network added further vulnerability. The observation of a judge in an earlier data breach case described that complexity and, implicitly, its consequent vulnerability:

> Every day, merchants swipe millions of customers' payment cards. In the seconds that pass between the swipe and approval (or disapproval), the transaction information goes from the point of sale, to an acquirer bank, across the credit-card network, to the issuer bank, and back. Acquirer banks contract with merchants to process their transactions, while issuer banks provide credit to consumers and issue payment cards. The acquirer bank receives the transaction information from the merchant and forwards it over the network to the issuer bank for approval. If the issuer bank approves the transaction, that bank sends money to cover the transaction to the acquirer bank. The acquirer bank then forwards payment to the merchant. (Rosenthal, 2011)

The judge described a four-party payment system: A credit-card network, usually Visa or MasterCard, is a network intermediary between the merchants' bank ("acquirer"), the merchant, and the customer's bank ("issuer"). The alternative, a three-party approach, links three participants: the card-carrying customer, the merchant, and the card issuer (e.g., American Express or Discover). In 2013, 82% of card payments went through the four-party system. To further the complexity, many merchants relied on outside payment processors for the link between their POS devices and acquiring banks. Two of these, Global Payments and Heartland Payments, had themselves been major victims of hackers.

Anatomy of the Target Breach

The first victim in the heist was not Target but Fazio Mechanical Services, a provider of refrigeration services to Target. The means of attack was uncertain but was likely executed via a bogus link or attachment as part of an e-mail "phishing" broadcast to multiple Target third-party vendors—a list of which was openly available on the Internet. To get inside the supplier's network, the attackers used a malware package called Citadel (Olavsrud, 2014) and then found and used Fazio's credentials to exploit its previously authorized access to Target's computer network. Fazio had access to several Target systems, including contract management, project management, and electronic billing. On November 12, 2013, the attackers gained access to Target's internal network, probably by uploading an executable file disguised as a legitimate document attachment through a web application. The name of the uploaded file was apparently chosen to be similar to that of other files commonly seen on the system.

Once inside Target's internal network, the attackers sought out logins, passwords, and network diagrams. Failing to find credit card credentials on Target servers, they instead, apparently patiently and successfully, penetrated Target's POS terminals. Harnessing a computer account they had created on Target's network, they deployed malware to POS terminals that the investigators named Kaptoxa (pronounced kar-toe-sha), available for about USD 2,000 on black market websites. The software then scraped each unencrypted card as it was read.

Between November 15 and 28, the attackers tested the malware[1] on a few of Target's POS devices. By November 30, the hack was fully installed on almost all POS devices and fully operational. That day, the attackers also installed malware to transfer the stolen data to an internal server. This data exfiltration malware,[2] the file name of which was disguised to look like a legitimate application, was updated twice: on December 2 and again on December 4. On December 2, the perpetrators began to transfer data to another Target server, one that was authorized for file transfers through Target's firewall. The data were moved from that server to servers outside the U.S., eventually ending up on a server in Russia. Data were moved during business hours to hide the illicit activity within an otherwise busy network traffic.

Stolen card numbers were almost immediately available on Internet black markets. One market, Rescator, had been described as "the Amazon.com of Stolen Credit Cards"

---

1 While not definitively linked to the Target data breach, in August of 2014 the U.S. Secret Service identified malware called "backoff" that was first detected in October of 2013 but not detectable by antivirus solutions until almost a year later. Backoff was estimated to have already affected more than 1,000 U.S. businesses. https://www.documentcloud.org/documents/1279345-secret-service-malware-announcement.html.

2 Data exfiltration is the transfer of stolen data from a compromised system within victims' network back to the attacker while attempting to remain undetected.

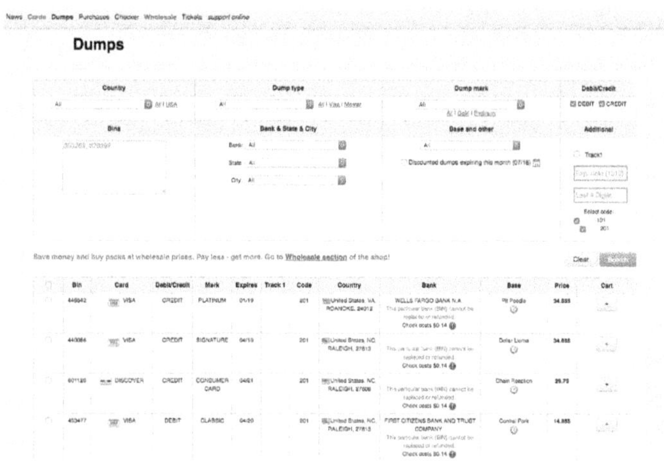

Figure 13.4. Rescator's efficient and user-friendly web shopping interface.

(Lawrence, 2014). Here, batches of credit cards could be purchased, sometimes for prices exceeding USD 100 (Figure 13.4). Cards data contained in the earliest batch released on Rescator sold for between USD 26.60 and USD 44.80 in the days before December 19 (Exhibit 13.3), when Target went public on the data breach (Krebs, 2014).

Failed Security Measures

Target's attackers exploited numerous security weaknesses. Target had publicly posted the names of its suppliers on the Internet. One of them, Fazio Mechanical Services, had relied on a free malware detection package, intended for use by individuals rather than for commercial use. The malicious detection package, installed at Fazio, probably captured login and password information during transactions. While two-factor authentication was required by PCI[1] for payment servers, it was not required, and from reports was rarely used, for non-payment related, externally accessible applications on Target's external network. Instead, Target relied on a scheme required by PCI policy: payment servers were segregated from the rest of the network. Indeed, PCI had recently given a clean audit of Target's network segregation—a segregation that subsequently proved inadequate.

Two different security packages triggered alarms as the data exfiltration malware was installed on November 30 and then again when it was updated. One of these packages, FireEye, installed at a cost of USD 1.6 million a few months earlier, recommended to its Target minders in Bangalore the deletion of the malware—a recommendation reportedly passed on to, but ignored by, the personnel in Target's security operations center in Minneapolis (Riley et al., 2014). Target also apparently did not maintain a "white list" of authorized processes, often used to ensure that malware is not allowed to run on a device or server. Neither did Target adequately monitor the creation of new accounts nor effectively block access to certain external file servers (e.g., servers in Russia).

Financial Consequences

The breach proved to be immediately costly, as reflected in the CEO's comments to analysts in a February 2014 earnings conference call.

> Target's fourth quarter financial results reflect better than expected US segments performance through the first three weeks of the holiday season, followed by meaningfully softer results following our December 19 [data breach announcement] . . . fourth quarter comparable sales decreased 2.5%, consistent with our updated guidance in January. (Target, 2014c, p. 3)

Target's cumulative stock return had beaten both the S&P 500 and Target's peer comparison group in February of 2013 but, by the following February, two months after the breach, had fallen precipitously behind both groups. Earnings per share had also fallen (Target, 2014a, pp. 15–16). Profits in the fourth quarter of 2013 were off 47% from the previous year, though the decline was partially attributed to poor performance at Target's Canadian stores.

Costs piled up. Eight months after the breach, the company reported USD 236 million in breach related costs, of which USD 90 million were covered by insurance (Target, 2014e, p. 9). One big expense was the cost to provide Target's customers with a year of credit screening services. Those reported expenses, coupled with a drop in expected earnings from 85 to 78 cents a share, stunned Wall Street; Target's stock price fell 4.4% the next day (Abrams, 2014).

John Kindervag, a vice president and principal analyst at Forrester Research, predicted that the eventual costs of the breach would be much higher:

> I don't see how they're getting out of this for under a billion, over time . . . One hundred fifty million in a quarter seems almost like a bargain. (Abrams, 2014)

Legal Consequences

In its 2014 second quarter earnings conference call (Target, 2014e, p. 9), Target trumpeted "dramatically lower"

---

1 The Payment Card Industry Security Standards Council (PCI SSC) was created in 2006 to develop security standards for the evolving Payment Card Industry (PCI). The resulting Payment Card Industry Data Security Standard (PCI DSS) is intended to ensure that participating companies that process, store, or transmit credit card information do so in a secure manner.

breach-related costs as compared to post-breach external estimates that had been more in line with Kindervag's billion-dollar estimate. But, three months later, in the risk assessment section of Target's November 2014 10-Q filing to the SEC (Target, 2014b, p. 9), Target identified many still unresolved potential sources for further costs and legal uncertainties.

> . . . more than 100 actions have been filed in courts in many states, along with one action in Canada, and other claims have been or may be asserted against us on behalf of guests, payment card issuing banks, shareholders or others seeking damages or other related relief allegedly arising out of the Data Breach. State and federal agencies, including State Attorneys General, the Federal Trade Commission and the SEC, are investigating events related to the Data Breach, including how it occurred, its consequences and our responses . . .

Target customers' numerous lawsuits were combined into a single class action suit to be adjudicated in a federal district court in Minnesota. One of nearly 100 customer reports included in the lawsuit described the damages and inconveniences suffered by one misfortunate Target customer:

> [A Target customer] used her Savannah State Bank Visa debit card to purchase goods at a Target store in Georgia during the period of the Target data breach. [The customer's] personal information associated with her debit card was compromised in and as a result of the Target data breach. [The customer] was harmed by having her financial and personal information compromised. She incurred multiple unauthorized charges totaling approximately $1900 in December 2013. [The customer] also experienced a loss of access to her funds, paid a replacement card fee for which she remains unreimbursed, and incurred late payment fees due to failed automatic payments. She also paid for credit monitoring services as a result of the Target data breach. (United States District Court: District of Minnesota, 2014, p. 23)

Estimates of the eventual total cost of fraudulent charges to customer cards ranged from USD 240 million to USD 2.2 billion (Weiss and Miller, 2015). Among the numerous damages enumerated by customers' lawyers were unauthorized charges to debit and credit card accounts, theft of personal and financial information, costs of detecting and protecting against identity theft and unauthorized use of accounts, lack of access to account funds, costs associated with that lack of access (e.g., late charges and fees, credit rating harm), and time and loss of productivity stemming from the need to deal with the challenges faced.

The customers' lawyers accused Target of:

> . . . failing to take adequate and reasonable measures to ensure its data systems were protected, failing to take available steps to prevent and stop the breach from ever happening, failing to disclose to its customers the material facts that it did not have adequate computer systems and security practices to safeguard customers' financial account and personal data, and failing to provide timely and adequate notice of the Target data breach (United States District Court: District of Minnesota, 2014, p. 4)

That same U.S. District Court in Minnesota would adjudicate another set of class action lawsuits, this time brought by banking institutions adversely impacted by their own customers' misfortune. Because of contracts with payment networks like Visa, historically the banks had shouldered the bulk of the losses for credit card breaches. This time they hoped, because of the retailers' alleged negligence, more of the responsibility would be assigned to Target. Estimates of the potential fines that might be levied on Target ranged from USD 71 million to USD 1.1 billion, numbers that represented anywhere from 2 to 37% of Target's net income for 2013 (Weiss and Miller, 2015). The American Bankers Association estimated that the data breach affected more than 8% of debit cards and nearly 4% of credit cards countrywide, with the average loss to banks of USD 331 per debit card and USD 530 per credit card (ABA, 2014).

Targeting Target: How to Steal a Hundred Million (B)

Everyone in this industry right now has to come together to make sure we're putting the right defense plans in place.

—Brian Cornell, CEO Target Stores (CBS News, 2014)

In May 2014, Greg Steinhafel resigned as Target's chairman, president, and CEO, a resignation partially attributed (Abrams, 2014) to a massive, criminal data breach suffered by Target during the 2013 holiday season. The breach had exposed over 100 million customer records; it depressed Target's holiday shopping revenues, increased administrative costs, and triggered legal liabilities. Moreover, the breach was a clear threat to Target's brand and reputation. In parallel with Steinhafel's May resignation, Institutional Shareholder Services, an overseer of corporate governance for institutional investors, recommended that shareholders reject the re-election of seven members of the board who served on Target's audit and corporate responsibility committee.

Following Steinhafel's resignation, John Mulligan, Target's CFO, took on the position of interim CEO. Three months later, in mid-August of 2014, Brian Cornell was named chairman and CEO. A previous CEO of PepsiCo Americas' Foods Division, Cornell brought extensive retail experience to Target; his impressive resume included CEO at Sam's Club, CEO at Michael's Craft Stores, and CMO at Safeway.

The breach foreshadowed a further shakeup in Target's management team. Prior to Steinhafel's resignation, and three months after the breach, Target's CIO resigned. The vice president of assurance risk and compliance, in keeping with his previously announced intention, also resigned.

Customer Communication

From its initial announcement of the breach on December 19 through January 15, Target sent six e-mails to its "guests" and a seventh to the holders of Target's proprietary REDcard payment card. Included among these were descriptions of what had happened, apologies, reassurances that the problem was being well taken care of and that the customer risk was small, advice about how recipients could protect themselves or what actions the customer should take (e.g., "Be wary of e-mails that ask for money or send you to suspicious websites.") or should not take (e.g., "Never share information with anyone over the phone, e-mail or text, even if they claim to be someone you know or do business with."), and explained how to take advantage of the year of free credit monitoring Target was providing. The company also quickly established, and continued to update, several web resources. One web page included links to the seven e-mails, related press announcements, and transcripts of CFO Mulligan's February 4 and March 26 testimony to congressional committees. A second web page included responses to 48 "frequently asked questions." The initial versions of these web resources were prominently displayed and accessible from Target's home page as of the announcement on December 19.

Rebuilding the Organization and Consumer Confidence

In April of 2014, Target hired a new CIO, Bob DeRoddes, who had served in a security advisory capacity to the U.S. Department of Homeland Security, the U.S. Secretary of Defense, the U.S. Department of Justice, and numerous multinational firms.

In the CIO announcement, Target also described its intention to move Target's "Red" branded credit and debit cards to a "chip-and-PIN-enabled technology" as well as accelerating a plan to install new payment devices in close to 1,800 stores (see Exhibit 13.4). Further, it identified a number of security enhancements already implemented (Target, 2014d). Among them were:

1. "Enhancing monitoring and logging [including] additional rules, alerts, centralizing log feeds and enabling additional logging capabilities."
2. "Installation of application whitelisting point-of-sale systems [including] deploying to all registers, point-of-sale servers and development of whitelisting rules."
3. "Implementation of enhanced segmentation [including] development of point-of-sale management tools, review and streamlining of network firewall rules and development of a comprehensive firewall governance process."
4. "Reviewing and limiting vendor access [including] decommissioning vendor access to the server impacted in the breach and disabling select vendor access points including FTP and telnet protocols."
5. "Enhanced security of accounts coordinated reset of 445,000 Target team member and contractor passwords, broadening the use of two-factor authentication, expansion of password vaults, disabled multiple vendor accounts, reduced privileges for certain accounts, and developing additional training related to password rotation."

In June of 2014, Brad Maiorino was appointed to a newly created position, that of senior VP and chief information security officer. Maiorino was previously with General Motors and, prior to that, General Electric. In those roles his responsibilities focused on information security. He would report to the CIO. Six months later, Target announced the appointment of Jacqueline Hourigan Rice to fill the roles of senior VP and chief risk and compliance officer. Hourigan Rice also came from GM, where she had spent 17 years, most recently as GM's chief compliance officer. According to the announcement, she would report to CEO Cornell. Her responsibilities would include "centralized oversight of enterprise risk management, compliance, vendor management and corporate security under her leadership" (Target, 2014f).

A Year Later

In a televised interview in November of 2014, a year after the breach and two days before "Black Friday,"[1] the semi-official start of the crucial holiday sales season, Cornell reassured customers, shareholders, and business partners that the Target leadership team was taking data security very seriously:

> We focus every day, every single day, not just during the holidays, but 52 weeks a year, on data security. Making sure we've the right team in place, to monitor, detect, contain. (CBS News, 2014)

These were confidence-building words, but even as he spoke, the perpetrator(s) had not been apprehended, the stolen credit card credentials were still for sale on Internet black markets, and a growing number of breach-related lawsuits still hung over Target.

Yet, the mood at Target seemed considerably more upbeat than a year earlier. So too were Target's financials. The 2014 fiscal year closed with sales up 1.3% and with digital channel

---

1 The first shopping day after Thanksgiving in the US, allegedly named because it was often the day when a retailer's profitability for the year went from red to black.

sales growth exceeding 30% (Target, 2015a), and by the first quarter of 2015, sales grew 2.3% from the same period in the prior year (Target, 2015b). Target's stock price, which had fallen to a low of USD 54.66 in February of 2014, had rebounded to over USD 75 in late January of 2015 (Exhibit 13.2). Target was confident that the data breach would not impact their reputation in the long term:

> We experienced weaker than expected sales immediately following the announcement of the Data Breach that occurred in the fourth quarter of 2013, and while we now believe the incident will not have a long-term impact to our relationship with our guests, it is an example of an incident that affected our reputation and negatively impacted our sales for a period of time. (Target, 2015a, p. 4)

The Target website, which had, until recently, prominently displayed links to information on the data breach, had returned to business as usual (Exhibit 13.5). By the end of 2015 the major lawsuits initiated by customers and credit card issuers were finally being settled. In March, Target agreed to pay USD 10 million to settle individual victims' damages up to USD 10,000 (Reuters and Fortune, 2015). In August, Visa issuers settled on up to $67 million in costs related to the data breach (Whipp, 2015). In December, an agreement was reached with MasterCard issuers for USD 19.11 million and banks and credit unions not covered in the other actions for up to USD 20.25 million (Stempel and Bose, 2015).

While the situation was increasingly back to normal, the company was still facing shareholder lawsuits as well probes by the Federal Trade Commission and State Attorneys General regarding the breach (Stempel and Bose, 2015).

The Broader Threat

Executives at other multinational companies were considerably more pessimistic than Cornell appeared to be, at least in his public pronouncements. Speaking at a panel at the 2015 World Economic Forum in Davos, Switzerland, several CEOs (Gelles, 2015) had expressed their apprehensions about data breaches. John Chambers, CEO of Cisco, predicted, "The number of security incidents this year will be exponentially greater than last year." Similarly, the CEO of Infosys, Visha Sikka, predicted "five times as many incidents as we did last year" (Figure 13.1). As vendors of IT and security solutions, Chambers and Sikka were perhaps predictably alarmist in their assessments. The comments of the CEO of IMAX, Richard Gelfond, probably better reflected the trepidation of many of Chambers' and Sikka's customers:

> The one thing that really scares me is that if someone wants to get into your system, they can get in. Almost no amount of money will keep them out.

Another vendor's study supported their pessimism (Riley et al., 2014), reporting that only 31% of companies had identified data breaches through their own monitoring. The percentage was far lower for retailers. As with Target, 95% of retail data breaches were not discovered by the retailer; one observer described retailers as "the wildebeests of the digital savannah."

Congressional Reactions to Target Breach

Compared to their European counterparts, U.S. retailers were particularly vulnerable, as Seth Berman, head of the London office of a risk management firm, observed:

> There's a fundamental flaw in the US credit card system in that they do not use chip and pin . . . The US is doing everyone a favor by acting as a honeypot for criminals, and in addition the country has more credit cards per head than anywhere else.

The growing, still seemingly uncontrollable, threat to U.S. firms posed by hackers was a growing concern in Washington D.C. Between February 3 and April 2, 2014, six congressional committees held seven different hearings related (Weiss and Miller, 2015, p. 2) to data breaches in general and the Target breach in particular. Among the options discussed were:

> Federal legislation to require notification to consumers when their data have been breached; legislation to potentially increase Federal Trade Commission (FTC) powers and authorities over companies' data security; and legislation that could create a federal standard for the general quality or reasonableness of companies' data security.

Study Questions

1. How was the attack on Target perpetrated? Can you identify its main phases?
2. Which weaknesses in Target security did hackers exploit?
3. Would you consider Target's data breach an information system failure? Why?
4. Who do you believe is to blame for the incident? Why? How did Target manage the situation when the breach was detected? Do you consider its reaction appropriate?
5. Do you believe it was the CEO's responsibility to inform customers about the data breach? What would you have done?
6. What lessons should a CEO learn from Target?
7. What lessons should a CIO learn?
8. What should Target do next?
9. Do you believe consumers are becoming tolerant of breaches?

References

ABA. (2014). *Target Breach Bank Impact*, American Bankers Association. Retrieved from https://silo.tips/download/target-breach-impact-survey.

Abrams, R. (2014, August 5). Target Puts Data Breach Costs at $148 Million, and Forecasts Profit Drop, *New York Times*. Retrieved from http://www.nytimes.com/2014/08/06/business/target-puts-data-breach-costs-at-148-million.html.

Bennett, B., Conover, D., O'Brien, S., and Advincula, R. (2014). Cash Continues to Play a Key Role in Consumer Spending: Evidence from the Diary of Consumer Payment Choice, *Federal Reserve Bank of San Francisco Fednotes (April 2014)*. Retrieved from https://www.frbsf.org/cash/files/FedNotes_Evidence_from_DCPC.pdf.

Breach Level Index. (2016). *2016 It's All About Identity Theft—First Half Findings from the 2016 Breach Level Index*, Gemalto. Retrieved from http://www.breachlevelindex.com/assets/Breach-Level-Index-Report-H12016.pdf.

CBS News. (2014, November 26). Target CEO on Black Friday: "We have to win that big playoff game," *CBS News*. Retrieved June 23, 2016, from http://www.cbsnews.com/news/target-ceo-brian-cornell-on-black-friday-data-security-free-shipping/.

Federal Reserve Board. (2014). *The 2013 Federal Reserve Payments Study—Recent and Long-Term Payment Trends in the United States: 2003–2012—Summary Report and Initial Data Release*, Federal Reserve System, p. 43. Retrieved from https://www.frbservices.org/binaries/content/assets/crsocms/news/research/2013-fed-res-paymt-study-summary-rpt.pdf.

Gelles, D. (2015, January 22). Executives in Davos Express Worries Over More Disruptive Cyberattacks, *New York Times' DealBook*. Retrieved June 23, 2016, from http://dealbook.nytimes.com/2015/01/22/in-davos-executives-express-worries-over-more-disruptive-cyberattacks/.

Geuss, M. (2015, November 26). Chip-Based Credit Cards Are Old News; Why Is the US Only Rolling Them Out Now?, *Ars Technica*. Retrieved May 13, 2016, from http://arstechnica.com/business/2015/11/chip-based-credit-cards-are-old-news-why-is-the-us-only-rolling-them-out-now/.

ITRC. (2015). *Data Breach Reports*, Identity Theft Resource Center, p. 197.

ITRC. (2016, January 25). ITRC Breach Statistics 2005–15. Retrieved May 13, 2016, from http://www.idtheftcenter.org/images/breach/DataBreachReports_2015.pdf.

Krebs, B. (2013, March 18). Sources: Target Investigating Data Breach—Krebs on Security, *Krebs on Security*. Retrieved from http://krebsonsecurity.com/2013/12/sources-target-investigating-data-breach/.

Krebs, B. (2014, February 19). Fire Sale on Cards Stolen in Target Breach, *Krebs on Security*. Retrieved from http://krebsonsecurity.com/2014/02/fire-sale-on-cards-stolen-in-target-breach/.

Langley, M. (2014, February 19). Inside Target, CEO Gregg Steinhafel Struggles to Contain Giant Cybertheft, *Wall Street Journal*. Retrieved from http://www.wsj.com/articles/SB10001424052702304703804579382941509180758.

Lawrence, D. (2014, September 4). The Amazon.com of Stolen Credit Cards Makes It All So Easy, *Bloomberg.com*. Retrieved May 13, 2016, from http://www.bloomberg.com/news/articles/2014-09-04/the-amazon-dot-com-of-stolen-credit-cards-makes-it-all-so-easy.

Olavsrud, T. (2014, September 2). 11 Steps Attackers Took to Crack Target, *CIO*. Retrieved May 13, 2016, from http://www.cio.com/article/2600345/security0/11-steps-attackers-took-to-crack-target.html.

PwC. (2016). *19th Annual Global CEO Survey*, PricewaterhouseCoopers, p. 44. Retrieved from http://www.pwc.com/gx/en/ceo-survey/2016/landing-page/pwc-19th-annual-global-ceo-survey.pdf.

Reuters and Fortune. (2015, March 19). Target will pay $10 million to settle data breach lawsuit, *Fortune*. Retrieved from http://fortune.com/2015/03/19/target-10-million-settle-data-breach/.

Riley, M., Elgin, B., Lawrence, D., and Matlack, C. (2014, March 17). Missed Alarms and 40 Million Stolen Credit Card Numbers: How Target Blew It, *Bloomberg.com*. Retrieved May 13, 2016, from http://www.bloomberg.com/news/articles/2014-03-13/target-missed-warnings-in-epic-hack-of-credit-card-data.

Rosenthal, L. H. In re: HEARTLAND PAYMENT SYSTEMS, INC. CUSTOMER DATA SECURITY BREACH LITIGATION, No. 834 F.Supp.2d 573 (United States District Court, SD Texas, Houston Division Dec. 1, 2011). Retrieved from http://www.leagle.com/decision/In%20FDCO%2020111202937/IN%20RE%20HEARTLAND%20PAYMENT%20SYSTEMS,%20INC.

Satanovsky, G. (2011, January 17). How Counterfeit Credit Cards Are Created from ATM Skimmers, *Fraud Fighter—Fraud Prevention Blog*. Retrieved May 12, 2016, from http://blog.fraudfighter.com/bid/52994/How-Counterfeit-Credit-Cards-are-Created-From-ATM-Skimmers.

Steinhafel, G. (2014, January 12). An Open Letter from CEO Gregg Steinhafel, *Target Corporate*. Retrieved April 26, 2016, from http://corporate.target.com/article/2014/01/target-ceo-gregg-steinhafel-open-letter-guests.

Stempel, J., and Bose, N. (2015, December 3). Target in $39.4 Million Settlement with Banks over Data Breach, *Reuters*. Retrieved from http://www.reuters.com/article/us-target-breach-settlement-idUSKBN0TL20Y20151203.

Target. (2014a). 2013 Annual Report, *Target.com*. Retrieved May 13, 2016, from https://investors.target.com/static-files/62bf2115-f6da-4e8d-b8da-a69fc2eb8c8b.

Target. (2014b). *Quarterly Report 10-Q, For the Quarterly Period Ended November 1, 2014* (SEC filing No. Commission File Number 1-6049). Retrieved from https://investors.target.com/static-files/6c3fddf5-ee47-4b4c-9ef0-41bf9d88ab6f.

Target. (2014c, February 26). Edited Transcript: TGT—Q4 2013 Target Corporation Earnings Conference Call, *Target.com*. Retrieved May 13, 2016, from http://phx.corporate-ir.net/External.File?item=UGFyZW50SUQ9MjIyNTE0fENoaWxkSUQ9LTF8VHlwZT0z&t=1.

Target. (2014d, April 29). Target Appoints New Chief Information Officer, Outlines Updates on Security Enhancements, *Target Corporate*. Retrieved June 23, 2016, from http://corporate.target.com/press/releases/2014/04/target-appoints-new-chief-information-officer-outl.

Target. (2014e, August 20). Edited Transcript: TGT—Q2 2014 Target Corporation Earnings Conference Call, *Target.com*. Retrieved May 13, 2016, from http://phx.corporate-ir.net/External.File?item=UGFyZW50SUQ9MjY0NDkzfENoaWxkSUQ9LTF8VHlwZT0z&t=1.

Target. (2014f, November 6). Target Names Jacqueline Hourigan Rice as Senior Vice President, Chief Risk and Compliance Officer, *Target Corporate*. Retrieved June 23, 2016, from http://

corporate.target.com/press/releases/2014/11/target-names-jacqueline-hourigan-rice-as-senior-vi.

Target. (2015a). *Quarterly Report 10-Q, For the Fiscal Year Ended January 31, 2015* (No. Commission File Number 1-6049). Retrieved from https://investors.target.com/static-files/5db8eda4-f87f-45c5-86de-2e6464139a9c.

Target. (2015b). *Quarterly Report 10-Q, For the Quarterly Period Ended May 2, 2015* (No. Commission File Number 1-6049). Retrieved from https://investors.target.com/static-files/5db8eda4-f87f-45c5-86de-2e6464139a9c.

United States District Court: District of Minnesota. In re: Target Corporation Customer Data Security Breach Litigation, No. 14-2522 (PAM/JJK) (2014, January 12). Retrieved from http://cdn.arstechnica.net/wp-content/uploads/2014/12/document4.pdf.

Wahba, P. (2014, October 21). Target Puts Focus Back on "Cheap-Chic" with Eye on Winning Back Holiday Shoppers. Retrieved April 26, 2016, from http://fortune.com/2014/10/21/target-holiday/.

Weiss, N. E., and Miller, R. S. (2015). The Target and Other Financial Data Breaches: Frequently Asked Questions, In *Congressional Research Service, Prepared for Members and Committees of Congress February*, Vol. 4, p. 2015.

Whipp, L. (2015, August 18). Target to Pay $67M over Visa Data Breach, *FT.com*. Retrieved July 31, 2016, from https://www.ft.com/content/a6b571d8-45c8-11e5-af2f-4d6e0e5eda22.

Zetter, K. (2014, September 30). How RAM Scrapers Work: The Sneaky Tools behind the Latest Credit Card Hacks, *WIRED*. Retrieved May 12, 2016, from https://www.wired.com/2014/09/ram-scrapers-how-they-work/.

Exhibit 13.1: Initial Notification to Target Customers on December 19th, 2013

Important Notice: Unauthorized access to payment card data in U.S. stores

We wanted to make you aware of unauthorized access to Target payment card data. The unauthorized access may impact guests who made credit or debit card purchases in our U.S. stores from Nov. 27 to Dec. 15, 2013. Your trust is a top priority for Target, and we deeply regret the inconvenience this may cause. The privacy and protection of our guests' information is a matter we take very seriously and we have worked swiftly to resolve the incident.

We began investigating the incident as soon as we learned of it. We have determined that the information involved in this incident included customer name, credit or debit card number, and the card's expiration date and CVV.

We are partnering with a leading third-party forensics firm to conduct a thorough investigation of the incident and to examine additional measures we can take that would be designed to help prevent incidents of this kind in the future. Additionally, Target alerted authorities and financial institutions immediately after we discovered and confirmed the unauthorized access, and we are putting our full resources behind these efforts.

We recommend that you closely review the information provided in this letter for some steps that you may take to protect yourself against potential misuse of your credit and debit information. You should remain vigilant for incidents of fraud and identity theft by regularly reviewing your account statements and monitoring free credit reports. If you discover any suspicious or unusual activity on your accounts or suspect fraud, be sure to report it immediately to your financial institutions. In addition, you may contact the Federal Trade Commission ("FTC") or law enforcement to report incidents of identity theft or to learn about steps you can take to protect yourself from identity theft. To learn more, you can go to the FTC's website, at www.consumer.gov/idtheft, or call the FTC, at (877) IDTHEFT (438–4338) or write to Federal Trade Commission, Consumer Response Center, 600 Pennsylvania Avenue, NW, Washington, D.C. 20580.

You may also periodically obtain credit reports from each nationwide credit reporting agency. If you discover information on your credit report arising from a fraudulent transaction, you should request that the credit reporting agency delete that information from your credit report file. In addition, under federal law, you are entitled to one free copy of your credit report every 12 months from each of the three nationwide credit reporting agencies.

Again, we want to stress that we regret any inconvenience or concern this incident may cause you. Be assured that we place a top priority on protecting the security of our guests' personal information. Please do not hesitate to contact us at 866-852-8680 or visit Target's website if you have any questions or concerns. If you used a non-Target credit or debit card at Target between Nov. 27 and Dec. 15 and have questions or concerns about activity on your card, please contact the issuing bank by calling the number on the back of your card.

Exhibit 13.2: Target data breach timeline (adapted Langley, 2014)

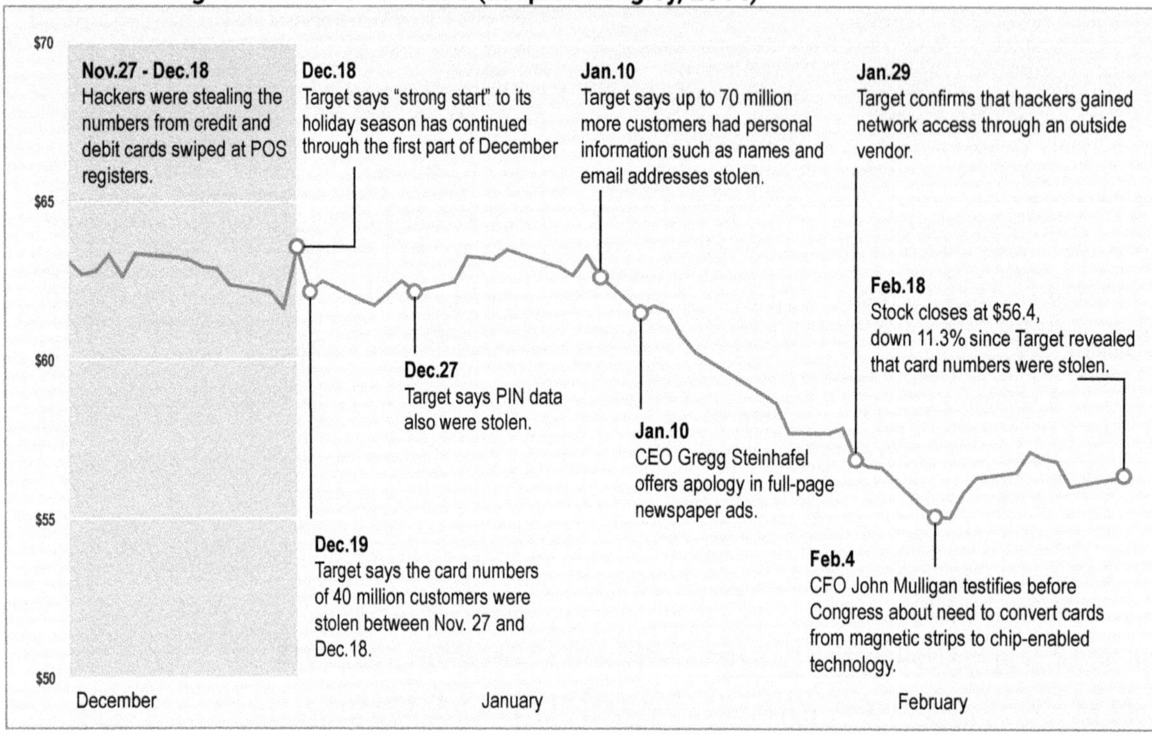

Exhibit 13.3: From hacking to monetization

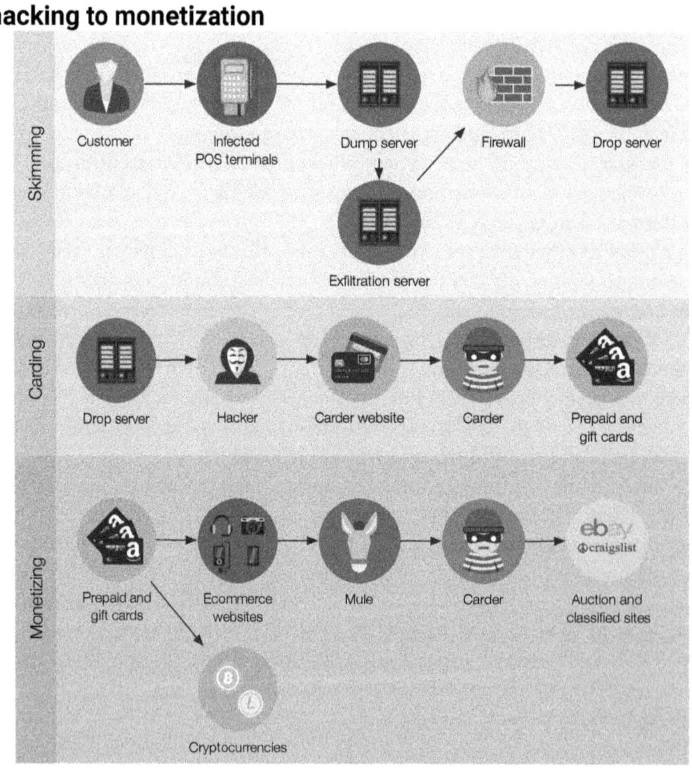

Exhibit 13.4: New MasterCard Initiative and Commitment to Chip-and-PIN

Today, Target also announced a significant new initiative as part of the company's accelerated transition to chip-and-PIN-enabled REDcards. Beginning in early 2015, the entire REDcard portfolio, including all Target-branded credit and debit cards, will be enabled with MasterCard's chip-and-PIN solution. Existing co-branded cards will be reissued as MasterCard co-branded chip-and-PIN cards. Ultimately, through this initiative, all of Target's REDcard products will be chip-and-PIN secured.

Earlier this year, Target announced an accelerated $100 million plan to move its REDcard portfolio to chip-and-PIN-enabled technology and to install supporting software and next-generation payment devices in stores. The new payment terminals will be in all 1,797 U.S. stores by this September, six months ahead of schedule. In addition, by early next year, Target will enable all REDcards with chip-and-PIN technology and begin accepting payments from all chip-enabled cards in its stores.

"Target has long been an advocate for the widespread adoption of chip-and-PIN card technology," said John Mulligan, executive vice president, chief financial officer for Target. "As we aggressively move forward to bring enhanced technology to Target, we believe it is critical that we provide our REDcard guests with the most secure payment product available. This new initiative satisfies that goal."

"Target and MasterCard are taking an important step forward in providing consumers with a secure shopping experience, and the latest in payments technology," said Chris McWilton, president, North American Markets for MasterCard. "Our focus, together with Target, is on safety and security."

Exhibit 13.5: Target Income Statement (adapted Target, 2014a, p. 63)

Quarterly Results (millions, except per share data)	First Quarter		Second Quarter		Third Quarter		Fourth Quarter		Total Year	
	2013	2012	2013	2012	2013	2012	2013	2012*	2013	2012*
Sales	16,706	16,537	17,117	16,451	17,258	16,601	21,516	22,370	72,596	71,960
Credit card revenues	–	330	–	328	–	328	–	356	–	1,341
Total revenues	16,706	16,867	17,117	16,779	17,258	16,929	21,516	22,726	72,596	73,301
Cost of sales	11,563	11,541	11,745	11,297	12,133	11,569	15,719	16,160	51,160	50,568
Selling, general and administrative expenses	3,590	3,392	3,698	3,588	3,853	3,704	4,235	4,229	15,375	14,914
Credit card expenses	–	120	–	108	–	106	–	135	–	467
Depreciation and amortization	536	529	542	531	569	542	576	539	2,223	2,142
Gain on receivables transaction	(391)	–	–	–	–	(156)	–	(5)	(391)	(161)
Earnings before interest expense and income taxes	1,408	1,285	1,132	1,255	703	1,164	986	1,668	4,229	5,371
Net interest expense	629	184	171	184	165	192	161	204	1,126	762
Earnings before income taxes	779	1,101	961	1,071	538	972	825	1,464	3,103	4,609
Provision for income taxes	281	404	350	367	197	335	305	503	1,132	1,610
Net earnings	498	697	611	704	341	637	520	961	1,971	2,999
Basic earnings per share	0.78	1.05	0.96	1.07	0.54	0.97	0.82	1.48	3.10	4.57
Diluted earnings per share	0.77	1.04	0.95	1.06	0.54	0.96	0.81	1.47	3.07	4.52
Dividends declared per share	0.36	0.30	0.43	0.36	0.43	0.36	0.43	0.36	1.65	1.38
Closing common stock price:										
High	70.67	58.86	73.32	61.95	71.99	65.44	66.89	64.48	73.32	65.44
Low	60.85	50.33	68.29	54.81	62.13	60.62	56.64	58.57	56.64	50.33

Note: Per share amounts are computed independently for each of the quarters presented. The sum of the quarters may not equal the total year amount due to the impact of changes in average quarterly shares outstanding, and all other quarterly amounts may not equal the total year due to rounding.

* The fourth quarter and total year 2013 consisted of 13 weeks and 52 weeks, respectively, compared with 14 weeks and 53 weeks in the comparable prior-year periods.

Index

Page numbers followed by *f* and *t* refer to figures and tables, respectively.

www.ingramcontent.com/pod-product-compliance
Ingram Content Group UK Ltd.
Pitfield, Milton Keynes, MK11 3LW, UK
UKHW051453290926
15471UKWH00008B/1190

9 781943 153848